AF556790

Advanced Hard and Soft Magnetic Materials

MATERIALS RESEARCH SOCIETY
SYMPOSIUM PROCEEDINGS VOLUME 577

Advanced Hard and Soft Magnetic Materials

Symposium held April 5–8, 1999, San Francisco, California, U.S.A.

EDITORS:

Michael Coey
Trinity College
Dublin, Ireland

Josef Fidler
Vienna University of Technology
Vienna, Austria

Laura H. Lewis
Brookhaven National Laboratory
Upton, New York, U.S.A.

Vincent G. Harris
Naval Research Laboratory
Washington, D.C., U.S.A.

Bao-Min Ma
Magnequench International, Inc.
Anderson, Indiana, U.S.A.

Ryusuke Hasegawa
AlliedSignal Inc.
Morristown, New Jersey, U.S.A.

Thomas Schrefl
Vienna University of Technology
Vienna, Austria

Akihisa Inoue
Tohoku University
Sendai, Japan

Ludwig Schultz
IFW Dresden
Dresden, Germany

Michael McHenry
Carnegie Mellon University
Pittsburgh, Pennsylvania, U.S.A.

Materials Research Society
Warrendale, Pennsylvania

Single article reprints from this publication are available through
University Microfilms Inc., 300 North Zeeb Road, Ann Arbor, Michigan 48106

CODEN: MRSPDH

Published by:

Materials Research Society
506 Keystone Drive
Warrendale, PA 15086
Telephone (724) 779-3003
Fax (724) 779-8313
Web site: http://www.mrs.org/

Library of Congress Cataloging-in-Publication Data

Advanced hard and soft magnetic materials / editors, Michael Coey, Laura H. Lewis, Bao-Min Ma, Thomas Schrefl, Ludwig Schultz, Josef Fidler, Vincent G. Harris, Ryusuke Hasegawa, Akihisa Inoue, Michael McHenry
p.cm.—(Materials Research Society symposium proceedings, ISSN 0272-9172 ; v. 577)
Includes bibliographical references and indexes.
ISBN 1-55899-485-8
1. Magnetic materials—Congresses. 2. Permanent magnets—Congresses. I. Coey, Michael II. Lewis, Laura H. III. Ma, Bao-Min IV. Schrefl, Thomas V. Schultz, Ludwig VI. Fidler, Josef VII. Harris, Vincent G. VIII. Hasegawa, Ryusuke IX. Inoue, Akihisa X. McHenry, Michael XI. Materials Research Society symposium proceedings ; v. 577
QC764.5 .A38 1999
538'.4—dc21 99-043779

Manufactured in the United States of America

CONTENTS

*Invited Paper

*Invited Paper

PART V: NANOSCALE HARD MAGNETISM III

PART VI: PERMANENT MAGNET APPLICATIONS

PART VII: MICROSTRUCTURE AND MICROMAGNETICS

*Invited Paper

PART VIII: THIN FILM PERMANENT MAGNETS

PART IX: FINE PARTICLE MAGNETS

*Invited Paper

*Invited Paper

*Invited Paper

PART XIV: NANOCRYSTALLINE MAGNETIC THIN FILMS

PREFACE

Many recent advances in magnetic materials have resulted from the ability to structure the materials on an appropriate magnetic length scale. This is typically the exchange length or the domain wall width of a hard phase, but in either case the characteristic length scale is a few nanometers. As the dimensions of the grains in a magnetic nanostructure approach this limit, the magnetic properties become significantly different from those in the bulk. Some examples are the following: remanence enhancement in isotropic, single-phase hard nanostructures, exchange spring behavior of nanostructures composed of hard and soft grains where the intrinsic magnetic properties such as magnetization and anisotropy are the volume-weighted averages of the two components, vanishing anisotropy in nanostructures composed of randomly-oriented soft grains and volume-averaging of magnetostriction in nanocomposites of two soft phases. In these examples, nanostructured materials significantly extend the range of available magnetic properties. A range of materials processing issues centers on the need to control nucleation and crystal growth on a very small length scale. Another focus is on the nature of the grain boundaries and the exchange coupling across them. The problems encountered here reappear when considering planar nanostructures such as multilayers or spin-polarized tunnel junctions, where the spin diffusion length is an additional length scale that has to be taken into consideration.

The papers in this proceedings volume were presented April 5–9 at Symposia H and I at the 1999 MRS Spring Meeting in San Francisco, California. Symposium H was on "Advanced Hard Magnets—Principles, Materials, and Processing," and Symposium I was on "Amorphous and Nanocrystalline Materials for Hard and Soft Magnetic Applications." In addition, several papers on magnetic thin films from Symposium L, "Polycrystalline Metal and Magnetic Thin Films," have been included to give the reader an overview of recent developments in the field.

Michael Coey
Laura H. Lewis
Bao-Min Ma
Thomas Schrefl
Ludwig Schultz
Josef Fidler
Vincent G. Harris
Ryusuke Hasegawa
Akihisa Inoue
Michael McHenry

June 1999

MATERIALS RESEARCH SOCIETY SYMPOSIUM PROCEEDINGS

Volume 535— III-V and IV-IV Materials and Processing Challenges for Highly Integrated Microelectonics and Optoelectronics, S.A. Ringel, E.A. Fitzgerald, I. Adesida, D. Houghton, 1999, ISBN: 1-55899-441-6
Volume 536— Microcrystalline and Nanocrystalline Semiconductors—1998, L.T. Canham, M.J. Sailor, K. Tanaka, C-C. Tsai, 1999, ISBN: 1-55899-442-4
Volume 537— GaN and Related Alloys, S.J. Pearton, C. Kuo, A.F. Wright, T. Uenoyama, 1999, ISBN: 1-55899-443-2
Volume 538— Multiscale Modelling of Materials, V.V. Bulatov, T. Diaz de la Rubia, R. Phillips, E. Kaxiras, N. Ghoniem, 1999, ISBN: 1-55899-444-0
Volume 539— Fracture and Ductile vs. Brittle Behavior—Theory, Modelling and Experiment, G.E. Beltz, R.L. Blumberg Selinger, K-S. Kim, M.P. Marder, 1999, ISBN: 1-55899-445-9
Volume 540— Microstructural Processes in Irradiated Materials, S.J. Zinkle, G.E. Lucas, R.C. Ewing, J.S. Williams, 1999, ISBN: 1-55899-446-7
Volume 541— Ferroelectric Thin Films VII, R.E. Jones, R.W. Schwartz, S.R. Summerfelt, I.K. Yoo, 1999, ISBN: 1-55899-447-5
Volume 542— Solid Freeform and Additive Fabrication, D. Dimos, S.C. Danforth, M.J. Cima, 1999 ISBN: 1-55899-448-3
Volume 543— Dynamics in Small Confining Systems IV, J.M. Drake, G.S. Grest, J. Klafter, R. Kopelman, 1999, ISBN: 1-55899-449-1
Volume 544— Plasma Deposition and Treatment of Polymers, W.W. Lee, R. d'Agostino, M.R. Wertheimer, B.D. Ratner, 1999, ISBN: 1-55899-450-5
Volume 545— Thermoelectric Materials 1998—The Next Generation Materials for Small-Scale Refrigeration and Power Generation Applications, T.M. Tritt, M.G. Kanatzidis, G.D. Mahan, H.B. Lyon, Jr., 1999, ISBN: 1-55899-451-3
Volume 546— Materials Science of Microelectromechanical Systems (MEMS) Devices, A.H. Heuer, S.J. Jacobs, 1999, ISBN: 1-55899-452-1
Volume 547— Solid-State Chemistry of Inorganic Materials II, S.M. Kauzlarich, E.M. McCarron III, A.W. Sleight, H-C. zur Loye, 1999, ISBN: 1-55899-453-X
Volume 548— Solid-State Ionics V, G-A. Nazri, C. Julien, A. Rougier, 1999, ISBN: 1-55899-454-8
Volume 549— Advanced Catalytic Materials—1998, P.W. Lednor, D.A. Nagaki, L.T. Thompson, 1999, ISBN: 1-55899-455-6
Volume 550— Biomedical Materials—Drug Delivery, Implants and Tissue Engineering, T. Neenan, M. Marcolongo, R.F. Valentini, 1999, ISBN: 1-55899-456-4
Volume 551— Materials in Space—Science, Technology and Exploration, A.F. Hepp, J.M. Prahl, T.G. Keith, S.G. Bailey, J.R. Fowler, 1999, ISBN: 1-55899-457-2
Volume 552— High-Temperature Ordered Intermetallic Alloys VIII, E.P. George, M.J. Mills, M. Yamaguchi, 1999, ISBN: 1-55899-458-0
Volume 553— Quasicrystals, J-M. Dubois, P.A. Thiel, A-P. Tsai, K. Urban, 1999, ISBN: 1-55899-459-9
Volume 554— Bulk Metallic Glasses, W.L. Johnson, C.T. Liu, A. Inoue, 1999, ISBN: 1-55899-460-2
Volume 555— Properties and Processing of Vapor-Deposited Coatings, R.N. Johnson, W.Y. Lee, M.A. Pickering, B.W. Sheldon, 1999, ISBN: 1-55899-461-0
Volume 556— Scientific Basis for Nuclear Waste Management XXII, D.J. Wronkiewicz, J.H. Lee, 1999, ISBN: 1-55899-462-9
Volume 557— Amorphous and Heterogeneous Silicon Thin Films: Fundamentals to Devices—1999, H.M. Branz, R.W. Collins, H. Okamoto, S. Guha, R. Schropp, 1999, ISBN: 1-55899-464-5

MATERIALS RESEARCH SOCIETY SYMPOSIUM PROCEEDINGS

Volume 558— Flat-Panel Displays and Sensors—Principles, Materials and Processes, F.R. Libsch, B. Chalamala, R. Friend, T. Jackson, H. Ohshima, 1999, ISBN: 1-55899-465-3

Volume 559— Liquid Crystal Materials and Devices, T.J. Bunning, S.H. Chen, L.C. Chien, S-C.A. Lien, T. Kajiyama, N. Koide, 1999, ISBN: 1-55899-466-1

Volume 560— Luminescent Materials, J. McKittrick, A. Kitai, K. Mishra, C. Ronda, B. DiBartolo, 1999, ISBN: 1-55899-467-X

Volume 561— Organic Nonlinear Optical Materials and Devices, B. Kippelen, H. Lackritz, R. Claus, 1999, ISBN: 1-55899-468-8

Volume 562— Polycrystalline Metal and Magnetic Thin Films, K.P. Rodbell, D. Laughlin, O. Thomas, B. Zhang, 1999, ISBN: 1-55899-469-6

Volume 563— Materials Reliability in Microelectronics IX, C.A. Volkert, A.H. Verbruggen, D. Brown, 1999, ISBN: 1-55899-470-X

Volume 564— Advanced Interconnects and Contacts, D.C. Edelstein, T. Kikkawa, M. Özturk, K-N. Tu, E. Weitzman, 1999, ISBN: 1-55899-471-8

Volume 565— Low-Dielectric Constant Materials V, J.P. Hummel, K. Endo, W.W. Lee, M.E. Mills, S-Q. Wang, 1999, ISBN: 1-55899-472-6

Volume 566— Chemical-Mechanical Polishing—Fundamentals and Challenges, S.V. Babu, S. Danyluk, M.I. Krishnan, M. Tsujimura, 1999, ISBN: 1-55899-473-4

Volume 567— Ultrathin SiO_2 and High-K Materials for ULSI Gate Dielectrics, H.R. Huff, M.L. Green, T. Hattori, G. Lucovsky, C.A. Richter, 1999, ISBN: 1-55899-474-2

Volume 568— Silicon Front-End Processing—Physics and Technology of Dopant-Defect Interactions, H-J.L. Gossmann, T.E. Haynes, M.E. Law, A.N. Larsen, S. Odanaka, 1999, ISBN: 1-55899-475-0

Volume 569— *In Situ* Process Diagnostics and Modelling, O. Auciello, A.R. Krauss, E.A. Irene, J.A. Schultz, 1999, ISBN: 1-55899-476-9

Volume 570— Epitaxial Growth, T.P. Pearsall, A-L. Barabasi, F. Liu, G.N. Maracas, 1999, ISBN: 1-55899-477-7

Volume 571— Semiconductor Quantum Dots, D. Ila, H. Lee, S. Moss, D. Norris, 1999, ISBN: 1-55899-478-5

Volume 572— Wide-Bandgap Semiconductors for High-Power, High-Frequency and High-Temperature Applications—1999, S. Binari, A. Burk, M. Melloch, C. Nguyen, 1999, ISBN: 1-55899-479-3

Volume 573— Compound Semiconductor Surface Passivation and Novel Device Processing, H. Hasegawa, M. Hong, Z.H. Lu, S.J. Pearton, 1999, ISBN: 1-55899-480-7

Volume 574— Multicomponent Oxide Films for Electronics, M.E. Hawley, D.H. Blank, C-B. Eom, S.K. Streiffer, D.G. Schlom, 1999, ISBN: 1-55899-481-5

Volume 575— New Materials for Batteries and Fuel Cells, D.H. Doughty, H-P. Brack, K. Naoi, L.F. Nazar, 1999, ISBN: 1-55899-482-3

Volume 576— Organic/Inorganic Hybrid Materials II, L.C. Klein, L. Francis, M.R. DeGuire, J.E. Mark, 1999, ISBN: 1-55899-483-1

Volume 577— Advanced Hard and Soft Magnetic Materials, L.H. Lewis, J.M.D. Coey, B-M. Ma, T. Schrefl, L. Schultz, M.E. McHenry, V.G. Harris, J. Fidler, R. Hasegawa, A. Inoue, 1999, ISBN: 1-55899-485-8

Prior Materials Research Society Symposium Proceedings available by contacting Materials Research Society

Part I

Permanent Magnet Processing

HYDROGENATION DISPROPORTIONATION DESORPTION RECOMBINATION PROCESSES APPLIED TO NdFeB-, SmFe- AND SmCo-TYPE ALLOYS

O. Gutfleisch, M. Kubis, A. Handstein, K.-H. Müller and L. Schultz
Institute of Solid State and Materials Research Dresden, P.O.Box 270016, 01171 Dresden, Germany

ABSTRACT

Recent aspects of hydrogenation disproportionation desorption and recombination (HDDR) phenomena in NdFeB-, SmFe- and SmCo-type alloys are reviewed in this paper. The effects of additives on hydrogen sorption and magnetic properties are discussed. The possibilities of producing magnetically anisotropic NdFeB HDDR magnets by hot deformation or alternatively pre-aligning and compacting anisotropic powders are described. Current models for the inducement of magnetic anisotropy in NdFeB HDDR powder are summarized and it is shown that anisotropic powder can be produced from a ternary compound when using a modified processing route. The application of ´extreme` hydrogenation conditions, namely high hydrogen pressure or reactive milling in a hydrogen atmosphere at enhanced temperatures, allows the disproportionation of thermodynamically very stable compounds such as $Sm_2Fe_{17-x}Ga_x$ ($x \geq 1$), $SmCo_5$ or Sm_2Co_{17}. Reactive milling and subsequent recombination in vacuum lead to structures, of both the disproportionated and the recombined type, which are on a remarkably finer scale than those commonly observed for standard HDDR procedures. Exchange coupling between the nano-scaled grains can result in magnetically single phase behavior despite a multiphase microstructure and in particular for the Sm_2Co_{17} alloy, a remanence enhancement was observed after recombination at temperatures ≤700°C.

INTRODUCTION - HDDR IN $Nd_2Fe_{14}B$ AND Sm_2Fe_{17}

A reversible hydrogen induced phase transformation, the so-called hydrogenation disproportionation desorption and recombination (HDDR) process [1,2], can be used to produce highly coercive $Nd_2Fe_{14}B$ powders, very suitable for the production of both, bonded and fully dense hot pressed, magnets. In the case of HDDR processed $Sm_2Fe_{17}N_3$-type magnets [3], the applications are restricted to bonded magnets because of the limited thermal stability of appr. 600°C [4]. However, the thermal stability can be improved when using $Sm_2Fe_{17-x}Ga_xC_y$ powders [5]. In the case of $Nd_2Fe_{14}B$, the HDDR process consists of two stages: firstly, the disproportionation into a finely divided mixture of neodymium hydride, iron and ferroboron typically at 800°C and 1bar hydrogen and secondly, the recombination on desorption to the original $Nd_2Fe_{14}B$ phase, but now with a much refined grain size. The HDDR reactions in the two systems can be described as:

$$Nd_2Fe_{14}B + (2\pm x)H_2 \Leftrightarrow 2NdH_{2\pm x} + 12Fe + Fe_2B \pm \Delta H \quad (1)$$

and

$$Sm_2Fe_{17} + (2\pm x)H_2 \Leftrightarrow 2SmH_{2\pm x} + 17Fe \pm \Delta H' \quad (2)$$

At lower temperatures the interstitial hydride is formed representing a metastable state (hydrogen-solid-solution) and the forward reaction as described in (1) and (2), i.e. the formation of the thermodynamically more stable binary rare earth-hydride, is prevented by the absence of sufficient kinetics at room temperature. The interstitial absorption, leading to a hydrogen-induced decrepitation [6], can be avoided by introducing hydrogen at elevated temperatures

Mat. Res. Soc. Symp. Proc. Vol. 577 © 1999 Materials Research Society

which results in the disproportionation without cracking, the so-called 'Solid-HDDR'-process [7].

Fig. 1 shows the typical absorption reactions of Sm_2Fe_{17} during heating in hydrogen atmosphere, namely the low temperature interstitial absorption by the Sm_2Fe_{17} phase (A) and the disproportionation of the Sm_2Fe_{17} phase (C) according to the equation (2) at around 550°C. The presence of minor amounts of $SmFe_3$ phase, as shown in Fig. 2, leads to an additional absorption (B) around 450°C corresponding to the disproportionation of this phase.

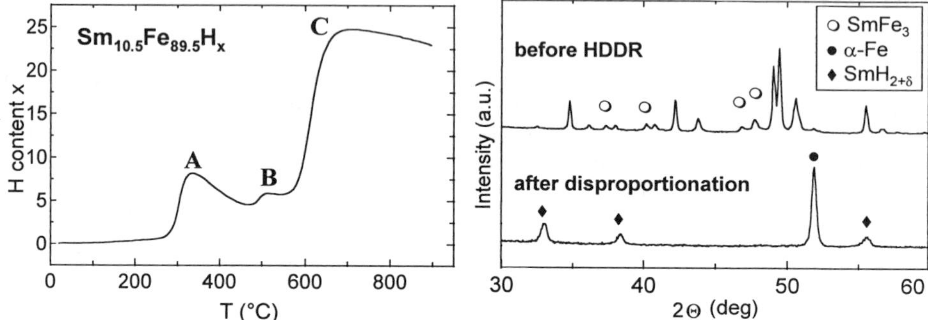

Fig. 1 (left): Temperature pressure analysis illustrating three exothermic hydrogen absorption reactions and the respective hydrogen uptakes on heating of $Sm_{10.5}Fe_{89.5}$ (A-interstitial absorption of Sm_2Fe_{17}, B-disproportionation of $SmFe_3$ and C-disproportionation of Sm_2Fe_{17}). Fig. 2 (right): XRD patterns of the alloy before and after disproportionation.

The hydrogen absorption and desorption characteristics are affected by alloy additions. In particular it has been found that Co and Ga accelerate the recombination reaction and shift its onset to lower temperatures and Zr makes the disproportionation reaction sluggish [8-10].

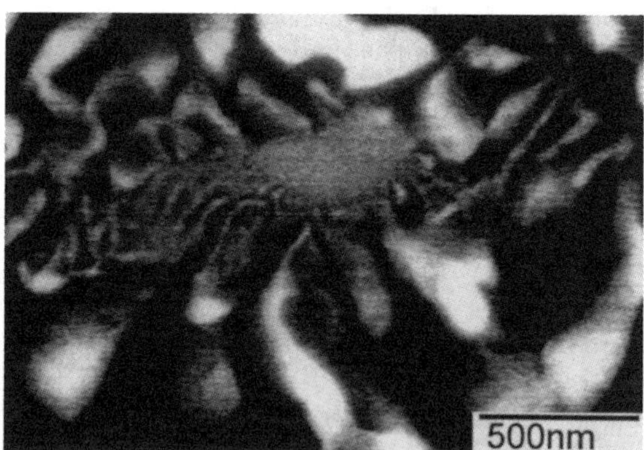

Fig. 3: Backscattered SEM image of a partially disproportionated $Sm_2Fe_{16}Ga$ illustrating the initial fine lamellar structure of Sm-hydride (white) and α-Fe (black) at the interface to the unreacted matrix phase (gray).

An example of a disproportionated structure is given in Fig. 3. The disproportionation products Sm-hydride and α-Fe (see also Fig. 2) form initially a fine lamellar type structure at the interface to the unreacted $Sm_2Fe_{16}Ga$ matrix phase.

Milling before HDDR did not improve the coercivity in NdFeB-type alloys, as observed for Sm_2Fe_{17}-type materials where coercivities as high as 3.5T were found after nitrogenation [11], but accelerated hydrogen absorption and desorption rates of the pre-milled material were found. An example of the magnetic properties of conventionally HDDR processed isotropic $Nd_{16.2}Fe_{78.2}B_{5.6}$ is given in Fig. 6.

MAGNETIC ANISOTROPY IN $Nd_2Fe_{14}B$

Nano-crystalline materials such as mechanically alloyed, intensively milled, rapidly quenched and HDDR powders require usually a hot compaction and deformation (die-upsetting) procedure in order to induce a texture leading to the high remanences necessary for high performance magnets [12]. The HDDR process offers the unique possibility and distinct advantage over e.g. the MQ (rapidly quenched) products, provided the right processing conditions and compositions are chosen, to produce powder in an anisotropic form [13] without, as in the case of anisotropic MQ powder (MQIII), having to apply the expensive route of die-upsetting and subsequent crushing.

A comparison of the magnetic properties of two anisotropic fully dense HDDR magnets with a similar composition produced by (a) hot compaction and subsequent deformation of isotropic powder and (b) pre-aligning and subsequent hot compaction of anisotropic HDDR powder (MQA-T) is given in Fig. 4.

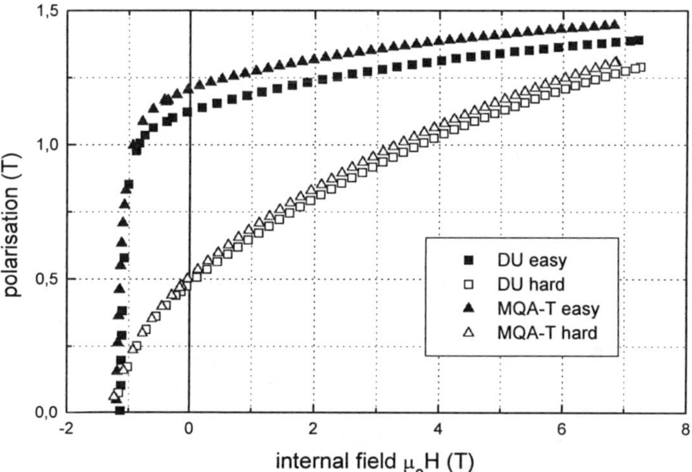

Fig. 4: Demagnetization curves of anisotropic HDDR magnets: (a) produced by hot compaction and deformation (DU) of isotropic $Nd_{13.9}Fe_{62.3}Co_{16.2}Al_{0.5}Ga_{0.3}B_{6.8}$ powder and (b) produced by pre-aligning anisotropic MQA-T powder before hot compaction.

A "memory-cell" model [14] has been proposed to explain the inducement of magnetic anisotropy and has been supported by thermodynamic arguments [15] and electron microscopy studies [16]: nano-scale $Nd_2Fe_{14}B$-type particles rich in the additives such as Co, Ga or Zr which remain unaffected by the disproportionation reaction and therefore retain the orientation of the original grains. Recently, an intermediate Fe_3B-type phase within the rod-like structure has been reported [17] for a solid-disproportionated Co and Ga containing alloy as an anisotropy mediating phase. On the other hand, it has been shown that it is possible to induce anisotropic features in a ternary alloy, however so far only combined with a comparatively low coercivity [18,19]. A modified HDDR procedure as shown in Fig. 5 can lead to a significant coercivity combined with an appreciable degree of anisotropy even for the ternary alloy. This procedure includes a solid-disproportionation step and a slow recombination under a partial hydrogen pressure. The latter step facilitates [20] the control of grain growth during the strongly endothermic desorption stage. Fig. 6 shows the demagnetization curve of a $Nd_{16.2}Fe_{78.2}B_{5.6}$ alloy to which this modified procedure has been applied. A remarkable degree of anisotropy (B_r=0.94T) and a high coercivity ($\mu_{0J}H_c$=1.11T) were obtained.

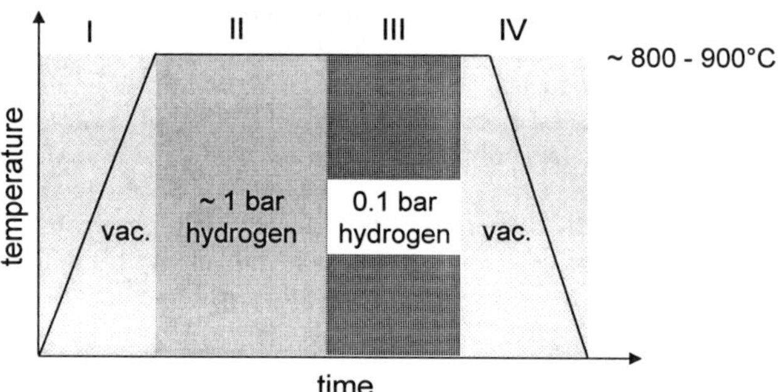

Fig. 5: Schematic presentation of the modified HDDR procedure applied to obtain anisotropic and highly coercive powders from a ternary NdFeB alloy: I - heating in vacuum, II - solid-disproportionation, III -slow recombination in a partial hydrogen pressure and IV - full desorption.

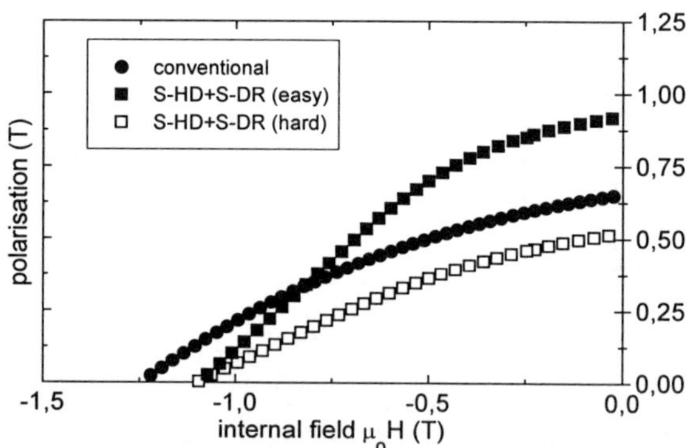

Fig. 6: Comparison of demagnetization curves along easy and hard directions of a $Nd_{16.2}Fe_{78.2}B_{5.6}$ alloy produced via the conventional method and the solid-disproportionation (S-HD) and slow recombination method (S-DR).

These observations for the ternary compound shed new light on the above ´memory cell´ model for the inducement of magnetic anisotropy as the coexistence of the disproportionated constituents and residual, nano-scale $Nd_2(Fe,Co,Ga)_{14}B$ no longer appears to be a feasible explanation for the phenomena observed. The key could lie in the type of disproportionated structure generated by the different procedures. Conventional processing leads to a granular structure [19] whereas solid-disproportionation results in a rod-like or lamellae-like structure as described in [21]. A slow recombination with controlled grain growth of a rod-like structure at temperatures as high as 900°C results in highly anisotropic and coercive material for ternary compounds. The driving force for the endothermic recombination reaction under a partial hydrogen pressure will be reduced compared to that of a vacuum recombination. This in turn could affect nucleation and growth of the $Nd_2Fe_{14}B$ phase which might be random in vacuum but preferential in a partial pressure. However, at the time of writing the final anisotropy mediating mechanism from the starting to the recombined alloy still needs to be identified.

´EXTREME` HDDR PROCESSING

HDDR Using High Hydrogen Pressures

Recently, it was shown that HDDR in thermodynamically stabilized compounds such as $Sm_2Fe_{16}Ga$ is successful when carried out under extreme conditions, namely under high hydrogen pressures [22] or by means of reactive milling in hydrogen [23]. High hydrogen concentrations in both, the ternary and binary hydride, can be achieved when using these extreme conditions. An example is given in Fig. 7 where the Curie temperatures of $Sm_2Fe_{15}Ga_2H_x$ hydrogenated in various hydrogen pressures p are shown. The increased Fe-Fe exchange interaction due to the lattice expansion on increasing hydrogen concentration causes the enhanced Curie-temperatures. The initial part of the curve can be described with Sievert´s law $p \propto c^2$ (c being the hydrogen concentration), i.e. the concentration c and in this case the Curie temperature increases linearly with the square root of p. With higher concentrations as all tetrahedral *18g* sites are filled a saturation is achieved and no further increase in Curie temperature is observed.

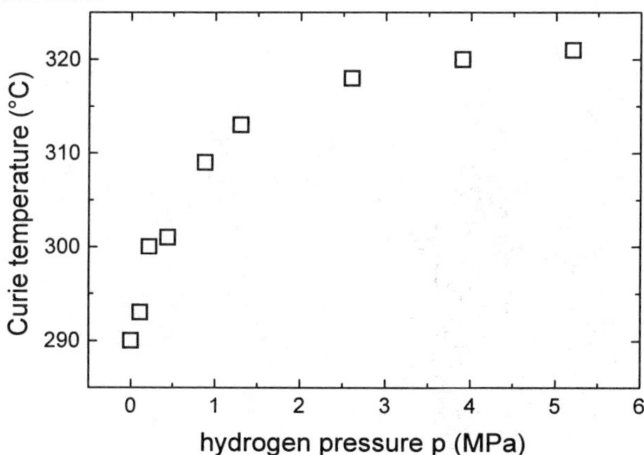

Fig. 7: Curie temperatures of hydrogenated $Sm_2Fe_{15}Ga_2H_x$ as a function of hydrogen pressure (the initial slope can be described with Sievert´s law $p \propto c^2$).

No successful application of the HDDR process to the Sm-Co system had been reported until recently [24,25] which is due to the high stability of the Sm-Co phases against the disproportionation by hydrogen. The DSC runs of $Sm_2(Co,Fe,Cu,Zr)_{17}$ in Fig. 8 show two main exothermic peaks corresponding to the interstitial absorption and the subsequent disproportionation above 500°C into $SmH_{2+\delta}$ ($1.9 < \delta < 2.8$) and Co-rich bcc α-(Co,Fe), as shown in the XRD pattern in Fig. 9. The application of different hydrogen pressures results in a variation of position and intensity of the exothermic peaks. With increasing hydrogen pressure, the exothermic peaks move to lower temperatures and become more dominant.

A microstructure generated by high pressure disproportionation using 40 bar starting pressure is shown in Fig. 10 and it is very interesting to note that a further refinement has taken place compared to a disproportionated mixture of e.g. Sm hydride and α-Fe induced by standard conditions [26].

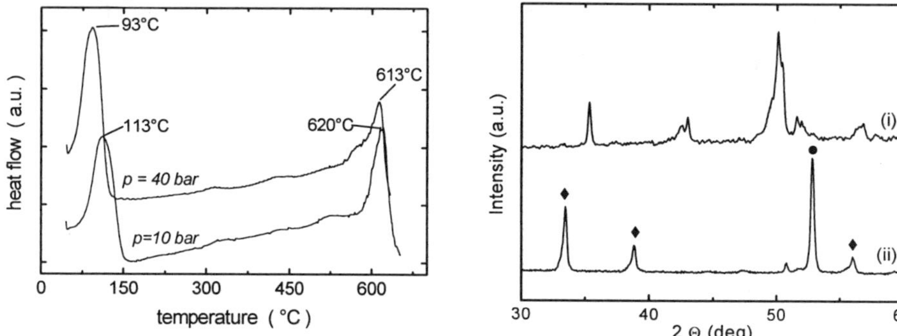

Fig. 8 (left): DSC curves for $Sm_2(Co,Fe,Cu,Zr)_{17}$ measured for starting pressures of 10 and 40 bar hydrogen, respectively showing two exothermic peaks related to the interstitial absorption of hydrogen and the disproportionation reaction.

Fig. 9 (right): XRD patterns before (i) and after (ii) a DSC run under 20bar hydrogen, the reflections of the starting material disappear almost completely and reflections of $SmH_{2+\delta}$ (♦) and Co-rich bcc α-(Co,Fe) phase (●) dominate after the high-pressure disproportionation.

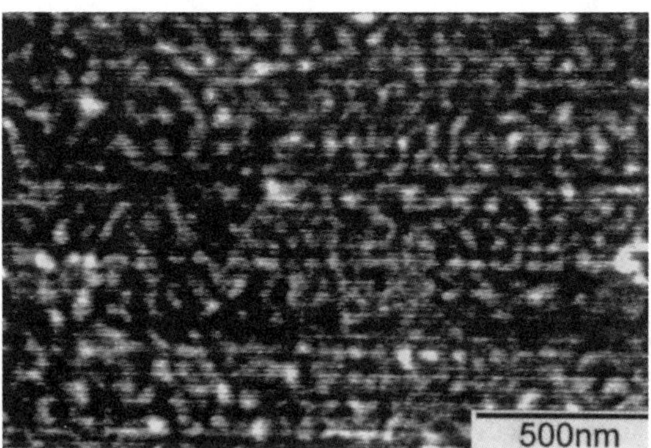

Fig. 10: Backscattered SEM image showing the very fine structure of $Sm_2(Co,Fe,Cu,Zr)_{17}$ disproportionated using 40bar starting pressure (Sm-hydride (white) and α-(Co,Fe) (black)).

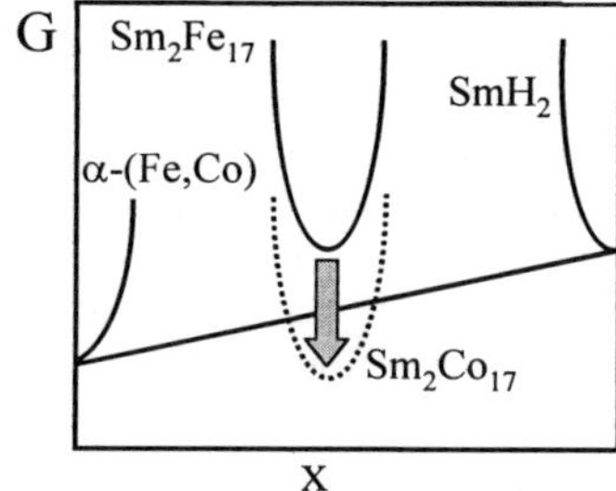

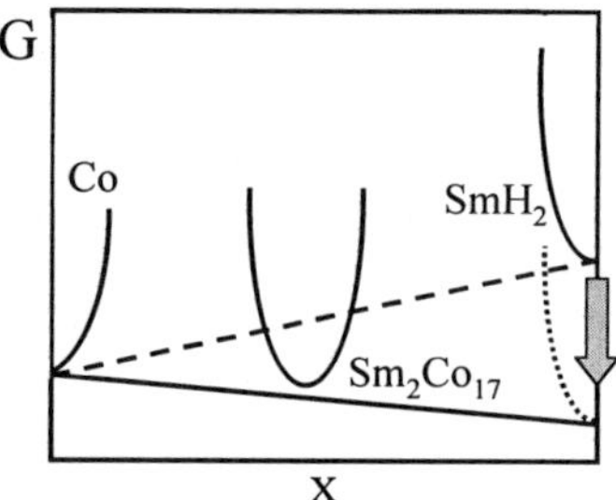

Fig. 11: Schematic free enthalpy diagram of $Sm_2(Fe,Co)_{17}$ alloys illustrating the decrease of *G* by the Co substitution making the disproportionation of Sm_2Co_{17} under standard condition impossible (left); the increased hydrogen pressure decreases the free enthalpy of the Sm-hydride and therefore the disproportionation can take place (right).

The effect of compositional changes such as the substitution of Fe by Co in $Sm_2(Fe,Co)_{17}$ on the thermodynamic stability of such compounds is described qualitatively in Fig. 11 (left). The Co substitution results in the situation that the tangent construction between the two disproportionation products is intersected by the new minimum in free enthalpy of the Sm_2Co_{17} phase. The favorable effect of high hydrogen pressures on disproportionation shown in Fig. 11 (right) could be explained in terms of a decrease of the free enthalpy of samarium hydride with increasing hydrogen pressure to that extent that the reaction is allowed to proceed.

HDDR Using Reactive Milling

Mechanical induced gas-solid reactions under non-equilibrium conditions can also be used to hydrogen disproportionate thermodynamically very stable compounds [25]. This reactive milling procedure in enhanced hydrogen pressures and temperatures and subsequent vacuum annealing to ensure full desorption and recombination is described schematically in Fig. 12.

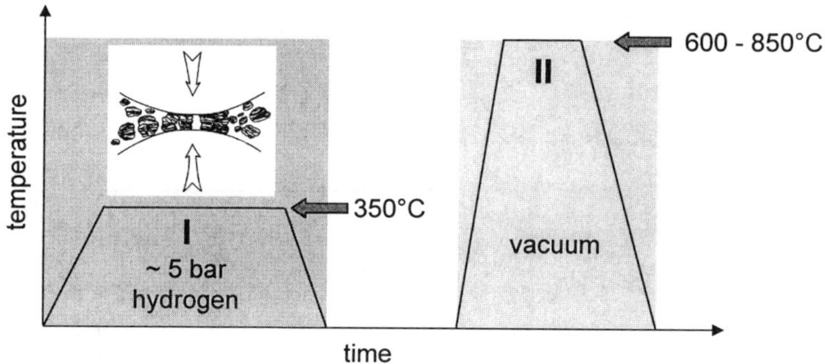

Fig. 12: Schematic presentation of the reactive milling procedure (I) and the subsequent recombination step in a vacuum furnace (II).

The effect of reactive milling on Sm_2Co_{17} alloys is illustrated in the XRD patterns of Fig. 13 and after milling the reflections of $SmH_{2+\delta}$ and fcc Co can be seen. The extremely broad reflections are caused by a very fine nano-crystalline structure of these phases and the average grain size of the $SmH_{2+\delta}$ phases was estimated to be around 9±2nm using the Williamson-Hall method [27]. This is a much finer structure than that of disproportionated Sm_2Fe_{17}-type or $Nd_2Fe_{14}B$-type materials prepared by the conventional HDDR method [21,26]. Fig. 13 also shows the XRD patterns of the Sm_2Co_{17}-type powders for different recombination temperatures. 1:5 type phase reflections due to Sm excess in the initial alloy can be seen in addition to the reflections of the 2:17 type phase. A decreasing mean peak width can be observed with increasing recombination temperatures due to larger average grain sizes. Grain sizes of 18±6nm (600°C), 25±8nm (650°C) and 29±4nm (700°C) were estimated from the XRD patterns for the recombined samples.

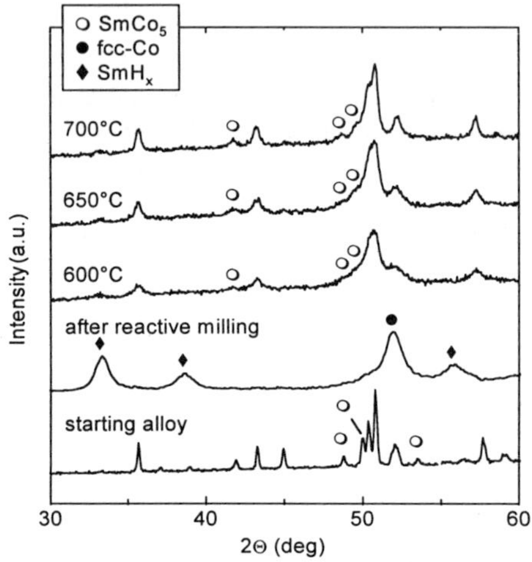

Fig. 13: XRD patterns of the starting Sm_2Co_{17} alloy, the reactively milled and various recombined powders.

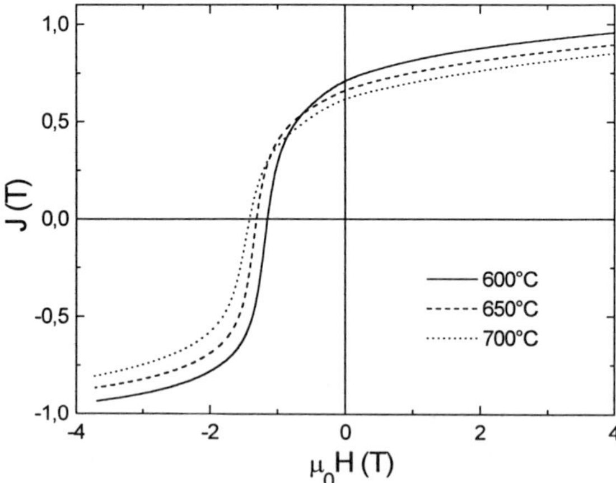

Fig. 14: Demagnetization curves of reactively milled Sm_2Co_{17} recombined at different temperatures.

Fig. 14 shows that despite the presence of two different phases, the demagnetization curves of the Sm_2Co_{17}-type powders have a magnetically single-phase like appearance for low recombination temperatures. The remanence J_r=0.71T for the sample recombined at 600°C is significantly higher than the theoretical maximum value $J_s/2$=0.65T [28] of isotropically distributed single-domain Sm_2Co_{17} particles which is attributed to the effect of exchange coupling between the nano-scaled grains. It has also to be taken into account that the amount of $SmCo_5$ phase within the sample should decrease the J_r value even further due to the lower value of the saturation magnetization compared to that of Sm_2Co_{17}.

HYDROGEN-INDUCED SYNTHESIS OF NANOSTRUCTURED MATERIALS

Gas-solid reactions of rare earth-transition metal compounds (R_nT_m) with hydrogen are an alternative route to mechanical alloying, intensive milling or rapid quenching for the preparation of amorphous or nano-crystalline materials. Depending on thermodynamics (stability of starting alloy and reaction products) and kinetics (temperature, hydrogen pressure and possibly mechanical activation) an interstitial modification resulting in a ternary hydride (c-$R_nT_mH_x$), an amorphous hydride (a-$R_nT_mH_x$) – the so-called hydrogen induced amorphisation (HIA) [29] or a disproportionation into a binary rare earth-hydride and the transition metal (nRH_x+mT) can occur (see Fig. 15). Metastable phases will be the reaction products when the activation, either thermally or mechanically, is insufficient to form the necessary concentration gradient of metal atoms and only hydrogen diffusion is possible ("chemical frustration" [29]).

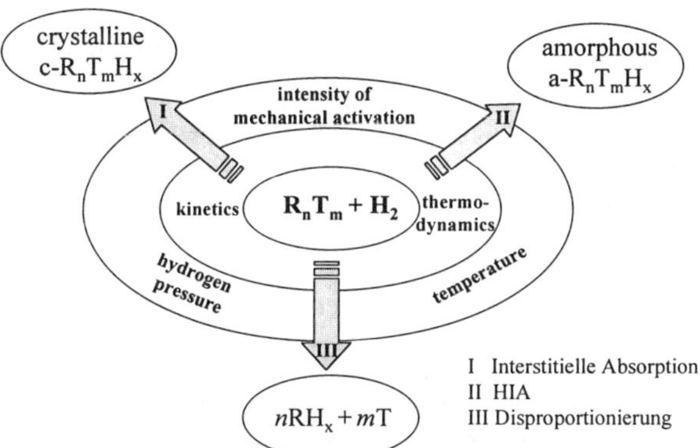

Fig. 15: Depending on thermodynamics and kinetics a ternary hydride (c-$R_nT_mH_x$), an amorphous hydride (a-$R_nT_mH_x$) or a binary rare earth-hydride and the transition metal (nRH_x+mT) is formed.

Fig. 16 illustrates that these various states can be achieved in a Sm_2Fe_{17} alloy by a variation of processing parameters. The XRD patterns show the amorphous state after milling (a), the HIA after milling at RT in hydrogen (b), the extremely finely grained disproportionated state after reactive milling under enhanced conditions (c), the standard disproportionated state after annealing in hydrogen (d) and (e) the interstitial ternary $Sm_2Fe_{17}H_x$ hydride after low temperature absorption.

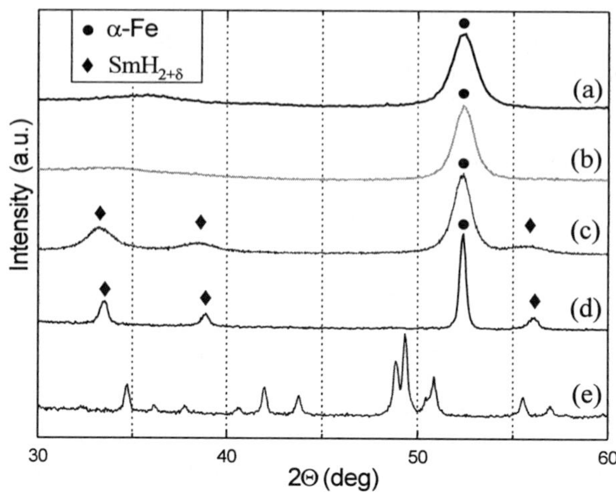

Fig. 16: XRD patterns of Sm_2Fe_{17} after (a) 63h milling in Ar at RT, (b) 30h milling in 1bar H_2 at RT, (c) 10h reactive milling in 5bar H_2 at 350°C, (d) 1h disproportionation in 0.6bar H_2 at 720°C and (e) after heating to 400°Cin 0.8bar H_2.

CONCLUSIONS

The differences in magnetic properties after conventional HDDR processing and after a modified HDDR procedure, i.e. a solid-disproportionation and a slow recombination in a partial hydrogen pressure, have been described. It was shown that the latter procedure induces magnetic anisotropy even in a ternary alloy and these findings have been discussed in terms of its implications on current models for the anisotropy mechanism in NdFeB-type alloys.

The application of 'extreme' hydrogenation conditions such as high hydrogen pressure or reactive milling in a hydrogen atmosphere at enhanced temperatures facilitates the disproportionation of thermodynamically very stable compounds such as $Sm_2Fe_{17-x}Ga_x$ ($x \geq 1$), $SmCo_5$ or Sm_2Co_{17} and thus extends the applicability range of the HDDR process. A simple qualitative thermodynamic description has been given in order to understand these phenomena.

The preparation of highly coercive magnet powders by reactive milling is a new processing route for the magnetic hardening of compounds which are highly stabilized towards the reaction with hydrogen and therefore disproportionate very sluggishly or, as in the case of the Sm-Co-type materials, do not react at all under standard HDDR conditions. Thus, an alternative approach for the magnetic hardening of Sm-Co-type materials, which normally require a time consuming and intricate annealing procedure using various alloy additions, is possible.

The full potential of this modified HDDR procedures regarding the tuning of one- or multi-phase nano-structures, possibly consisting of normally metastable phases, suitable for example for exchange-coupled coupling is yet to be explored.

ACKNOWLEDGMENTS

Thanks are due to Prof. IR Harris for stimulating discussions and the use of the HRSEM facility at the University of Birmingham and to Dr. V. Panchanathan (Magnequench) for the provision of MQA-T material. The support of this work by the Deutsche Forschungsgemeinschaft and Sächsisches Minsterium für Wissenschaft und Kunst is also acknowledged.

REFERENCES

[1] T. Takeshita and R. Nakayama, 10th Intl. Workshop on RE Magnets and their Applications, Kyoto, Japan, 551 (1989)

[2] P. J. McGuiness, X. J. Zhang, X. J. Yin and I. R. Harris, J. Less-Common Metals 158, 379 (1990)

[3] H. Nakamura, S. Sugimoto, M. Okada and M. Homma, Materials Chemistry and Physics 32, 280 (1992)

[4] J.M.D. Coey, Proc. of the Intl. Conf. on Ferrites (ICF96), Kyoto, Japan, 1992, p. 1064

[5] L. Cao, K.H. Müller, A. Handstein, W. Grünberger, V. Neu and L. Schultz, J. Phys. D: Appl. Phys. 29, 271 (1996)

[6] I. R. Harris, C. Noble and T. Bailey, J. Less-Common Met. 106, L1 (1985)

[7] O.Gutfleisch, N. Martinez, M. Verdier and I.R. Harris, J. of Alloys and Comp. 215, 227 (1994)

[8] A. Fujita and I.R. Harris, IEEE Trans. Magn., vol.30, no.2, 860 (1994)

[9] H. Nakamura, S. Sugimoto, T. Tanaka, M. Okada and M. Homma, J. Alloys and Comp. 222, 136 (1995)

[10] S. Sugimoto, O. Gutfleisch and I.R. Harris, J. of Alloys and Comp. 260, 284 (1997)

[11] N.M. Dempsey, P.A.P. Wendhausen, B. Gebel, K.H. Müller and J.M.D. Coey, Proc. of 14th Intl. Workshop on Rare-Earth Magnets and their Applications, Sao Paolo, Brasil 1996 p.349

[12] A. Kirchner, W. Grünberger, O. Gutfleisch, V. Neu, K.-H. Müller and L. Schultz, J. Phys. D: Appl. Phys. 31, 1660 (1998)

[13] T. Takeshita and R. Nakayama, 12th Intl. Workshop on RE Magnets and their Applications, Canberra, 670 (1992)

[14] I.R. Harris, Proc. 12th Intl. Workshop on Rare-Earth Magnets and their Appl., Canberra, Australia, 347 (1992)

[15] K.H.J. Buschow, IEEE Trans. Magn., vol. 30, no.2, 565 (1994)

[16] T. Tomida, N. Sano and M. Uehara, J. Appl. Phys. 81 (11), 7170 (1997)

[17] T. Tomida, N. Sano, K. Hanafusa, H. Tomizawa and S. Hirosawa, Acta mater. 47, no.3, 875 (1999)

[18] H. Nakamura, K. Kato, S. Sugimoto, M. Okada and M. Homma, Proc. of 15th Int. Workshop on RE Magnets and their Appl., Dresden, Germany, vol. 1, p. 507 (1998)

[19] O. Gutfleisch and I.R. Harris, Proc. of 15th Int. Workshop on RE Magnets and their Appl., Dresden, Germany, vol. 1, p. 487 (1998)

[20] J. Gao, X. Song and X. Wang, J. Alloys and Comp. 267, 270 (1998)

[21] O. Gutfleisch, M. Matzinger, J. Fidler and I.R. Harris, J. Magn. Magn. Mat. 147, 320 (1995)

[22] M. Kubis, O. Gutfleisch, K.H. Müller, I.R. Harris and L. Schultz, J. Appl. Phys. 83 (11), 6905 (1998)

[23] M. Kubis, L. Cao, K.H. Müller and L. Schultz, J. Phys. D: Appl. Phys. 30, L51 (1997)

[24] A. Handstein, M. Kubis, O. Gutfleisch, B. Gebel and K.-H. Müller, J. Magn. Magn. Mat. 192, 73 (1999)
[25] O. Gutfleisch, M. Kubis, A. Handstein, K.-H. Müller and L. Schultz, Appl. Phys. Lett. 73 (20), 3001 (1998)
[26] M. Okada, K. Saito, H. Nakamura, S. Sugimoto and M. Homma, J. Alloys and Comp. 231, 60 (1995)
[27] G. K. Williamson and W. H. Hall, Acta Metall. **1** (1953), 22
[28] J. M. D. Coey (Editor), *Rare Earth Iron Permanent Magnets*, Clarendon Press Oxford, UK (1996), p.35
[29] X.L. Yeh, K. Samwer and W.L. Johnson, Appl. Phys. Lett. 42, 242 (1983)

ISOTROPIC AND ANISOTROPIC Nd-Fe-B MAGNET DEVELOPMENTS

V. Panchanathan
Magnequench International, 6435 Scatterfield Road, Anderson, IN 46013

ABSTRACT

This paper deals with the magnetic and technological properties of isotropic Nd-Fe-B magnets over a wide range of Nd and B contents. They can be used to determine the optimum composition based on requirements. In anisotropic magnets, Nd and B ranges for production of high energy product rings was determined in Nd-Fe-Co-Ga-B system. This is also used in production.

INTRODUCTION

The processing of Nd-Fe-B magnets has been carried out by two different methods. One is the conventional powder sintering process and the other is the rapid solidification process. The sintering process produces only fully dense anisotropic magnets. The rapid solidification process makes both bonded isotropic magnets and fully dense isotropic and anisotropic magnets. The rapid solidification is carried out by melt spinning producing flake like particles having a highly stable and magnetically hard microstructure. They are ground into powder before being processed into bonded and fully dense magnets.

Bonded Magnets

The bonded Nd-Fe-B market is a significant growing fraction of total permanent magnet market worldwide. The reasons for the growth are the emergence of large developing economies, growth of PC and its related technologies, and government regulation for efficiency in appliances and automobiles. Bonded Nd-Fe-B magnets being an engineered material can be tailored to specific application needs [1]. Its flexibility in processing, shape, and magnetic properties contribute to its success in the market place. Improvements of hard magnetic properties, ease of processing and cost reduction have been the motivating factors for much of the research of Nd-Fe-B permanent magnet materials since they were discovered. The technical development activities in isotropic powders for bonded magnets have been concerned with improvements in remanence, reducing the losses and the squareness of the demagnetization curve. The improved squareness of the demagnetization curve is made possible by suitable process control. In this process the heat extraction from quench wheel is controlled to have uniform fine grain size (~ 30 nm) across the thickness of the ribbon. The aging losses in magnets can be reduced by the addition of Nb; however it slightly reduces the remanence. In fact additions of 2% Nb has been used in production of commercially available powders, with aging losses at 180° C being less than 5% in injection molded magnets of load line 2.0 [1-4]. The recent studies on the improvement of hard magnetic properties include a mixture of two different phases (a hard and a soft phase) with the aim of enhancing magnetic properties while reducing the overall Nd content. Compared to the standard magnet the microstructure consists of $Nd_2Fe_{14}B$ with either Fe_3B or α-Fe. In a magnetized state these materials have an enhanced remanence due to higher intrinsic magnetization of α-Fe or Fe_3B phases. Due to an exchange interaction with the neighboring magnetically hard $Nd_2Fe_{14}B$ grains, these phases retain their magnetized state in the presence of a reverse field. Various researchers have reported remanence values as high as 12 kG with intrinsic coercivity values of

Mat. Res. Soc. Symp. Proc. Vol. 577 © 1999 Materials Research Society

~3kOe [5-9]. A low coercivity powder (~4kOe) with easy magnetization behavior is now being produced commercially [10]. There is still scope for improved remanence and good coercivity material with less than stoichiometric composition. Hence a detailed investigation is carried out in Nd-Fe-B systems over a wide range of Nd and B contents.

Anisotropic Magnets

Anisotropic magnets are produced by hot pressing the melt spun ribbons followed by hot deformation. The processing occurs quite rapidly and at low pressures at temperatures above 725°C. The alignment mechanism is purely crystallographic and no alignment filed is applied. These magnets find applications in voice coils motors, speakers and industrial servo motors where high flux densities are required [11, 12]. Other benefits include improved response time, higher efficiency and improved performance in smaller package size. The applications for servo motors are wide and varied. Some examples are heavy industrial application, factory automation, machine tools, process equipment etc. Servo motors are the best method for accurate motor control for use in appliances, recreational products, etc.

The energy products of hot deformed magnets in production are lower than the sintered magnets. However there are several offsetting factors which favor the use of hot deformed magnets. It is known that sintered magnets are easier to magnetize compared to MQ3 magnets. However it shows constricted curve at low magnetizing field, which causes a phenomenon called wiggles in multi-pole magnetization. In fact sintered magnets need higher magnetizing voltage to decrease wiggles. Further the surface flux density distribution along the circumference and height is uniform in a hot deformed magnet compared to sintered magnet having similar remanence and coercivities. The losses are high in sintered magnets compared to hot deformed magnets for similar coercivities.

Rings of OD 10-225 mm width minimum wall thickness of 1.5 mm and height less than 1.5 times the OD can be made by hot extrusion process. In sintered process the ring OD is generally up to 150 mm with maximum wall thickness restricted to 5 mm and height less than 0.5 times OD. Radially oriented rings with energy product of 30-35 MGOe have been routinely used in application. The industry demands higher magnetic properties for further improvement in application. With that in mind a detailed investigation was carried out in Nd-Fe-Co-Ga-B system by varying Nd and B contents.

EXPERIMENT

Bonded Magnets

Alloys having varying amounts of Nd and B as shown in figure 1 were melt spun in a laboratory melt spinner. The Nd content was varied from 9% to 27% and B content from 0.7% to 4%. The compositions are in weight %. The ribbons were crushed into powder and annealed at 600 to 700°C for four minutes for optimum magnetic properties. The magnetic properties were determined using a VSM at room temperature with 0.333 demag correction factor. The properties of the powder were related to the Nd and B contents using Sigma plot. From selected annealed powders, compression bonded magnets were made using 2% epoxy as the binder. The following properties were evaluated for the bonded magnets:

- Recoil permeability values
- Magnetization behavior
- Aging behavior

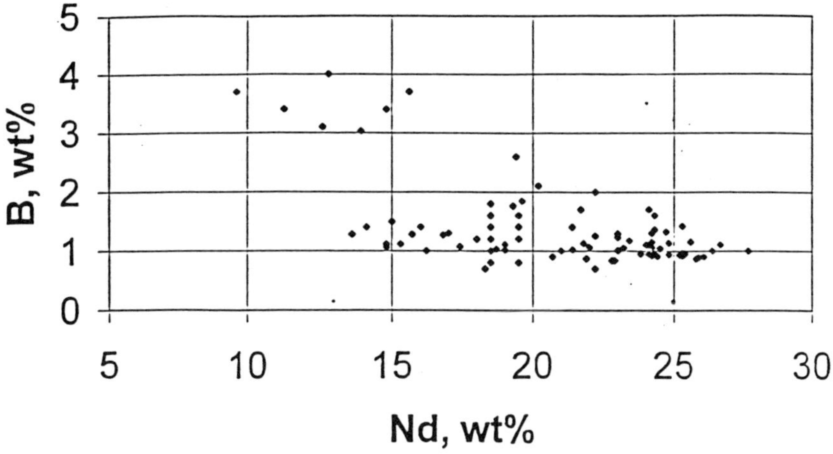

Fig. 1. Range of Nd and B contents studied.

The relationship between bonded magnet and powder properties was determined. The effect of Co, Ga and Nb on magnetic properties was studied.

Anisotropic Magnets

For this study Nd-Fe-Co-Ga-B system was used since the rings are normally made using this type of material. Nd was varied from 29-32% and B from 0.8 to 1.0%. Co and Ga additives were maintained constant respectively at 6% and 0.5% in all the trials. The alloy was melt spun, crushed into powder, compacted and hot pressed at 750°C. Rings with radial orientation of OD 18 mm and ID 13.5 mm were prepared by extruding at 800°C. The hot workability was evaluated from the extrusion length.

RESULTS

Magnetic Properties

The variation in magnetic properties (Br and Hci) for various Nd and B contents is shown in figures 2 and 3. From these figures the following general remarks can be made for varying B contents:

Nd, wt%	Br, kG	Hci, kOe
<14	~11	~2
14-18	~10	~2
18-22	~9	~3-5
22-24	~8-9	~5-7

Only at higher Nd contents, B content greater that 1.5% decreases both Br and Hci. In view of this a simplified relationship is drawn relating Nd content to Br and Hci as shown in figures 4 and 5. In these figures the B content was kept less than 1.5% especially for Nd contents exceeding 24%. The relationship can be expressed thus:

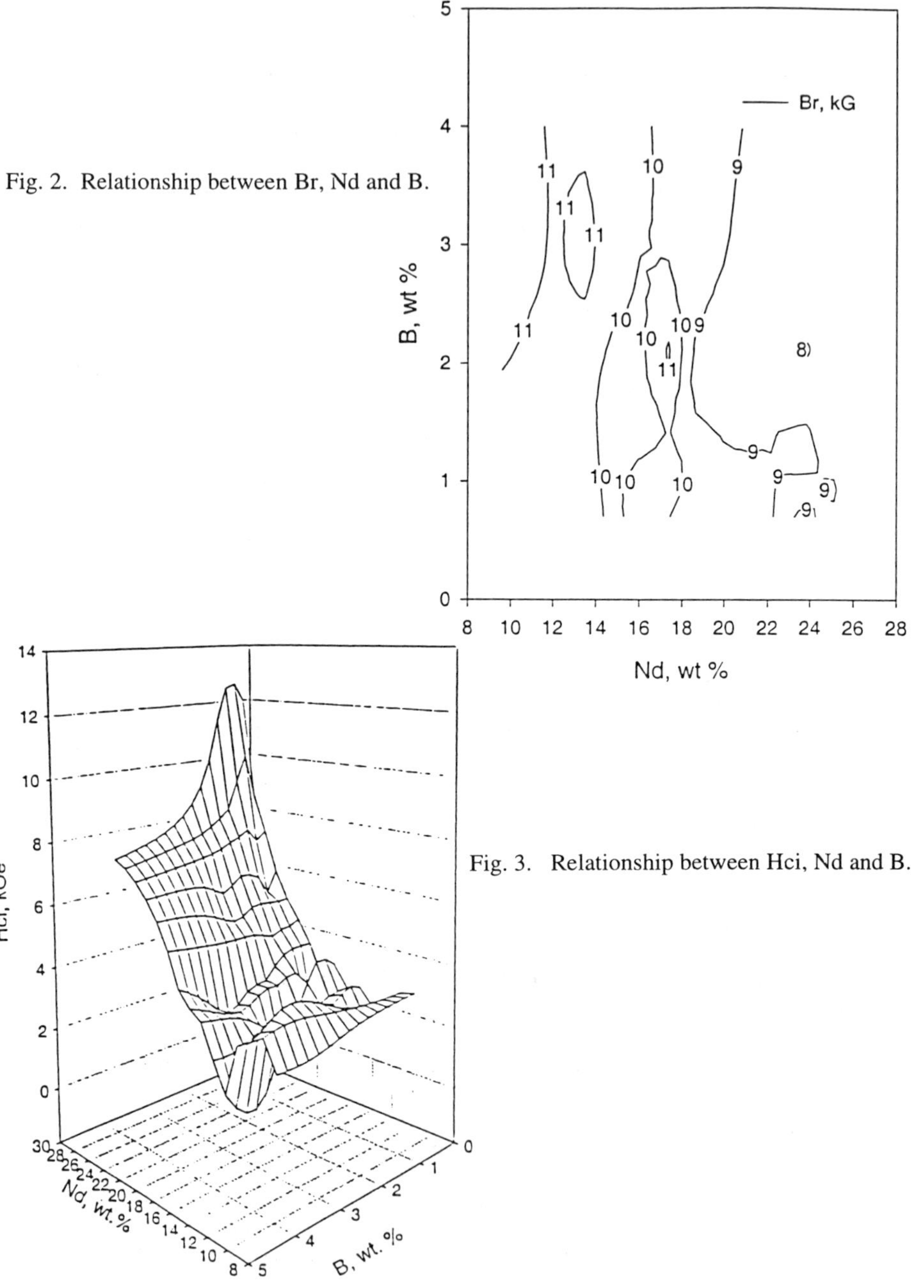

Fig. 2. Relationship between Br, Nd and B.

Fig. 3. Relationship between Hci, Nd and B.

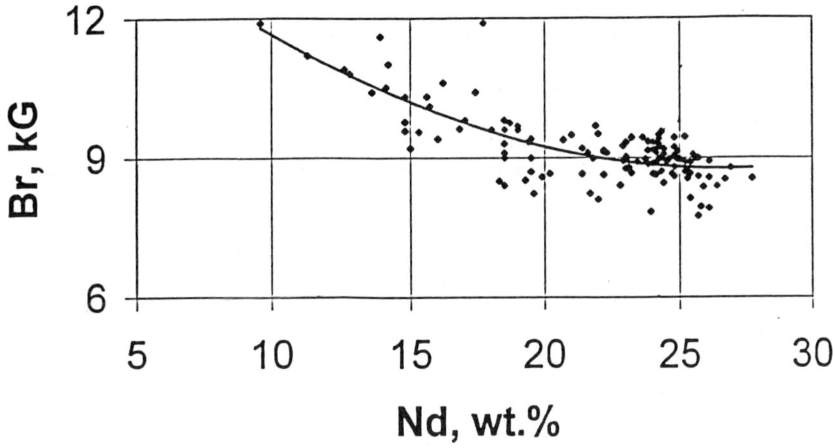

Fig. 4. Br variation with Nd.

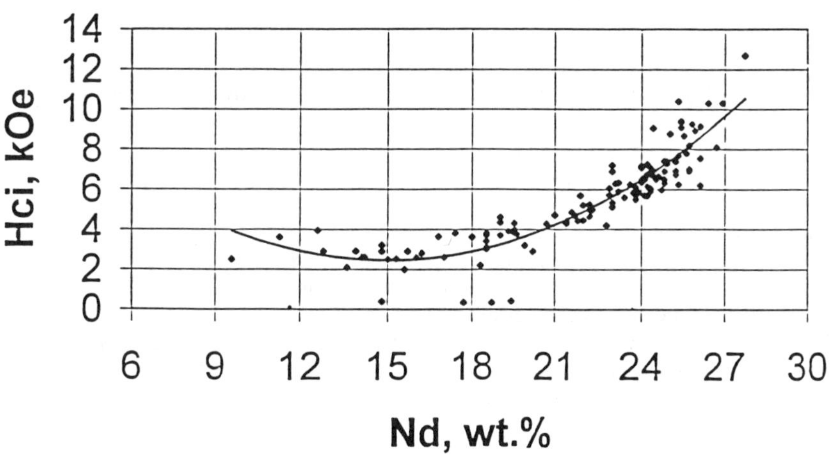

Fig. 5. Hci variation with Nd.

$$Br = 0.01\ (Nd)^2 - 0.55\ (Nd) + 16.1 \quad (1)$$
$$Hci = 0.05\ (Nd)^2 - 1.51\ (Nd) + 13.8 \quad (2)$$

Where Br is in kG, Hci is in kOe, Nd is amount of Nd in wt%. The R^2 value for Hci is 0.85 indicating that Hci is mainly dependent on Nd content; however for Br it is less at 0.65 indicating besides Nd content, the process variables may be responsible to its variability. These two figures can be used to narrow in on the final composition based on the requirements.

The effect of additives on magnetic properties was studied for Nd content 23-25%. The additives include Co (2%), Ga (0.5%) and Nb (1.0, 1.5 and 2.0%). Co and Ga did not appreciably affect the Br and Hci of the powders; however Nb especially at 2% considerably increased the Hci and decreased Br. As an example Hci increased from 7 to 9 kOe and Br dropped from 9.1 to 8.2 kG at 24% Nd. Nb additives are especially useful in reducing the aging losses as discussed later.

Technological Properties

Bonded magnets were made from powders having various Hci values. These magnets were evaluated for the following technological properties:

- Recoil permeability
- Saturation behavior
- Aging losses

Figure 6 shows the variation in recoil permeability values for various Hci ranges in systems having α-Fe or Fe_3B soft phases. The values for various Hci ranges are as follows:

Hci, kOe	Soft Phase	Recoil Permeability
2-4	Fe_3B	3.3 - 2.5
1-5	α-Fe	1.6 – 1.5
5-8	α-Fe	~1.4

Recoil permeability value is very important in application since lower values help in having higher flux in the reverse field.

The α-Fe containing material shows an improved saturation behavior compared to Fe_3B containing material [10]. The good saturation behavior is needed in applications where magnetization is very difficult, e.g. small multi-pole rings. Figure 7 shows the saturation behavior at 15 kOe for magnets having various Hci values in closed circuit. In general up to 95% saturation is possible with Hci values up to 6 kOe and it drops to 90% at 8 kOe. In an open circuit magnetization the values will be lower by about 5-10%.

Table 1 list the aging behavior of magnets having various Hci values; also included is the data for a magnet having 2% Nb. A study of the table indicates that in general higher the Hci lower the losses and vice versa. However it can be seen that 2% Nb additive in a magnet with Hci value of 6.4 kOe reduces the losses to about 7% at 125°C compared to 11% without Nb additives.

Powder – Magnet Br

Usually the powders are used to make bonded magnets. The Br of bonded magnet is related to the amount of binder. For compression molded magnet the amount of binder is ~2% by weight. One would expect the Br of magnet to be ~90% of the Br of the powder. Many other factors come into play. The two major factors are pore volume and inner magnetic shear

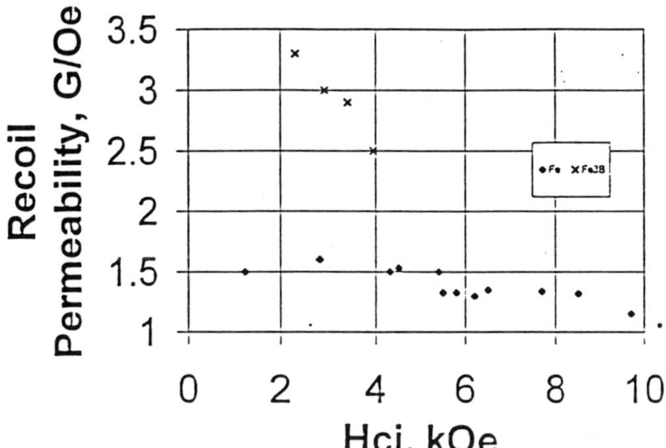

Fig. 6. Relationship of recoil permeability with Hci.

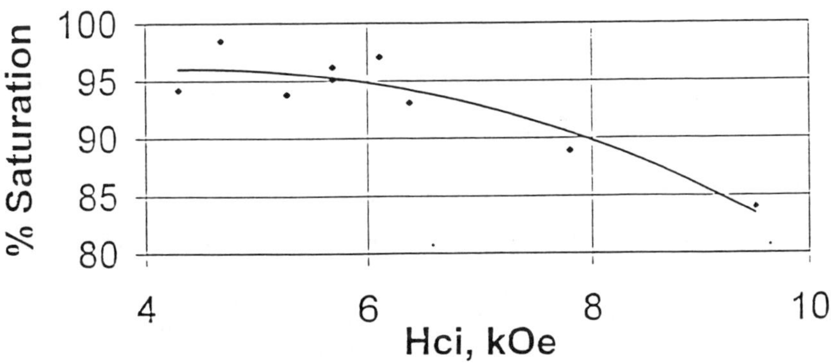

Fig. 7. % saturation of Br variation with Hci.

TABLE 1

AGING LOSSES OF MAGNETS AFTER 1000 HOURS

LOAD LINE 2.0

SYSTEM →	Nd-Fe-B				Nd-Fe-Nb-B
Hci, kOe →	5.0	5.8	6.6	7.2	6.4
TEMPERATURES, ^{O} C	LOSSES IN %				
80	10.3	7.8	7.5	5.0	4.0
100	13.6	11.3	9.1	8.2	6.8
125			11.0	11.0	7.2

loss. Pores reduce density and hence lead to low Br values. The other major weakening effect is due to "inner shear loss" [13]. The powders of the bonded magnets are diluted with binder and magnetic particles are insulated from each other by plastic. Every particle is in magnetically sheared condition. The sum of the shear conditions of the powder particles in bonded magnets result in flux loss of the magnet as a whole and is referred as "inner shear loss". This loss is high with higher binder content and higher value of (Br/coercivity). From this point low rare earth materials having $Nd_2Fe_{14}B$ with α-Fe or Fe_3B is likely to have lower Br value as a magnet. To study this aspect bonded magnets were made using powders having $Nd_2Fe_{14}B$ with both α-Fe and Fe_3B soft phases. Figure 8 shows the relationship between magnet Br and powder Br in compression molded magnets. There are clearly three regions; Br of powder is < 9 kG, between 9 and 9.7 kG and > 10 kG. They nearly correspond to soft phases being respectively 20 – 30%, 40 – 50%, and > 60%. A linear relationship between powder Br (Brp) and magnet Br (Brm) is given below:

$$8.5 < \text{Brp} < 9.0 \qquad \text{Brm} = 1.6\ \text{Brp} - 7.2 \qquad (3)$$

$$9.0 < \text{Brp} < 9.7 \qquad \text{Brm} = 1.6\ \text{Brp} - 8.0 \qquad (4)$$

Where Brp is greater than 10 kG there is no relationship and the properties are no better than those having Brp of ~ 9 kG. For the magnets made from powders having 2-14-1 phase only the properties are very nearly the same as those of 8.5 < Brp < 9.

Anisotropic Magnets

The hot workability of the material is an important criteria of hot deformed/extruded magnets. In the case of ring magnets it can be done by measuring the extruded length of a magnet as shown in figure 9. It can be seen that higher the extrusion length the better the hot workability and vice versa. The hot deformation/workability of the material is very much influenced by the amount of low melting point grain boundary phase which plays an important role in alignment [14]. The amounts of Nd and B influence the amount of grain boundary phase and the magnetic properties. In view of the very complex interrelationship a contour plot is obtained for extrusion length and energy product for various Nd and B contents respectively as shown in figures 10 and 11. From figure 10 it can be seen that rings can be extruded over a wide range of Nd and B contents. The energy product shows a wide variation from 272 to >320 kJ/m^3 depending on the Nd and B content. In fact energy product of >320 kJ/m^3 is

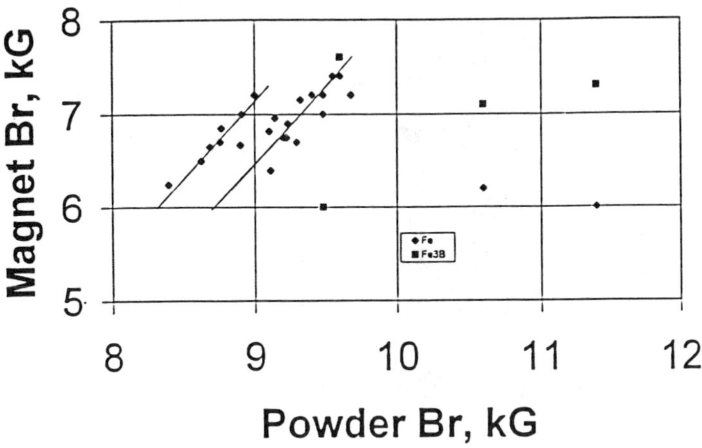

Fig. 8: Relationship between Magnet and Powder Br Values

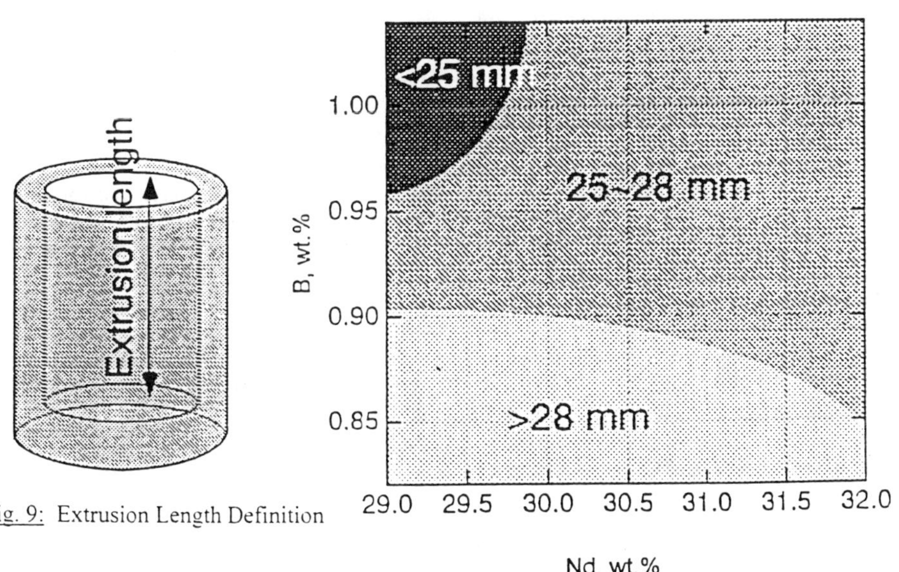

Fig. 9: Extrusion Length Definition

Fig. 10: Relationship of Extrusion Length with Nd and B

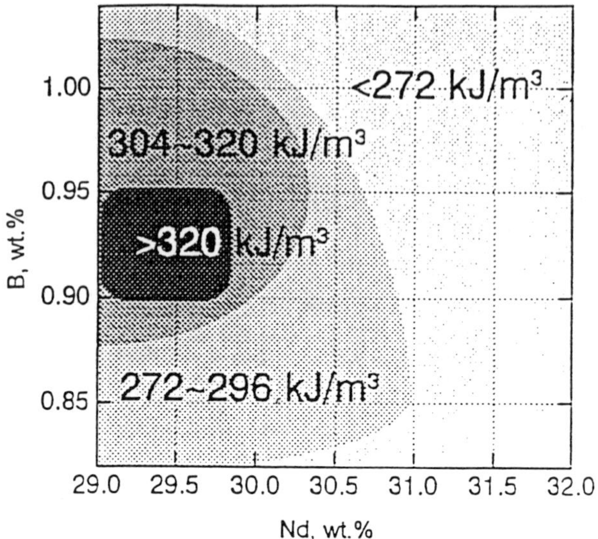

Fig. 11: Relationship of Energy Product with Nd and B

obtained only with Nd 29.0 – 29.8% and B 0.90 – 0.95% in the Nd-Fe-Co-Ga-B system. This is now used in production of ring magnets for industrial servo motors. The performance was improved by 5-6% compared to using a ring magnet with energy product of 280 kJ/m^3.

CONCLUSIONS

The magnetic properties (Br and Hci) of isotropic powders were related to varying Nd and B contents in Nd-Fe-B system. They can be used to narrow in on the final composition based on the requirements. Technological properties were evaluated over a wide range of Hci values. Nb addition at 2% was found to be effective in reducing the aging losses. The magnet Br value was related to powder Br values in systems having $Nd_2Fe_{14}B$ with either α-Fe or Fe_3B.

The energy product and extrusion length of anisotropic ring magnets were related to varying Nd and B contents in Nd-Fe-Co-Ga-B system. Specific composition for production of high energy product ring magnet was determined.

ACKNOWLEDGEMENTS

The author thanks Mr. B. Edwards, MQI, Anderson, USA and Dr. T. Nishio, Mr N. Yoshikawa and Dr. T. Iriyama of Daido Steel Co. Ltd, Nagoya, Japan for their help in the preparation of the manuscript.

REFERENCES

1. V. Panchanathan, Rare Earth Science, Technology and Applications III, Edited by R. Bautista, C.O. Bounds, T.W. Ellis and B.T. Kilbourn, The Minerals, Metals and Materials Society, p. 137 (1997).

2. S. Koyama, T. Nishio and V. Panchanathan, Denki–Seiko 69 (4), p. 229 (1998).

3. V. Panchanathan and K. Sparwasser, Proceedings of the 15th International Workshop on Rare Earth Magnets and Their Applications, Dresden, p. 671 (1998).

4. L.H. Lewis, K. Gallagher and V. Panchanathan, Paper #GD-12, MMM Conference, Miami, (1998).

5. T. Schrefl, H. Kronmuller and J. Fidler, J. Magn. Mag. Mat. 127, p. 273 (1993).

6. J. Ding, P.G. McCormick and R. Street, J. Magn. Mag. Mat. 124, p. 1 (1993).

7. S. Hirosawa and H. Kanekiyo, J. Appl. Phys. 73 (10), p. 6,488 (1993).

8. L. Withanawasam, G.C. Hadjipanayis, and R. Krause, J. Appl. Phys. 75, p. 6,646 (1994).

9. L.H. Lewis, K. Gallagher, B. Hoermann and V. Panchanathan, J. Alloy Compd. 270, p. 265 (1998).

10. V. Panchanathan, IEEE Trans. Mag. 31 (6), p. 3,605 (1995).

11. V. Panchanathan, Proceedings EIC/EMCWA Conference, Cincinnati, p. 173 (1998).

12. T. Iriyama, N. Yoshikawa, H. Yamada, Y. Kasai and V. Panchanathan, Denki–Seiko 69 (4), p. 219 (1998).

13. W. Ervens, Private Communication.

14. R.K. Mishra, J. Appl. Phys. 62, p. 967 (1987).

PRASEODYMIUM AND NEODYMIUM-BASED NANOCRYSTALLINE HARD MAGNETIC ALLOYS

H.A. DAVIES *, C.L. HARLAND * , J.I. BETANCOURT R.*, G. MENDOZA S.*+
*Department of Engineering Materials, University of Sheffield, Sheffield S1 3JD, UK. h.a.davies@sheffield.ac.uk
+Present address: CINVESTAV-Saltillo, Carr. Saltillo-Mty, Apdo. Postal 663, Mexico.

ABSTRACT

The magnetic properties of nanophase exchange enhanced RE-Fe-B based alloys ribbons prepared by melt spinning are presented and discussed. The alloy systems investigated include $Nd_yFe_{94-y}B_6$ and $Pr_yFe_{94-y}B_6$, for y within the range 6-20 at%, and the data are compared in the context of the larger anisotropy field for the $Pr_2Fe_{14}B$ phase in comparison with $Nd_2Fe_{14}B$. Experimental studies are also reported for nanophase (Nd-Pr)-Fe-B alloys with various proportions of Nd:Pr, to investigate the increase in coercivity of NdFeB by partial substitution of Nd by Pr, and for (Nd, Pr)-(Fe-Co)-B-based systems, to study the degree to which the remanence enhancement can be maintained for alloys in which Co substitutes for Fe to increase the thermal stability.

INTRODUCTION

Much of the research on melt spun nanophase RE-Fe-B hard magnetic alloys thus far has focused on Nd-based alloys [1-6], partly because these have been commercialised and are widely used in microcrystalline form [7]. Neodymium is more abundant than praseodymium in rare earth ores by a factor typically in the range 3-4 and, in principle, should be correspondingly cheaper. However, in practice, the relative cost is dependent on demand and, currently, Pr is actually less costly than Nd because it is not being exploited as effectively.

The nanophase NdFeB-based alloys exhibit, for mean $Nd_2Fe_{14}B$ crystallite diameters d_g smaller than ~ 45nm [4], exchange enhancement of remanent polarisation J_r relative to the expected value of $J_s/2$ for uniaxial, randomly oriented and non-interacting particles [8] (where J_s is the saturation polarisation). Generally, the smaller is d_g the greater is the degree of exchange interaction and enhancement of J_r and hence of $(BH)_{max}$. The nanocomposite alloys of sub-stoichiometric compositions (i.e., Nd concentrations < 11.76 at% and containing an increasing volume fraction of soft magnetic α-Fe phase with decreasing Nd content) manifest additional J_r enhancement because of exchange coupling also between $Nd_2Fe_{14}B$ and α-Fe crystallites which have a high J_s (2.2T) in comparison with the hard magnetic phase (1.6T) [5,6]. The lower rare earth content also reduces the alloy cost. A corollary of the increased J_r is a reduction in intrinsic coercivity ${}_iH_c$ because the resistance to reverse magnetisation is reduced by the exchange coupling effects at the $Nd_2Fe_{14}B$ crystallite boundaries [4-6]. This becomes especially important for low Nd contents (< 9at%) for which the volume fraction of soft phase is large since ${}_iH_c$ then becomes so small ($\leq$500kAm^{-1}) that it has a detrimental effect on $(BH)_{max}$ [6,9].

Mat. Res. Soc. Symp. Proc. Vol. 577 © 1999 Materials Research Society

It is known that $Pr_2Fe_{14}B$ has a significantly larger anisotropy field H_A than $Nd_2Fe_{14}B$ [10] and hence the coercivity is expected to be consistently larger, for a given d_g and level of J_r enhancement, providing the grain morphologies are the same in the two systems . This would be especially beneficial for the sub-stoichiometric nanocomposite compositions for which ${}_iH_c$ is attenuated by the presence of the soft phase. Thus, the magnetic properties of nanocomposite melt spun Pr-Fe-B alloys have been the subject of recent investigations [11-13]. There have also been studies of mechanically alloyed material. On the other hand, a disadvantage of the Pr-Fe-B alloys is that the Curie temperature T_c for $Pr_2Fe_{14}B$ is ~ 25 K lower than for $Nd_2Fe_{14}B$ which reduces the thermal stability and maximum useful operating temperature. This reduction in T_c can be compensated for through partial substitution of Fe by Co [15-17], providing it does not also result in a reduction in ${}_iH_c$ through a decreased H_A. Experimental studies indicate that nanophase Nd-FeCo-B alloys manifest levels of J_r and $(BH)_{max}$ enhancement, compatible with the corresponding ternary NdFeB alloys, in combination with substantially increased T_c [15-17].

It is also of interest to study nanophase Didymium-Fe-B alloys in which Nd is partly substituted by Pr to establish if increases in ${}_iH_c$ can be achieved, consistent with the concentration of Pr but without compromise of the J_r enhancement. Initial studies, albeit for samples with rather variable d_g between various alloys, suggest that this is achievable [18,19].

This paper compares the magnetic properties of nanocrystalline NdFeB and PrFeB alloy ribbons, in particular the influence of RE:Fe ratio spanning the composition range 6 to 20 at% RE. Similarly, Nd(Fe-Co)B and Pr(Fe-Co)B nanophase alloys are compared, both for single phase and nanocomposite variants. The influence of Nd-Pr mixtures in various proportions on the magnetic properties of nanophase Didymium-Fe-B alloys are also outlined. The results are discussed in terms of the anisotropy field, for the Nd and Pr 2/14/1 phases and of the relative magnitudes of the effects of Co substitution for Fe on T_c.

EXPERIMENTAL PROCEDURES.

Rare earth-iron-boron alloys of the type $RE_yFe_{94-y}B_6$ (at%), based on Nd, Pr or on Nd-Pr (Didymium) combinations of various atomic fractions and with various values of y within the range y = 6-20, were prepared by co-melting the pure constituents in an argon arc cold-hearth melting unit. Additionally, NdFeB and PrFeB based alloys, in which the Fe was partly substituted by Co, were prepared in similar way. The alloy ingots were turned over on the hearth and remelted up to six times to promote increased compositional homogeneity. Fragments of the ingots, typically of mass 5-10g, were then chill block melt spun onto a copper wheel rotating at speeds V_r, in the range 16-25m/s under an argon atmosphere, usually at 1 atm pressure. However, for the PrFeCoB and NdFeCoB alloys series, the alloys were spun in argon at 0.5 atm pressure in order to minimise the area fraction of entrapped gas pockets on the roll contact surface of the ribbons and hence to improve the uniformity of the grain diameters. This optimisation of the process was recently studied for NdFeB [20] and PrFeB ribbons [21]. It was shown to lead to improved magnetic properties and has now been implemented for all alloy series. For quartz nozzles with an orifice diameter of 0.6 mm and an argon gas propelling pressure of

0.2-0.25 atm, the ribbon thicknesses were typically within the range 20-35 μm, depending on V_r and alloy composition. The roll speed influenced the scale of the microstructure, notably d_g for the $RE_2Fe_{14}B$ and other phases, (d_g varying inversely with V_r) which, in turn, partly determined the degree of J_r enhancement for RE concentrations ≤14 at%.

The values of d_g for the $RE_2Fe_{14}B$ phase in each case were estimated by x-ray diffraction (XRD) line broadening analysis of non-overlapping diffraction peaks [22] and the size and distribution of primary and secondary phase particles were also observed, for selected samples using transmission electron microscopy (TEM). The XRD and TEM observations indicated that the crystallites were randomly oriented.

The magnetic properties (J_s, J_r and $_iH_c$) of individual pre-weighed pieces of ribbon, typically 3-5mm long and 1-2mm wide, were determined from J-H loops recorded using a vibrating sample magnetometer (VSM) coupled with a superconducting magnet having a maximum field of 5T. The ribbons were magnetised across their widths so that no corrections for self-demagnetisation were required. $(BH)_{max}$ was derived for the computed second-quadrant B-H loop in each case. T_c for the $RE_2Fe_{14}B$ phase was determined from the maximum in the ferro-/para-magnetic transition on the specific heat-temperature trace, recorded by differential scanning calorimetry, for alloys in the RE-(Fe-Co)-B and (Nd-Pr)-Fe-B series.

EFFECTS OF RE/Fe RATIO ON THE MAGNETIC PROPERTIES OF $Nd_yFe_{94-y}B_6$ AND $Pr_yFe_{94-y}B_6$ ALLOYS

The dependence of J_r, $_iH_c$ and $(BH)_{max}$ on RE concentration for the $Nd_yFe_{94-y}B_6$ [6] and $Pr_yFe_{94-y}B_6$ [11] alloys (for d_g corresponding to the maximum J_r in each case) are plotted in Fig 1 for RE concentrations y within the range 6-20 at%. The d_g, for the $RE_2Fe_{14}B$ phase ranged between 25 and 30 nm for the NdFeB series and somewhat wider, between 20 and 30 nm, for the PrFeB series. These data show clearly that for very similar J_r over the whole range of RE:Fe ratios, for the two systems and allowing for experimental scatter, $_iH_c$ is systematically larger for the Pr series than its Nd counterpart. The difference averages 20-30% and increases in absolute terms to ~400 kAm^{-1} at 14 at% RE although it narrows progressively for higher RE content. This difference in $_iH_c$ up to ~15 at% RE is consistent with the 30% larger H_A for the $Pr_2Fe_{14}B$ phase at room temperature (8.7 T vs 6.7 T)[10].

The J_r are also slightly larger for the Pr system for RE contents ≤ 13at% (although the values are not significantly different for larger y). The deviation increases progressively down to 8 at% RE, which was the measurement limit for the Nd system, although, in both cases, J_r rises continuously and is close to 1.3 T at y = 6 at% Pr for the PrFeB system. The larger J_r for PrFeB is reflected also in the $(BH)_{max}$ data for y down to 8 at%. However, the benefit in increased $(BH)_{max}$ arising from enhanced J_r is eventually limited by the decreasing $_iH_c$ at low RE contents and, below y ~ 8 at%, $(BH)_{max}$ drops precipitously as a result of a non-linear second quadrant B-H characteristic [9].

For a NdFeB non-stoichiometric alloy (with a minor amount of silicon dopant), the critical d_g corresponding to the appearance of significant J_r enhancement was found to be approximately 40 nm [4]. It is interesting to observe that the degree of J_r enhancement for similar d_g values is virtually identical for the two stoichiometric alloys (Fig.1)

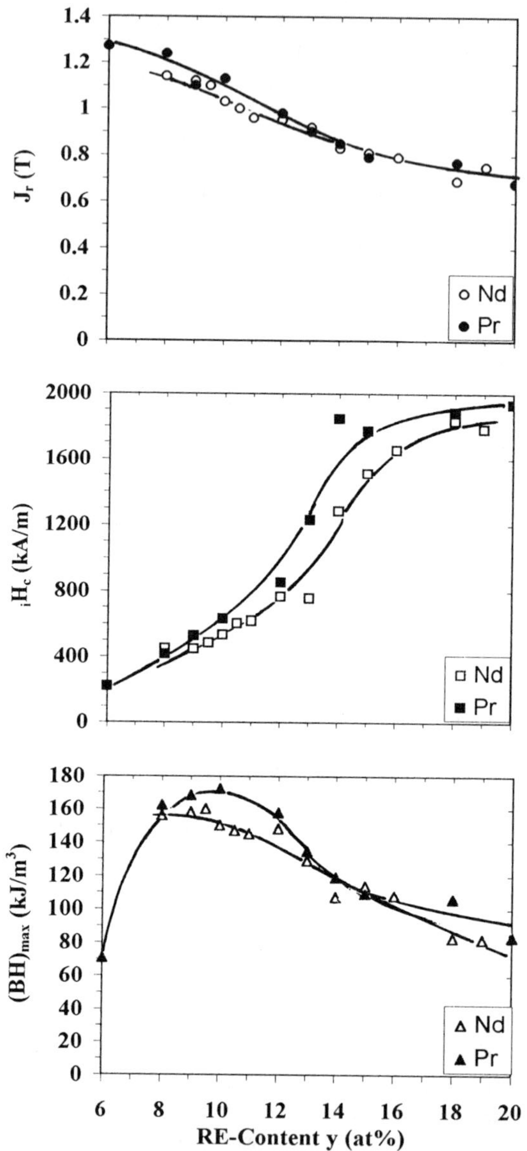

Fig 1. Effect of rare-earth metal concentration on the magnetic properties of nanocrystalline $Pr_yFe_{94-y}B_6$ and $Nd_yFe_{94-y}B_6$ alloy ribbons with d_g for $RE_2Fe_{14}B$ phase ~ 20-30 nm and ~ 25-30nm, respectively.

The enhancement of J_r above the respective Stoner-Wohlfarth values of $J_s/2$ [8] for the near-stoichiometric compositions $RE_{12}Fe_{82}B_6$ are due to exchange coupling between 2/14/1 unit cells in neighbouring crystallites, which extends over a length L_e , the exchange length.The effect on J_r should become noticeable when d_g is such that the exchange volume in unfavourably oriented crystallites, with respect to the magnetising direction, becomes a significant fraction of the grain volume.

The values of the exchange constants A are similar for the two 2/14/1 phases while H_A is about 30% larger for $Pr_2Fe_{14}B$ [10]. Thus, it would be expected that L_e for the $Pr_2Fe_{14}B$ phase would be approximately 15 % smaller than for $Nd_2Fe_{14}B$ (which is about 4nm) since $L_e \propto (A/K)^{1/2}$, where K is the anisotropy constant. Hence, on this basis, for a given d_g, J_r enhancement would be predicted to be smaller for the Pr 2/14/1 phase in contrast to the present experiemental observation.

The larger enhancement for the PrFeB system at low RE contents can be ascribed to a slight reduction in d_g with decreasing Pr [11], whereas it was independent of Nd concentration for NdFeB [6]. As has been noted previously, the increasing J_r with diminishing y can be ascribed to a progressively increasing volume fraction of soft magnetic α-Fe phase [5,6].The 2/14/1 crystallites exchange couple also to the α-Fe grains and, since L_e is substantially larger (~ 35 nm) for the latter than for the former and the α-Fe crystallites are smaller than the $Pr_2Fe_{14}B$ grains, all the moments in the α-Fe grains are aligned parallel by the exchange process. This results in the so-called 'exchange spring' phenomenon, a marke decrease in the recoil permeability [23].

The systematically higher ${}_iH_c$ for the PrFeB alloys (at least up to 15 at% RE) is consistent with the 30% larger H_A for the $Pr_2Fe_{14}B$ [8]. However, the overall shape of the ${}_iH_c$-y relationship is broadly similar in the two systems with a roughly linear decrease below y = 12 at% RE, reflecting the increasing volume fraction of the α-Fe phase and consequent reduction in the effective overall H_A.

Between 12 and ~15 at% RE, the rate of increase of ${}_iH_c$ with y accelerates in both systems, because of the presence of an increasing fraction of the paramagnetic RE-rich phase ($Nd_{73}Fe_{27}$ for the NdFeB system) which tends to decouple the $RE_2Fe_{14}B$ crystallites and thus hinders the process of domain reversal. The rate of increase of ${}_iH_c$ levels off sharply at ~ 15 at% Pr in the case of the PrFeB system, suggesting that the Pr-rich phase efficiently wets the 2/14/1 grain surfaces and that coverage is almost complete at 15at% Pr. The continuing decrease in J_r for y > 12 at% eventually to below $J_s/2$ reflects both the exchange decoupling of the nanocrystallites and the volume dilution of the 2/14/1 phase by the paramagnetic RE-rich phase.

The data for PrFeB alloys are broadly in agreement with those of Goll and Kronmuller [13]. In the latter case J_r are generally smaller than the present values over the range 8-12 at%Pr, in spite of systematically smaller d_g, while Goll's values of ${}_iH_c$ are also generally somewhat smaller.

It is clear from the comparison in Fig.1 that the nanophase PrFeB melt spun alloys give significantly better overall combinations of magnetic properties for a given mean diameter of the 2/14/1 crystallites and, between ~9 and 12 at% Pr, excellent combinations of $(BH)_{max}$ and ${}_iH_c$ are obtained. This has significant implications for the exploitation of the nanophase ribbon in polymer bonded magnets for technological applications, since the nanocomposite NdFeB alloys, especially, tend to have rather low coercivities.

EFFECTS OF COBALT SUBSTITUTION ON THE MAGNETIC PROPERTIES OF NdFeB AND PrFeB ALLOYS

Curie Temperature Enhancement

Substitution of Co for Fe in REFeB alloys is known to enhance the Curie temperature T_c. The effects on T_c of Co fraction x up to 0.2 in nanostructured stoichiometric $Nd_{11.7}(Fe_{1-x}Co_x)_{82.4}B_{5.9}$ and in composite $Nd_{10}(Fe_{1-x}Co_x)_{84.1}B_{5.9}$ alloys [16], with mean $Nd_2Fe_{14}B$ crystallite diameters d_g of 25-30 nm, are shown in Fig 2. The overall trend indicates ~100°C increase in T_c for each 0.1 fraction substitution of Fe by Co. The T_c values for the two alloy series are in very close agreement. This suggest that the Co is partitioning uniformly between the 2/14/1 hard phase and the α soft phase in the nanocomposite alloys. Thus, J_s for the α-FeCo phase will be expected to increase progressively with increasing x and to contribute to the J_r enhancement for the two-phase nanocrystalline alloys, at least up to x ~ 0.4.

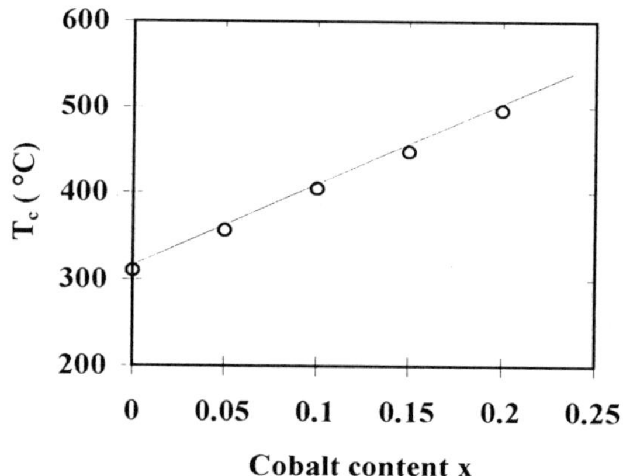

Fig. 2. Curie temperature for $Nd_2Fe_{14}B$ phase as a function of Co concentration x in nanocrystalline $Nd_{11.7}(Fe_{1-x}Co_x)_{82.4}B_{5.9}$ (continuous line) and $Nd_{10}(Fe_{1-x}Co_x)_{84.1}B_{5.9}$ (o symbols) alloys with d_g = 25-35 nm.

Magnetic Properties of Co-Substituted Nanophase NdFeB Alloys

The magnetic properties of nanostructured $Nd_{11.7}(Fe_{1-x}Co_x)_{82.4}B_{5.9}$ and $Nd_{10}(Fe_{1-x}Co_x)_{84.1}B_{5.9}$ alloys, with d_g in the range 25-35 nm [16], are plotted against x up to x = 0.24, in Figs 3(a) and 3(b), respectively. For the stoichiometric Nd series (11.7 at%

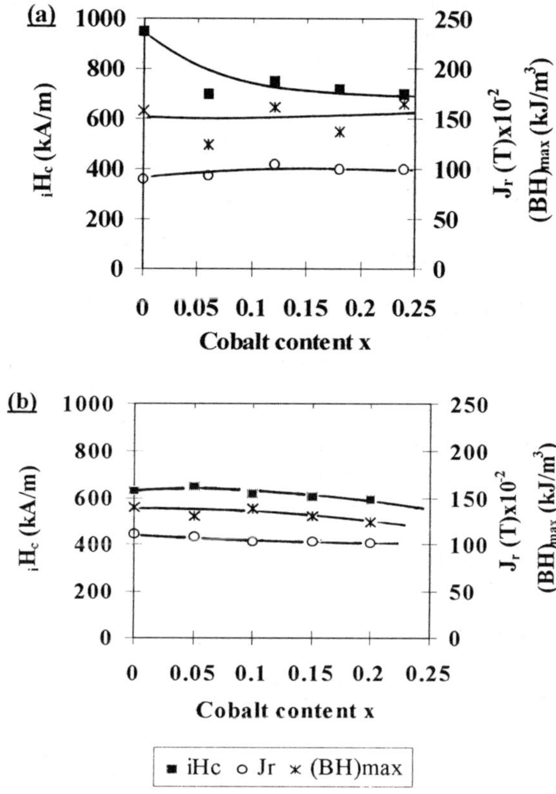

Fig 3. Effect of Co substitution for Fe on magnetic properties in nanocrystalline (a) $Nd_{11.7}(Fe_{1-x}Co_x)_{82.4}B_{5.9}$ and (b) $Nd_{10}(Fe_{1-x}Co_x)_{84.1}B_{5.9}$ alloy ribbons with d_g for $Nd_2Fe_{14}B$ phase ~ 25-30 nm

Nd), J_r enhancement (with values ~ 1.0T) was observed for all x. A shallow increase in J_r with x is apparent. The $(BH)_{max}$ data show considerable scatter but there is evidence that it also increases with x, consistent with the slightly rising J_r. The coercivity is relatively very high for the x = 0 base ternary alloy, the value being close to 1000 kAm^{-1}. It decreases quite sharply to ~ 750kAm^{-1} for x = 0.05 but remains approximately constant for higher x. Thus, the x = 0.24 alloy has an excellent combination of properties: $(BH)_{max}$ ~ 160kJm^{-3}, $_iH_c$ ~ 700kAm^{-1}, Tc ~ 550°C.

In the sub-stoichiometric alloy series, containing an estimated 15 vol % of α-(Fe-Co) phase, in addition to the hard magnetic $Nd_2(FeCo)_{14}B$ phase, J_r enhancement was systematically larger than for the stoichiometric series, consistent with the additional enhancement arising from the high J_s soft phase. However, J_r appeared to decrease slightly with increasing x so that the increase that was expected from the rising J_s of the soft-phase was apparently more than compensated for by a decrease in J_r of the hard phase, as Co enters the 2/14/1 lattice. This diminishing J_r was also reflected in a decreasing $(BH)_{max}$. The coercivity was much smaller for the base x = 0 alloy than for its stoichiometric counterpart but the overall decrease with increasing x was much shallower. The reasons for these differences are not yet clear but it is evident that the coercivities of the nanophase Co-containing alloys, beyond x = 0.1, are much less sensitive to a 2% decrease in RE concentration than is the case for the ternary alloy (x=0).

Magnetic Properties of Co-Substituted Nanophase PrFeB Alloys

The magnetic properties of nanostructured $Pr_{12}(Fe_{1-x}Co_x)_{82}B_6$ and $Pr_{10}(Fe_{1-x}Co_x)_{84}B_6$, alloys with d_g in the range 35 - 45 nm [23], are plotted against x up to x = 0.4 in Figs 4(a) and 4(b), respectively. For the near-stoichiometric Pr series (12at% Pr) strong J_r enhancement, (with values in the range 1.02 - 1.07 T) was observed for all x. J_r increased to ~ 1.07 T at x = 0.3 and then appeared to decrease for higher x. Excellent coercivities were also observed for all Co concentrations; ${}_iH_c$ increased from ~ 900 to ~ 1000kAm^{-1} between x = 0 and 0.1 and then showed a steady decrease for further increase of x, down to ~ 650kAm^{-1} at x = 0.4. The $(BH)_{max}$ values for alloys in the range x = 0.1 - 0.3 are ~ 190kJm^{-1} which reflect the large J_r enhancement and good J-H second quadrant loop shapes. Hence, the combination of $(BH)_{max}$, ${}_iH_c$ and T_c for the x = 0.1 and 0.2 alloys, in particular, are excellent, viz. ~ 195kJm^{-3} / 1050kAm^{-1} / 385°C and ~ 190kJm^{-3} / 950kAm^{-1} / 485°C for x= 0.1 and 0.2, respectively. Such large $(BH)_{max}$ are rather surprising considering that the level of grain refinement achieved was less than for the other alloy series discussed in the previous sections (i.e. d_g was larger).

For the nanocomposite series (10 at% Pr) (Fig 4(b)) [23], Jr enhancements are systematically larger than for the 12 at% series, owing to the presence of the α-(Fe-Co) phase (with J_r being 1.20T for x = 0), but, in contrast, decreased gradually with increasing x (to ~1.12T at x = 0.4). The ${}_iH_c$ were, as expected, substantially smaller than for the corresponding 12 at% Pr series (Fig 4(a)), due to reduced overall H_A resulting from the exchange coupling between the hard and soft phases. Both ${}_iH_c$ and $(BH)_{max}$ decrease with increasing x, that of $(BH)_{max}$ reflecting the declining J_r. Nevertheless, the combination of properties, for instance for the x = 0.2 alloy, are again excellent with $(BH)_{max}$ ~ 180kJm^{-3}, ${}_iH_c$ ~ 575kAm^{-1} and T_c = 485°C.

NANOPHASE DIDYMIUM (Nd-Pr)-Fe-B ALLOYS

Stoichiometric $(Nd_{1-x}Pr_x)_{12}Fe_{82}B_6$ Alloys

The Curie temperature T_c of the $RE_2Fe_{14}B$ phase for $(Nd_{1-x}Pr_x)Fe_{14}B$ single phase alloys, with d_g in the range 26-32 nm is shown in Fig 5 as a function of fractional substitution of Nd by Pr_x. The decrease from 311°C to 287°C between $Nd_2Fe_{14}B$ and $Pr_2Fe_{14}B$ is non-linear, the initial rate of decrease up to x ~ 0.25 being less steep. This is broadly in agreement with previous T_c data for a series of nanocomposite PrNdFeB ribbons containing 10.7 at% RE [18].

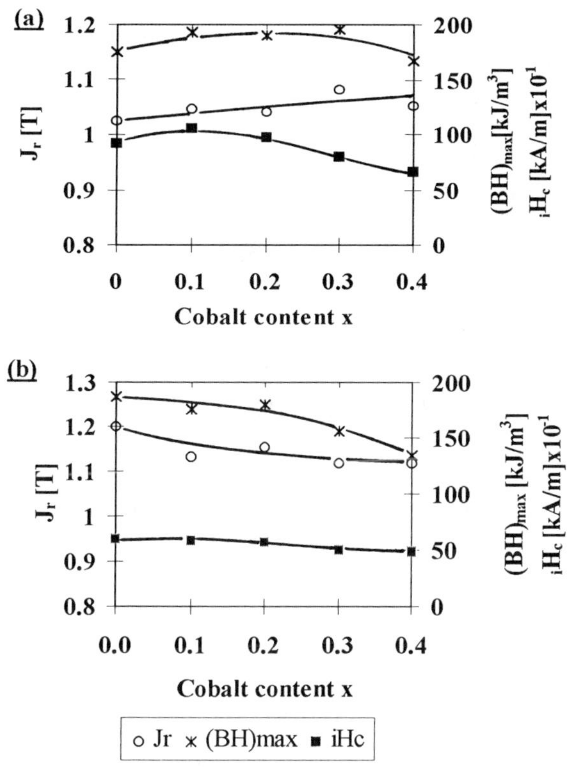

Fig 4. Effect of Co substitution for Fe in nanocrystalline $Pr_{12}(Fe_{1-x}Co_x)_{82}B_6$ and $Pr_{10}(Fe_{1-x}Co_x)_{84}B_6$ alloy ribbons with d_g for $Pr_2Fe_{14}B$ phase 35-45 nm and 35-40 nm, respectively.

The magnetic properties J_r, $_iH_c$ and $(BH)_{max}$ for the $(Nd_{1-x}Pr_x)Fe_{14}B$ series of stoichiometric alloys are plotted in Fig 6 for an approximately constant d_g within the range 26-32 nm. In a previous paper, data were presented for alloys having more widely varying d_g between the various compositions [19] which resulted in substantial scatter. The narrowing of the grains size range has resulted in smoother composition dependencies for all three properties. The $_iH_c$, as expected, increases progressively and roughly linearly between the Nd and Pr end points, reflecting the rising H_A with increasing Pr substitution into the 2/14/1 unit cell. The enhanced J_r is roughly constant at ~ 0.95 T, across the system, consistent with the constant d_g, except for the 0.25Nd / 0.75Pr alloy for which it is substantially larger at 1.05T. The reason for this is unclear as the estimated d_g was nominally the same for all alloys. This larger J_r is reflected also in a large $(BH)_{max}$ of 180kJm^{-3}. Since this alloy has $_iH_c$ of 800 kAm^{-1}, it shows an outstanding combination of magnetic properties. $(BH)_{max}$ for the 0 and 0.25 Pr fraction alloys are < 150kJm^{-3}, principally because of inferior loop shape, probably a result of a greater variation of grain size within the ribbon section.

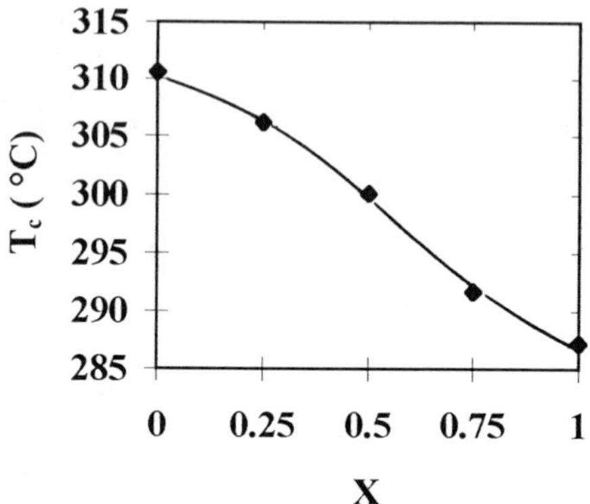

Fig 5. Effect of Pr substitution for Nd on the Curie temperature for nanocrystalline $(Nd_{1-x}Pr_x)_{12}Fe_{82}B_6$ alloys with $d_g \sim$ 30nm.

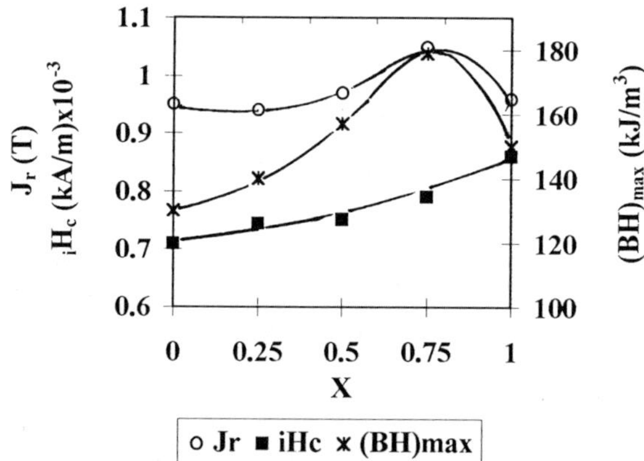

Fig 6. Influence of Pr fraction x on the magnetic properties of nanophase $(Nd_{1-x}Pr_x)_{12}Fe_{82}B_6$ alloy ribbons with d_g for the $RE_2Fe_{14}B$ phase in the range 26-32 nm.

$(Nd_{0.75}Pr_{0.25})_yFe_{94-y}B_6$ alloy series

Nanophase alloys, again with d_g for the $RE_2Fe_{14}B$ phase in the range 26-32 nm, were investigated for the series $(Nd_{0.75}Pr_{0.25})_yFe_{94-y}B_6$ for four values of y: 8, 10, 12 and 14 at% RE (Fig. 7). The 0.75 Nd: 0.25 Pr ratio was chosen for initial study on the basis that this corresponds roughly with the ratio in RE ores and also that the decrease in T_c , compared with NdFeB, is relatively small. $_iH_c$ increases progressively more steeply with increasing RE:Fe ratio, especially beyond 12 at% RE where the second phase becomes the paramagnetic rare earth-rich phase located around the $RE_2Fe_{14}B$ grain boundaries. This decouples the 2/14/1 crystallites and the behaviour is analogous to that of the PrFeB and NdFeB systems (Fig.1). Similarly, the increasing enhancement of J_r above $J_s/2$ between 12 and 8 at% RE shows the effect of the growing volume fraction of α-Fe, by analogy with its NdFeB and PrFeB counterparts (Fig.1).

Although J_r is generally intermediate between those for the Nd and Pr variants for all RE concentrations studied, $_iH_c$ somewhat surprisingly, is generally lower than for the corresponding Nd and Pr alloys, in spite of the higher predicted H_A for the $(Nd\text{-}Pr)_2Fe_{14}B$ than for the $Nd_2Fe_{14}B$. The $(BH)_{max}$ are slightly inferior to both the Nd counterparts and more so in comparison with the Pr counterparts (Fig 1) for 8, 10 and 12 at% RE alloys, in spite of the intermediate J_r enhancements.

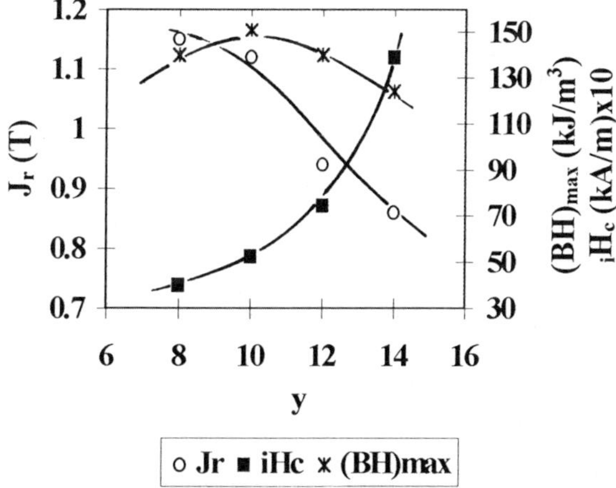

Fig 7. Influence of RE (Pr+Nd) concentration y on the magnetic properties of nanophase $(Nd_{0.75}Pr_{0.25})_yFe_{94-y}B_6$ alloy ribbons with d_g for $RE_2Fe_{14}B$ phase in the range 26-32 nm.

CONCLUSIONS

1. Substantial enhancements of J_r and $(BH)_{max}$ can be induced in melt spun REFeB alloys of stoichiometric composition by refining the mean $RE_2Fe_{14}B$ grain size d_g below ~ 45 nm. This results from intergrain exchange coupling and the effect increases with decreasing d_g due to an increasing volume fraction of coupled material in the grains.
2. Further enhancement of J_r and $(BH)_{max}$ occurs for sub-stoichiometric alloys (RE contents < 11.7 at%) owing to additional exchange coupling between the $RE_2Fe_{14}B$ and α-Fe soft magnetic grains, having a higher J_s. The effect on J_r increases progressively with decreasing RE content as the volume fraction of α-Fe increases. The J_r enhancement is accompanied by a reduction in ${}_iH_c$ due to reduced resistance to domain reversal. However, ${}_iH_c$ for PrFeB alloys is systematically larger (by 20 - 30%) than for corresponding NdFeB alloys having the same d_g, due to a larger H_A for the $Pr_2Fe_{14}B$ phase.
3. Substitution of Fe by Co leads to increased T_c in all cases without major adverse affects on the other magnetic properties. For nanophase PrFeCoB alloys, in particular, outstanding combinations of $(BH)_{max}$, ${}_iH_c$ and T_c are obtained.
4. Substitutions of Nd by Pr increase ${}_iH_c$ for the nanophase alloys without substantial adverse effects on the J_r / $(BH)_{max}$ enhancements.

ACKNOWLEDGEMENT

The research was supported in part by the Engineering and Physical Sciences Research Council through the Advanced Magnetics Programme. JIBR and GMS acknowledge the award of scholarships by DGAPA, UNAM and by CONACYT, respectively.

REFERENCES

1. J.E. Keem, G.B. Clemente, A. M. Kadin and R.W. McCallum, in: Hard an Soft Magnetic Materials with Applications including Superconductivity, Proc. Conf. ASM Materials Week 87, ed. J.A. Salsgiver (American Society for Metals 1987), p. 87
2. R. Coehoorn, D. B. DeMooij, J.P.W.B. Duchateau and K.H.J. Buschow, J. Physique **49**, p. 669 (1988).
3. G.C. Hadjipanayis and W. Gong, J. Appl. Phys. **64**, p. 5559 (1988)
4. A. Manaf, R.A. Buckley, H.A. Davies and M. Leonowicz, J. Mag. Magn. Mater. **101**, p. 360 (1991)
5. A. Manaf, R.A. Buckley and H.A. Davies, J. Magn. Magn. Mater. **128**, p. 302 (1993)
6. H. A. Davies, J. Magn. Magn. Mater. **157/8**, p. 11 (1996)
7. J.J Croat, J.F. Herbst, R.W. Lee and F.E. Pinkerton, J.Appl. Phys. **55**, p.2078 (1984)
8. E.C Stoner and E.P. Wohlfarth, Phil. Trans. Roy. Soc. **A 240**, p. 599 (1948)
9. A. Manaf, P.Z. Zhang, I. Ahmad, R.A. Buckley and H. A Davies, IEEE Trans. Magn. **29**, p. 2866 (1993).
10. S. Hirosawa, Y. Matsuura, H. Yamamoto, S. Fujimura, M. Sawaga and H. Yamauchi. J. Appl. Phys. **59**, p. 873 (1986)

11. G. Mendoza and , H.A.Davies, J. Alloys and Compounds. **281**, p. 17 (1988).
12. D. Goll, M. Seeger and H. Kronmuller, J. Magn. Magn. Mater.**185**, p. 49 (1998)
13. D. Goll and H. Kronmuller, Proc. 15th Int. Workshop RE Magnets and their Applications , eds. L. Schultz and K-H Muller. (MATINFO, Frankfurt 1998), p. 189.
14. J. Ding, P.G. McCormick and R. Street, J. Magn. Magn. Mater. **124**, p. 1 (1993).
15. J.F. Liu and H.A. Davies, J. Magn. Magn. Mater. **157-158**, p. 29 (1996).
16. R.Yapp, H. A. and Davies, Proc. 14th Int.Workshop on Rare-Earth Magnets and their Applications, eds. F.P. Missel et. al. (World Scientific Singapur 1996), p. 563..
17. R. Yapp and H. A. Davies, Proc. 15th Int. Workshop RE Magnets and their Applications , eds. L. Schultz and K-H Muller. (MATINFO, Frankfurt 1998), p. 315
18. W.P. Sequeira, S.J. Dobson, R.A. Buckley, and H.A. Davies, Mat. Lett. **15**, p. 376 (1993).
19. J. I. Betancourt R. and H. A. Davies, J. Appl. Phys. **85**, (1999) in the press.
20. R. Yapp, H. A. Davies, F. Leccabue and B.E. Watts, Mater. Lett. **38**, p. 33 (1999).
21. C.L. Harland, H.A. Davies, B.E. Watts and F. Lecabbue, 15th Int. Workshop RE Magnets and their Applications , eds. L. Schultz and K-H Muller. (MATINFO, Frankfurt 1998), p. 263.
22. G.E. Carr, H.A. Davies and R.A. Buckley, Mater. Sci. Eng. **99**, p. 147 (1988)
23. F.F. Kneller and R. Hawig, IEEE Trans. Magn. **27**, p. 3588 (1991).
24. C.L. Harland and H.A. Davies, to be published.

SURFACE COATING OF HDDR PROCESSED AND MECHANICALLY ALLOYED POWDER BASED ON RE-TM

*S. KOBE, *S. NOVAK, *P.J. McGUINESS, **Đ MANDRINO, **M. JENKO
*Jožef Stefan Institute, 1001 Ljubljana, Slovenia, spomenka.kobe@ijs.si
**Institute of Metals and Technology

ABSTRACT

Chemical surface modification can be used as a method for corrosion protection of sensitive powders based on intermetallic alloys between rare earth and transition metals. Surface coating is used for preventing fine powders, based on Nd-Fe-B, or $Sm_2Fe_{17-x}Ta_xN_{3-\delta}$ prepared by HDDR processing and mechanically alloying, from hydrolysis. Powders coated by chemisorbed organic substance, after exposing to a humid atmosphere, do not show any chemical or physical change.
Different coating agents were used and the sufficient amount of various materials was optimised with the emphasis on minimising their quantity. Simple experiment shows that the surfactant is successfully chemisorbed on the powder surface and that the coated powders are hydrophobic indefinitely. Magnetic properties were measured on samples after they were exposed to the same corrosion tests. Measurements on coated and bonded samples were compared with the measurements of non-coated and bonded samples. By using Auger electron spectroscopy the thickness of the coating was controlled. In order to distinguish the nature of the bonding between the powder surface and the surface-active substance FT-IR spectroscopy in absorbency and diffuse reflection modes was used. The protection of the fine particles is based on the formation of a covalent bond between the hydroxyl groups at the particle surface and the surface-active substance. The monomolecular layer of organic substance does not damage the magnetic properties of the powder, but successfully protect the powder against humidity.

INTRODUCTION

Several routes can be used to prepare high coercivity powders of Nd-Fe-B or $Sm_2Fe_{17-x}Ta_xN_{3-\delta}$, which can be used as a basic material for bonded magnets. For achieving high coercivities, in addition to composition the average particle size, grain size and its distribution are the most important parameters. Single domain sized grains (~300 nm) are preferable. For the production of nanocrystalline powders, in addition to melt spinning, the hydrogenation, disproportionation, desorption, recombination (HDDR) process (1,2) and mechanical alloying (3) can be used successfully as preparation methods for obtaining nanocrystalline powders without any further milling.

It is well known that magnetic powders based on rare-earth and transition metals are very sensitive to the atmospheric conditions and one of the tasks of many researchers was, and still is, to improve the properties of the basic material by lowering the sensitivity of the bulk powder itself. The addition of small quantities of Co and Al improved the corrosion resistance of Nd-Fe-B magnets (4). Also refractory elements like Zr (5,9) and V (6) or ZrO_2 can achieve this in sintered magnets (7,8) and HDDR processed powders (10). The addition of Zr or ZrO_2 has been the subject of our further research (11), together with the study on additional improvement of corrosion resistivity in powders, which is the basis of this paper.

Based on the results achieved with other materials (12) and our preliminary experimental results, the idea was to improve the basic resistivity of Nd-Fe-B powder not only by the additives, but also by coating individual particles with a very thin protective layer of organic molecules. It is important that the magnetic properties of the powders processed either by the HDDR processing route or mechanical milling (MM) should not be damaged by the coating

Mat. Res. Soc. Symp. Proc. Vol. 577 © 1999 Materials Research Society

process. The resistivity against corrosion should be improved to the extent that very fine powders can be handled without any protective atmosphere on the production level and that this procedure also should not significantly increase the production costs.

Surface modification of powders is based on the formation of a covalent bond between the hydroxyl groups that are usually present at metal or metal oxide surfaces, and surface modifying agents, such as long-chain carboxylic acids or silanes. The surface-active agent is first physically adsorbed onto the powder surface while during heating, condensation takes place, resulting in the formation of a covalent bond. After the reaction, the powder surface is covered with a mono- molecular layer of organic molecules. According to the literature data this method was successfully used to prevent powder agglomeration and to improve suspension stability as well as to prevent the chemical reaction of different powders with environmental humidity. It was reported (13) that the pressing behaviour of the alumina powder was improved by surface modification with carboxylic acids, and that the effect increases with the acid chain length. Coating with carboxylic acids was also used for hydrophobization of alumina powder in a paraffin suspension in order to prevent humidity-induced destabilisation of the suspension for low-pressure injection moulding (12). Zhang and co-authors (14) used n-octyltriethoxysilane to significantly improve the fluidity of a silicon and silicon nitride injection moulding blend. One of the most significant results of coating was obtained with aluminium nitride, which is extremely sensitive to water. The protection of AlN with stearic acid even enables the processing to take place in water since, as reported by Gorter (15) no reaction was detected for at least 72 hours. Different methods were used to control the reactivity of as -received and coated AlN powder: measurement of pH of the aqueous suspension (during the hydrolysis NH_3 forms) (16-18), thermogravimetric analysis (16), X-ray analysis (17), XPS and FTIR (12,18). In general, the results confirmed the presence of a chemical bond between the surface-active agent and the powder surface, which suggests that the presence of hydroxyl groups on the powder surface is a pre-requisite for a successful treatment.

EXPERIMENT

Different powders: $Nd_{16}Fe_{76}B_8$, $Nd_{15}Dy_1Fe_{76}B_8$, $Nd_{15}Dy_1Fe_{75.3}Zr_{0.7}B_8$ or $Sm_2Fe_{17-x}Ta_xN_{3-\delta}$, were produced from 5 kg batches by HDDR processing or high energy milling. The compositions were chosen on the basis of our previous work (10, 21). For HDDR processing the alloys were exposed to a hydrogen atmosphere in a specially constructed rotating furnace. Detailed processing parameters were published previously (19). As a parallel method for preparing nanocrystalline powders high energy milling in a planetary mill and an Anutech high-energy mill were used. After the milling process the powders were annealed at different temperatures (650°C to 850°C) to optimise the magnetic properties of crystalline powders. After HDDR processing the resulting powder was hand crushed. Both powders HDDR processed and mechanically milled were coated with a thin organic layer. Silanes were used for this purpose. The experimental procedure is published elsewhere (20). Coated and non-coated powders prepared by different routes and with different compositions were exposed to atmospheric conditions as well as more severe corrosive tests in a humidity chamber (Weiss) (95 % relative humidity and 40°C, 8 hours; ~ 40 % relative humidity and 25°C, 16 hours) in accordance with the DIN 50017 standard. The weight increase was followed and after the corrosion tests were completed, magnetic properties were measured and compared with the properties of the basic powders. The morphology of the powders was observed using a JEOL 840A SEM. The surface of the coated powders was analysed using high-resolution field emission AES (Microlab 310-F VG Scientific) combined with XPS. AES analysis was performed under standard conditions: a primary electron beam of 5 keV and the current between 1 and 10 nA. AES point analyses were performed on different grains of powder. Points for AES analysis were selected from SEM images with magnifications between 200 and 1000. XPS (X-ray Photoelectron Spectroscopy) of images with magnifications between 200 and 1000. XPS (X-ray Photoelectron Spectroscopy) of the same sample was used to determine the chemical state of the silicon. For XPS analysis MgKα radiation was used. In order to determine the nature of the bonding between the powder surface and the surface-active substance, FT-IR spectroscopy (Digilab FTS-80 spectrometer) in absorbency and diffuse reflection modes was used.

RESULTS AND DISCUSSION

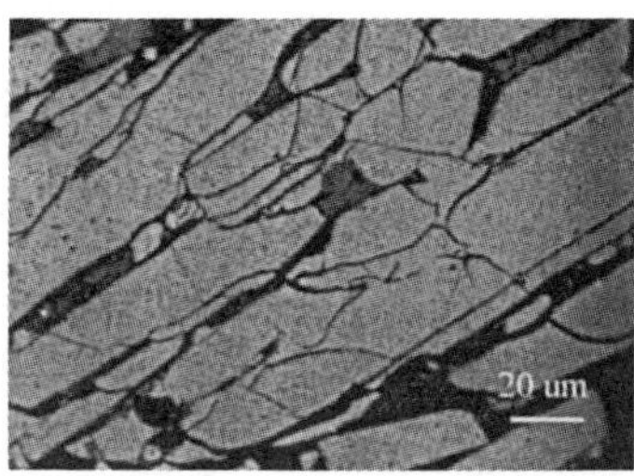

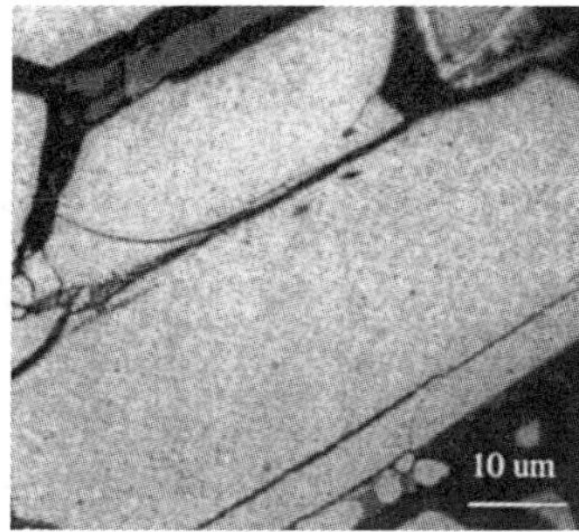

Fig. 1. Image from the optical microscope in a polarised light showing the polished surface of the HDDR processed powder ($Nd_{15}Dy_1Fe_{76}B_8$)) with a visible domain structure at higher magnification.

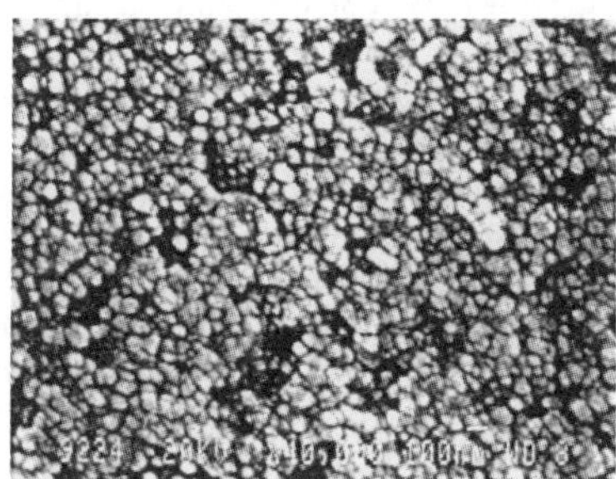

Fig. 2. SEM image of HDDR processed powder ($Nd_{15}Dy_1Fe_{75.3}Zr_{0.7}B_8$).

Fig. 5. FT-IR spectrum of the coated powder.

Figure 1 shows the image of the powder observed by optical microscopy in polarised light showing the polished surface of the HDDR processed powder with a visible domain structure at higher magnification. A SEM image of the powder is shown In Figure 2. The thin organic layer cannot be seen from this image, but a simple experiment shows that the surfactant is successfully chemisorbed on to the powder surface. If the powder is mixed with water, uncoated powder sinks immediately, but coated powder is hydrophobic and therefore floats on the water. Powders coated by chemisorbed silanes are hydrophobic indefinitely, while powders with only physisorbed organic molecules are not sufficiently resistant to the water and sink after a finite time. It is known from previous work that if the temperature used for the chemical reaction is too low, the water molecules replace the organic molecules (12). With AES analysis the presence of C and Si on the powder surface was detected. Adsorbed carbon from the air was removed from the sample surface by Ar ion sputtering. XPS analysis of the same sample showed that peaks of Si 2s at 153.1 eV and Si 2p at 101.4 eV correspond to Si in silanes. Si (2s 153.1 eV) is shown in Figure 3. Further analyses (Figure 4) showed besides Si (2p 101.4 eV) also Fe (3s 92.6 eV),

powder. AES and XPS analyses show that the coated powder is covered with an ultra thin layer of organic material and one can estimate the thickness of this coating as a monolayer.

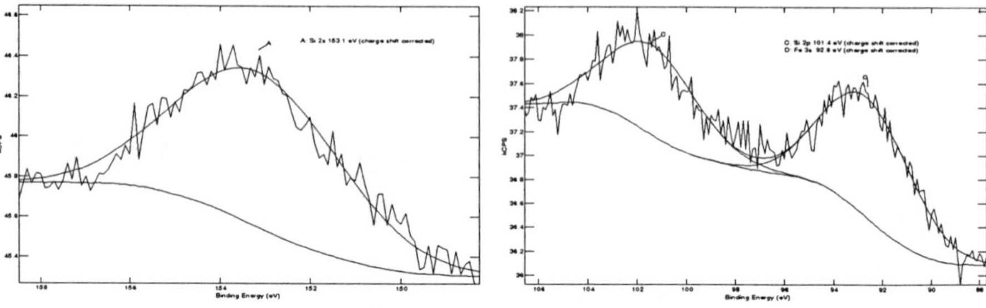

Fig. 3. XPS spectra of Si (2s 153.1 eV) which corresponds to Si in silanes.

Fig. 4. XPS spectra of Si (2p 101.4 eV) and Fe (3s 92.6 eV) measured after Ar ion sputtering.

By using FT-IR spectroscopy on the powders observed in absorbency and diffuse reflection modes, we expected that the vibrational modes observed in the measured spectra could help to differentiate between species which are present on the surface and in the bulk of the crystallites. In order to distinguish between the surface of the powders with respect to the bulk, we measured the diffuse reflectance spectra of the powders. Polycrystalline samples were simply mixed with KBr without applying pressure, which may cause grinding and crushing of the powder particles. In such a way, the surface of the particles remained unaffected and the inner part of the bulk of the crystallites did not become exposed to infrared radiation when examined in the diffuse scattering mode. The spectrum obtained is shown on Figure 5. We can estimate the characteristic peaks for Si-O-Si groups (γ and δ), as well as CH_2 groups, which must be the consequence of some remaining organic surface modifying agent's groups not used for the reaction. This can only indicate that the number of OH^- groups on the powder surface was a minimum, but enough for the reaction and consequently for the protection of the powder with a monomolecular layer. The weight gain during the corrosion test of the HDDR processed powders with two different compositions; uncoated and coated with silane is shown in Figure 6. Uncoated powders show a very sharp increase in the first 50 hours of exposure to the severe humid conditions with no further progress in the corrosion after 70 hours.

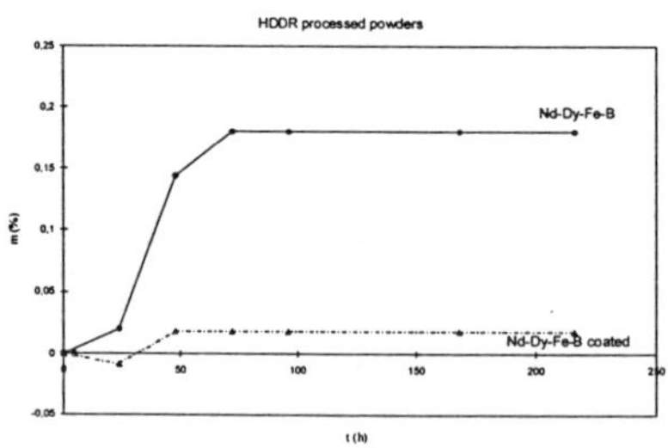

Fig. 6. The weight gain during the corrosion of HDDR processed powder $Nd_{15}Dy_1Fe_{75.25}Zr_{0.75}B8$: uncoated coated with silane.

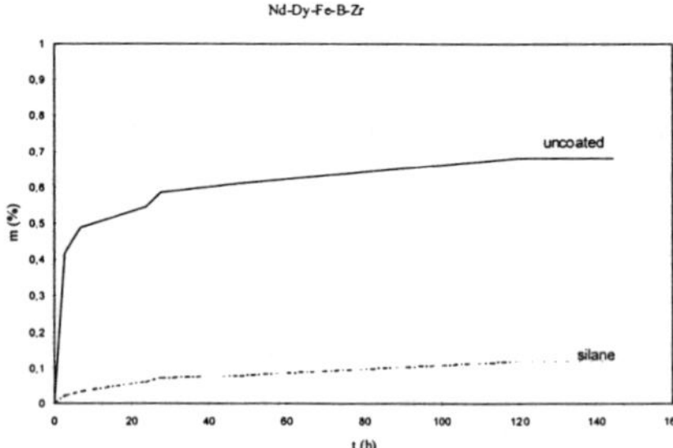

Fig. 7: Weight gain during the corrosion test of $Nd_{15}Dy_1Fe_{75.25}Zr_{0.75}B8$ MM powders: uncoated and coated with silane

Nanocrystalline powders with highly reactive surfaces react with the atmosphere immediately and most likely form a passivated surface of a thin NdO_x or/and Nd_2O_3 layer, corrosion does not proceed any further. Coated powders with silane show a marginal weight gain (less than 0.01 %) in the first 50 hours of exposure to the severe corrosion conditions with complete passivation afterwards. One can conclude that not a 100 % coating of particles were achieved and those uncoated particles oxidise, which contributes to the observed weight gain.

Magnetic properties of the HDDR processed powders, before the corrosion tests compared to the properties of the powders coated with silane and measured after the test are shown in table 1.

Table.1. Magnetic properties of the basic powders processed by HDDR measured immediately on uncoated powders and after the corrosion test on coated powders.

Composition and processing conditions - Disproportionation T and t (°C/minutes)	Hc (**uncoated** powder) (kA/m)	Hc (**coated** powder) (kA/m)
$Nd_{15}Dy_1Fe_{76}B_8$ (820/H_2/1000; 790/vac./60)	**844**	**860**

Results in Table 1. show that coercivity did not change as a result of the chemical treatment, which makes the use of these organic substances suitable for the coating of nanocrystalline Nd-Dy-Fe-B powders with a very thin (~10 nm) protective layer and without damaging the magnetic properties.

Results in Figure 7 show the corrosion rate of mechanically milled powders exposed to the atmospheric conditions of the laboratory. Corrosion of MM powders proceeds much faster than with HDDR processed powders due to the higher reaction surface and the difference between coated and uncoated powders is remarkable. With the very high resistance to corrosion of the coated surface in the first 20-hours free handling of the powder in laboratory atmospheric conditions (~ 40 % R.H.) is possible.

This is of great importance in the production where the handling of sensitive powders in the air without any consequences would contribute to the lowering of production costs. The above results imply that the starting powder is covered with a thin layer of hydroxyl groups that enable the adhesion of silane molecules to the powder's surface. Presumably the Nd-rich phase has to be the reacting precursor for hydrolysis of the particle's surface. Even with very careful handling of the powder in a protective atmosphere and/or liquid, a very small number of OH^- groups in a protective gas or organic liquid are always present. It is not clear yet, if these groups adhere onto the extremely sensitive powder from the surrounding in spite of the very careful handling in the protective atmosphere throughout the processing. It seems to be more likely, that these groups proceed from the organic liquid used as a solvent for silane. According to the proposed reaction schemes, during the subsequent heating in vacuum, condensation takes place (water or alcohol molecules are eliminated) and the covalent bond forms.

CONCLUSIONS

Coating achieved by the chemical reaction between the Nd-Fe-B powder's surface and silane molecules is a very powerful method for the protection of sensitive nanocrystalline powders. HDDR processed powders coated with stearic acid show a marginal weight gain (less than 0.01%) in the first 50 hours of exposure to the severe corrosion conditions. The weight increase after the corrosion test of the more sensitive mechanically milled powders coated with silanes shows a weight increase in the first few hours which is lower than 0.05%, suggesting that this coating has excellent prospects. The magnetic properties are not damaged and the coating enables free handling of the powder for at least 20 hours in air. The results achieved could be of great importance in production where the handling of sensitive powders in air without any significant degradation would contribute to a lowering of production costs.

Acknowledgements. Prof. Boris Orel is gratefully acknowledged for providing us with FT-IR analyses and fruitful discussions. The Ministry of Science and Technology of Slovenia is gratefully acknowledged for providing the financial support necessary to carry out this programme of research.

REFERENCES

1. T. Takeshita and R. Nakayama: 10th Int. Workshop on RE Magnets and their Applications, Kyoto, Japan (1989), 551
2. P.J. McGuiness, X.J. Zhang, X.J. Yin and I.R. Harris: J. Less-Common Metals, 158 (1989), 359
3. L. Schultz, K. Schnitzke, J. Wecker, M. Katter, and C. Kuhrt: J. Appl. Phys., 70 (1991), 6339
4. M. Sagawa: Japanese Patent No. 63-38555 (1988)
5. C.H. Allibert: CEAM, eds. I.V. Mitchell, J.M.D. Coey, D. Givord, I.R. Harris and R. Hanitch, Elsevier Appl. Science, Barking (1989), 358
6. P. Tenaud, F. Vial and M. Sagawa: J. IEEE Trans. Magn., 26, 5 (1990),1930
7. S. Kobe-Beseničar, J. Holc, G. Dražič, B. Saje: J. IEEE Trans. Magn., 30, 2 (1994), 693
8. S. Beseničar, B. Saje, G. Dražič and J. Holc: JMMM, 104-107 (1992), 1175
9. C. Burkhardt, I.R. Harris, S. Kobe, L. Vehovar and Steinhorst: 14th Int. Workshop on RE Magnets and their Applications, Sao Paulo, Brazil (1996), 689
10. S. Kobe, A.J. Williams, C. Burkhardt, F. Dimc and B. Saje: 14th Int. Workshop on RE Magnets and their Applications, Sao Paulo, Brazil (1996), 213
11. P.J. McGuiness, I. Škulj, A. Porenta and S. Kobe: Proc. Of the 15th Inter. Workshop on RE Magnets & Appl., Dresden, Germany (1998) 553-563
12. S. Novak, K.Vidovič, M.Sajko, T.Kosmač: J.Europ.Cer.Soc., 17 (1997) 217-223
13. S. M. Wolfrum, J.J. Ponjee: J.Mat.Sci.Letters, 8 (1989) 667-669
14. J.G.Zhang, M.J. Edirisinghe, J.R.G.Evans: J.Mat.Sci., 23 (1989)2115-2120
15. H. Gorter, J. Gerretsen, R.A. Terpstra: Proc. Third Euroceramics V.1, pp. 615-620 (1993) ed. P.Duran, J.F. Fernandez
16. M. Egashira, Y. Shimizu, Y. Takao, R. Yamaguchi, Y. Ishikawa: J.Am.Ceram.Soc. 77(7) (1994) 1793-98
17. M. Egashira, Y. Shimizu, S. Takatsuki: J.Mat.Sci.Letters, 10 (1991) 2115-2120
18. R. Metselaar, R. Reenis, M. Chen, H. Gorter,H.T. Hintzen: J.Europ.Cer.Soc., 15 (1995) 1079-1085
19. P.J. McGuiness, I. Škulj, A. Porenta and S. Kobe, JMMM188 (1998) 119-124
20. S. Kobe, S. Novak, I. Škulj, P.J. McGuiness, Proceedings of the Fifteenth International Workshop on Rare-Earth Magnets and their Application, Ed. L. Schultz, K.-H. Müller, 1998, 897-90
21. K. Žuzek, P.J. McGuiness, and S. Kobe: this conference

MICROSTRUCTURAL AND MAGNETIC STUDIES OF HDDR MAGNETS FROM HIGH BORON NdFeB(Zr) ALLOYS

DN Brown *, AJ Williams *, O Gutfleisch **, M Strangwood * and IR Harris *.
* The University of Birmingham, School of Metallurgy and Materials, Elms Road, Edgbaston, Birmingham. B15 2TT. UK
** Institute of Solid State and Materials Research Dresden, PO Box 270016, Dresden, Germany.

ABSTRACT

Some boron-rich (with respect to the $Nd_2Fe_{14}B$ composition) NdFeB alloys (with and without small additions of zirconium) have been investigated in an attempt to understand the influence of zirconium in the subsequent magnetic properties of HDDR hot pressed magnets. It was found that zirconium additions had a marked influence on the as-cast microstructures with a general refinement of the grain size and evidence of a zirconium-rich phase (probably ZrB_2) at the grain boundaries. The magnetic property measurements of fully dense, isotropic HDDR hot pressed magnets indicated that those containing zirconium maintained their coercivity to significantly higher hot pressing temperatures and this was attributed to the pinning of grain boundaries by the zirconium rich phase.

INTRODUCTION

A number of processing routes applied to NdFeB-based alloys are being investigated in the search for new, cost effective methods of manufacture and improved magnetic properties. The HDDR (Hydrogenation-Disproportionation-Desorption-Recombination) process offers great potential to produce highly coercive NdFeB powder from conventionally cast alloys [1,2]. However, to improve upon the energy product achieved from current isotropic HDDR material the average recombined grain size needs to be reduced from the 300 nm level to something comparable with that of the MQ material [3]. In optimising HDDR processing with respect to grain size, the stages involved in grain size modification within the process and their dependence on composition and process parameters need to be determined.

In recent years, there have been a number of studies into the effects of zirconium additions to NdFeB alloys. Early investigations with sintered NdFeB magnets found that small additions of zirconium precipitated additional Zr-rich phases in $Nd_{14}Fe_{78}B_7Zr_1$ and $Nd_{16.5}Fe_{74.7}B_{7.8}Zr_1$ microstructures, which reduced grain growth and hence increased coercivity [4,5]. Further work with HDDR treated NdFeB bonded magnets showed that a $Nd_{16}Fe_{75.5}B_8Zr_{0.5}$ composition produced anisotropic magnets with remanences over 1 T and coercivities over 700 kAm^{-1} ($BH_{max} \approx 150$ kJm^{-3}) [6,7].

Further studies have associated zirconium additions with the elimination of free iron from small arc melted castings, inhibition of grain growth during annealing treatments and increased magnetic anisotropy induced through localised alignment of HDDR grains [8]. TEM investigations of zirconium-modified alloys have identified ZrB_2 platelets (0.3-2 μm diameter and 50–300 nm thickness) within intergranular Nd-rich areas and linked these to grain growth inhibition [9,10]. It has been suggested that these zirconium precipitates induce a memory-effect between the cast alloy and subsequent HDDR material. This memory-effect has also been proposed to originate from zirconium-stabilised $Nd_2Fe_{14}B$ cells, which nucleate HDDR grains in the orientation of the original cast grains [11].

Mat. Res. Soc. Symp. Proc. Vol. 577 © 1999 Materials Research Society

A more fundamental X-ray absorption study has shown Zr additions, below the solid solubility limit in NdFeB alloys, occupy a Fe(j_2) site in the $Nd_2Fe_{14}B$ structure [12]. This substitution can act to stabilise the $Nd_2Fe_{14}B$ structure and could allow parts of the original cast structure to be preserved during disproportionation. On recombination these stable regions could then seed new HDDR grains, and lead to magnetic anisotropy in the subsequent material.

In term of processing, zirconium additions can extend the temperature and time processing windows in the HDDR treatment, via increased grain growth control [13]. Increased corrosion resistance of NdFeB magnets has also been observed in Zr-modified material [10].

In this study a number of NdFeB-based high boron, high and low neodymium compositions, with and without zirconium additions have been cast and subsequently HDDR treated and then hot pressed over a range of temperatures. The microstructures of these alloys were studied in relation to the magnetic properties with an aim of throwing further light on the effects of zirconium additions. High boron compositions were studied in order to promote Zr-B interactions and identify the role of particles, such as ZrB_2 on microstructural development.

EXPERIMENTAL

Casting.

The compositions of the alloys investigated are presented in Table 1. These were prepared by melting 1.5 kg of iron, ferro-boron and neodymium constituents for 2 minutes at 1500°C in a refractory crucible, and casting into a chill-mould. This method differed from conventional mould casting practice, as a Durville pouring technique [14], a 2 bar argon atmosphere and a zirconia-coated cast-iron mould were employed.

Table 1: Compositions of Cast Alloys with $Nd_{16}Fe_{76}B_8$ for Reference.

Alloy Type	Weight % ± 0.1%				Atomic % ± 0.1%			
	Nd	: Fe	: B :	Zr	Nd	: Fe	: B :	Zr
Reference Neomax Alloy	34.7	64.0	1.3		16	76	8	
Nd-Rich	34.7	62.7	2.6		15.0	70	15	
Nd-Rich + Zr	34.9	60.7	2.9	1.5	15.0	67.3	16.6	1.1
Nd-Lean	29.0	68.8	2.2		12.3	75.2	12.5	
Nd-Lean + Zr	28.2	65.5	2.1	4.2	12.2	72.8	12	3

In the Durville casting technique the furnace chamber is rotated through 180° on pouring, such that the melt passes into the mould without disturbing any surface debris. A diagram of this apparatus is shown in Figure 1. The cast-iron mould was designed to produce a series of cylindrical and rectangular bars from the cast, with a range of diameters from 5 mm to 30 mm, as shown in Figure 2. Cracking can be seen across these cast bars, which is due to the brittle alloy being sensitive to thermal shock on cooling.

Metallographic samples were prepared from section XX (Figure 2) of the 20 mm diameter rods from each of the four cast compositions. The microstructures of these samples were investigated by optical and scanning electron microscopy, and the area analysis of phases was measured on a Leica Quantimet 500 image analyser. Phase compositions, in terms of elements with atomic numbers >12, were determined using a LINK QX2000 energy dispersive X-ray (EDX) spectrometer system fitted to a JOEL 6300 SEM.

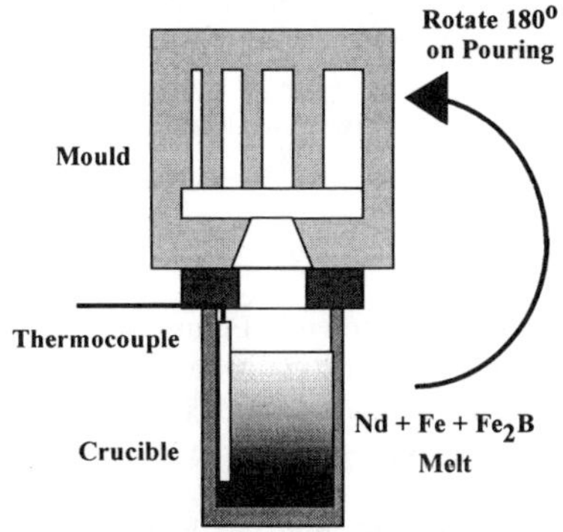

Figure 1: Schematic of furnace

Figure 2: Example cast bars in mould

HDDR Treatment.

The 20 mm diameter rods from each cast composition were lightly crushed, and 40 g of powder was placed into a static laboratory-scale reactor and HDDR treated by means of the cycle shown in Figure 3. This cycle began with the reactor being evacuated to 10^{-2} mbar, and subsequently purged with 1bar of hydrogen. This atmosphere was maintained during the 10°C/minute heating stage to 860°C, and through a 15 minute disproportionation dwell period. This was followed by a desorption stage for approximately 10 minutes, where the reactor pressure was reduced to 10^{-1} mbar, and the sample recombined. At this pressure, the system was quenched with an external copper water-cooled jacket being placed around the Inconel reactor vessel. The magnetic properties of the four HDDR powder compositions were determined by means of a Vibrating Sample Magnetometer (VSM).

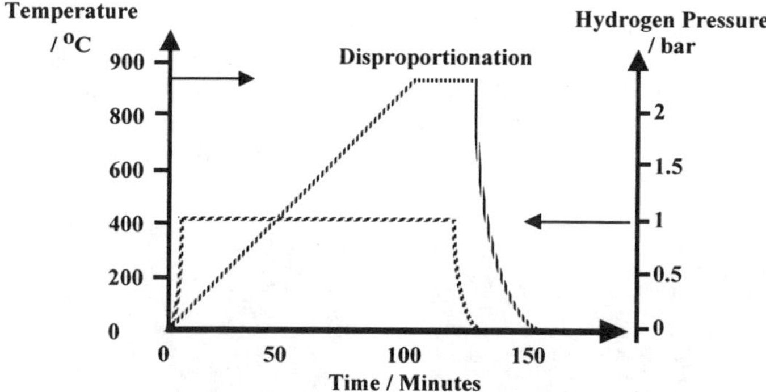

Figure 3: A schematic of the HDDR cycle.

Hot Pressing.

Isotropic hot pressed magnets were produced from the HDDR powders at IFW, Dresden. Initially the powders were ground in an inert atmosphere with a mortar and pestle and sieved to produce particles with size ≤106 μm. 3.2 g batches of the sieved powders were then pressed into 8 mm diameter cylinders at 150 MPa over a range of temperatures between 685°C and 900°C. Pressing took place for a two minute period from the point the sample reached the particular pressing temperature. The magnetic properties of each hot pressed magnet were determined by a permeameter technique.

Both the HDDR powder and hot pressed material (after a cross-sectional fracture) were studied as above on an Hitachi S-4000 FEG-SEM.

RESULTS and DISCUSSION

As-Cast Material.

The as-cast microstructures of the four compositions are shown in Figures 4 to 7, and a phase balance is presented in Table 2. Phase identification was made on the basis of EDX-determined Nd:Fe:Zr atomic ratios, as noted in Table 2. The relative volume percentages of each phase in the four microstructures were obtained from image analysis area fractions of optical micrographs, assuming isotropy on a random plane.

All the alloys exhibit $Nd_2Fe_{14}B$ dendrites (magnetic domains revealed by Kerr effect in these grains) with interdendritic phases rich in iron (Nd:Fe = 20:80) and neodymium (Nd:Fe = 98:2). EDX atomic ratios indicate these interdendritic phases to be $Nd_1Fe_4B_4$ and Nd-rich. The $Nd_{12.3}Fe_{75.2}B_{12.5}$ composition did not exhibit any Nd-rich phase but a further intergranular iron-rich phase (**D**). Zirconium-rich precipitates (**E**) were observed in both Zr-containing alloys (Figures 5 and 7). These were of such a fine morphology (1-5 μm long needles) that TEM observations will be necessary for further detailed investigation.

There are striking differences in grain morphology and phase distribution between the two neodymium-rich alloys. The $Nd_{15}Fe_{70}B_{15}$ composition shows a microstructure (Figure 4) with fine columnar grains of the $Nd_2Fe_{14}B$ matrix phase (**A**), and intergranular Nd-rich (**C**) and $Nd_1Fe_4B_4$ (**B**) phases. The $Nd_{15}Fe_{67.3}B_{16.6}Zr_{1.1}$ microstructure (Figure 5) shows a matrix of $Nd_2Fe_{14}B$ columnar and rosette grains (**A**). The intergranular Nd-rich (**C**) and $Nd_1Fe_4B_4$ (**B**) phases appears more blocky and less continuous in the Zr containing alloy.

The neodymium-lean alloys show a similar change in microstructure with zirconium addition. The $Nd_{12.3}Fe_{75.2}B_{12.5}$ microstructure (Figure 6) shows a matrix of $Nd_2Fe_{14}B$ columnar grains (**A**) and a feathery intergranular $Nd_1Fe_4B_4$ phase (**B**). There was no Nd-rich phase present but an occasional, dark, iron-rich phase (**D**) was observed at the grain boundaries. With zirconium additions ($Nd_{12.2}Fe_{72.8}B_{12}Zr_3$) (Figure 7) rosettes were again seen in the columnar dendritic $Nd_2Fe_{14}B$ matrix (**A**) with feathery intergranular $Nd_1Fe_4B_4$ phases (**B**) and blocky Nd-rich regions (**C**), that replace the iron-rich phase. The interdendritic phases are more blocky and less continuous.

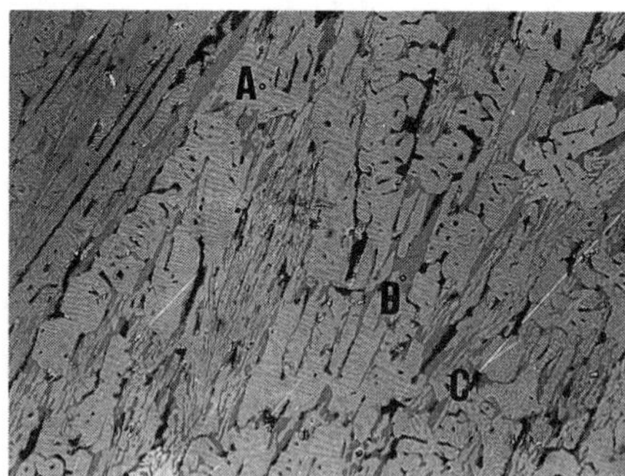

Figure 4: Optical microstructure of as-cast $Nd_{15}Fe_{70}B_{15}$

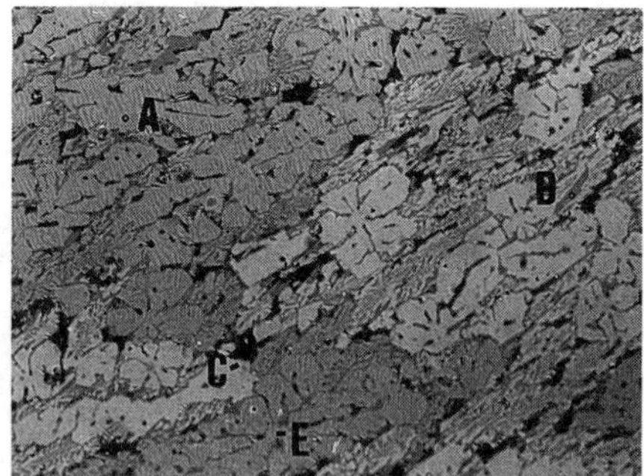

Figure 5: Optical microstructure of as-cast $Nd_{15}Fe_{67.3}B_{16.6}Zr_{1.1}$

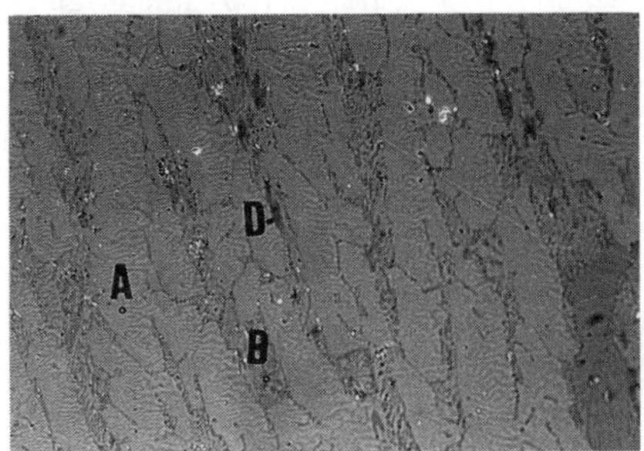

Figure 6: Optical microstructure of as-cast $Nd_{12.3}Fe_{75.2}B_{12.5}$

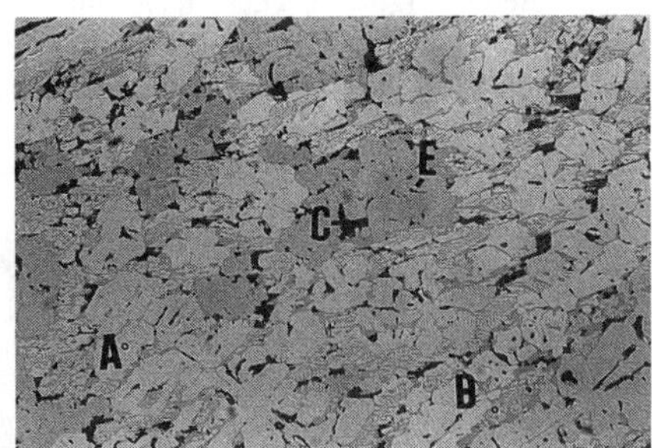

Figure 7: Optical microstructure of as-cast $Nd_{12.2}Fe_{72.8}B_{12}Zr_3$

In terms of phase balance, the four alloys have between 59% and 66% $Nd_2Fe_{14}B$ matrix phase and between 21% and 30% $Nd_1Fe_4B_4$ phase. The most significant observations were the presence of a fine-scale Zr-rich phase in $Nd_{15}Fe_{67.3}B_{16.6}Zr_{1.1}$ and $Nd_{12.2}Fe_{72.8}B_{12}Zr_3$ alloys, and the appearance of the Nd-rich phase in $Nd_{12.2}Fe_{72.3}B_{12}Zr_3$ and not in $Nd_{12.3}Fe_{75.2}B_{12.5}$. The latter observation is consistent with zirconium combining with some of the boron from neodymium-containing phases to form ZrB_2.

Table 2: Phase balance of as-cast alloys.

Alloy	**^{A}Matrix Phase** $Nd_2Fe_{14}B$	**^{B}Dark Phase** $Nd_1Fe_4B_4$	**^{C}White Phase** Nd-Rich	**Other Phases, possibly** ^{D}Fe-rich, ^{E}Zr-rich
$Nd_{15}Fe_{70}B_{15}$ (Nd-Rich)	**62** ± 7	**29** ± 7	**7** ± 2	**2** (Porosity and oxides)
$Nd_{15}Fe_{67.3}B_{16.6}Zr_{1.1}$ (Nd-Rich+Zr)	**59** ± 2	**26** ± 2	**7** ± 3	**8** (Zr-rich grain boundary phases, porosity and oxides)
$Nd_{12.3}Fe_{75.2}B_{12.5}$ (Nd-Lean)	**65** ± 5	**30** ± 3	-	**2** (Fe-rich phases, porosity and oxides)
$Nd_{12.2}Fe_{72.8}B_{12}Zr_3$ (Nd-Lean+Zr)	**66** ± 2	**21** ± 3	**6** ± 2	**7** (Zr-rich grain boundary phases, porosity and oxides)

Phases identified by EDX atomic ratio Nd:Fe
^{A}$Nd_2Fe_{14}B$~12.5/87.5 ^{B}$Nd_1Fe_4B_4$~20/80 ^{C}Nd-rich→98/2 ^{D}Fe→100%Fe ^{E}ZrB_2~100%Zr

HDDR Material.

From the FEG-SEM study of the HDDR powders, fine faceted $Nd_2Fe_{14}B$ grains (200 – 500 nm in diameter) were observed in all compositions, as illustrated in Figure 8. The Nd-lean compositions also had significant amounts of large grains (approximately 100 μm) present. This phase is likely to be $Nd_1Fe_4B_4$, as it is known that this phase does not readily disproportionate [15]. Despite all compositions having similar amounts of $Nd_1Fe_4B_4$ (21% to 30%) in their as-cast state, the Nd-rich alloys ($Nd_{15}Fe_{70}B_{15}$ and $Nd_{15}Fe_{67.3}B_{16.6}Zr_{1.1}$) have more capacity to react $Nd_1Fe_4B_4$ with the Nd-rich grain boundary phase and produce additional $Nd_2Fe_{14}B$ on HDDR treatment. Hence the large unreacted particles are less obvious in the Nd-rich HDDR powders.

A further difference between the two Nd-rich compositions was the presence of rod shaped structures (approximately 20 μm in length) in the zirconium-modified material, as illustrated in Figure 9. These have only been observed on fracture surfaces (possible original as-cast grain boundaries), and thus may reflect segregation or columnar boundary features in the as-cast structure. The feature could be ZrB_2 needles with fine $Nd_2Fe_{14}B$ HDDR grains coating their surface.

Figure 8: HDDR microstructure of Nd-rich material.

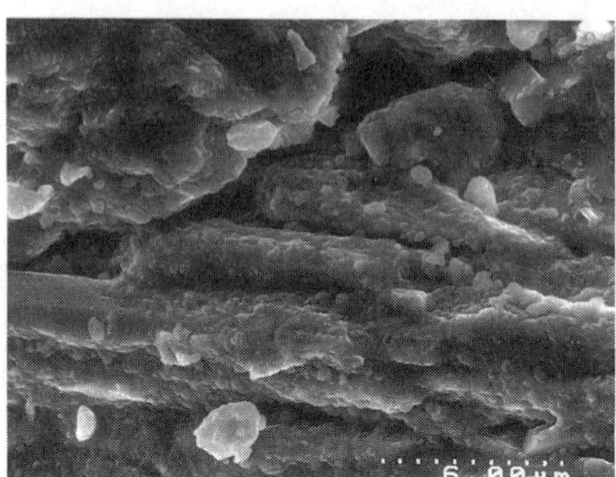

Figure 9: HDDR microstructure of Nd-rich + Zr material.

The magnetic properties of the HDDR powders are presented in Table 3. Generally, these compositions have lower remanences when compared to that of a more conventional $Nd_{15}Fe_{77}B_8$ composition [16]. This difference can be explained by the high proportion of the non-magnetic $Nd_1Fe_4B_4$ phase present (21% to 30%) in these alloys.

The two Nd-rich alloys have the highest remanences and reasonable coercivities. The effect of zirconium is slight in comparison to the effects of different neodymium levels, although Table 3 shows a consistent downward trend in magnetic properties with the introduction of zirconium.

Table 3. HDDR Powder Properties (VSM)

Alloy	Remanence / mT	$_iH_c$ / kAm^{-1}
$\mathbf{Nd_{15}Fe_{77}B_8}$ (Reference alloy [16])	740	980
$\mathbf{Nd_{15}Fe_{70}B_{15}}$ (Nd-Rich)	590	990
$\mathbf{Nd_{15}Fe_{67.3}B_{16.6}Zr_{1.1}}$ (Nd-Rich+Zr)	575	985
$\mathbf{Nd_{12.3}Fe_{75.2}B_{12.5}}$ (Nd-Lean)	275	50
$\mathbf{Nd_{12.2}Fe_{72.8}B_{12}Zr_3}$ (Nd-Lean+Zr)	100	10

Hot Pressed HDDR Material.

As the Nd-lean powders had poor magnetic properties at the HDDR powder stage, only the two Nd-rich compositions were hot pressed. A summary of the magnetic properties achieved from hot pressing at temperatures between 685°C and 900°C is presented in Table 4. All hot pressed magnets exhibited a density of 7.4 ± 0.1 gcm^{-3}, indicating they were virtually fully dense.

Table 4 shows that the properties of the two alloys are similar at pressing temperatures up to 825°C. However at the higher hot pressing temperature of 900°C, the zirconium modified material maintains its properties, whereas the other alloy experiences a significant reduction. This feature is illustrated further in Figure 10, where the coercivity of each material is plotted against the hot pressing temperature.

Table 4: Magnetic properties of HDDR powder hot pressed.

Hot Press Temperature	$Nd_{15}Fe_{70}B_{15}$ (Nd-Rich)		$Nd_{15}Fe_{67.3}B_{16.6}Zr_{1.1}$ (Nd-Rich+Zr)		$Nd_{15}Fe_{77}B_8$ Reference alloy [16]	
	B / mT	$_iH_c$ / kAm^{-1}	B / mT	$_iH_c$ / kAm^{-1}	B / mT	$_iH_c$ / kAm^{-1}
685°C	636	1039	647	987	780	940
725°C	651	1015	605	983	710	660
750°C	642	1020	624	933	710	580
775°C	644	975	637	948	685	500
825°C	627	888	642	937	n/a	n/a
900°C	573	530	597	945	n/a	n/a

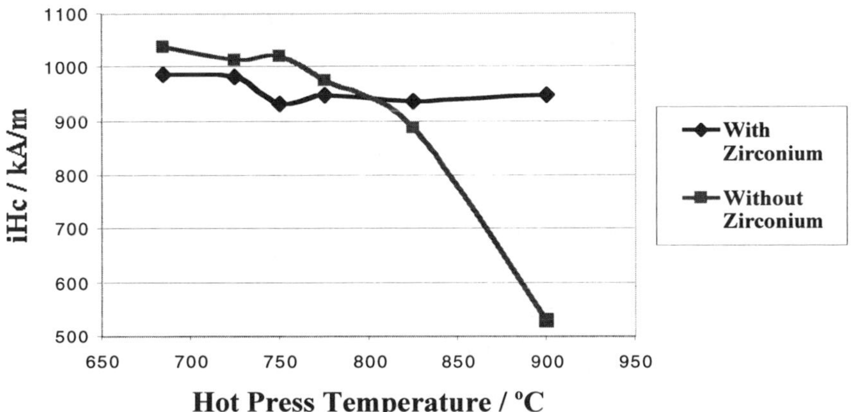

Figure 10: Coercivity of Nd-rich magnets as a function of hot pressing temperature.

The much greater thermal stability of the coercivity of the $Nd_{15}Fe_{67.3}B_{16.6}Zr_{1.1}$ magnets can be explained by comparing the microstructures of the hot pressed magnets from the two compositions. Cross-sectional fracture surfaces of the magnets pressed at 900°C were studied using a FEG-SEM. These microstructures are compared in Figure 11 ($Nd_{15}Fe_{70}B_{15}$) and Figure 12 ($Nd_{15}Fe_{67.3}B_{16.6}Zr_{1.1}$).

The zirconium-free magnet exhibits large facetted $Nd_2Fe_{14}B$ grains, indicating extensive grain growth. These grains are also randomly distributed, which suggests such grain growth is not uniform. This is in contrast with the zirconium-modified sample, which shows a much more uniform distribution of smaller $Nd_2Fe_{14}B$ grains. Furthermore, the matrix of grains appears finer in the zirconium-modified sample, indicating a possible effect of zirconium on grain boundary mobility. This pinning of $Nd_2Fe_{14}B$ boundaries during hot pressing could arise from a uniform, fine dispersion of Zr-rich precipitates (possibly ZrB_2), or from interfacial zirconium segregation.

This observation is in agreement with other studies, which have found similar effects of reduced grain growth kinetics and extended processing windows with the addition of zirconium to NdFeB alloys [6,13]. To investigate these effects further, TEM studies are being undertaken.

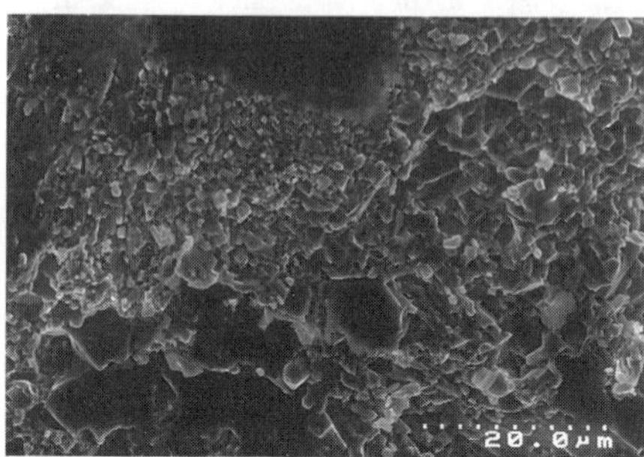

Figure 11: Microstructure of 900°C hot pressed Nd-rich sample without Zr.

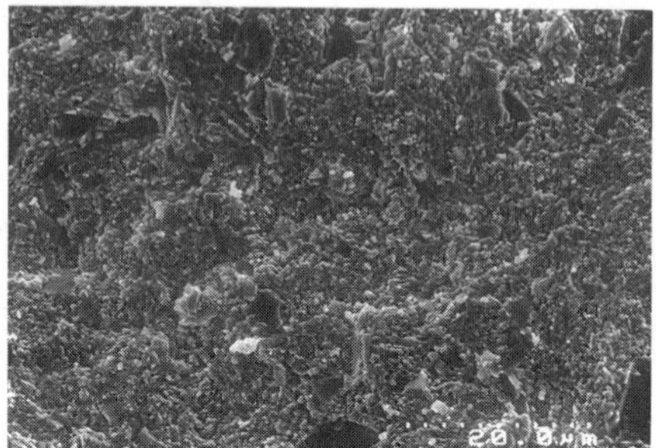

Figure 12: Microstructure of 900°C hot pressed Nd-rich sample with Zr.

CONCLUSIONS

1. Zirconium can significantly change the cast structure of NdFeB alloys. Grains with rosette morphology, and fine Zr-rich precipitates were observed. With Nd-lean compositions, zirconium has the additional effect of promoting the grain boundary Nd-rich phase, possibly by the consumption of boron from phases with Nd, for the formation of ZrB_2.
2. Reasonable coercivities (990 kA/m) have been achieved with both types of Nd-rich HDDR powder.
3. Hot pressed isotropic Nd-rich HDDR powders with zirconium maintain their coercivity at higher processing temperatures in comparison to corresponding HDDR powders without zirconium additions. This feature is possibly associated with zirconium additions giving a uniform distribution of grain boundary pinning sites, either by precipitation or in solution.

ACKNOWLEDGMENTS

The authors would like to thank EPSRC and the British Council (Germany) for their financial support, and Rare Earth Products Ltd, Widnes for the supply of material. In addition, the help and support of Prof. John Campbell and Mr. Adrian Caden with foundry activities is appreciated.

REFERENCES

1. T. Takeshita and R. Nakayama, Proc. 10th Int'l Workshop on RE Magnets & Their Applications, Kyoto, (1989), pp. 551-557.
2. P.J. McGuiness, X. J. Zhang, X. J. Yin and I.R. Harris, J. Less-Common Met., Vol. 158, pp. 359-365.
3. R.K. Mishra, J. Magn. and Magn. Mat., Vol. 54-57, (1986), p. 450.
4. R.J. Pollard, P.J. Grundy, S.F.H. Parker and D.G. Lord, IEEE Trans. Magn., Vol. 24, No. 2, (1988), pp. 1626-1628.
5. C.H. Allibert, J. Less-Common Met., Vol. 152, (1989), pp. L1-L4.

6. P.J. McGuiness, C.L. Short and I.R. Harris, IEEE Trans. Magn., Vol. 28, No. 5 (1992), pp. 2160-2162.
7. C.L. Short, P. Guegan, O. Gutfleisch, O.M. Ragg and I.R. Harris, IEEE Trans. Magn., Vol. 32, No. 5 (1996), pp. 4368-4370.
8. A. Fujita and I.R. Harris. IEEE Trans. Mag., Vol. 29, No. 6 (1993), pp. 2803-2805.
9. M. Matzinger, J. Fidler, A. Fujita and I.R. Harris, J. Magn. and Magn. Mat., Vol. 157/158 (1996), pp. 54-56.
10. S. Kobe-Besenicar, J. Holc, G. Drazic and B. Saje, IEEE Trans on Magn., Vol. 30, No. 2, (1994), pp. 693-695.
11. I.R. Harris, Proc. 12th Int'l Workshop on RE Magnets & Their Applications, Canberra, (1992), pp. 347-371
12. A. Ashfaq, M. Matsuura and M. Sakurai, Appl. Phys. Letters, Vol. 73, No. 17, (1998), pp. 1-3.
13. P.J. McGuiness, I. Skulj, A. Porenta and S. Kobe, Proc. 15th Int'l Workshop on RE Magnets & Their Applications, Dresden, (1998), pp 553-562.
14. J. Campbell, *Castings*, Butterworth-Heinemann Ltd, Oxford, 1993 p. 70.
15. O. Gutfleisch, I.R. Harris and V.A. Yartys, Proc. 14th Int'l Workshop on RE Magnets & Their Applications, San Paulo, (1996), pp. 268-277.
16. O.Gutfleisch, A. Kirchner, W. Grünberger, D. Hinz, R. Schäfer, L. Schltz, I.R. Harris and K. H. Müller, J. Magn. and Magn. Mat., Vol. 183 (1998), pp. 359-364.

EFFECTS OF QUENCHING ENVIRONMENT ON THE STRUCTURE OF MELT- SPUN $Nd_2Fe_{14}B$

M. J. KRAMER[1,2], YALI TANG[1], K.W. DENNIS[1], R. W. MCCALLUM[1,2]

[1]Ames Laboratory, [2]Department of Materials Science and Engineering, Iowa State University, Ames, IA 50011-3020

ABSTRACT

Melt-spun $Nd_2Fe_{14}B$ (2-14-1) ribbons were produced under active vacuum and different partial pressures of inert gases of Ar and He. Microstructure and thermal analyses were performed to understand the microstructural evolution and glass formability (GF) of the ribbons. He atmosphere enhances the quenchability of the ribbons over Ar and vacuum. Ribbons made under 250 Torr He have more uniform microstructure and smoother surfaces than those under 760 Torr He. The higher quenchability induced by He, which increases the interfacial heat transfer coefficient between the melt and rotating wheel during melt spinning, is due to its higher thermal conductivity compared to Ar. The lower pressure stabilizes the turbulence between the melt-pool and Cu wheel, enhancing the heat transfer resulting in a more uniform quench. As a result, a more uniform ribbon microstructure can be obtained at relatively low wheel speeds.

INTRODUCTION

Melt–spinning is a widely used production method of rapidly solidifying (RS) materials [1,2]. Processing parameters, such as the thermal conductivity of the rotating quench wheel, wheel speed, ejection pressure, thermal history of the melt before ejection, distance between nozzle of crucible and rotating wheel as well as processing atmosphere, have sensitive influence on the microstructure and properties of melt-spun ribbons [3-12]. There have been numerous studies concerned with the effect of processing parameters on the microstructure and properties of materials [6]. While these studies give direction to the control of the microstructure of melt-spun ribbon, the effect of atmosphere on the microstructure during melt spinning has rarely been studied [7]. The thin boundary layer of gas between the melt and quench wheel gives rise to an impedance layer through which the heat must flow. As will be discussed in this paper, this is a significant parameter in the initial quench. Another crucial aspect is the transient nature of the quench process. With increasing wheel speeds for a given ejection rate, the increase in extraction rate results in thinner ribbons. The thinner ribbons have higher dT/dz, but shorter contact time with the quench wheel. Thus the amount of heat conduction away from the ribbon during its free flight phase can vary with wheel speed.

In addition to steady state effects, there are non-steady state or chaotic effects. One critical aspect is melt pool stability. Instability in the melt pool can result in non-uniform extraction rate. Another aspect is the pockets of gas entrapped between the quench wheel and the melt pool. In the former case, the thickness of the ribbon varies along its length while in the latter the gas pockets give rise to regions of poor thermal conductivity. Both of these cases result in non-uniformity of microstructure along the length of the ribbon. Our research has shown that

Mat. Res. Soc. Symp. Proc. Vol. 577

the pressure and composition of the processing atmosphere affect both chaotic and steady state processes mentioned above.

EXPERIMENTAL

The compositions of the alloys in present study are a stoichiometric $Nd_2Fe_{14}B_2$ alloy. Alloys were prepared with an arc-melter from pure elements of 99.95 Nd, 99.99Fe and 99.5B. All arc melting was done under ultra high purity (UHP) argon. Each ingot was melted, flipped and remelted a minimum of four times to minimize macrosegregation in the as-cast ingot. RS alloys were obtained through melt-spinning. The melt-spinning chamber was pumped down and backfilled several times with UHP Ar or He before operation. In order to study the effects of atmospheric pressure, the melt spinning was done at either full or one-third atmosphere (250 or 750 torr). X-ray diffraction (XRD) was done on a Phillips PW-1830 Diffractometer with Cu K_α radiation. Both wheel and free side of the ribbon were analyzed to check the uniformity of the alloys. Optical microscopy (OM) was performed to study the surface morphology of the ribbons. Transmission electron microscopy (TEM) was performed on a Phillips CM-30 analytical electron microscope. TEM samples were prepared by thinning preferentially from either wheel or free side of the ribbon. Some select ribbons were analyzed along their cross section. Differential thermal analysis (DTA) was performed on a Perkin Elmer DTA 7 at a scanning rate of 10 °C/min. Differential scanning calorimetry (DSC) was done on a Perkin Elmer DSC 7 at a scanning rate of 40 °C/min. Samples for thermal analysis were ground powders from melt-spun ribbon.

RESULTS AND DISCUSSION

A qualitative measure of the crystallinity of the samples can be determined from XRD. Since the quench is not uniform through the thickness of the ribbon, about 15 ribbon fragments were placed free side up and an XRD scan taken, then the ribbons were flipped over so the

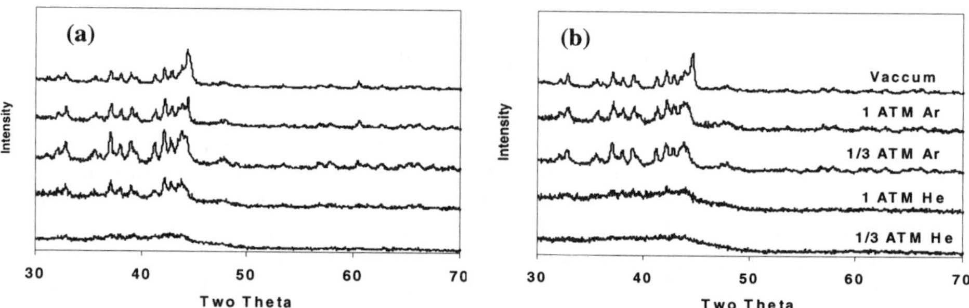

Figure 1 X-ray diffraction patterns obtained from both free (a) and wheel (b) side of ribbons made under different quenching atmospheres at wheel speed of 20m/s

wheel side was exposed and then re-scanned. The ribbons melt-spun at 20 m/s showed a considerable crystalline fraction except samples produced using 250 and 750 Torr He chamber gas (Fig. 1). When only the free surface of the ribbons are examined, only ribbons produced in the 250 torr He pressure are predominately amorphous (Fig. 1a). At higher wheel speeds the samples produced in He were nearly all amorphous as measured by XRD. However, XRD is not sensitive to small amounts of nanocrystalline material. To better characterize the degree of crystallinity, TEM and DSC measurements were performed. A full cross-section through the thickness of the 22.5 m/s 750 torr He ribbon obtained for TEM analysis (Fig. 2) confirms the XRD results. The region closest to the wheel is fully amorphous with gradual increase in density of "nuclei" of 2-14-1 with increasing distance from the wheel side. A broad diffuse selected area diffraction patterns (SADP) is indicative of an amorphous alloy. As the density of nuclei increases, they also become larger, with discrete diffraction spots visible in the SADP. The free side has a truly nanophased 2-14-1 microstructure with no evidence of an amorphous fraction. The metastable mixture of fine grains of 2-14-1 in glass has implications for understanding the DSC results to be discussed below. TEM performed on a sample melt-spun at 22.5m/s in 250 torr He is shown to be nearly entirely amorphous with only a few nuclei observed on the free side (Fig. 3).

Table 1. For all runs the ejection temperature was 1395°C, with an ejection pressure of 125 torr with the nozzle 5 mm from the quench wheel. Estimated glass content in melt-spun ribbons from DSC measurements under the various conditions tested. Where multiple testes were performed the average is reported.

Wheel Speed (m/s)	Glass Fraction (%)			
	Ar (torr)		He (torr)	
	0	750	250	750
17.5			34.5	13.4
20.0	0.0	0.0	80.0	45.1
22.5		0.0	100.	82.0
25.0	11.9	15.3	99.4	90.8
30.0		51.8	100.	93.2
35.0		54.2	94.4	82.0
40.0		57.2	93.4	82.9

Given the spatial inhomogenity of the ribbons, the DSC is a more quantifiable measure of the glassy fraction. The exothermic energy released during devitrification should scale with the

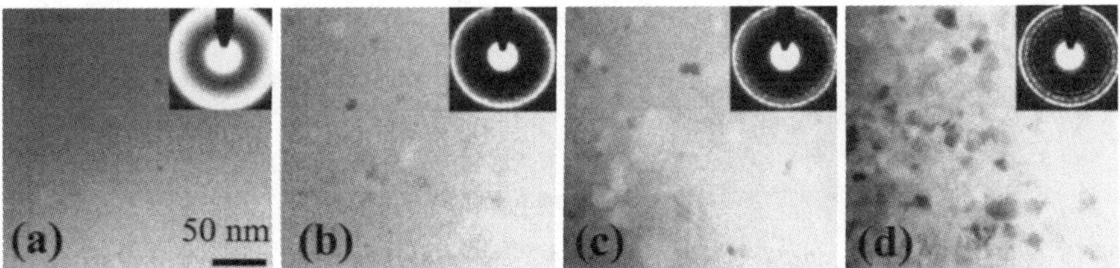

Figure 2. TEM photographs of cross section in melt-spun ribbon under full atmosphere He at wheel speed of 22.5m/s. (a), (b), (c) and (d) indicate microstructure evolution from the wheel side to the free side of the ribbon, respectively

mass fraction of glass (Table 1). Two measurements were performed on each sample (Fig. 4). The effect of He versus Ar on the exothermic energy is quite dramatic. Not only does the He spun samples always have a higher exotherm compared to Ar spun samples for a given wheel speed, but the exothermic energy for the nominally amorphous melt spun alloy in 250 torr He is 50% higher than Ar. The samples produced under lower He pressure always have a slightly higher exotherm than the 750 torr He spun samples. The peak in the exothermic energy occurs between 20 and 35 m/s.

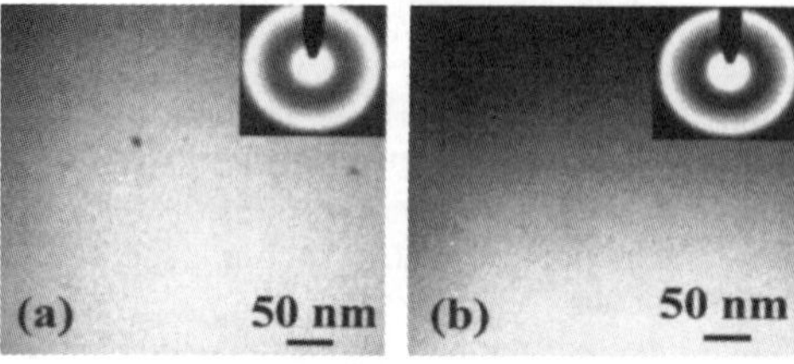

Figure 3 TEM photographs from free (a) and wheel side (a) of melt-spun ribbon under one third atmosphere helium at wheel speed of 22.5m/s, showing very uniform structure

This is the same wheel speed regime where longer, more uniform ribbons are produced. At higher wheel speeds (> 30 m/s), the ribbons are fine fragments of irregular thickness. By placing a longer catch basin with some fine silica wool in place, we were able to determine that these finer ribbon fragments were formed during the melt spinning and not breaking up due to colliding with the walls. We suspect that the slight lowering of the exothermic energy at the highest wheel speeds is due to melt-pool instability on the wheel. This instability could be due to mechanical vibrations or turbulence due to higher gas flow.

The DSC measurement are clearly in agreement with the XRD results that He does provide a more efficient quench of the ribbon and that a lower pressure is more effective. There are enigmas that need explaining. If lower chamber pressure alone provides a better coupling of the liquid to the quench wheel, why are the samples produced in an active vacuum so poorly quenched. And why should the same chamber pressure give so large a difference between Ar

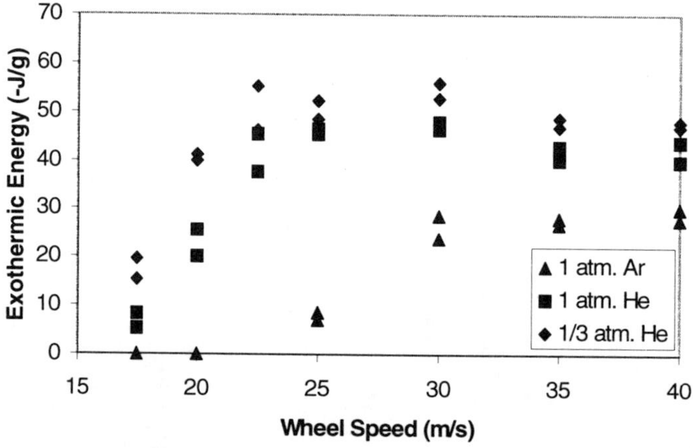

Figure 4. Exothermal Energy of crystallization through DSC measurement on melt-spun ribbons made under different atmospheres and wheel speeds. Every ribbon has been tested two times at heating rate of 40K/s

and He. The surface morphology of the sample and heat capacity of the gasses are clues. Ribbons spun using either an active vacuum of 250 torr He had a much smoother wheel surface than 750 torr Ar or He (Fig. 5c and 5d). This is attributed to fewer gas pockets trapped between the wheel and liquid. These gas pockets are significant since they result in regions of poor heat transfer. These poorly quenched regions propagate through the ribbon and result in porous crystalline pockets on the free surface (Fig. 5a and 5b).

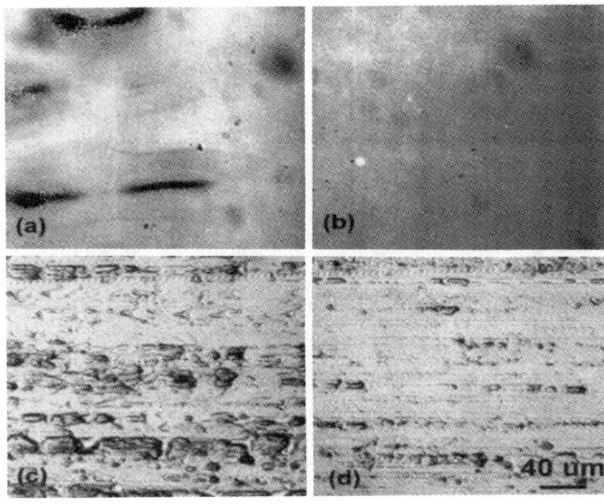

Fig. 5 Optical micrographs showing the surface structure of melt-spun ribbons prepared under different partial pressure. (a) and (c): free and wheel side of ribbon under full atmosphere helium, respectively. (b) and (d): free and wheel side of ribbon under one third atmosphere helium, respectively

Reduced gas pockets does not explain the He effect. He has higher thermal conductivity and lower density than that Ar. This increases the interface heat coefficient between the ribbon and rotating wheel and may reduce the turbulence around the wheel.

In a quasi-steady condition, the average Nussel Number at the surface between the melt and rotating wheel can be estimated as [13]:

$$N_{Nu} = 2+0.6N_{Re}^{1/2}N_{pr}^{1/3} \quad (1)$$

Where $N_{Nu} = h_i t/\lambda$ is the Nussel Number (h_i: interface heat transfer coefficient, t: thickness of ribbon and λ_i: the heat conductivity of materials). $N_{re} = \rho v t/\mu$ is Reynolds Number (ρ: density of matter, v: relative velocity between the melt and gas layer or gas layer and rotating wheel, μ viscosity of the gas, C: Heat capacity of the melt). $N_{pr} = C\mu/\lambda\rho$ is the Prandtl Number. We can rewrite (1):

$$h_I = (\lambda/t)[2+0.6(\rho v^3 t^3 C^2/\mu\lambda)^{1/6}] \quad (2)$$

The heat transfer coefficient between the melt and the wheel can be expressed as:

$$h_I = 1/(1/h_1+t'/\lambda+1/h_2) \quad (3)$$

Where h_1 and h_2 are interface heat transfer coefficient in melt/gas layer and gas layer/wheel interface, while t' is the thickness of gas layer which should get thinner with increasing wheel speeds (in present calculation we assume t' is much less than t).

It can be seen clearly that the higher thermal conductivity of He has positive effect on enhancing the heat transfer coefficient, leading to a direct increase of cooling rate of RS ribbons. Assuming the relative velocity is zero, h_i can be calculated with equation (2) using data from [14] (Fig.

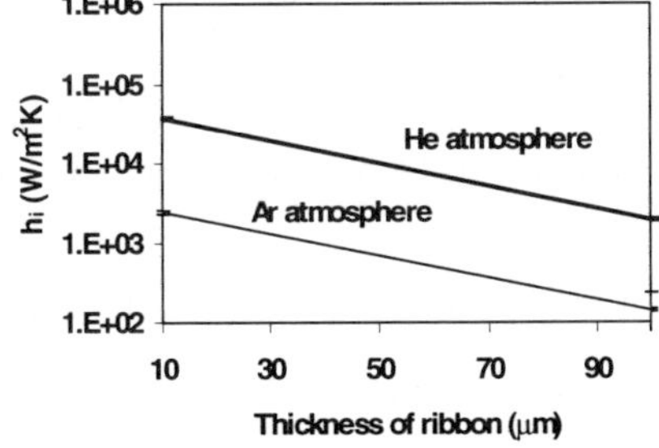

Figure 6 Effect of quenching atmosphere on the heat transfer coefficient between wheel and alloy

6). Because the heat is transferred through the extremely thin layer of gas between the melt and the rotating wheel, h_i can be increased by using atmosphere with higher λ. According to equation (2), ρ has little influence on heat transfer coefficient but may affect the stability of the flow of the liquid metal from the crucible to the melt pool. The lower pressure in the chamber also reduces the size of the gas pockets. This improves the overall uniformity of the microstructure.

CONCLUSIONS

The use of He as the chamber gas enhances the quenching by increasing the heat transfer through the gas between the liquid metal and the quench wheel. Lowering the partial pressure improves the uniformity of the microstructure by reducing the size of the gas pockets, which forms between the ribbon and wheel. The lower pressure and reduced density of the gas may also reduce effects of turbulence on the melt pool stability.

ACKNOWLEDGMENT

Ames Laboratory is operated for U.S. Department of Energy by Iowa State University under Contract No. W-7405-Eng-82.

REFERENCES

1. R. Pond and R. Maddin, Transaction of the Metal Society AIEM, 245, (1969), p.2475-2476.
2. M. C. Narasimham, U.S. Patent No. 4212343, 1980.
3. F. Fogiku, et al., Rapidly Solidified Alloys and Their Mechanical and Magnetic Propertied. MRS, B.C. Giessen et al. (eds.) Vol.58,(1986),p.15
4. S. Krishnamurthy, P.R.Smith and J.G. Paine, Proc. Of Enhanced Propertied in Structural Metals via Rapid Solidification, F.H. Froes(ed.) (ASM. Orlando, FL. Oct. 1986), p.67
5. Y.L.Tang, S. K. Guan, N. F. Shen and D. S. Zhao, J. Mater. Sci. Lett.,12 (22) p.1749(1993)
6. H. Jones, Proc. of Enhanced Propertied in Structural Metals via Rapid Solidification, F.H. Froes(ed.) (ASM. Orlando, FL. Oct. 1986), p.77
7. T.H. Kanekiyo, M. Uehara and S. Hirosawa, Abstract in 7th Joint MMM-Intermagnetics Conference, San Francisco, CA (1998), p.20.
8. D. J. Branagan, Ph.D. Thesis, Iowa State University, 1995
9. M.J. Kramer, C.P. Li, K.W. Dennis and R.W. McCallum, J. Applied Physics, 83(11) p.6631(1998)
10. D. J. Branagan, M. J. Kramer, R. W. McCallum J. Alloys Compd.,244, p.27-39(1996)
11. D.J. Branagan, and R.W. McCallum, J. Alloys and Compd., 218, p.143(1995)
12. M. J. Kramer, C.P. Li, K. W. Dennis, R.W. McCallum et al. J. Applied Phys. 81, p.4459(1997)
13. B. Abramzon and C. Elta, Numerical Method in Thermal Problems, Vol. 2, R.W. Lewis, K. Morgan and B.A. Shrefler (eds.), Swansea, Pineridge: 1981,pp.1145
14. H. B. Crawford and R. J. Kepler (eds.) Chemical Engineer's Handbook, McGraw-Hill Inc.,1973, pp.3-244

Processing Sm-Fe(Ta)-N hard magnetic materials

K ŽUŽEK, PJ MCGUINESS and S KOBE
Institute Jožef Stefan, Ljubljana, Slovenia, paul.mcguiness@ijs.si

Abstract

SmFe based alloys interstitially modified with nitrogen are potential candidates for high energy permanent magnets. In order to obtain the optimum properties a thorough understanding of the starting material and processing parameters is required. The microstructures of two cast alloys of composition $Sm_{13.8}Fe_{82.2}Ta_{4.0}$ and $Sm_{13.7}Fe_{86.3}$ were carefully examined with a SEM equipped with EDX and the exact stoichiometries of the phases were determined. The SmFeTa material was found to contain significant amounts of $TaFe_2$ as well as the Sm_2Fe_{17}, $SmFe_2$, $SmFe_3$ phases observed in the SmFe material but without the α-iron dendrites which are characteristic of the latter material. The optimum conditions necessary to provide the highest coercivities using the conventional HDDR process, and for the HDDR process combined with pre-milling were investigated. The coercivities obtained after using the HDDR process and subsequent nitriding were 680 kA/m for the SmFeTaN and 360 kA/m for the SmFeN samples. Significantly higher coercivites of 1000 kA/m for SmFeN and 1275 kA/m for SmFeTaN were achieved by reducing the particle size with milling prior to the HDDR process.The better coercivities obtained with the Ta containing sample were found to be due to the presence of a much smaller amount of α. The milling prior to the HDDR treatment improves the magnetic properties because of the small particle size which prevents the grains growing too large, with their consequent very negative effect on the coercivity.

Introduction

SmFeN magnets have received considerable attention since their discovery in 1991 [1]. Their intrinsic properties are comparable with, or better than, those of magnets based on $Nd_2Fe_{14}B$, and so they have the potential of taking a significant share of the rare earth permanent magnet market. SmFeN magnets are open to a number of possible processing routes, for example, mechanical alloying [2,3], melt spinning [4,5] and conventional powder metallurgy [1,6] however the HDDR process [7,8] appears to be the most promising.

The addition of a third element, in this case Ta, is primarily intended to reduce the incidence of α-iron which typically constitutes some 25% of the phases present [9] in the as-cast binary alloy. The α-iron can be removed by an extended vacuum or inert gas heat treatment, however this is expensive, environmentally detrimental and can cause problems with maintaining precise composition as a result of samarium evaporation. The addition of 4-5% Nb [10] or Ta [11] or 1% Zr [12] to the SmFe material allows the primary crystallization of α-iron to be avoided.

Effects of reducing the particle size in combination with the HDDR process have been noted by other workers, Müller et al [13] found that by milling after the HDDR process was completed a remarkable increase in coercivity could be observed. This increase was attributed to the well known effects of size reduction as a result of milling. They also found milling prior to the HDDR process had a minimal effect unless an additional 5 wt% of samarium was added. In this case the improvement was attributed to a more even microstructure not possible with larger particles due to having grain formation begin earlier in the centre of the large particles. Gebel et

Mat. Res. Soc. Symp. Proc. Vol. 577 © 1999 Materials Research Society

al [14] also used vibration milling under toluene to produce fine powders prior to applying the HDDR process as part of their experimental procedure.

In this paper we have investigated the effects of a Ta addition on the magnetic properties and microstructure of a conventional SmFe cast alloy as well as the influence of attritor milling, prior to the HDDR process. The effect of the Ta substitution and the pre-milling of the material on the magnetic properties, will be discussed.

Experimental

Both the SmFe and SmFeTa cast alloys were produced by conventional induction melting methods in 5 kg batches by Less-Common Metals Ltd. The ingot material was then crushed to a particle size of less than 1 mm and then reduced further by milling. The milling was carried out in an attritor mill for times of 1 to 60 minutes, under hexane, in the inert atmosphere of a glove box. The particle size of the attritor milled powder was determined using a Fisher sub sieve sizer, investigation of the size distribution of the powders was undertaken with a scanning electron microscope. The HDDR processing was carried out in a rotating furnace capable of operating between 1 bar over pressure and a vacuum of 10^{-2} mbar. Two forms of sample were introduced into the rotating furnace; crushed <1mm powder and the attritor milled powders. The initial stages of the HDDR processing were carried out in pure H_2, samples were heated at 5°C/min to 750°C and then held at this temperature for 60 mins. The second, recombination, stage was carried out under vacuum for a further 60 mins using temperatures between 700 and 900°C. All samples were subsequently nitrided at 450°C for 4 hours in a flow of nitrogen gas. Permanent magnet bonded samples were produced by mixing the powder with epoxy resin, these samples were measured at room temperature in a conventional permeameter after pulsing the magnets in a field of 5 T.

Results

Small pieces of approximately 0.5 cm^3 were cut from similar parts of each alloy and mounted and polished for metallographic examination. The resulting microstructures of the SmFe and SmFeTa materials in the as-cast condition can be seen in Fig 1.

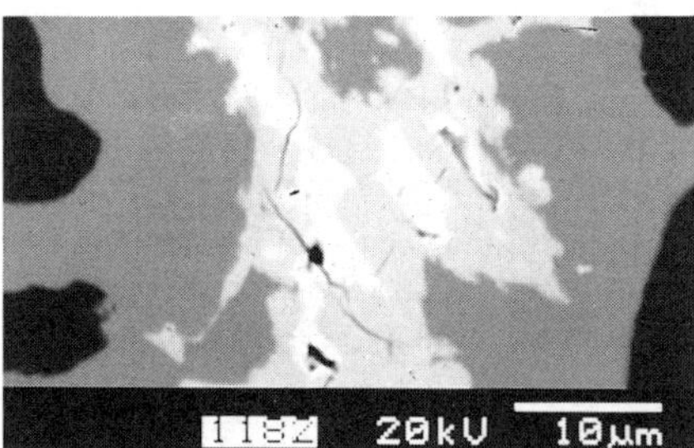

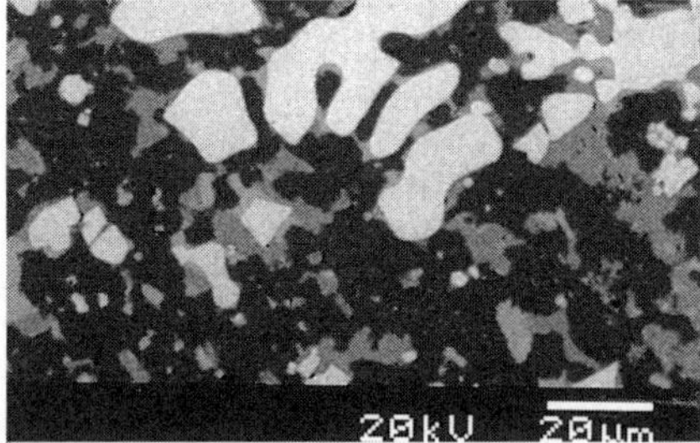

Figure 1. Microstructures of as cast SmFe and SmFeTa alloys

The conventional SmFe binary alloy exhibits the characteristic microstructure with dendrites of peritectically formed α-Fe, a Sm_2Fe_{17} phase which forms the majority phase, with $SmFe_2$ and $SmFe_3$ observed to form at the grain boundaries. The results of the analysis confirm the existence of Sm_2Fe_{17}, $SmFe_2$, $SmFe_3$ and α-Fe. The SmFeTa alloy is characterised by a very different microstructure. No α-Fe is observed. Rather, the primary phase in this case is $TaFe_2$ which forms throughout the alloy in grains of 1-10 μm. Sm_2Fe_{17} is formed as the major phase and $SmFe_2$ and $SmFe_3$ are also observed to exist. The detected phases and their composition are given in Table 1 and have been confirmed by XRD studies[15].

SmFe		SmFeTa	
Sm_2Fe_{17} *phase*	$Sm_{11}Fe_{89}$	**Sm_2Fe_{17}** *phase*	$Sm_{10}Fe_{88}Ta_2$
$SmFe_3$ *phase*	$Sm_{25.2}Fe_{74.8}$	**$SmFe_3$** *phase*	$Sm_{23}Fe_{75}Ta_2$
$SmFe_2$ *phase*	$Sm_{32.2}Fe_{67.8}$	**$SmFe_2$** *phase*	$Sm_{30}Fe_{69}Ta_1$
αFe *phase*	$Sm_{0.1}Fe_{99.9}$	**$TaFe_2$** *phase*	$Ta_{70}Fe_{30}$

Table 1. Observed phases and compositions in as cast SmFe and SmFeTa alloys

Looking first at the material introduced to the rotating furnace as coarse un-milled powders. Both SmFe and SmFeTa samples exhibit peak coercivities at recombination temperatures of 750°C (Fig.2), but the SmFeTa sample (680kA/m) has almost twice the coercivity of the Ta-free, SmFe binary alloy (360kA/m). XRD and metallographic examination of the resulting materials indicates that the SmFe binary material still contains very significant quantities of iron even after the HDDR process. The presence of this iron has a very negative effect on the material's coercivity. In both materials the coercivity decreases at processing temperatures higher than 750°C.

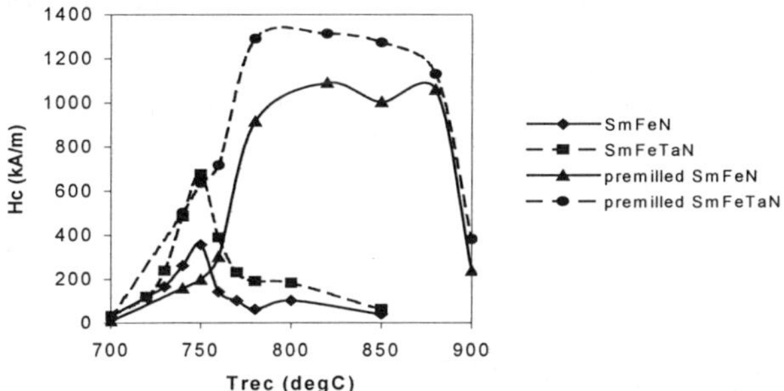

Figure 2. Variation of coercivity with recombination temperature for crushed (<1mm) and 30 min premilled Sm-Fe and SmFeTa materials

In the case of the 30 min pre-attritor milled material, similar coercivities to those obtained with the coarse, <1mm starting material are obtained in the recombination temperature range 700-750°C. Above this temperature however, whereas a fall off in coercivity occurs for the coarse material, the 30 min pre-attritor milled material continues to show an increase in the coercivity. The peak coercivity of the SmFeN and SmFeTaN prepared by the HDDR process are increased by this pre-milling from 360kA/m to 1075kA/m for SmFeN and from 680kA/m to 1280kA/m for SmFeTaN (Fig.2). At very high recombination temperatures a decrease in the coercivity of the 30 min pre-milled material is observed for both SmFe and SmFeTa materials.

The results of a FSSS particle size analysis on the SmFeTa materials milled for between 1 and 60 mins can be seen in Fig 3. Measurements indicated that the average size of the particles decreases rapidly with milling time at least for the first 10-15 mins before settling at size of approximately 3-4 μm. SEM micrographs of the SmFe and SmFeTa powders (Fig.4) show that the size distribution for the two materials is basically similar for the same milling times.

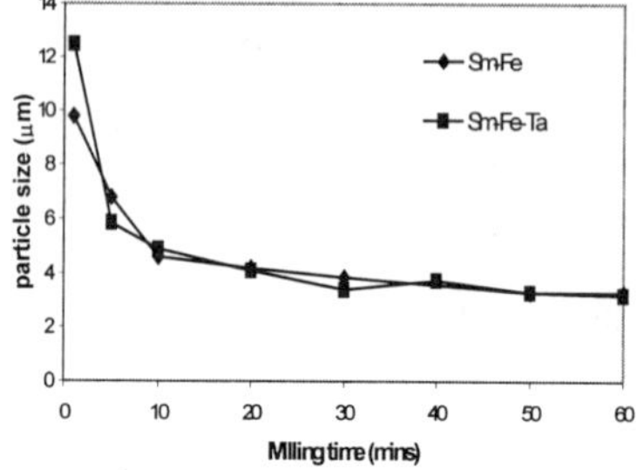

Figure3. Variation of particle size with attritor milling time

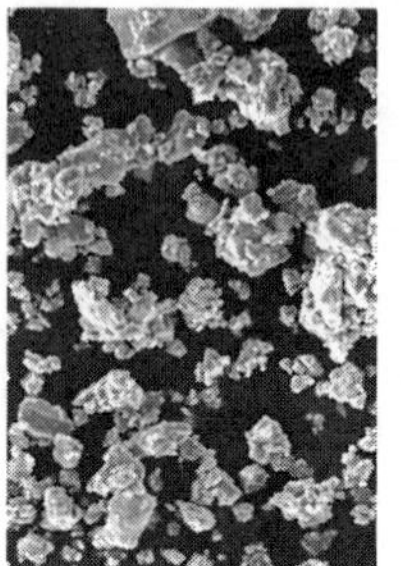

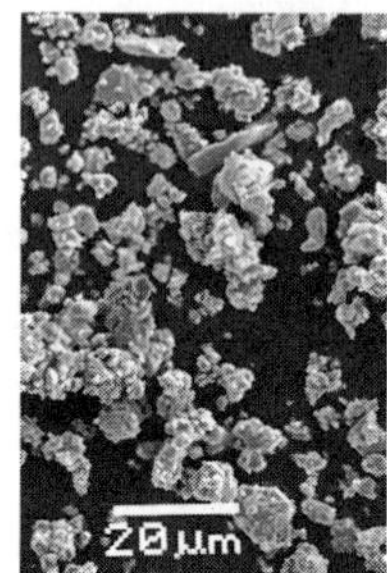

Figure 4. Pre-milled SmFe and SmFeTa material (40 mins)

Figure 5 shows the effect of pre-milling time on SmFe and SmFeTa processed at recombination temperatures of 740 and 850°C. Both materials show a sharp increase in coercivity with relatively short milling times when a recombination temperature of 850°C is used. This is particularly so for the SmFeTa material. In contrast milling time appears to have little effect when the lower recombination temperature is used.

A non pre-milled sample of SmFe desorbed and recombined at 850°C was mounted, polished and etched for examination on the SEM. The microstructure in Fig. 6. indicates the presence of grains significantly larger than 5μm.

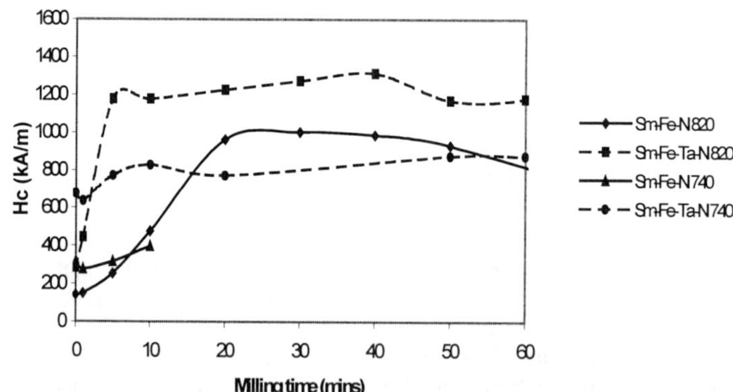

Figure 5. Variation of coercivity with milling time for SmFe and SmFeTa materials recombined at 740 and 820°C

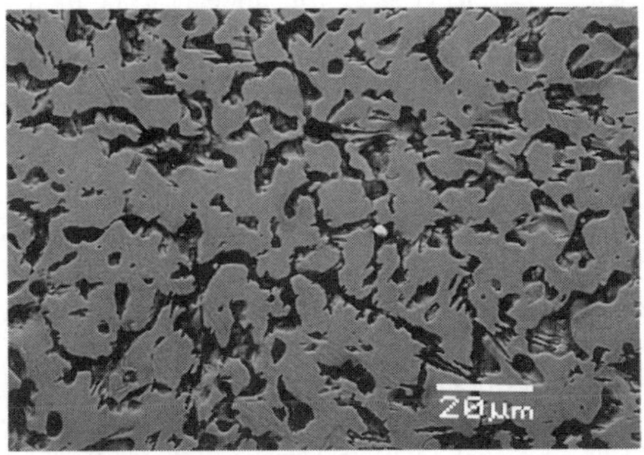

Figure 6. SEM micrograph showing big grains (>20 µm) of Sm_2Fe_{17} phase (grey)

Discussion

The addition of Ta, or a similar element is clearly important to developing the best coercivities from SmFe based materials. Heat treatments to homogenise the cast materials are a potential alternative but are less attractive for reasons of cost and ease of manufacture. The use of pre-milled material in the HDDR process and the resulting ability to process at much higher temperatures to some extent reduces the benefits of Ta additions because the higher temperatures cause a homgenisation of the material and a decrease in the α-iron content, nevertheless, ternary materials produce consistently higher coercivities. This beneficial homgenising effect of the HDDR process has been previously observed for near stoichiometric NdFeB based materials [16]

Samples pre-milled for 30 mins prior to the HDDR process show a very substantial improvement over the coarse starting material. At low recombination temperatures the coercivities for both types of starting material are essentially the same, only at temperatures above 750°C do differences in properties become apparent. The fall off in coercivity at higher recombination temperatures for the non pre-milled material is usually ascribed to the increase in grain size, the importance of achieving grain sizes comparable with that of the single domain size of ~0.3µm is often pointed out, for example [17], however it seems unlikely that if the critical grain size is ~0.3µm then it should be important whether the particle size is 5µm or 500µm since both these particle sizes are very substantially larger than the single domain size. The results in Fig.5 further emphasise this. The low recombination temperature samples are largely particle size independent. The grains in these samples are not growing very quickly because of the relatively low temperature and so the size of the particles which contain these grains is relatively unimportant. In the case of the high temperature processing, particle size is critical. The elevated temperatures cause a considerable degree of grain growth and the resulting coercivities in non pre-milled, or 1 and 5 mins pre-milled, material are low, but with a particle size reduction to ~5µm (for the SmFeTa) material a considerable increase in the coercivity is observed. The SEM micrograph in Fig. 6 shows the presence of grains substantially larger than 5µm in the coarse material for samples recombined at 850°C, such grains could not exist in the pre-milled material simply because of its physical dimensions. This being the case it would seem that in order to achieve the highest coercivities for SmFe based HDDR materials it is critical to restrict the size of the largest grains rather than attempt to produce an even distribution of smaller grains. At much higher recombination temperatures the coercivity of the pre-milled material does begin to fall off (Fig.2), this appears to be the result of partial sintering of the material which allows the possibility of the growth of some much larger grains as the particles begin to fuse together.

Conclusions

The introduction of Ta has a very significant and beneficial impact on both the cast material and the final nitrided HDDR product. The enhanced values of coercivity in the SmFeTa material indicate the critical importance of Ta in reducing as far as possible any free iron in the material. This is of lesser importance in the pre-attritor milled material because of the higher processing temperatures which enable a greater degree of homogenisation to take place during the HDDR process than with low temperature processing associated with coarse starting material.

Milling of the material before the HDDR treatment, reduces the particle size and prevents the grains from growing to a size which influences negatively the coercivity. At elevated temperatures the coercivity drops because of sintering and agglomeration, which allows larger grains to form.

Acknowledgement

The Ministry of Science and Technology of Slovenija is gratefully acknowledged for the provision of research funds.

References

1. JMD Coey and H Sun, *J. Magn. Magn. Mater.*, 87 (1991) L251
2. K Schnitzke, L Schultz, J Wecker and M Katter, *Appl. Phys. Lett.*, 57 (1990) 2853
3. M Endoh, M Iwata and M Tokunaga, *J. Appl. Phys.*, 70 (1991) 6030
4. M Katter, J Wecker and L Schultz, *J. Appl. Phys.*, 70 (1991) 3188
5. C N Christodoulou and T Takeshita, J. Alloys & Comp., 196 (1993) 161
6. MQ Huang, LY Zhang, BM Ma, Y Zbeng, JM Elbicki, WE Wallace and SG Sankar, *J. Appl. Phys.*, 70 (1991) 6027
7. T Takeshita and R Nakayama, *Proc. 11th Int. W'shop on RE Magnets & their Appl.* (1991) 49
8. PJ McGuiness, XJ Zhang, H Forsyth and IR Harris, *J. Less-Common Met.*, 162 (1990) 379
9. SA Sinan, DS Edgely, IR Harris, J. Alloys. Comp. 226(1995)170
10. AE Platts, IR Harris and JMD Coey, J. Alloys. Comp. 185(1992)251
11. B Saje, AE Platts, S Kobe Besenicar, IR Harris and D Kolar, IEEE Trans. Magn. 30(1994)690
12. B Gebel, M Kubis and K-H Muller, J. Magn. Magn. Mat. 174(1997)L1-L4
13. K-H Müller, L Cao, NM Dempsey and PAP Wendhausen, J. Appl. Phys. 79(8)(1996)5045
14. B Gebel, M Kubis and K-H Müller, J. Magn. Magn. Mater. 174(1997)L1-L4
15. K Žužek, PJ McGuiness and S Kobe, accepted by J. Alloys & Compounds
16. XJ Zhang, XJ Yin, PJ McGuiness and IR Harris, J. Mat. Proc. Tech., 48(1995) 461
17. X Zhao, Z Zhang, W Lei, Q Xiao and XK Sun, J. Magn. Magn. Mater., 148(1995)419

MICROSTRUCTURAL AND MAGNETIC IMPROVEMENTS IN TUBE-CAST Pr-Fe-B-Cu ALLOYS VIA HEAT TREATMENTS AND RAPID UPSET FORGING

G P Hatch *, A J Williams and I R Harris

School of Metallurgy and Materials, University of Birmingham, Birmingham B15 2TT, UK
* Now at Dexter Magnetic Technologies, 1050 Morse Avenue, Elk Grove Village, IL 60090

ABSTRACT

Alloys of Pr-Fe-B-Cu were cast into Cu and stainless steel tubes. It could be seen that a preferred crystallographic orientation was obtained after cooling. Fine platelets of $Pr_2Fe_{14}B$ matrix phase were formed, surrounded by various grain boundary phases, and there was a significant reduction in the amount of free Fe present, in comparison to conventional slab cast alloys. In order to improve the magnetic properties in the as-cast state, two alternative routes were used. The first involved a series of two step heat treatments to develop an improved microstructure. After an optimum heat treatment of 1000 °C for 24 hours + 500 °C for 3 hours, significantly improved magnetic properties were obtained for a 17.5 % Pr alloy; B_r = 752 mT, H_{ci} = 613 kAm^{-1} and BH_{max} = 96 kJm^{-3}. The second route involved a rapid upset forging [RUF] process, with a strain rate of 11.5 s^{-1}, to hot deform the alloys. Following post-forging heat treatments, properties of B_r = 966 mT, H_{ci} = 780 kAm^{-1} and BH_{max} = 160 kJm^{-3} were obtained for a 15% Pr alloy. That nature of the improvements in properties as a result of heat treatments and RUF were investigated and are discussed in the present work.

INTRODUCTION

In recent years there have been several studies of the microstructural characteristics of Pr-Fe-B-Cu alloys in the as-cast state. Previous workers, most notably Shimoda et al [1], were able to show that in the as-cast state, Pr-Fe-B-Cu alloys exhibit an appreciable coercivity. This was significantly improved with subsequent anneals; workers in this laboratory have shown that this also occurs in similar alloys containing Nd as well as Pr [2, 3].

The present study examined the microstructural characteristics of alloys cast into metal tubes. Casting directly into a tube, it was thought, would induce a preferred crystallographic orientation within the alloys, due to the radial nature of the heat flow in the cylinders during solidification. Alloys were also cast in slab form for comparison.

Much work has also been done on developing a viable hot deformation process. Shimoda et al [4-7] looked in some detail at Pr-Fe-B alloys. These workers found that Cu additions improved the magnetic properties. Ohki et al [8] investigated the deformation of Pr-Fe-B-Cu alloys further. Further work by Shimoda et al [9] compared the upset forging process to hot rolling. Annealing these alloys was found to enhance their magnetic properties. Hot rolling was found to lead to a 'sea of grains' microstructure associated with high strain rate deformation. Other workers [10, 11] continued investigations on hot rolled materials, concluding that post-deformation heat treatments were critical in obtaining good magnetic properties. Akioka et al [12] used an iron ring to constrain Pr-Fe-B-Cu alloys during upset forging to prevent cracking and excessive discharge of the Pr-rich liquid.

Chen et al [13] did further work, on Pr-Fe-B-Cu alloys and observed that the $Pr_2Fe_{14}B$ matrix grains were crushed during forging. Kwon et al [14, 15], also looked at the upset forging of high Pr-based alloys. These workers theorised that the removal of free Fe during forging was due to the cracking of the $Pr_2Fe_{14}B$ grains, which caused the free-Fe and Pr-rich liquid to react peritectically. Rotation of the platelet-like grains caused magnetic alignment with the process being aided by a significant volume fraction of liquid phase. Other workers have undertaken further work on upset forging [16-21].

As these previous workers have shown, the hot deformation of cast Pr-Fe-B-Cu alloys leads to significant improvements in magnetic properties, especially when combined with post-forging heat treatments.

Mat. Res. Soc. Symp. Proc. Vol. 577 © 1999 Materials Research Society

In addition to looking at as-cast alloys, the present work looked at these tube-cast alloys after upset forging, as part of a preliminary study of the process [19, 20, 22]. It was thought that using tubes to contain such alloys would aid in reducing oxidation and would facilitate more stable deformation.

EXPERIMENTAL

All of the alloys investigated in the present work were supplied by REP of Widnes (UK). Molten alloys were poured into tubes of Cu or stainless steel. These alloys were then allowed to cool. The compositions of the alloys were $Pr_{15}Fe_{78.5}B_5Cu_{1.5}$, $Pr_{17.5}Fe_{76}B_5Cu_{1.5}$ and $Pr_{20}Fe_{73.5}B_5Cu_{1.5}$ [atomic %].

The final alloys were cylindrical, approximately 300 mm in length and 32 mm in diameter. After heat treatments in a vacuum furnace, cross-sectional slices was taken from each cylinder. Cross-sections of each of the slab-cast alloys were cut into small cubes.

Once cut, all samples were mounted and then ground and polished. All samples were examined optically; some samples were also examined in a Scanning Electron Microscope [SEM], switched to the Back Scattered Electron [BSE] mode. Energy Dispersive X-ray (EDX) measurements were used to identify the compositions of the phases within each system.

For rapid upset forging, a muffle furnace was located close to a 200 Tonne mechanical press. Sample tubes with an aspect ratio (l:d) of 1.5:1 were used. The platens of the mechanical press were coated with graphite to further aid the deformation process. The platens were then heated to reduce thermal shock. The machine produced an average strain rate of approximately 11.5 s^{-1}, comparable to the strain rate found during hot rolling. Samples were then heat treated in a vacuum furnace and samples taken as outlined above.

RESULTS

The effects of heat treatments on pre-forged alloys

Typical optical micrographs of the tube-cast alloys are shown in Figure 1. Each alloy contained

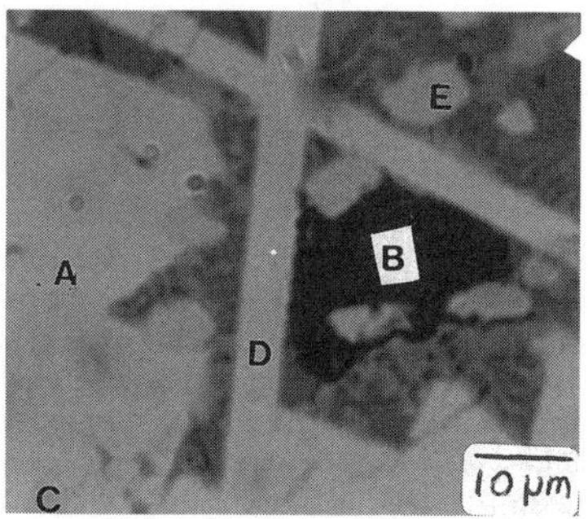

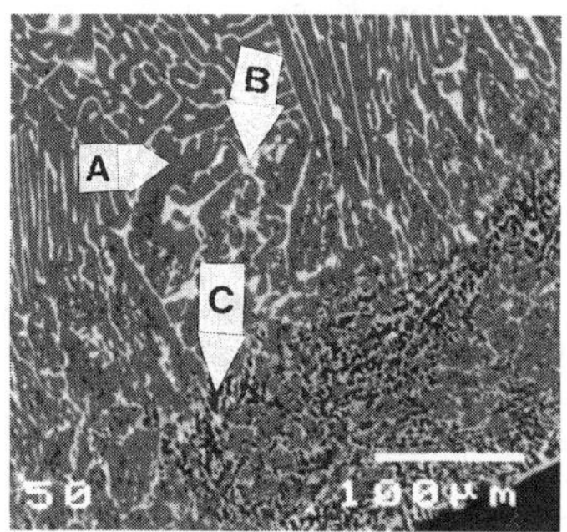

Fig 1a) Tube-cast $Pr_{15}Fe_{78.5}B_5Cu_{1.5}$ **Fig 1b) Tube-cast $Pr_{20}Fe_{73.5}B_5Cu_{1.5}$**
[A=$Pr_2Fe_{14}B$ B=Pr-rich C= Free Fe D=$Pr_6Fe_{13}Cu$ E=Boride]

several phases; the largest volume fraction was the $Pr_2Fe_{14}B$ matrix phase. Free Fe was also present, along with various grain boundary phases including $Pr_6Fe_{13}Cu$ (δ phase).

A phase distinct from those identified by EDX was also found. The appearance of this phase was similar to that previously reported [23, 24] as a boride phase. The needle-like $Pr_6Fe_{13}Cu$ phase was found at the grain boundaries, along with the Pr-rich - PrCu eutectic and primary Pr-rich phase.

Each of the tube-cast alloys exhibited noticeable directional solidification. This was particularly evident in the 20% Pr alloy as seen in Figure 1b. It could be seen that there were long columnar grains emanating from the centre of the tube. At the edge of the alloy, a fine grain structure could be seen.

Within these grains of $Pr_2Fe_{14}B$, free Fe could be seen. This was most likely as a result of the incomplete peritectic reaction

$$L + Fe \rightarrow Pr_2Fe_{14}B \quad (1)$$

Free Fe was absent from the long columnar grains present further into the alloy. The difference between these grains, and the much finer grains that contain free Fe, is explained in terms of a grain refinement mechanism discussed in other studies by workers in this laboratory [2,3].

The bulk of the alloy consisted of platelet grains with a high aspect ratio. Although there was free Fe present within the matrix grains, there was very little present compared to that of grains in the slab-cast alloys. It seems that the process of direction solidification that occurs with the tube-cast alloys leads to less free Fe remaining after cooling.

A two step heat treatment regime was used to improve the magnetic properties of these alloys. A homogenisation of 1000 °C for the first step, and an anneal of 500 °C for the second was used. The alloys were subjected to these temperatures for various times. Each heat treatment took place under vacuum; after each step, the furnace was rolled off the furnace tube and the samples, remaining in the evacuated tube, were left to cool in air.

Figure 2 shows the typical microstructure of a heat treated 15% alloy. The heat treatments had number of effects. The first was an increase in the amount of the $Pr_6Fe_{13}Cu$ present at the grain boundaries. The morphology of the phase was different too; broad bands of $Pr_6Fe_{13}Cu$ could be seen, instead of the previous needle-like morphology.

A second important effect was the virtual elimination of free Fe from the matrix grains; even after as short a heat treatment as 1000 °C / 2 h + 500 °C / 1 h [the heat treatment applied to the alloy shown in figure 2], there was virtually no free Fe left. There was also a noticeable increase, in certain regions of the grain boundaries, of the PrCu phase, normally found within the Pr-rich - PrCu eutectic. There was little evidence of the Pr-rich phase at the grain boundary of these alloys after heat treatment.

The other important effect, was the apparent coarsening of the Pr-rich - PrCu eutectic present. There also seemed to be a decrease in the amount of this eutectic present - possibly due to the increase on PrCu and $Pr_6Fe_{13}Cu$ phases.

Other observations included a generally smoothening of matrix grains after heat treatment. The apparent boride phase within the grain boundary region was absent after treatment.

Figure 3 shows the second quadrant demagnetisation curve for the as-cast 15%, 17.5% and 2-% Pr Cu tube-cast alloys. It can be seen that the as-cast remanence was highest for the 15% Pr alloy but

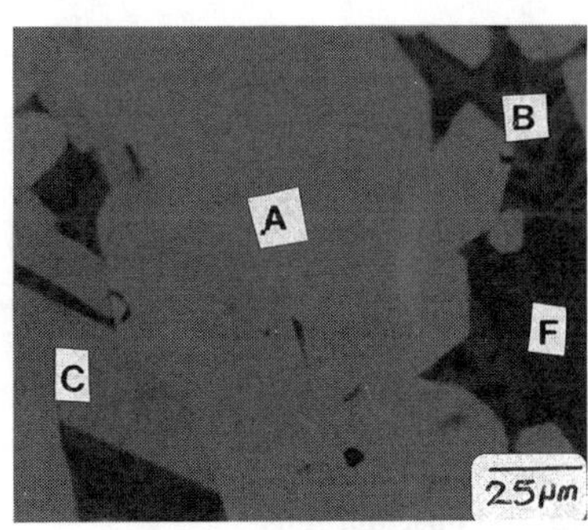

Fig 2) Tube-cast 15% Pr alloy after heat treatments [A=$Pr_2Fe_{14}B$; B=Pr-PrCu eutectic; C=$Pr_6Fe_{13}Cu$; F=PrCu]

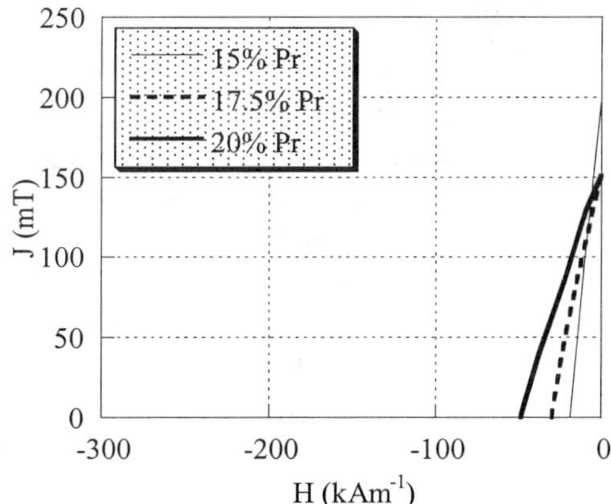

Fig 3) second quadrant demagnetisation curves for 15%, 17.5% and 20% Pr alloys [as-cast].

that this alloy had the lowest coercivity. This can be explained by the larger volume fraction of $Pr_2Fe_{14}B$ matrix phase being present.

Figure 4 shows the second quadrant demagnetisation curves for the optimum heat

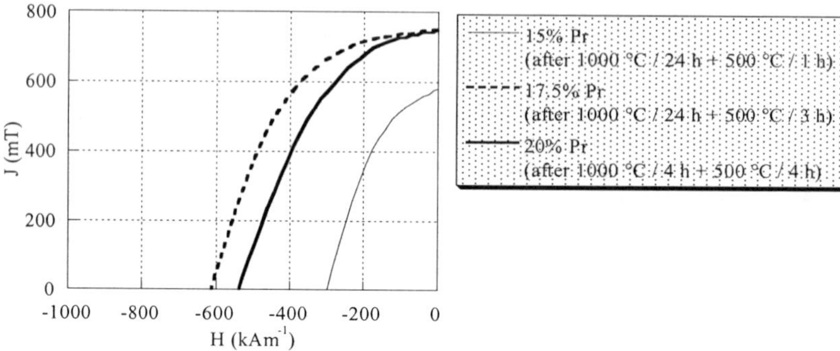

Fig 4) Second quadrant demagnetisation curves for optimally heat treated alloys.

treatments of each of the alloys. All measurements were taken parallel to the axis of the tube. The heat treatments significantly improved the magnetic properties of the alloys. For almost all of the samples, the best properties were obtained after heat treating for 24 hours.

The effects of heat treatments on rapidly upset forged tube-cast alloys

Samples of steel tube-cast alloys were cut and upset forged in the mechanical press, to deformations of up to 84 %. The samples were forged without the aid of an inert atmosphere. Once forged and cooled, the samples were sectioned.

Figure 5 shows the bulk microstructure of the 20% Pr alloy after rapid upset forging. Free Fe was still present in some of the grains. There was some evidence of cracking in the bulk

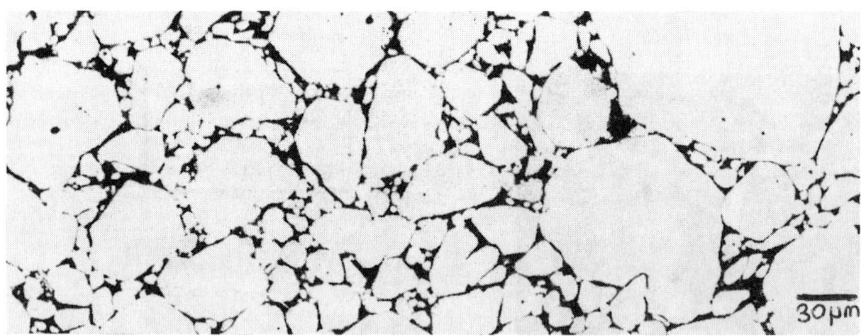

Fig 5) Bulk grains in 20% Pr alloy after rapid upset forging.

grains, which may have allowed most of the free Fe to react with the liquid grain boundary phase. The speed of the process probably prevented this reaction from going to completion. Specimens of the alloys were heat treated for various times using a similar regime to the heat treatments applied to the as-cast alloys. The best second quadrant demagnetisation curves were obtained for the 15% Pr alloy and are shown in Figure 6. The microstructure of this optimally heat treated alloy was not dissimilar to those of

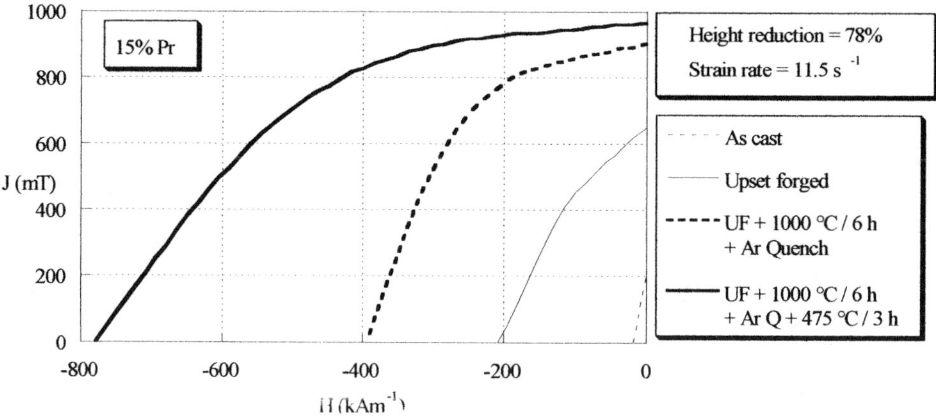

Fig 6) Best second quadrant demagnetisation curves for 15% Pr alloy after processing.

the un-heat treated alloys. The heat treatment appeared to have virtually eliminated all the remaining free Fe, though some did remain.

The microstructure of these alloys, both after rapid upset forging (RUF) and heat treatments, consisted of the classic 'sea of grains' microstructure reported by other workers in hot rolled materials [9, 25]. It would seem that this structure is a characteristic of the high strain rate used, as opposed to the particular deformation mechanism used.

The optimum properties of B_r = 966 mT, $_iH_c$ = 780 kAm^{-1} and BH_{max} = 160 kJm^{-3}, were obtained for the 15% Pr alloy rapidly upset forged to 78% height deformation, at 1000 °C and with a strain rate of 11.5 s^{-1}, and then heated treated at 1000 °C / 2 h + 500 °C / 1 h. It is possible that pre-forging heat treatments, similar to those that were applied to the as-cast alloys, would improve the properties even further.

CONCLUSIONS

Casting Pr-Fe-B-Cu alloys into tubes induced a preferred crystallographic orientation in the final alloy. Alloys cast into tubes had less free Fe present within the matrix grains than in similar slab cast alloys. The magnetic properties of tube-cast Pr-Fe-B-Cu alloys were significantly improved after heat treatments. The best magnetic properties obtained were $\mathbf{B_r}$ = 752 mT, $_i\mathbf{H_c}$ = 613 kAm^{-1} and $\mathbf{BH_{max}}$ = 96 kJm^{-3}, obtained for a 17.5% Pr alloy, after a heat treatment of 1000 °C / 24 h + 500 °C / 3 h. It could be seen in the tube-cast alloys that free Fe could be eliminated after very short homogenisation times.

Upset forged alloys were successfully processed without the need for an inert atmosphere. A two step heat treatment was found to improve significantly the magnetic properties. Heat treating the alloys led to significant improvements in the magnetic properties. The optimum properties of B_r = 966 mT, $_iH_c$ = 780 kAm^{-1} and BH_{max} = 160 kJm^{-3}, were obtained for a 15% Pr alloy, rapidly upset forged to 78% height deformation, at 1000 °C and with a strain rate of 11.5 s^{-1}, and then heated treated at 1000 °C / 2 h + 500 °C / 1 h.

REFERENCES

[1] T Shimoda, K Akioka, O Kobayashi and T Yamagami, J. Appl. Phys., **64**, p 5290, 1988

[2] G W Delamore, G P Hatch, P J McGuiness, G J Mycock and I R Harris, Proc. 8th International Symposium on Magnetic Anisotropy and Coercivity in Rare Earth Transition Metal Alloys, pp 253-260, Birmingham, 1994

[3] G P Hatch, G W Delamore and I R Harris, submitted to J. All. Comp., 1999

[4] T Shimoda, K Akioka, O Kobayashi and T Yamagami, J. Appl. Phys., **64**, pp 5290-5292, 1988a

[5] T Shimoda, K Akioka, O Kobayashi and T Yamagami, J. de Phys., **49**, pp 631-632, 1988b

[6] T Shimoda, 'The Global Business and Technical Outlook for NdFeB Magnet Markets', Gorham Advanced Marketing Institute, Monterrey, 1989a

[7] T Shimoda, K Akioka, O Kobayashi and T Yamagami, Proc. 10th. International Workshop on Rare Earth Magnets and their Applications, pp 389-398, Kyoto, 1989c

[8] T Ohki, T Yuri, M Miyagawa, Y Takahashi, C Yoshida, S Kambe, M Higashi and K Itayama, Proc. 10th. International Workshop on Rare Earth Magnets and their Applications, pp 399-407, Kyoto, 1989

[9] T Shimoda, K Akioka, O Kobayashi, T Yamagami, T Ohki, M Miyagawa and T Yuri, IEEE. Trans. Magn., **25**, pp 4099-4104, 1989b

[10] W C Chang, C R Paik, H Nakamura, N Takahashi, S Sugimoto, M Okada and M Homma, IEEE. Trans. Magn., **26**, pp 2604-2606, 1990

[11] T Shimoda, K Akioka, O Kobayashi, T Yamagami and A Arai, Proc. 11th. International Workshop on Rare Earth Magnets and their Applications, pp 17-28, Pittsburgh, 1990

[12] K Akioka, O Kobayashi, T Yamagami, A Arai and T Shimoda, J. Appl. Phys., **69**, pp 5829-5831, 1991

[13] Z Chen, F Xie, Z Shi, L Wang and H Fu, J. Appl. Phys., **70**, pp 2868-2870, 1991

[14] H W Kwon, P Bowen and I R Harris, J. All. Comp., **189**, pp 131-137, 1992a

[15] H W Kwon, P Bowen and I R Harris, Proc. 12th. International Workshop on Rare Earth Magnets and their Applications, pp 705-719, Canberra, 1992b

[16] E Iwamura, T Yuri, A Hanaki, H Mitani and K Itayama, Proc. 12th. International Workshop on Rare Earth Magnets and their Applications, pp 384-391, Canberra, 1992

[17] Y Luo and T-S Xia, Proc. 12th. International Workshop on Rare Earth Magnets and their Applications, pp 473-481, Canberra, 1992

[18] G P Hatch, A J Williams, G J Mycock and I R Harris, J. Mag. Magn. Mats., **157/158**, pp 69-70, 1996

[19] G P Hatch, A J Williams, I R Harris and G J Mycock, Proc. 14th International Workshop on Rare Earth Magnets and their Applications, pp 501-510, Sao Paulo, 1996

[20] G P Hatch, A J Williams and I R Harris, submitted to J. Mag. Magn. Mats., 1999

[21] M Ferrante, V Sinka, O B G Assis, I de Oliveira and E de Freitas, Proc. 14th. International Workshop on Rare Earth Magnets and their Applications, pp 255-263, Sao Paulo, 1996

[22] G P Hatch, A J Williams and I R Harris, submitted to J. Mag. Magn. Mats., 1999

[23] H W Kwon, P Bowen and I R Harris, J. All. Comp., **182**, pp 233-242, 1992

[24] R N J Faria, PhD thesis, University of Birmingham, 1993

[25] M Leonowicz, P J McGuiness, G P Hatch, H A Davies and I R Harris, J. Mag. Magn. Mats., 166, pp 165-171, 1997

High Performance, Light weight Thermoplastic/Rare Earth Alloy Magnets

Jun Xiao[1], Joshua U. Otaigbe[1, 2]
[1] Department of Materials Science and Engineering
[2] Department of Chemical Engineering
Iowa State University of Science and Technology, Ames, IA 50010

ABSTRACT

We report progress on our exploratory research on surface modification of magnetic NdFeB fillers, characterization of suitable magnetic rare earth alloy powders and high-performance polymer matrices, processability, and properties of novel thermoplastic/NdFeB magnets. The results suggest that blending liquid crystal polymer (LCP) with a high-thermoplastic polymer such as polyphenylene sulfide (PPS) provides the required balance of properties. These properties include superior magneto-mechanical performance, minimal melt viscosity at optimal NdFeB volume loading, enhanced thermal stability, high stiffness, high strength, improved dimensional stability, and excellent chemical resistance; making the thermoplastics magnets suitable for use in high temperature and aggressive environments where commercial polymer-bonded magnets are not useable.

INTRODUCTION

Neodymium-iron-boron(NdFeB), commonly known as "Neo" magnetic materials, is being widely used because of the combination of superior magnetic properties and low cost. However, the inherently poor thermal stability of NdFeB limits both its processability and potential automobile applications in "under-the-hood" environment (normally exceeding 180 °C). This limitation is due to the fact that the magnetic powder is readily oxidized especially at elevated temperature [1, 2]. Further, exposure to humid environments deteriorates the performance of NdFeB magnets because the presence of water corrodes the material [3, 4]. Consequently, the magnetic materials degrade and eventually the magnet is destroyed due to excessive corrosion in the magnets. To enlarge the processing window and prevent the poor performance caused by oxidation and corrosion, we performed chemical surface modification on the NdFeB magnets to improve the thermal and environmental stability [5].

Due to their superior mechanical damping characteristics, corrosion resistance, and facile processing, thermoplastic polymer bonded rare earth alloy magnets are used extensively in electromechanical devices such as motors and actuators. In these applications, injection-moldable thermoplastic/NdFeB alloy magnets offer a number of opportunities for component design of parts with intricate and complex shapes. However, the high injection molding temperature (350 ± 50°C) required by the high-performance thermoplastic polymers causes thermal oxidation of the NdFeB alloys, unless the alloys are coated with an impervious coating or passivation layer prior to incorporating them into the thermoplastic polymer.

The purpose of this paper is to describe the effect of the coating treatment on the high temperature stability and moisture resistance of NdFeB powder. The effects of the coating agent on magnetic performance are characterized by short and long term irreversible loss test (STILT and LTILT, respectively). PPS and LCP were

Mat. Res. Soc. Symp. Proc. Vol. 577

investigated as potential binders or matrices for the coated NdFeB fillers for making useful thermoplastic polymer bonded magnets with desirable magneto-mechanical properties. The mechanical properties of the polymer bonded magnets were characterized under varying test conditions that they are likely to encounter during use.

EXPERIMENT

Materials and Sample Preparation

Commercial NdFeB magnetic powder (MQP-O, Magnequench International, Inc.) was used for this study. The powder is a multimodal mixture of plate-like particles. The surface modification of the NdFeB was performed following the method developed by Otaigbe and coworkers [6]. Both polyphenylene sulphide (Ryton P-6, Philips Petroleum Company) and liquid crystal polymer (Vectra 900, Ticona plastics) were used as binders. A dry blend of the PPS, LCP and magnetic NdFeB powders was mixed by melt blending according to the desired composition. Before blending, the neat polymers were dried in an oven at 120°C for at least 36 hours and the mixing was performed for 20 min. at 320 °C in a roller rotor mixer (HAAKE Rheocord90) operating at 10 rpm. These mixing conditions were found to give reproducible results.

Compression moldings were performed in a lab-scale compression molding press. Small disks (25 mm diameter × 3 mm height) were compression molded under a pressure of 7 MPa at 290 ± 5 °C using the premixed PPS+LCP+NdFeB batch in a standard lab-scale compression molding press. A mold residence time of about 15 minutes was used and the molded sample was allowed to cool in air to ambient temperature under the molding pressure of 7 MPa. The specimens used for the dynamic mechanical measurements were cut from the molded disks.

Thermogravimetric analysis

TGA tests were performed on all the samples using a Perkin-Elmer TGA-7, under identical test conditions that were found to give comparative reproducible data. The samples were heated from ambient temperature to 400 °C at a heating rate of 40 °C/min using air as the purge gas. A calibrated humidity controller was used to control the humidity of the purge gas. The weight changes of the samples were recorded dynamically. The TGA runs for each sample were performed in triplicate and the mean values of the data obtained are reported in this paper.

Dynamic Mechanical Analysis

A dynamic mechanical analyzer (Perkin-Elmer DMA 7) was used to perform the dynamic mechanical testing in the 3-point bend configuration. Rectangular DMA test specimens (length = 19 mm, width = 2 mm and depth = 1.5 mm) were cut from the compression molded samples using Buehler slow speed diamond saw. The specimen span length was 15 mm. The experiments were performed at a scanning rate of 5°C/min from 25 to 260 °C and a frequency of 1 Hz. All experiments were conducted in triplicates and the results averaged.

Magnetic properties measurements

The samples were magnetized by using Walker Scientific® magnetizer at a voltage of 1,400 V, which generates 4,500 kOersted field strength to saturate the PPS bonded NdFeB magnets. The samples used for STILT test were saturated at room temperature and their flux density (B) were obtained by using Helmholtz coil (Magnetic Instrumentation Inc. ML-40D). After that all the samples were heated in a convection air oven (Tenney Jr.) at 50 °C for an hour. After heating, the samples were cooled on a metal grid in air for an hour and the flux density was measured again. The above procedure was repeated at 90, 130 and 180 °C and the corresponding flux density was recorded. The samples were saturated again after all the above measurement and the flux density was obtained for computing the irreversible flux density loss. The LTILT test was performed in a similar way except that it was done at 180 °C for 168 hours.

Morphology Analysis

Scanning electron micrographs of the coated NdFeB magnetic powders were obtained with a JEOL 6100 Scanning Electron Microscope with an accelerating voltage of 10 kV. An EDS detector was used to detect the generated x-ray in windowless mode.

RESULTS

The purpose of the surface modification is to passivate the NdFeB powder by encapsulating the magnetic filler with a protective coating layer. Due to their well known properties such as improved filler dispersion and water resistance [7, 8], organofunctional silane coupling agents are used widely in fiberglass industry as adhesion promoters between organic polymers and glass fibers. In this study, the surface modification was performed with silanes because silane coupling agents contain chemical functional groups that can react with -OH groups on NdFeB powders. By examining the treated NdFeB powder using x-ray mapping, it can be seen in Fig. 1 that the Si element (from the silane) is uniformly distributed on the magnetic filler surface.

Thermogravimetric analysis was used to study the oxidation and corrosion behavior of NdFeB powder dynamically at elevated temperature and in different environments. Since the processing temperature of the thermoplastics used is around 400 °C, the treated and untreated NdFeB powder were heated to 400 °C and held there for 40 minutes under air atmosphere. It has been reported that the humidity reduces the thermal stability of rare earth magnetic filler [3, 4], therefore, the TGA experiments were also performed at 400 °C with 100% humid air as purge gas to understand the effect of coating

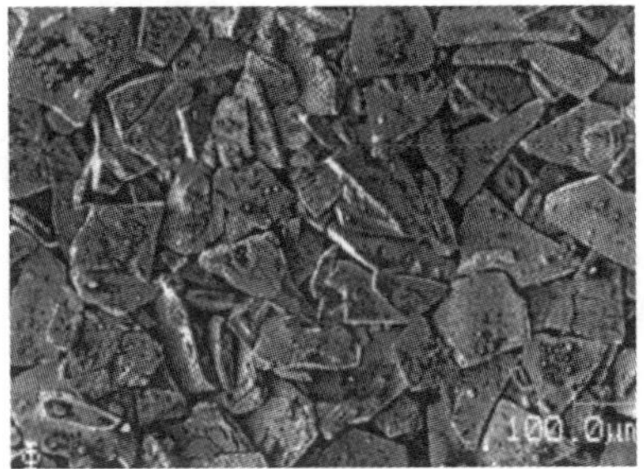

(a)

(b)

Fig.1. Morphology of coated magnetic NdFeB powder:
(a) Secondary electron image and (b) Si map of (a) after sputtering

in a humid environment. The TG curves obtained are shown in Figures 2 and 3.

It can be seen that the untreated powder showed a more severe oxidation or corrosion at elevated temperature compared to the treated powder in both dry and humid environment (see Figs. 2 and 3). It has also been observed that the humidity of purge gas has a negative effect on the thermal stability of NdFeB powder, which limits the application of polymer bonded NdFeB magnets.

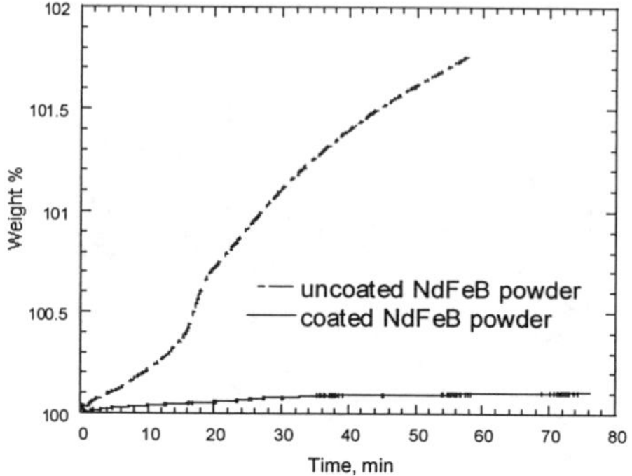

Fig. 2. TG curves of coated (solid line) and uncoated (dashed line) NdFeB powder heated in dry air for 40 min. at 400 °C

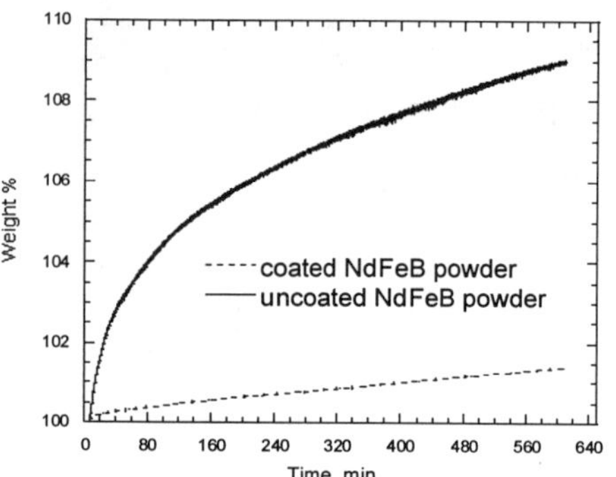

Fig. 3. TG curves of coated (dashed line) and uncoated (solid line) NdFeB powder heated in humid air for 10 hrs. at 400 °C

The surface modification was found to reduce the weight gain of the treated NdFeB powder by 1/18 (dry air, 40 min. at 400 °C) and 1/6 (100% humid air, 10 hrs at 400 °C) relative to that of the untreated powder. It should be emphasized that the surface modification also served as corrosion barrier in the humid environment in the long term test discussed later. It is believed that the passivation layer provides the observed improved thermal stability to the rare earth magnetic fillers. Consequently, polymer bonded magnets made with treated powder as in this study will exhibit improved magnetic performance for new applications where the untreated commercial polymer bonded magnets are not usable.

The coating on the magnetic fillers protects them from the moisture and oxygen (from air) attack at elevated temperature. This corrosion and oxidation protection are expected to provide enhanced magnetic properties at high temperature. The results of short term irreversible loss test (STILT) and a portion of the long term irreversible loss test (LTILT) confirm the expectation of corrosion and oxidation protection (see Figs. 4 and 5).

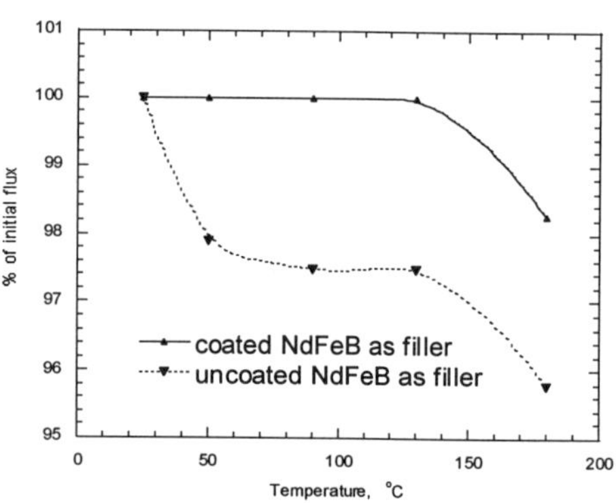

Fig. 4. STILT curves of PPS bonded NdFeB magnets (62 vol %, compression molded at 320 °C for 15 min.)

The STILT results showed that the loss in flux density (B) of PPS bonded treated powder was much smaller relative to that of untreated powder. After cooling and re-saturating the magnets, it was found that the irreversible loss of flux density of the treated sample was zero while that of untreated sample was 2.53%. The LTILT results showed that the decrease in flux density of the treated sample was slower than that of the untreated sample (Fig. 5). It is proposed that while NdFeB magnets are in the hot and humid environment, the corrosion mechanism start with a surface reaction of water vapor with the Nd-rich grain boundary phase, resulting in a volume expansion of the grain boundary phase. In turn, the volume expansion leads to the decrepitation of the magnetic materials. The hydrogen reaction with the Nd-rich grain boundary phase is the predominant mechanism in the humidity-related corrosion of NdFeB magnets [9]. Even in this corrosive environment, the passivation layer on the surface of the powder is still effective in preventing the corrosion. These results further confirm that the coated NdFeB powder has excellent oxidation and corrosion resistance in aggressive and corrosive environments, especially at elevated temperatures.

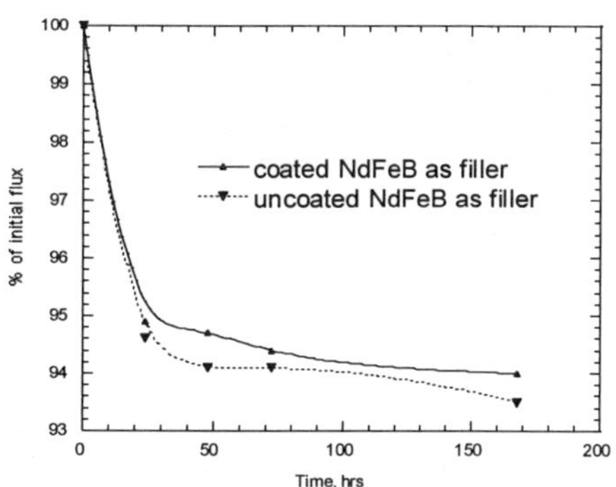

Fig. 5. LTILT curves of epoxy bonded magnets at 180 °C in air

The mechanical properties of polymer bonded magnets depend on properties of the matrix, magnetic filler and the interfacial strength between the two components. The concept of tailoring the polymer matrix's properties through blending different polymers to enhance the mechanical properties has been a target of the permanent

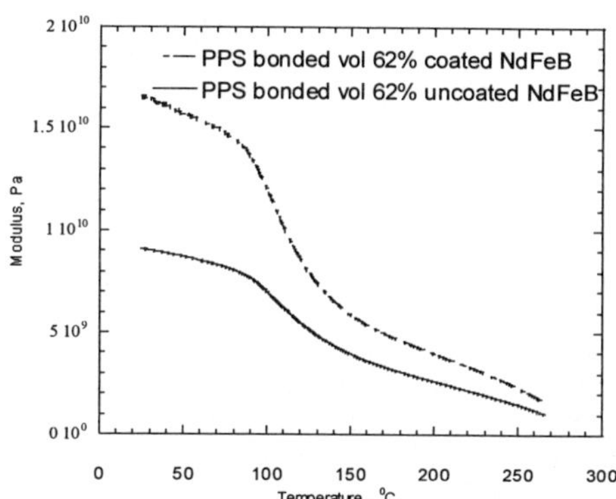

Fig. 6. Temperature dependence of modulus of PPS bonded magnets showing the effect of coating

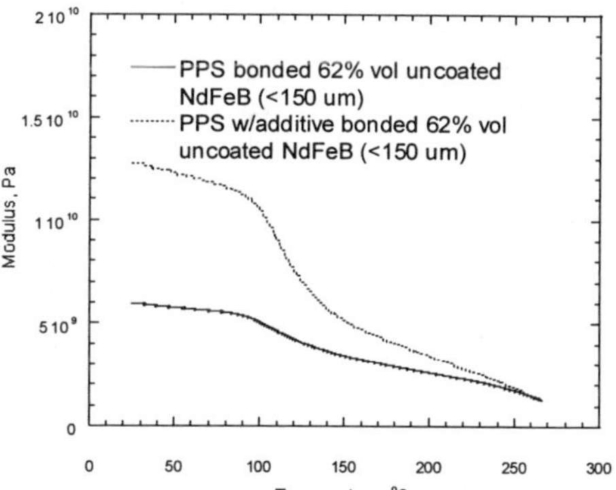

Fig. 7. Temperature dependence of modulus of PPS bonded magnets showing the effect of LCP additives

magnet industry. In this study, a liquid crystalline polymer (LCP) were used as an additive to the PPS matrix because of its high strength and stiffness, excellent chemical resistance, low thermal expansion and low melt viscosity [10]. Blending LCP with other thermoplastics offers the possibility of exploiting the desirable characteristics of the LCP.

We found that with a very small amount of LCP additives to the PPS matrix, the modulus of the bonded magnet is improved.

Figs. 6 and 7 show that the performance of coated samples was superior to the uncoated samples, and the small amount of LCP addition gave further improvement performance of the bonded magnets.

The improvement in mechanical properties by the coupling agent is ascribed to the fact that the coupling agent may act as a bridge to bond the NdFeB powder to the polymer. This improved interfacial bonding between the polymeric matrix and magnetic fillers results in the observed increase in modulus of polymer bonded magnets (see Figs. 6 and 7).

The morphology studies of the fracture surfaces showed that the dispersed LCP droplet domains hinder the crack propagation, leading to improved mechanical properties [11]. This observation is consistent with the expectation of enhanced dimensional stability and strength by applying a coating treatment and LCP additives to polymer bonded magnets during processing. It is believed that by combining these two techniques, i.e. surface modification and adding optimum amount of LCP during processing, bonded magnets with optimum mechanical properties can be obtained with beneficial magnetic properties for many uses.

CONCLUSIONS

(1) At elevated temperature (400 ^{o}C) and high humidity conditions, the surface modification applied to NdFeB powders improved the oxidation and corrosion resistance by passivating the surfaces of active sites against oxidation and corrosion. The high temperature and environmental stability of coated NdFeB powders allowed the bonded magnets to be injection molded at elevated temperature to produce parts with enhanced magneto-mechanical properties for use in high temperature and humid environments.

(2) Short term irreversible loss test results of the polymer bonded magnets containing coated NdFeB magnetic filler with high loading (62 vol%) showed zero irreversible loss in magnetic flux density after exposure to temperatures up to 180 ^{o}C. The use of silane coupling agent is recommended for processing of polymer bonded magnets that are designed to be exposed to aggressive environments such as high humidity and high temperature.

(3) The use of liquid crystal polymers as additives in the processing of polymer bonded magnets improved dimensional stability and stiffness of the polymer bonded magnets. Therefore, employing both the coating treatment prior to incorporating NdFeB into polymer and LCP additives should provide high performance, light weight thermoplastics magnets with optimum magnetic and mechanical properties for the targeted applications.

ACKNOWLEDGEMENT

The authors wish to thank Dr. Hyun Seog Kim, Mr. Steve Constantinides, Dr. Tho Bui of Group Arnold, and Dr. Barbara Lograsso of Ames Laboratory. Without their assistance this work would have been impossible. This work was supported by the National Science Foundation with grant No. NSF DMR - 9712688.

REFERENCE

[1] J. Jacobson and A. Kim, J. Appl. Phys. 61(8), 3763-3765 (1987)

[2] F. E. Camp and A. S. Kim, J. Appl. Phys. 70(10), 6348-6350 (1991)

[3] C. J. Willman and K. S. V. L. Narasimhan, J. Appl. Phys. 61(8), 3766-3768 (1987)

[4] D. F. Cygan and M. J. McNallan, Journal of Magnetism and Magnetic materials, 139, 131-138 (1995)

[5] J. Xiao, H. Kim, J. Otaigbe and T. Tacke, SPE-ANTEC '99 Technical papers, Vol. 55, 1999 (in press)

[6] J. Otaigbe, J. Xiao, H. Kim and S. Constantinides, J. Mater. Sci. Letts. Vol. 18, No. 4 1999 (in press)

[7] Edwin P. Plueddemann, *Silane Coupling Agents* (Plenum Press, New York 1982) p.4
[8] Joel R. Fried, *Polymer Science and technology* (Prentice Hall PTR, New jersy 1995), p. 281
[9] A. S. Kim, F. E. Camp and T. Lizzi, J. Appl. Phys. 79(8), 4840-4842 (1996)
[10] Woo Nyon Kim and Morton M. Denn, Journal of Rheology, 36(8), 1477-1499 (1992)
[11] J. Xiao and J. Otaigbe, Polymer Composites Journal (1999, submitted)

Part II

Intrinsic Properties of Permanent Magnetic Materials

NEUTRON DIFFRACTION CHARACTERIZATION OF PERMANENT MAGNET PHASES

WILLIAM B. YELON
University of Missouri Research Reactor, Columbia MO
Yelon@reactor.murr.missouri.edu

ABSTRACT

Neutron diffraction has been invaluable in the characterization of intermetallic magnetic phases due to the unique properties of the neutron and the availability of high performance powder diffractometers that allow large numbers of samples to be investigated. The structural data derived from these studies serves as fundamental input for theory, and constraints on analyses of data from other methods, such as Mossbauer spectroscopy. The empirical structure-property relations derived from these studies serves to guide synthesis of new compounds and may help to define the next generation of high performance permanent magnets.

INTRODUCTION

Neutron scattering is an immensely powerful and versatile tool in modern condensed matter science. Inelastic scattering is used to measure the forces between atoms, including magnetic exchange interactions, while quasielastic scattering is used to study single particle motion such as hydrogen diffusion in solids. Small angle scattering is used to examine the size (and shape) distribution of macroscopic density fluctuations, while reflectometery is used to measure the density profile of scattering close to surfaces and interfaces. In contrast, neutron powder diffraction seems almost primitive in its simplicity, and yet it has proven to be invaluable in the study of magnetic alloys. The magnetic interaction of the neutron with ordered moments in solids allows for determination of the arrangement and magnitudes of the moments, which is especially valuable for the study of antiferromagnetic and ferrimagnetic materials. In ferromagnetic materials the moment direction can usually be determined by other means, but even here powder diffraction has been an important tool in the investigation of new materials. Ironically, it is the use of this method as a simple structural tool that has proven most useful. This paper will review the use of powder diffraction to help understand the behavior of compounds, based on the $CaCu_5$ structure and similar structures, that have been modified in various ways in order to change their magnetic properties.

NEUTRON DIFFRACTION

Neutron diffraction is, of course, governed by Bragg's Law and can be probed in either of two basic ways. The first, like its typical x-ray analogue, uses a monochromatic beam and measures the scattering as a function of angle. The second takes advantage of the energy velocity relation for heavy particles and uses time-of-flight methods at fixed angle to determine the wavelength of the scattered neutrons. In either case intensities are measured as a function of d-spacing and these data are used for structural modeling in which the observed intensities are compared to calculated intensities. The calculated intensities are adjusted through the structural parameters, in a least-squares method until satisfactory agreement is reached. Because neutron sources are weak compared to x-ray sources, various strategies are used to enhance the rates of

Mat. Res. Soc. Symp. Proc. Vol. 577

data acquisition. Large samples, coupled with multi-detectors, each with its individual Soller slit collimators, has been the most popular method for monochromatic experiments, while area detectors are employed in time-of-flight, with special methods employed to account for resolutions that vary with wavelength and angle. At the University of Missouri Research Reactor (MURR) a different strategy has been employed (Fig. 1). Focusing neutron optics (using a bent perfect crystal) is combined with small sample and position sensitive detection. The sample "back selects" the wavelength band that passes and defines the resolution [1]. The gain in data acquisition, through the open optics and area detector, more than compensates for the loss in sample volume. The gain factor is about x 40 compared to a 64 element multidetector and sample ten times larger, and typical measurements on 1 gm. samples are completed in a few hours. This high throughput has been a vital aspect of the large-scale program at MURR.

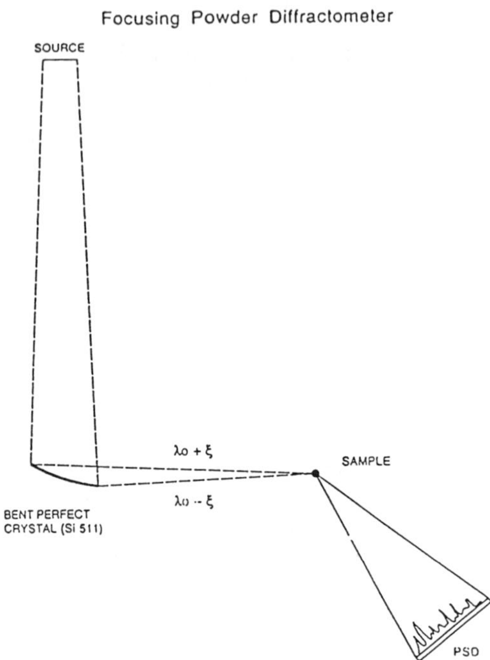

Figure 1:Schematic of MURR diffractometer.

The neutron possesses several unique attributes that render it invaluable for these studies. As a neutral particle it is highly penetrating. This simplifies sample preparation since surface effects can generally be neglected and makes the use of peripheral devices like furnaces and cryostats a rather simple matter. The (non-magnetic) scattering is a neutron-nuclear property that does not show the simple z-dependence exhibited by x-ray scattering. Of special interest is the strong scattering by light elements and the large contrast that often is seen between adjacent elements on the periodic chart (Fig. 2). Because the nucleus is point-like on the scale of the neutron wavelength, the scattering length shown in Fig. 2 is a scalar quantity, independent of scattering vector. Consequently, the neutron diffraction diagram damps out at high angle much more slowly than does x-ray scattering, where the form factors lead to a rapid decrease in scattering power. Furthermore, the neutron diffraction peaks (for monochromatic scattering) are typically Gaussian, an ideal shape for peak fitting. These aspects lie at the heart of the studies carried out at MURR.

The Rietveld method, embodied either in the GSAS [2] or FULLPROF [3] programs is used for data reduction. These programs are both capable of multiphase refinements. Typical refinements include lattice parameters, position parameters and magnetic moments and isotropic temperature factors. In general these are stable and robust, and give unique structural results.

Figure 2: Thermal Neutrons Nuclear Scattering Lengths

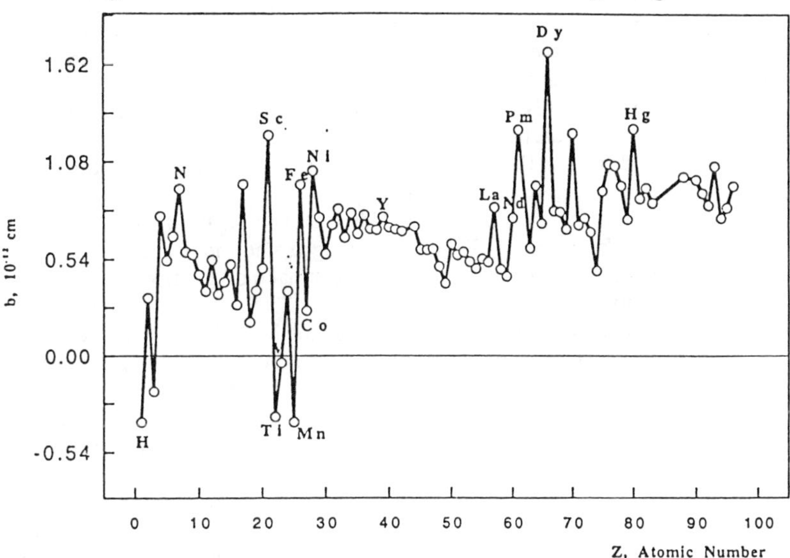

Samples for these studies were prepared either by induction or arc-melting and annealed for extended periods to achieve homogeneity. A large number of laboratories have engaged in sample preparation, including the Materials Research Center at the University of Missouri-Rolla, National Cheng-Chung University, in Taiwan, the University of Sao Paulo and the University of Delaware.

RESULTS

It is not possible, in a short paper, to summarize more than 20 years of research on magnetic materials. Consequently, this paper will focus on the improvements that have been made in methodology, and in our understanding of the magnetic properties of these important classes of materials. When the first experiments were carried out on the R_6TM_{23} phase (where R is a rare-earth or Y and TM is a transition metal) it was unknown if the replacement of one TM element by another (Fe and Mn) would occur in a random fashion on the crystallographically distinct TM sites [4]. The results showed that site preference was a strong effect with some sites holding significantly higher (lower) concentrations of one species than expected from random. Similar findings were reported for the $Nd_2Fe_{17-x}Co_x$ system [5]. At the time of those studies, however, theoretical modeling of magnetic properties was not well developed, nor was their any theory to explain the observed site preferences. Although size considerations clearly play a role in this phenomenon, it was clear from the first studies that this was not sufficient to explain the totality of the results. It was also noted, in the 6:23 phase, that the minority species was not magnetically coupled to the majority transition metal sublattice, leading to a rapid decrease in the Curie temperature compared to either end member. For significant mixing of Fe and Mn spin glass behavior was observed. This was attributed to antiferromagnetic coupling of the Fe and Mn atoms.

These early studies were carried out with a conventional single detector diffractometer. In spite of the poor resolution of the system, definitive results were produced. With the development of the first-generation position sensitive detector diffractometer, a wider range of studies were possible with significantly higher resolution. A major accomplishment of this period was the structure solution of $Nd_2Fe_{14}B$ [6]. Although structure solution from powder data is uncommon, simple trial and error, based on the observed space group, led to rapid convergence of the Rietveld fits and the full structure was extracted. At the time, this structure was the largest ever directly solved from powder data.

The observation of good magnetic properties in $Nd_2Fe_{14}B$ and the similarity of that structure to the binary R-Fe compounds (especially R_2Fe_{17}) encouraged a new look at those compounds as possible permanent magnet phases. It was found that interstitial modification of these binaries, by C or N could substantially increase the Curie temperature, T_c, [7,8]. The early reports of these compounds gave a wide range of stoichiometry, with the interstitial content determined from TGA or similar weight change analysis. These results, though, were compromised by the decomposition of the alloys through the formation of α-Fe and rare-earth nitrides or carbides. Neutron diffraction was used to directly observe the interstitial atoms and to quantify the changes in bond lengths associated with the interstitial atoms. The rhombohedral R_2Fe_{17} structure, which forms with the light rare-earths, contains two distinct sites potentially capable of holding interstitial N or C. The larger of these has an octahedral environment, while the smaller is tetrahedral. Studies by several groups showed that fully nitrided material contained only octahedral interstitials, corresponding to a stoichiometry of $R_2Fe_{17}N_3$. However, diffraction diagrams for some samples suggested a variable stoichiometry as manifested in broadened Bragg reflections. Careful analysis of the neutron data reveals that two nitride

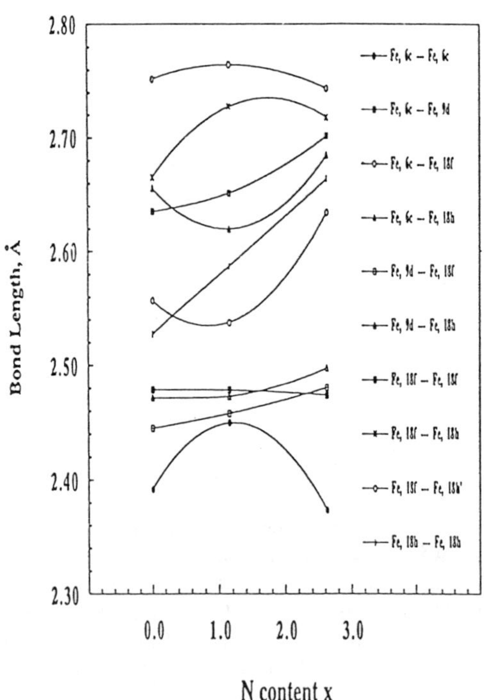

Figure 3: Fe-Fe bond lengths in $Nd_2Fe_{17}N_x$

phases were present in one of these samples [9]. One of these is fully nitrited with only octahedral occupation by N, while the second phase contains less nitrogen but these atoms are found on both octahedral and tetrahedral sites. A model for the nitrogenation process emerges from this result. It appears that nitrogen can enter both sites (at low concentration) but once a nitrogen atom fills an octahedral site it remains fixed at that position. Nitrogens in the tetrahedral positions are more mobile and continue to move until trapped on the octahedral sites. However, as these fill the lattice can less easily expand to accommodate the tertrahedral occupancy and eventually only the octahedral sites remain filled. Interestingly, the Fe-Fe bond lengths are distinctly different in the fully and partially occupied structures (Fig. 3).

It was widely assumed that the increase in T_c was the result of magnetovolume effects. That is, the exchange between Fe near neighbors was assumed to increase as the near neighbor distance increased. Nitrogenation of the R_2Fe_{17} phase is accompanied by a 7% volume increase, but this is not a uniform expansion. Bonds close to the interstitial nitrogen expand strongly, while others expand less so or may even show slight contraction. Band theory has been applied to examine the spin-polarized density of states in order to model the changes that occur. The detailed neutron structure is obviously an essential starting point for this analysis [10].

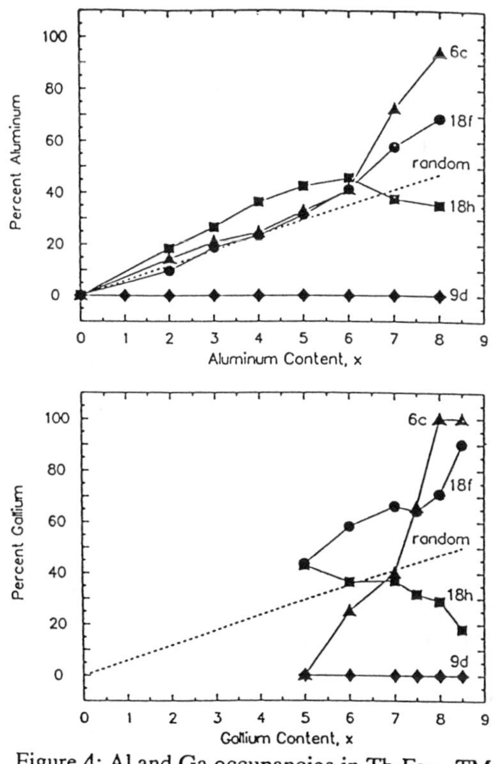

Figure 4: Al and Ga occupancies in $Tb_2Fe_{17-x}TM_x$

The magnetovolume assumption appeared to gain further support with the observation that Al or Ga substitution of Fe in the R_2Fe_{17} phase also led to a volume expansion and increase in T_c at low substituent concentration [11,12]. The changes in bond lengths with these substitutions are easily determined by neutron Rietveld analysis, and the location and concentration of the Al atoms are particularly easily determined [13]. Because the neutron scattering contrast between Fe and Ga is only modest the refinement of the Ga occupancy is somewhat less certain, and combining the neutron data with Mossbauer spectroscopy on the same samples leads to an improved precision [14]. It is interesting that the Ga occupancy does not increase monotonically on all sites. In fact, the 18h site has a high initial affinity for Ga, but the occupancy of this site actually decreases at high Ga content ($x > 7$ in $Tb_2Fe_{17-x}Ga_x$) (Fig. 4). This appear to be related to a tendency to form a new ordered structure, and the Curie temperature is also seen to increase modestly around $x = 8$. Apparently the presence of one nearly full and one nearly empty Fe sublattice gives more effective magnetic coupling than two partially full sublattices.

In light of the observations on Al and Ga substitution, it was surprising to find that Si substitution also leads to an increase in T_c, since the unit cell contracts as Si is added [15]. The occupancy pattern of the Si substitution is similar to that seen for Al and Ga, with the highest initial affinity of the Si for the 18h site [16]. Obviously, though, the pattern of bond length changes must be quite different, due to the overall lattice contraction. Thus, the enhancement in T_c must be explained by electronic effects rather than by simple Neel-Pauling-Slater type arguments.

The R_2Fe_{17} phase proves to be quite unusual. It can accommodate a large number of substituent species, including the metalloids already discussed and most row one and row two transition elements, as well as forming with all of the rare earths except Eu, and with Y. Furthermore, we have found that every substituent increases the Curie temperature, at least at concentrations around $x = 1$, and in many cases at higher concentrations, despite the dilution of the Fe-Fe exchange. This might be less surprising if all substituents initially occupied the same sites. This is, in fact, not the case, most transition elements prefer the 6c "dumbbell" site rather than the 18h site preferred by the metalloid elements. Although this site has one short bond, to the adjacent dumbbell atom, it also includes 12 long bonds to other Fe atoms, and the average Fe-Fe bond length to this site is substantially longer than any of the other sites. Thus, one might expect that such substitution would strongly suppress the Curie temperature. Band theory for these substituents has, in general, not been carried out. However, significant progress has been made in understanding the site preferences in terms of elastic and chemical effects [17]. For example, the metalloid occupancy appears to be controlled by the affinity of those elements to rare-earth near neighbors.

Although the magnetic properties of the carbides and nitrides, produced by gas phase insertion, are suitable for permanent magnet application, these compounds are thermodynamically unstable and decompose at the temperatures required for processing into dense magnets. Substituted compounds are stable, but none of them have sufficiently high Curie temperatures, and their remanance is degraded by the replacement of magnetic Fe by non-magnetic species. Carbon can be introduced into these alloys during phase formation from the melt, but only in limited quantities. It was found, however, that replacement of some of the Fe by Si or Ga allows higher concentrations of C to be incorporated into the lattice. These resulting compounds, while not achieving the high Curie temperature of the fully carbided or nitrided compounds, do represent possible practical systems for certain limited applications. Neutron studies of these combined substitutional/interstitial alloys shows that the occupancy of the metalloid atoms is unaffected by the incorporation of C. However, similar attempts at enhancing the magnetic properties have been made on transition metal substituted interstitial alloys. The neutron results (Fig. 5) show that the occupancy of the substituents is dramatically affected by the presence of C in the lattice [18]. Instead of predominantly occupying the dumbbell sites, the transition elements such as Ti are found in the Fe sites adjacent to the octahedral interstitial site. Interestingly, though, the C content of that site is lower than expected if every Ti atom was bonded to a C. It appears that the Ti enters the structure with an accompanying C, but that these are later lost during the extended annealing. It would be

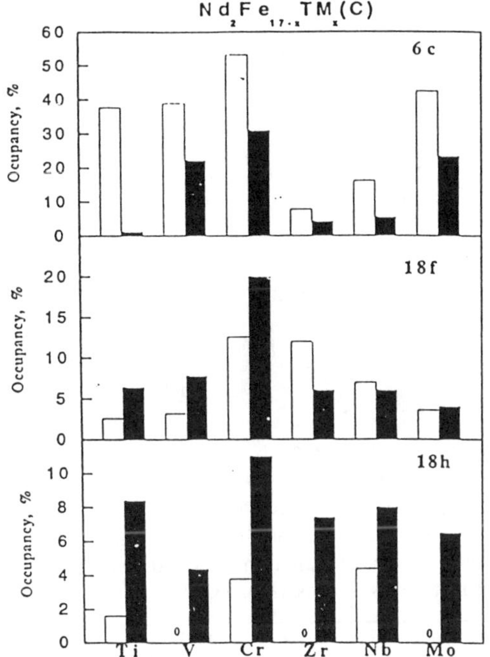

Figure 5: Transition metal occupancies in $Nd_2Fe_{17-x}TM_x$ (open bars) and $Nd_2Fe_{17-x}TM_xC$ (solid bars).

interesting to study these samples as a function of annealing time to see if such a speculation can be verified.

The R_2Fe_{17} phase exists in two forms. The rhombohedral structure forms with the light rare-earths. This structure is fully ordered and stoichiometric. It consists of an abc stacking of $CaCu_5$ layers, with 1/3 of the rare earths replaced by transition metal dumbbell pairs. For the heavy (and smaller) rare-earths a hexagonal form is found. The stacking is abab and it is found that the ideal structure shown in Fig. 6 is not realized. Instead, some of the dumbbell sites shown actually contain rare-earth atoms, while Fe-dumbbells can be found at some of the ideal rare earth positions. Thus, instead of containing two rare-earth sites and four Fe sites, the true structure contains three R sites and five Fe sites. Table I lists the crystallographic information for this structure. In addition to the disorder represented by the dumbbell-rare-earth interchange, the material can also support a range of stoichiometry through the substitution of more or less than the ideal 1/3 of the rare-earth atoms in the parent $CaCu_5$ phase. These complexities have been thought to limit the ability of neutron diffraction to extract the site occupancy of atoms replacing the Fe in the lattice. However, we realized that if the possible rare-earth-dumbbell sites are assumed to be fully occupied, either by a rare-earth or a dumbbell pair of transition or metalloid atoms, then an unambiguous determination can be carried out. The result of these studies indicate that the occupancy patterns for substituents in the

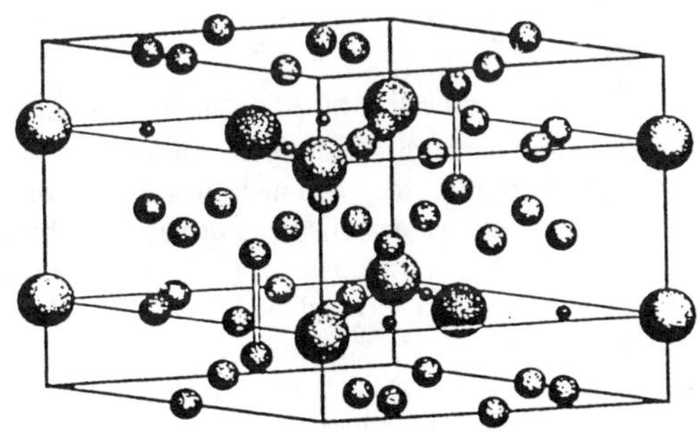

Figure 6: Ideal structure of hexagonal R_2Fe_{17} phase.

Table I: crystallographic sites in hexagonal R_2Fe_{17} compounds.

Multiplicity	Site	Atom	Position
2	d	RE	1/3,2/3,3/4
2	c	RE	1/3,2/3,1/4
2	b	RE	0,0,1/4
12	k	Fe	x,2x,z $x \approx 0.17$, $z \approx 0.98$
12	j	Fe	x,y,1/4 $x \approx 0.33$, $z \approx -0.03$
6	g	Fe	1/2,0,0
4	f	Fe	1/3,2/3,z $z \approx 0.10$
4	e	Fe	0,0,z $z \approx 0.10$
6	h	C	x,2x,1/4 $x \approx 0.83$

hexagonal R_2Fe_{17} compounds is substantially the same as observed for the rhombohedral compounds, i.e. the metalloids prefer those sites with high rare-earth coordination, and the concentrations at other sites resemble those seen for the equivalent rhombohedral sites. The effect of rare-earth on the occupation of the substituents is weak (Fig. 7) [19].

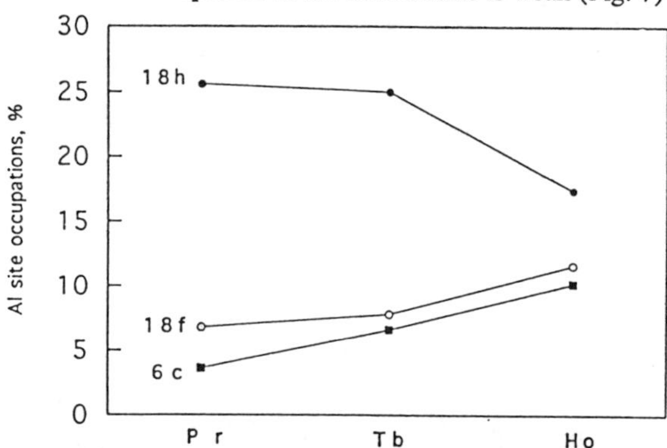

Figure 7: Rare-earth effect on Al occupancy in $R_2Fe_{15}Al_2$

An interesting observation concerns the transformation from the hexagonal crystal structure to the rhombohedral. Addition of C (from the melt) or substitution of Fe by Al or Ga, all of which cause the lattice to expand, trigger this transformation. This is consistent with the observation that the rhombohedral form exists for the larger rare-earths, but further analysis is needed to establish whether there is a "critical" value for cell volume or for the volume of a particular site which determines the crystal structure.

The studies discussed above primarily concern the R_2Fe_{17} phase, but similar results are obtained for the $RFe_{12-x}TM_x$ structure, which is derived from the $CaCu_5$ structure through the replacement of one-half of the rare earth atoms by transition metal dumbbells.

CONCLUSIONS

Modification of the $CaCu_5$ derived phases (2:17 or 1:12) can lead to significant enhancement in magnetic properties. Gas phase incorporation of N or C can lead to Curie temperatures superior to those of the equivalent $R_2Fe_{14}B$ compounds (Fig. 8), but these materials are not thermodynamically stable. This makes the production of dense magnets impossible and often leads to the presence of soft phases (α-Fe) in powders, degrading the magnetic performance for bonded magnets, which can be produced. Stable compounds with Curie temperatures above those of the parent R-Fe compounds can be produced through substitution or combined substitution and interstitial incorporation in the phase formation from the melt. None of these compounds appear to offer significant improvement over the readily available $R_2Fe_{14}B$ alloys. Nevertheless, neutron diffraction studies of these materials have yielded improved insight into the structures and the factors that contribute to the magnetic coupling in these compounds. In addition, the neutron results have provided essential information for band theoretical modeling of the spin-polarized density of states, and for analysis of the preferred occupancy patterns. The neutron data has also served as constraints on other experimental studies of these materials, especially Mossbauer spectroscopy.

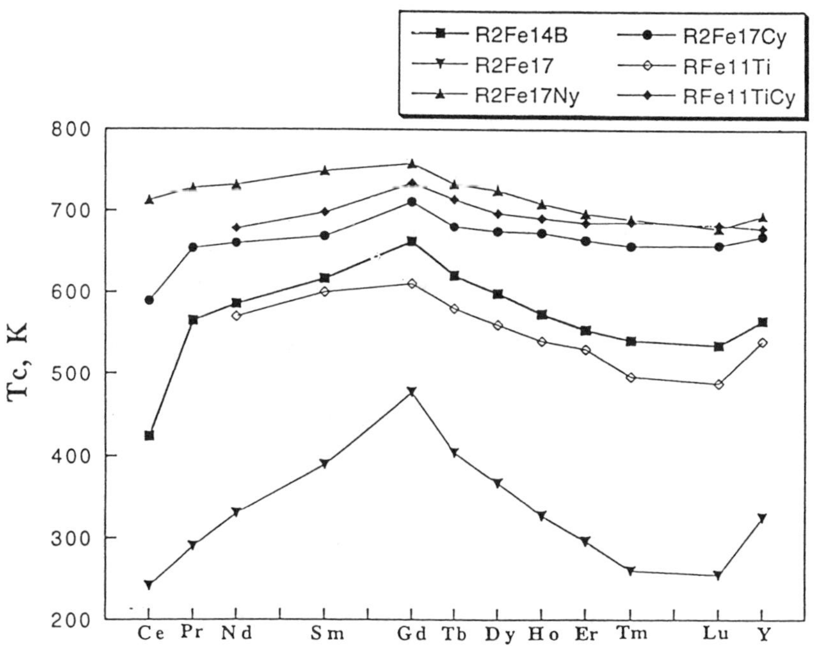

Figure 8: Curie temperatures for various R-Fe based series.

It appears unlikely that the $CaCu_5$ derived materials will provide further opportunities for optimization of magnetic properties, and so our attention has shifted to other families of compounds. Included as potential candidates are Nd_5Fe_{17} and a variety of Mn based compounds that exhibit ferromagnetic coupling, such as MnBi. The full power of neutron diffraction will be brought to bear on these in order to correlate the changes in magnetic properties with modifications to the crystal structures. It is expected that these results will provide new insights and perhaps will lead to the formulation of magnetic compounds with optimum properties.

ACKNOWLEDGMENTS

The author gratefully acknowledges the contributions of his collaborators including W.J. James and H. Luo and others too numerous to name individually. Support of the National Science Foundation through grant DMR-9614596, and its predecessor DMR-9305782, and DARPA through contract DAAG55-98-1-0267 is also gratefully acknowledged.

REFERENCES

1. W. B. Yelon, R. Berliner and M. Popovici, Physica **B241**, 216 (1997)

2. A. C. Larson and R. B. Von Dreele, *GSAS: General Structure Analysis System,* LANSCE, MS-H805, Los Alamos National Laboratory, (1985-1994)

3. J. Rodriguez-Carvajal, *FULLPROF, Version 2.4.2 DEC93-ILLJRC,*Institute Laue-Langevin, (1993)

4. K. Hardman, W.J. James and W. B. Yelon, J. Phys. Chem. Solids **41**, 1105 (1980)

5. J. Herbst, J. Croat and W. B. Yelon, J. Appl. Phys. **53**, 250 (1982)

6. J. Herbst, J. Croat, F. Pinkerton and W. B. Yelon, Phys. Rev. **B29**, 4176 (1984)

7. K. H. J. Buschow, R. Coehoorn, D.B. deMooij, K. de Waard and T.H. Jacobs, J. Mag. Mag. Mat. **92**, L35 (1990)

8. J. M. D. Coey and H. Sun, J. Mag. Mag. Mat. **87**, L251 (1990)

9. Z. Hu and W. B. Yelon, J. Appl. Phys. **80**, 6175 (1996)

10. S. S. Jaswal, W. B. Yelon, G. C. Hadjipanayis, Y. Z. Wang and D.J. Sellmyer, Phys. Rev. Lett. **67**, 644 (1991)

11. P. C. M. Gubbens, A. M. van der Kraan, T. H. Jacobs, and K. H. J. Buschow, J. Mag. Mag. Mat. **80**, 265 (1989)

12. W. B. Yelon, Z. Hu, S. Mishra, O. A. Pringle, F. Grandjean, D. P. Middleton and K. H. J. Buschow, J. Appl. Phys. **76**, 5383 (1994)

13. W. B. Yelon, H. Xie, G. J. Long, O. A. Pringle, F. Grandjean and K. H. J. Buschow, J. Appl. Phys. **73**, 6029.(1993)

14. O. A. Pringle, G. J. Long, S. R. Mishra, D. Hautot, F. Grandjean, D. P. M. Middleton, K. H. J. Buschow, Z. Hu, H. Luo and W. B. Yelon, J. Mag. Mag. Mat. **183**, 81 (1998)

15. E. E. Alp, A. M. Umarji, S. K. Malik, G. K. Shenoy, M. Q. Huang, E.B. Bolthich and W. E. Wallace, J. Mag. Mag. Mat. **68**, 305 (1987), R. Van Mens, J. Mag. Mag. Mat. **61**, 24 (1986)

16. G. J. Long, G. K. Marasinghe, S. Mishra, O. A. Pringle, F. Grandjean, K. H. J. Buschow, D. P. Middleton, W. B. Yelon, F. Pourarian and O. Isnard, Solid. State Comm. **88**, 761 (1993)

17. Er. Girt and Z. Altounian, Phys. Rev. **B57**, 5711 (1998)

18. G. K. Marasinghe, P. C. Ezekwenna, W. J. James, G. J. Long, O. A. Pringle, Z. Hu, W. B. Yelon and F. Grandjean, J. Appl. Phys. **79**, 4587 (1995)

19. W. C. Chang, S. H. Tsai, W. B. Yelon, Z. Hu, H. Luo and M. Chen, J. Appl. Phys. **81**, 4548 (1997)

MAGNETIC CHARACTERISTICS OF $RCo_{7-x}Zr_x$ ALLOYS (invited) (R = Sm, Pr, Er, Gd, and Y)

M.Q. HUANG [a,b], W.E.WALLACE [b], M.E.McHENRY [b], and S.G.SANKAR [a]
a: Advanced Materials Corporation, Pittsburgh, PA 15230
b: Department of Materials Science and Engineering, Carnegie Mellon University Pittsburgh, PA 15213
QUN CHEN and B.M.MA
Rhodia Inc/Chem Dist. Center, CN7500, Cranbury, NJ 08512

ABSTRACT

A brief review is presented of recent work dealing with the structure and magnetic properties of $RCo_{7-x}Zr_x$ alloys (R=Sm, Pr, Er, Gd, and Y). The experimental results obtained are consistent with a model in which Zr atoms partly replace dumbbell Co atoms and play an important role in stabilizing the $TbCu_7$ structure while significantly increasing the anisotropy field (H_A). For example, when R=Sm, H_A increases from 90 kOe for x=0 to 130 kOe for x=0.2 at 300 K, and from 140 kOe for x=0 to 220 kOe for x=0.2 at 10 K. In the case of R=Y and Gd, H_A is mainly contributed by the Co sublattice. For R=Y alloys, H_A increases from 18 kOe for x=0 to 74 kOe for x=0.2 at 300 K and from 20 kOe for x=0 to 82 kOe for x=0.2 at 10 K. For R=Gd alloys, H_A shows the largest enhancement. It increases from 35 kOe for x=0 to 140 kOe for x=0.2 at 300 K and from planar for x=0 to uniaxial with 182 kOe for x=0.2 at 10 K. In general, experimental results are in accord with the theory of Greedan and Rao for anisotropies of R-Co alloys. The magnetic moments for cobalt and rare earth in $RCo_{7-x}Zr_x$ compounds (x=0~0.2) have been estimated from the experimental values. The results show that they are nearly the same as those in RCo_5 or R_2Co_{17}. The Co moment is 1.5~1.6 μ_B. Some phase transition phenomenon between RCo_5, RCo_7, R_2Co_{17} and R_2Co_7 at different heat treatment conditions will also be discussed.

INTRODUCTION

For many years there have been a vigorous on-going effort in the international magnetic community to produce new or improved permanent magnetic materials. The premier materials at present are $SmCo_5$, Sm_2Co_{17} based alloys and $Nd_2Fe_{14}B$. The latter material is deficient in respect to Curie temperature and resistance to corrosion. $SmCo_5$ has very high magnetocrystalline anisotropy (H_A >200 kOe), but is lacking in regard to magnetic moment (7.3 μ_B/f.u). Sm_2Co_{17} has higher moment (23.2 μ_B/f.u) than that of $SmCo_5$, but does not possess the high anisotropy exhibted by $SmCo_5$. H_A is reduced from 200 kOe to 65 kOe. This is considered to be mainly caused by the partial replacing of Sm ions with Co dumbbells. Since the composition of $SmCo_7$

Mat. Res. Soc. Symp. Proc. Vol. 577

lies between that of $SmCo_5$ and Sm_2Co_{17}, it may retain the benefits of both materials: reasonably high moment and anisotropy. Therefore, it may become a useful high temperature permanent magnetic material. $SmCo_7$ is a metastable phase at room temperature with the $TbCu_7$ structure. The structure of $SmCo_7$ and Sm_2Co_{17} are both derived from the structure of $SmCo_5$. The structure of Sm_2Co_{17} can be viewed [1] as one in which 1/3 of the Sm atoms in the $SmCo_5$ are replaced by dumbbells of Co in an ordered fashion. The structure of $SmCo_7$ is one in which 22% of the Sm atoms in the $SmCo_5$ are replaced by Co dumbbells randomly distributed over two of the five sublattices [2]. This metastable phase can be obtained by splat cooling [3], melt spinning [2] or mechanical alloying [4]. Some interesting magnetic properties have been reported by H. Saito et al [2]. They reported that $Sm_2(Co,Mn)_{17}$ with the $TbCu_7$ structure exhibits an anisotropy of 105 ~ 140 kOe , which is 1.2 -1.4 times larger than that of $Sm_2(Co,Mn)_{17}$ with the Th_2Zn_{17} structure. This work shows the extreme structure sensitivities between the 2-17 and 1-7materials. In regard to magnet fabrication, it was found that the 1-7 phase plays an important role for developing the cellular structure and the high coercivity in 2-17 type magnets [5,6]. Interestingly, in 1982, S.Liu and K.J.Strnat observed that the $Sm(Co, Fe, Cu, Zr)_{7+x}$ magnets with x=0, exhibited less temperature dependence of coercivity than that of magnets with $x > 0$ [7].

Currently, there is, as noted above, an intense search to find new magnets or to improve the magnetic properties of existing magnets for high temperature applications. A systematic study of the RCo_7 systems will be of use in this regard. In 1977 T. Ojima et al. [8] have observed that Zr doping enhances the magnetic properties of the alloys used to produce Sm_2Co_{17} type magnets. Wallace et al [9] in 1982 reported that Zr is the best additive element for enhancement of H_A in R_2Co_{17} systems. Because of these features, it seems desirable to study the ternary systems $RCo_{7-x}Zr_x$ in hopes of clarifying the role of Zr doping. In our previous work [10,11] on the $RCo_{7-x}Zr_x$ alloys (R=Sm, Pr and Er), it has been reported that partly replacing Co dumbbells by Zr atoms plays an important role for stabilizing the $TbCu_7$ structure and also significantly increases the magnetocrystalline anisotropy. The earlier work has been extended in the present study with R =Y, and Gd, in which the magnetic anisotropy is derived mainly from the Co sublattices.

In this paper, a brief review is presented of our recent work dealing with the structure and magnetic properties of $RCo_{7-x}Zr_x$ alloys (R=Sm, Pr, Er, Gd, and Y). R values have been chosen to exemplify cases in which R is a light rare earth (R = Pr, Sm), or a heavy rare earth (Er), or a spin-only with zero orbital moment rare earth (Gd) and non-magnetic R (Y).

EXPERIMENTAL DETAILS

The experimental procedures employed are the same as those used in the previous studied in this series [10,11].

RESULTS AND DISCUSSION

I. Structure and phases present in as-cast alloys.

Information in regard to the structure and phases present obtained by XRD (random and aligned powder samples), TMA and SEM measurements is shown in Fig. 1-3 and summarized in Tables I and II.

It was found that the crystal structure of the as-cast alloys $RCo_{7-x}Zr_x$ varies significantly with the Zr content. The structures of the main phase were observed to change from ordered rhombohedral (Th_2Zn_{17}-type) or hexagonal (Th_2Ni_{17}-type) to disordered hexagonal ($TbCu_7$-type) when x increases to 0.1 ~ 0.2. With further increase in Zr content, other R-rich phases, such as 2-7 (Ce_2Ni_7 or Er_2Co_7 - type) and 1-3 ($PuNi_3$ - type) phases also form.

(a). Alloys With $0 \leq x \leq 0.1$ or 0.2

The XRD patterns of the random powder samples show that the crystal structure of as-cast alloys $RCo_{7-x}Zr_x$ varies significantly with Zr content. For R = Sm or Pr alloys, the typical superlattice reflections for the Th_2Zn_{17} structure (015 and 204) disappear when $x \geq 0.2$ (Fig.1a) For R =Y, Gd or Er alloys, the typical superlattice lines for the Th_2Ni_{17} structure (203) also disappear when $x \geq 0.1$ or 0.2 (Fig.1b and 1c). This indicates that the structure of the main phase of $RCo_{7-x}Zr_x$ alloys changes from ordered rhombohedral (Th_2Zn_{17} -type) or hexagonal (Th_2Ni_{17}-type) to disordered hexagonal ($TbCu_7$-type) with a small amount of Zr doping (x =0.1~ 0.2). This is also evident in the XRD of aligned samples. As shown in Fig. 1, the two pairs of field-strengthened reflections at x = 0 became a single pair strong reflections (110, 200), which only belong to 1-7 phase when $x \geq 0.1$ ~ 0.2. Nearly single phase materials in the $TbCu_7$ structure were formed for R=Sm or Pr alloys when x =0.2~0.3 and for R=Y, Gd or Er alloys when x =0.1~0.2.

The above results have also been confirmed by an examination of the microstructure employing SEM. For example, as shown in Fig. 2, for $SmCo_{6.8}Zr_{0.2}$ alloy, its composition was determined to be Sm ~ 12.5, Co~ 8.5, Zr~2.5 (at%) for the main phase (in gray domain), which is very close to $Sm(Co,Zr)_7$ and Sm~12.7,Co~8.4, Zr~ 2.9 (at%) for the minor phase (In the white boundary part). Clearly, a small amount of Zr doping can significantly increase the stability of the $TbCu_7$ structure phase in the RCo_7 alloys.

As listed in Table I, the lattice parameters a and c change slightly as the Zr content increases. This can be ascribed to a substitution of Zr for Co dumbbells. Zr substitution plays an important role for improving the anisotropy of RCo_7 system, as will be discussed later.

(b). Alloys with x > 0.3 or 0.4

When the Zr content further increases (x >0.3 or 0.4), the XRD intensity of line 101 decreases and other extra lines appear. Other R-rich, (R, Zr)-(Co, Zr), phases, such as 2-7 (Ce_2Ni_7 or Er_2Co_7 -type) and 1-3 ($PuNi_3$- type) phases were observed (see Fig. 1). The above mentioned magnetic phases were also identified by TMA measurements. The four Curie temperatures (T_{c1}, T_{c2}, T_{c3} and T_{c4}) correspond to four magnetic phases, i.e. the 2-17, 1-7, 2-7 and 1-3 phases respectively (see Figs. 3a, 3b, 3c and Table II).

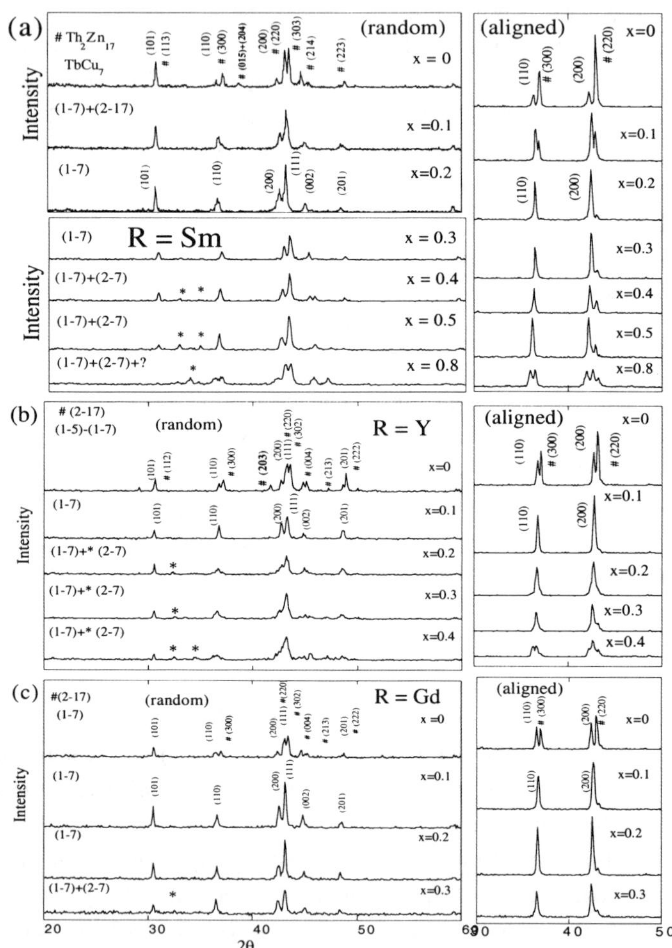

Fig.1. XRD of $RCo_{7-x}Zr_x$ as-cast alloys (x=0 ~ 0.8) in random or magnetically aligned powder (H ~ 1.6T). (a) R=Sm (b) R=Y and (c) R=Gd (* -- for 2-7 phase).

Table I. Phases Present and Crystal Structure of $RCo_{7-x}Zr_x$ as-cast alloys (x=0-0.2)

R	x	Phases Present (main)	(minor)	a (Å) for (1-7)	c(Å)	c/a
Sm	0	2-17 a	1-7 c	4.856	4.081	0.84[1]
	0.1	1-7	2-17 a	4.869	4.079	0.83
	0.2	1-7		4.916	4.049	0.82
Pr	0	2-17 a	1-7	4.873	4.087	0.84[1]
	0.1	1-7	2-17 a	4.947	4.066	0.82
	0.2	1-7		4.966	4.040	0.82
Gd	0	2-17 b	1-7	4.837	4.066	0.84[1]
	0.1	1-7		4.904	4.046	0.83
	0.2	1-7		4.910	4.035	0.82
Er	0	2-17 b	1-7	4.798	4.065	0.84[1]
	0.1	1-7		4,864	4.057	0.83
	0.2	1-7	2-7	4.867	4.038	0.83
Y	0	2-17 b	1-7	4.824	4.064	0.84[1]
	0.1	1-7		4.886	4.040	0.83
	0.2	1-7	2-7	4.894	4.019	0.83

a. Th_2Zn_{17} –type structure, b. Th_2Ni_{17} -type structure, c. $TbCu_7$ –type structure
d. Ce_2Ni_7 or Er_2Co_7 type structure, only can be detected by TMA when x<0.2
[1]data were taken from [1] or calculated as: c(Th_2Zn_{17}-type)=3/2c(Th_2Ni_{17}-type)= 3c($TbCu_7$-type); a(Th_2Zn_{17}-type)= a(Th_2Ni_{17}-type)=$3^{1/2}$ a($TbCu_7$-type)[1].

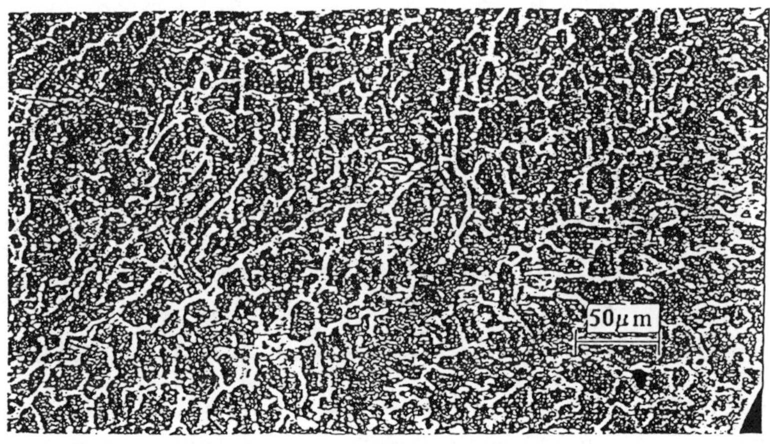

Fig. 2 Microstructure of $SmCo_{6.8}Zr_{0.2}$ as-cast alloy

Table II. Magnetic Properties of $RCo_{7-x}Zr_x$ as-cast alloys (x = 0–0.8)

		T_c (°C)				M (300 K)	M (10 K)			H_A (kOe)- DoM	
R	x	Tc1	Tc2	Tc3	Tc4	emu/g	emu/g	μ_B/f.u		300K	10K
Sm	0	920	780			104.0	114.0	11.5		90	140
	0.1	905	770			97.0	106.5	10.8		106	190
	0.2	(*)	762			90.8	131.0	10.5		131	220
	0.3	(*)	760			85.3	93.9			166	240
	0.4		752	439		78.6	85.6			176	292
	0.5		736	428		70.1	73.7			183	312
	0.8			420	157		43.8			195	380
Pr	0	897	692			126.1	139.5	13.8		EB	EB
	0.1	877	689			116.9	127.6	12.7		EB+EA	EC
	0.2	(*)	683			111.0	120.5.	12.1		94	EC
	0.3	(*)	670	368		99.0	109.6			97	EC
	0.4		660	350		90.9	100.3			100	EC
	0.8			312		49.2	67.4			100	EC
Gd	0	928	879			54.3	44.1	4.5		35	EB
	0.1	(*)	872			49.7	40.9	4.2		113	148
	0.2	(*)	840			46.6	35.3	3.6	35	140	182
	0.5		766	437		30.4	19.6			162	190
Er	0	893	863			77.3	15.1	1.56		50	85
	0.1		849			68.7	18.4	1.92		95	168
	0.2	(*)	835	417		66.9	25.6			92	150
	0.3	(*)	830	416		61.0	27.4			73	130
	0.4	(*)	830	414	207	56.1	33.1			78	140
	0.8				206	48.2	38.7			83	200
Y	0	894	850			125.1	128.1	11.2		18	20
	0.1		841			114.9	116.6	10.5		50	60
	0.2	(*)	780	414		109.3	112.5	10.3		74	82
	0.3		730	375		117.7	116.2			74	85
	0.4		720	375		90.9	93.6			75	85
	0.8			367		37.3	53.5			75	105

(*) related to (2-17) phase, which was formed after heating up to 750 ~ 800 C due to a phase transformation from 1-7 to 2-17 and 1-5. DoM-Easy direction of Magnetization, EB- Easy planar, EC - Easy conical, EA - Easy uniaxial.

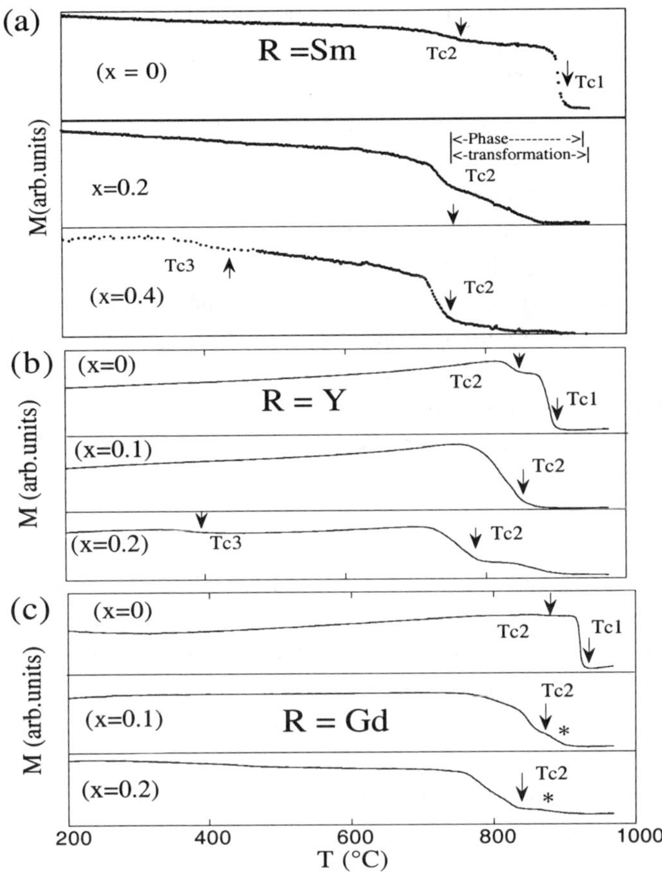

Fig .3. M vs T of $RCo_{7-x}Zr_x$ as-cast alloys (R =Sm, Gd and Y) (H= 500 Oe)

It is to be noted that the above information about the phases present in $SmCo_{7-x}Zr_x$ as-cast alloys match well with S. Derkaoui et al's phase diagram of Sm-Co-Zr system at 1150°C [12].

II. Magnetic Properties of as-cast alloys

As noted above, by substitution of Co with a small amount of Zr in the $RCo_{7-x}Zr_x$ systems, nearly single phase materials with the $TbCu_7$ structure were obtained. Investigation was made of their magnetic properties and the effects of Zr doping on their magnetic properties.

(a). Curie Temperature T_c

Curie temperatures of $RCo_{7-x}Zr_x$ are given in Table II and Fig. 3. As mentioned before, for x=0, two magnetic phases were observed, Tc_1 for the 2-17 phase and Tc_2 for the 1-7 phase. As the Zr content increases, the amount of 1-7 phase increases and that of the 2-17 phase decreases and finally disappears. Almost 1-7 single magnetic phase was formed for R=Sm and Pr at x=0.2~0.3 and for R=Y, Er and Gd at x=0.1~0.2. The Tc's are in the range 960 ~ 1130 K, which are ~ 50 K lower than that of the 2-17 phase and ~ 50 K higher than that of the 1-5 phase. With increased Zr content, the Tc_2 for the 1-7 phase slightly decreases. For x>0.3 ~0.4, other magnetic phases, such as 2-7 phase (Tc_3) and 1-3 phase (Tc_4) were detected.

As in many other transition metal-rich RE-TM compounds, the T_c of $YCo_{7-x}Zr_x$ compounds is contributed by only the Co-Co magnetic interaction. The varieties of Tc with other R are due to the different R-Co magnetic interactions.

(b). Magnetocrystalline Anisotropy field H_A

As shown in Table II , the anisotropy fields H_A, at both 300 K and 10 K, were significantly increased by a small amount of Zr substitution.

In the case of the Sm alloys, H_A increases from 90 kOe for x=0 to over 130 kOe for x=0.2 at 300 K, and from 140 kOe for x=0 to over 220 kOe for x=0.2 at 10 K. For Pr alloys, the anisotropy changes from basal plane for x=0 to uniaxial (EA) with H_A ~ 94 kOe for $x \geq 0.2$ at 300 K. A spin reorientation from easy uniaxial to easy conical was observed in the Pr alloys during heating from 10 to 300 K. The spin reorientation temperature (Tsr) for the $PrCo_{6.8}Zr_{0.2}$ alloy is around 135 K and shifts slightly to higher temperature as the Zr content increases (as shown in Fig. 4a). This behavior is similar to that observed in $PrCo_5$ alloys (Tsr ~ 105 K) [14].

In the case of R=Er alloys, as x increases from 0 to 0.1, H_A increases from 50 to 95 kOe at 300 K and from 85 to168 kOe at 10 K. For R = Y or Gd alloys, the anisotropy is mainly contributed by the 3d (Co) sublattice, because Y is nonmagnetic and Gd^{+3} has an S ground state with zero orbital moment. Even so, large increases in H_A due to Zr doping are observed in these two systems. For R=Y alloys, H_A increases from 18 kOe for x=0 to 74 kOe for x=0.2 at 300 K and from 20 kOe for x=0 to 82 kOe for x=0.2 at 10 K. For R=Gd alloys, H_A increases from 35 kOe for x=0 to 140 kOe for x=0.2 at 300 K, and to ~ 182 kOe for x=0.2 at 10 K.

The above information shows that strong uniaxial anisotropy was developed by a small amount of Zr substitution for Co in the $RCo_{7-x}Zr_x$ systems. As we pointed out earlier [10,11], this can be ascribed to the replacement of the Co dumbbells by Zr atoms. This restores the anisotropy of the Co (2c sites) sublattice , which was reduced when Co was replaced by R atoms (i.e. as the Th_2Zn_{17} or Th_2Ni_{17}-type structure were formed from the $CaCu_5$ -type structure).

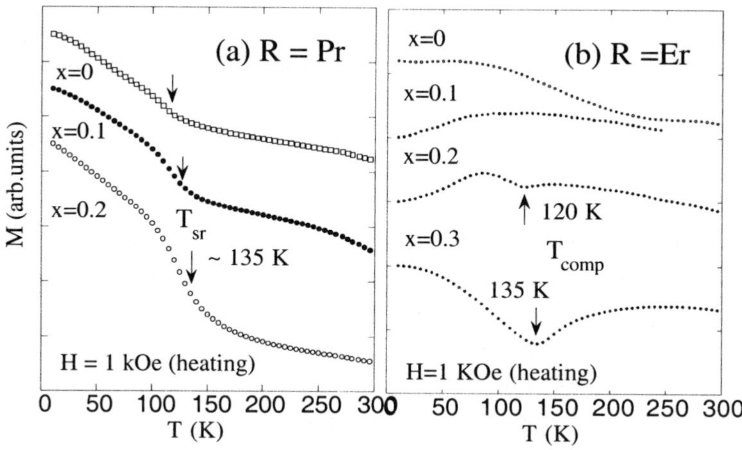

Fig.4. M vs T (10-300K) at H~1000 Oe (chunk samples). (a) Spin reorentation temperature T_{sr} ~135 K for $PrCo_{7-x}Zr_x$. (b) Moment compensation temperature T_{comp} for $ErCo_{7-x}Zr_x$

For R = Sm, Pr or Er alloys, H_A also contains a contribution by the R sublattice. Using quantum mechanics, Greedan and Rao [15] showed that the sign of α_J, the Stevens factor, determined whether the anisotropy of the R sublattice is planar (α_J <0) or axial (α_J >0). Sm^{+3} and Er^{+3} have positive α_J's. Hence they reinforce that of the Co sublattice anisotropy which is known to be axial from work on YCo_5. In contrast, the Pr and Nd anisotropies are in opposition to that of Co. Hence their overall anisotropies are expected to be either weakly positive or negative. For Gd^{+3}, α_J = 0 and Gd systems are general to be isotropic. Experimental results obtained in the present study are in general accord with theory.

When x ≥ 0.3~0.4, other R-rich phases were formed. Some high anisotropy 2-7 phases [13] can make a further increase in H_A

(c). Magnetic Moment M (at 5T, and 10 K)

The moment M, measured at 10 K in a field up to 5 T and in a loose powder form (≤ 38 μm) are listed in Table II. As with the other R-3d systems, the M in the $RCo_{7-x}Zr_x$ systems should contain contributions from both R^{+3} ion and 3d (Co) atom. From the experimental moment of $YCo_{7-x}Zr_x$ (x=0~0.1 and 0.2), we can estimate the Co moment, M(Co). It decreases slightly with increased Zr from ~1.6 to 1.5 μB, which is nearly the same as that in YCo_5 and Y_2Co_{17}.

It was found that the experimental moments for R=Pr, Sm, Er or Gd alloys (x=0~ 0.1-0.2) also obey the general rule for the 4f-3d exchange interaction. The light R couple ferromagnetically with Co, while the heavy R couple antiparallel. This results in a very low saturation for the heavy R, which has high moment. For example, in the Er alloys, as shown in Fig .4b, the partial moment cancellation, resulting from this antiparallel coupling, was observed around 120 ~ 135 K when x=0.2~0.4.i.e. T_{comp}- compensation point.

The estimated moments for R^{+3} ion in $RCo_{7-x}Zr_x$ systems are similar to those in RCo_5 and R_2Co_{17} systems. For example, M(Pr) ~ 2.2 μ_B and M(Er) ~ 8.6 μ_B. With an increase in x, a decrease in M for all systems is reasonable, because of an increase in the amount of the non-magnetic element Zr.

III. Information about the phase transformation.

Fig. 5a shows the XRD of aligned powder (H~1.6T) of annealed $SmCo_{6.8}Zr_{0.2}$ alloys. Two strong lines, 110 and 200, which belong to the 1-7 phase, begin to broaden after annealing

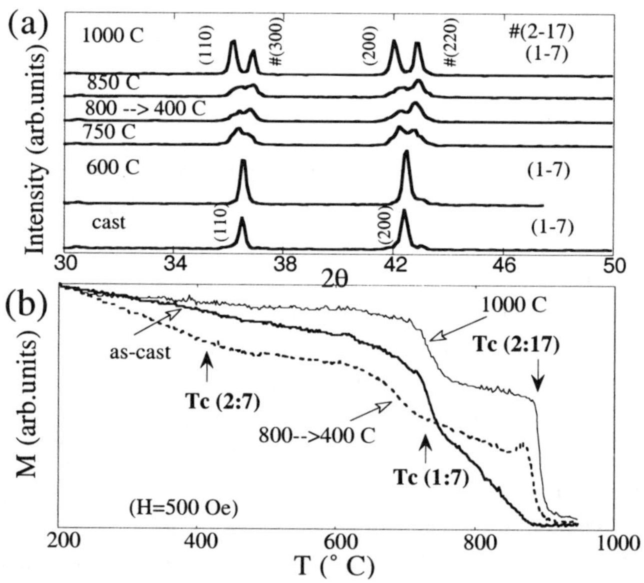

Fig. 5. (a) XRD of aligned powders (H ~ 1.6T) of annealed $SmCo_{6.8}Zr_{0.2}$

(b) M vs T of annealed $SmCo_{6.8}Zr_{0.2}$

above 750 °C. Thus the 1-7 phase begins to decompose. When the annealing temperature increased to 1000°C, two other strong lines, 300 and 220, which belong to the 2-17 phase, clearly show up. Thus the 2-17 phase precipitates out. The above phase transformation was also observed by TMA measurement as seen in Fig.5b. The T_c for 1-7 phase varies slightly at different annealing temperatures, which may be associated with a shift in composition. It was found that H_c (loose powder) at 300 K increases from 1.5 kOe for as-cast sample, to 2.5 kOe when the sample was heated at 800 C followed by a slow cooling to 400 C and decreases to 1.1 kOe after annealing at 1000 C. The increase of H_c may be related to precipitation of a strong anisotropy 2-7 phase, as seen in Fig. 5b . The decrease of H_c can be ascribed to the formation of a weak anisotropy 2-17 phase.

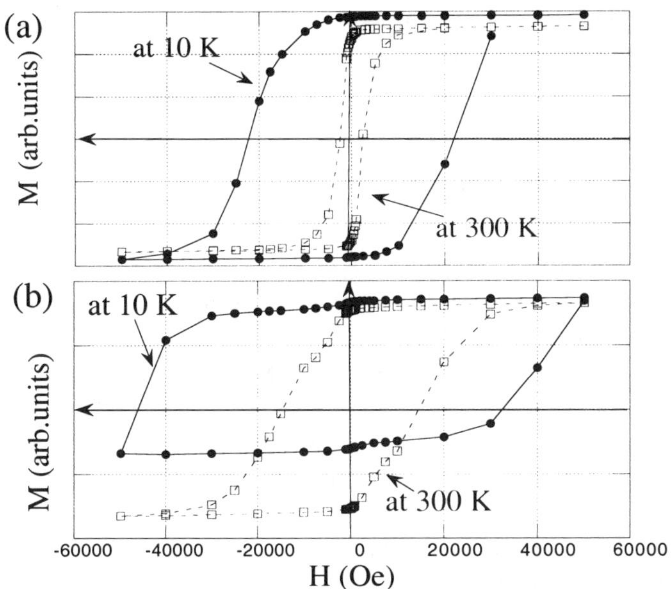

Fig.6. BH loops of Sm (Co,Fe,Cu.Zr)$_7$ aligned powders (~ 38 μm) at 10K and 300 K (a) as-cast (b) after annealing at 800 °C followed by a slow cooling to 400 °C.

In particular, we have noted that, for Sm(Co,Fe,Cu,Zr)$_7$ alloys, after heat treatment, the aligned powders (~ 38 μm) show a large increase of coercivity, H_c - from 1.9 kOe to near 15 kOe at 300 K and from 21 kOe to near 50 kOe at 10 K. (as seen in Fig 6a and 6b)

CONCLUSIONS:

(1) Zr plays an important role in stabilizing the $TbCu_7$ structure in $RCo_{7-x}Zr_x$ alloys. (2) Zr replaces the Co dumbbells and significantly increases anisotropy. (3) Experimental anisotropies are in keeping with expectations derived from the theoretical treatments of anisotropies by Greedan and Rao. (4) $RCo_{7-x}Zr_x$ or R $(Co,Fe,Cu,Zr)_7$ may be a new candidate for permanent magnetic materials and hopefully will play an important role in efforts to improve the high temperature magnetic properties of current 2-17 magnets.

ACKNOWLEDGMENT

This work was supported by the Air Force Office of Scientific Research, Air Force Material Command and USAF under Grant No. F49620-96-1-0454) and by DARPA through the office of Naval Research under contract No. N00014-98-C-0268.

REFERENCE

1. Y. Khan, Acta Cryst. **B29**, 2501 (1973)
2. H. Saito, M. Takahashi, T. Wakiyama, G. Kodo, and H. Nakagawa, J. Magn. Magn. Mater. **82**, 322 (1989)
3. K. H. J. Buschow and F. J. A. den Broeder, J. Less-Common Met. **3**, 191 (1973)
4. J. Yang, O. Mao and Z. Altounian, IEEE Trans. Magn. **32,** 4390 (1987)
5. Y. Mortia, T. Umeda and Y. Kimura, IEEE Trans. Magn. **23**, 2702 (1987)
6. Y. Fukui, T. Nishio and Y. Iwama, IEEE Trena. Magn. **23**, 2705 (1987)
7. S. Liu, H.F. Midrum and K.J. Strnat, J.Appl.Phys. **53** (3) 2383 (1982)
8. T. Ojima, S. Tomizawa, T. Yoneyama, and I.Hari, IEEE Trans. Magn. **13**, 1317 (1977)
9. H. Fujii, M. V. Satyanaryana and W.E. Wallace, Solid State Comm., **41**, 445 (1982)
10. M.Q. Huang, W.E. Wallace, M.E. McHenry, Q. Chen and B. M. Ma, J. Appl. Phys., 83,6718 (1998)
11. M.Q. Huang, M. Drennan, W.E. Wallace, M.E. McHenry, Q. Chen and B.M. Ma, J.Appl. Phys., (1999) (in press)
12. S. Derkaui, N. Valignat and C.H. Alibert, J. Alloy, Comm. **232**, 296(1996)
13. K.H. J. Buschow, J.Less-Common Met. **33**, 311 (1973)
14. W.E. Wallace, Prog. Solid St. Chem, **16**, 127 (1985)
15. J.E. Greedan and V.U.S. Rao , J. Solid State Chem., **6**, 387 (1973).

FIRST-PRINCIPLE STUDIES OF PERMANENT-MAGNET MATERIALS

S.S. JASWAL, R.F. SABIRYANOV
Behlen Laboratory of Physics and Center for Materials Research and Analysis, University of Nebraska-Lincoln, Lincoln, NE 68588-0111

ABSTRACT

First-principle electronic structure studies complement experimental research on hard-magnet materials. Since the discovery of $Nd_2Fe_{14}B$ in 1984, the research in this area has been concentrated on T(Fe,Co)-rich rare-earth compounds such as RT_{12} and R_2T_{17} and exchange coupled hard/soft phases. Self-consistent spin-polarized electronic structure calculations are carried out for the sequence $YFe_2 \rightarrow YFe_3 \rightarrow Y_2Fe_{17} \rightarrow YFe_{12}$ to study the variation of the magnetization and Curie temperature as a function of the Fe concentration. Calculations are performed for R_2T_{17} systems which show significant improvements in their Curie temperatures with interstitial and substitutional modifications. The calculated results are compared with the available experimental data.. Computer simulations are carried out for FePt/Fe and $SmCo_5/Co_{1-x}Fe_x$ hard/soft multilayers.

INTRODUCTION

A good permanent magnet must have large saturation magnetization (M) and high Curie temperature (T_c) and coercivity (H_c). Fe and Co have very good M and T_c values but have low H_c due to their low magnetic anisotropies. Because of the high anisotropies of the rare-earths (R) , transition metal-rich rare-earth (T-R) compounds have been the prime candidates for permanent-magnet research. Most of the studies in this field deal with light rare earths because of their ferromagnetic coupling with Fe and Co. Unfortunately, rare earths lead to a dramatic reduction in T_c in Fe compounds such as R_2Fe_{17} and $RFe_{12-x}M_x$ (M is a stabilizing metal) compared to bcc Fe. The lowering of T_c in Co-rich compounds such as $SmCo_5$ and Sm_2Co_{17} is a relatively small expected dilution effect and must be tolerated to gain in anisotropy. Thus, a tremendous amount of research has been devoted to improve T_c of Fe-rich compounds by substitutional or interstitial modifications during the last decade. We have performed first-principle studies in this area to complement the experimental work.

As mentioned above, R-Fe compounds have T_c considerably lower than that of bcc Fe and T_c tends to decrease as the Fe concentration increases in these compounds ($YFe_2 \rightarrow YFe_3 \rightarrow Y_2Fe_{17} \rightarrow$ Fe(fcc)) [1]. We have carried out electronic structure calculations to understand this unusual variation in T_c with Fe concentration in these compounds [2]. This is related to a very large magnetovolume effect in Fe-rich compounds similar to that in fcc Fe. It is shown how the magnetovolume effect leads to very large increases in T_c of Fe-rich compounds with interstitial and substitutional modifications [3,4].

A way to improve the properties of permanent-magnet materials is by exchange coupling of hard and soft magnetic phases as proposed by Kneller and Hawig [5]. The basic idea here is to combine the large saturation magnetization and T_c of the soft phase with the large magnetic anisotropy of the hard phase to produce a composite with superior hard-magnet properties [6]. We have performed calculations for FePt/Fe and $SmCo_5/Co_{1-x}Fe_x$ multilayers and the results look quite promising [7,8].

Mat. Res. Soc. Symp. Proc. Vol. 577 © 1999 Materials Research Society

METHOD

Self-consistent spin-polarized electronic structure calculations are carried out using the linear-muffin-tin-orbitals method in the atomic sphere approximation [9]. The calculated electronic structure is used to find densities of states (DOS), local magnetic moments (μ_i) and the exchange parameters (J_{ij}) in the Heisenberg Hamiltonian. The Green function formalism to calculate interatomic exchange parameters from the electronic structure has been developed by Liechtenstein *et al.* [10]. Monte Carlo simulations, based on the Heisenberg Hamiltonian, are used to find the thermodynamic average of M^n (n = 1 to 4) as functions of temperature for different cell sizes. The T_c for each cell size is calculated from the locations of the extrema of the quantities such as susceptibility and third and fourth order cumulants. Finite-size scaling is used to derive T_c for the infinite size of the sample using a procedure due to Chen *et al.* [11].

RESULTS AND DISCUSSION

A. Y-Fe Compounds [2]

The details of this study are reported in Ref. 2. Briefly, the changes in the magnetic properties of Y-Fe compounds as a function of the Fe concentration depend on the chemical composition and the local environment of the various Fe sites. The important environment-related properties and the calculated magnetic properties are listed in Table I. The calculated average saturation magnetization for Fe atom is compared with the experimental data in Fig. 1. We see from Table I that the total number of near neighbors that determine the Voronoi polyhedron (VP) is essentially the same for Fe sites in all the compounds. However, the

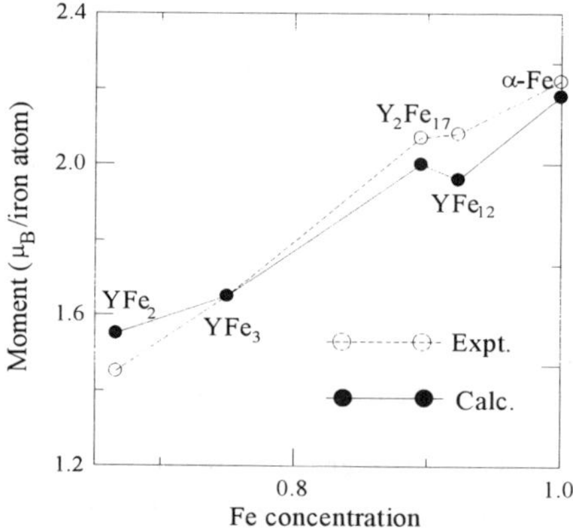

Fig. 1. Magnetization as a function of the Fe concentration. The experimental data from references in R. Coehoorn, Phys. Rev. B **39**, 13072 (1989).

number of Fe neighbors increases and the VP volume decreases with increasing Fe concentration and these changes are qualitatively responsible for the increasing magnetization and decreasing T_c as shown in Fig. 1 and Table I. As more and more non-magnetic Y neighbors are replaced with Fe atoms with increasing Fe concentration, the changes in the interatomic hybridization leads to an increase in the intratomic exchange splitting and hence the magnetization.

Table. I. Numbers of first neighbors (m), Voronoi polyhedron volumes (V), exchange parameters (J_0), magnetic moments (M_i) of Fe atoms and Curie temperatures of yttrium-iron compounds.

		m (all/Fe)	V(Å^3)	J_0(meV)	$M_i(\mu_B)$	T_c^{calc}(K)	T_c^{expt}(K)
YFe_2	1Fe(d)	12/6	12.21	105	1.89	650	570[a]
YFe_3	1Fe(b)	12/6	13.64	120	1.73	592	580[a]
	2Fe(c)	12/9	13.95	133	2.10		
	6Fe(h)	12/7	13.83	96	1.77		
Y_2Fe_{17}	2Fe(c)	14/13	14.02	73	2.51	262	300[b]
	3Fe(d)	12/10	12.67	67	1.50		
	6Fe(f)	12/10	13.47	30	2.10		
	6Fe(h)	12/9	13.78	11	2.10		
YFe_{12}	4Fe(j)	12/10	11.21	79	1.55	486	~500[c]
	4Fe(i)	14/13	12.60	125	2.33		
	4Fe(f)	12/10	11.86	48	2.09		

[a]J. P. Gavigan *et al.*, Physica B **149**, 345 (1988).
[b]H. Fuji *et al.*, J. Alloys and Compounds **219**,10 (1995).
[c]An extrapolated value of T_c for the hypothetical structure YFe_{12} from H. Sun *et al.*, Physica B **183**, 33 (1993).

The exchange interactions (J_{ij}) are fairly long range in these compounds. As an example we list J_{ij} for the Fe-Fe pairs for three shells in Y_2Fe_{17} in Table II. One can see that the near-neighbor exchange interactions in Y_2Fe_{17} are ferromagnetic for all Fe-Fe pairs except for two f-h bonds. Also the J_{ij} for the c-c pair has a large positive value. The results are similar for the other compounds. Thus our calculated exchange parameters do not fit a Bethe-Slater curve where the exchange interactions change smoothly from antiferromagnetic to ferromagnetic with increasing bond length.

In fitting exchange parameters to their hyperfine-field (H_{hf}) data for Sm_2Fe_{17} and $Sm_2Fe_{17}N_y$ with a mean-field model, Li *et al.*[12] make the following assumptions:
(i) Only near-neighbor exchange interactions are considered. Our calculated J_{ij} (Table II) are long range.
(ii) J_{ij} are functions only of the bond length and not the environment. Our calculations show that J_{ij} are strongly influenced by the environment (Table II).
(iii) H_{hf} for different Fe sites are scaled by a constant factor to get the magnetic moments. A comparison of Mössbauer and neutron scattering data for R-Fe compounds shows that this is not a good approximation [13].

Their fitted J_{ij} follow a Bethe-Slater curve with a large negative value for the c-c bond. Unfortunately they replace this curve with a step function (J_{ij}= - 115 K for bond lengths less than 2.45Å and J_{ij}= 55 K for longer bonds) to fit T_c with different models for interstitial and substitutional impurities. Using their values of J_{ij} for Sm_2Fe_{17} and site occupancies for $Sm_2Fe_{15}Si_2$, we find that the mean-field value of the Fe contribution to T_c of $Sm_2Fe_{15}Si_2$ is

Table II. Fe-Fe pair exchange parameters (meV) for three shells in Y_2Fe_{17}. The numbers in the brackets are the number of neighbors and their distance (Å) respectively.

	J_{01}	J_{02}	J_{03}
Fe(c)	62.1 (1x2.355)	1.44 (3x4.043)	1.21 (3x5.183)
	2.1 (3x2.591)	-0.39 (3x4.140)	1.10 (6x5.237)
	7.6 (3x2.622)	-0.52 (6x4.202)	-0.63 (3x5.413)
	3.6 (6x2.706)	-2.28 (6x4.293)	
Fe(d)	7.2 (4x2.403)	0.69 (4x4.010)	0.39 (4x4.653)
	4.3 (4x2.436)	1.44 (2x4.043)	2.11 (4x4.761)
	2.1 (2x2.591)	-0.13 (2x4.223)	-0.88 (2x4.867)
		-0.37 (4x4.230)	2.38 (4x4.895)
		0.03 (4x4.577)	-1.34 (4x5.137)
Fe(f)	7.2 (2x2.403)	6.23 (1x3.587)	-1.64 (2x4.650)
	12.4(2x2.436)	0.69 (2x4.010)	5.12 (2x4.806)
	4.0 (2x2.516)	-1.92 (2x4.069)	0.72 (1x4.873)
	-2.4 (2x2.620)	0.81 (2x4.134)	3.88 (4x5.089)
	3.6 (2x2.706)	-3.69 (2x4.142)	-1.34 (2x5.137)
		-0.09 (2x4.153)	-1.55 (2x5.175)
		-0.52 (2x4.202)	-0.52(4x5.248)
		-3.83 (2x4.220)	
		0.03 (2x4.577)	
Fe(h)	4.3 (2x2.436)	5.28 (1x3.819)	-1.64 (2x4.650)
	8.3 (2x2.476)	-1.92 (2x4.069)	0.39 (2x4.653)
	4.0 (2x2.516)	0.81 (2x4.134)	-1.94 (4x4.763)
	-2.4 (2x2.620)	-0.39 (1x4.140)	-0.88 (2x4.867)
	7.6 (1x2.622)	-0.09 (2x4.153)	3.94 (1x4.873)
		0.84 (2x4.196)	2.38 (2x4.895)
		-0.13 (1x4.223)	-2.87 (2x4.930)
		3.42 (2x4.264)	1.75 (2x4.986)
		-2.28 (2x4.293)	-1.55(2x5.165)

lower than the corresponding value for Sm_2Fe_{17} by 10% (Since the volume of $Sm_2Fe_{15}Si_2$ is lower than that of Sm_2Fe_{17} by only about 1%, it seems reasonable to use the J_{ij} of Sm_2Fe_{17} in this calculation). This is contrary to the results of their model fitted to the T_c of $Sm_2Fe_{17-x}Si_x$ which increases with x for low values of x.

Due to the oscillatory and decaying nature of J_{ij}, the largest contribution to $J_o = \Sigma J_{ij}$ come from the first neighbors. The average values of the sum of the first-neighbor exchange interactions over inequivalent Fe sites ($<J_{01}>$) for YFe_2, YFe_3, Y_2Fe_{17} and YFe_{12} are 91, 77, 52 and 81 meV respectively. In general, an increase in the first-neighbor Fe atoms randomizes their distribution into a broad group which leads to this decrease in $<J_{o1}>$. However, in the case of YFe_{12}, the first neighbor group breaks up into two distinct subshells for the i site similar to bcc Fe which leads to a very large value (123 meV) of its J_{01}.

For a low first-neighbor compound (YFe_2 and YFe_3 with 6 to 8 first-neighbor Fe atoms), the DOS peaks are sharper than those of a high first-neighbor compound and the Fermi energy (E_f) is not located next to any major peak. Therefore, like bcc Fe, such a system is not expected to show any appreciable magnetovolume effect. On the other hand, a high first-neighbor compound like Y_2Fe_{17} has the E_f located on the sharp slope of the large peak in the majority

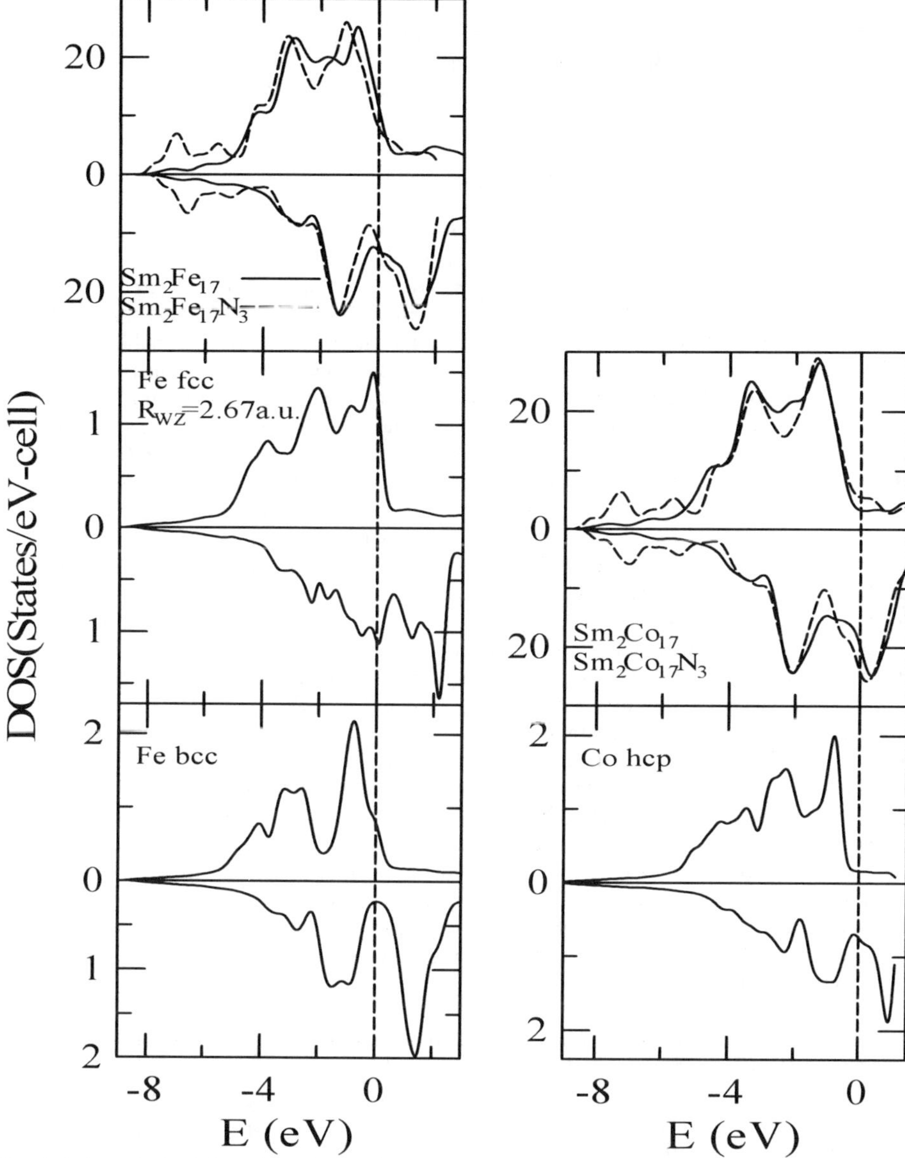

Fig. 2. Spin-polarized densities of state for Sm_2Fe_{17}, $Sm_2Fe_{17}N_3$, and bcc and fcc Fe.

Fig. 3. Spin-polarized densities of state for Sm_2Co_{17}, $Sm_2Co_{17}N_3$, and hcp Co.

DOS and this, as in the case of fcc Fe [14], is responsible for the well known large magnetovolume effect in this system. YFe_{12} has two fcc-like sites and a bcc-like site. As a result, the magnetovolume effect in this system is less pronounced than in Y_2Fe_{17}.

The calculated results for magnetization and T_c are in good agreement with the experimental data as shown in Fig. 1 and Table I. It is interesting to note that Monte Carlo values of T_c are proportional to J_o averaged over Fe sites ($<J_o>$) but the proportionality constant is ~0.76 of that of the mean-field value of 2/3 k_B ($T_c^{MF} = 2/3\ k_B <J_o>$) for these compounds.

B. Interstitial Modification of R_2Fe_{17} and R_2Co_{17} [3]

We consider here the effect of interstitial N on the magnetic properties of 2-17 compounds with Sm. As mentioned in section A, Sm_2Fe_{17} is expected to have a large magnetovolume effect. The calculated densities of states of Sm_2Fe_{17} are compared with those of fcc and bcc iron in Fig. 2. It is interesting to note that Fe d bands of Sm_2Fe_{17} are closer to those of fcc Fe than those of bcc Fe. This is due to the fact that the close-packed environment of Fe atoms in 2-17 compounds is similar to that in fcc Fe. Because of the similarity in their electronic structure, the magnetic properties of 2-17 compounds are closer to those of fcc Fe [14]. For example, both R_2Fe_{17} and fcc Fe show strong dependence of the magnetization on the volume. On the other hand, Co d bands in R_2Co_{17} are similar to those of hcp Co because of the close-packed environment of Co in both the systems (see Fig. 3). As a result, the rare earths in Co-rich compounds have much smaller effect on the magnetic properties of pure Co.

The nitrogenation with 3N atoms per formula unit produces a volume expansion of 6.2% in both Sm_2Fe_{17} and Sm_2Co_{17} [15]. It can be seen from Fig. 2 that the changes in the electronic structure due to the addition of nitrogen to Sm_2Fe_{17} correspond essentially to a rigid downward shift of the majority band. Thus the main consequence of nitrogenation of R_2Fe_{17} compounds is the magnetovolume effect. In contrast, Fig. 3 shows no rigid-band shift and hence nitrogen mainly produces small changes in hybridization in R_2Co_{17} compounds.

The exchange parameters increase in Sm_2Fe_{17} and decrease in Sm_2Co_{17} upon nitrogenation. These changes are reflected in the values of J_i for various sites in Fe and Co compounds as shown in Table III. The average $<J_i>$ increases by ~156% for Sm_2Fe_{17} and decreases by ~33% for Sm_2Co_{17} upon nitrogenation, qualitatively reflecting the changes in T_c of the two compounds. The physical origin of the large increase in T_c of Sm_2Fe_{17} is the magnetovolume effect mentioned above with the changes in hybridization playing a smaller role. The moderate decrease in $<J_i>$ of the Co compound is related to the location of E_f in a valley in the majority DOS (Fig. 3). There is practically no rigid-band shift in this system and the band structure changes come primarily from the changes in the hybridization. The Monte Carlo results for T_c for these compounds are in very good agreement with the experimental data as shown in Table III.

Table III. Exchange parameters J_i (meV) for different sites, average exchange parameters $<J_i>$, and calculated and experimental T_c(K).

	J_c^{Sm}	J_c^{M}	J_d^{M}	J_f^{M}	J_h^{M}	$<J_i>$	T_c^{calc}	$T_c^{expt.,a}$
Sm_2Fe_{17}	-44	67	52	49	40	48	287	389
$Sm_2Fe_{17}N_3$	-77	106	128	128	136	123	768	752
Sm_2Co_{17}	-90	228	219	185	192	187	1169	1189
$Sm_2Co_{17}N_3$	-63	171	138	109	141	125	776	783

[a]Reference 15.

From the results of sections A and B we conclude that Fe-rich R-Fe compounds have large magnetovolume effect which can be used to enhance the permanent-magnet properties of these compounds by weakly hybridizing interstitials such as N and C. Similar improvements are not possible in Co compounds because of the weak magnetovolume effect in the opposite direction.

C. Substitutional Modification of R_2Fe_{17} [4]

Substitution of Al, Ga and Si for Fe in R_2Fe_{17} compounds leads to an increase in their T_c which peaks around three substitutions per formula unit. For example, the substitution of three Si or Ga atoms per formula unit in $SmFe_{17}$ leads to a substantial enhancement of T_c [16,17]. We have calculated T_c for $Sm_2Fe_{16}Ga$ and $Sm_2Fe_{16}Si$ and the results are in very good agreement with the experimental data as shown in Table IV.

Table IV. Curie temperatures (K) of $Sm_2Fe_{16}Ga$, $Sm_2Fe_{16}Si$, and Sm_2Fe_{17}.

	T_c^{MC}	T_c^{expt}
$Sm_2Fe_{16}Ga$	476	490[a]
$Sm_2Fe_{16}Si$	495	468[b]
Sm_2Fe_{17}	282	389[b]

[a]Reference 16.
[b]Reference 17.

The changes in T_c are due to the magnetovolume effect and changes in hybridization around the impurities. The volume expansion and reduction in the hybridization around the impurity lead to narrowing of the d band which is responsible for T_c enhancement with Ga impurity. Even though Si produces a small volume contraction, the reduction in hybridization around Si more than compensates for the contraction to produce a large increase in T_c.

D. Exchange-Coupled Multilayers [7,8]

Exchange-spring magnets are composites of hard and soft phases with strong exchange coupling where the hard phase provides the uniaxial anisotropy and the soft phase the large saturation magnetization. Recently, Liu *et al.* reported large energy products and coercivities in FePt and $SmCo_x$/Co nanoscale multilayers and composites [18]. We have carried out model studies for hard/soft multilayers of two-systems.

The large magnetization and anisotropy of FePt makes it an attractive hard magnetic phase. We performed a first-principle study of the magnetic moments, exchange interactions, T_c and magnetocrystalline anisotropy in layered structures of tetragonal FePt and bcc Fe. FePt and Fe lattice parameters have a mismatch of about 3.5%. The calculations were carried out for 5 bilayers of the hard phase and different bilayers of the soft phase where a bilayer of the FePt phase represents one layer of Fe and one adjacent layer of Pt atoms and for the soft phase it is just two layers of Fe atoms.

The Fe sites maintain their high local moments both in hard and soft phases. The theoretical $(BH)^{th}_{max} = (2\pi M)^2$ as a function of the soft phase is plotted in Fig. 4 along with the anisotropy field. For approximately equal thickness of hard and soft phases (FePt(5)/Fe(8)) the present calculations given $(BH)^{th}_{max} \sim 90$ MGOe and anisotropy field $H_a \sim 60$ kOe. Both the soft and hard phases basically preserve their exchange interactions but the interface atoms have slightly lower values. The mean-field value of T_c based on these interactions is ~1100 K. Since the mean-field approximation usually overestimates T_c by 20-30%, a realistic value of T_c for this system is expected to be 800-900 K. Thus, this composite system is a promising hard-magnet candidate for experimental studies.

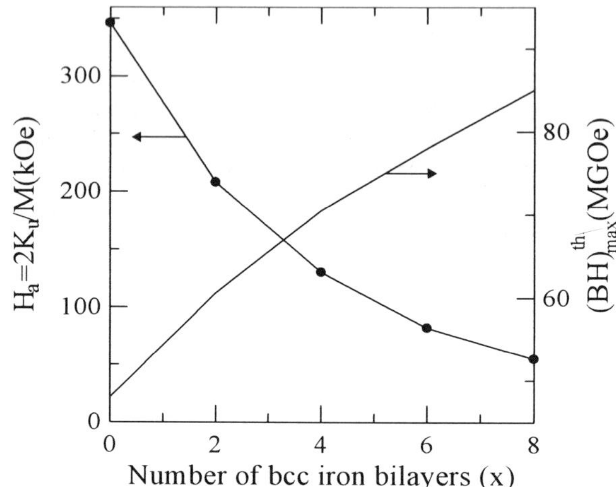

Fig. 4. Energy product $(BH)^{th}_{max}$ and anysotropy field as functions of the soft-phase thickness in FePt(5)/Fe(x) composites.

The second system we have modeled is a composite of $SmCo_5/Co_{1-x}Fe_x$ multilayers. $SmCo_5$ has the largest anisotropy field among the rare earth-transition metal compounds and also a high T_c (1000 K). However, it has a relatively low theoretical value of $(BH)^{th}_{max}$ = 33 MGOe. Among the fcc phases Co has the highest T_c (1388 K) and large magnetization. An addition of 10-15% Fe stabilizes the fcc phase and increases the magnetization to 1.8 T ($(BH)^{th}_{max}$ = 82 MGOe) making it an excellent candidate for the soft phase. Also, the lattice mismatch is only about 1% when the (111) direction of fcc Co is aligned with the c-axis of $SmCo_5$ as shown in Fig. 5.

Calculations were carried out for three supercells with 3, 6 and 12 layers of Co and four units of $SmCo_5$ in each case. The results for magnetization and exchange parameters are almost the same in each case. The local magnetic moments in two phases are quite similar as expected. The exchange parameters in the soft phase are slightly larger than those in the hard phase reflecting the difference in their T_c. The exchange parameters at the interface are in-between those of the two phases implying strong exchange coupling of the two phases. It is clear from the calculated exchange parameters that T_c for the composite is expected to be in-between the values of T_c for $SmCo_5$ and Co depending on the relative amount of the two phases.

These results show that $SmCo_5$ and Co are strongly exchange coupled with excellent hard-magnet properties. As mentioned above, a small amount of Fe is necessary to stabilize Co in the fcc phase. We find that the exchange stiffness parameters for Co-rich $Co_{1-x}Fe_x$ are almost independent of Fe concentration. Therefore we have chosen fcc Co_3Fe as a representative of the soft phase in the following micromagnetic calculations.

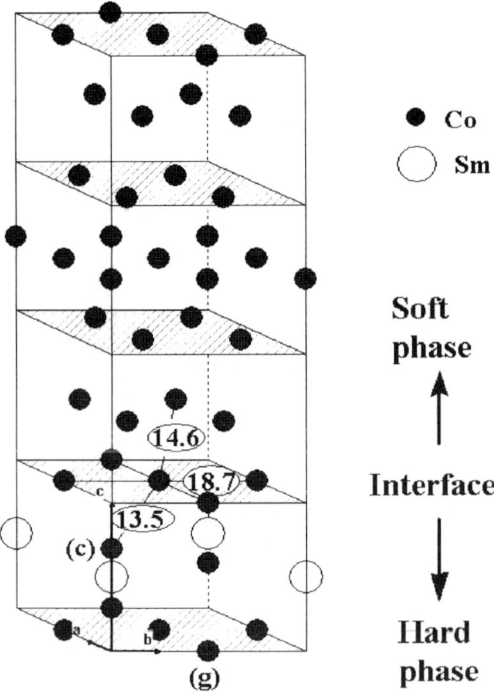

Fig. 5. Crystal structure near the interface of hexagonal $SmCo_5$ and fcc Co with the c axis of $SmCo_5$ and the (111) direction of Co aligned normal to the interface. The interface pair-exchange parameters (meV) are shown in the figure.

Skomski and Coey minimize the magnetic energy of a periodic layered structure in the presence of a nucleation field to arrive at a micromagnetic relation among the various parameters of the hard and soft phases [6]. We solve this equation numberically to find the optimum thickness of two phases for a given $(BH)_{max}^{th}$ and the results are shown in Fig. 6. We see that the thickness of the soft phase remains constant (~13 nm) while the thickness of hard phase decreases with increasing $(BH)_{max}^{th}$ the soft layer thickness begins to decrease when the hard layer thickness is about 2 nm or less. The micromagnetic solutions based on the continuum model are questionable in this region. One needs to use discrete models to deal with such structures. We see from Fig. 6 that $(BH)_{max}^{th}$ is ~65 MGOe for hard and soft layer thicknesses of 5 and 13 nm respectively. Such a multilayer is expected to have a T_c of ~1200 K making it a good candidate for a high-temperature permanent-magnet material. Stacking of hcp Co and $SmCo_5$ in the (1100) direction also has a very good match at the interface and such a layered structure may form with easy axis perpendicular to the layering direction.

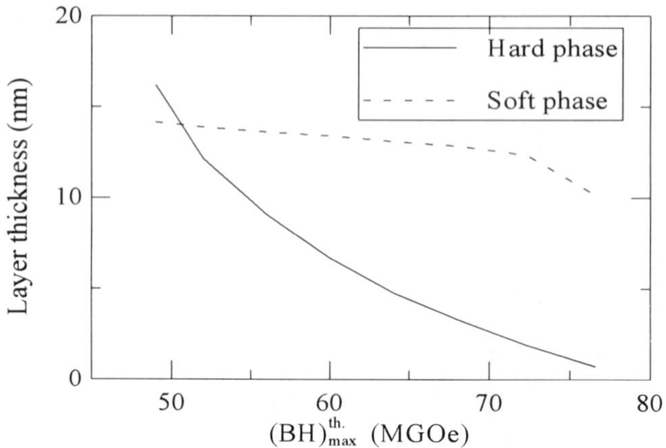

Fig. 6. The optimal thicknesses of hard and soft phases versus the theoretical energy product $(BH)^{th}_{max}$ for $SmCo_5/Co_3Fe$ multilayers.

CONCLUDING REMARKS

Given the computational power, first-principle self-consistent calculations can be carried out on very complex magnetic materials. The results for the magnetization and Curie temperature are quite good now. The same cannot be said of the anisotropy calculations but progress is being made in this extremely difficult and computationally intensive area of study.

Calculated exchange parameters for R-Fe compounds are fairly long range and do not fit a Bethe-Slater curve where exchange interactions change smoothly from antiferromagnetic to ferromagnetic with increase in the bond length. Our calculations show that the dumbbell exchange interactions for Y_2Fe_{17} are strongly ferromagnetic.

The magnetovolume effect is a complicated function of changes in hybridization due to the changes in volume as discussed in the last section. This effect is large in Fe-rich compounds and the lattice expansion leads to a considerable improvement in their hard-magnet properties. The hybridization changes produced by the substitutional impurities like Si can simulate the magnetovolume effect producing large enhancements in T_c even though the lattice contracts slightly.

Since the Co-rich compounds show a small magnetovolume effect in the direction opposite to that of Fe-rich compounds, the interstitial or substitutional modifications of the former are not helpful in improving their magnetizations or T_c. Such impurities might improve their anisotropies.

Finally, we have presented model calculations for two exchange coupled hard/soft multilayers (FePt/Fe and $SmCo_5/Co_{1-x}Fe_x$) as very promising hard-magnet materials for experimental studies.

ACKNOWLEDGMENTS

This work is supported by NSF (Grant No. DMR-9705044), DOE (Grant No. DE-FG2-86ER45262), AFOSR (Grant No. F496209810098), DARPA (Grant No. DAAG-55-98-1-0268) and the National Computational Science Alliance.

REFERENCES

1. K.H.J. Buschow, *Phys. Stat. Sol. (a)* **7**, 199 (1971).
2. R.F. Sabiryanov and S.S. Jaswal, *Phys. Rev. B* **57**, 7767 (1998).
3. R.F. Sabiryanov and S.S. Jaswal, *Phys. Rev. Lett.* **79**, 155 (1997).
4. R.F. Sabiryanov and S.S. Jaswal, *J. Appl. Phys.* **81**, 5615 (1997).
5. E.F. Kneller and R. Hawig, *IEEE Trans. Magn.* **27**, 3588 (1991).
6. R. Skomski and J.M.D. Coey, *Phys. Rev. B* **48**, 15812 (1993).
7. R.F. Sabiryanov and S.S. Jaswal, *J. Magn. Magn. Mater.* **177-181**, 989 (1998).
8. R.F. Sabiryanov and S.S. Jaswal, *Phys. Rev. B* **58**, 12071 (1998).
9. O.K. Andersen, *Phys. Rev. B* **8**, 3060 (1975).
10. A.I. Liechtenstein, M.I. Katsnelson, V.P. Antropov, and V.A. Gubanov, *J. Magn. Magn. Mater.* **21**, 35 (1988).
11. K. Chen, A.M. Ferrenberg, D.P. Landau, *Phys. Rev. B* **48**, 3249 (1993).
12. Z.W. Li, X.Z. Zhou, and A.H. Morrish, *Phys. Rev.B* **55**, 2891 (1995).
    Z.W. Li, A. H. Morrish, *Phys. Rev. B* **55**, 3670 (1997).
13. G.J. Long, G.K. Marasinghe, S. Mishra, O.A. Pringle, Z. Hu, W.B. Yelon, D.P. Middleton, K.H.J. Bushow, and F. Grandjean, *J. Appl. Phys.* **76**, 5383 (1994);
14. R.F. Sabiryanov, S.K. Bose, and O.N. Mryasov, *Phys. Rev. B* **51**, 8958 (1995).
15. M. Katter, J. Wecker, C. Kuhrt, and L. Shultz, *J. Magn. Magn. Mater.* **114**, 35 (1992).
16. Z.W. Li, X.Z. Zhou, and A.H. Morrish, *J. Magn. Magn. Mater.* **150**, 57 (1995).
17. Z. Wang and R.A. Dunlap, *Philos. Mag. B* **69**, 103 (1994).
18. J.P. Liu, Y. Liu, C. Luo, Z. Shan, and D.J. Sellmyer, *J. Appl. Phys.* **81**, 5644 (1997);
    J.P. Liu, C. Luo, Y. Liu, and D.J. Sellmyer, *Appl. Phys. Lett.* **72**, 483 (1998);
    J.P. Liu, Y. Liu, R. Skomski, and D.J. Sellmyer, *J. Appl. Phys.* (in press).

ELECTRONIC STRUCTURE AND MAGNETIC PROPERTIES OF Nd_5Fe_{17}

R. F. SABIRYANOV AND S. S. JASWAL
Behlen Laboratory of Physics and Center for Materials Research and Analysis, University of Nebraska, Lincoln, NE68588-0111

ABSTRACT

The Nd_5Fe_{17} compound has attracted some attention lately as a possible permanent-magnet material. This compound has a very complicated crystal structure (264 atoms per unit cell with 14 and 7 different Fe and Nd sites respectively!) Using linear-muffin-tin-orbital method, self-consistent spin-polarized studies of Nd_5Fe_{17} have been carried out to determine the electronic structure, magnetic moments, and the Curie temperature. The results show strong effect of the local environment on the magnetic properties of individual Fe sites. The calculated moments are compared with the magnetization data. The Curie temperature estimated for the iron sublattice using the mean-field approximation is in a reasonable agreement with the experimental result.

INTRODUCTION

Nd_5Fe_{17} reported first by Landgraf *et al.* [1] has a very complicated structure refined recently by neutron diffraction measurements [2,3]. Nd_5Fe_{17} belongs to the hexagonal space group $P6_3/mcm$, a=20.2Å, c=12.322Å. It has 12 formula units per unit cell with a total of 264 atoms per unit cell. Its magnetization is comparable to that of $Nd_2Fe_{14}B$ (162 emu/g at 4.2K [1], 48.2 μ_B/f.u. = 1232 emu/cm^3 at 1.5K [2]). The Mössbauer measurements give an estimation of the average moment on the iron sublattice of $\langle M^{Fe}\rangle$=2.12 μ_B/atom [4]. Combining the magnetization and Mössbauer measurements gives the average neodymium moment $\langle M^{Nd}\rangle$=2.43 μ_B/atom.This is somewhat lower than the value obtained using Russel-Saunders coupling ($\langle M_f^{Nd}\rangle$=3.27 μ_B) which consists of an orbital moment contribution of 5.72 μ_B and a spin moment of 2.45 μ_B pointed in opposite direction to the orbital moment. The deviations from Russel-Saunders coupling is also found in other Nd-transition metal compounds such as $NdCo_5$ (see review by J.J.M. Franse and R.J. Radwanski in *Handbook of magnetic materials*, ed. by K.H.J. Bushow, North-Holland, v. 7, p. 442). Finally, room temperature neutron diffraction measurements give iron moments ranging from 1.99 to 2.21 μ_B and $\langle M^{Nd}\rangle$=1.2 μ_B [2,3]. The Curie temperature (T_c) obtained from magnetic measurements is 507-509K [2,3].

We present in this paper a detailed study of the magnetic properties of Nd_5Fe_{17} using first-principle self-consistent spin-polarized electronic structure calculations. The site projected densities of states (DOS) and magnetic moments are determined. The exchange parameters calculated by Green function method are used to estimate T_c. The calculated results are compared with the available experimental data.

RESULTS AND DISCUSSION

The top view of the unit cell, as one looks down the c-axis, can be seen in Fig. 1. The number of near neighbors of each site are listed in Table I. There are three well

Mat. Res. Soc. Symp. Proc. Vol. 577 © 1999 Materials Research Society

defined regions in the cell. One region is a column of Nd atoms of about 9Å in diameter. The central atoms of this column have no iron in their environment . The second region

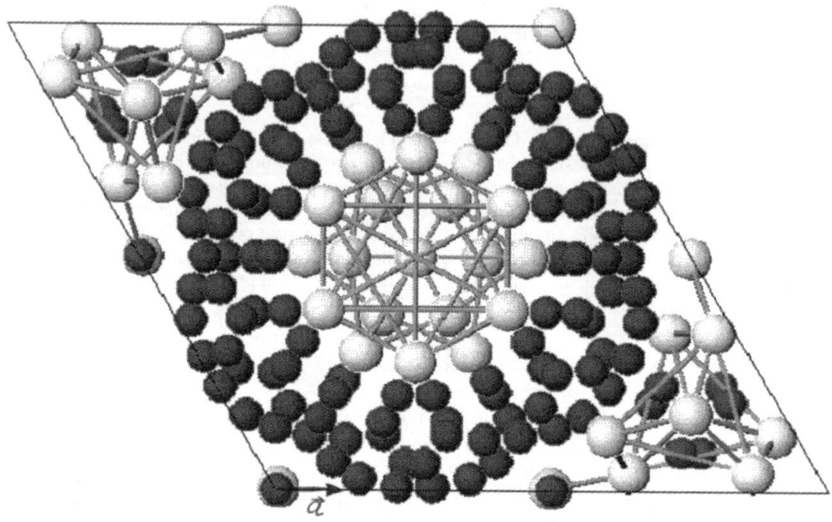

Fig. 1. Projection in the (a,a) plane of the unit cell of Nd_5Fe_{17} with 264 atoms (black balls -Fe, white balls - Nd.)

consists of only iron atoms and it surrounds the Nd column. The width of this region is also about 9Å. The central atoms in this region have about two Nd atoms as near neighbors. The third region with both iron and neodymium atoms is somewhat triangular in shape. We can view this region as a compound. Iron sites in this "compound" have 6 Nd neighbors and 6 Fe neighbors. This spatial distribution in Nd_5Fe_{17} makes it similar to a nano-composite material with the size of each component close to 1nm. Thus Nd_5Fe_{17} is like an exchange-coupled system with hard, soft and neodymium phases. This unusual structure leads to a large variation in the local environment of various atoms as shown in Table I. The Voronoi polyhedron was constructed with near-neighbor Nd-Fe bonds partitioned such that $r_{Nd}:r_{Fe}=1.4:1$ to account for the larger Nd atom radius. The Wigner-Seitz radii were derived from Voronoi polyhedron such that Wigner-Seitz sphere volume is equal to the corresponding Voronoi polyhedron volume. Radii are given in Table I. It is known that magnetism of an iron atom is very sensitive to its environment and can show strong-magnet behavior with high Curie temperature and almost no volume dependence (bcc Fe) or weak-magnet behavior with strong magnetovolume effect (fcc Fe) [5]. This behavior is closely connected with the symmetry and strength of hybridization.

We use linear-muffin-tin-orbitals method in atomic-sphere and frozen-core approximations to derive the electronic structure and magnetic properties of Nd_5Fe_{17}. The core spin densities for Nd f electrons are calculated using Russel-Saunders

Table I. The number of neighbors (N_{Fe}, N_{Nd}), Wigner-Seitz sphere radii (R), spin moments (M), and total exchange parameters (J_0) of various sites.

Site	Type	N_{Fe}	N_{Nd}	R(Å)	M(μ_B)	J_0(mRy)
Nd1	2a	0	12	2.171	0.48*	
Nd2	4d	12	6	1.974	-0.43*	
Nd3	6g1	4	8	2.065	-0.07*	
Nd4	6g2	13	4	1.887	-0.16*	
Nd5	6g3	9	6	1.975	-0.11*	
Nd6	12i	10	7	2.027	-0.24*	
Nd7	24l	13	4	1.903	-0.23*	
Fe1	6f	6	6	1.426	2.02	6.3
Fe2	6g	10	2	1.437	2.25	6.4
Fe3	12k1	9	3	1.419	2.47	3.4
Fe4	12k2	9	3	1.401	2.51	7.7
Fe5	12k3	9	3	1.446	2.28	8.3
Fe6	12j1	8	4	1.433	2.21	12.7
Fe7	12j2	9	3	1.396	1.97	6.3
Fe8	12j3	10	2	1.392	2.82	5.4
Fe9	12j4	10	2	1.401	2.63	3.5
Fe10	12j5	6	6	1.464	2.61	10.6
Fe11	24l1	12	2	1.450	1.99	8.7
Fe12	24l2	11	3	1.402	2.28	8.9
Fe13	24l3	8	4	1.421	1.89	10.5
Fe14	24l4	10	2	1.408	2.46	8.8

* spin-moments of non-4f valence electrons

coupling spin moment of -2.44 μ_B (population of 4f majority is 2.72 *e* and minority 0.28 *e*).The calculated DOS of Fe atoms are presented in Figures 2 and 3. We see that the DOS of Fe atoms show strong dependence on the local environment. The degree of spin polarization is high with the majority DOS main peaks below the Fermi level. The minority DOS have deep minimum at the Fermi energy which makes the position and shape of its main peaks similar. The shape of majority DOS is more sensitive to the volume and hybridization with neighboring atoms. For atoms with similar environment there is an increase of the moment with the volume (f $\rightarrow$j5 and j2,l3 $\rightarrow$ l2, k1, k2, k3).

The local moments are given in Table I. There is a large variation in the local moments of Fe atoms (1.9 to 2.8 μ_B). The average magnetization of iron sublattice ($\langle M^{Fe} \rangle$=2.28 μ_B/Fe-atom) is in a good agreement with the room temperature Mössbauer data [4]. Magnetization of the Nd sublattice derived from the difference between the experimental magnetization and the calculated iron sublattice data gives $\langle M^{Nd} \rangle$=1.89 μ_B/Nd-atom which is lower than the Hund's rule value of 3.27 μ_B. This may be related to the positive value of non-4f moment on Nd1 (Table I) which implies an antiferromagnetic ordering of the Nd1 atom with respect to the Fe atoms. Also pure Nd has a non-collinear structure with ordering temperature of 19K [6]. Similar ordering may be present in the Nd region of Nd_5Fe_{17}.

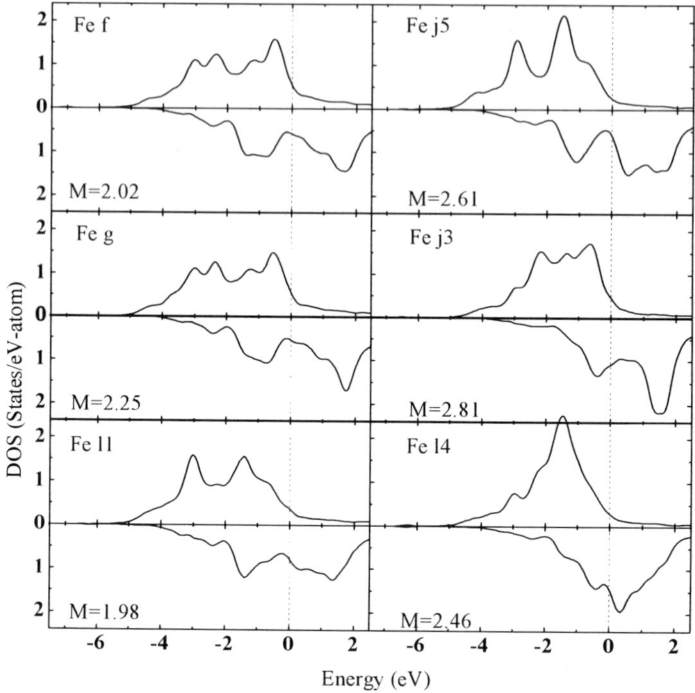

Fig. 2. Density of states of Fe atoms with 6 Nd neighbors (f, j5) and 2 Nd neighbors (g, j3, l1, l4). Zero energy corresponds to Fermi energy.

The modelling of the exchange interactions and Curie temperature of Nd_5Fe_{17} is very complicated because of its structure and the presence of Nd atoms in two different environments. We performed calculations of the total exchange parameter J_0 for each iron atom. The results given in Table I show that there is no simple correlation between the magnitude of J_0 of a given Fe site and its near-neighbor environment. The average total exchange parameter of the iron atoms is $\langle J_0 \rangle = 7.4$mRy. Ignoring the Nd atoms this gives a mean-field Curie temperature T_c^{MF}=789K. It is well known that mean field calculations overestimates T_c by 20-30%. This qualitatively accounts for the difference between the calculated T_c and the experimental value of T_c^{expt}=509K.

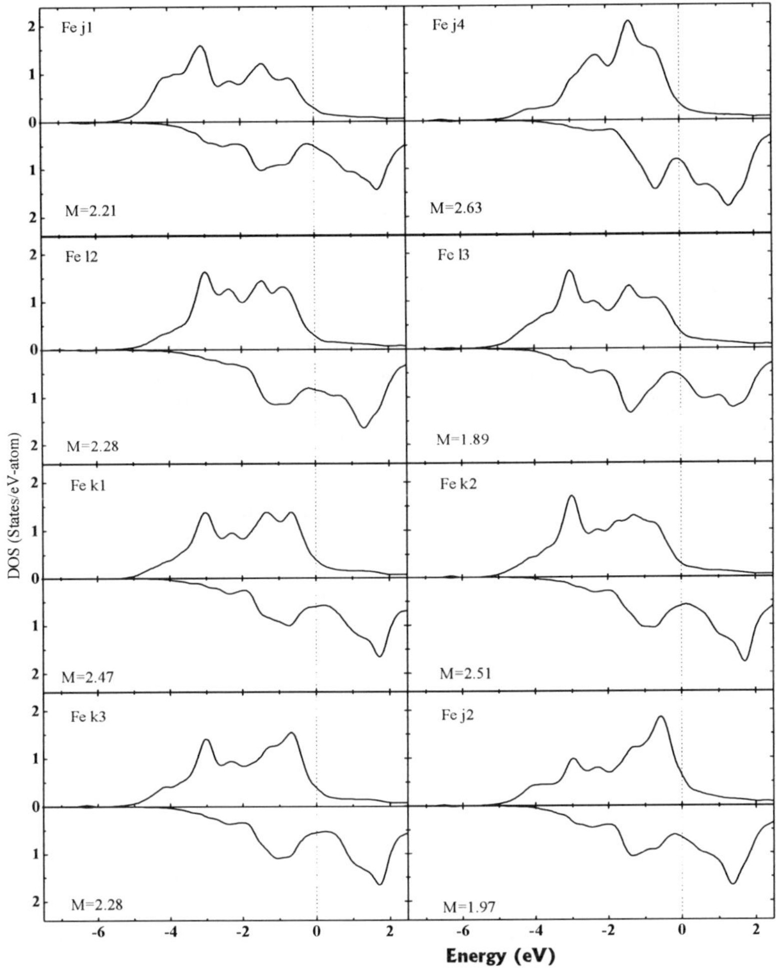

Fig. 3. Density of states of Fe atoms with 3 and 4 Nd neighbors. Zero energy corresponds to Fermi energy.

CONCLUSIONS

The Nd_5Fe_{17} compound can be viewed as a three part columnar structure (Nd, Fe and NdFe) with the lateral dimension of each column being about 1nm. This unusual structure gives rise to large variations in the local environment of various Fe atoms. As a result, the magnetic moments of the 14 different Fe sites obtained from our first-principle electronic structure calculations vary over a wide range (1.9-2.8 μ_B). The calculated values for the magnetization and the Curie temperature are in reasonable agreement with available experimental data.

ACKNOWLEDGMENTS

This work is supported by NSF (Grant No. DMR-9705044), DOE (Grant No. DE-FG2-86ER45262), AFOSR (Grant No. F496209810098), DARPA (Grant No. DAAG-55-98-1-0268) and the National Computational Science Alliance.

REFERENCES

1. F.J.G. Landgraf, G. Schneider, V. Villas-Boas, F.P. Missel, J. Less-Common Met. **163**, 209 (1990); F.J.G. Landgraf, F.P. Missel, H.R. Rechenberg, G. Schneider, V. Villas-Boas, J.M. Moreau, L. Piccard, J.P. Nozieres, J. Appl. Phys. **70**, 6125 (1991).
2. C. Lin, C.-X. Liu, Y.-X. Sun, Z.-X. Liu, D.-F. Chen, C. Gou, K. Sun, J.-L. Yang , J. Magn. Magn. Mater **186**, 129 (1998).
3. W.B. Yelon, H. Luo, M. Chen, F.P. Missel, J. Appl. Phys. (in press).
4. J.P. Nozieres, H.R. Rechenberg, Solid State Commun. **79**, 21 (1991).
5. R.F. Sabiryanov, S.K. Bose and O.N. Mryasov, Phys. Rev. B **51**, 8958 (1995).
6. W.C. Koehler, R.M. Moon, H.R. Child, AIP Conf. Proc. **10**, 1319 (1972).
7. R.F. Sabiryanov and S.S. Jaswal, Phys. Rev. B **57**, 7767 (1998).

MAGNETIC ANISOTROPY IN RARE-EARTH - TRANSITION METAL INTERMETALLIC COMPOUNDS

NGUYEN HOANG LUONG *, NGUYEN PHU THUY *,** , PHAM HONG QUANG *, PHAM DUC THANG **
* Cryogenic Laboratory, Faculty of Physics, National University of Hanoi
334 Nguyen Trai, Thanh Xuan, Hanoi, Vietnam
** International Training Institute for Materials Science, 1 Dai Co Viet St., Hanoi, Vietnam

ABSTRACT

In this work the rare-earth anisotropy in the rare-earth - transition metal intermetallic compounds is investigated by studying the temperature-induced spin reorientation phenomena on aligned powder samples and by analyzing the temperature dependence of the magnetization measured along different principal axes of single crystals. From such investigations the crystalline-electric-field parameters of the rare-earth ions in these compounds have been derived. The results on different hard magnetic materials of the compositions $R_2Fe_{14}B$, $RFe_{11}Ti$ and their substituted compounds as well as R_2Co_{17} are summarized.

INTRODUCTION

The rare earth (R) - transition metal (T) compounds are important intermetallic compounds because of favorable properties associated with the R elements and the 3d elements. These compounds have led to important industrial applications, especially as permanent magnets. The magnetic properties of the R-T intermetallic compounds are governed by the interplay between crystalline electric field (CEF) and exchange interactions. The origin of the CEF is the electrostatic interaction of the aspherical orbital of the 4f electrons of the R ions with the aspherical electric field of the surroundings. Among the different types of anisotropy, the magnetocrystalline anisotropy (MCA) originating from the CEF is dominant. In this work we focus our discussion on the spin-reorientation phenomena in some hard magnetic R-T intermetallics and on the analysis of the temperature dependence of the magnetization in some R-T single crystals. Such considerations give valuable information on the CEF parameters.

THEORETICAL OUTLINE

In the phenomenological description, for crystals with tetragonal symmetry the MCA energy can be expressed as [1]:

$$E_A = K_1\sin^2\theta + K_2\sin^4\theta + K_3\sin^6\theta + (K_4\sin^4\theta + K_5\sin^6\theta)\cos4\phi + \ldots \tag{1}$$

whereas for hexagonal or rhombohedral symmetry it is given by

$$E_A = K_1\sin^2\theta + K_2\sin^4\theta + K_3\sin^6\theta + K_4\sin^6\theta\cos6\phi + \ldots, \tag{2}$$

where θ is the angle between the (uniaxial) *c*-axis and the magnetization vector, ϕ is the angle between the component of the magnetization in the *ab*-plane and the *a*-axis and K_i (i = 1, 2,

Mat. Res. Soc. Symp. Proc. Vol. 577 © 1999 Materials Research Society

3...) are the anisotropy constants. Minimizing the anisotropy energy, E_A, with respect to θ gives the orientation of the magnetization. For the uniaxial R-T compounds, usually it is sufficient to consider only the K_1 and K_2 terms. The spin reorientation starts at the temperature where K_1 changes sign [2,3]. If K_1 dominates and $K_1 > 0$, the preferred magnetization direction is along the *c* axis, which corresponds to the case of easy-axis anisotropy. If K_1 dominates and $K_1 < 0$, the preferred magnetization direction is perpendicular to the *c* axis, which corresponds to easy-plane anisotropy. When K_1 is not dominant, the preferred magnetization may be in other directions.

For a deeper understanding of the phenomena, one has to consider a microscopic model. As mentioned in the introduction, the magnetic properties of 3d-4f compounds are governed by a combination of the 3d-4f exchange and crystalline-electric-field interactions. The Hamiltonian of a 4f ion in 3d-4f compounds is usually given by

$$H_R = H_{CEF} + H_{ex} , \tag{3}$$

where H_{CEF} and H_{ex} are the crystalline-electric-field and exchange Hamiltonians, respectively. The CEF Hamiltonian is expressed by

$$H_{CEF} = \sum_{n=0}^{\infty} \sum_{m=-n}^{n} B_n^m O_n^m , \tag{4}$$

where B_n^m are called the crystalline-electric-field parameters and O_n^m are the Stevens equivalent operators. The parameters B_n^m can be written as:

$$B_n^m = \theta_n \langle r_{4f}^n \rangle A_n^m , \tag{5}$$

where the terms related to the 4f ion, $\theta_n \langle r_{4f}^n \rangle$, and those related to the surrounding charges, A_n^m, are separated. The coefficients A_n^m are known as the crystalline-electric-field coefficients. θ_n is the appropriate Stevens factor of order n which represents the proportionality between the operator functions of x, y, z and the operator functions of J_x, J_y, J_z. The parameter θ_n is denoted as α_J, β_J, γ_J for n = 2, 4, 6. $\langle r_{4f}^n \rangle$ is the mean value of the n-th power of the 4f radius and taken from [4].

The exchange Hamiltonian is given by

$$H_{ex} = g\mu_B \mathbf{J} \mathbf{B}_m = 2(g-1)\, \mu_B \mathbf{J} \mathbf{B}_{ex} , \tag{6}$$

where $\mathbf{B}_m$ is the molecular field acting on the rare-earth magnetic moment, which is related to the exchange field $\mathbf{B}_{ex}$ acting on the rare-earth spin by $\mathbf{B}_m = [2(g-1)/g]\mathbf{B}_{ex}$. J and g are the total angular momentum and the Landé factor of the R^{3+} ion, respectively. The field $\mathbf{B}_m$ is related to the exchange constant n_{RT}, which links both rare earth (R) and transition metal (T) sublattices by

$$B_m = [2(g-1)/g]\, n_{RT} \langle M_T \rangle , \tag{7}$$

where $\langle M_T \rangle$ is the average transition-metal sublattice magnetization.

The relation between K_1^R and B_n^m is [5]

$$K_1^R = -(3/2)\,B_2^0 <O_2^0> - 5\,B_4^0 <O_4^0> - (21/2)\,B_6^0 <O_6^0>, \tag{8}$$

where $<O_n^m>$ are the thermal averages of the Stevens operators.

The ground state of the 4f ion is calculated by diagonalizing the Hamiltonian (3). The temperature dependence of the rare-earth sublattice anisotropy, K_1^R, which normally dominates, is then calculated according to Eq. (8). In case of the $R_2Fe_{14}B$ compounds, for example, K_1^R has been calculated in [6] as $K_1^R = E_{\perp c} - E_{\parallel c}$ where $E_{\perp c}$ and $E_{\parallel c}$ represent the ground state energy of the Hamiltonian (3) for B_m being perpendicular and parallel to the *c*-axis, respectively. To calculate the temperature dependence of the anisotropy energy of the rare-earth ion, the Boltzmann distribution function is used. The temperature dependence of the transition metal sublattice anisotropy, K_1^T, is obtained from the study of the isostructural compound, in which R is non-magnetic.

If the data on the temperature dependence of the angle θ in the aligned powder samples are available, the analysis proceeds as follows.

For the aligned powder samples of the material with axial anisotropy, the crystallites (powder particles) are oriented so that their *c*-axes are parallel to each other, whereas the *a*- and *b*-axes are randomly distributed in the plane perpendicular to the alignment direction. Therefore, it is appropriate to confine the exchange field to the x-z plane [7]. This approximation leads to the following exchange Hamiltonian

$$H_{ex} = g\mu_B B_m\,(J_z\cos\theta + J_x\sin\theta). \tag{9}$$

The rare-earth energy is obtained by diagonalization of the Hamiltonian (3) and by calculating the partition function Z(θ,T) and is given by

$$F_R(\theta,T) = -k_B T \sum_{i=1}^{n} \ln Z_i(\theta,T). \tag{10}$$

Here, the contributions for n crystallographically inequivalent rare-earth sites have been taken into account.

For a mixed system such as $R_{1-x}R'_xFe_{11}Ti$, the total free energy is then expressed as

$$F(\theta,T) = (1-x)\,F_R(\theta,T) + x\,F_R'(\theta,T) + E_A^T(\theta,T), \tag{11}$$

where the last term represents the contribution from the transition-metal sublattice ($=K_1^T \sin^2\theta$). To calculate F(θ,T) we need to know the CEF parameters, B_n^m, and molecular-field parameter, $g\mu_B B_m$. The temperature dependence of transition-metal sublattice anisotropy, K_1^T, is again obtained from the study of the isostructural compound, in which R is non-magnetic. The angular dependence of the total free energy is then calculated and the minimum in the free energy gives the orientation of the total magnetization vector. In this way, the temperature dependence of the magnetic structure is determined.

In an external magnetic field, **B,** we should add to the Hamiltonian of the R ion the term describing the magnetostatic energy. Thus, the Hamiltonian for the lowest multiplet J of a R ion becomes:

$$H_R = \sum_{n=0}^{\infty} \sum_{m=-n}^{n} B_n^m O_n^m + g\mu_B \mathbf{J}\mathbf{B}_m + g\mu_B \mathbf{J}\mathbf{B}. \tag{12}$$

The total free energy of the R-T compound at the angle θ corresponding to the minimal free energy can now be obtained as:

$$F(\mathbf{B},T) = -k_B T \sum_{i=1}^{n} \ln Z_i(T) + E_A^T - \mathbf{M}_T\mathbf{B}, \tag{13}$$

where the first term is the free energy of the R sublattice and the last two terms are the MCA and magnetostatic energy of the T sublattice, respectively. By minimizing the free energy at a given temperature and external field, the magnetization process as well as the temperature dependence of the magnetization can be described.

RESULTS AND DISCUSSION

$R_2Fe_{14}B$ compounds

In the tetragonal $R_2Fe_{14}B$ compounds, the rare-earth sublattice consists of two inequivalent rare-earth sites and its anisotropy at room temperature is uniaxial in the compounds with rare-earth elements for which the Stevens factor α_J is negative, while it is planar in the compounds with rare-earth elements for which α_J is positive [8]. In the compounds with Sm, Er, Tm, and Yb for which α_J is positive, a competition between rare-earth and iron sublattice anisotropies is expected and may lead to a spin reorientation. Such reorientations were observed in $Er_2Fe_{14}B$, $Tm_2Fe_{14}B$ [9] and $Yb_2Fe_{14}B$ [10]. In $Sm_2Fe_{14}B$ a spin reorientation has not been observed. $Nd_2Fe_{14}B$ undergoes a different type of spin reorientation, namely from a conical to an axial arrangement with increasing temperature.

We have calculated the temperature dependence of the rare-earth contribution to the MCA energy of the $R_2Fe_{14}B$ compounds (R = Sm, Er, Tm, Yb), using the Hamiltonian (3) with only one lowest-order CEF parameter, B_2^0, [6]. In our calculations, B_2^0 and $g\mu_B B_m$ are adjustable parameters to get the best fits to the experimental data. We could obtain satisfactory fits to the experimental rare-earth anisotropy constant, K_1^R, for Sm, Er, Tm compounds in a wide temperature range and, in combination with the data on the iron sublattice anisotropy obtained in[11], reproduce the correct values for the spin-reorientation temperature in $R_2Fe_{14}B$ with R = Er, Tm and Yb [6]. We showed that a spin reorientation can not be expected in $Sm_2Fe_{14}B$, in agreement with experiment. It turned out, furthermore, from our analysis [6] that the CEF parameter, A_2^0, is not constant for the $R_2Fe_{14}B$ series. A_2^0 in $Tm_2Fe_{14}B$ is somewhat smaller than that in $Er_2Fe_{14}B$, while this parameter in $Yb_2Fe_{14}B$ is significantly smaller than that in $Er_2Fe_{14}B$. Such a behavior of A_2^0 in the $R_2Fe_{14}B$ series is also reported by Cadogan et al. [12].

$RFe_{11}Ti$ compounds

We have studied the effect of Y substitution on the spin reorientation in $NdFe_{11}Ti$ by magnetization measurements on bulk and oriented powder samples and by measurements of the *ac* susceptibility on bulk samples as a function of temperature [13]. The spin-reorientation temperatures, T_{SR}, were determined by ac-susceptibility method for different $Nd_{1-x}Y_xFe_{11}Ti$ compounds. It was shown that above T_{SR} the anisotropy is axial, but below T_{SR} the easy magnetization direction deviates from the *c*-axis. The method of analysis, based on anisotropy constants (Eq. (8)), was applied to discuss the spin-reorientation phenomenon in the $Nd_{1-x}Y_xFe_{11}Ti$ compounds [14]. For calculating K_1^R according to Eq. (8) we have chosen the set of coefficients A_n^m presented in table I. Value of 161 K for $g\mu_B B_m$, derived from the value of n_{RT} obtained by Hu et al. [15], has been used. The temperature dependence of the iron sublattice anisotropy, K_1^{Fe} (=K_1^T), is obtained from the study of $YFe_{11}Ti$ [15]. The calculated values of the spin reorientation temperature are in good agreement with the experimental ones (see [13,14]).

In the case of $Tb_{1-x}Y_xFe_{11}Ti$ and $Dy_{1-x}Y_xFe_{11}Ti$, we have carried out a more detailed experimental study of the spin-reorientation [16,17]. We have obtained the temperature dependence of the cone angle θ [16,17], from the polar plots of the magnetization parallel and perpendicular to the applied magnetic field direction measured on the aligned samples, using an extraction magnetometer equipped with four Helmholtz coils in the temperature range 77-300 K. We analyzed the experimental results with a method based on the calculation of the free energy (Eq. (11)). In our analysis of the $Tb_{1-x}Y_xFe_{11}Ti$ and $Dy_{1-x}Y_xFe_{11}Ti$ data, we assume that the exchange interaction stabilizes the ferrimagnetic collinearity between the magnetic moments of the transition-metal and rare-earth sublattices. The CEF coefficients, A_n^m, which we proposed for fitting the experimental results for the $Tb_{1-x}Y_xFe_{11}Ti$ and $Dy_{1-x}Y_xFe_{11}Ti$ compounds, are presented in Table I. The value of 128.7 K for the molecular-field parameter, $g\mu_B B_m$, for $Tb_{1-x}Y_xFe_{11}Ti$ was derived from the exchange constant, n_{RFe}, obtained by Hu et al. [15]. For $Dy_{1-x}Y_xFe_{11}Ti$, the value of 85.6 K for $g\mu_B B_m$ was deduced from the value for n_{RFe} obtained by Hu et al. [18]. The temperature dependence of the iron sublattice anisotropy, K_1^{Fe}, was again taken from the study of $YFe_{11}Ti$ [15]. The calculated temperature dependencies of the angle θ for the $Dy_{1-x}Y_xFe_{11}Ti$ compounds are shown together with the experimental ones in Fig. 1 as an example [16]. As can be seen from this figure, a good agreement between calculation and experiment has been obtained. For the $Tb_{1-x}Y_xFe_{11}Ti$ compounds the calculated magnetic structure is also in agreement with the experimental one [17].

Table I shows that the leading CEF coefficients A_2^0 obtained for the three systems investigated herein have the same sign and are of the same order of magnitude. However, to reproduce the experimental results, higher-order CEF coefficients also have to be included.

Table I. Crystalline-electric-field coefficients A_n^m (in unit of Ka_0^{-n}) for the $R_{1-x}Y_xFe_{11}Ti$ compounds.

R	A_2^0	A_4^0	A_6^0	A_4^4	A_6^4
Nd	-57	-14.3	0	0	0
Tb	-54.1	-0.9	0	0	0
Dy	-32	-9	3.2	105	0

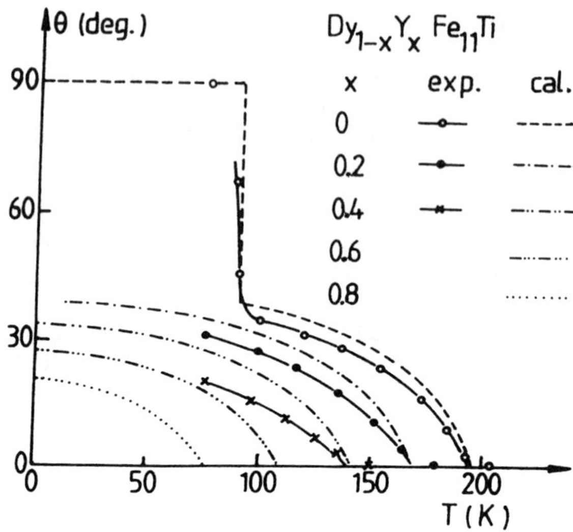

Fig. 1. Experimental and calculated thermal dependence of the cone angle θ between the magnetization and the *c* axis for the $Dy_{1-x}Y_xFe_{11}Ti$ compounds [16].

From an analysis of the single-crystal data of $DyFe_{11}Ti$, Hu et al. [18] deduced a set of A_n^m with $A_2^0 = -32.3\ \mathrm{K}a_0^{-2}$, $A_4^0 = -12.4\ \mathrm{K}a_0^{-4}$, $A_4^4 = 118\ \mathrm{K}a_0^{-4}$, $A_6^0 = 2.56\ \mathrm{K}a_0^{-6}$ and $A_6^4 = 0.64\ \mathrm{K}a_0^{-6}$. We have used this set of CEF coefficients to study the spin-reorientation phenomenon in the systems considered herein. For the $Nd_{1-x}Y_xFe_{11}Ti$ compounds, the calculated values for the spin-reorientation temperature are somewhat different from those obtained in our experiments [13]. Moreover, the calculations do not predict a spin reorientation in the Y-substituted compounds with $x > 4$, which is not consistent with our experimental data. Our set of CEF coefficients given in Table I predicts that a spin reorientation occurs in the compounds with x values up to 0.6 [14] (in agreement with experiments, see [13]), and that no spin reorientation in the compound with $x = 0.8$ is expected since for this compound the iron anisotropy is dominant at all temperatures. Measurements on $Nd_{0.2}Y_{0.8}Fe_{11}Ti$ [14] indeed support this expectation. For the $Dy_{1-x}Y_xFe_{11}Ti$ compounds, calculations using the set of A_n^m from Hu et al. [18] gave spin-reorientation temperatures which are again somewhat different from our experimental values. In addition, this set of A_n^m [18] does not reproduce the spin-reorientation behavior observed in $TbFe_{11}Ti$.

Very recently Ivanova et al. [19] have discussed the magnetic anisotropy and spin-reorientation transitions in $TbFe_{11-x}Co_xTi$ single crystals. These authors have shown the temperature dependence of K_1 for $TbFe_{11}Ti$. From these results and from the data on the iron sublattice anisotropy, K_1^{Fe}, derived by Hu et al. [15], we extract the magnetic anisotropy of the terbium sublattice, K_1^{Tb}, whose temperature dependence is shown in Fig. 2. Using the set of CEF parameters in Table I for the $TbFe_{11}Ti$ compound, we have calculated the temperature dependence of K_1^{Tb} which is also plotted in Fig. 2. The agreement between the calculated $K_1^{Tb}(T)$ curve and the experimental data is satisfying in a wide temperature range.

The spin reorientation in the $Tb_{1-x}Sm_xFe_{11}Ti$ compounds, in which the Stevens factor α_J of Tb is negative and α_J of Sm is positive, has been studied by *ac*-susceptibility and

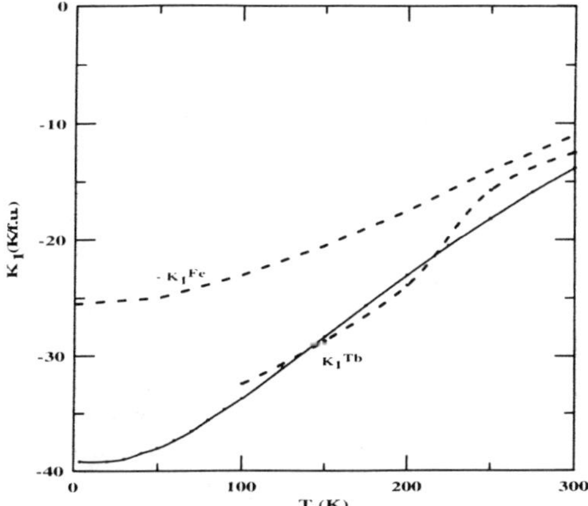

Fig. 2. Calculated temperature dependence of K_1^{Tb} (solid line) of the terbium sublattice in $TbFe_{11}Ti$. Experimental K_1^{Fe} values quoted from Hu et al. [15] and experimental K_1^{Tb} (dashed line) values extracted from Ivanova et al. [19] are also shown.

magnetization measurements on bulk and aligned powder samples [20]. Substitution of Tb by Sm reduces considerably the spin-reorientation temperature. T_{SR} for the compounds with x = 0, 0.1, 0.2 and 0.3 is 325, 285, 210 and 130 K, respectively. In the compound with x = 0.4, no spin reorientation has been observed down to 77 K. We have measured the temperature dependence of the cone angle θ (see Fig. 3) [20]. The analysis of the spin reorientation in these compounds may need a more sophisticated model than that described above because of a possible mixing of the ground state and the first excited state.

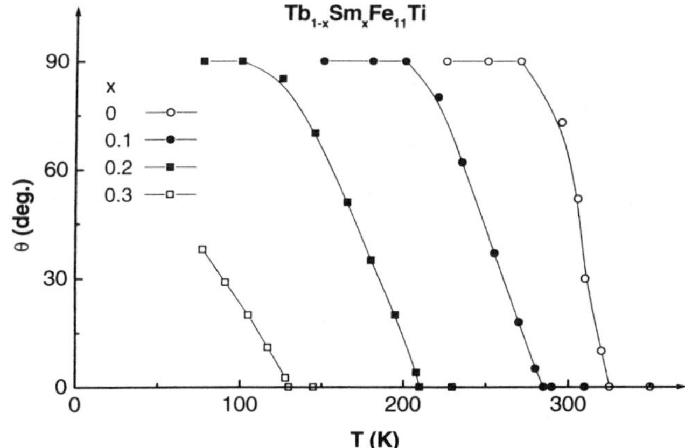

Fig. 3. Temperature dependence of the cone angle θ between the magnetization and the *c*-axis for the $Tb_{1-x}Sm_xFe_{11}Ti$ compounds [20].

R_2Co_{17} compounds

The magnetic properties of the R_2T_{17} intermetallic compounds have been investigated in much more details because of the availability of single crystals of high quality [21]. The isothermal magnetization measured along different crystallographic directions in high magnetic field has been well explained for several series of binary and ternary compounds based on microscopic models. Valuable information on the CEF and exchange field parameters has been obtained from these analyses. The temperature dependence of the magnetization along the principal axes of the crystals has also been satisfactorily explained by these models. However, the differences in the temperature dependencies of the magnetization along the principal axes within the basal plane of the single crystals of R-T compounds received less attention. Quite recently, we have carried out a detailed analysis of such experimental results, namely, the temperature dependence of the magnetization along the *a*- and *b*-axis in the R_2Co_{17} compounds (R= Nd, Pr, Ho and Dy) [22]. By fitting the experimental magnetization curves, the high-order CEF parameter, B_6^6, and a precise value of coupling parameter, J_{RT}, have been derived.

The computed results for the thermomagnetic curves along the principal axes of the R_2Co_{17} single crystals (R= Nd, Pr, Ho and Dy) [22] are shown in Fig. 4. The experimental results are taken from Sinnema [20]. In our calculations, all the CEF parameters, except B_6^6, are from Refs. [23-25], whereas B_6^6 and B_m are adjustable parameters to get the best fits to the experimental data. The anisotropy constant, K_4^R, which determines the MCA term in the basal plane (Eq. (2)), can be expressed via the CEF parameters as $K_4^R = B_6^6(<O_6^0> + <O_6^6>/2)/16$. It is clear from Fig. 4 that the calculations reproduce rather well the complicated temperature dependencies along the *a*- and *b*-axis in the basal plane.

To investigate the different roles of B_6^6 and B_m in determining the behavior of the thermomagnetic curves, we have studied the dependence on B_6^6 and B_m of the difference in magnetization along the *a*- and *b*-axis at 4.2 K, $\Delta M = M_b - M_a$, and of the value of the critical temperature, T_{CR}, above which the sixfold anisotropy in the basal plane disappears. It turned out from this analysis that the strength of B_m has a crucial effect on the T_{CR} value, whereas B_6^6 plays a decisive role in determining the ΔM value.

In Table II, ΔM values measured by Sinnema [21] are listed along with our fitting values of B_6^6 for the R_2Co_{17} compounds. It can be clearly seen that the larger the absolute value of B_6^6, the larger the magnitude of ΔM at 4.2 K.

It can be noted from our analysis that:

1. Precise values of the high-order CEF parameter, B_6^6, which plays a decisive role in determining the sixfold anisotropy and was neglected or roughly evaluated in previous studies, have been obtained (see Table II). Nevertheless, the simple relation between easy magnetization direction and the sign of B_6^6 can not be derived and still needs more investigation.

2. Precise values of the molecular field, B_m, and exchange parameter, J_{RT}, have been derived (see Table II). Up to now, several methods have been applied for determining J_{RT} such as fitting the high field magnetization curves, analyzing the Curie or compensation temperatures. Our calculation can be considered as a new method for obtaining precise values of the R-T exchange interaction.

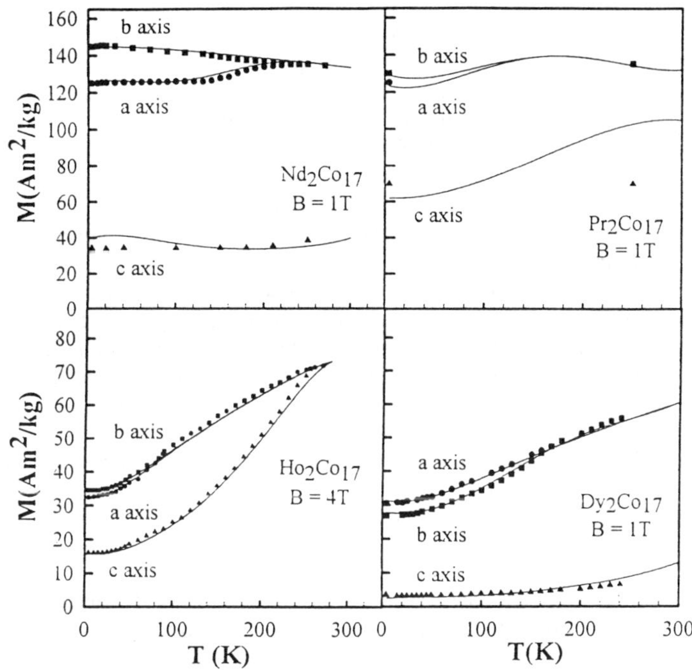

Fig. 4. The calculated results (solid lines) for Nd_2Co_{17}, Pr_2Co_{17}, Ho_2Co_{17} and Dy_2Co_{17} single crystals with the fitting parameters B_m and B_6^6 as listed in Table II [22]. Other CEF parameters used in the fitting procedure were taken from Refs. [23-25]. The experimental results (points) are taken from Ref. [21].

Table II. Values of B_m, J_{RT} and B_6^6 obtained from the fitting procedure of Fig. 4 [22]. Values of $\Delta M = M_b - M_a$ measured at 4.2 K in R_2Co_{17} compounds (applied field is indicated in brackets) [21] are also listed.

R	B_m (T)	$J_{RT}(10^{-23}J)$	$B_6^6(10^{-3}K)$	ΔM (Am^2/kg)	
Pr	280	33.80	-1.466	10	(1 T)
Nd	320	25.86	-8.0	18.7	(1 T)
Dy	91.3	11.04	-0.284	-3.6	(1 T)
Ho	63.5	9.40	0.137	1.9	(4 T)

CONCLUDING REMARKS

The above analysis of the microscopic origin of the magnetocrystalline anisotropy in the R-T compounds shows that the magnetocrystalline anisotropy of the R sublattice is fully determined by the values of the CEF parameters and exchange field of the R ion. Our results show that the study of spin-reorientation phenomena in intermetallic compounds with substitution of a non-magnetic rare earth for a magnetic one is very useful in obtaining information on crystalline-electric-field parameters. This is particularly valuable when data on single crystals are lacking. By fitting the temperature dependencies of the magnetization along the principal axes in the basal plane of single crystals of R-T compounds, the high-order CEF parameter, B_6^6, and the value of the coupling parameter, J_{RT}, have been derived. It should be noted, however, that the fitting procedures used to reproduce the magnetization curves of the (in best case single-crystalline) R-T compounds can not give a unique set of CEF parameters and exchange field without additional information obtained from other studies like neutron diffraction, inelastic neutron scattering, specific heat, Mössbauer experiments etc. Besides, due to the complexity of the microscopic nature of the magnetocrystalline anisotropy of the 3d sublattice, this contribution to magnetocrystalline anisotropy is usually treated in a phenomenological way on the basis of the data obtained on the isostructural R-T compounds in which R are non-magnetic (R = Y, La, Lu).

ACKNOWLEDGMENTS

The authors would like to thank Prof. Dr. T.D. Hien, Dr. L.T. Tai and all colleagues of the Cryogenic Laboratory of National University of Hanoi and Prof. Dr. J.J.M. Franse and Dr. F.F. Bekker of the University of Amsterdam for collaboration and discussion over a number of years. One of us (N.H.L) wishes to thank the National Science Foundation, USA, for financial support.

REFERENCES

1. R.F. Pearson, in *Experimental Magnetism*, Vol. 1, edited by G.M. Kalvius and R.S. Tebble (Willey, New York, 1979), p. 107.
2. K.P. Belov, A.K. Zvezdin, A.M. Kadomseva and R.Z. Levitin, *Orientation Transitions in Rare Earth Magnetics* (Nauka, Moskva 1979), in Russian.
3. J. Bartolome, in *Supermagnets, Hard Magnetic Materials*, edited by G.J. Long and F. Grandjean (Kluwer Academic, Dordrecht, 1991), p. 261.
4. A. J. Freeman and J.P. Desclaux, J. Magn. Magn. Mater. **12**, 11 (1979).
5. P.A. Lingard and O. Danielson, Phys. Rev. B **11,** 351 (1975).
6. Nguyen Hoang Luong, Nguyen Phu Thuy, Luu Tuan Tai and Than Duc Hien, Phys. Stat. Sol. (a) **111**, 591 (1989).
7. M.R. Ibarra, P.A. Algarabel, C. Marquina, J.I. Arnauda and A. del Moral, Phys. Rev. B **39**, 7081 (1989).
8. S.Sinnema, R.J. Radwanski, J.J.M. Franse, D.B. de Mooij and K.H.J. Buschow, J. Magn. Magn. Mater. **44**, 333 (1984).
9. S. Hirosawa and M. Sagawa, Solid State Commun. **54**, 335 (1985).

10. P. Burlet, J.M.D. Coey, J.P. Gavigan, D. Givord and C. Meyer, Solid State Commun. **60**, 723 (1986).
11. M. Sagawa, S. Fujimura, H. Yamamoto, Y. Matsuura and S. Hirosawa, J. Appl. Phys. **57**, 4094 (1985).
12. J.M. Cadogan, J.M.D. Coey, J.P. Gavigan, D. Givord and H.S. Li, J. Appl. Phys. **61**, 3974 (1987).
13. N.H. Luong, N.P. Thuy, L.T. Tai and N.V. Hoang, Phys. Stat. Sol. (a) **121**, 607 (1990).
14. N.H. Luong, N.P. Thuy and J.J.M. Franse, J. Magn. Magn. Mater. **104-107**, 1301 (1992).
15. B.H. Hu, H.S. Li, J.P. Gavigan and J.M.D. Coey, J. Phys.: Condens. Matter. **1**, 755 (1989).
16. P.H. Quang, N.H. Luong, N.P. Thuy, T.D. Hien and J.J.M. Franse, J. Magn. Magn. Mater. **128**, 67 (1993).
17. P.H. Quang, N.H. Luong, N.P. Thuy, T.D. Hien and J.J.M. Franse, IEEE Trans. Magn. **30**, 893 (1994).
18. B.H. Hu, H.S. Li, J.M.D. Coey and J.P. Gavigan, Phys. Rev. B **41**, 2221 (1990).
19. T.I. Ivanova, Yu. G. Pastushenkov, K.P. Skokov, I.V. Telegina and I.A. Tskhadadze, J. Alloys and Compounds **280**, 20 (1998).
20. Pham Hong Quang, Nguyen Hoang Luong, Nguyen Phu Thuy and Tran Le Anh, Proc. of 4th National Conference on Physics, Hanoi, Oct. 1993, p. 482, in Vietnamese.
21. S. Sinnema, PhD Thesis, University of Amsterdam, 1988.
22. P.D. Thang, N.P. Thuy, L.V.D. Khuong, C.V. Thang, J.J.M. Franse and S. Sinnema, J. Magn. Magn. Mater. **177-181**, 1099 (1998).
23. J. Colpa, P.H. Frings, R.J. Radwanski, J.J.M. Franse and M. Alba, J. Magn. Magn. Mater. **131**, 167 (1994).
24. R. Radwanski, J.J.M. Franse, S. Sinnema, H.J.M. Heerooms and J.H.P. Colpa, J. Magn. Magn. Mater. **76-77**, 1982 (1988).
25. J. Colpa, S. Sinnema, J.J.M. Franse, R.J. Radwanski and P.H. Frings, Physica B **156**, 731 (1989).

Part III

Nanoscale Hard Magnetism I

MATERIALS PROPERTIES AND UTILIZATION OF $Fe_3B/Nd_2Fe_{14}B$-TYPE NANOCOMPOSITE PERMANENT MAGNETS BASED ON Nd-Fe-Cr-Co-B

S. HIROSAWA, H. KANEKIYO, and Y. SHIGEMOTO
R&D Division, Sumitomo Special Metals Co. Ltd., 2-15-17 Egawa, Osaka 618-0013, Japan, HIROSAWA.S@ssmc.co.jp

ABSTRACT

Recent progresses of materials development and their utilization of Cr-and-Co-doped $Fe_3B/Nd_2Fe_{14}B$-type nanocomposite permanent magnets are reported. A practical balance between enhanced intrinsic coercivity and temperature stability of magnetic properties has been realized by simultaneous additions of Cr and Co. This type of magnets are less susceptible to oxidation because of the less rare-earth content. Consequently, even fine powders (e. g., under 38 micrometer sieve) are physically and chemically stable in contrast to the conventional $Nd_2Fe_{14}B$-type melt-spun materials. The stability in terms of the structural losses is also superior to that of the $Nd_2Fe_{14}B$-type melt-spun materials. Utilization of the Cr-and-Co-doped $Fe_3B/Nd_2Fe_{14}B$ nanocomposite permanent magnets as a hard magnetic component of injection-molded resin-bonded magnets seems promising because of the excellent stability of the magnetic properties of fine powders. Direct quenching of a melt into the nanocomposite structure has become possible recently in addition to the conventional processing route of crystallization of an amorphous precursor, opening up the possibility of less-expensive production of the material.

INTRODUCTION

The multiphase nanocrystalline permanent magnet materials based on Fe_3B and $Nd_2Fe_{14}B$ were briefly reported by Buschow et al.[1] and, more fully, by Coehoorn et al. in 1988[2]. Owing to their relatively large remanence of about 1.2 T for an isotropic material and low rare earth content of only about 4 at. %, these nanocomposite permanent magnets attracted considerable attention. However, the lack of a large-enough coercivity retarded utilization of these materials for practical applications. On the other hand, the theoretical prediction by Skomski and Coey[3] in 1993 for giant magnetic energy in textured nanocomposite permanent magnets has guided intense research into systems such as $Fe/Nd_2Fe_{14}B$[4, 5, 6], Fe/(Sm-Zr)Fe_7N_x[7] and $Fe/Sm_2Fe_{17}N_3$[8].

Theoretically, a nanocomposite permanent magnet with a large energy product may be fabricated by combining a ferromagnetic phase having a high magnetic ordering temperature and a large magnetic polarization with a hard magnetic phase having a large magnetocrystalline anisotropy in a nanometer-scaled structure in which both phases are ferromagnetically coupled via exchange interactions through grain boundaries. Practically, however, realization of such a ferromagnetic nanostructure is possible only in a system where the two phases can coexist at least in a metastable equilibrium. Also important is the requirement that the hard magnetic phase needs to be crystalline with few defects in order to generate a large magnetocrystalline anisotropy.

Techniques to produce nanocomposite permanent magnets include melt-spinning[1,2, 4-7, 9-16], mechanical alloying[8, 17, 18], and thin-film fabrication techniques such as sputter deposition[19-21], and ultra-high vacuum deposition[22]. In these techniques, formation of the nanocomposite structure relies on heterogeneous crystallization of the composite phases from an

Mat. Res. Soc. Symp. Proc. Vol. 577 © 1999 Materials Research Society

either glassy or highly defected nanocrystalline matrix into a mixture of the nanometer-sized composite phases which are stable or kinetically favored metastable phases.

To date, utilization of the nanocomposites has been limited practically to the application as magnetic components of resin-bonded magnets because the nanometer scale structure is not stable at temperatures at which atomic diffusion becomes possible to accomplish self sintering, although efforts using elaborated techniques such as spark-plasma sintering have been proposed to produce consolidated nanocomposite magnets[23-25].

One of the advantageous features of the $Fe_3B/Nd_2Fe_{14}B$-type nanocomposites over the α-Fe-based ones is that a certain coercivity (H_{cJ}) value can be obtained with a significantly smaller rare earth content, which potentially means low material cost and less susceptibility to oxygen or humidity. Another consequence is that the Fe_3B-based permanent magnetic materials contains a larger fraction of the soft magnetic phase (in some cases, over 85 percent) than in the bcc-Fe-based nanocomposites if H_{cJ} values are same. This leads to an obvious manifestation of the "exchange-spring" behavior, which was discussed by Kneller and Hawig in 1991[9]. One of consequences of such a behavior is a large recoil permeability. Improvements of the hard magnetic properties, particularly coercivity, of the $Fe_3B/Nd_2Fe_{14}B$-based nanocomposites have been achieved by present authors by means of modifications of alloy compositions[11-16]. Additions of small amounts of elements such as Al, Si, V, Cr, and Ga were found to be effective in this respect. Among these, Cr is unique because it can be added to replace a large portion of Fe and enhance H_{cJ} significantly. Mechanism of coercivity enhancement by Cr addition has been investigated by microscopic techniques such as Mössbauer spectroscopy[26] and atom probe analysis[27].

Purpose of this paper is to provide a brief description of magnetic properties of Cr-Co-doped $Fe_3B/Nd_2Fe_{14}B$ type nanocomposites and to examine practical aspects such as stability against oxidation, long term stability, and possibility of low-cost processing of nanocomposite permanent magnets.

EXPERIMENTAL PROCEDURES

Rapidly solidified flakes were produced by single roller melt-spinning apparatuses using pre-alloyed mother ingots. Crystallization heat treatment was performed using a continuous furnace under a protective Ar gas flow. Powders of the crystallized alloys were obtained by milling with a screened knife-blade mill and a pin-arrayed disc mill. Magnetic properties of the powders were measured by a vibrating sample magnetometer.

Specimens of compaction-molded magnets were made using 2 weight percent epoxy resin and 0.1 weight percent Ca-stearic acid powder. Magnetic properties of the bonded magnets were measured using a conventional *B-H* looper. Open magnetic flux of the bonded magnets was measured by the extraction method using a closely fit coil and a flux meter. For a long-term exposure test under the 80 degrees Celsius, 90 relative humidity environment, a programmable chamber was used with a particular caution so as to preheat specimens under dry air in order to avoid condensation of water on the surfaces of the specimens.

MAGNETIC PROPERTIES OF Cr-Co-DOPED $Fe_3B/Nd_2Fe_{14}B$

In contrast to other additives such as Ga, Cu, Ag, Au, which are effective when used by a small amount up to about one atomic percent[11-14], Cr can be used to replace Fe up to about 20 atomic percent. Simultaneous addition of Co with Cr is required, however, in order to

Table 1. Typical magnetic properties of powders of Cr-Co-doped $Fe_3B/Nd_2Fe_{14}B$ nanocomposites.

Code	Composition	$(BH)_{max}$ (kJ/m^3)	B_r (T)	H_{cJ} (kA/m)
A	$Nd_{3.8}Fe_{74}B_{18.5}Cr_{0.7}Co_3$	86	1.05	300
C	$Nd_{4.5}Fe_{73}B_{18.5}Cr_2Co_2$	77	0.95	370
D	$Nd_{5.5}Fe_{66}B_{18.5}Cr_5Co_5$	69	0.76	570

compensate a large decrease in the Curie temperature caused by the Cr addition. Consequently, a wide range of magnetic properties can be covered by Cr-Co-doped $Fe_3B/Nd_2Fe_{14}B$ nanocomposites as is shown in Fig. 1, in which B_r and H_{cJ} of flakes with density of 7.53×10^3 kg/m$_3$ (the same value is assumed for all cases) of typical $Fe_3B/Nd_2Fe_{14}B$ nanocomposites are plotted. Shown in Table 1 are magnetic properties of powders obtained by crashing the flakes down to 150 micron mesh size. A, C, and D are three grades of Cr-Co-doped powders for which stability tests have been performed in this paper. Remanence and maximum energy products of the powder materials are inferior to those of flakes due to demagnetization fields in the powder specimens.

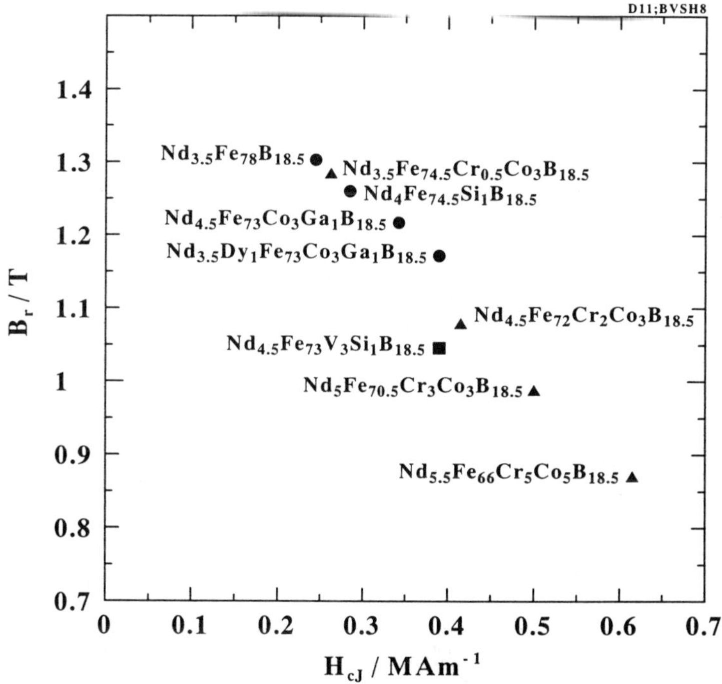

Fig. 1 Remanence (B_r) and intrinsic coercivity (H_{cJ}) of typical $Fe_3B/Nd_2Fe_{14}B$ nanocomposites with the Cr-Co-addition (▲) and with or without other additives (● and ■).

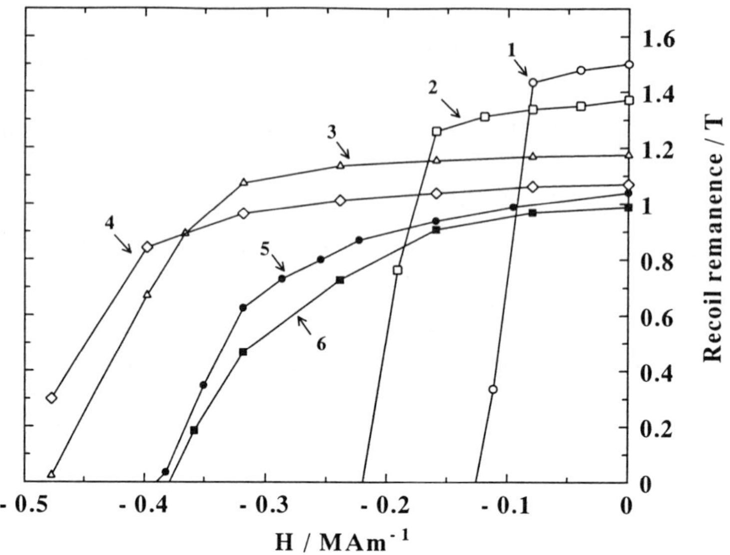

Fig. 2 Dependence of recoil remanence on demagnetizing force in several nanocomposite magnet ribbons. Details of specimens are shown in Table 2.

Recoil remanence defined as magnetic flux density measured under zero field after application and removal of a certain demagnetizing force is plotted in Fig. 2 for several nanocomposite permanent magnet flakes, compositions of which are shown in Table 2. Also shown in this table are critical fields H_{crit} at which irreversible demagnetization begins to take place and the intrinsic coercivity of the specimens. H_{crit} is defined as demagnetization field above which recoil remanence decreases by more than 5 % of the intrinsic remanence.

In Fig. 2 and Table 2, specimens of Nos. 5 and 6 are of the α-Fe/$Nd_2Fe_{14}B$ type. Comparison of H_{crit} vs. H plots of the Fe_3B/$Nd_2Fe_{14}B$ and the α-Fe/$Nd_2Fe_{14}B$ types of nanocomposites reveals another characteristics of the former; that is, an Fe_3B/$Nd_2Fe_{14}B$ type shows sharper transition from exchange-spring behavior to irreversible demagnetization than the

Table 2 H_{crit} and H_{cJ} values of several nanocomposite magnet flakes.

No.	Composition	H_{crit} (kA/m)	H_{cJ} (kA/m)
1	$Nd_2Fe_{79.5}B_{18.5}$	83	106.4
2	$Nd_3Fe_{78.5}B_{18.5}$	130	196.3
3	$Nd_5Fe_{70.5}Co_5Ga_1B_{18.5}$	280	366.1
4	$Nd_{4.5}Fe_{72}Cr_2Co_3B_{18.5}$	c. 340	404.4
5	$Nd_6Fe_{87}Nb_1B_6$	c. 100	283.3
6	$Nd_8Fe_{86}B_6$	c. 130	379.7

α-Fe/$Nd_2Fe_{14}B$ type nanocomposites. Probably, the sharp transition from reversible to irreversible demagnetization reflects a collective magnetic reversal in $Fe_3B/Nd_2Fe_{14}B$.

STABILITY OF MAGNETIC PROPERTIES

Oxidation behavior of the $Fe_3B/Nd_2Fe_{14}B$ nanocomposite magnetic powders under a hot and humid environment was investigated by observing weight changes (*dW*) during exposition to an 80 degrees Celsius, 90 % relative humidity atmosphere. The results are shown in Fig. 3, which clearly indicates that the nanocomposites with less than half Nd content than the $Nd_2Fe_{14}B$-type isotropic powder are significantly stable under a hostile atmosphere. The sample powders of $Fe_3B/Nd_2Fe_{14}B$ nanocomposites are those listed in Table 1.

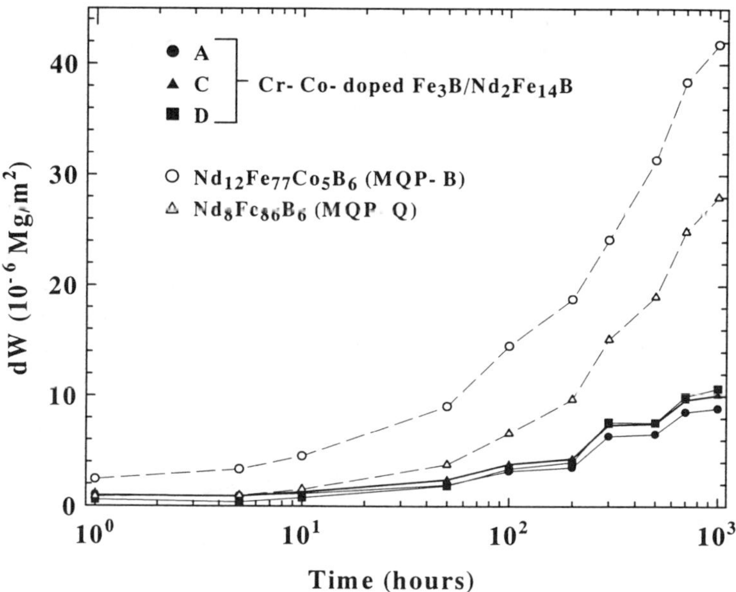

Fig. 3 Weight change (*dW*) of compaction-molded resin-bonded magnets made of $Fe_3B/Nd_2Fe_{14}B$ nanocomposite and isotropic $Nd_2Fe_{14}B$ powders caused by exposure to an 80 degrees Celsius, 90 % relative humidity environment. $Fe_3B/Nd_2Fe_{14}B$ powders are A, C, and D (■, ▲, and ◆, respectively). Powder sizes are less than 150 micrometers sieve.

Stability of magnetic properties of the nanocomposite magnets were examined by measuring open flux of the compaction-molded resin-bonded magnets as a function of exposure time. The results are shown in Fig. 4. Magnetic properties of the specimens will be described later (Table 3). The initial losses are thermal losses which include both reversible and irreversible losses. Once the thermal losses occurred, the nanocomposite powders show only slow decay which is approximately linear with logarithmic time whereas the $Nd_2Fe_{14}B$-type powder shows accelerated decay after about 100 hours exposure.

Thermal losses of the bonded $Fe_3B/Nd_2Fe_{14}B$ nanocomposite are shown in Fig. 5. The

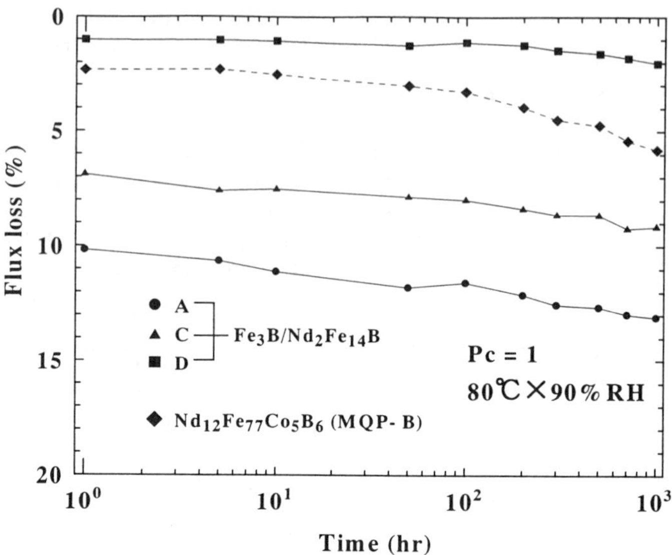

Fig. 4 Flux changes of compaction-molded resin-bonded magnets made of $Fe_3B/Nd_2Fe_{14}B$ nanocomposite and $Nd_2Fe_{14}B$ isotropic powders as a function of exposure time under an 80 degrees Celsius, 90 % relative humidity atmosphere. No surface coating was applied on the specimens. See Table 3 for magnetic properties of the specimens.

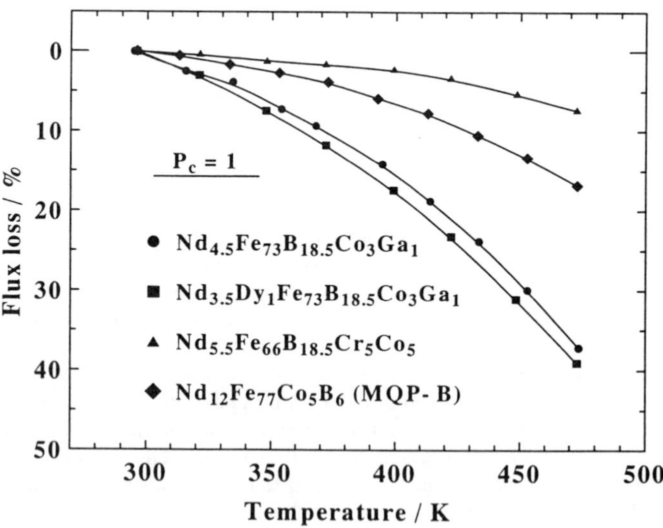

Fig. 5 Thermal losses of compaction-molded resin-bonded magnets.

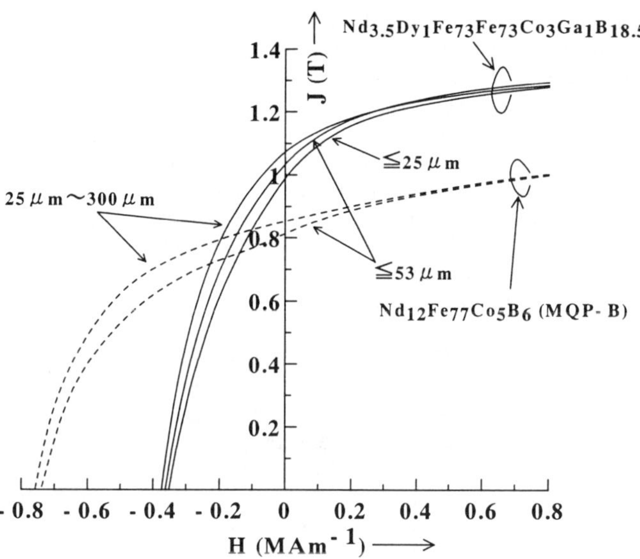

Fig. 6 Deterioration of demagnetization curves of a $Fe_3B/Nd_2Fe_{14}B$ nanocomposite $Nd_{3.5}Dy_1Fe_{73}Co_3Ga_1B_{18.5}$ and $Nd_2Fe_{14}B$-type $Nd_{12}Fe_{77}Co_5B_6$ caused by milling.

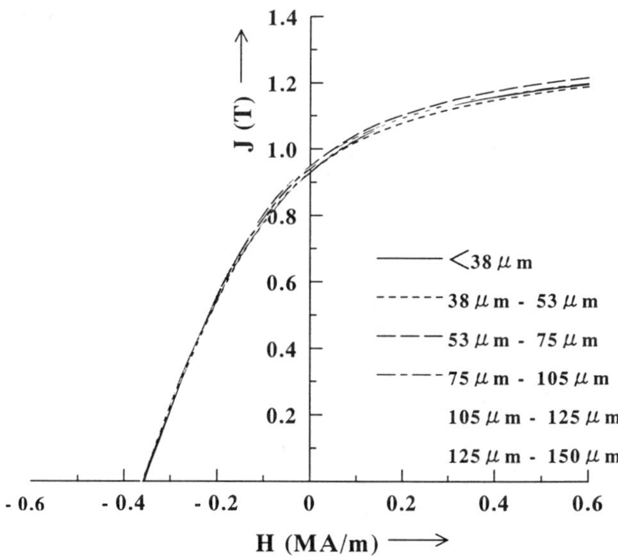

Fig 7 Demagnetization curves of Cr-Co-doped $Fe_3B/Nd_2Fe_{14}B$ nanocomposite powder (type C) of different particle sizes obtained by milling.

low-coercivity grades of Co-Ga-doped $Fe_3B/Nd_2Fe_{14}B$ nanocomposites show larger thermal losses because of the low coercivity. However, it should be noted that the $Nd_{5.5}Fe_{66}B_{18.5}Cr_5Co_5$ powder exhibits excellent thermal stability. This is due, al least partially, to its small temperature coefficient of H_{cJ} which is approximately -0.32 %/K.

Fig. 6 shows demagnetization curves of powders of conventional $Nd_2Fe_{14}B$ powder (MQP-B) and $Nd_{3.5}Dy_1Fe_{73}Co_3Ga_1B_{18.5}$ powder of different particle sizes. It is notable that the demagnetization curves deteriorate as particle size becomes small. On the other hand, the Cr-Co-doped $Fe_3B/Nd_2Fe_{14}B$ nanocomposite powder shows negligible dependence of demagnetization curves on particle size, as is demonstrated in Fig. 7. This stability of magnetic properties during milling is practically quite useful for injection-molding processes and also for compaction-molding of small magnetic parts.

LOW-SPEED QUENCHING AND AS-SPUN $Fe_3B/Nd_2Fe_{14}B$ NANOCOMPOSITES

Since the chemical composition of the $Fe_3B/Nd_2Fe_{14}B$ nanocomposites falls on the deep valley of liquidus of the Fe-B-Nd alloy system, this type of alloys exhibits great amorphous formability. Consequently, in contrast to $Nd_2Fe_{14}B$ and α-Fe/$Nd_2Fe_{14}B$ nanocomposites alloys, optimum surface velocity of quenching roller is significantly slower for $Fe_3B/Nd_2Fe_{14}B$ than for the these types of alloys. As an example, Fig. 8 shows the dependence of H_{cJ} of as-spun flakes of $Nd_xFe_{78.5-x}B_{18.5}Cr_3$ alloys produced by the single roller method under an Ar gas atmosphere of 13 kPa pressure on the chill substrate velocity. The reduced pressure was essential to obtain consistently repeatable results. The optimum velocity is 3 to 5 m/s for the alloy with x = 4.5. These velocities are significantly smaller than the optimal velocity of the $Nd_2Fe_{14}B$-type alloys which is around 20 m/s[26]. It is possible to anneal back over-quenched alloys to obtain best magnetic properties, which has been the original method adopted in former researches.

Adoption of slow quenching results in several advantageous features which are unique to

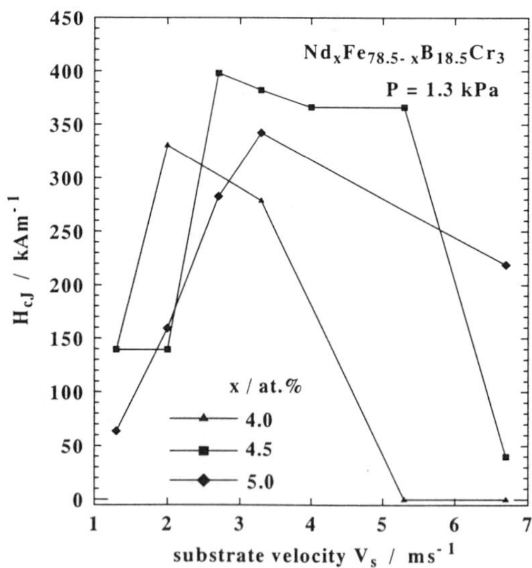

Fig. 8 Dependence of H_{cJ} of as-spun flakes of $Nd_xFe_{78.5-x}B_{18.5}Cr_3$ alloys.

$Fe_3B/Nd_2Fe_{14}B$ nanocomposites:

1) flakes are thick, around 100 micrometers,
2) consequently, aspect ratio of crashed powder particles is close to unity, see Fig. 9,
3) design and operation of liquid solidification apparatus become simple.

Fig. 9 Scanning electron micrograph of crashed powder of thick flakes of $Fe_3B/Nd_2Fe_{14}B$ nanocomposite magnet (C powder).

The second and third features are particularly favorable for industrial production of powders for resin-bonded magnets. The possibility of fabrication of thick platelet magnets stems from the first feature also[16]. The thickness range around 100 micrometers has been difficult to achieve by either vapor deposition processes such as sputtering or mechanical thinning of bulk sintered magnets.

UTILIZATION OF $Fe_3B/Nd_2Fe_{14}B$ NANOCOMPOSITES

Most realistic way of utilization of $Fe_3B/Nd_2Fe_{14}B$ nanocomposite permanent magnet powders is to use them as magnetic component of resin-bonded magnets. The stability of fine powders of this type of materials is particularly favorable for injection-molding process in which powders are exposed to high temperatures near 300 degrees Celsius. For instance, $Nd_{3.5}Dy_1Fe_{73}Co_3Ga_1B_{18.5}$ powder of $(BH)_{max}$ = 109 kJ/m^3, B_r = 1.07 T, H_{cJ} = 366 kA/m could make an injection-molded magnets of $(BH)_{max}$ of approximately 42 kJ/m^3 and density of approximately 4.7×10^3 kg/m^3 without particular efforts to optimize compounding recipe.

Magnetic properties of compaction-molded bonded magnets made from the Cr-Co-doped

Table 3 Magnetic properties of compaction-molded resin-bonded magnets

Powder Used	$(BH)_{max}$ (kJ/m^3)	B_r (T)	H_{cJ} (kA/m)	D (10^3 kg/m^3)
A	62	0.84	285	5.9
C	56	0.73	370	5.9
D	48	0.62	540	6.1

powders are listed in Table 3. The $Fe_3B/Nd_2Fe_{14}B$ nanocomposites possess a unique position as the high-remanence, low-coercivity magnetic components in comparison to the commercially available $Nd_2Fe_{14}B$ type isotropic powder. However, utilization of the low-coercivity grades require careful consideration of magnetic operation points in order to fully utilize the high-remanence characteristics. Namely, the operation line ($-B/H$) of magnetic circuits should be large (preferably, larger than *c.a.*, 4) and maximum demagnetization force should be smaller than a certain value. Recoil remanence of ESM-A, -C, and -D powders is plotted against demagnetizing field in Fig. 10. Because of a distribution of the demagnetizing fields, transition from reversible to irreversible magnetization process is less clear for the powders than for flakes (Fig. 2). The largest allowable demagnetization forces are estimated as approximately 160, 230, and 360 kA/m for ESM-A, -C, and -D powders on the same criteria as used for defining H_{crit} (namely, 5 % irreversible loss in the recoil remanence).

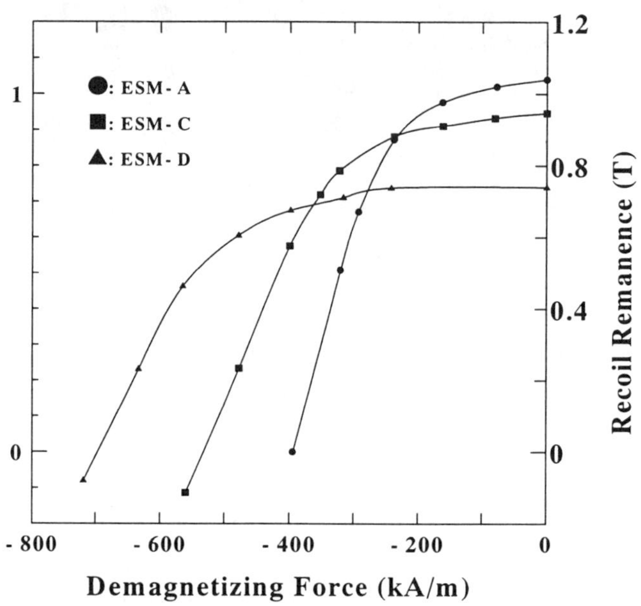

Fig. 10 Dependence of recoil remanence of $Fe_3B/Nd_2Fe_{14}B$ nanocomposite magnetic powders on demagnetizing field.

Efforts to utilize this type of nanocomposite permanent magnet powders for motor applications are reported recently. A prominent example is an internal permanent magnet (IPM) type of rotor in which a magnetic compound comprising of an $Fe_3B/Nd_2Fe_{14}B$ nanocomposite magnetic powder is injected into slots of a laminated steel core in order to achieve a high efficiency[27]. Another example is a multi-pole rotor magnet used in a stepping motor. In this case, the good magnetizability proved to be favorable because of the lack of sufficient magnetizing force to saturate isotropic magnets[28]. In order to illustrate this point, dependence of relative magnetic flux on magnetizing force is shown in Fig. 11. To construct this figure, remanent magnetic flux density of an $Fe_3B/Nd_2Fe_{14}B$ resin-bonded magnet (CA in Table 3) obtained after magnetizing by

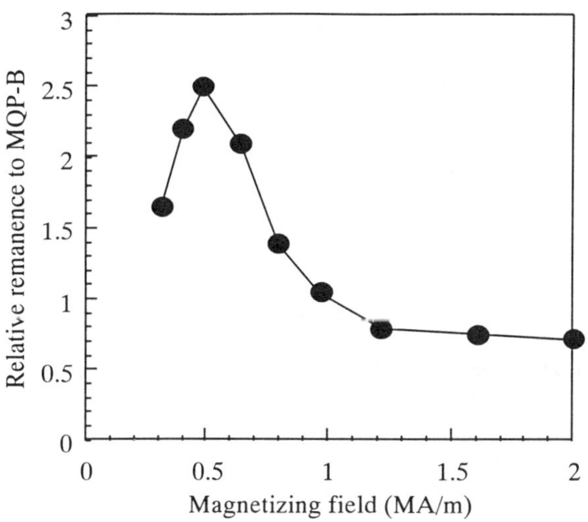

Fig. 11 Relative remanence of unsaturated $Fe_3B/Nd_2Fe_{14}B$ nanocomposite magnets as a function of magnetizing field. Remanent magnetic flux density of an $Fe_3B/Nd_2Fe_{14}B$ resin-bonded magnet (CA in Table 3) obtained after magnetizing by a given field is divided by that of isotropic $Nd_2Fe_{14}B$ type resin-bonded magnet (powder used is MQP-B) obtained after magnetizing by the same field and is plotted as a function of magnetizing field.

a given field is divided by that of isotropic $Nd_2Fe_{14}B$ type resin-bonded magnet (powder used is MQP-B) and is plotted as a function of magnetizing field. For magnetizing fields smaller than approximately 1 MA/m, significantly greater magnetic flux can be obtained with the low-coercivity grade of the $Fe_3B/Nd_2Fe_{14}B$ nanocomposite in comparison to the isotropic $Nd_2Fe_{14}B$ type resin-bonded magnet. However, caution should be reminded concerning the thermal losses of magnetic flux of this type of low coercivity material (Fig. 5).

CONCLUDING REMARKS: FUTURE OF $Fe_3B/Nd_2Fe_{14}B$ NANOCOMPOSITES

It has been demonstrated in this article that the Cr-Co-doped $Fe_3B/Nd_2Fe_{14}B$ nanocomposite magnetic powders have unique advantages over conventional isotropic $Nd_2Fe_{14}B$ powders in spite that magnetic performance of the former, particularly $(BH)_{max}$, of fully saturated magnets is not superior to the latter. The advantageous aspects of the $Fe_3B/Nd_2Fe_{14}B$ nanocomposite magnetic powders may vary from case to case according to particular interests in each application:

1) Stability of the powders both in physical and chemical aspects which ensures safe operation in powder handling,
2) Availability of relatively fine powder sizes for molding small parts and for injection-molding process,
3) Negligible long-term structural losses,
4) Relatively spherical (or non-flake-like) particle shapes,
5) Good magnetizability of the high remanence-low coercivity type powder which may be useful for applications which require fine magnetization patterns on small magnets.

Injection-molded $Fe_3B/Nd_2Fe_{14}B$ nanocomposite magnets have $(BH)_{max}$ significantly larger than those of hard ferrites. When the slowly quenched powders having the above-mentioned characteristics and novel injection-insertion technologies are combined, the $Fe_3B/Nd_2Fe_{14}B$ nanocomposite magnets may become well competitive even with the hard ferrites in certain application areas where difficulty of assembling or insertion of pre-shaped bulk magnets is causing restrictions on magnetic circuit designs and eventually preventing the maximum performance of the product from realization.

REFERENCES

[1] K. H. J. Buschow, D. B. de Mooij, and R. Coehoorn, J. Less-Common Met. 145 (1988) 601-611.

[2] R. Coehoorn, D. B. de Mooij, J. P. W. B. Duchateau, and K. H. J. Buschow, J. de Phys. Colloque C8, 49 (1988) C8-669-670.

[3] R. Skomski and J. M .D. Coey, Phys. Rev. B, 48 (1993) 15812-15816.

[4] A. Manaf, R. A. Buckley, and H. A. Davies, J. Magn. Magn. Mater. 128 (1993) 302-306.

[5] L. Withanawasam, G. C. Hadjipanayis, and R. F. Krause, J. Appl. Phys. 75 (1994) 6646.

[6] J. Bauer, M. Seeger, A. Zern, and H. Kronmüller, J. Appl. Phys. 80 (1996) 1667-1673.

[7] T. Yoneyama, T. Yamamoto, and T. Hidaka, Appl. Phys. Lett. 67 (1995) 3197-3199.

[8] J. Ding, P. G. McCormick and R. Street, J. Magn. Magn. Mater. 124 (1993) 1.

[9] E. F. Kneller and R. Hawig, IEEE Trans. Magn. 27 (1991) 3588-3600.

[10] H. Kanekiyo and S. Hirosawa, J. Magn. Soc. Jpn. 17 (1993) 2863 (text in Japanese).

[11] S. Hirosawa, H. Kanekiyo, and M. Uehara, J. Appl. Phys. 73 (1993) 6488.

[12] H. Kanekiyo, M. Uehara, and S. Hirosawa, IEEE Trans. Magn. 29 (1993) 2863.

[13] S. Hirosawa and H. Kanekiyo, Trans. Mat. Res. Soc. Jpn. 14B (1994) 965.

[14] H. Kanekiyo, M. Uehara, and S. Hirosawa, Mater. Sci. Eng. A181/A182 (1994) 868.

[15] S. Hirosawa and H. Kanekiyo, Proc. 13th Int. Workshop on Rare Earth Magnets and Their Applications, Birmingham, UK (Univ. Birmingham, 1994), pp. 87.

[16] H. Kanekiyo and S. Hirosawa, Proc. 15th Int. Workshop on Rare Earth Magnets and Their Applications, Dresden, Germany,(German Metallurgical Society, 1998)

[17] W. Gong, G. C. Hadjipanayis, and R. F. Krause, J. Applied Phys. (1994) 6649.

[18] W. F. Miao, J. Ding, P. G. McCormick, and R. Street, J. Alloys. Comp. 240 (1996) 200.

[19] S. Parhofer, C. Kurt, J. Wecker, G. Gieres, and L. Schultz, J. Appl. Phys. 83 (1998) 2735.

[20] E. E. Fullerton, J. S. Jiang, C. H. Sowers, J. E. Pearson, and S. D. Bader, Appl. Phys. Lett. 72 (1998) 380.

[21] J. P. Liu, Y. Liu, C. P. Luo, Z. S. Shan, and D. L. Sellmyer, J. Appl. Phys. 81 (1997) 5644.

[22] K. Mubu, T. Nagahama, and T. Shinjo, J. Magn. Magn. Mater. 163 (1996) 75.

[23] M. Jurczyk and P. B. Gwan, J. Alloys. Comp. 230 (1995) L1.

[24] R. K. Mishra, and V. Panchanathan, J. Appl. Phys. 75 (1994) 6652.

[25] A. Kojima, A. Makino, A. Inoue, Nanostruct. Mater. 8 (1998) 1015.

[26] J. J. Croat, J. F. Herbst, R. W. Lee, and F. E. Pinkerton, J. Appl. Phys. 55 (1984) 2078.

[27] F. Yamashita and S. Hashimoto, Matsushita Technical J. 44 (1998) 190 (text in Japanese).

[28] K. Ohara, H. Ueda, F. Yamashita, Digests of 19th Ann. Conf. Magn. Jpn.(1995) 127 (text in Japanese).

EMERGE

European **M**agnets with **E**nhanced **R**emanence for **G**reater **E**fficiency
: a new and cost effective magnetic material for use in mid-priced applications

V. ARCHAMBAULT, D. PERE
Rhodia Research Centre, 52 Rue de la Haie Coq, 93308 Aubervilliers, France

ABSTRACT

EMERGE is an European project based on the collaboration between three industrial partners (Rhodia, Philips and Magnetfabrik Bonn) and three research laboratories (Centre National de la Recherche Scientifique CNRS (Laboratory Louis NEEL), Trinity College of Dublin TCD, Vienna University IATP). The aim of this project is to improve the coercivity in enhanced remanence material. This three years project used statistical set-ups to reduce the number of samples to be realised and tested. The complete description of the project is developed in this paper and the main conclusions on sample composition and process issues are also discussed.

INTRODUCTION

In the field of permanent magnets there is a need for a magnet whose performance and price is intermediate between low-cost low-performance ferrites and high-cost high-performance Nd-Fe-B sintered magnets. Such a mid-range magnet would find applications in many sectors including consumer electronics, computer peripherals and the automobile industry. Surveys by Gorham and Wheeler [1], [2] indicate that the market for such permanent magnets is expected to grow from 83 MECU in 1993 to 1200 MECU by the year 2000.

Philips possesses the patent rights [3] for a low Nd concentration material with a J_r of 0.8 T and a ${}_JH_c$ of 0.2 MA / m. From a technical point of view this material has a surprisingly large remanence which is 1.5 times larger than MQI. However, it has one technical shortcoming, which is, its coercivity is too low for most applications.

The surprisingly large "enhanced-remanence" (increased magnetic strength) of the Philips material is a consequence of the so-called exchange coupling between hard and soft grains in a heterogeneous material. This phenomenon was discovered by Coehoorn et al. [4] in a boron-rich, neodymium-poor composition $Nd_{4.5}Fe_{77.5}B_{18.5}$. The initial composition was melt-spun and crystallised to form an intimate mixture of nanoscaled phases, mainly $Nd_2Fe_{14}B$ (hard) and Fe_3B (soft) which were exchange coupled. This material was found to have a $J_r = 1.2$ T and ${}_JH_c = 0.24$ MA / m.

From the outset, it was obvious that the coercivity of this material needed to be significantly increased for applications. To this end, the obvious approach is to try substituting or adding elements to the parent composition. This has been studied by

Mat. Res. Soc. Symp. Proc. Vol. 577 © 1999 Materials Research Society

Sumitomo [5], by a Chinese group [6] and by Philips [7]. Their results are shown in Table 1. One of the outcomes of these studies is that the introduction of only one additional element is not sufficient to induce a significantly high coercivity. The results are however encouraging, for example a $_JH_c$ of 0.48 MA / m has been obtained (material A) although at the expense of a reduction in J_r compared with the parent composition. These results suggest that there is scope for improvement by performing more comprehensive additions and substitutions.

Material	Composition	J_r (T)	$_JH_c$ (MAm^{-1})	BH_{max} (kJm^{-3})	Ref.
Parent	$Nd_{4.5}Fe_{78}B_{18.5}$	1.20	0.24	95	[4]
A	$Nd_3Dy_2Fe_{70.5}B_{18.5}Co_5Ga_1$	0.98	0.48	108	[5]
B	$Nd_5Fe_{70.5}B_{18.5}Co_5Ga_1$	1.18	0.34	121	[5]
C	$Nd_{4.0}Fe_{72.9}B_{18.5}Co_{4.6}$	n/p	0.24	108	[6]
D	$Pr_{4.0}Tb_{0.4}Fe_{77.8}B_{17.8}$	1.07	0.33	100	[7]

Table 1 - Influence of substitutions and additives (A - D) to the parent composition on room temperature magnetic properties of powders.

To understand the phenomenon at the origin of remanence enhancement and exchange hardening, Kneller and Hawig [8] discussed the elementary theory of the isotropic "exchange spring" magnet. Computer simulations by Schrefl and Fidler [9] and calculation for the oriented case by Skomski and Coey [10] have helped to clarify the optimum nanostructure for exchange hardening. The size of the soft regions is the most critical parameter as soft regions act as nucleation centres for magnetisation reversal. Ideally they should be no wider than twice the domain wall width of the hard phase. For $Nd_2Fe_{14}B$, this yields an optimum size of the soft magnetic phase of 10 nm.

The EMERGE project was set to understand the mechanism of enhanced remanence in NdFeB/FeB type materials, to study and propose new compositions around the Philips composition (Nd4.5Fe77B18.5) with better coercivity, to analyse and adjust the process parameter to produce this kind of powders and finely to produce application demonstrators from this new materials. The aim of the EMERGE project is to produce powders with room temperature magnetic properties ranging from J_r = 1.2T with $_JH_c$ = 0.4 MA / m (target 1) to J_r = 1.0T with $_JH_c$ = 0.6 MA / m (target 2) and to use these powders to make bonded magnets with J_r = 0.8 T with $_JH_c$ = 0.4 MA / m to J_r = 0.7 T with $_JH_c$ = 0.6 MA / m. Fig 1 represents the magnetic properties of the states of the art, the micro magnetic limits of this type of products and the targets of the EMERGE project when it started in 1996.

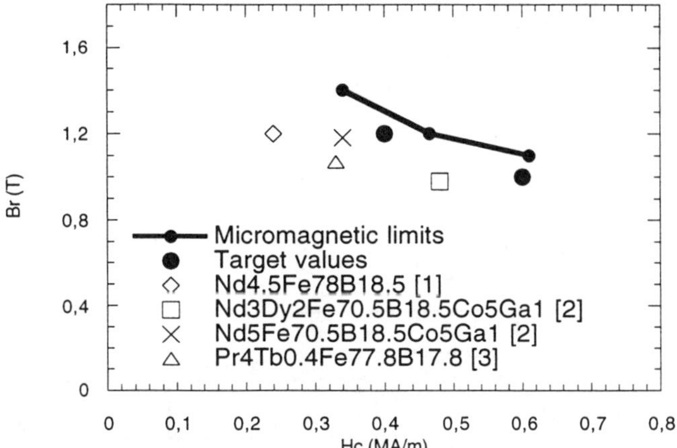

Fig.1 : Magnetic properties of the states of the art, the micro magnetic limits of this type of products and the targets of the EMERGE project

This project, founded by the European Commission, is based on the collaboration between different companies and research laboratories in Europe. An overview of the partners involved in the project is shown in the table 2. The project was co-ordinated by Rhodia.

Organisation name	Function
RHODIA (France) V. Archambault, D. Péré	Powder supplier
PHILIPS (Netherlands) K. de Kort, T. Rijks	Developer and end-user
MAGNEFABRICK BONN (Germany) M. Grönefeld, U. Czernik	Permanent magnet manufacturer
TCD (Ireland) J.M.D. Coey, J.F. O'Sullivan, P.A.I. Smith	Developer (speciality : materials - intrinsic properties)
CNRS/LLN (France) D. Givord, S. David	Developer (speciality : materials - coercivity mechanisms)
IATP (Austria) J. Fidler, T. Schrefl, J. Bernardi	Developer (speciality : micro structural analysis - modelling)

Table 2 - Overview of the consortium

A wide variety of techniques were used during this project to make and to study remanence-enhanced nanostructured materials.

Statistical set-ups were used in order to select most effectively the compositions to be made and the process parameters to be tested. A two step procedure was followed : coarse selection (Phase 1) followed by fine selection (Phase 2). Phase 3 corresponds to the realisation at the pilot scale of the project targets. A schematic representation of the work plan is given in Fig 2.

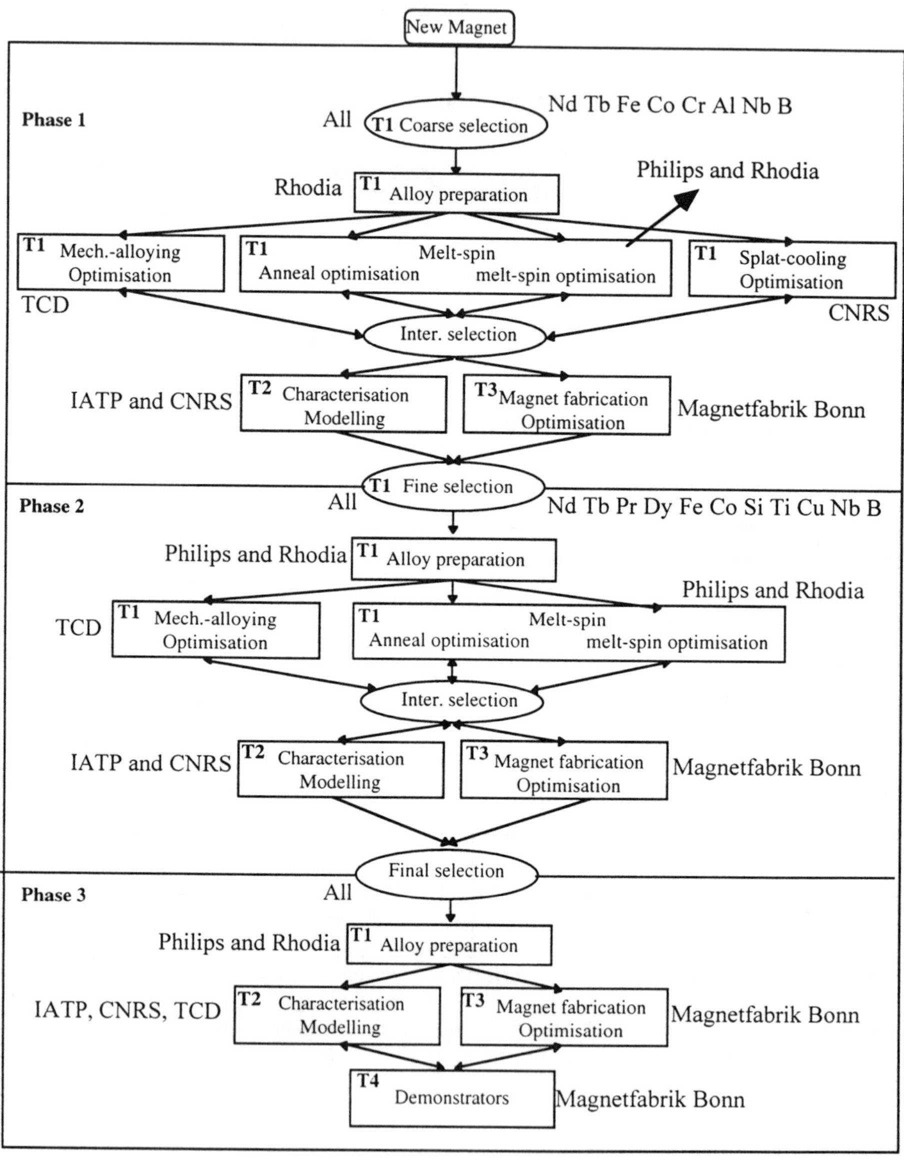

Fig. 2 - Schematic representation of the EMERGE work plan

This work plan led to the preparation and the study of some 130 different compositions around the parent composition. The two step composition selection allowed adjustment, during the project, of the list and amount of additives to be studied. We were able to stop investigations on non attractive additives to focus on attractive additions.
The large amount of samples and technics used during this project allowed the study of the coercivity mechanism in such enhanced remanence materials. For details about experiments, modelling and coercivity mechanism, please refer to the paper [11] where the results from the EMERGE project on this topics are discussed.

PROCESS OPTIMISATION

This part on the paper is more related to the specific work done by Rhodia on the optimisation of the preparation conditions of EMERGE type powders.

The magnetic properties of such enhanced remanence powders strongly depend on the structure of the material. To obtain the optimum structure it is required to start with totally amorphous ribbons or flakes. Fig 3 shows a typical X-Ray diffraction pattern of such amorphous product produced byRhodia.

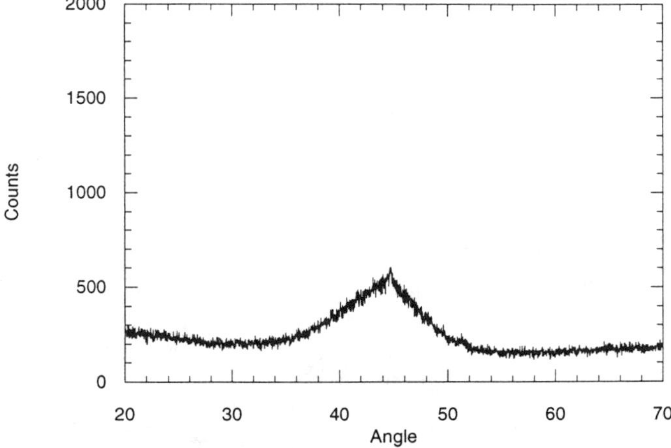

Fig 3 : XRD pattern of as melt-spun flakes showing the high amorphousness of Rhodia powder.

A high cooling rate ($\approx 10^6$ K/s) of the melted material leads to the formation of amorphous product. No crystallized phase can be detected. The broad pick between 35 and 50 ° corresponds to the Nd-Fe-B. Depending on the concentration of certain additives and also Boron, partial crystallization may occur during melt-spinning. This partial crystallization is not wanted as it may affect the controlled crystallization step needed to generate the right nanostructure. To avoid grain growth and to achieve the best nanostructure, fast annealing of the amorphous sample is needed.

To perform fast annealing, Rhodia developed a rotary Infrared furnace to study the annealing parameters (heating rate, cooling rate, sample size, sample morphology...). Fig 4 gives a description of this furnace. This furnace is composed of a rotary quartz tube, to let the infrared radiation goes through, surrounded by 12 Infrared lamps (1 kW each). This tube is air tight so different sample surrounding atmosphere can be tested (vacuum, Helium, Argon...). The powder is introduced in a quartz case built with internal lifters to keep the powder in motion. The temperature control is realised by a thermocouple directly in contact with the powder to monitor the heating or the cooling rate of the powder itself.

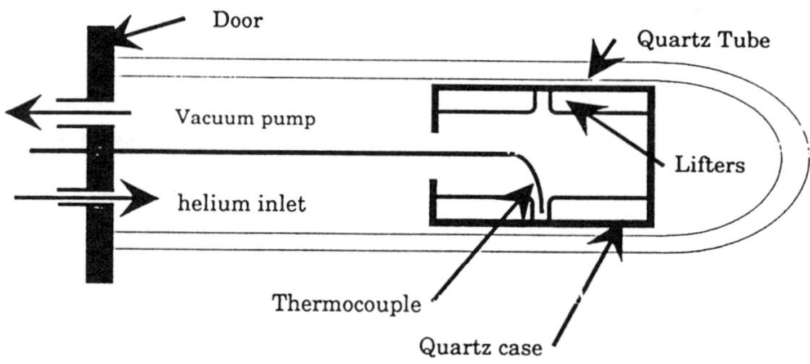

Fig 4 : Drawing of the rotary Infrared furnace

During the annealing treatment, both heating rate and cooling rate are important to get the optimum nanostructure. Fig 5 and fig 6 show the furnace performances.

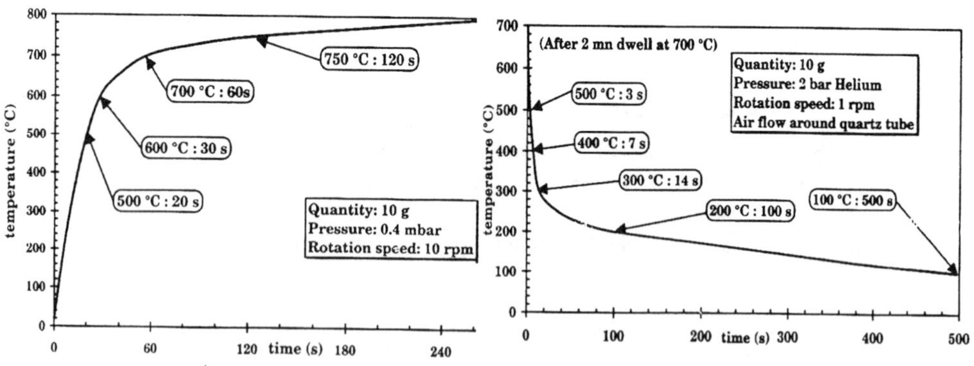

Fig 5 : IR furnace heating rate *Fig 6 : IR furnace cooling rate*

The heating rate is important to generate the fine nanostructure. Rotary infrared heating allows an homogeneous and rapid heat treatment of 10 to 50 grams of powder.
After optimisation of the annealing treatment to obtain the best magnetic properties, XRD pattern and TEM study of the powder show that annealed Rhodia powder are usually composed of very small (from 20 nm to 40 nm according to composition) and homogeneous grains.
TEM investigations were performed by Pr Fidler's group on selected samples. The optimally treated powders are homogeneous and the average grain size is around 20 to 30 nm. The TEM micrograph (fig 7) is typical of a good sample with the optimum annealing treatment. Very small grains can be seen and the picture shows a good homogeneity of the grains size distribution.

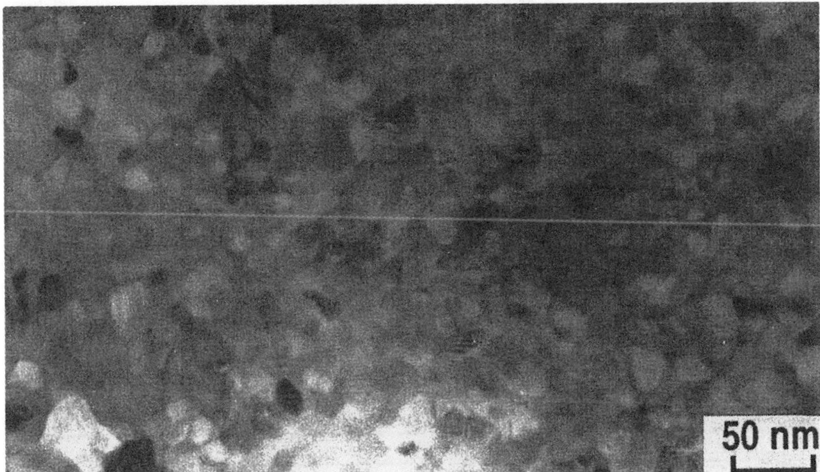

Fig 7 : TEM micrograph of annealed melt-spun

The grain size is also determine by the analysis of the XRD pattern and the use of the Sherrer formula to determine the average grain size of the main phases present in the sample. This analysis confirmed the grain size determined by TEM on a larger amount of grains. Only the combination of the size and the homogeneity can lead to good remanence enhancement. The cooling rate is crucial to reduce the grain growth after annealing. The most important parameter as far as cooling is concerned is that the sample temperature should be lowered to below 300°C as fast as possible. Below this temperature no grain growth was detected. Figure 8 presents the XRD pattern of an annealed sample. This sample is mainly composed of $Nd_2Fe_{14}B$, Fe_3B, Fe_2B and α-Fe.

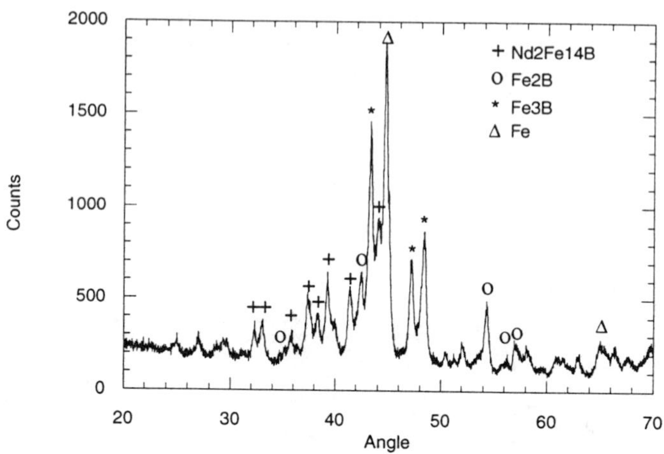

Fig 8 : XRD pattern of optimized annealed melt-spun sample

COMPOSITION OPTIMIZATION :

Table 3 gathers the main results obtained during phase 1 and phase 2 of the project.

Target	Composition	Jr (T)	jHc (MA/m)	BHmax (kJ/m3)
1	Nd3.25Tb1FebalCo5B18.5	1.08	0.394	100
2	Nd4.5Tb1FebalCr5B18	0.77	0.723	
1	Nd3.25Tb1FebalCo5Si1B18.5	1.1	0.386	105
1	Nd3.25Tb0.5FebalCo5Si1B18.5	1.17	0.327	111
1	Nd3.25Tb1FebalCo5Nb0.5Cu0.5B18.5	1.04	0.419	89
1	Nd3.25Tb0.5FebalCo5Si1Nb0.4Cu0.4B19	1.13	0.349	97
1	Nd3.05La0.2Tb1FebalCo5B18	1.11	0.363	88

Table 3 : Results of the EMERGE project composition optimization

The first two results were obtained at the end of the phase 1. Good magnetic properties and amelioration of the coercivity were already obtained at this moment of the project. During this first phase, it has been clearly showed that high coercivity can only be obtained with significant amount of Cr but this is at the expense of the remanence. Tb addition has a good effect on the remanence and is always needed for large remanence properties. Good coercivity can be found in Tb contening sample. Tb is a good compromize to get both remanence and coercivity.

During phase 2 of the project, we decided to focus on target 1 for the composition optimization. The effect of Si addition to improve the grain refinement can be seen in table 3.

This effect is confirmed in Nd3.25Tb0.5FebalCo5Si1B18.5 where large enhanced remanence is measured. NbCu addition gives a large increase in the coercivity without distroying the remanence. For details about composition effect on the magnetic properties, please refer to the paper [12] where results from the EMERGE project on non Cr containing sample are discussed.
Results from table 3 are plotted in the fig 9 which shows the results of the EMERGE project versus the targets.

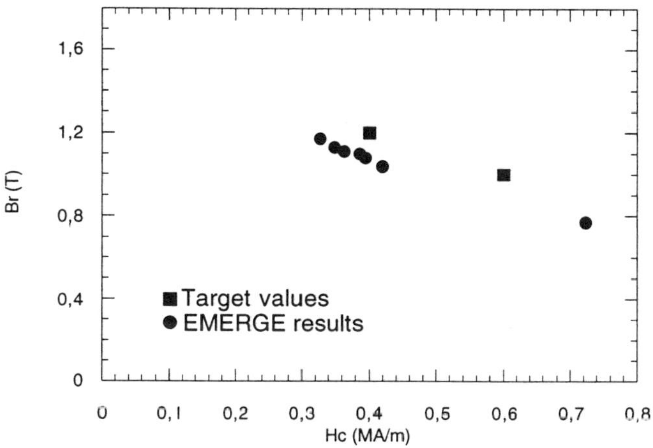

Fig 9 : EMERGE results summary

CONCLUSION :

Powders with high enhanced remanence and acceptable coercivity were obtained .
The process to obtain the right nanostructured material is a critical issue. The melt-spinning parameters should be adjusted to get the more amorphous melt-spun product. This is highly composition dependent. Crystallization of the amorphous material needs to be well controlled to generate the appropriate nanostructure.
It has been showed that judicious combinaison of additives to the parent composition can lead to large improvement of the magnetic properties of the starting material. Cr leads to high coercivity but this is at the expense of the remanence. Tb was found to be a good compromize to get high remanence and acceptable coercivity.

ACKNOWLEDGEMENTS :

The authors wish to acknowledge the EMERGE partners for the hard work performed during this project and for the fruitful exchanges. They also want to thank the European Commission for supporting this project. They are thankful to the MRS team for the invitation to the MRS 1999 congress.

REFERENCES :

[1] Gorham, Proc; of Nd-Fe-B permanent magnets conference, Orlando, Florida, July 12-14, 1993.

[2] Wheeler Associates, 93 Update on rare-earth permanent magnets

[3] Philips, European patent EP 0 264 153 B1
United States patent 4935074

[4] R. Coehoorn, D. B. de Mooij, J.P.W.B. Duchateau and K.J.H. Buschow, J. de Phys. C8, 669 (1988).

[5] S. Horosawa, H. Kanekiyo and M. Uehara, J. Appl. Phys. 73, 6488 (1993).

[6] Shen Bao-Gen, Yang Lin-yuan, Liang Jian-zhen, Zjang Kim-xian and Yang Fu-ming, J. Magn. Magn. Mater. 104, 1088 (1992).

[7] R. Coehoorn and C. de Waard, J. Magn. Magn. Mater. 83, 228 (1990).

[8] E. F. Kneller and R. Hawig, IEEE Trans. Mag. MAG 27, 3588 (1991).

[9] T. Schrefl, J. Fidler and H. Kronmüller, Phys. Rev. B49, 6100 (1994).

[10] R. Skomski and J.M.D. Coey, IEEE Trans. Magn. MAG 29, 2860 (1993) ; Phys. Rev. B48, 15812 (1993).

[11] S. David, D. Givord, "Coercivity and Modelling", Journal of Magnetism and Magnetic Materials, to be published.

[12] J. Bernardi, J. Fidler, "non Cr containing samples", Journal of Magnetism and Magnetic Materials, to be published.

MICROMAGNETICS OF NANOCRYSTALLINE PERMANENT MAGNETS

T. SCHREFL, J. FIDLER
Institute of Applied and Technical Physics, Vienna University of Technology,
Wiedner Hauptstr. 8-10, A-1040 Vienna, Austria, schrefl@email.tuwien.ac.at

ABSTRACT

Micromagnetic finite element calculations provide the theoretical limits for the remanence, the coercive field, and the coercive squareness of nanocomposite $Nd_2Fe_{14}B/\alpha$-Fe,Fe_3B,$Fe_{23}B_6$ magnets. The influence of the intrinsic magnetic properties and the microstructure were investigated using an energy minimization technique. The coercive field reaches a maximum as a function of the average grain size at about 15 nm - 20 nm. The replacement of α-Fe with Fe_3B improves the coercive field but deteriorates the loop shape. The reduction of the magnetization and the exchange constant in the soft magnetic Fe_3B phase by 20% improves the coercive field without a significant loss in the remanence. An increase of the hard phase anisotropy by 8% enhances the coercive field by more the 100 kA/m in two-phase α-Fe/$Nd_2Fe_{14}B$ and by 60 kA/m in two-phase $Fe_3B/Nd_2Fe_{14}B$ magnets. Dynamic simulations for magnetostatically interacting particles of a bonded magnet show a slight influence of the particle arrangement on magnetization reversal. The interaction field in the range of 100 kA/m to 200 kA/m rapidly decreases the distance from the particle.

INTRODUCTION

Single-phase nanocrystalline permanent magnets [1-2] show a decrease of the coercive field and an increase of the remanence with respect to the limits given by the Stoner-Wohlfarth theory [3]. The increase of the remanence and decrease of the coercive field with decreasing grain size can be attributed to the competitive effects of intergrain exchange interactions and of the magnetocrystalline anisotropy [4-6]. The magnetization significantly deviates from the local easy axis near the grain boundaries. The width of nonuniform states are restricted by the domain wall width of the material. Thus the volume fraction, where the magnetization is not parallel to the anisotropy direction, enlarge with decreasing grain size. The inhomogeneous magnetic state near the grain boundary enhance the remanence but act as nucleation site for reversed domains [5]. A further increase of the remanence is possible, if the hard magnetic phase is partially replaced by a soft magnetic phase [7-8]. These so-called nanocomposite magnets show reasonable hard magnetic properties, provided that both phases are sufficiently exchange coupled [9].

The magnetic properties of nanocomposite magnets sensitively depend on the microstructure such as the size and the shape of the grains. Recently, the effects of the microstructure on the remanence and the coercive field of nanocrystalline permanent magnets were investigated using micromagnetic calculations [10-12]. Experimentally, the magnetic properties of the nanocomposite magnets are improved, changing the alloy composition [13-14]. The substitution of Nd by Dy or Tb or the addition of Cr changes the intrinsic magnetic properties of the magnetically hard and soft phase, respectively. This work numerically treats the influence of the intrinsic magnetic properties on the remanence, the coercive squareness, and the coercive field of nanocomposite magnets.

Mat. Res. Soc. Symp. Proc. Vol. 577

MICROMAGNETIC FINITE ELEMENT SIMULATIONS

The successive calculation of equilibrium states of a ferromagnetic structure for decreasing external field provides the demagnetization curve of the model system. The equilibrium states follow either from the numerical minimization of the total magnetic Gibb's free energy or the time integration of the Gilbert equation of motion [15]. The finite element techniques which are applied to solve the equations numerically are described in previous work [5,12,16]. The use of tetrahedral finite elements allows to model grains with realistic shape. A Voronoi construction was applied to construct the grain structures used for the micromagnetic simulations of magnetization reversal. The grains are assumed to grow with constant velocity in each direction, starting from randomly located seed points.

The coercive field and the remanence strongly depends on the magnetic interactions between the grains. Therefore the local arrangement of grains significantly affects the magnetic properties. Figure 1 describes a simple microstructure which was used to investigate the effects of grain size and phase distribution on the coercive field. A single grain of polyhedral shape is surrounded by 20 $Nd_2Fe_{14}B$ grains. The phase of the interior grain is either α-Fe, or Fe_3B. The anisotropy directions of the surrounding hard magnetic grains are randomly oriented. The exterior grains are truncated in order to obtain a volume fraction of 30% α-Fe or Fe_3B. The microstructure given in fig. 2 was used, in order to calculate the remanence and the coercive field as a function of the phase distribution and different intrinsic magnetic properties. The intrinsic properties of the different phases are listed in table I.

NUMERICAL RESULTS

Exchange Hardening and Remanence Enhancement

The coercive field of a nanocomposite magnet was studied assuming a α-Fe grain or Fe_3B grain surrounded by randomly oriented $Nd_2Fe_{14}B$ grains. The volume fraction of the magnetically soft phase is 30%. After the magnet has been saturated and the external field has been decreased to zero, the magnetization within the hard magnetic grains is parallel to the local anisotropy direction. The magnetization points upwards towards the original saturation direction. Provided that the soft magnetic grain is small enough, the exchange energy becomes the dominating energy term. The exchange field overrides the demagnetizing field and the anisotropy field. Therefore the magnetization within the soft magnetic grain becomes aligned parallel to the average

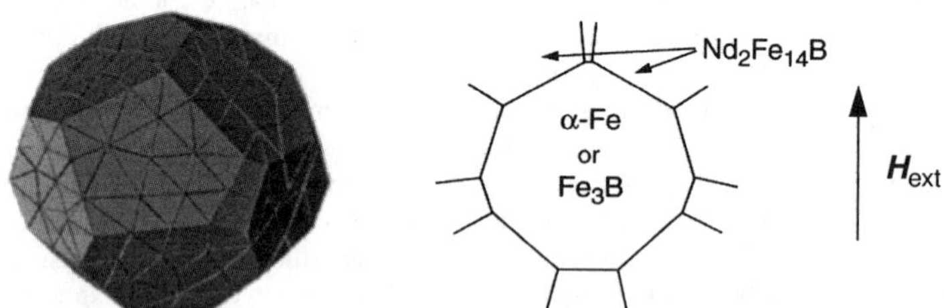

Fig. 1. Model structure consisting of a soft magnetic grain surrounded by hard magnetic neighboring grains. Left hand side: Finite element mesh at the surface. Right hand side: Slice through the structure.

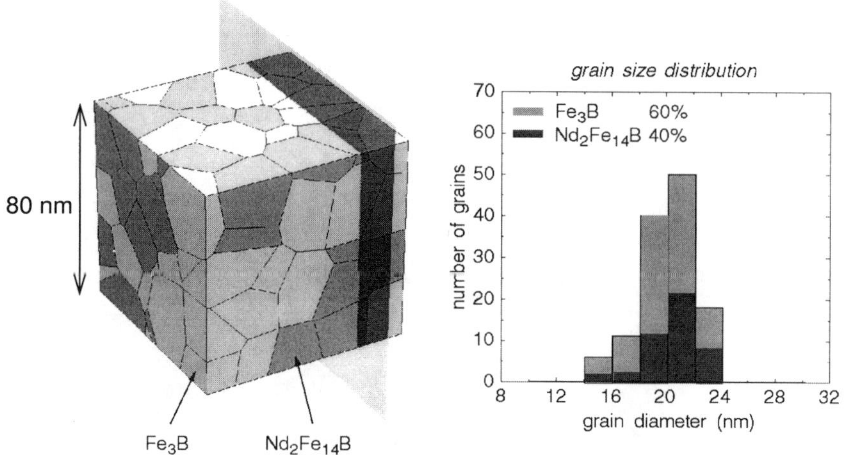

Fig. 2. Model of a nanocomposite magnet consisting of 125 polyhedral grains. The volume fraction of the hard magnetic phase is 40%. Left hand side: Grain structure and a slice plane used as probe for the magnetization and demagnetizing field; right hand side: grain size distribution.

Table I: Intrinsic magnetic properties. K_1 and K_2 (MJ/m^3), the anisotropy constants; J_s(T), the spontaneous magnetic polarization and the exchange constant A (pJ/m).

	anisotropy	K_1	K_2	J_s	A	
$Nd_2Fe_{14}B$	uniaxial	4.5	0.66	1.61	12.5	[17]
$(NdTb)_2Fe_{14}B$	uniaxial	4.85	0.72	1.38	12.5	[18]
α-Fe	cubic	0.046	0.015	2.15	25	[8]
$Fe_{23}B_6$	cubic	0.01		1.7	17	[8]
Fe_3B	easy plane	-0.32		1.62	12.5	[19]
Fe_3B (reduced J_s)	easy plane	-0.32		1.3	12.5	
Fe_3B (reduced A)	easy plane	-0.32		1.62	8	
Fe_3B (reduced A and J_s)	easy plane	-0.32		1.3	8	

direction of the magnetization of the neighboring hard magnetic grains. Owing to the magnet's history, this direction is nearly parallel to the original saturation direction. Therefore intergrain exchange interactions between the soft and the hard phase cause remanence enhancement in nanocomposite magnets. Figure 3 shows the calculated demagnetization curves for increasing diameter of the magnetically soft grain. The solid line denotes the total magnetic polarization which is the sum of the contribution of $Nd_2Fe_{14}B$ (long-dashed line) and of α-Fe (dotted line) or Fe_3B (dashed line).

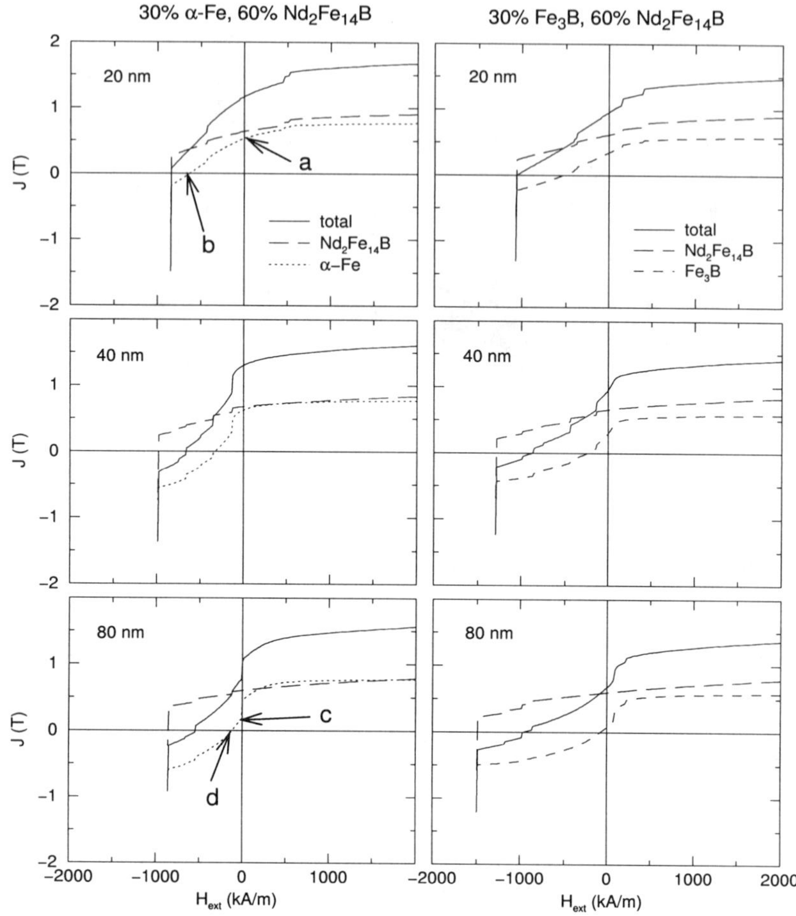

Fig. 3. Grain size dependence of the magnetic properties. Left hand side: Two-phase $Nd_2Fe_{14}B$/α-Fe magnet. Right hand side: Two-phase $Nd_2Fe_{14}B/Fe_3B$ magnet. The letters refer to the magnetization distributions given in fig. 4.

The magnetization starts to rotate reversibly within both phases, when a small opposing field is applied. The hard magnetic grains behave like a Stoner-Wohlfarth particle under an oblique applied field. In the soft magnetic grain, the magnetization rotates towards the direction of the applied field. However, the exchange interactions between the α-Fe grain and the $Nd_2Fe_{14}B$ grains hinders the rotation of the magnetization. The exchange interactions resemble a spring that tries to keep the magnetization within the α-Fe grains parallel to the magnetization of the neighboring hard magnetic grains. When the external field is switched off, the magnetization rotates back towards its original direction. Figures 4a and 4b show the rotation of the magnetization within the α-Fe grain for zero applied field and for H_{ext} = - 640 kA/m.

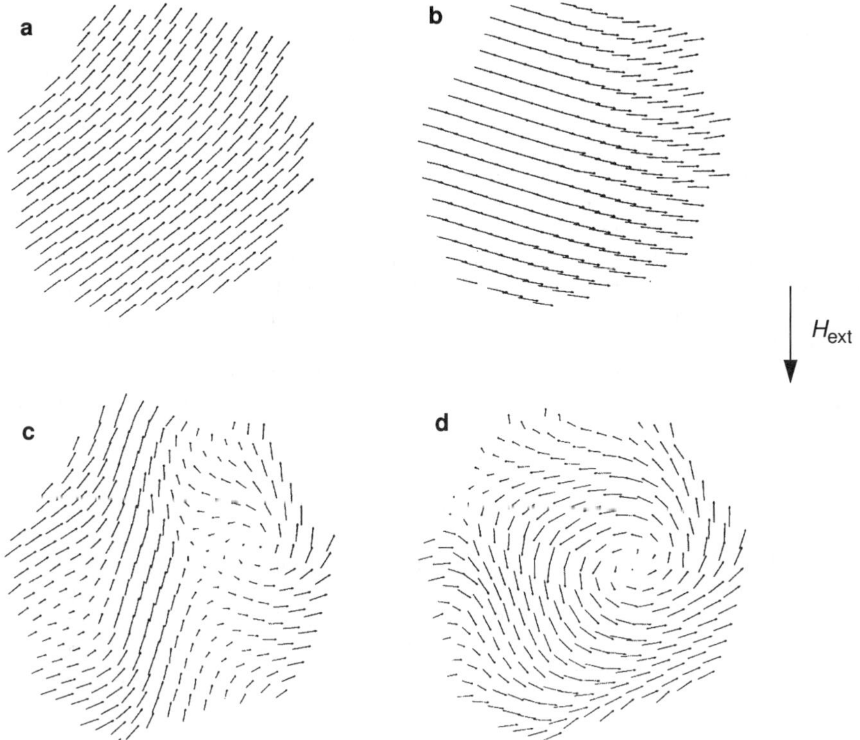

Fig. 4. Magnetization distribution within the α-Fe grain magnetic grain. Top: D = 20 nm, H_{ext} = 0, - 640 kA/m; bottom: D = 80 nm, H_{ext} = -10, - 130 kA/m. The magnetization is projected onto a slice plane.

The reversible rotation of the magnetization within the α-Fe grain is clearly visible in the demagnetization curve. An irreversible change of the magnetization occurs, if one of the hard magnetic grains reverses its magnetization [12]. For D = 20 nm, irreversible switching of the α-Fe/$Nd_2Fe_{14}B$ magnet occurs at an external field of H_{ext} = - 840 kA/m. Both phases reverse their magnetization together at the same critical value of the external field. The effective exchange coupling between α-Fe and $Nd_2Fe_{14}B$ becomes less significant for D = 40 nm. The α-Fe grain starts to reverse at an external field of H_{ext} = -140 kA/m, clearly visible as kink in the demagnetization curve. Both phases reverse their magnetization at different values of the external field for $D \geq 40$ nm. The magnetization within the α-Fe grain may reverse without inducing the reversal of the magnetization in neighboring hard magnetic grains. Magnetostatic effects become dominant for D ≥ 80 nm. Figures 4c and 4d show the formation of vortex within the α-Fe grain as the external field is decreased from -10 kA/m to -130 kA/m.

Soft phase properties

Two phase $Fe_3B/Nd_2Fe_{14}B$ magnets show a higher coercive field than α-$Fe/Nd_2Fe_{14}B$ magnets with the same microstructure. The coercive field increases by more than 200 kA/m if α-Fe is replaced by Fe_3B. This behavior was found for the simple model structure given in fig. 1 and for the microstructure consisting of 125 grains. The comparison of the demagnetization curves for α-$Fe/Nd_2Fe_{14}B$ and $Fe_3B/Nd_2Fe_{14}B$ (figures 3 and 5), shows that Fe_3B deteriorates the loop shape. This effect has to be attributed to the lower exchange constant of Fe_3B as compared to the exchange constant of α-Fe. The Fe_3B grain may reverse while the neighboring hard magnetic grains remain stable up to higher values of the opposing field. In the α-$Fe/Nd_2Fe_{14}B$ system the nucleation field of the hard magnetic phase remains nearly constant when the grain diameter is increased from 20 nm to 80 nm, whereas in the $Fe_3B/Nd_2Fe_{14}B$ the nucleation field of the hard phase is shifted from H_{ext} = - 1060 kA/m to H_{ext} = - 1500 kA/m with an increase of the grain diameter in the same range.

In addition figure 5 gives the demagnetization curves of the three-phase system α-$Fe/Fe_3B/Nd_2Fe_{14}B$ and the two-phase magnet containing the metastable $Fe_{23}B_6$ phase as magnetically soft phase. The results show that the coercive field increases at the expense of remanence, as α-Fe is replaced by Fe_3B. The magnetic properties of the two-phase $Fe_{23}B_6/Nd_2Fe_{14}B$ magnet are intermediate between the properties of α-$Fe/Nd_2Fe_{14}B$ and $Fe_3B/Nd_2Fe_{14}B$ magnets.

A further improvement of the coercive field in two-phase $Fe_3B/Nd_2Fe_{14}B$ magnets becomes possible reducing the magnetization and the exchange constant of Fe_3B by the addition of Cr. The corresponding change of the coercive field observed experimentally [14] can be qualitatively explained using micromagnetic simulations. Figure 6 compares the calculated demagnetization curve for a two phase $Nd_2Fe_{14}B/Fe_3B$ with a mean grain size of 20 nm for different assumptions of the effect of Cr on the intrinsic magnetic properties of Fe_3B. Either the magnetic polarization or the exchange constant, or both were reduced by 20% with respect to the pure Fe_3B phase.

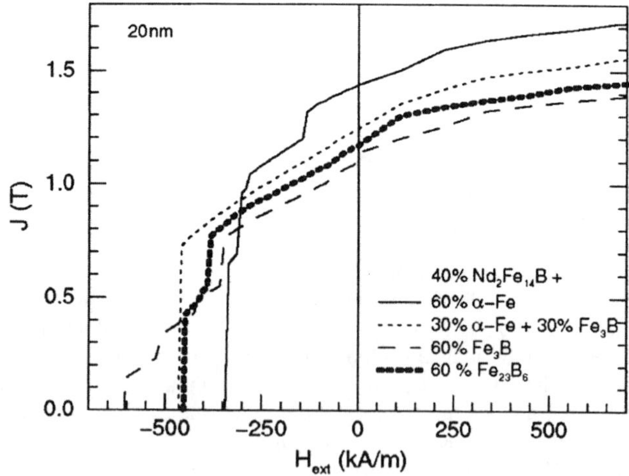

Fig. 5. Demagnetization curves as a function of the volume fraction of the different phases. 125 polyhedral grains with a mean diameter of 20 nm.

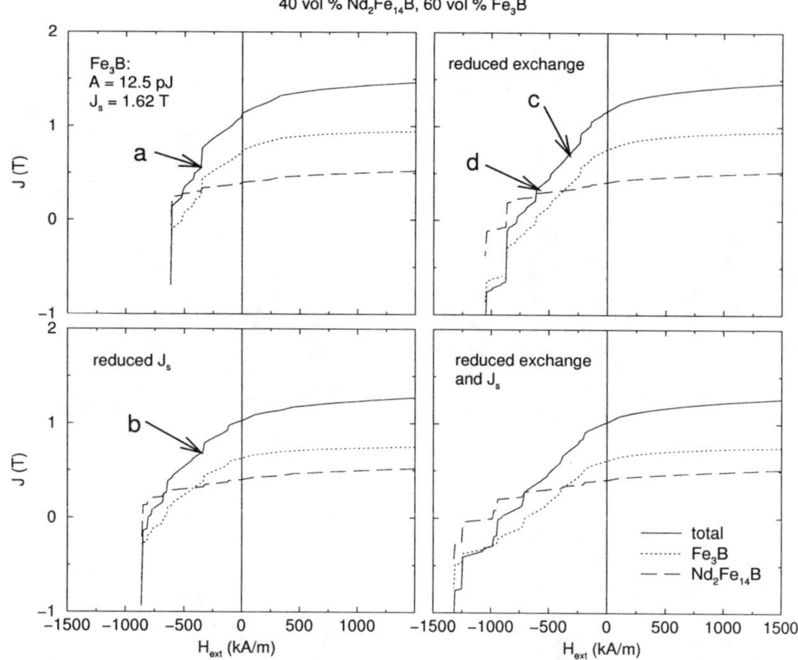

Fig. 6. Numerically calculated demagnetization curves for a nanocomposite Fe_3B/ $Nd_2Fe_{14}B$ magnet assuming different intrinsic properties of the Fe_3B phase. The letters refer to the magnetization distribution given in fig. 7 and 8.

A smaller magnetization results in a lower demagnetizing field which in turn increases the coercive from 600 kA/m to 800 kA/m while keeping a remanence of about 1.1 T. A reduction of the exchange constant within Fe_3B deteriorates the loop shape but increases the coercive field. Figure 7 shows the demagnetizing field for zero applied field and the magnetization distribution of an applied field of H_{ext} = - 340 kA/m. The comparison of the magnetization configurations a and b show that a high self-demagnetizing field causes the reversal of Fe_3B grains which remain stable, if J_s is reduced. In what follows, the exchange interactions between the different phases causes the reversal of neighboring hard magnetic grains when the opposing field is increased. A reduced exchange constant within the soft magnetic phase decreases the exchange energy density stored at the interface between the soft and the hard magnetic phase. The magnetization distributions given in fig. 8 reveal that the magnetic polarization within the soft phase may reverse without the immediate nucleation of a reversed domain in neighboring hard magnetic grains. The coupling between the different phases remains large enough to obtain a reasonable loop shape.

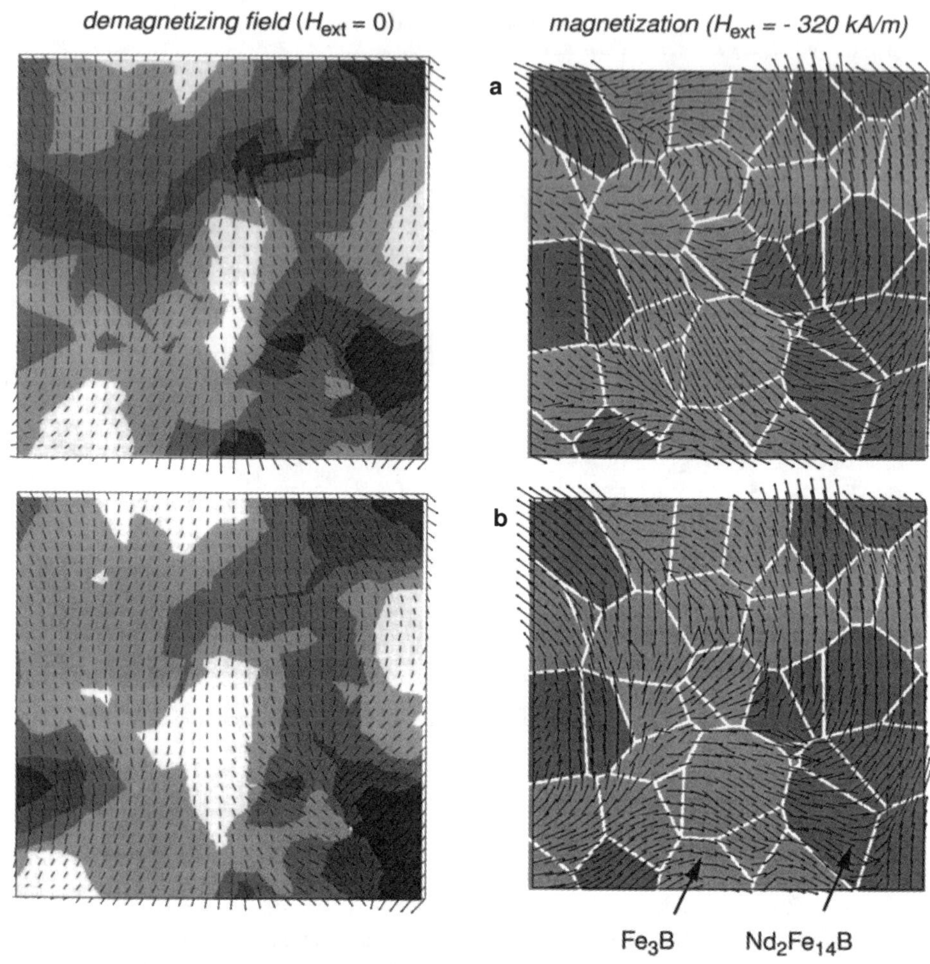

Fig. 7. Magnetization reversal in two-phase $Fe_3B/Nd_2Fe_{14}B$ magnets. Demagnetizing field for zero applied field (left hand side) and magnetization distribution in a slice plane for H_{ext} = -340 kA/m (right hand side). The arrows indicate the direction of the demagnetizing field. The grey scale gives the component of the demagnetizing field parallel to the external field, ranging from 240 kA/m (white) to 640 kA/m (dark grey). Top: Reference calculation; bottom: J_s of Fe_3B reduced by 20%.

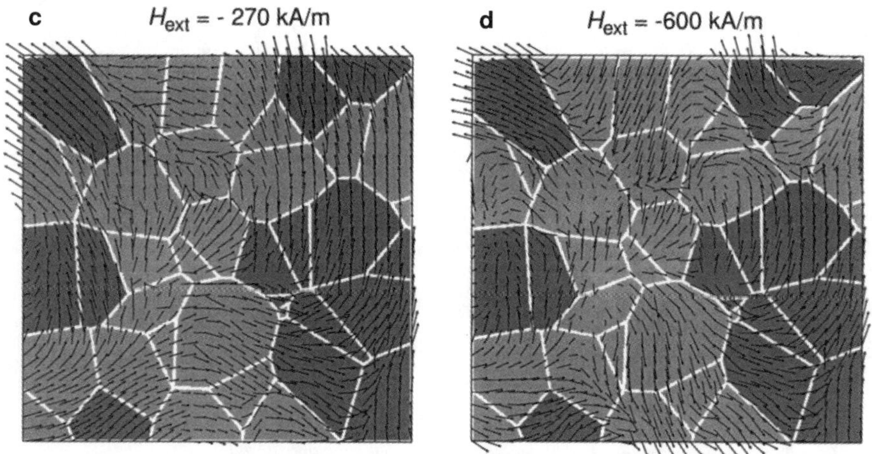

Fig. 8. Magnetization distribution of a nanocomposite $Fe_3B/Nd_2Fe_{14}B$ magnet at different applied fields. The exchange constant of Fe_3B was reduced by 20%.

Hard Phase Anisotropy

The substitution of Nd by Tb or Dy modifies the hard phase properties. An increase of the hard phase anisotropy by 8% and a corresponding reduction of J_s by 14% enhances the coercive field without a significant loss in remanence. Figure 9 compares the effect of hard phase anisotropy as a function of the mean grain size and as a function of the Fe_3B content. The mean grain size was varied from 10 nm to 30 nm. The increase of the anisotropy of only 8% increases the coercive field from 340 kA/m to 450 kA/m in the two phase $Nd_2Fe_{14}B/\alpha$-Fe magnets. The improvement of H_c decreases with increasing Fe_3B content. The grain size dependence of the increase of H_c owing to the substitution of Nd by Tb shows a maximum for $D = 20$ nm.

Interaction of magnetic particles in bonded magnets

In bonded magnets, the magnetostatic interactions between the particles may influence the magnetic properties. The strength of the magnetostatic interactions field and the possible effects on the remanence and coercivity were estimated using a hybrid finite element / boundary element method [16]. A particle of a bonded magnet was assumed to consist of several hard ($Nd_2Fe_{14}B$) and soft (Fe_3B) magnetic grains. Figure 10 compares the demagnetization curves calculated for a single particle and for a particle under the influence of the demagnetizing field of its neighbors with the particles arranged parallel or in a T-shaped form. The magnetostatic interactions decrease the remanence and the coercive field. Figure 11 shows that the interaction field in the range of 100 kA/m to 200 kA/m rapidly decreases with the distance from the particle. Fields of this magnitude cover about on half of a grain of the neighboring particle.

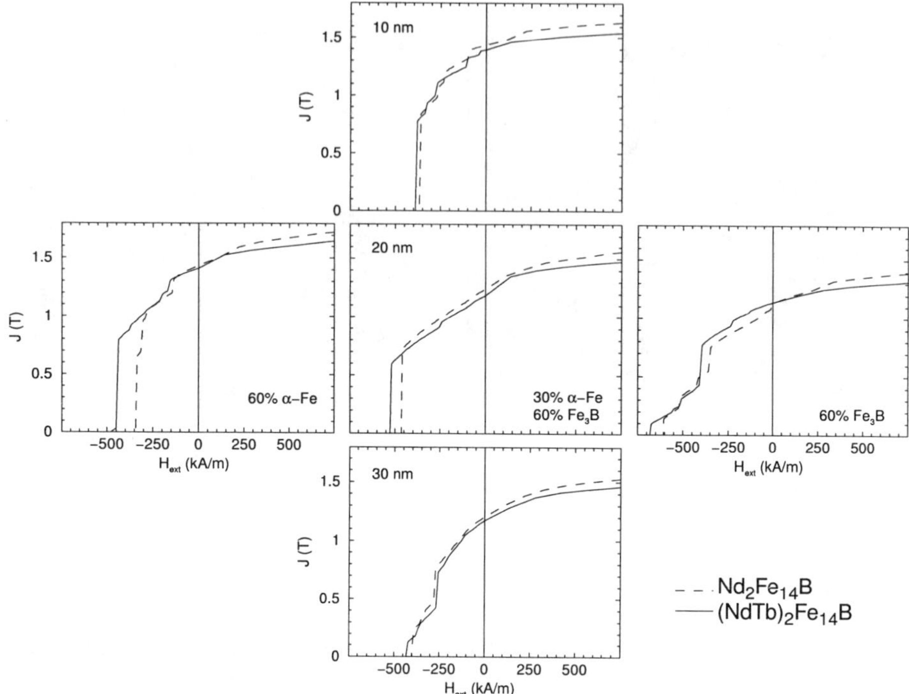

Fig. 9. Effect of the hard phase anisotropy on the demagnetization curve of nanocomposite magnets. From top to bottom the grain size is increased from 10 nm to 30 nm; from left to right the α-Fe is replaced by Fe_3B.

CONCLUSIONS

Numerical micromagnetic simulations provide guidelines for the optimization of magnetic properties of nanocomposite $Nd_2Fe_{14}B$ based permanent magnets:

The increase of the Fe_3B content at the expense of α-Fe increases the coercive field but deteriorates the loop shape. The reduction of the exchange constant and/or the magnetic polarization of the soft phase by about 20% increases the coercive field of two-phase $Fe_3B/Nd_2Fe_{14}B$ magnets without a significant loss in the remanence. The partial substitution of Nd by Dy or Tb improves the coercive field of two-phase α-$Fe/Nd_2Fe_{14}B$ magnets, while keeping the remanence and the coercive squareness.

ACKNOWLEDGMENTS

This work has been supported by the Austrian Science Foundation FWF (Grant No.P10511-NAW) and the EC-BRITE/EURAM project BRPR-CT95-0097.

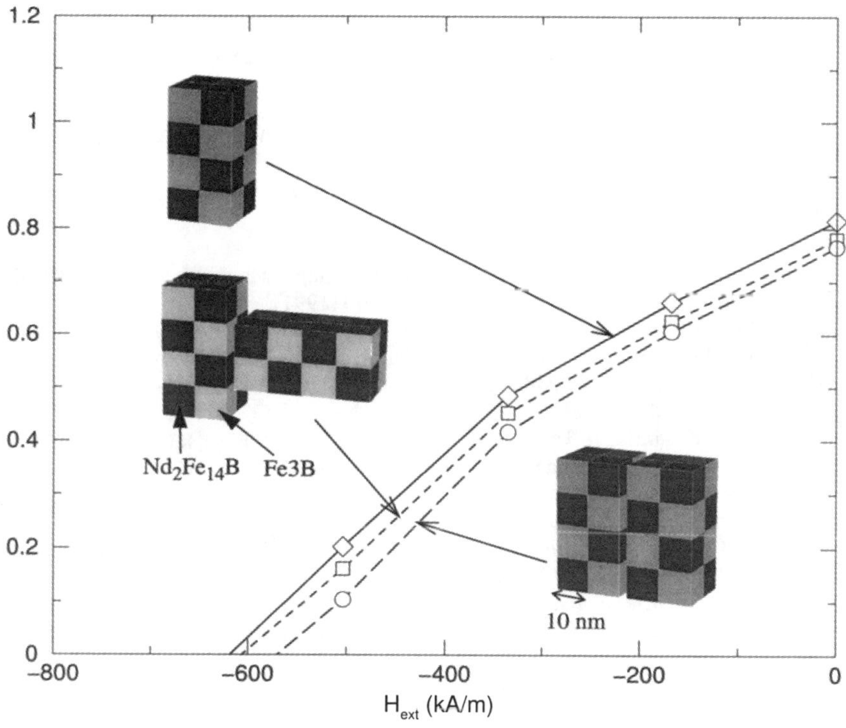

Fig. 10. Demagnetization curve of the left-most particle for different particle arrangements. The gap between the particles is 2.5 nm.

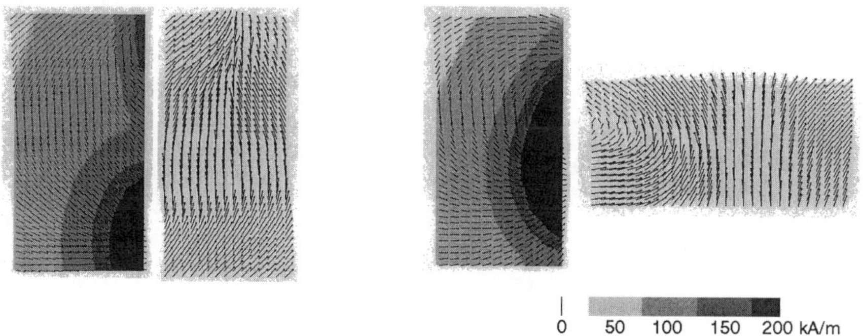

Fig. 11. Magnetostatic interaction field acting on the left particle caused by the magnetization of the right particle for parallel and T-shape arrangement of the particles for zero applied field.

REFERENCES

1. R. W. McCallum, A. M. Kadin, G. B. Clemente, and J. E. Keem, J. Appl. Phys. **61**, p. 3577-3579 (1987).
2. G. C. Hadjipanayis and W. Gong, J. Appl. Phys. **64**, p. 5559-5561 (1988).
3. E. C. Stoner and E. P. Wohlfarth, Philos. Trans. R. Soc. **240**, p. 599-642 (1948).
4. H. Fukunga and H. Inoue, Jpn. J. Appl. Phys. **31**, p. 1347 (1992).
5. T. Schrefl, J. Fidler, and H. Kronmüller, Phys. Rev. B **49**, p. 6100-6110 (1994).
6. M.K. Griffiths, J.E.L. Bishop, J. W. Tucker, and H.A. Davies, J. Magn. Magn. Mater. 183, p. 49 (1998).
7. R. Coehoorn, D. B. De Mooij, and C. De Waard, J. Magn. Magn. Mater. **80**, p. 101-104 (1989).
8. E. F. Kneller, IEEE Trans. Magn. *27*, p. 3588-3600 (1991).
9. H. Kanekiyo and S. Hirosawa, J. Appl. Phys. **83**, p. 6265-6267 (1998).
10. M. Bachmann, R. Fischer, and H. Kronmüller, in *Magnetic Anisotropy and Coercivity in Rare-Earth Transition Metal Alloys,* edited by L. Schultz, K.-H. Müller (Werkstoff-Informationsgesell-schaft, Dresden, 1998), p. 217-236.
11. J. Kuma, N. Kitajima, Y. Kanai, and H. Fukunaga, J. Appl. Phys. **83**, p. 6623-6625 (1998).
12. T. Schrefl and J. Fidler, J. Magn. Magn. Mater. **177**, p. 970-975 (1998).
13. S. Hirosawa, H. Kanekiyo, and M. Uehara, J. Appl. Phys. **73**, p. 6488-6490 (1993).
14. P. A. I. Smith, J. F. O'Sullivan, and J. M. D. Coey, in *Magnetic Anisotropy and Coercivity in Rare-Earth Transition Metal Alloys,* edited by L. Schultz, K.-H. Müller (Werkstoff-Informations-gesellschaft, Dresden, 1998), p. 205-214.
15. W.F. Brown, Jr., *Micromagnetics*, Wiley, New York, 1963.
16. T. Schrefl, J. Fidler, K. J. Kirk, and J. N. Chapman, IEEE Trans. Magn. **33**, p. 4182-4184 (1997).
17. M. Sagawa, S. Fujimura, H. Yamamoto, Y. Matsuura, and S. Hirosawa, J. Appl. Phys. **57**, p. 4094-4096 (1985).
18. S. Hirosawa, Y. Matsuura, H. Yamamoto, S. Fujimura, and M. Sagawa, J. Appl. Phys. **59**, p. 873 (1986).
19. W. Coene, F. Hakkens, R. Coehoorn, D. B. de Mooij, C. de Waard, J. Fidler, and R. Grössinger, J. Magn. Magn. Mater. **96**, p. 189-196 (1991).

ANALYSIS OF MAGNETIZATION PROCESSES IN NANOCOMPOSITE HARD MAGNETIC MATERIALS FROM MACROSCOPIC MAGNETIC MEASUREMENTS AND X-RAY MAGNETIC CIRCULAR DICHROISM

Stéphane David, Kenneth Mackay, Marlio Bonfim, Stefania Pizzini,
Alain Fontaine and Dominique Givord
Laboratoire Louis Néel-CNRS, BP 166, 38042 Grenoble Cedex 9, France
david@labs.polycnrs-gre.fr

Abstract

A series of $RE_xTM_{bal}B_y$ (3≤x≤5.5 , 15≤y≤25 at.%) nanocomposite magnets were prepared using the classical method of alloy rapid quenching and subsequent devitrification. Appropriate elemental additions and substitutions allowed the technical magnetic properties to be optimized. The analysis of the both hysteresis cycles and static susceptibility measurements in the range 20K-300K reveals that magnetization reversal in these materials can be interpreted schematically as a two-step process : the initial magnetization reversal is dominated by reversible processes while irreversible processes, which characterize the coercivity mechanism, occur at larger field values. To complement this study, X-Ray Magnetic Circular Dichroism (XMCD) spectroscopy was applied as an element-selective probe of magnetization reversal in nanocomposite samples. This allowed the hysteresis cycles of both hard and soft phases to be separated from the total cycle. Magnetization processes in these bulk heterogeneous systems are discussed in the light of information brought by the complementary applied experimental techniques.

1-Introduction

In the search for hard isotropic magnetic materials with higher remanence, the concept of nanocomposite materials with reduced rare-earth content has emerged as a highly promising alternative. These systems consist of an ultrafine mixture of magnetically hard and soft crystallites, both hard and soft materials being exchange-coupled at the nanometer scale. Magnetization reversal in the soft phase is hindered by this coupling, thus leading to macroscopically homogeneous magnets, whose coercivity is intermediate between those of the hard and soft phases. In addition, the coupling between the nanosize crystallites gives rise to a large remanence-enhancement (up to 0.75 M_s) . This attractive specific property leads to a remarkably high remanent magnetization in the two-phase isotropic nanocomposite magnets.

The melt-spun ternary alloys with typical composition $RE_xFe_{bal}B_y$ (3≤x≤5.5 , 15≤y≤25 at.%) , which were discovered by Coehoorn *et al.* [1] , constituted the first category of nanocomposite materials that came to light. In these systems, the hard uniaxial $RE_2Fe_{14}B$ phase is formed in association with Fe_3B as main soft magnetic phase. Later on, another prominent family of nanocomposites, with α-Fe as main soft phase, was reported [2] in the hypostoichiometric (6≤RE≤12 at.%) RE-Fe-B systems with lower boron content (2≤B≤10 at.%) . Subsequently, several compounds and different preparation techniques were discovered, allowing nanocomposite materials to be obtained [3-5] .

Up until now, the industrial exploitation of nanocomposite materials has been waiting for a significant improvement in their technical properties - mainly coercivity - to make them economically viable as medium-cost permanent magnets. This objective has stimulated the formation of a consortium, supported by the E.U. , bringing together industrial and academic partners in a 3-year collaboration [6] . The work realized within this project concerned i) the

Mat. Res. Soc. Symp. Proc. Vol. 577 © 1999 Materials Research Society

preparation of nanocomposites of the $RE_2Fe_{14}B/Fe_3B$ type with improved properties ii) the investigation of magnetization processes in these systems, with the aim in particular to improve our understanding of the relationship between microstructure and coercivity mechanisms.

The main trends and results found during the optimization phase of the materials are presented in the section 2 of the present paper; a more thorough description of the output can be found in ref. [7] . Here however, the focus is mainly on the characterization and analysis of the magnetization processes. Usual macroscopic magnetization measurements were applied for this purpose (section 3) . To complement this study, XMCD experiments were performed at the European Synchrotron Radiation Facility (ESRF) , Grenoble, France. This is presented in section 4. The element-selectivity of the XMCD probe was used to separate contributions of both hard and soft phases to the total magnetization. These measurements provided a unique insight into the characteristic magnetization reversal mechanisms occurring in the nanocomposite systems (section 5) .

2-Optimization of the technical coercive field of the system $RE_2TM_{14}B/Fe_3B$

2.1 Sample preparation and characterization

The typical composition of the starting alloys was $RE_x(Fe_{1-t}T_t)_{bal}B_yA_z$, where T = Co, Cr and A = Al, Si, Cu, Nb, ... A small piece of alloy (typically 100 mg) was placed in a levitating RF coil operated at 350 kHz and melted. When the RF power was turned off, the molten alloy dropped into the gap of a splat-cooling quenching device incorporating an anvil-hammer arrangement [8] . In this device, the synchronous closure of a low-loss magnetic circuit allows a violent motion to be transmitted to the hammer. The hammer intercepts the molten droplet which is quenched between the hammer and the anvil. With the alloys used in the present study, the cooling rate reached was about 10^7K/s, and the resulting flakes were 20-40 mm thick. They were found to be amorphous by X-ray and electron diffraction.

The crystallization process of the quenched amorphous flakes was studied by DTA. In general, two exothermal peaks were observed at about 870K and 920K with a 20K/min. scanning rate. In agreement with previous studies [9,10] , these crystallization events were associated with the initial formation of the Fe_3B and $RE_2Fe_{23}B_3$ phases, followed by the transformation of the latter metastable $RE_2Fe_{23}B_3$ phase into $RE_2Fe_{14}B$, Fe_3B and α-Fe at higher temperature. However, the sequence of phase formation and their nature were found to depend critically on the initial alloy composition. In particular, the formation of the $Nd_{1.1}Fe_4B_4$ phase, instead of $RE_2Fe_{14}B$, was found to be detrimental to coercivity.

Annealing of the quenched flakes was performed in a high-vacuum furnace (10^{-7} Torr) equipped with two Kyocera heating elements [8] . In this furnace, the heating volume is small and the thermal inertia is low, allowing a 20 K/sec. heating rate, up to 1070 K, to be attained. In most cases, optimum magnetic properties were obtained at maximum heating and cooling rates. Under such conditions, the crystallization temperatures, as obtained by DTA slow heating, were shifted by approximately 100 K, in agreement with values deduced from the Kissinger approach [11, 12] . In practice, the annealing temperature for each alloy was determined empirically by scanning the temperature from 970 K to 1100 K, to achieve optimum magnetic properties.

The technical magnetic properties of each annealed alloy were deduced by recording the room temperature demagnetization curve, using the magnetometer described in section 3. For these measurements, a 8T magnetic field was initially applied. The field was applied in the plane of the flakes allowing the demagnetizing field to be neglected.

The relevance and the reliability of the present protocol were established by reproducing the earlier results of Coehoorn and co-workers [1] (see Table 1 and Figure 1) . The same protocol was then repeated to prepare materials based on more than 120 different compositions. Structural characterization was carried out by XRD and TEM on a restricted number of samples. This allowed to unambiguously associate coercivity with the presence of the $RE_2Fe_{14}B$ hard phase. Depending on the starting composition, the soft phase was found to be composed of either Fe_3B and $Fe_{23}B_6$ in association with α-Fe [7] . A direct relationship between microstructure and technical magnetic properties was observed : the best samples present a homogeneous finely-grained (10-30 nm) nanostructure [14] .

2.2 Search for optimum technical magnetic properties

The alloys composition was varied with a view toward the optimization of remanence M_r and coercivity ${}_MH_c$, which constitute the technical magnetic properties of magnets. A series of alloys was prepared to sweep a broad portion of the $Nd_xFe_{bal}B_y$ ternary phase diagram, in the vicinity of the reference composition [1] (3□ x□ 5.5 at. %, 15□ y□ 25 at %) .

Three different RE contents were investigated, namely 3, 4.25 and 5.5 at. %. The associated hard/soft volume fractions, in percent, were approximately 30/70, 45/55 and 55/45. In general, it was found that the lower the RE content, the lower the coercivity and the higher the remanence. In agreement with previous studies [15] , the optimum B concentration was found to be close to 18% . In agreement with Hirosawa [16] , a few percent of Cr addition (1 to 5 %) was required at higher RE content (5.5 %) to form the $RE_2Fe_{14}B$ phase at the expense of the $RE_{1.1}Fe_4B_4$ phase. Recently, Atom-Probe Field-Ion Microscopy [17] and calorimetric measurements [18] have revealed that Cr addition leads to thermodynamic and kinetic modifications of the crystallization process in $RE_5Fe_{77}B_{18}$ alloys. In addition, we found that in alloys containing more than 5 % Cr, Fe-B-Cr paramagnetic phases are formed which lead to decoupling of the hard grains. This explains the two-step hysteresis loops then obtained, associated with high coecivity but low magnetization values.

Some of our best results are presented in Table I and Figure 1.

Composition	$\mu_0{}_MH_c$[T]	μ_0M_r[T]	μ_0M_s[T]
Ref: $Nd_{4.25}Fe_{77.75}B_{18}$	0.35	1.2	1.65
$Pr_{4.25}Fe_{77.75}B_{18}$	0.34	1.1	1.54
$Nd_3Fe_{73}Co_5Nb_1B_{18}$ I	0.29	1.31	1.67
$Nd_{3.25}Tb_1Fe_{77.75}B_{18}$	0.47	1.17	1.53
$Nd_{3.25}Tb_1Fe_{72.75}Co_5B_{18}$	0.51	1.05	1.44
$Nd_{3.25}Tb_1Fe_{70.75}Co_5Cu_{0.5}Nb_{0.5}Si_1B_{18}$ II	0.53	1.06	1.38
$Nd_{4.5}Tb_1Fe_{70.75}Co_{2.5}Cr_3Si_{0.5}B_{18}$ III	0.7	0.94	1.27

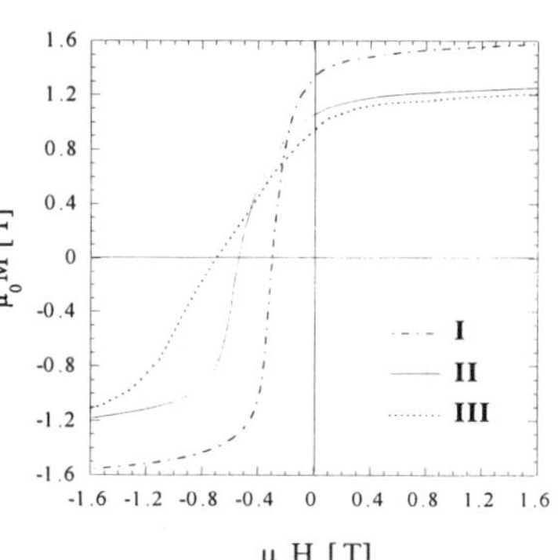

Table I : Technical magnetic properties of the optimized $RE_2TM_{14}B/Fe_3B$ materials. μ_0M_s refers to the magnetization of the samples at 8T.

Figure 1: Demagnetization curves of the three samples (I, II, III) indicated in Table 1. Each of these samples is representative of the best technical properties we obtained for the different families: RE 3% (I) , 4.25% (II) and 5.5% (III) .

In general, the main conclusions which can be drawn from these studies are the following : -i) Tb substitution is effective in improving $_{M}H_c$. This can be attributed to hardening of the hard phase, through the large enhancement of the magnetocrystalline anisotropy when Tb is substituted for Nd [19] . - ii) 5% Co addition is associated in most cases with high technical coercivities. By itself the substitution of Co for Fe in $RE_2Fe_{14}B$ is known to decrease M_s but substantially increase T_c . In usual sintered NdFeB magnets, the substitution of Co is not favorable. In the present case, a structural influence can be inferred, such as grain refinement. -iii) Optimum properties were achieved by a combined addition of Cu and Nb (approximately 1%) and/or Si. These elements are known to induce grain refinement in Fe-based nanostructured soft magnetic materials [20] .

Nanostructure refinement and homogenization appear from the fundamental study of the magnetization processes, to be a stringent requirement to optimize the technical magnetic coercivity in nanocomposite systems [21, 26] .

3. Macroscopic magnetic measurements: characterization of reversible and irreversible processes.

The discussion in the rest of the paper concerns the properties observed in the reference alloys with compositions $Nd_{4.25}Fe_{77.75}B_{18}$ and $Pr_{4.25}Fe_{77.75}B_{18}$, simply referred as NdFeB and PrFeB. The experimentally observed behavior was found to be qualitatively similar in all investigated samples [12] and the discussion applies essentially to all of them.

Magnetization measurements were performed using a high-sensitivity axial-extraction magnetometer equipped with a superconducting coil supplying fields up to 12T, with a range of accessible temperatures between 4.2K and 600K. Hysteresis cycles were measured at different temperatures from 300K down to 20K (Figure 2) .

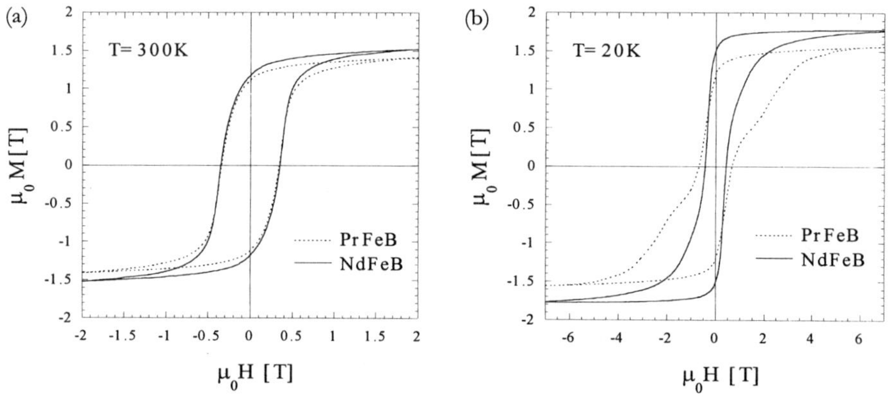

Figure 2: Hysteresis cycles of NdFeB and PrFeB at 300K (a) and 20K (b)

At 300K, the hysteresis cycles apparently indicate homogeneous magnetization reversal (Figure 2a) . As the temperature is decreased, a two-phase process becomes apparent (see data at 20K in Figure 2b) as a shoulder develops in the third quadrant. The effect is more visible in the case of PrFeB. In NdFeB, the easy-magnetization tetragonal axis is not unique below the spin reorientation temperature of 135K, and this tends to mask the two-phase reversal. This is why the PrFeB has been preferentially used to study properly the low-temperature magnetization processes.

As discussed in reference [21] , the analysis of magnetization processes in nanocomposite materials requires reversible and irreversible processes to be separated in the experimental hysteresis cycles. In order to extract the reversible susceptibility (χ_{rev}) contribution, a series of truncated recoil loops (ΔH=30mT) were measured at 20mT field intervals along the major hysteresis cycle. In general, time-dependant magnetization variations were found to occur at the field on the major cycle where the recoil loop was started; these viscous effects were particularly significant in the third quadrant. The measurements were started after a 100s waiting time, when the viscous effects had become negligible. From χ_{rev} and the total susceptibility χ_{tot} , the quantity χ_{tot}-χ_{rev} can be deduced (Figure 3) . The change in magnetization represented by χ_{tot}-χ_{rev} is predominantly associated with irreversible reversal phenomena and it is a meaningful quantity to characterize reversal. The field at which the maximum of χ_{tot}-χ_{rev} occurs may thus be defined as the coercive field H_c of the system. In usual single-phase magnets, H_c is often taken to be equal to the technical coercive field $_{M}H_c$ at which magnetization vanishes. As shown in Figure 3, this assumption is not valid in the case of nanocomposite systems.

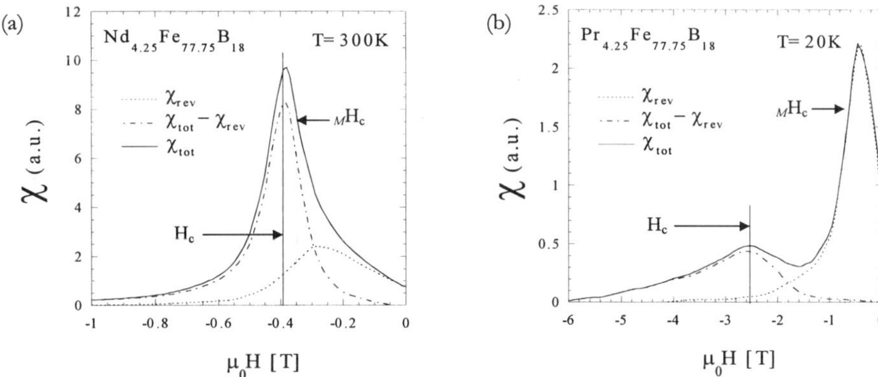

Figure 3: Static susceptibilities (total, reversible, difference) as a function of the reverse applied field in NdFeB at 300K (a) and PrFeB at 20K (b) .

Both low-temperature cycles and static susceptibility measurements (Figure 3) indicate that : (i) the mechanism of magnetization reversal in nanocomposites involves two-steps ; (ii) the two steps are different in nature: reversal is mainly reversible at low field, while irreversible processes take place at higher field values. In a first approach, it has been assumed that irreversible processes correspond to the reversal of the hard magnetization: the thermal variations of H_c , have been analyzed semi-quantitatively using the "Nucleus expansion model" [21] . On the other hand, the relation between the soft magnetization reversal and reversible processes appeared to be less straightforward, because of the coupling between the hard and soft grains (section 5.2) .

To elucidate the contributions of the hard and soft magnetization to the reversal of the total magnetization, we have developed an original approach based on X-ray Magnetic Circular Dichroism (XMCD) spectroscopy.

4 XMCD element-selective magnetometry

4.1 XMCD: element-selectivity and magnetism sensitivity

XMCD takes advantages of the energy and polarization tunability of the synchrotron radiation provided by the relativistic electrons accelerated in a storage ring. It is an element-selective X-ray Absorption Spectroscopy (XAS) which measures the difference between the absorption of right and left circular polarized X-ray light by a ferro (ferri) magnetic materials. The absorption cross-section of the photons σ(E) is given by the matrix element of the Hamiltonian – which can be limited to the dipolar term -which couples the initial and final states (Fermi Golden Rule). Thanks to the dipolar interaction between the X-ray photon and the excited core electron of a specific atom, the selection rules make this helicity-dependent spectroscopy, sensitive to the orbital momentum of the band hosting the photoelectron. In addition, the spin-orbit interaction makes this orbital momentum-sensitive spectroscopy also spin–sensitive. It can be demonstrated that thanks to spin-orbit interaction, the photoelectron emitted in the absorption process is spin-polarized. In magnetic samples, the XMCD becomes sensitive to the asymmetry in vacant states above the Fermi level in the majority and minority valence bands [22] . By taking σ^- (σ^+) as the absorption cross-section for the right (left) polarization, the dichroic cross-section is given by:

$$\sigma^{XMCD} = \sigma^- - \sigma^+ \qquad (1)$$

Sum rules have established that, at least for $L_{2,3}$ absorption edges of transition metals, integrated XMCD spectra can be related quantitatively to the expectation values of spin and orbital moments in the valence band [23] . XMCD probes the selected magnetic moments along the direction of propagation of the X-rays (**z**) . As a consequence, the signal measured on a macroscopic sample is proportional to the projection of the selected magnetization along **z**:

$$\sigma^- - \sigma^+ \propto \langle \mathbf{M} \rangle_{\mathbf{z}} \qquad (2)$$

In our approach, we focus only on the relative change of the XMCD signal as a function of the applied field, to probe the element-specific hysteresis of the hard phase. In contrast with usual XMCD experiments, no quantitative information of the value on the magnetic moments is sought .

The element-selectivity of the X-ray dichroism is the key-advantage over classical visible light dichroism. In addition, the penetration depth of the X-rays allows the bulk of samples to be probed.

4.2 Configuration of the hard phase magnetization as a function of the applied field.

For the present measurements, we selected the L_2 absorption edge of the rare earth element (Nd or Pr) contained in our samples. The energy involved in this transition (6-7 keV) allows a sufficient penetration of the X-rays, to probe 20μm-thin samples throughout the whole thickness. Room-temperature measurements were carried out on the energy dispersive beamline (ESRF, ID 24) in transmission geometry. This line is equipped with a planar undulator which delivers linearly polarized incident light. Circular polarization was obtained using a diamond Quarter Wave Plate [24] placed upstream the sample. Measurements in the temperature range 20K-300K were performed on a beamline equipped with scanning optics (ESRF, ID12A) . Spectra were measured in fluorescence detection. On this beamline, the X-ray circular polarization was achieved by an helical undulator, the polarization being reversed by changing the phase of the undulator.

The L_2 absorption edge corresponds to the $2p_{1/2} \rightarrow 5d$ transition, and it is therefore linked to the 5d polarization. Through 5d-4f polarization, it is an indirect probe of the 4f RE magnetization. Thus, the signal measured at the RE L_2 edge is representative of the hard phase magnetization.

Prior to the experiments, the samples were saturated *ex-situ* in a 8T field. They were mounted afterwards in an Helmoltz coil, whose axis is collinear with the X-Ray beam, in a quasi-planar (30°) configuration to allow the beam to pass trough. A static external field was applied and swept in the range [-1.5T; +1.5T] ([-7T; +7T]) at room temperature (low temperatures) experiments.

For each applied field value, an XMCD spectrum was measured. A typical XMCD spectrum and the variations for different field values are given in Figure 4 and 5 respectively.

The maximum amplitude of the XMCD signal as a function of the applied field is reported in Figure 6. Assuming that the RE ions are distributed exclusively in the $RE_2Fe_{14}B$ hard grains, it is inferred that these graphs represent the specific hysteresis cycle of this phase in both samples.

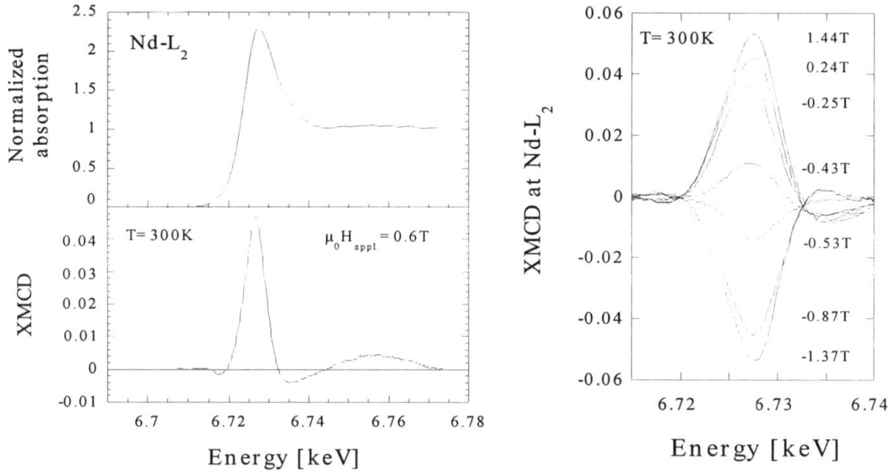

Figure 4: XAS spectrum at the Nd-L2 absorption edge (top) and typical XMCD spectrum at 300K (bottom) in the NdFeB nanocomposite sample.

Figure 5: Field dependence of the XMCD signal in NdFeB at 300K.

4.3 Separation of the hard phase and soft phase contributions to the total magnetization.

In order to be related to magnetization absolute values, the XMCD data need to be normalized to some reference value. For this purpose, it was assumed that the magnetization variation at high fields (>2T) , as obtained in macroscopic magnetization measurements, is due to moment rotation in the hard grains only, the magnetization in the soft grains being saturated. Thus, the law of approach to saturation may be expressed as :

$$M(H) = M^{soft} + M^{hard}(1 - \tfrac{a}{H} - \tfrac{b}{H^2}) \qquad (3)$$

The various parameters in expression (3) were obtained from fit of the magnetization curves. The value of the hard phase magnetization under 2 T (or any other higher field value) can thus be deduced and related to the XMCD signal under the same field.

Note also that in the XMCD measurements, the field was applied at 30° from the sample plane. The applied field must thus be corrected for demagnetizing field effects. The demagnetizing field was determined experimentally by comparing bulk magnetization measurements obtained for H applied in-plane and at 30° from the plane respectively.

Finally, from the phase volume composition and assuming that:

$$M^{soft}(H) = M(H) - M^{hard}(H) \tag{4}$$

the magnetization hysteresis cycles of both hard and soft phases were separated (Figure 7) .

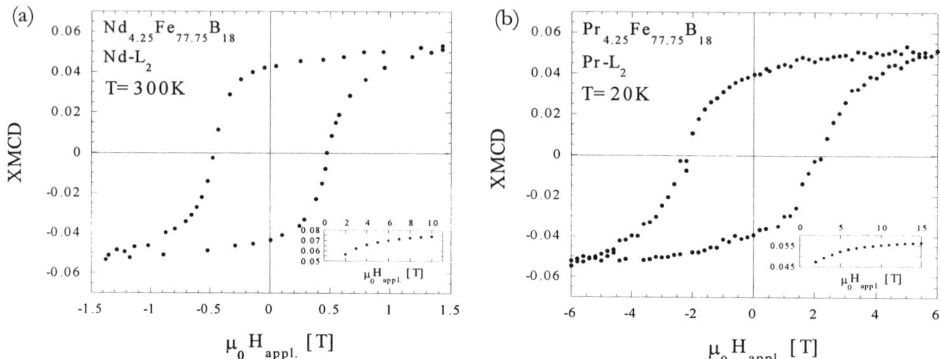

Figure 6: XMCD maximum amplitude signal as a function of the applied field at the Nd-L_2 (a) and Pr-L_2 (b) edges, respectively at 300K and 20K. The inserts show the XMCD values at high field, which are extrapolated from a law of approach to saturation (see 4.3).

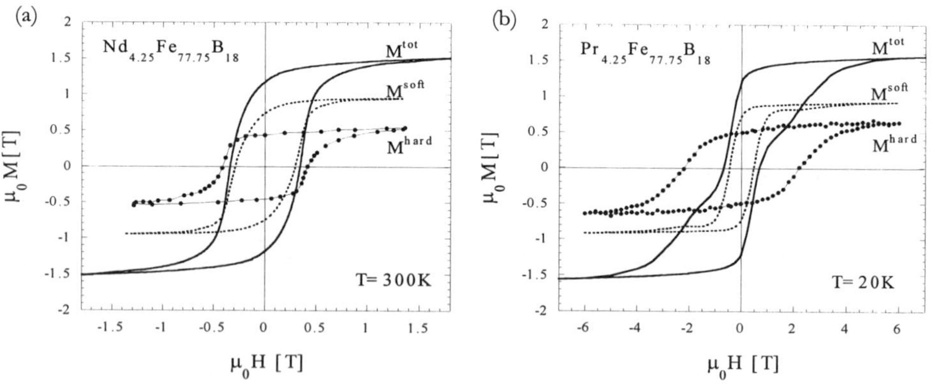

Figure 7: Specific hysteresis cycles of the hard and soft phase in NdFeB (a) and PrFeB (b) nanocomposite materials at 300K and 20K respectively.

5. Discussion

As shown on Figure 7, XMCD selective measurements reveal unambiguously the nature of the two-step reversal mechanism: soft phase magnetization reverses first and is followed by the reversal of the hard magnetization. Due to the nanometer-size of the constituent crystallites, the transition between the two steps is continuous at 300K [25] . At 20K, the enhanced anisotropy of the hard phase considerably increases the value of the hard magnetization switching field. This phenomenon leads to a de-coupled reversal mechanisms [26] , hence the shoulder in the third quadrant.

The static susceptibilities of both hard and soft magnetization (Figures 8, 10) are deduced by taking the derivative of each specific cycle. This allows us to compare the reversible and irreversible processes (obtained from bulk magnetization measurements) to soft and hard phase contributions (deduced from XMCD data) . This comparison provides a quantitative basis for a discussion of the interrelation between these different components.

5.1 Soft phase reversal processes

The maximum in χ_{soft} (Figure 8) , which occurs at a field value much higher than the value expected for a soft magnetic phase, directly indicates that exchange interactions exist between the hard and the soft phases. Moreover, Figure 7 shows clearly that the hard phase magnetization varies only slightly at low reverse field values (in the field range where soft magnetization reverses predominantly) and hence can be assumed to be constant in first approximation. Reversal may thus be described in a semi-quantitative approach by considering that the moment configuration in a soft grain is determined by the competition between Zeeman energy which tends to align the moments along H and exchange interactions which favor their alignment along the moment direction in the adjacent hard grains. Assuming that the anisotropy in the hard grains is very large, it may be assumed that under H, a rotation of the moments in the soft phase occurs only. The resulting moment configuration consists in a progressive rotation from an angle of maximum deviation, θ_{max} , in the center of the soft grains, to alignment with the hard phase easy direction at the grain boundary with adjacent grains. To first approximation, the stored exchange energy in a soft grain, exchange coupled with two hard grains with parallel magnetization, may be expressed as:

$$\gamma = \frac{4A\theta_{\max}^2}{d_{soft}} \tag{5}$$

where A is the exchange constant and d_{soft} is the soft grain size. The Zeeman energy is :

$$E_Z = \mu_0 M_s H\left(\frac{\sin\theta_{\max}}{\theta_{\max}}\right)d_{soft} \tag{6}$$

Minimization yields:

$$H(\theta_{\max}) = \frac{8A\theta_{\max}^3}{\mu_0 M_s d_{soft}^2}\frac{1}{\left[\theta_{\max}\cos\theta_{\max} - \sin\theta_{\max}\right]} \tag{7}$$

The exchange constant A varies with temperature as M_s^2 and thus, from relation (7) , the applied field value $H(\theta_{max})$ giving rise to the magnetization configuration determined by θ_{max} should vary as M_S as a function of temperature.

In reference [21] , magnetization processes were analyzed by simply attributing the reversible processes to the soft phase, taking $\theta_{max}= \pi$ in relation (7) (corresponding approximately to the reversal of half the soft phase, *i.e.* to the maximum of χ_{rev}) . The comparison of χ_{soft} and χ_{rev} in Figure 8 shows that this assumption is valid at low temperature, but not at 300K. In fact, only half the soft grain magnetization can be associated with reversible processes at 300K, the remaining half remaining coupled to the hard neighbors and reversed irreversibly at a higher field value. In this case, the actual maximum of reversible reversal is reached for a lower field value than predicted by (7) : it corresponds approximately to the reversal of 25% (half the half) of the soft magnetization ($\theta_{max} \approx \pi/2$) . Consequently, in the analysis based on the above crude assumption, the proportion of soft magnetization involved in reversible process is progressively overestimated as the temperature is increased from 20K to 300K. This is the cause of the observed discrepancy between the thermal variation of $M_s(T)$ and the field $H_{c,rev}(T)$ at which χ_{rev} is maximum (Figure 9) . The new insight provided by selective magnetometry shows that reversible processes and soft magnetization reversal can be associated qualitatively, but not quantitatively across the entire temperature range 20K-300K.

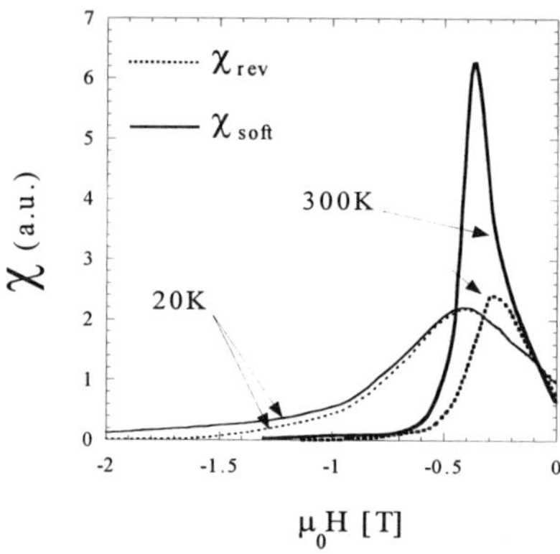

Figure 8: Comparison of χ_{soft} and χ_{rev} as a function of the applied field at 300K and 20K. At 20 K, the two curves overlap, thus showing that reversible reversal can be fully identified to the reversal of the soft phase. At 300K, integrated area of χ_{rev} and χ_{soft} show that reversible process involves only half the soft magnetization.

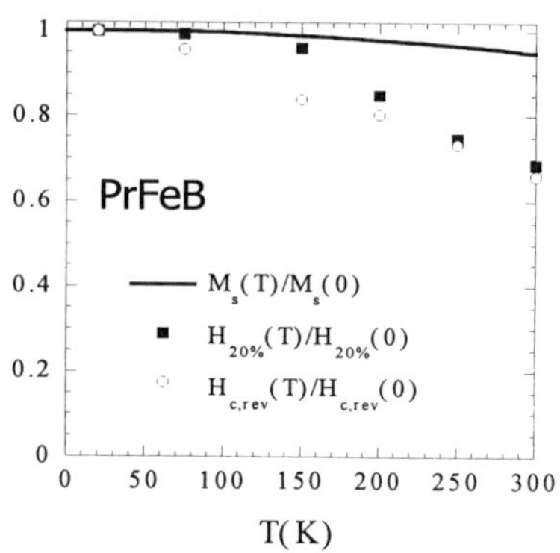

Figure 9: Reduced thermal variation of $H_{c,rev}$ and $H_{20\%}$, compared to the variation of M_s in PrFeB. The quantity $H_{20\%}$ corresponds to the field value at which 20% of the macroscopic magnetization is reversed (see text) .

However, at fields lower than $H_{c,rev}$, the above hypothesis becomes valid again : at 300 K, in a field of 0.2 T, the total magnetization reversal is due almost exclusively to the soft phase reversal and is essentially reversible. Integrating χ_{soft} (or χ_{rev} or χ_{tot}) from H =0 to 0.2 T, shows that 20% of the magnetization is reversed in this field. At lower temperatures, under such experimental conditions, the association between reversible processes and soft phase contribution is even more justified. The field at which 20% of the soft phase is reversed was then deduced at all temperatures from macroscopic magnetic measurements. It is plotted as a function of T in Figure 9. Again the experimental thermal variation does not follow that of the spontaneous magnetization but it is significantly faster. Three arguments may be proposed for this unexpected behavior. On the one hand, it may be that the magnetocrystalline anisotropy of the soft Fe_3B phase is not completely negligible. The anisotropy is expected in general to decrease rapidly when the temperature increases and thus may explain the observed behavior. Another possibility is that the exchange constant A at the interface between the soft and the hard phases is altered. In this case, its thermal decrease is likely to be larger than in the bulk. This would again account for the observed behavior, because the grains are exchange-coupled through their interfaces. Finally, it may be an indication that a non-uniform magnetization state (rotational) is created in the hard grains due to the exchange-coupling. The increase with temperature of the typical length of such configuration –typically δ_0^{hard} - can be invoked. This last possibility is compatible with the remanence-enhancement of the hard magnetization which has been experimentally observed and which is discussed in section 5.3.

5.2 Hard phase reversal

The clear identification of the hard phase properties in the present measurements, establishes that magnetization reversal in the hard grains occurs at higher fields than in the soft grains. The field at which χ_{hard} is maximum approximately coincides with the one at which $\chi_{tot}-\chi_{rev}$ is maximum (Figure 10) . In reference [21] , an analysis of coercivity in REFeB nanocomposites was presented which is based on this hypothesis. These results are thus strengthened by the present study.

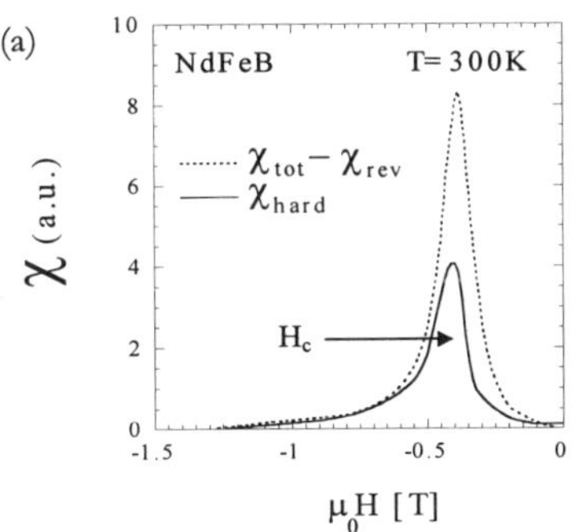

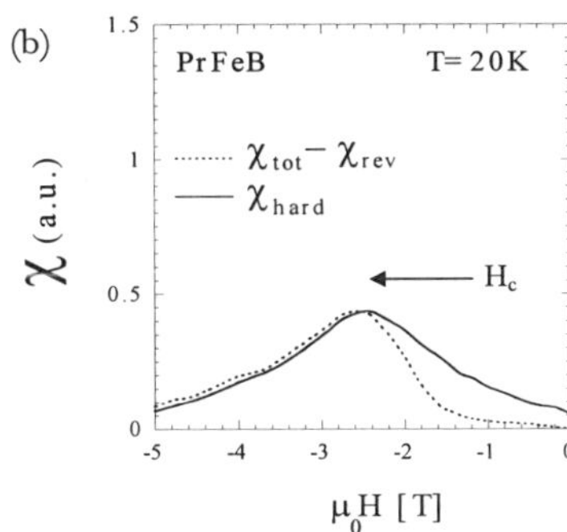

Figure 10: Comparison of χ_{hard} and $\chi_{tot}-\chi_{rev}$ in NdFeB at 300K (a) and PrFeB at 20K (b) . H_c characterizes the reversal of the hard phase in nanocomposites.

5.3 Remanence enhancement

It is generally accepted that the large remanent magnetization in NdFeB nanocomposites results from the large remanence-enhancement (up to $0.75.M_s$) due to the soft phase which is exchange coupled to the hard phase. From XMCD measurements, the ratio, r, of remanence to saturation, in the $Nd_2Fe_{14}B$ hard grains, reaches 0.6 at 300K . For non-interacting uniaxial grains, the theoretical value of r is 0.5. This value of r, larger than 0.5, is a direct experimental evidence for remanence-enhancement in the hard phase. Thus remanence-enhancement in these materials is not restricted to the soft phase only.

The remanent magnetization enhancement in the nanocomposite systems is of the same order of magnitude as in single-phase NdFeB nanostructured magnets which consist of exchange coupled hard grains only. In general, one may expect that each constituent grain is exchange-coupled to 12 other grains. In the nanostructured magnets, a given hard grains is coupled to 12 other hard grains ; in the soft/hard nanocomposites, a given hard grain is coupled approximately to 6 hard grains and 6 soft grains. From the higher number of hard neighbors, one would *a priori* expect that the remanence-enhancement should be higher in the nanostructured systems. However, remanence-enhancement is due to the moments of atoms close to the grain surface which tend to align to the moments in a neighboring grain. In each grain, the easy magnetization direction differs from the one in the others. A certain competition may exist when a grain is coupled to more than one other grain. Then, the remanence-enhancement cannot fully develop. It may be expected that this frustration effect is more important in nanostructured magnets than in nanocomposites, due to the higher number of neighboring hard neighbors. The similar remanence enhancement values observed in both nanostructured magnets and nanocomposites may thus be ascribed to the fact that frustration compensates for the higher number of hard neighbors.

Acknowledgement

The support of the European Union within the Brite-Euram project EMERGE (n° BRPR-CT95-0097) is gratefully acknowledged. The authors are thankful to S. Pascarelli, T. Neisius (ESRF, ID 24 Beamline) and A. Rogalev (ESRF, ID12A Beamline) for their assistance during the XMCD experiments.

REFERENCES

1. R. Coehoorn, D.B. de Mooij, J.P.W.B. Duchateau, K.H.J. Bushow, J. de Phys **49**-C8 , 669 (1988); R. Coehoorn, D.B. de Mooij, C. de Waard, J. Magn. Magn. Mater. **80**, 101 (1989).
2. A. Manaf, R.A. Buckley, H.A. Davies, J. Magn.Magn.Mater. **128**, 302 (1993); H. A. Davies, J. Magn. Magn. Mater. **157/158**, 11 (1996).
3. J. Ding, Y. Liu, P.G. McCormick, R. Street, J. Magn. Magn. Mater **123**, L239 (1993).
4. K. O'Donnell, C. Kuhrt, J.M.D. Coey, J. Appl. Phys. **76** (10) , 7068 (1994).
5. P. Smith, J.F. O'Sullivan, J.M.D. Coey, *in* Proceedings of the 15th International Symposium on Rare-Earth Magnets and their Applications, Dresden 1998, Vol.1, Werkstoff-Informationgesellschaft, Francfort, L. Schutz & K.H. Müller editors, 205 (1998).
6. V. Archambault, D. Peré, these Proceedings.
7. J. Bernardi *et al.* , submitted to J. Magn. Magn. Mater. (1999).
8. S. David, R. Raphel, D. Givord (unpublished) .
9. Y. Khan, Phys. Stat. Sol. (a) **122**, K1 (1990).
10. T. Hinomura, S. Nasu, H. Kanekiyo, S. Hirosawa, J. Japan. Inst. Metals, **61**-n°2, 184 (1997)
11. H.E. Kissinger, Analyt. Chem. **29**, 1702 (1957).
12. S. David, PhD. Thesis, Université Joseph Fourier, Grenoble (June 1999).
13. The invaluable help of J. Bernardi and J. Fidler from IATP at TUWien in TEM characterization is gratefully acknowledged.
14. J. Bernardi, S. David, G.F. Soto, J. Fidler, D. Givord, J. Appl. Phys. **85** (8) , 5905 (1999).
15. Z.W. Li, X.Z. Zhou, A.H. Morrish, Hyperfine Interactions **72**, 111 (1992).
16. S. Hirosawa,..;
17. N. Sano, T. Tomida, S. Hirosawa, M. Hueara, H. Kanekiyo, Mater. Sci.& Eng. **A250**, 146 (1998).
18. K. Suzuki, J.M. Cadogan, M. Uehara, S. Hirosawa, H. Kanekiyo, J. Appl. Phys., **85** (8) , 5914 (1999)
19. L. Pareti, F. Bolzoni, M. Solzi, J. of Less-Com. Met. **132**, L5 (1987).
20. K. Hono, J.-L. Li, Y. Ueki, A. Inoue, T. Sakurai, Appl. Surf. Science **67**, 398 (1993).
21. S. David, D. Givord, J. Alloys compd. **281**, 6 (1998)
22. J. Stöhr, Y.Wu, New Directions in Research with3rd Generation Soft X-Ray Synchrotron Radiation Sources, NATO Advanced Study Institute (Kluwer Academic Publishers, Dordrecht, 1993).
23. B.T. Thole, P. Carra, F. Sette, G. Van der Laan, Phys. Rev. Lett. **68**, 1943 (1992).
24. S. Pizzini, M. Bonfim, F. Baudelet, H. Tolentino, A. San Miguel, K. MacKay, C. Malgrange, H. Hagelstein, J. Synchrotr. Rad. **5**, 1298 (1998).
25. E. Kneller and R. Hawig, IEEE Trans. Mag. 27, 3588 (1991)
26. S. David, K. MacKay, M. Bonfim, S. Pizzini, A. Fontaine, D. Givord, (Submitted)

RESIDUAL AMORPHOUS PHASE IN α Fe/NdFeB NANOCOMPOSITE HARD MAGNETS

M. HAMANO*, M. YAMASAKI*, H. MIZUGUCHI*, T. KOBAYASHI**,
H. YAMAMOTO***, A.INOUE****
*AMC Div., Toda Kogyo Corp., Otake, Hiroshima, 739-0652 Japan
**Dept. of Phys., Siga Univ. of Medic. Sci., Shiga, 520-2192 Japan
***Dept. of Electr. Eng., Meiji Univ., Kawasaki, 214-8571 Japan
****Inst. for Mater. Res., Tohoku Univ.,Sendai, 980-8577 Japan

ABSTRACT

A residual amorphous phase can be observed when Nd content is less than 8 at% in a α Fe/NdFeB nanocomposite with various additional elements which was prepared by melt-spinning and heat treatment. The surface velocity of the single-roller melt-spinner for optimal coercive force (Hcj) was 10-15 m/s. By adding Nb, Hcj increases monotonically. Nearly homogeneous crystal size of around 5 nm is observed in a sample composition of $Nd_8Fe_{76}Co_8Nb_2B_6$. With Nd=8 at%, the maximum Hcj is as large as 575kA/m (7.22kOe) at Nb=2.5 at%, and the best (BH)max is 154kJ/m^3 (19.4MGOe) at Nb=0.5 at% in optimally heat-treated and pulse-current magnetized samples. By high-resolution transmission electron microscopy, a residual amorphous phase is observed in all alloy ribbons with Nd=8 at%, which shows soft magnetic properties and is thought to act as a crystal growth inhibitor. To estimate the existing ratio of the amorphous phase and crystallized phases, the Mössbauer effect has been studied. It was found that the ratio of the crystallized $Nd_2Fe_{14}B_1$ phase increases with increasing Nb content and that several wt% of the residual amorphous phase is certified. Epoxy-resin bonded magnets were prepared by compression molding, and revealed that, comparing to the MQP bonded magnet using conventional MQP-B powder, their (BH)max are in the same order of 72kJ/m^3 (9MGOe), and magnetization at lower magnetic field than 800kA/m (10kOe) is superior. Initial flux loss at 100 ℃ for 1.8ks, however, is in the range of 3-8% depending on Nb contents.

INTRODUCTION

As a next generation magnet, nanocomposite alloys, that is so-called exchange spring magnets, have been interested since they were reported firstly [1]. Especially Fe_3B/NdFeB and α Fe/NdFeB alloys have been investigated earnestly by many researchers worldwidely [2-11]. Recently, amorphous-phase remaining α Fe/NdFeB alloys with Nd content of less than 8 at% were found to show a good hard magnetic properties by Inoue et al.[12-15]. The role of amorphous phase is analyzed and revealed to be a grain growth inhibitor.

In this paper, the effect of adding small amount of elements to the fundamental composition of the above Nd-lean Nd-Fe-B ternary alloys on coercivity increment is reported and magnetic properties of melt-spun and heat-treated alloy ribbons and their bonded magnets are discussed.

EXPERIMENT

After various experiments as to additional elements such as Ti, V, Cr, Zr, Al, Si and Ga, we have made choice of the composition of $Nd_8Fe_{bal}Co_8Nb_xCu_yB_z$, where x=0-2.5, y=0-0.5, z=6-7. The ingots were prepared by arc-melting the constituent elements in an argon atmosphere. Melt-spun alloy ribbons were obtained by using a single-roller melt-spinner having a copper roller of 200 mm in diameter and a quartz tube with an orifice diameter of about 0.5 mm. The surface velocity of the roller was determined to be in the range of 10-15 m/s for getting higher Hcj. Resulting ribbons were of 30-55 μ m in thickness and about 1 mm in width. The amorphization of ribbons was examined by X-ray diffractometry (XRD) and transmission electron microscopy (TEM). Heat-treatment of ribbons was made in vacuum in sealed quartz tubes in a temperature range of 700-780℃ for 180-300 s. The heat-treatment mode was that a quartz tube was inserted in a furnace already heated up and after a certain

Mat. Res. Soc. Symp. Proc. Vol. 577 © 1999 Materials Research Society

duration the tube was picked out and dropped into a bucket filled with water. The crystallization behavior was examined by differential scanning calorimetry (DSC), XRD and high-resolution TEM (HR-TEM).

Magnetization curves of ribbons of about 10 mm in length were measured with a vibrating sample magnetometer (VSM) in field up to 1.6 MA/m (20 kOe) at room temperature. No correction of the demagnetization field was made. In some cases, magnetization was done using a pulse current field up to 4.8 MA/m (60 kOe). The Mössbauer study were carried out in the conventional transmission mode at room temperature with a ^{57}Co source of about 740MBq in order to estimate the existing ratio of the amorphous phase and two crystallized phases, α Fe and $Nd_2Fe_{14}B_1$, in some powdered ribbons of melt-spun and heat-treated alloys. Observed Mössbauer spectra were analyzed with the non-linear least-square method.

Bonded magnets were made from powders of the above ribbons pulverized in the range of 53-500 μ m in size using a ball-milling machine. An epoxy resin of 1.5 wt% was mixed with powders and resulting compounds were compression molded with pressure of 980MPa. After curing treatment at 150 °C for 3.6 ks, bonded magnets with a size of 10 mm in diameter and 7 mm in height, that is nearly equal to a permeance coefficient Pc=2, were obtained. For the sake of comparison, MQP bonded magnets using MQP-B powder available from the MQI Corp. were prepared using the same procedure. Magnetic properties of bonded magnets magnetized with a pulse current in field up to 4.8 MA/m (60 kOe) were measured with a B-H curve tracer in field up to 1.6 MA/m (20 kOe) at room temperature or at higher temperatures for estimating temperature-coefficients of Br and Hcj. Initial flux losses were measured with a flux meter at room temperature after magnets had been exposed at 100 °C for 3.6 ks in air. Magnetization at various magnetization fields were examined by controlling magnetizing pulse currents and by measuring open flux values at room temperature.

RESULTS

On melt-spun ribbons

First of all, the dependence of Hcj on Nd content in melt-spun and heat-treated $Nd_wFe_{bal}Co_8Nb_2B_6$ alloy ribbons, where w=6.0-8.0, is shown in Table I. Values of Hcj are top data at each Nd content, which depend on preparing conditions shown in the table, especially on the roller-surface velocity (Vsurf). It should be emphasized that, in the alloys prepared at such a velocity, XRD patterns from α Fe and $Nd_2Fe_{14}B_1$ crystals already appear even in an as-quenched (melt-spun) state. Examples of the XRD patterns are shown in Fig.1. A hysteresis loop of a $Nd_8Fe_{77.5}Co_8Nb_{0.5}B_6$ melt-spun and heat-treated alloy ribbon which has the best (BH)max of 154 kJ/m^3 (19.4 MGOe) in our study is shown in Fig. 2.

Table I. Top data on Hcj of melt spun $Nd_wFe_{bal}Co_8Nb_2B_6$ alloys (w=6.0～8.0at%).

Nd (at%)	HcJ (kA/m)	Br (T)	J1.2MA/m (T)	$(BH)_{max}$ (kJ/m^3)	Vsurf. (m/s)	Heat Treat. (℃-sec)
8.0	512	1.12	1.36	143	15	760-180
7.5	438	1.09	1.33	122	15	740-180
7.0	403	1.09	1.35	116	15	760-180
6.5	331	1.11	1.40	103	20	760-180
6.0	341	1.19	1.50	111	20	760-180

Pulse magnetized: 4.8MA/m

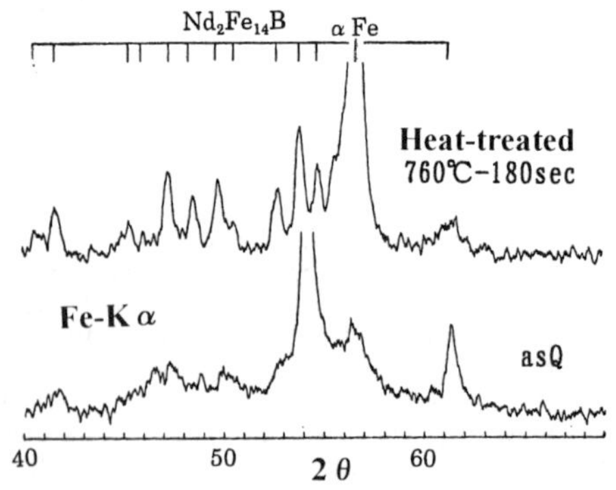

Fig. 1. X-ray diffraction patterns of a melt spun $Nd_8Fe_{76}Co_8Nb_2B_6$ ribbon.

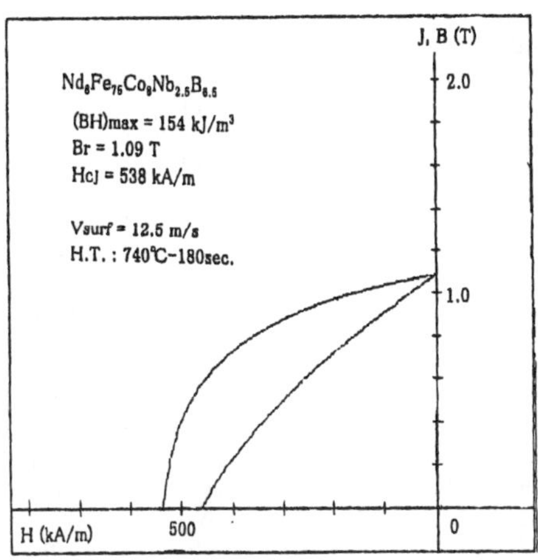

Fig. 2. A demagnetization curve of a melt spun $Nd_8Fe_{75}Co_8Nb_{2.5}B_{6.5}$ ribbon. (Max Happl.: 4.8MA/m)

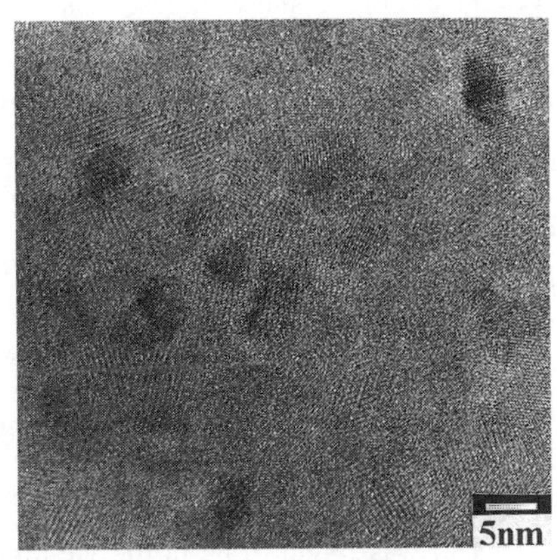

Fig. 3. Bright-field electron microscopy of a $Nd_8Fe_{76}Co_8Nb_2B_6$ melt-spun and heat-treated alloy.

Fig. 4. The electron-beam diffraction pattern of a $Nd_8Fe_{76}Co_8Nb_2B_6$ melt-spun and heat-treated alloy

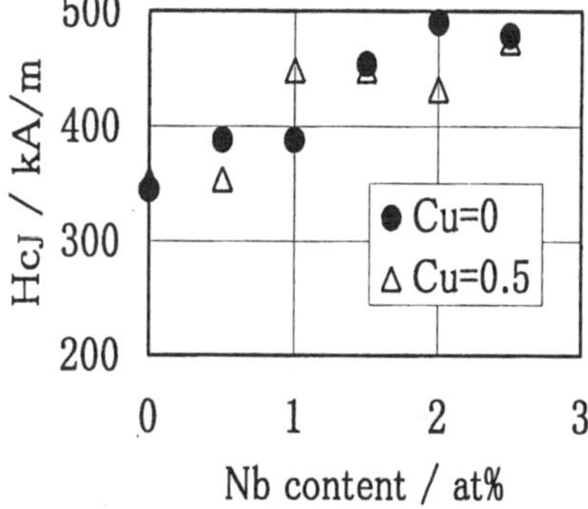

Fig. 5. The Nb content dependence of Hcj in $Nd_8Fe_{bal}Co_8Nb_xCu_yB_6$ alloy ribbons. (Max Happl.: 1.2MA/m)

A photograph of the finest microstructure that was taken using the HR-TEM on a $Nd_8Fe_{76}Co_8Nb_2B_6$ melt-spun and heat-treated alloy is shown in Fig. 3. It can be seen that crystal sizes after heat-treatment is almost uniform of being about 5 nm. Although crystals are not identified in the figure, the XRD analysis shows two patterns from α Fe and $Nd_2Fe_{14}B_1$ phases. The average crystal size in the heat-treated alloys studied was found to be ranging from 5 to 50 nm in diameter. It should be noted that these fine structure may be realized with the choice of optimum alloy compositions and heat-treatment conditions. In Fig. 4, an electron-beam diffraction pattern is shown, which proves the coexistence of the amorphous phase with crystallized α Fe and $Nd_2Fe_{14}B_1$ phases. From these data, it can be said that, in a alloy with a residual amorphous phase, crystal sizes are kept small enough even after heat-treatment, and as a result the stronger exchange interaction between soft and hard phases can be realized, as is predicted by the micromagnetic theory [16].

Fixing Nd content to 8 at%, effects of Nb, Cu and B contents to Hcj have been investigated. Fig. 5 shows the Nb content dependence of Hcj. As Nb increases, Hcj increases monotonically. By analogizing to soft magnetic materials of Fe-Nb-Cu-Si-B Finemet alloys which were prepared as to be fully amorphous, as reported by Hono et al.[17], Nb atoms are observed to be enriched in the residual amorphous phase after optimum heat-treatment and as a result very fine crystal grains are realized. In the present alloys, therefore, the residual amorphous phase with enriched Nb atoms is thought to act as a crystal growth inhibitor as well, referring to the fundamental study of amorphous-phase remaining Nd-Fe-B ternary alloys reported by Inoue et al.[12,13]. Thus, Hcj increases with increasing Nb content according to the above referred micromagnetic theory [16]. For example, the best Hcj attained in this study is as large as 575 kA/m (7.22 kOe) in a $Nd_8Fe_{bal}Co_8Nb_{2.5}B_{6.5}$ alloy with the largest Nb content.

Fig. 6 shows the Cu content dependence of Hcj. The role of Cu in magnetic properties is not clear in our data. However, Zhang et al.[18] have analyzed solute Cu clusters in the Fe-Zr-Cu-B melt-spun alloy which were formed at the first stage of crystallizing heat-treatment, by means of a 3-dimensional atom probe equipment, and reported that Cu clusters should play as a inhomogeneous nucleation site of a semi-stable crystal phase. The behavior of Cu in our alloys, therefore, should be worth to examine in further detail. The optimum B content for higher Hcj was found to be in the vicinity of 6.5 at% in these alloys as shown in Table Ⅱ. Concerning optimization of melt-spinning conditions, examples of the roller-surface velocity dependence of Hcj are shown in Fig. 7. In as-spun state, the optimum velocity is 10 m/s. In as heat-treated state, however, the optimum velocity moves to higher, that is 15-20 m/s, and values of Hcj increases a little bit comparing to those of as-spun state. This suggests that by heat-treatment crystal sizes may become more uniform but do not grow so large in these alloys. Fig. 8 shows an example of the heat-treating temperature dependence of Hcj. The optimum temperature range is found to be wide enough fortunately.

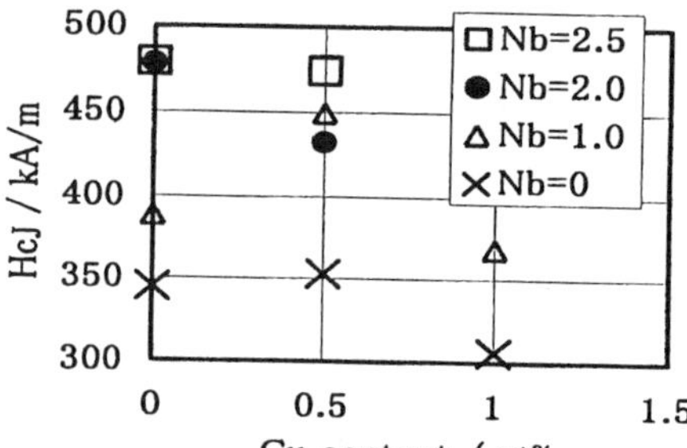

Fig. 6. The Cu content dependence of Hcj in $Nd_8Fe_{bal}Co_8Nb_xCu_yB_6$ alloy ribbons. (Max Happl.: 1.2MA/m)

Table Ⅱ. The B content dependence of Hcj in $Nd_8Fe_{bal}Co_8Nb_{1.5}B_z$ (z=6~7) alloy ribbons. (Max Happl.: 1.2MA/m)

B (at%)	HcJ (kA/m)	Br (T)	(BH)max (kJ/m^3)
6.00	481	0.97	111
6.25	470	1.01	120
6.50	505	1.01	127
7.00	513	0.96	116

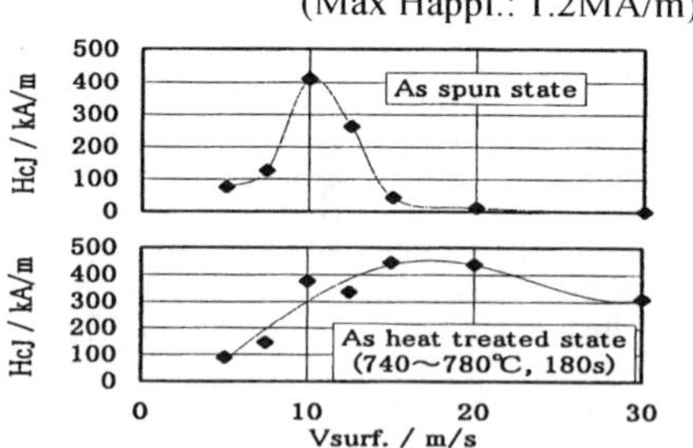

Fig. 7. The roller-surface velocity dependence of Hcj in $Nd_8Fe_{76.5}Co_8Nb_1Cu_{0.5}B_6$ alloy ribbons. (Max Happl.: 1.2MA/m)

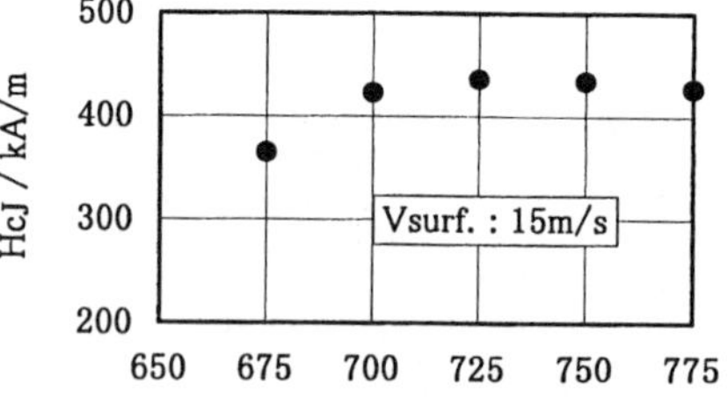

Fig. 8. The heat treating temperature dependence of Hcj in $Nd_8Fe_{76.5}Co_8Nb_1Cu_{0.5}B_6$ alloy ribbons. (Max Happl.: 1.2MA/m)

Mössbauer spectroscopic study

Mössbauer study was made on powdered alloy samples of $Nd_8Fe_{77.5}Co_8B_{6.5}$ (M1), $Nd_8Fe_{76}Co_8Nb_{1.5}B_{6.5}$ (M2), $Nd_8Fe_{75.5}Co_8Nb_{1.5}Cu_{0.5}B_{6.5}$ (M3), and $Nd_8Fe_{75}Co_8Nb_{2.5}B_{6.5}$ (M4) in order to know the existing ratio of the amorphous phase and two crystallized phases, α Fe and $Nd_2Fe_{14}B_1$. All samples had been melt-spun and heat-treated in the same way as to have optimal Hcj. For the sake of comparison, just an as-quenched (melt-spun) alloy powder with the same composition as M2 were prepared with the roller-surface velocity of 30 m/s, which shows only amorphous-like pattern by XRD.

The Mössbauer spectrum of the as-quenched alloy can be fitted satisfactorily to a superposition of two components with positive and negative isomer-shifts and the hyperfine field of which are broadly distributed as is generally seen in an amorphous materials. By analyzing the spectra of the heat-treated alloys M1-M4 , it is found that some components with a lower hyperfine field which should not be derived from both α Fe and $Nd_2Fe_{14}B_1$ crystal phases are existing. In Fig. 9, an example of the analyzed result for the alloy M2 is shown. Furthermore, these distribution of the components are unlike to those of the above mentioned typical amorphous phase. It is then considered that there exist some discrete intermediate phases whose structures are gradually changing from the crystal state to the typical amorphous state. In this paper, however, these intermediate phases should be certified as the amorphous phase in the sense of the XRD or electron beam diffraction. In Table III, Mössbauer parameters of the alloys M1-M4 are shown, in which the existing ratio (wt%) of the intermediate phases (A1-A3) are summarized and listed as the single amorphous phase for the sake of convenience. The detailed discussion on each intermediate phase will be done in elsewhere [19].

From the data on existing ratio, it can be said that the ratio of the $Nd_2Fe_{14}B_1$ phase increases with increasing Nb content in spite of no change in the Nd content. These data is in good agreement to the fact that Hcj increases with increasing Nb content. Further investigation on the mechanism why Nb can increase $Nd_2Fe_{14}B_1$ phase is necessary.

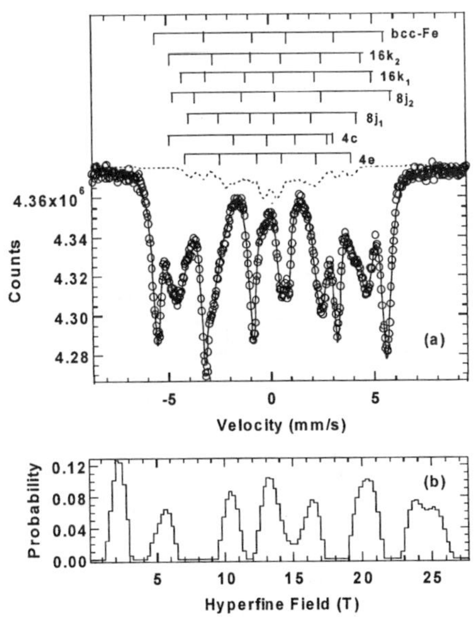

Fig. 9. The analyzed Mössbauer spectrum for the $Nd_8Fe_{76}Co_8Nb_{1.5}B_{6.5}$ alloy M2.

Table III. Mössbauer parameters of the alloys M1-M4 at room temperature.

Alloy	Phase	Site	d (mm/s)	D (mm/s)	Hyperfine Field (T)	Relative Intensity (%)	Existing Ratio(wt%)
M1	bcc-Fe		0.00	0.00	34.3	45.5	38.7
	$Nd_2Fe_{14}B$	$16k_2$	−0.13	−0.02	29.3	13.2	54.3
		$16k_1$	−0.09	0.45	29.0	13.2	
		$8j_2$	0.05	0.54	33.6	6.6	
		$8j_1$	0.00	0.22	26.5	6.6	
		4c	−0.05	−0.44	25.9	3.3	
		4e	−0.19	0.02	26.3	3.3	
	A1		0.03	0.13	24.8	1.4	7.0
	A2		–	–	–	–	
	A3		0.00	−0.05	–	6.8	
M2	bcc-Fe		0.00	0.00	34.4	38.1	31.8
	$Nd_2Fe_{14}B$	$16k_2$	−0.10	−0.04	28.6	14.7	59.3
		$16k_1$	−0.03	0.44	28.4	14.7	
		$8j_2$	0.07	0.59	33.3	7.3	
		$8j_1$	−0.00	0.24	25.1	7.3	
		4c	−0.22	−0.76	24.7	3.7	
		4e	−0.16	0.03	24.8	3.7	
	A1		0.12	0.55	22.1	5.4	8.9
	A2		0.15	0.07	13.5	2.8	
	A3		−0.00	0.02	–	2.5	
M3	bcc-Fe		0.00	0.00	34.6	29.5	24.3
	$Nd_2Fe_{14}B$	$16k_2$	−0.12	−0.01	29.1	16	63.7
		$16k_1$	−0.06	0.47	29.0	16	
		$8j_2$	0.12	0.42	33.7	8	
		$8j_1$	−0.04	0.23	26.5	8	
		4c	−0.07	−0.83	25.7	4	
		4e	−0.11	0.06	26.1	4	
	A1		−0.05	0.21	24.5	7.4	12.0
	A2		0.33	−0.39	15.0	4.4	
	A3		0.04	−0.03	–	2.8	
M4	bcc-Fe		0.00	0.00	34.5	21	16.9
	$Nd_2Fe_{14}B$	$16k_2$	−0.09	0.00	28.3	18.2	70.8
		$16k_1$	−0.03	0.50	28.3	18.2	
		$8j_2$	0.06	0.61	33.2	9.1	
		$8j_1$	−0.01	0.24	24.8	9.1	
		4c	−0.25	−1.11	25.5	4.5	
		4e	−0.14	0.18	25.1	4.5	
	A1		0.06	0.41	21.4	7.2	12.3
	A2		−0.01	0.10	15.0	4.6	
	A3		0.01	0.04	–	3.5	

On bonded magnets

Magnetic properties of bonded magnets made from the above alloy ribbons measured at room temperature are listed in Table IV. Typical demagnetization curves of the best magnet studied and the MQP bonded magnet are shown in Fig. 10. For practical use of bonded magnets, most important three factors are a) magnetic properties such as (BH)max at room temperature, b) irreversible flux losses at higher temperatures, and c) magnetization at lower applied field. Therefore, we have set the corresponding aim as; a) larger than the MQP bonded magnet, b) less than 5% loss in initial flux loss at 100℃ for 3.6ks, and c) more than 5% larger in open flux value comparing to the MQP bonded magnet at magnetizing field lower than 800kA/m (10kOe).

Fig. 11 shows an example of data on the initial flux loss. Some samples with higher Nb content show flux losses of less than 5%. However, they are still larger than that of the MQP bonded magnet. Although the time depending loss from 3.6ks to 3.6Ms of nanocomposite type alloys is generally smaller than that of the MQP alloy [20], it is desirable to lower the initial loss as small as possible. Efforts on this point have being made in our laboratory. For example, an approach on the basis of the micromagnetic theory is conducting by measuring relation between variation of working points field Hd with temperatures and initial flux losses.

Table IV. Magnetic properties of some bonded magnets. (Pulse magnetized at 4.8MA/m)

Alloy composition	Br (T)	HcJ (kA/m)	(BH)max (kJ/m^3)	Hk/HcJ (%)
$Nd_6Fe_{bal}Co_8Nb_2B_6$	0.756	297	45.8	15.9
$Nd_7Fe_{bal}Co_8Nb_2B_6$	0.720	368	52.4	18.4
$Nd_8Fe_{bal}Co_8Nb_2B_6$	0.765	465	64.1	18.0
$Nd_8Fe_{bal}Co_8Nb_{0.5}B_{6.5}$	0.756	347	55.9	19.2
$Nd_8Fe_{bal}Co_8Nb_{1.0}B_{6.5}$	0.777	426	68.4	21.7
$Nd_8Fe_{bal}Co_8Nb_{1.5}B_{6.5}$	0.776	487	74.7	23.6
$Nd_8Fe_{bal}Co_8Nb_{2.0}B_{6.5}$	0.731	473	60.6	17.9
$Nd_8Fe_{bal}Co_8Nb_{2.5}B_{6.5}$	0.680	483	59.0	21.5
$Nd_8Fe_{bal}Co_8Nb_{1.0}Cu_{0.5}B_6$	0.765	441	71.9	25.3
$Nd_8Fe_{bal}Co_8Nb_{2.0}Cu_{0.5}B_6$	0.810	386	73.2	25.2
$Nd_8Fe_{bal}Co_8Nb_{1.0}Cu_{0.5}B_{6.5}$	0.728	356	52.0	18.2
$Nd_8Fe_{bal}Co_8Nb_{2.0}Cu_{0.5}B_{6.5}$	0.717	398	50.9	16.1
MQP-B	0.690	690	72.6	26.2

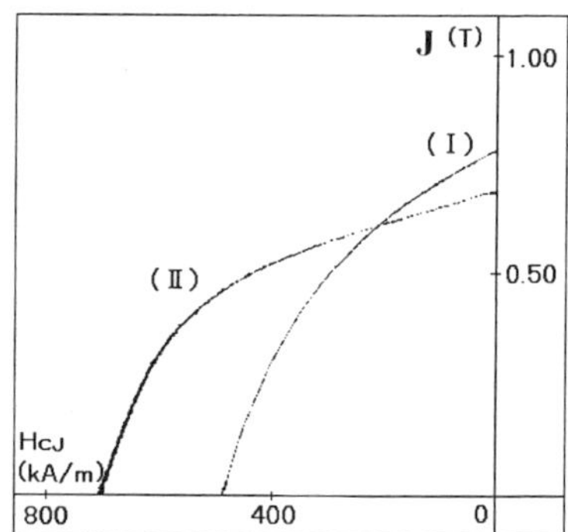

Fig. 10. Demagnetization curves of the $Nd_8Fe_{76}Co_8Nb_{1.5}B_{6.5}$ (I) and MQP-B (II) bonded magnets. (Pulse magnetized at 4.8MA/m)

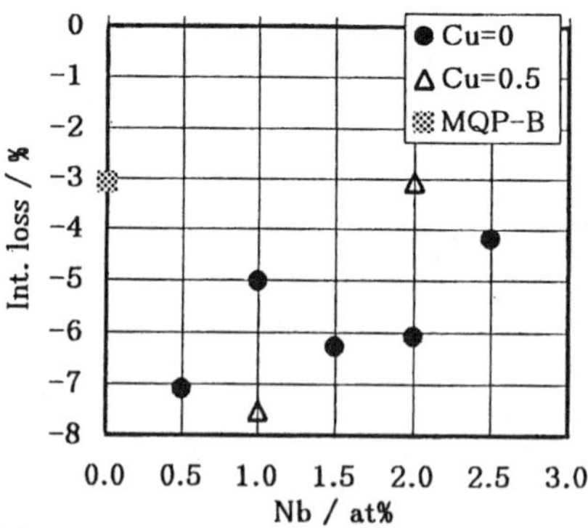

Fig. 11. The initial flux loss of $Nd_8Fe_{bal}Co_8Nb_xB_{6.5}$ bonded magnets (Pc=2), at 100℃ for 3.6ks.

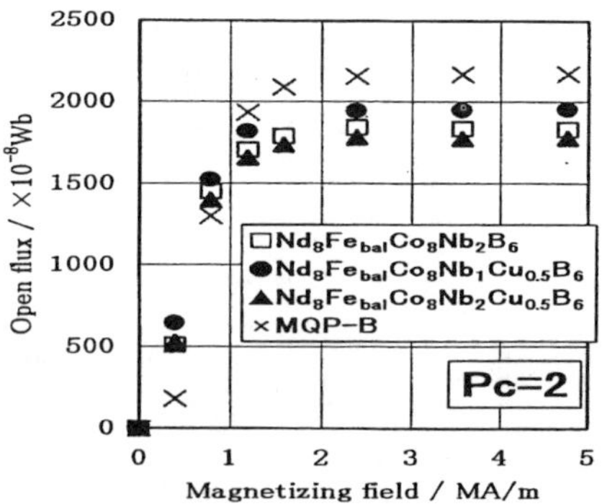

Fig. 12. Magnetization (Open flux value) of some bonded magnets.

Table V. Temperature coefficients, α (Br) and β (Hcj), of some bonded magnets.

Alloy composition	α %/℃	β %/℃	Int.-loss %	(BH)max kJ/m^3	Br T	HcJ kA/m
$Nd_8Fe_{bal}Co_8Nb_2B_6$	-0.06	-0.30	-4.9	64.1	0.765	465
$Nd_8Fe_{bal}Co_8Nb_{1.5}B_{6.5}$	-0.04	-0.42	-6.3	74.7	0.776	487
$Nd_8Fe_{bal}Co_8Nb_2B_{6.5}$	-0.08	-0.36	-6.1	60.6	0.731	473
$Nd_8Fe_{bal}Co_8Nb_{2.5}B_{6.5}$	-0.07	-0.39	-4.2	59.0	0.680	483
MQP-B	-0.08	-0.33	-3.1	72.6	0.690	690

Table V shows temperature coefficients, α (Br) and β (Hcj), measured using the B-H curve tracer. It was found that α is slightly smaller and β is a little larger than those of the MQP bonded magnet. Roughly speaking, they are in the same level in both types of magnets. Fig. 12 shows magnetization of some bonded magnets studied. Comparing to the MQP bonded magnet, the values of open flux of nanocomposite alloys exceed more than 5% at applied field less than 800 kA/m (10 kOe). This may be mostly attributed to lower coercivity of nanocomposite alloys. This character is, however, very important and useful when magnets are multi-polarly magnetized or magnetization is done after magnets are assembled.

Further investigations of attaining balanced magnetic properties among higher (BH)max, lower initial flux losses, and better magnetization by means of selecting appropriate alloy compositions are necessary for actual application of the present type of α Fe/NdFeB nanocomposite magnets.

CONCLUSIONS

Magnetic properties of melt-spun and heat-treated alloy ribbons and their bonded magnets with composition of $Nd_8Fe_{bal}Co_8Nb_xCu_yB_z$, where x=0-2.5, y=0-0.5, z=6-7, have been studied. By selecting the heat-treatment mode as rapid heating and rapid cooling, a residual amorphous phase coexists with α Fe and NdFeB crystalline phases. Fine and uniform crystal grains of about 5 nm in diameter is observed in the ribbon of $Nd_8Fe_{bal}Co_8Nb_2B_6$, and this is attributed to the residual amorphous phase which may act as a grain growth inhibitor as well, as have been discussed in referenced papers. The Hcj increases monotonically with increasing Nb content. The best magnetic properties of ribbons are (BH)max=154 kJ/m^3 (19.4 MGOe), Br=1.16 T, and Hcj=405 kA/m, which has prepared with a roller-surface velocity of 15 m/s and an annealing condition at 740℃ for 180 s. The highest Hcj attained is 575 kA/m (7.22 kOe) in another ribbon.

Mössbauer study have been made on some ribbon powders in order to estimate the existing ratio of the amorphous phase and the crystallized α Fe and $Nd_2Fe_{14}B_1$ phases. It is found that the ratio of the NdFeB phase increases with increasing the Nd content and that the residual amorphous phase should be thought as some discrete intermediate phases whose structure are gradually changing from the crystal state to the real amorphous state.

On epoxy-bonded magnets studied, the best magnetic properties are (BH)max=74.7 kJ/m^3 (9.39 MGOe), Br=0.776 T, and Hcj=487 kA/m, whereas those of the MQP bonded magnet is 72.6 kJ/m^3 (9.12 MGOe), 0.690 T, and 690 kA/m, respectively. However, initial flux losses at 100℃ for 3.6ks are a little bit larger than that of the MQP bonded magnet. Magnetization at lower applied fields is better comparing to the MQP bonded magnet.

REFERENCES

[1] K.H.J. Buschow, D.B. DeMooij and R. Coehoorn: J. Less-Common Met., **145**, 601 (1988).

[2] E.F. Kneller and R. Hawig: IEEE Trans. Magn., **27-4**, 3588 (1991).

[3] T. Schrefl,H. Kronmüller and J. Fidler: J. Magn. Magn. Mater., **127**, L273 (1993).

[4] R. Skomsky and J.M.D. Coey: Phys. Rev., B **48**, 15812 (1993).

[5] A. Manaf, R.A. Buckly and H.A. Davis: J. Magn. Magn. Mater., **128**, 302 (1993).

[6] S. Hirosawa, H. Kanekiyo and M. Uehara: J. Appl. Phys., **73**, 6488 (1993).

[7] R. Skomsky: J. Appl. Phys., **76**, 7059 (1993).

[8] J.M. Yao, T.S. Chin and S.K. Chen: J. Appl. Phys., **76**, 7071 (1994).

[9] L. Withanawasam, A.S. Murthy, G.C. Hadjipanayis and R.F. Krause: J. Appl. Phys., **78**, 7065 (1994).

[10] V. Panchanathan: IEEE Trans. Magn., **31-6**, 3605 (1995).

[11] M. Hamano, M. Yamasaki, H. Mizuguchi, H. Yamamoto and A. Inoue: Proc. 15th Int'l Workshop on Rare Earth Magnets, Dresden, Germany. 30 Aug.-3 Sep., Vol.**1**, 199 (1998).

[12] A. Inoue, A. Takeuchi, A. Makino and T. Masumoto: IEEE Trans. Magn., **31-6**, 3626 (1995).

[13] A. Inoue, A. Takeuchi, A. Makino and T. Masumoto: Mater. Trans., JIM, **36-7**, 962 (1995).

[14] A. Inoue, A. Takeuchi, A. Makino and T. Masumoto: Sci. Rep. RITU, A **42-1**, 143 (1996).

[15] A. Takeuchi, A. Inoue and A. Makino: Mater. Sci.and Engin., A **226-228**, 458 (1997).

[16] J. Kuma, N. Kitajima, Y. Kanai and H. Fukunaga: J. Appl. Phys., 83, 6623 (1998).

[17] K. Hono, J.L. Li, Y. Ueki, A, Inoue and T. Sakurai: Appl. Surf. Sci., **67**, 398 (1993).

[18] Y. Zhang, K. Hono, A. Inoue and T. Sakurai: Appl. Phys. Lett., **69**, 2128 (1996).

[19] T. Kobayashi, M. Hamano and M. Yamasaki: to be presented at 44th Annual Conf. Magn. Magn. Mater., San Jose, CA, Nov. 15-18, (1999).

[20] H. Fukunaga, S. Hayashida, Y. Kanai and F. Yamashita: Proc. of Magnetism of Nanostructured Phase (MNP) Conf., San Sebastian, Spain, September 1-6, (1998).

Part IV

Nanoscale Hard Magnetism II

HARD MAGNETIC NANOPARTICLES AND NANOCOMPOSITES

ANIT GIRI, KRISHNA CHOWDARY, SARA A. MAJETICH*,
Department of Physics, Carnegie Mellon University, Pittsburgh, PA 15213
* Author to whom correspondence should be addressed. email: sm70@andrew.cmu.edu.

ABSTRACT

Many high performance permanent magnets are nanostructured materials. The magnetic properties of nanoparticles are discussed in terms of characteristic length scales, including the maximum monodomain size and the exchange length. Experimental results for ball milled SmCo5 nanoparticles are presented, showing deviations from idealized behavior. Because of the short exchange length, this can be understood in terms of independent nucleation of reverse domains in grains within larger particles. With a much longer exchange length, FeCo alloy nanoparticles show reduced coercivity in a high density compact, in accordance with the random anisotropy model. The SmCo5 and FeCo nanoparticles were mixed and compacted in an attempt to make an exchange spring nanocomposite. However, significant exchange between the hard and soft phases was not observed because the sample density was too low. Processing considerations for improved co-compaction of these nanoparticles are discussed.

INTRODUCTION

Nanoparticle-based magnetic materials have potential advantages over existing materials in numerous applications. Because nanoparticles have a size dependent magnetization reversal field, or coercivity H_c [1], nanocrystalline materials can have either very hard or very soft magnetic materials [2]. Many permanent magnet materials such "Sm_2Co_{17}"[3] and Alnico9 [4] and soft magnet materials such as finemet [5] actually contain multiple phases and distinct structures on a submicron scale. Unfortunately it is difficult to extrapolate from theories for isolated particles, such as the Stoner-Wohlfarth coherent rotation model [6], to the exact behavior of nanocrystalline solids where particles or grains are coupled.

To expand and exploit these capabilities, it is important to experimentally determine how the structure and chemical composition affect the coercivity , H_c, and the energy product, $(BH)_{max}$. The potential uses of nanocomposites as either hard or soft magnets will depend on the type and extent of interactions between particles or grains. Here we consider the particular case of nanocomposites made from the compaction of nanoparticles, and the properties of the nanoparticles as well.

Coercivity in Nanoparticles

The coercivity of a perfect isolated nanoparticle can be calculated exactly, but the behavior of real nanoparticles and nanocomposites is more complex. In the Stoner-Wohlfarth model for an ellipsoidal monodomain particle of a uniaxial material, all spins within the particle are aligned, and magnetization reversal occurs by coherent rotation [6]. The maximum coercivity is given by

$$H_c = \frac{2K}{M_s} - N_{eff} M_s \ , \qquad (1)$$

where N_{eff} is the demagnetization factor, K is the magnetocrystalline anisotropy, and M_s is the saturation magnetization. The largest monodomain size d_{cr} can be estimated from the material

Mat. Res. Soc. Symp. Proc. Vol. 577 © 1999 Materials Research Society

parameters:

$$d_{cr} = \frac{72(AK)^{1/2}}{M_s^2} \ , \qquad (2)\ [7].$$

where A is the exchange stiffness. Below d_{cr}, the coercivity drops, reaching zero at the superparamagnetic diameter, d_{sp}:

$$d_{sp} = \left[\frac{6kT\ln(\tau / \tau_o)}{\pi K}\right]^{1/3} , \qquad (3)\ [8]$$

where T is the temperature, τ is the magnetization measurements time, and τ_o is the Larmor precession frequency. Here thermal fluctuations are sufficient to reverse the magnetization direction of individual particles within the measurement time, and the average magnetization of an assembly of particles is zero.

Even with the nanoparticles, there are significant deviations between theory and experiment. Brown's Paradox [9] states that the measured coercivities are always less than the theoretically predicted values. In many high performance magnets, the actual value is only 20% of the ideal [10]. In addition to ignoring particle interaction effects, most fine particle magnet theories assume perfect particles which contain no defects such as grain boundaries, no internal strains, and no chemical variations such as the presence of a surface oxide phase. The experimental values of critical sizes for monodomain and superparamagnetic behavior also deviate from the theoretical estimates. In many cases particles of hard magnet materials behave as if they are monodomain when nearly an order of magnitude above the predicted d_{cr} [11]. In very soft magnetic materials, the predicted d_{sp} can be larger than d_{cr}, indicating that the model used to derive these formulas omits important factors such as magnetostriction and surface anisotropy.

Nanocomposite Hard Magnets: The Exchange Spring

Bulk solids are necessarily too large to be monodomain, and they reverse their magnetization direction incoherently. In this process a small region with a reverse magnetization direction nucleates, forming a domain wall around it. Under a force from an external field, this region expands as the domain walls sweep through the solid until the magnetization direction is reversed everywhere. As the domain walls move, they may become pinned at low energy sites, requiring higher fields and therefore greater forces to move them further. The coercivity of hard magnets may be dominated by either high nucleation fields ($SmCo_5$ and $Nd_2Fe_{14}B$) or strong pinning sites ("Sm_2Co_{17}"), and pinning by domain walls is desirable. In soft magnets the coercivity is often pinning dominated, but pinning is undesirable.

Opportunities for nanocomposite hard magnets arise from the limitations of single phase magnets. The standard figure of merit for comparison of permanent magnets is the energy product, $(BH)_{max}$. For an ideal single phase magnet,

$$(BH)_{max} \le \mu_o M_r^2/4 \le \mu_o M_s^2/4, \qquad (4)\ [12]$$

where M_r is the remanent magnetization, or magnetization at zero field. The maximum *useable* coercivity of a permanent magnet is given by

$$H_c = M_s/2. \qquad (5)\ [12]$$

The energy product of a single phase material such as $SmCo_5$ is limited not by its coercivity, but by its low saturation magnetization M_S. While H_C is sensitive to grain or particle size and to details of the microstructure, M_S depends mainly on the magnitude of the atomic magnetic moments, and therefore the chemical composition of the material. Adding iron will increase M_S, since Fe has a larger moment than Co, but the uniaxial crystal structure must be preserved in order to retain the high coercivity. One was to do this is to make an exchange spring magnet [13, 14].

In an ideal exchange spring magnet the hard and soft phases are uniformly distributed so that the soft phase is always within a magnetic exchange length,

$$L_{ex} = \left(\frac{A}{K}\right)^{1/2}, \tag{6}$$

of the hard phase. The exchange interaction prevents the soft phase from reversing its magnetization direction at low applied fields (Fig. 1), and the presence of the soft phase raises the overall value of M_S. For optimal performance, the easy axes of the grains of the hard phase should be magnetically aligned. Maximum coercivity arises when the thickness of the soft phase is given by

$$\delta_{soft} = \pi\left(\frac{A_{soft}}{2K_{hard}}\right)^{1/2}, \qquad (7)\ [12]$$

which is on the order of 5 nm for typical materials. If the hard phase grains are uniform in size and uniformly dispersed, a volume fraction of only 10% of the far more expensive rare earth-containing hard phase is feasible.

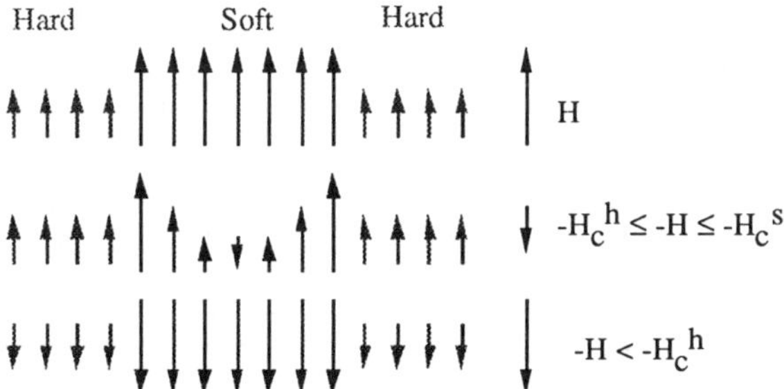

Fig. 1. Schematic of an exchange spring magnet, with a region of a soft phase surrounded by the hard phase to the right and left. Following saturation (top), application of a reverse field about the coercivity of the soft phase starts to reverse its magnetization direction, but regions exchange coupled to the hard phase require higher fields to reverse. Above the coercivity of the hard phase, all spins reverse their direction.

When materials with large and small coercivities are mixed, and the soft phase is exchange coupled to the hard phase, the exchange spring effect is anticipated. Here a small reverse field on the order of H_C^{soft} is insufficient to reverse the magnetization direction in the soft phase because it is coupled to the hard phase. If the soft phase has a large M_S, and if it's inexpensive, then the composite magnets have potential advantages over existing permanent magnet materials.

There has been considerable recent work on the growth and characterization of exchange-coupled thin films [15, 16]. This area is very interesting because clean and simple systems are ideal for studying the fundamental physics of this effect. While necessarily messier and more complex, bulk exchange spring magnets are also of interest since they would be more economical to make in large quantities.

There have been two approaches: 1) annealing an alloy until phase segregation occurs between hard and soft phases, typically $Nd_2Fe_{14}B$ and an iron-rich phase [13, 17-19], and 2) compaction of mixed hard phase and soft phase particles [20]. Each has its advantages and difficulties. Exchange coupling an dispersion of the hard and soft phases is not a problem with the former, but is a major challenge in compacted systems. Particle compaction could be used for a wider choice of materials, with possibilities for better control of the grain size and phases. Both merit further exploration. While enhancements in the energy product and improved reversibility have been observed, the theoretical values remain much larger.

NANOPARTICLES

$SmCo_5$ was chosen as the model hard magnetic material because it has a very high magnetocrystalline anisotropy, and because it is uniform on a nanometer scale. It is important to optimize the starting materials for the compacted nanocomposites. Here in order to maximize the coercivity the main factors investigated were the particle and grain size, oxygen exposure, and strain. The $SmCo_5$ nanoparticles were made by ball milling, since gas phase and chemical preparation methods either do not make single phase materials, or cannot make submicron particles. Ball milled $SmCo_5$ has been studied by several groups [21-23] as well as our own [24].

Oxidation

The samarium oxidizes preferentially, leaving behind a Co-rich alloy with a lower nucleation field than $SmCo_5$. ΔG for oxidation of samarium is higher than for most other elements [25]. All ball milling was therefore done under an argon atmosphere in a glove box. A vacuum furnace (base pressure 5×10^{-6} Torr) was directly connected, so that annealing and controlled atmosphere experiments could be done with minimal O_2 exposure. Samples for magnetization measurements were prepared in the glove box, but powders had to be removed for x-ray diffraction and electron microscopy. Oxidation of the unannealed nanoparticles was not observed by x-ray diffraction, where an estimated 4 wt % is typically required for detection. However, once the sample is removed from the glove box, surface oxides are present, and in the annealed samples oxides were more abundant, as seen from x-ray diffraction (Fig. 2).

For the initial annealing experiments, portions of this powder were transferred directly to a vacuum furnace attached to the glove box (base pressure 5×10^{-6} Torr), and heat treated at different temperatures (200 °C, 400 °C, 600 °C, 800 °C, and 1000 °C) for 6 hours. In a separate experiment, a sample was heat treated in H_2 at 600 °C for 12 h. A sample was also sealed in a quartz ampoule evacuated with a mechanical pump (base pressure 4×10^{-3} Torr) for rapid heating and cooling; it was exposed to 950 °C for only 30 min.

Portions of the cooled samples were removed from the glove box for x-ray diffraction, to determine whether decomposition had occurred, and to find how annealing affected the average strain and grain size. Debye-Scherrer powder diffraction measurements were performed with a Rigaku diffractometer. Instrumental broadening was determined using a silicon standard.

The effect of annealing on the composition was found from the x-ray diffraction patterns shown in Fig. 2. There is a eutectoid at in the Sm-Co phase diagram at approximately 800 °C at which $SmCo_5$ decomposes into Sm_2Co_{17} and Sm_2Co_7 [26]. Other researchers have noted that decomposition is relatively slow because it involves long range diffusion [22, 27, 28]. The

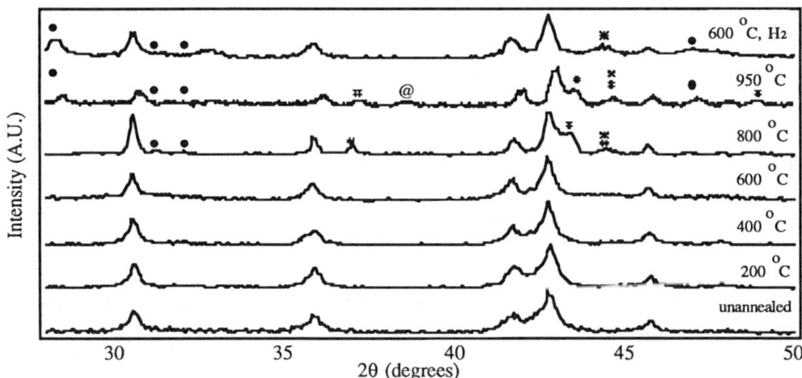

Fig. 2. Powder x-ray diffraction of ball milled $SmCo_5$ nanoparticles as a function of annealing temperature. Unless otherwise indicated, samples were annealed in vacuum. Peaks of phases other than $SmCo_5$ are indicated: Sm_2O_3 (•), FCC Co (*), Sm_2Co_{17} (#), and Co_2O_3 (@).

Sm_2Co_7 byproduct has frequently been undetected by x-ray diffraction [26, 27, 29], perhaps because it is highly disordered.

The decomposition occurred at an even lower temperature in the presence of H_2, and yielded different products, FCC Co and Sm_2O_3. Differential scanning calorimetry indicates diffusion into the lattice to form a hydride, beginning at 200°C, and oxide removal at ~400 °C [30]. Though large amounts of H_2 are reversibly absorbed at room temperature without affecting the magnetic properties of bulk $SmCo_5$ [31, 32], at elevated temperature in N_2, the nitrided samarium cobalt decomposes into samarium nitride and Co [32]. A similar process may be responsible for the high temperature decomposition to form Co and unstable Sm hydride, which oxidizes after the sample is reexposed to air for x-ray diffraction.

Both the quench rate and the heating rate were important, since the sample could decompose whenever it was in the eutectoid decomposition temperature region, and cooling to room temperature in an alumina boat in vacuum can take several hours. The ampoules were used to make rapid heating and cooling possible, but even so some decomposition occurred. Comparison with samples slowly heated in a better vacuum show a similar but more extensive decomposition. The difference in oxygen appears to be less important than the heating and cooling rate.

Strain and Grain Size

For $SmCo_5$ nanoparticles described in the previous section, the effect of annealing on the average strain and grain size was found from analysis of the diffraction peak profiles. Comparison of the (101) ($2\theta = 30.53°$) $SmCo_5$ peaks shows that as the annealing temperature is increased, the profile sharpens and becomes more Lorentzian in nature. Random strains lead to a Gaussian lineshape, while a finite diffracting grain is generally modeled by a Lorentzian lineshape [33, 34]. To a good approximation, the size and strain contributions to the total full width at half maximum (FWHM) linewidth $\delta 2\theta$ are related by

$$\frac{\delta 2\theta_{strain}}{\delta 2\theta} = 1 - \left(\frac{\delta 2\theta_{size}}{\delta 2\theta} \right)^2 . \qquad (8)\ [34]$$

Substituting the theoretical expressions for the size and strain broadening,

$$\delta 2\theta_{size} = \frac{K\lambda}{D\cos\theta} \quad \text{and } \delta 2\theta_{strain} = 4e\tan\theta, \tag{9}$$

where K=0.9, λ is the x-ray wavelength, D is the average grain size, and e is the average local strain, an expression,

$$\frac{(\delta 2\theta)^2}{\tan^2\theta} = \frac{K\lambda}{D}\left(\frac{\delta 2\theta}{\tan\theta\sin\theta}\right) + 16e^2 \qquad (10)[16]$$

is obtained. Plotting $(\delta 2\theta/\tan\theta)^2$ versus $(\delta 2\theta/\tan\theta\sin\theta)$, for the linewidths found from fitting the experimental data, The average grain size and strain were found by plotting $(\delta 2\theta/\tan\theta)^2$ versus $(\delta 2\theta/\tan\theta\sin\theta)$, for the linewidths found from fitting the experimental data.

The average size and strain as a function of the annealing temperature are plotted in Figure 3. Size broadening dominates the total linewidth δ2θ at all temperatures, unlike for $SmCo_5$ nanoparticles made by more vigorous milling [21], where the x-ray profiles are more Gaussian in lineshape. Annealing reduces the relative amount of strain, with a significant drop between 400 °C and 600 °C. Unfortunately, while the strain relaxes, part of the $SmCo_5$ decomposes into other phases, suggesting higher temperatures and/or faster heating and cooling times are needed. The average grain size increases slightly with the annealing temperature. The error bars in Fig. 3 refer to the fitting procedure. In addition, Eq. 8 systematically overestimates $\delta 2\theta_{strain}/\delta 2\theta$ by an amount ranging from 2 % to 4 %, which may vary with the annealing temperature [33]. SEM results indicated the particle size distribution was not changed notably by annealing at 600 °C, though most particles were multi-grain, and showed no evidence of sintering. Above 950 °C, however, the sharp increase in the average grain size indicates atomic diffusion is significant and sintering is likely to occur.

Annealing up to 800 °C lowers the coercivity, relative to the unannealed sample, because the eutectoid decomposition dominates the strain reduction. Above 950 °C the coercivity rises again, but because of the soft phases is still lower than H_c for the unannealed sample. In bulk SmCo5 annealing at 1050°C leads to complete recovery of the coercivity, but in fine particles either disorder or trace oxygen amounts at the grain boundaries may interfere, and optimal coercivities are found in unannealed particles.

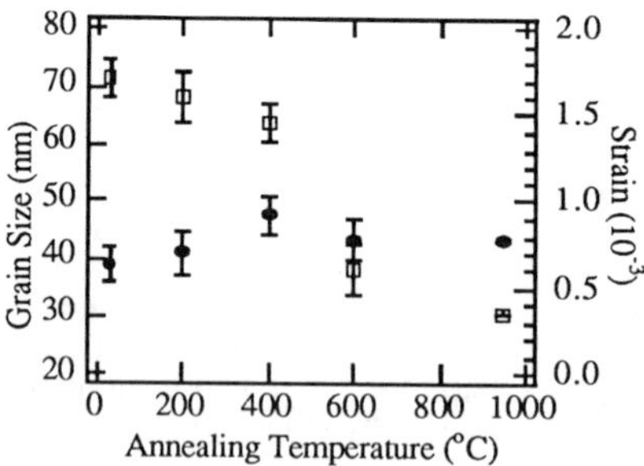

Fig. 3. Average size (•) and strain (◻) as a function of annealing temperature.

Particle Size and Grain Size

Nanoparticles were prepared by ball milling commercially available alloy powder with initial average size of 10 μm. As the micrometer-sized $SmCo_5$ grains collide with the zirconia balls, they fracture into smaller particles. At first the particles have irregular shapes, but when generating nanoparticles this fracturing occurs many times and with random orientations, so normally the resulting particles are spherical [24]. Powder samples were added to a container along with zirconia balls of varying sizes and milled for different lengths of time. The results presented here are for particles gently milled for 15 days, but the times are very sensitive to the type of mill used. High energy milling can achieve the same average grain size in a few hours [21]. Ball milling always yields a large size distribution. The average grain size was determined by removing small amounts of powder periodically and using Scherrer analysis on the $SmCo_5$ (110) x-ray diffraction peak at $2\theta = 35.922°$. The average particle size was estimated from electron microscopy. For magnetic measurements a series of samples were prepared. The remaining portions of the powders were dispersed in epoxy (mass fraction of 10 % $SmCo_5$) within the glove box for magnetic measurements. Remanent magnetization curves were recorded using a Quantum Design MPMS SQUID magnetometer at 300 K between ± 50 kOe.

The coercivity of the 10 μm $SmCo_5$ powder as received was on the order of 3000 Oe at room temperature. Ball milling this powder until the average diameter of the particles is 25 nm raises the coercivity to over 13,000 Oe, showing the significant improvement in H_c which can be achieved by reducing the particle size. Following the milling the saturation magnetization is unchanged, but saturation is approached more slowly with increasing field, indicating the presence of a small amount of paramagnetic material such as a thin oxide coating or superparamagnetic material such as very small $SmCo_5$ nanoparticles. In earlier measurements of remanent magnetization hysteresis loops [35], this amount was estimated to be less that 6 wt.% .

Theories of fine particle magnets predict size dependent coercivities for defect-free ellipsoidal particles with uniform phase, and no interparticle interactions. However experiments find significantly lower coercivities. A small amount of a soft magnetic phase can lead to low nucleation fields for reverse domains [26, 27]. Defects such as grain boundaries and even an irregular surface can lower the anisotropy field H_A by as much as an order of magnitude [36]. Finally, local strains can also cause low energy nucleation sites [37]. The large surface area and the strain induced by ball milling should make $SmCo_5$ nanoparticles particularly vulnerable to these coercivity reducing effects. Ideally H_c is maximized at the largest monodomain size (~2 μm for $SmCo_5$). However the observed maximum occurred at an *average* particle and grain size of ~25 nm [24], as seen in Figure 4.

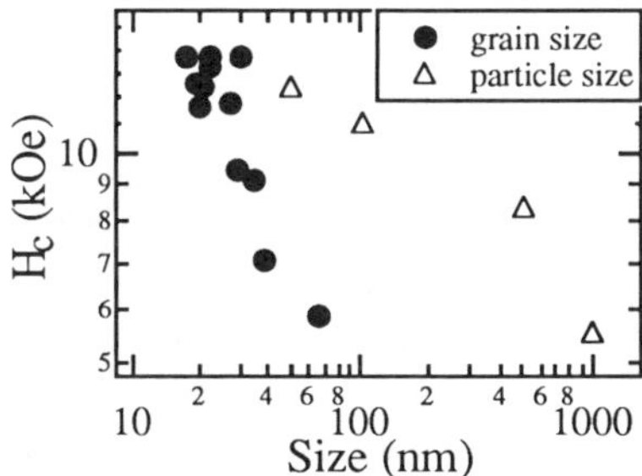

Fig. 4. Dependence of the $SmCo_5$ nanoparticle coercivity on the average particle and grain size.

NANOCOMPOSITES

Our previous work on nanoparticles lead to interest in particle-based nanocomposites. Here we discuss compacts made from ball milled $SmCo_5$ nanoparticles and FeCo nanoparticles made by the polyol method.

$SmCo_5$

For comparison with multiphase composites, a pellet was made from unannealed ball milled $SmCo_5$ nanoparticles by uniaxial compaction, without heating. A cold compaction method was chosen because temperatures between 400 and 800 °C cause eutectoid decomposition, and above 950 °C there is rapid grain growth. In addition, increased temperatures enhance the oxidation rate. The density of the pellet was ~ 60 %, assuming no oxygen content. Figure 5 shows magnetization curves (half loops) for the powder and the resulting pellet. Because of error involved in determining the exact amount of powder dispersed in the epoxy sample, the difference in the magnetization may not be significant. The coercivity of the pellet is slightly smaller, but the general shape of the curve is unchanged. The coercivity is still dominated by nucleation in individual grains.

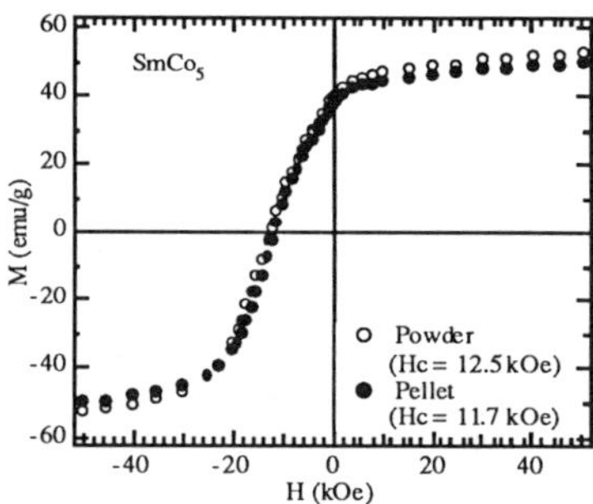

Fig. 5. Comparison of magnetization curves for $SmCo_5$ powder, dispersed in epoxy, and compacted pellet made from the powder.

FeCo Nanoparticles and Nanocomposites

FeCo nanoparticles were chosen as the soft phase, since by varying the composition the magnetocrystalline anisotropy can be minimized [38]. Here the alloy nanoparticles were synthesized by the polyol method [39], a chemical route which leads to highly monodisperse particles with a thin carbonaceous coating. Unlike for $SmCo_5$, the exchange length in soft materials like FeCo is large, and often comparable to the grain or particle size. Adjacent grains can then be exchange coupled to each other, leading to a further reduction in the anisotropy and coercivity, as explained by the random anisotropy model [2, 40].

Here an alternative compaction method was investigated in order to achieve higher density. Plasma pressure compaction was performed by Mr. S. Yoo and Dr. S. Sudarshan of Materiasl Modification, Ind., of Fairfax, VA. Here an electric arc is passed through the powder both preceding and during uniaxial compaction. A density of over 90 % was achieved, assuming only Fe and Co in the sample. Figure 6 shows a comparison of the hysteresis loops for the powder and compacted pellet. Compaction significantly reduces the coercivity, indicating that the exchange coupling is increased, as predicted by the random anisotropy model.

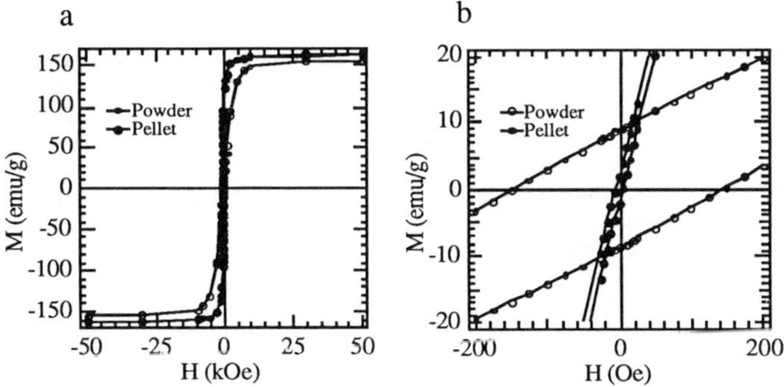

Fig. 6. a.) Comparison of $Fe_{25}Co_{75}$ nanoparticles made by the polyol method, and a compact made from them; b.) An expanded region shows the reduced coercivity of the pellet.

$SmCo_5$/FeCo Nanocomposites

A rolling mill was used to disperse a 50 wt % mixture of the SmCo5 and FeCo powders. For a preliminary test, a pellet was compacted by cold isostatic pressing by Dr. S. Sankar of Advanced Materials Corp., using a pressure of 40,000 psi. The resulting density was ~ 60%. In addition to the potential problems associated with heating SmCo5 nanoparticles by themselves, this alloy can react with Fe to form softer phases. Figure 7 shows the magnetization curves (half loops) for the separate SmCo5 (a) and FeCo (b) powders, and for the nanocomposite (c). Unfortunately the composite behaves as if the nanoparticles have almost no exchange coupling between the hard and soft phases. On a magnified scale, small changes were observed near the coercivity of the FeCo powder. However, recoil measurements for the pellet and powder indicated that only 7% of the recoil magnetization could be attributed to exchange or possibly dipolar coupling, rather than to rotational effects.

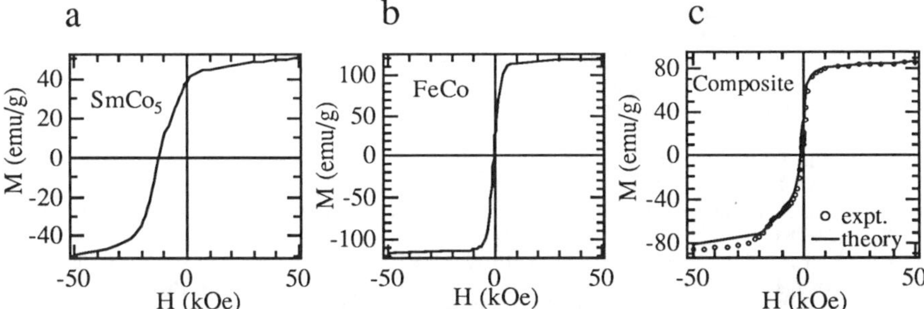

Fig. 7. Comparison of magnetization curves (half loops) for a.) ball milled $SmCo_5$ nanoparticles, b.) $Fe_{50}Co_{50}$ nanoparticles made by the polyol method, and c.) the composite made from them. Due to its low density, the overall magnetization curve is nearly equal to the weighted sum of the individual contributions.

CONCLUSIONS

In the relatively porous pellet, only regions where the hard and soft phases are directly in contact can be exchange coupled. Higher densities are clearly needed to observe significant exchange spring behavior in compacted nanocomposites. For composites based on SmCo5 and FeCo nanoparticles, these are additional restrictions. Both materials, but particularly the samarium alloy, are sensitive to oxygen contamination. Grain growth occurs within a few hours for the iron alloy above 700 °C and for the samarium alloy above 950 °C. The temperature region of 400-800 °C should also be avoided to prevent decomposition of the hard phase. A rapid, relatively low temperature compaction method is most likely to be successful in making a particle-based exchange spring magnet.

ACKNOWLEDGMENTS

This work was supported in part by the National Science Foundation under Grant numbers DMR-9500313 and ECD-8907068, and by the Air Force Office of Scientific Research under grant number F49620-96-1-0454.

REFERENCES

1. F. E. Luborsky, J. Appl. Phys. **32**, 171S (1961).
2. G. Herzer, IEEE Trans. Mag. **25**, 3327 (1989); IEEE Trans. Mag. **26**, 1397 (1990); J. Magn. Magn. Mater. **112**, 258 (1992).
3. G. C. Hadjipanayis, J. Appl. Phys. **55**, 2091 (1984).
4. R. A. McCurrie, *Ferromagnetic Materials: Structure and Properties*, Academic Press, NY, 1994, p. 210.
5. Y. Yoshizawa, S. Oguma, and K. Yamauchi, J. Appl. Phys. **64**, 6044 (1988).
6. E. C. Stoner and E. P. Wohlfarth, Phil. Trans. Roy. Soc. London **A240**, 599 (1948).
7. S. Chikazumi, *Physics of Magnetism*, Wiley, N. Y., 1964.
8. E. Kneller, in *Magnetism and Metallurgy*, eds. A. Berkowitz and E. Kneller, Academic Press, New York, 1969, p.366.
9. W. F. Brown, Jr., J. Appl. Phys. **29**, 470 (1958);J. Appl. Phys. **30**, 130S (1959).
10. H. Kronmüller, in *Science and Technology of Nanostructured Magnetic Materials*, G. C. Hadjipanayis and G. A. Prinz, Plenum Press, NY, 1991, p. 657.
11. R. A. McCurrie and G. P. Carswell, J. Mater. Sci. **5**, 825 (1970).
12. J. M. D. Coey, *Rare Earth Iron Permanent Magnets*, Oxford Univ. Press, Oxford, 1998, p. 8.
13. E. Kneller and R. Hawig, IEEE Trans. Mag. **27**, 3588 (1991).
14. T. Schrefl and J. Fidler, J. Magn. Magn. Mater. **177-181**, 970 (1998).
15. J. S. Jiang, E. E. Fullerton, M. Grimsditch, C. H. Sowers, and S. D. Bader, J. Appl. Phys. **83**, 6238 (1998).
16. J. P. Liu, Y. Liu, Z. S. Shan, and D. J. Sellmyer, IEEE Trans. Mag. **33**, 3709 (1997).
17. R. Coehoorn, D. B. de Mooij, and C. De Ward, J. Magn. Magn. Mater. **80**, 101 (1989).
18. M. Sagawa, S. Hirosawa, H. Yamamoto, S. Fujimura, and Y. Matsura, Jpn. J. Appl. Phys. **26**, 785 (1987).
19. R. W. McCallum, A. M. Kadin, G. B., Clemente, and J. E. Keem, J. Appl. Phys. **61**, 3577 (1987).
20. P. G. McCormick, W. F. Miao, P. A. I. Smith, and R. Street, J. Appl. Phys. **83**, 6256 (1998).
21. R. L. Schalek, D. Leslie-Pelecky, J. Knight, D. J. Sellmyer, and S. C. Axtell, IEEE Trans. Mag. **31**, 3772 (1995).
22. J. Wecker, M. Katter, and L. Schultz, J. Appl. Phys. **69**, 6058 (1991).

23. J. Ding, P. A. I. Smith, P. G. McCormick, R. Street, J. Magn. Magn. Mat. **161**, 303 (1996).
24. S. A. Majetich, K. M. Chowdary, and E. M. Kirkpatrick, IEEE Trans. Mag. **34**, 985 (1998); S. A. Majetich and E. M. Kirkpatrick, IEEE Trans. Mag. **33**, 3721 (1997).
25. *Handbook of Chemistry and Physics*, Chemical Rubber Co. Cleveland, 1972, p. D-49.
26. K. Kumar, J. Appl. Phys. **63**, R13 (1988).
27. R. W. Bartlett and P. J. Jorgensen, J. Less-Common Metals, **37**, 21 (1974).
28. M. G. Benz and D. L. Martin, J. Appl. Phys. **43**, 3165 (1972).
29. K. J. Strnat, in *Ferromagnetic Materials*, eds. E. P. Wohlfarth and K. H. J. Buschow, (Elsevier, Amsterdam and New York, 1988), p. 131.
30. M. V. Susic, J. Serb. Chem. Soc. **62**, 349 (1997).
31. H. Zijlstra and F. F. Westerdorp, Solid State Commun, **7**, 857 (1969).
32. X. G. Li, A. Chiba, and S. Takahashi, J. Appl. Phys. **81**, 2895 (1997).
33. H. P. Klug and L. E. Alexander, *X-ray Diffraction Procedures for Polycrystalline and Amorphous Materials*, (J. Wiley and Sons, New York, 1974), pp. 661-665.
34. G. K. Williamson and W. H. Hall, Acta Metallurgica **1**, 22 (1953).
35. E. M. Kirkpatrick, Ph. D. thesis, Carnegie Mellon University, 1997.
36. A. Aharoni, J. Appl. Phys. **30**, 705 (1959); Phys. Rev. **119**, 127 (1960); J. H. Wernick, Ann. Rev. Mat. Sci. **2**,607 (1972).
37. Y. Shimada and H. Kojima, J. Appl. Phys. **44**, 5125 (1973).
38. F. Pfeifer and C. Radeloff, J. Magn. Magn. Mat.. **19**, 190 (1980).
39. G.Viau, F. Fievet-Vincent, and F. Fievet, J. Mater. Chem.. **6**, 1047 (1996).
40. R. Alben, J. J. Becker and M. C. Chi, J. Appl. Phys. **49**, 1653 (1978).

A STUDY ON THE PHASE TRANSFORMATION AND EXCHANGE-COUPLING OF $(Nd_{0.95}La_{0.05})_{9.5}Fe_{bal}Co_5Nb_2B_{10.5}$ NANOCOMPOSITES

Q. Chen*, B. M. Ma*, B. Lu**, M. Q. Huang**, D. E. Laughlin**

*Rhodia Inc., Rare Earths and Gallium, CN 7500, Cranbury, New Jersey 08512
**Department of Materials Science and Engineering, Carnegie Mellon University, Pittsburgh, Pennsylvania 15213

ABSTRACT

The phase transformation and the exchange coupling in $(Nd_{0.95}La_{0.05})_{9.5}Fe_{bal}Co_5Nb_2B_{10.5}$ have been investigated. Nanocomposites were obtained by treating amorphous precursors at temperatures ranging from 650°C to 950°C for 10 minutes. The magnetic properties were characterized via the vibrating sample magnetometer (VSM). X-ray diffraction (XRD), thermomagnetic analysis (TMA), and transmission electron microscopy (TEM) were used to perform phase identification, measure grain size, and analyze phase distribution. The strength of the exchange coupling between the magnetically hard and soft phases in the corresponding nanocomposite was analyzed via the ΔM-versus-H plot.
It was found that the remanence (B_r), coercivity (H_{ci}), and maximum energy product (BH_{max}) obtained were affected by the magnetic phases present as well as the grain size of constituent phases and their distribution. The optimal magnetic performance, BH_{max}, occurred between 700°C to 750°C, where the crystallization has completed without excessive grain growth. TMA and TEM indicated that the system was composed of three phases at this point, $Nd_2(Fe\ Co)_{14}B$, α-Fe, and Fe_3B. The exchange coupling interaction among these phases was consistently described via the ΔM-versus-H plot up to 750°C. The B_r, H_{ci}, and BH_{max} degraded severely when the thermal treatment temperature increased from 750°C. This degradation may be attributed to the grain growth of the main phases, from 45 to 68nm, and the development of precipitates, which grew from 5nm at 750°C to 12nm at 850°C. Moreover, the amount of the precipitates was found to increase with the thermal treatment temperatures. The precipitates, presumably borides, may cause a decrease in the amount of the α-Fe and Fe_3B and result in a redistribution of the Co in the nanocomposites. The increase of the Co content in the $Nd_2(Fe\ Co)_{14}B$ may explain the increase of its Curie temperature with the thermal treatment temperatures. In this paper, we examine the impacts of these factors on the magnetic properties of $(Nd_{0.95}La_{0.05})_{9.5}Fe_{bal}Co_5Nb_2B_{10.5}$ nanocomposite.

INTRODUCTION

NdFeB nanocomposite intermetallic compounds have been of great interest of both the scientific community and bonded magnet industry [2-6, 11, 12]. Composed of both magnetically hard and soft phases, nanocomposites usually possess enhanced remanence due to the exchange coupling interaction[2-5]. The overall magnetic performances, namely, remanence (B_r), coercivity (H_{ci}), and maximum energy product ($(BH)_{max}$) are dictated by the strength of the exchange coupling as well as by the magnetic phases present in the nanocomposites.
Using a three dimensional modeling, Schrefl et. al.[5] estimated that the ideal grain size should be approximately 20 nm for the hard phases and 10 nm for the soft phases. Nano-scaled grain size can be achieved by optimal quenching or by over-quenching followed by a crystallization

Mat. Res. Soc. Symp. Proc. Vol. 577 © 1999 Materials Research Society

treatment. It is generally recognized that an increase in the thermal processing temperature results in grain growth. The grain growth usually becomes more rapid at higher temperatures. An excessive grain growth may weaken the exchange coupling and lowers the overall magnetic performance of nanocomposites. However, the impact of the magnetic phase transformation under different thermal processing conditions on the exchange coupling and subsequent magnetic properties have not received the same attention.

The ΔM-versus-H plot has been widely used in the recording media to describe the interacting particle systems [7-9]. This technique can also provide a quantitative means to describe the exchange coupling in the newly developed nanocomposites for permanent magnets [10]. However, due to other factors such as the phase content in the system, the strength of the exchange coupling reflected from the maximum value of ΔM may not be consistently associated with the maximum magnetic performance.

In this work, using a rare-earth-lean and boron-rich NdFeB nanocomposite, $(Nd_{0.95}La_{0.05})_{9.5}Fe_{bal}Co_5Nb_2B_{10.5}$, the phase transformation and the exchange coupling were studied as a function of thermal treatment temperature. Their impact on the magnetic performances, namely, B_r, H_{ci}, $(BH)_{max}$, and $M_{H=18koe}$ (magnetization at 18kOe, an estimation of the saturation magnetization) is reported. The microstructures, including the grain size distribution and development of the intergranular phases, were studied via transmission electron microscopy (TEM). The exchange coupling of the system was investigated via the ΔM-versus-H plot. The main purpose of the study is not to obtain the highest $(BH)_{max}$, but to understand the impact of phase transformation and the strength of the exchange coupling interaction on the magnetic performance from the microstructure point of view.

EXPERIMENTAL

$(Nd_{0.95}La_{0.05})_{9.5}Fe_{bal}Co_5Nb_2B_{10.5}$ ribbons were prepared via the conventional melt-spinning technique. The as-melt-spun materials were isothermally treated in a tube furnace under vacuum (20mtorr) at 650°C, 700°C, 750°C, 850°C, and 950°C. The magnetic properties at room temperature were measured by an EG&G Model 4500 Vibration Sample Magnetometer (VSM) with a maximum applied magnetic field of 18 kOe. All samples were pulse-magnetized at 100 kOe prior to each measurement to mimic hysteresis loop measurement with higher maximum applied field. A Perkin-Elmer Model 7 Thermogravimetric Analyzer (TGA) was modified to perform thermomagnetic analysis (TMA). An X-ray diffractometer (XRD) (Siemens) with Co radiation (λ=1.78897Å) was used to determine the amorphocity of the as-spun materials and phases developed after the thermal treatments.

The exchange interactions of the system were characterized via the ΔM-versus-H plot, in which ΔM (=$m_d(H)$-(1-2$m_r(H)$, where m_d is the reduced DC magnetization at a given applied magnetic field and m_r is the reduced isothermal remanence) is plotted against the applied magnetic field [7-9].

To carry out microstructure study, as-spun and all thermally treated samples were studied using Phillips-420 transmission electron microscope (TEM) [1]. Samples were prepared by ion milling on a cold stage of a GATAN 600 ion miller. Grain size measurement and phase identification using electron diffraction pattern (EDP) were performed on samples treated with temperatures ranging from 650°C from 850°C.

RESULTS AND DISCUSSIONS

Shown in Figure 1 are the variations of B_r, H_{ci}, BH_{max} and $M_{H=18koe}$, obtained with various thermal processing temperatures. H_{ci} increased from 0.1 kOe in the as-spun state to 1.5 kOe when treated at 650°C, peaked to 11.7 kOe at 700°C then decreased to 8.8 and 7.0 kOe when treated at 750°C and 850°C, respectively. For 950°C, our test resulted in a more dramatic decrease in H_{ci} (to3.3kOe). Unlike H_{ci}, B_r increases from 1.0 kG in the as-spun state to 4.2, 6.8, and 7.0 kG when treated at 650°C, 700°C and 750 °C, respectively. A slight decrease in B_r (from 7.0 to 6.9 kG) was noticed when treated at 850°C. At 950°C, B_r decreased to 5.8 kG. It is of interest to note that H_{ci} reaches the highest value at 700°C, while the highest B_r was obtained when treated between 700°C to 750°C. This difference in optimum values may arise from the fact that two different mechanisms, the exchange coupling, and the phase content, govern H_{ci} and B_r. While exchange coupling is important for both H_{ci} and B_r, the amount of the soft phase competes with the exchange interaction in its effect on B_r. During this competition, B_r can be increased at the expense of reduced exchange interaction and therefore lowered H_{ci}. At 850°C and 950°C, the desired hard and soft magnetic phases may transform into undesired phases and result in the decrease in both H_{ci} and B_r. The trend of BH_{max} is a combined effect of H_{ci} and B_r. BH_{max} begun at 0.0 MGOe in the as-spun state, reached a maximum value of 10.0 MGOe at 700°C and 750°C, and then decreased when treated at a temperature higher than 750°C. Over the course of the study, $M_{H=18kOe}$ exhibited an initial decrease (at 650°C and 700°C) as a result of the crystallization of the amorphous phase. Nearly constant $M_{H=18koe}$ were observed for temperatures higher than 700°C, suggesting although phase transformation exist at higher temperatures, the change in the relative phase content might not be large enough to create observable changes in the overall magnetization at 18kOe.

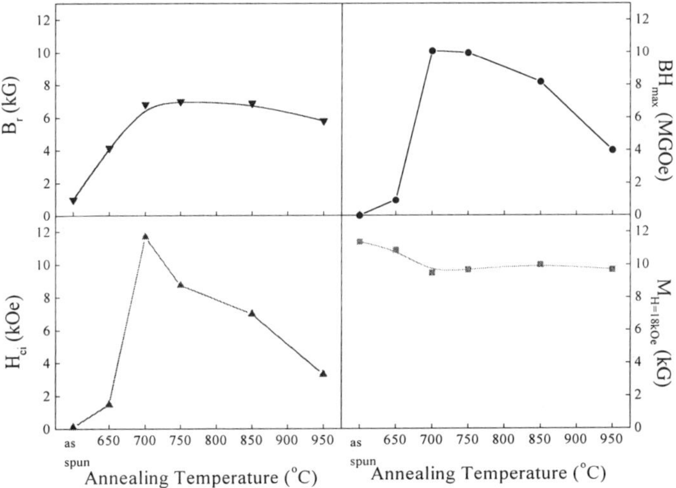

Figure 1. Magnetic properties of as-melt-spun and thermally processed $(Nd_{0.95}La_{0.05})_{9.5}Fe_{bal}Co_5Nb_2B_{10.5}$.

Shown in Figure 2 are the TMA scans of the samples studied. The as-spun and the 650°C treated materials show a large amount of the unknown phase with a T_c of 170°C. The low H_{ci} values obtained on these two samples suggest that this amorphous phase is indeed a soft magnetic phase. During crystallization, this soft and amorphous phase transformed into phases dictated by the nominal alloy composition and the phase transformation paths. Two magnetic phases, namely, $Nd_2(Fe\ Co)_{14}B$ and α-(Fe Co) (for α-Fe, T_c~770°C) were detected on samples treated at 700°C. When treated at 750°C, the amount of α-Fe increased slightly and a traceable amount of Fe_3B (T_c~510°C) was also detected. This increase in the amount of α-Fe is consistent with the slight increase in B_r observed. At 850°C, the amount of Fe_3B decreased significantly suggesting that Fe_3B may transform into some other, possibly non-magnetic, phases. Additional test at 950°C exhibited consistent trend of decrease of Fe_3B with the thermal treatment temperatures. It should be noted that the T_c of $Nd_2(Fe\ Co)_{14}B$ increased as a function of the thermal processing temperatures. When processed at 650, 700, 750, 850°C, and 950°C, the T_c of $Nd_2(Fe\ Co)_{14}B$ was 343°C, 351°C, 367°C, 395°C, and 410°C, respectively. The cause of this change in T_c is not clear at this point. Phase transformation from $Nd_2(Fe\ Co)_{14}B$ is unlikely based on the XRD data shown in Figure 3. However, the decrease in the amount of the α-Fe and Fe_3B may result in a redistribution of the Co in the nanocomposites. The increase of the Co content in the $Nd_2(Fe\ Co)_{14}B$ may therefore in turn explain the increase of its Curie temperature with the thermal treatment temperatures.

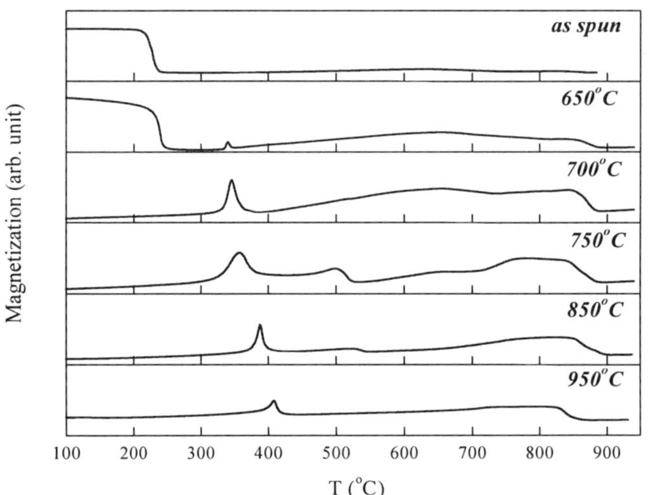

Figure 2. TMA scans of the as-melt-spun and thermally processed $(Nd_{0.95}La_{0.05})_{9.5}Fe_{bal}Co_5Nb_2B_{10.5}$.

Shown in Figure 3(a) are the XRD scans of the samples treated with various thermal processing temperatures. Evolution of the crystallization from amorphous material to a composite of $Nd_2(Fe\ Co)_{14}B$ and α-Fe were demonstrated. To compare the growth rate of $Nd_2(Fe\ Co)_{14}B$ and α-Fe, the intensities of the (410) peak of the $Nd_2(Fe\ Co)_{14}B$ phase and the (110) peak of the α-Fe phase were plotted against the temperature in Figure 3(b). It can be seen that most of the crystallization of the $Nd_2(Fe\ Co)_{14}B$ phase occurred at temperature lower than 700°C, while the growth of the α-Fe phase accelerated as the temperature rose.

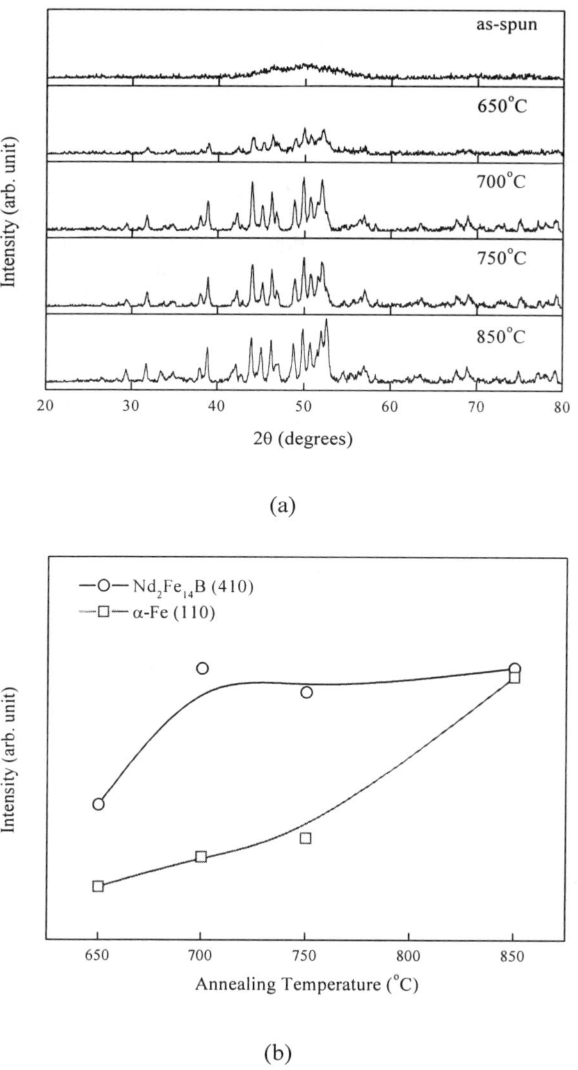

Figure 3. (a) XRD of as-spun $(Nd_{0.95}La_{0.05})_{9.5}Fe_{bal}Co_5Nb_2B_{10.5}$. (b) Intensities of the (410) peak of $Nd_2(Fe\ Co)_{14}B$ and the (110) peak of α-Fe versus the temperature.

Shown in Figure 4 are the ΔM-versus-H of samples included in this study. To illustrate the effect of grain size on this type of plot, a ΔM versus H curve of a sintered NdFeB with an average grain size of 10 μm, pulverized to an equivalent powder size, was also included for comparison. As expected, the crushed sintered NdFeB showed mostly negative ΔM, suggesting a magnetostatic-dominated particle interaction[7,8]. The ΔM of the as-melt-spun materials and samples heat treated at 650°C exhibited similar behavior. For samples treated above 650°C, a positive ΔM was observed indicating the existence of exchange coupling between phases[7,8,10]. The height of the ΔM peak reached a maximum when materials were treated at 700°C, suggesting the strongest exchange coupling among samples studied. At 750 and 850°C, the peaks of ΔM decreased significantly, an indication of a weakened exchange interaction. Microstructural examinations, both on average grain size and phase distribution, are necessary to quantify this change in exchange coupling. We should also point out that unlike the classical exchange coupling case where the system is composed of 60% hard magnetic phases and 40% of soft magnetic phases [4], the system involved in this study is composed of more than 90% of the hard magnetic phase, determined by the nominal composition used in this study. The purpose is to maintain relatively higher H_{ci}, while still being able to have an exchange-coupled nanocomposite system to study.

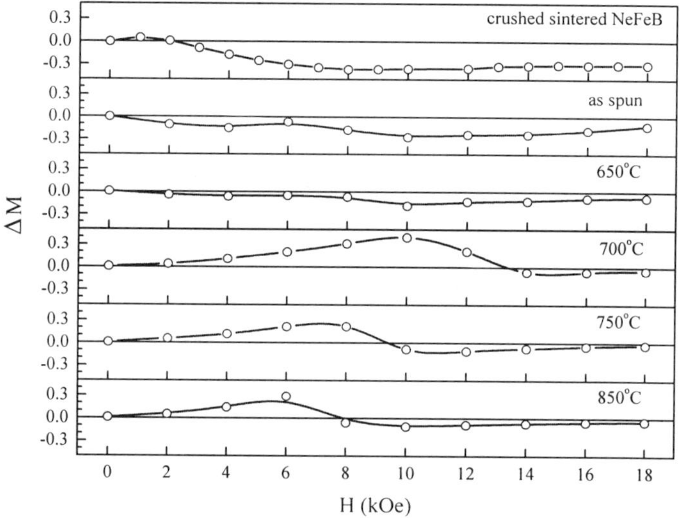

Figure 4. The ΔM-versus-H plots of the sintered $Nd_{15}Fe_{79}B_6$ and as-melt-spun and thermally processed $(Nd_{0.95}La_{0.05})_{9.5}Fe_{bal}Co_5Nb_2B_{10.5}$.

Table I. The average grain sizes obtained from TEM images.

sample	average grain size (nm)	
	primary phase	secondary phase
as-spun	n/a	n/a
650°C	38	n/a
700°C	40	n/a
750°C	45	5
850°C	68	12

Shown in Figure 5 are the TEM images and the inset EDP of samples in the as-spun state and after thermal treatments at 650°C, 700°C and 750°C. The as-spun sample, shown in Figure 5(a), appeared to be homogenous and showed no crystal grain. The amorphous material observed agreed well with our results from TMA and XRD reported earlier in this paper. After thermal treatment at 650°C, we observed $Nd_2(Fe\ Co)_{14}B$ grains embedded in a large amount of amorphous residual, indicating a partial crystallization occurred at 650°C (see Figure 5(b)). At 700°C, shown Figure 5(c), more well-defined grains could be observed and the elimination of the amorphous phase was noticed. The development of soft phases and some secondary phases became more prevalent in samples treated at temperatures higher than 700°C. TEM images at 750°C are shown in Figure 6(a) and (b). Some observable small precipitates started to develop at the grain boundaries as intergranular phases (black arrows in Figure 6(b)). At 850°C, as these intergranular phases grew, they were able to be identified as either α-Fe or Fe_3B, using micro-diffraction technique. See Figure 7(a) and (b). Interestingly, new phase(s) started to precipitate within the $Nd_2(Fe\ Co)_{14}B$ grains at 850°C. Due to the small grain size of these grains, they have not been characterized by either electron diffraction or elemental analysis at this point. Since no additional ferromagnetic phase was observed in the TMA scan for 850°C, we suspected that the new precipitate may actually be non-magnetic phases, such as borides.

The development of the primary and secondary phases as well as the grain growth can be reflected by the change in the grain size distributions. Descriptive statistical data on the grain sizes were obtained from the TEM images. Shown in Figure 8 are the grain size distribution of samples thermally treated from 650°C to 850°C. At 650°C, a symmetrical single peak distribution was observed. The distribution became biased to the smaller grain size at 700°C, which could be an indicator of the formation of the bi-modal distribution found in samples at 750°C. Listed in Table I are the average grain sizes of the primary and secondary phases determined [1]. It suggests that the as-spun materials are fully amorphous. Grain growth from 650°C (38nm) to 700°C (40nm), was not significant. At 750°C, the grain size of the primary phase increased to 45nm and newly developed intergranular phases resulted in a second peak averaged at 5nm. At 850°C, the primary phase and the secondary phase peaks shifted to the larger grain size, i.e., 68nm, and 12nm, respectively. It is of interest that the grain size of the primary phase, $Nd_2(Fe\ Co)_{14}B$ grains, did not increase significantly until 850°C. It is then suspected that the intergranular precipitates may serve as barriers at the grain boundaries to prevent the $Nd_2(Fe\ Co)_{14}B$ grain coarsening [1].

Our microstructure analyses agree well with the magnetic performances as well as the strength of the exchange coupling interaction between the $Nd_2(Fe\ Co)_{14}B$ and α-Fe, Fe_3B phases characterized by the ΔM-versus-H plot. The optimal magnetic performance occurred between 700°C and 750°C, where no amorphous phase is left, no excessive grain growth has started, and reasonable fraction of soft phase is present. The highest ΔM was found a 700°C where the crystallization was just completed and smallest average grain size was observed. The highest ΔM, representing the strongest exchange interaction, occurred at the same temperature as did H_{ci}. However, the optimal magnetic performance, namely, BH_{max}, appeared to stay at the same level between 700°C and 750°C, where B_r increased at the expense of H_{ci}. This resulted from the fact that the optimal magnetic properties are not only impacted by the strength of the exchange interaction, but also dictated by the relative content of various phases of the composite.

The 2% Nb addition in our composition is also of our interest. While we found that Nb is very effective in improving the H_{ci} of the system, the actual location of Nb in terms of microstructure is not clear at this point. We suspect Nb may be present in forms of various borides and serve as pinning centers for the domain walls during the demagnetizing process. However, we were unable to identify them via XRD due to their extremely small volume fractions, if any was present. The precipitation developed within the grains of primary phase was also suspected to be related to the Nb addition. Works on identifying them are currently underway and we will report our progress in the near future.

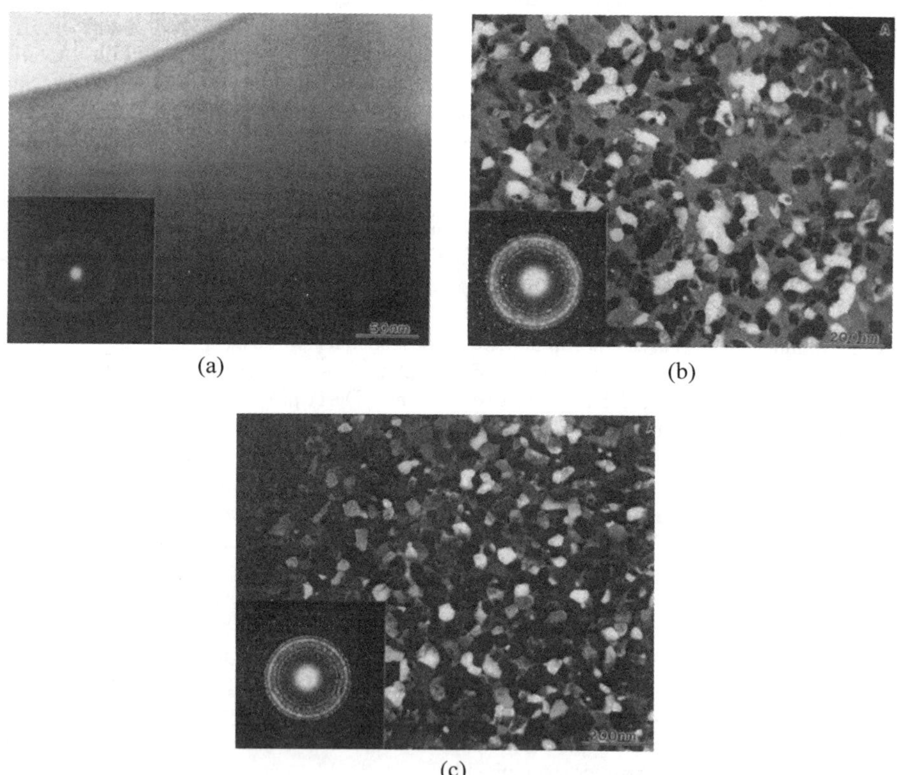

(a) (b) (c)

Figure 5. TEM images of ribbons in as-spun state and after thermal treatments. (a) As-spun. (b) 650°C. (c) 700°C.

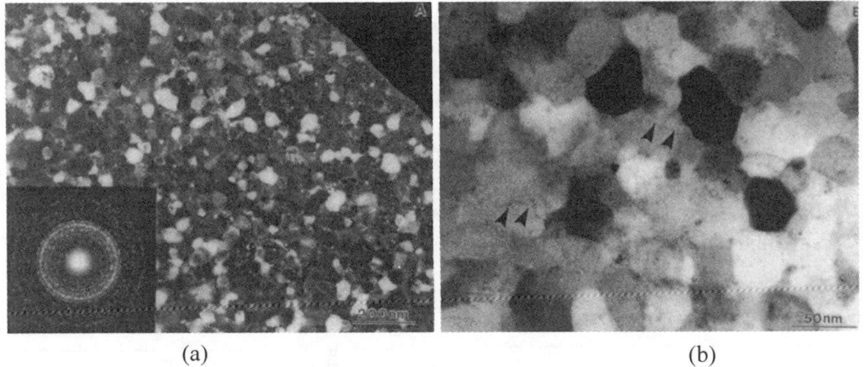

Figure 6. TEM images of ribbons after thermal treatments at 750°C. (a) Lower magnification. (b) Higher magnification. Black arrows point to intergranular phases.

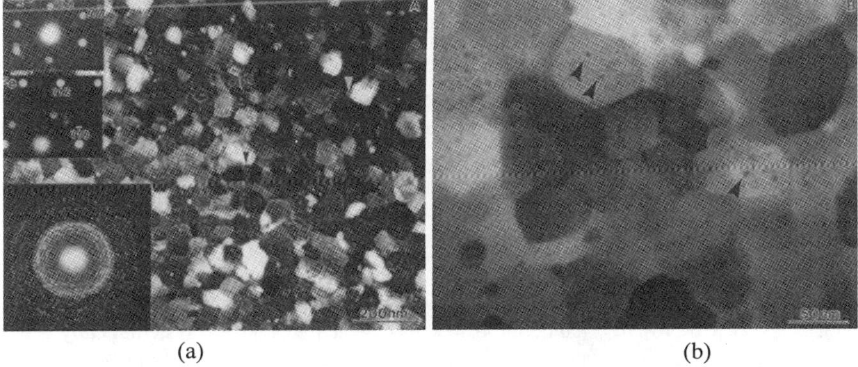

Figure 7. TEM images of ribbons after thermal treatments at 850°C. (a) Lower magnification. (b) Higher magnification. Black arrows point to new phase(s) precipitated within $Nd_2(Fe\ Co)_{14}B$ grains.

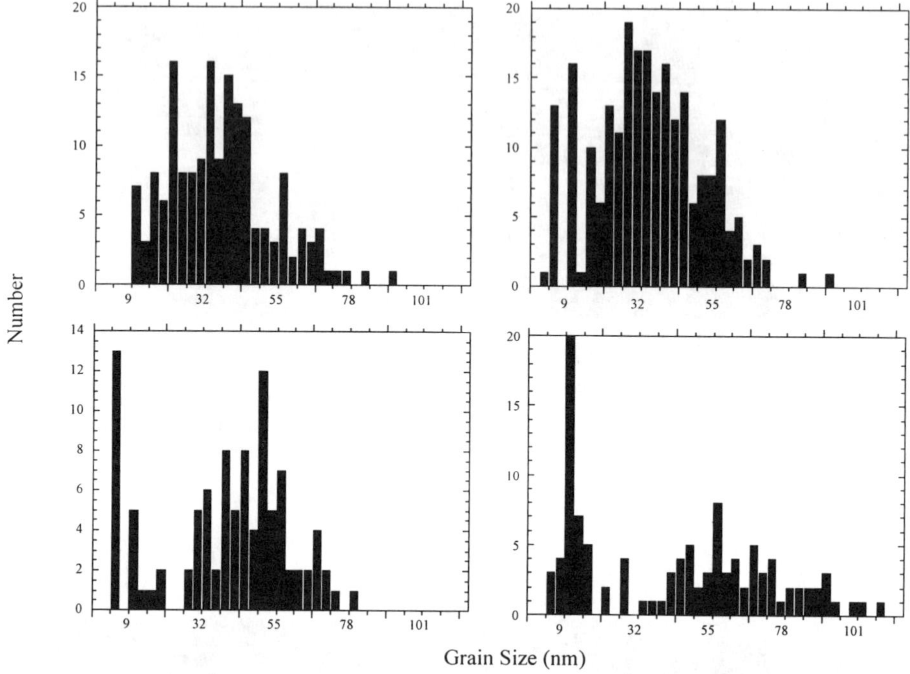

Figure 8. Grain size analyses from TEM images for thermally treated ribbons. Top left: 650°C. Top right: 700°C. Bottom left: 750°C. Bottom right: 850°C.

CONCLUSIONS

The effects of phase transformations and microstructures on the magnetic properties of $(Nd_{0.95}La_{0.05})_{9.5}Fe_{bal}Co_5Nb_2B_{10.5}$ nanocomposites has been investigated systematically. It was found that B_r, H_{ci}, and BH_{max} obtained were affected by the magnetic phases present as well as the strength of the exchange coupling. The optimal magnetic performance, BH_{max}, occurred between 700C to 750°C, where the crystallization has completed without excessive grain growth. TMA and TEM indicated that the system was composed of three phases at this point, Nd_2(Fe Co)$_{14}$B, α-Fe, and Fe_3B. The ΔM-versus-H plot successfully interpreted the effect of the exchange coupling on B_r, H_{ci}, and BH_{max} obtained for samples treated below 750°C. While similar ΔM-H curves were obtained for samples treated at higher than 750°C, degradation in B_r, H_{ci}, and BH_{max} was found. This degradation may be attributed to the grain growth of the main phases, from 45 to 68nm, and the development of precipitates developed within the Nd_2(Fe Co)$_{14}$B grains. It was hypothesized that the precipitated phases could be finely dispersed borides. The formation of these borides, in conjunction with the mass balance, may explain the increase in the T_c of Nd_2(Fe Co)$_{14}$B and the decreased amount of α-Fe and Fe_3B when thermally processed at temperatures above 700°C.

REFERENCES

[1] B. Lu, M.Q. Huang, Q. Chen, B.M. Ma, and D.E. Laughlin, Proc. Of 43rd Annual Conference on Magnetism and Magnetic Materials, Miami, Florida (1998), in press.

[2] E. F. Kneller and R. Hawig, IEEE Trans. Mag. **27**, p.3588 (1991).

[3] R. Skomski and J. M. D. Coey, *Phys. Rev. B* **48**, p.1581 (1993).

[4] T. Schrefl, J. Fidler, and H. Kronmuller, *Phys. Rev. B*, **49**, p.6100 (1994).

[5] R. Fischer, T. Schrefl, H. Kronmuller, and J. Fidler, *J. Magn. Magn. Mater.* **153**, p.35 (1996).

[6] J. Ormerod, S. Constantinides, J. Appl. Phys. **81**, p.4816 (1997).

[7] E. P. Wohlfarth, J. Appl. Phys. **29**, p.595 (1958).

[8] P. E. Kelly, K. O. Grady, P. I. Mayo, and R. W. Chantrell, IEEE Trans. Mag. **25**, p.3881 (1989).

[9] M. Fearon, R. W. Chantrell, E.P.Wohlfarth, *J. Magn. Magn. Mater.* **86**, p.197 (1990).

[10] W. C. Chang, S. H. Wu, B. M. Ma, C. O. Bounds, S. Y. Yao, J. Appl. Phys. **83**, p.2147 (1998).

[11] Manaf, R. A. Buckley, and H. A. Davies, J. Magn. Magn. Mater. **128**, p.302 (1993).

[12] L. Withanawasam, A.S. Murphy, G. C. Hadjipanayis, K. R. Lawles, R. F. Krause, J. Magn. Magn. Mater. **140-145**, p.1057 (1995).

MAGNETIC PROPERTIES OF POLYMER BONDED EXCHANGE-SPRING NdFeB MAGNETS

B.MOKAL, N.A.SMITH, A.J.WILLIAMS and I.R.HARRIS
School of Metallurgy and Materials, The University of Birmingham, Birmingham B15 2TT, U.K.

ABSTRACT

The production of remanence enhanced powders, using a method of mechanical alloying of free iron to melt spun ribbon, has been studied. The final composition is $Nd_{8.1}Fe_{77.1}Co_{10.9}B_{3.9}$ and both hard and soft materials possess an initial fine grain size; approximately 30nm for the $Nd_2Fe_{14}B$ phase in the melt spun ribbon and 1μm for the α-Fe. The magnetic properties of resin bonded magnets fabricated using such material have been measured and found to exhibit exchange-spring behaviour. The experiments performed in this work are described with respect to the volume fraction of the soft phase and subsequent annealing temperature of the as-milled material. Remanence enhancement has been observed and this can be attributed to the exchange interactions between the ferromagnetic soft and hard phases. For a 30% by volume addition of α-Fe, typical values of remanence were found to be ~880mT with a coercivity of ~250 kA/m. The reversible and irreversible components of the magnetisation have been determined, to demonstrate the exchange coupling and coercivity behaviours respectively.

Keywords: Permanent magnets, exchange coupling

INTRODUCTION

Rapidly solidified NdFeB materials have long been used in the production of permanent magnets; one of the main methods of production was that of Croat et al in 1984, whereby a randomly oriented crystal structure of mean crystallite size, d=60nm was produced (1). High coercivity could be acheived as the anisotropy constant of the tetragonal $Nd_2Fe_{14}B$ phase is high and each grain is single domain. However, the maximum remanence Br was limited by this random grain orientation to $M_s/2$, according to the Stoner -Wohlfarth model (2), where M_s is the measured saturation magnetisation. Later it was shown that the Br and BHmax of NdFeB melt spun alloys could be enhanced by reducing d to ~40nm (3). Enhancement was found to occur because of the exchange coupling that exists between the nanocrystals.

Mechanical alloying has also been used extensively in the manufacture of permanent magnets. This technique allows the production of very fine composite particles during milling which leads to good magnetic properties after heat treatment of this material (3 to 10). Grain sizes of ~20nm for heat treated powder have been reported for α-Fe/$Nd_2Fe_{14}B$ nanocomposite structures as compared to 5-10 nm in the as-milled powder.

Remanence can be increased in these materials by using a mixture of hard and soft magnetic phases and the composite magnets produced may exhibit high remanence and coercivity when the two phases become exchange coupled across their interfaces. Exchange coupling causes the magnetisation vector of the soft phase to be rotated toward that of the hard phase and increases the remanent magnetisation in the direction of initial magnetisation. Thus, although the material is still crystallographcially isotropic, the remanence can be much higher than the isotropic value of 0.5Ms for uniaxial magnetocrystalline materials, without crystallographic alignment. Good coercivities are difficult to attain as coercivity is a sensitive function of the dimension of the soft phase (11 to 15). The amount of iron present, as well as its mean grain diameter, has been found to influence the magnetic interactions (16). It has been shown that the grain size of the soft phase should be less than the exchange length of the soft magnetic phase (17).

The mixture of magnetically hard and soft phases decreases the rare-earth content whilst maintaining reasonable hard magnetic properties (18,23). The magnetisablity of the nanocomposite $Nd_2Fe_{14}B$ magnets is improved when compared to conventional resin bonded magnets (19). Work performed by Panchanathan (20) compared the Ms behaviour of single phase $Nd_2Fe_{14}B$, two phase $Fe_3B/Nd_2Fe_{14}B$ and two phase α-Fe/ $Nd_2Fe_{14}B$; the composite magnets were found to show a saturation of about 80% with an applied field of only 960kA/m and magnets from these materials had a coercive field of 320 kA/m.

EXPERIMENTAL

The powders were produced by the addition of fine grain sized (d=1μm) iron powder, in various amounts, to MQ-type melt spun ribbon, composition $Nd_{12.3}Fe_{65.1}Co_{16.6}B_6$ with subsequent milling in a planetary mill for 80hours under an argon atmosphere. It was necessary subsequently to crush the as-milled material and then heat treat to crystallise the amorphous phase formed during milling (in the temperature range 640 -720°C). Powder samples were set in wax and pulse magnetised in a 4.5 Tesla field; subsequent magnetic measurements were performed using vibrating sample magnetometers capable of applying a maximum field of up to 2387 kA/m Remanence enhancement was calculated using the 'saturation magnetisation' value obtained at maximum field and Br values corrected for self-demagnetisation fields (self demagnetising factor, N=0.3 (24).

Epoxy bonded magnets were fabricated using the 30% Fe containing powder which was previously heat treated at 680°C for 30 minutes. The magnets contained 15% by volume of epoxy resin and were pressed at 5 Ton/cm^2 without a magnetic field. The magnets were characterised using a permeameter or VSM.

The reversible and irreversible components of the magnetisation were evaluated using recoil measurements, on the permeameter, and the relation $M_{tot}(Hi)=M_{rev}(Hi) +M_{irr}(Hi)$ where $M_{tot}(Hi)$ is the drop in magnetisation from the remanence, measured at a reverse field, Hi. $M_{rev}(Hi)$ is the reversible change in magnetisation on application of a demagnetising field, Hi and is measured as the increase in magnetisation observed on decreasing the field from Hi to zero. $M_{irr}(Hi)$ is the irreversible drop in magnetisation on application of a demagnetising field, Hi and is measured as the drop in magnetisation compared to the remanence after decreasing the field from Hi to zero. The total and irreversible susceptibilities (κ_{tot} and κ_{irr}) are then calculated by differentiating the $M_{tot}(Hi)$ and $M_{irr}(Hi)$ curves with respect to Hi, respectively.

RESULTS AND DISCUSSION

Powders

VSM measurements were carried out and are presented in table 1 for milled as-received MQ material and table 2 for milled MQ material with 30% volume fraction additions of soft phase. The materials were subsequently heat treated at temperatures of 640, 660, 680, 700 and 720°C for 30 minutes. Heat treatments were undertaken under a diffusion pump vacuum (~10^{-5} mbar) and then quenched with a water cooled coil, around the Inconel furnace tube containing the samples.

Table 1: Samples containing 0% Fe

HT temp (°C)	Br (J/Tkg)	Corrected Br* (J/Tkg)	iHc (kA/m)	Br*/Ms
640	64.8	73.5	657.2	0.57
660	90.5	102.5	669.2	0.80
680	86.8	99.8	656.0	0.78
700	87.7	102.6	656.5	0.80
720	74.4	83.3	629.4	0.65

* Corrected for self demagnetising fields using N=0.3.

Table 2: Samples containing 30% Fe

HT temp (°C)	Br (J/Tkg)	Corrected Br* (J/Tkg)	iHc (kA/m)	Br*/Ms
640	80.8	117.8	302.5	0.76
660	79.8	116.2	305.7	0.75
680	90.0	126.9	324.9	0.82
700	80.88	124.7	302.2	0.81
720	60.9	104.3	237.5	0.67

* Corrected for self demagnetising fields using N=0.3.

Bonded Magnets

The finished magnet, fabricated from the 30% Fe containing powder, heat treated at 680°C for 30 minutes, had a density of 5.28g/cm^3, with a remanence, corrected for density, of 876 mT (92.9 J/Tkg) and an intrinsic coercivity of 254 kA/m. The coercivity is slightly lower than that of the VSM measurement and may be due to the different density of the samples and/or the application of a load during the magnet fabrication. The value of remanence is similar to the uncorrected VSM measurement, which could indicate that although the permeameter does not have a bulk self demagnetisation field, the low density bonded magnet may also suffer from internal self demagnetisation.

Figures 1 shows the reversible component of the magnetisation. The high values for M_{rev} observed for these samples at values of Hi greater than the coercive field of α-Fe indicate that there is exchange coupling between the hard and soft phases (21).

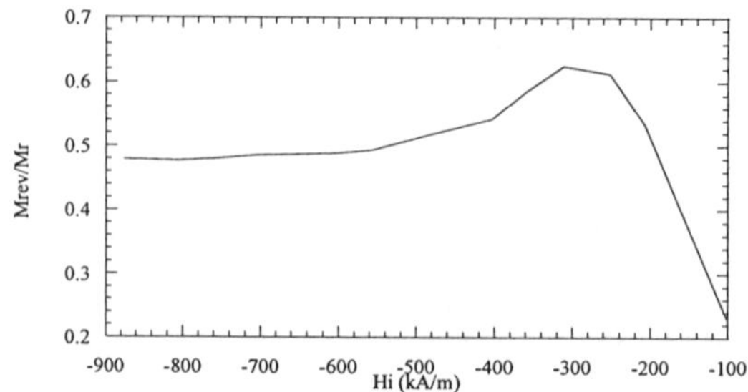

Figure 1 Measurements of the reversible component Mrev(H)/Mr showing a peak at around 310 kA/m.

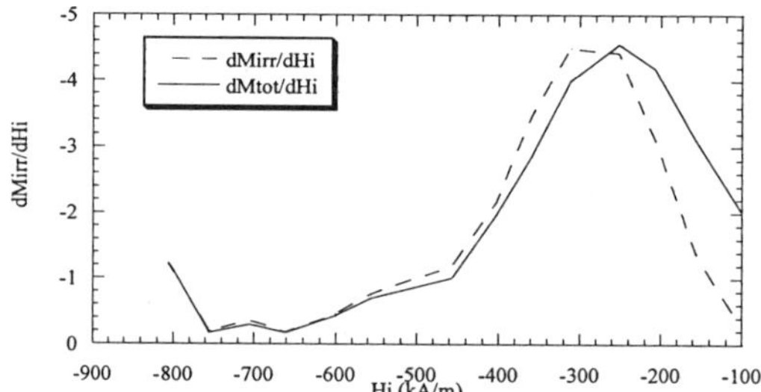

Fig.2 Susceptibility plots κ_{tot}, and κ_{irr} as a function of Hi.

Figure 2 shows measurements of total and irreversible susceptibilities as a function of Hi for these bonded samples. The two main single peaks in both κ_{irr} and κ_{tot} shown in figure 2 imply that they originate from the reversal of the hard phase.

The irreversible susceptiblity κ_{irr} represents the degree of collectiveness of the reversal process (22), narrow peaks indicating a highly collective reversal occurring over a narrow range, broad ones indicating a less collective process extending over a large field range, it would appear that the latter process is in play. In addition the peak of the κ_{irr} shown in figure 2 occurs at a very similar field to that of the peak in the M_{rev} as seen in figure 1, implying that the two phases are exchange coupled and undergo coherent reversal from an initial magnetically aligned state. The nucleation field in the bonded sample, as measured on the permeameter, (310 kA/m) is lower than that of the unbonded case (385 kA/m), measured on the VSM. This compares well with values reported in the literature (20).

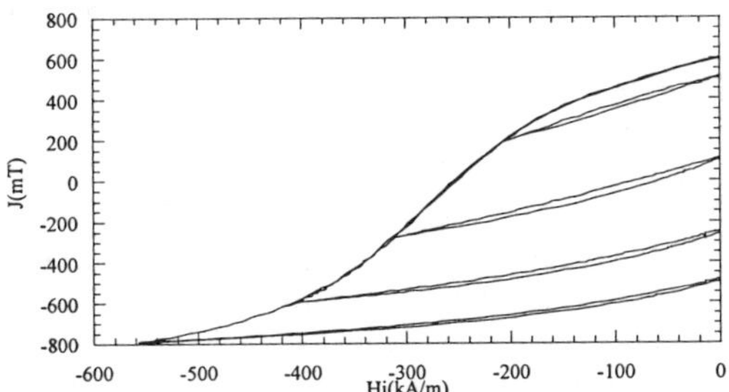

Figure 3 Part of the demagnetising curve and some recoil measurements for the epoxy bonded magnet.

Figure 3 shows that the bonded magnet exhibits reversible recoil curves typical of exchange spring behaviour (21) with two phases acting collectively.

In agreement with previous work (22), the separation between the recoil curves on increasing and decreasing the field, as shown in figure 3, is significantly greater than that in non-exchanged coupled material. In the present work the maximum vertical separation in these curves was found to be approximately 4 % of Ms, compared with 0.2 % of Ms for standard MQP-D (22). In the former the vertical separation seems to decrease as higher negative fields are applied.

CONCLUSIONS

Although the processing parameters have not yet been optimised, it is possible that these materials would make good candidates for bonded magnets. Coercivity is lower in these materials but certain advantages have been outlined (22) in that as these materials have a remanence above $0.5M_s$, the need to induce anisotropy by magnetic/mechanical means has been eliminated; lowered coercivity may even be an advantage in certain applications as these magnets may be magnetised by significantly lower fields than conventional MQI materials. Stepper motors were one of the applications mentioned where the available magnetising force can be limited.

Corrosion resistance of these magnets may also be better than that of conventional resin bonded materials, firstly as the Nd content in these materials is low and secondly as the free iron may assist in forming a protective 'envelope' surrounding the corrosion susceptible $Nd_2Fe_{14}B$ phase (21).

REFERENCES

1. J.J.Croat, J.F.Herbst, R.W.Lee and F.E. Pinkerton Appl. Phys lett. **44** (1984) 148
2. E.C. Stoner and E.P. Wohlfarth *Phil.Trans. R.Soc. A-240 (1948) 599*
3. J.E. Keem, G.G. Clemente, A.M.Kadim and R.W.Mccallum in: *Hard and Soft Magentic Materials with Applic. Inc. Superconductivity,* Proc.Conf.ASM Materils Week 87 (1987) p87
4. L.Scultz, J.Wecker and E.J. Hellstern J.Appl.Phys. **61** 3583 (1987)
5. G.C.Hadjipanayis and W. Gong J.Appl.Phys **64** 5559 (1988)

6. K.Schitzne, L.Schultz, J.Wecker amd M.Katter Appl.Phys.Lett **57** 2853 (1990)
7. P.G. McCormick, R.Street,J.Ding, Y.Liu and P.A.I Smoth in 12th Int. Workshop on RE Magnets and their Applic. (1992) 329
8. J.Ding, R.Street and P.G. McCormick JMMM **115** 211 (1992)
9. W.Gong and G.C. Hadjipanayis IEEE Trans Mag. Mater. MAG-28 2563 (1992)
10. W.Gong and G.C. Hadjipanayis J.Appl.Phys **73** 6245 (1993)
11. G.C.Hadjipanayis, L.Withanawasam and R.F. Krause IEEE Trans Magn. **31** 3596 (1995*)*
12. J.M.Yao, T.S.Chin and S.K.Chen, J.Appl.Phys **76** 7071 (1994)
13. R.Fischer, T.Schrefl, H.Kronmuller and J.Fidler *JMMM 150 329 (1995)*
14. E.H.Feutrill, P.G.McCormick and R.Street *J.Appl.Phys 75 5701 (1994)*
15. H.Fukunaga and H.Inoue *Jp.J.Appl.Phys 1 31 1347 (1992)*
16. L.H. Lewis, D.O.Welch and V.Panchanathan *J.Appl.Phys 81 8 (1997) p4422*
17. R.Fischer, T.Schrefl, H.Kronmuller and J.Fidler *JMMM 153 35 (1996)*Skomski 1993 and
18. T.Screfl, R.Fischer, J.Fidler and H.Kronmuller *J.Appl.Phys 76 7053 (1994)*
19. J.J.Croat *J.Appl.Phys 81 4804 (1997)*
20. V.Panchanathan *IEEE Trans.Magn. 31 3605 (1995)*
21. E.F. Kneller and R.Hawig *IEEE Transactions on Magnetics 27 4 (1991) p3588*
22. L. Folks, R.Street, R.C. Woodward, S.Hirosawa and H.Kanekiyo *JMMM 147 (1995) p360-366*
23. R.Skomski and J.M.D. Coey *Phys.Rev. B 48 1581 (1993)*
24. D.Jiles *Introduction to Magnetism and Magnetic Materials* Chapman & Hall 1996

Part V

Nanoscale Hard Magnetism III

MAGNETIZATION AND INTERACTIONS OF SINGLE DOMAIN Sm_2Co_{17} PARTICLES EMBEDDED IN CaO MATRIX

Wenge Liu and Paul G. McCormick
Special Research Center for Advanced Mineral and Materials Processing
The University of Western Australia, Perth, Australia.

ABSTRACT

Investigation of the interactions in two magnetic composite systems composed of ~8 vol% isolated Sm_2Co_{17} nanoparticles of different size ranges in a CaO matrix has been made by comparing the isothermal remanent magnetization curves (IRM) with the dc demagnetization curves (DCD) using Wohlfarth remanence relation. The presence of a weak demagnetizing field was evidenced in the sample with smaller magnetic particles (7-90nm), indicating the magnetostatic interaction between the magnetic particles was dominant in this sample. The large positive deviation from Wolhfarth relation in the sample with larger particles (10-250nm) suggests the presence of multigrain particles in which neighboring grains interact strongly with each other. The hysteresis curves of the two composite systems were measured at temperatures between 5 and 350K. A nearly linear temperature dependence of the coercivity was observed in both samples up to 350 K. From these results it is shown that although the sizes of the Sm_2Co_{17} particles are well below the critical size for single domain particle, there is no evidence that the magnetization reversals in these particles are coherent rotation.

INTRUDUCTION

Magnetic composites consisting of fine ferromagnetic particles dispersed in a non-magnetic matrix have been the subject of intensive investigation due to their interesting properties and potential technological applications [1-4]. While extensive work has been done on the preparation and magnetic properties of the magnetic composites containing soft magnetic nanoparticles, such as Fe (e.g., for a review see [5]), the study on the nanoparticles of magnetically hard rare-earth intermetallic compounds in a non-magnetic matrix has been lacking partly due to the difficulty in making the composites.

Magnetic composites have been made by a variety of methods, including sputter depositing [3] and mechanical milling [4]. Recently, a new technique, known as mechanochemical processing, has been developed for the fabrication of a wide range of metallic and ceramic nanoparticles from their respective oxides or chlorides [6,7]. The process involves the mechanical milling of a mixture of precursor powders (e.g., $FeCl_3$ + Na) to form a nanostructured composite structure (e.g., Fe +NaCl). The starting powders react either during milling or subsequent heat treatment so that separated nanocrystals of the desired phase embedded in a salt or oxide matrix can be obtained. The size and volume fraction of the desired particles can be adjusted by adding an appropriate amount of an inert diluent into the starting materials prior to milling.

Magnetic interactions are known to have strong influence on the magnetic properties of an assembly of ferromagnetic magnetic particles. Remanent magnetization of magnetic materials provides useful information relating to the effect of particle interactions. For an assemble of non-interacting uniaxial single domain particles Wohlfarth derived the well-known remanence relation [8]: $M_d(H) = M_{r,\,max} - 2M_r(H)$, where $M_r(H)$ is the isothermal remanence magnetization (IRM) obtained after progressively magnetizing an initially demagnetized sample, $M_{r,\,max}$ is the saturation remanence and $M_d(H)$ is the dc demagnetization (DCD) obtained after demagnetizing an initially

Mat. Res. Soc. Symp. Proc. Vol. 577 © 1999 Materials Research Society

positively satuarted sample. It has been shown by Gaunt et al [9] that the Wohlfarth relation still holds for incoherent or multidomain behavior provided that magnetization reversal occurs by an identical wall pinning mechanism during both magnetization and demagnetization. Therefore any deviations from the Wohlfarth relation can be interpreted as either magnetic interaction between single domain particles or a difference in the initial and demagnetizing processes. Following Wohlfarth's theory, Kelly et al [10] have developed the so-called δM technique which uses the difference between the measured $M_d(H)$ values and the $M_d(H)$ values predicted by the Wohlfarth relation to give a quatitative indication of the deviations from the case of noninterating, single domain particles.

In this work, two magnetic composite systems containing ~8 vol% of nanoscale Sm_2Co_{17} particles of different size ranges in a non-magnetic matrix have been prepared by mechanochemical processing. The details concerning the preparation of the magnetic composites have been reported elsewhere [11]. The δM technique has been used to examine the magnetic interactions among the particles in the two systems. In conjunction with these measurements the temperature dependence of the magnetic properties has been determined to provide further information on the magnetization processes in the Sm_2Co_{17} particles.

EXPERIMENT

The composite materials studied in this work were prepared by mechanochemical processing which involves mechanical milling and subsequent heat treatment of a mixture of Sm_2O_3, CoO and Ca with 200 wt% extra CaO as diluent. The starting materials were Sm_2O_3 (99.9%), CoO (99.5%), CaO (>96%) powders and small Ca pieces (≤ 1 mm). The reactant powders, mixed for the reaction $Sm_2O_3 + 17CoO + 20Ca = Sm_2Co_{17} + 20CaO$, were sealed in a hardened steel vial with 6.4 mm diameter steel balls. The vial was loaded and sealed under a high-purity argon atmosphere. To compensate for evaporation and oxidation of Sm during subsequent heat treatment, an additional 20 wt% Sm_2O_3 over the stoichiometric amount was added. An excess of 20 wt% Ca above the amount needed for full reduction of the starting CoO and Sm_2O_3 was added to the mixture to ensure a full reaction. To suppress the occurrence of mechanochemical combustion and to separate the product particles, 200wt% extra CaO was added to the reactants prior to milling. Milling was carried out in a Spex 8000 mixer/mill for 12 hours. The as-milled samples were heat treated at 600°C (**Sample A**) and 800°C (**Sample B**), respectively, for 30 min in silica tubes under a continuously pumped vacuum of less than $5x10^{-7}$ Torr.

X-ray diffraction (XRD) patterns were measured using a Siemens D5000 x-ray diffractometer with monochromatic CuK_α radiation. The microstructure studies were carried out on a Philips EM430 transmission electron microscope (TEM) equipped with energy dispersive spectroscopy (EDS) for composition analysis. Special sealed sample holders were used for both the measurements mentioned above. The hysteresis loops were measured with a Quantum Design SQUID magnetometer between $\pm$70kOe and 5-350K. The IRM and DCD were measured at room temperature (~298K) on cold-pressed cylindrical samples using an Aerosonics VSM3001 vibrating sample magnetometer with a maximum applied field of 50 kOe. Corrections for demagnetisation were made using the actual magnetization of the specimen and a demagnetization factor taken from [12].

RESULTS

Microstructure characterization

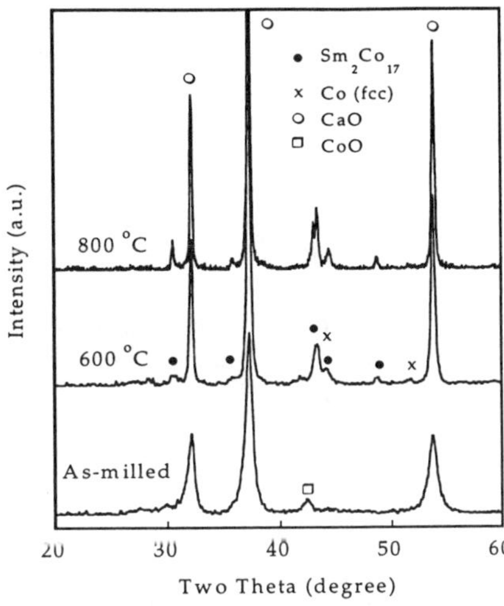

Fig. 1 X-ray diffraction patterns for as-milled and heat treated samples

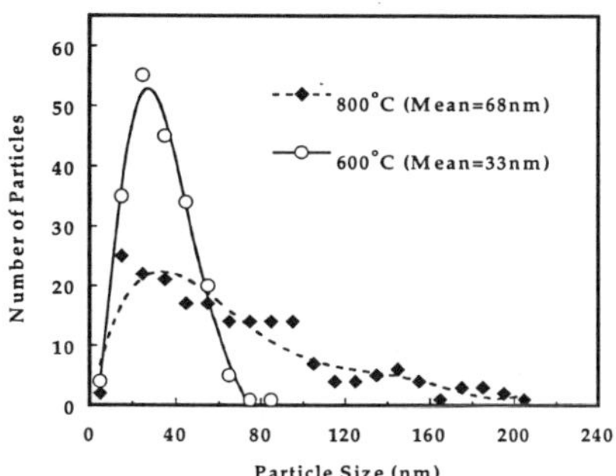

Fig. 2 Size distribution of the Sm_2Co_{17} particles measured on TEM micrographs of washed samples

Fig. 1 shows the XRD patterns for the as-milled and heat treated sample. Mechanical milling produced a nanocrystalline CaO supersaturated with Sm_2O_3, CoO and Ca as indicated by the broadened and shifted CaO peaks in the XRD pattern of the as-milled sample. Magnetic measurements revealed the presence of a soft magnetic phase in the as-milled sample, suggesting the partial reduction of CoO to Co by Ca during milling. Heat treatment at 600°C resulted in a composite structure consisting of CaO, Sm_2Co_{17} and a small amount of fcc Co. The Co phase disappeared in sample B which was heat treated at 800°C. It was also noted that the peaks corresponding to the Sm_2Co_{17} phase in sample B were sharper than those in sample A due to the grain growth at higher heat treatment temperature.

TEM observations revealed that most of the isolated Sm_2Co_{17} particles in sample A were spherical in shape and had particle sizes ranging from 7-90nm. The complex composite structure and fine particle size prevented the differentiation of the Sm_2Co_{17} particles from the Co particles. As shown in Fig. 2 the Sm_2Co_{17} particles in sample B were much coarser and had a wider size distribution (10-250nm) compared with their counterpart in sample A. In addition, the particle shape changed from spherical in sample A to equiaxed in sample B. It should be noted that the particle sizes in both samples are well below the critical size for single domain particles which has been estimated to be ~500nm for Sm_2Co_{17} alloys.

Magnetic measurements confirmed the XRD results. The hysteresis loop of sample A showed two phase behavior as given in Fig. 3. Susceptibility analysis suggested that the hard phase, i.e. Sm_2Co_{17}, in sample A had a coercivity value of 11.1 kOe while the soft phase, presumably being fcc Co, showed a coercivity of ~400Oe. Sample B exhibited single hard phase behavior with a coercivity value of 13.7kOe as shown in Fig. 3 and 4. These coercivity values are less than half of the predicted values (~35 kOe) for an assembly of randomly oriented non-interacting Sm_2Co_{17} single domain particles coherent rotation process in which the magnetization reversal occurs by coherent rotation.

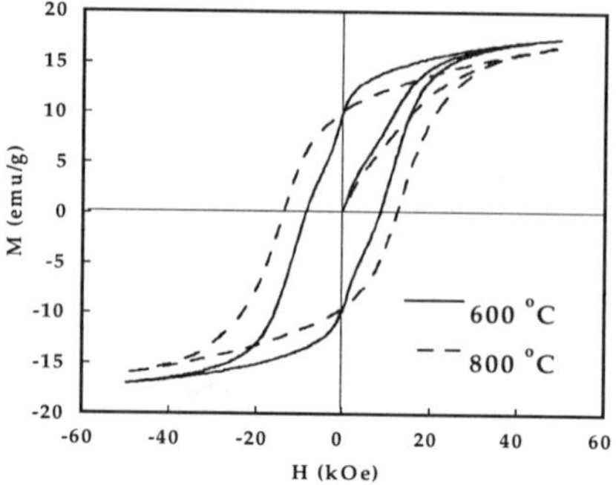

Fig. 3 Hysteresis curves of the samples heat treated at 600°C and 800°C as measured on VSM

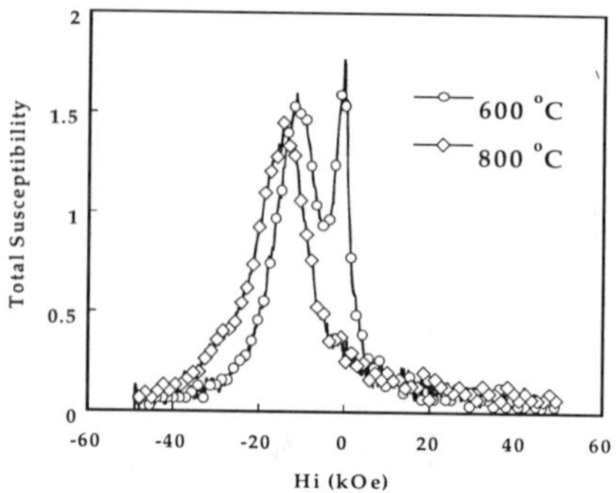

Fig. 4 Total susceptibility as a function of internal field along the demagnetization curve in Fig. 3

Magnetic measurements

The coercivity, the maximum magnetization and the remanent magnetization in an applied field of 70kOe are plotted as a function of temperature in Fig. 5 for both samples. The predicted $T^{1/2}$ dependence of coercivity for single domain particles was not observed. Instead, the coercivity decreases almost linearly with increasing temperature for both samples. This deviation from the theoretical behavior is attributed to the non-coherent magnetization processes and magnetic interactions among the particles.

Maximum magnetization generally increased with increasing temperature in both samples, indicating that the maximum applied field of 70kOe in SQUID was not enough to saturate the samples, especially at low temperature regime. The remanence of sample A decreased monotonically with rising temperature while that of sample B remained largely unchanged.

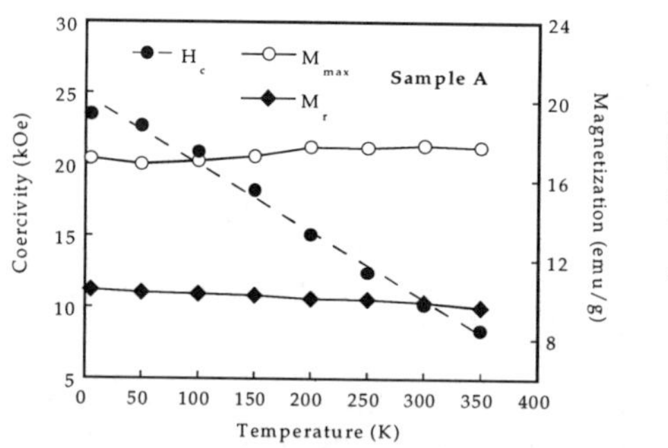

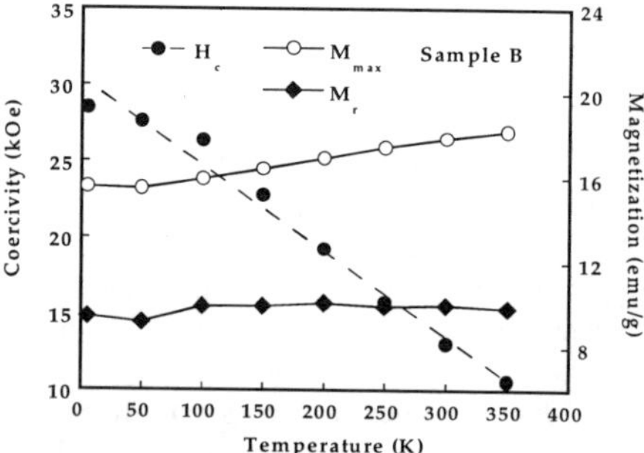

Fig. 5 Coercivity (H_c), maximum magnetization (M_{max}) and remanent magnetization (M_r) measured in an applied field of ±70kOe are plotted as a function of temperature for samples A and B

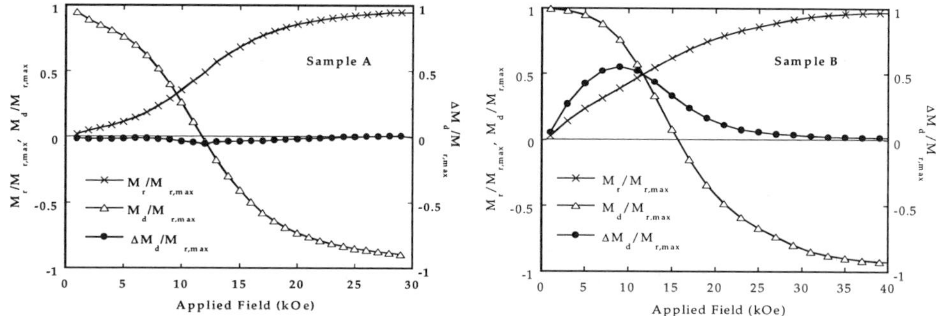

Fig. 6 Normalized initial remanence ($M_r/M_{r,max}$), normalized demagnetization remanence ($M_d/M_{r,max}$) and normalized difference ($\delta M_d/M_{r,max}$) between the measured M_d value and the calculated $M_d(H)=M_{r,max}-2M_r(H)$ as a function of applied field for samples A and B

The IRM and DCD data are summarized in Fig 6 for samples A and B. For sample A the normalized δM_d had small negative values up to an applied field of 30 kOe with a peak value of ~-0.06 at an applied field slightly higher than the coercivity of the sample. This indicates that the net effect of interparticle interactions is demagnetizing in nature, suggesting the presence of a weak magnetostatic interaction field in the sample. For sample B the normalized δM_d reached a peak at an applied field of ~7kOe, much lower than the coercivity of the sample. The relative difficulty in demagnetizing the sample is attributed to exchange interactions between neighboring grains, indicating the existence of multigrain particles in the sample. Similar positive interactions were observed in nanostructured Nd-Fe-B and Sm-Co materials [13, 14].

CONCLUSIONS

Composite systems consisting of ~8 vol% isolated Sm_2Co_{17} particles embedded in a non-magnetic matrix were prepared by mechanochemical processing. Higher heat treatment temperature yelded larger particles with broader size distribution. At room temperature, the Sm_2Co_{17} particles in sample A, which have a smaller average size and a narrow size distribution, shows slightly lower coercivity than the Sm_2Co_{17} particles in sample B. However, both coercivity values are less than half of the values calculated for a coherent rotation process, indicating other mechanisms must be responsible for the magnetization reversals in the particles. The coercivity of both samples showed nearly linear temperature dependence in the temperature range of 5K to 350 K.

Isothermal remanent magnetization (IRM) and dc demagnetization (DCD) measurements revealed the nature of interactions among the magnetic particles in the composites. A weak demagnetizing field was present in sample A, indicating that the interactions among Sm_2Co_{17} nanoparticles interact was magnetostatic in nature. The large positive deviation from Wolhfarth relation in sample B suggests the presence of the multigrain particles in which strong exchange interactions exist between the neighboring grains.

ACKNOWLEDGEMENTS

The authors would like to thank Prof. R. Street, Dr. A.W.S. Johnson and R. Woodward for helpful discussions.

REFERENCES

1. M. Pardave-Horvath and L. Takacs, Scripta Metall. Mater., **33**, 1731 (1995).
2. S.A. Majetich, J. H. Scott, E.M. Kirkpatrick, K. Chowdary, K. Gallgher, M.E. McHentry, Nanostructured Mater., **9**, 291 (1997).
3. I. Panagiotopoulos, X. Meng-Burany, and G.C. Handjipanayis, J. Magn. Magn. Mater., **172**, 225 (1997).
4. R.L. Schalek, D.L. Leslie-Pelecky, J. Knight, D.L. Sellmyer, and S.C. Axtell, IEEE Trans. Mag., **31**, 3772 (1995).
5. C.L. Chien in *Science and Technology of Nanostructured Magnetic Materials*, edited by G.C. Hadjipanayis and G.A. Prinz, Plenum Press, New York, 1991, 477.
6. J. Ding, W.F. Miao, P.G. McCormick, and R. Street, Appl. Phys. Lett., **67**, 3804 (1995).
7. J. Ding, T. Tsuzuki, P.G. McCormick, and R. Street, J. Am. Ceram. Soc., 79, 2956 (1996).
8. E.P. Wohlfarth, J. Appl. Phys., **29**, 595 (1958).
9. P. Gaunt, G. Hadjipanayis, and D. Ng, J. Magn. Magn. Mater., **54-57**, 841 (1986).
10. P.E. Kelly, K. O'Grady, P.I. Mayo, and R.W. Chantrell, IEEE Trans. Mag., **25**, 3881 (1989).
11. W. Liu and P.G. McCormick, J. Magn. Magn. Mater., **195**, 279 (1999).
12. D.X. Chen, J.A. Brug, and R.B. Goldfarb, IEEE Trans. Mag., **27**, 3601 (1991).
13. L. Folks, R. Street, and R. Woodward, J. Appl. Phys., **75**, 6271 (1994).
14. J. Ding, P.A.I. Smith, P.G. McCormick, and R. Street, J. Magn. Magn. Mater., **161**, 303 (1996).

MAGNETIC PROPERTIES OF α Fe/NdFeB NANOCOMPOSITE ALLOYS PREPARED BY TWO-STEP RAPID SOLIDIFICATION

M. YAMASAKI*, H. MIZUGUCHI*, H MORIOKA*, M. HAMANO*, A. INOUE**
*AMC Div, Toda Kogyo Corp, Hiroshima, 739-0652 JAPAN
**Inst. for Materials Research, Tohoku Univ., Sendai, 980-8577 JAPAN

ABSTRACT

In order to obtain directly powders of nanocomposite α Fe/NdFeB alloys with the composition of $Nd_xFe_{bal}Co_8Nb_yB_z$, where x: 8.0-8.5, y: 1.5-2.5, z: 6.0-6.5 in at%, have been prepared using the two-step rapid solidification apparatus, which is composed of a gas atomizing system at the upper side and a collision-and-solidificating rotor system at the lower side. It is found that the coercivity (HcJ) of flat-shaped powders obtained depends on powder sizes. The best HcJ in each sieved powder is 525 kA/m (6.60 kOe) for a size more than 500 μ m, 449 kA/m (5.64 kOe) for 150-500 μ m, and 391 kA/m (4.91 kOe) for 50-150 μ m. These values are not so high compared to those of powders of ribbons prepared using a single-roller melt spinning apparatus. This should be attributed to the powder size distribution that causes fluctuation of the quenching rate. Thickness of the flat powder is found to be controllable up to 70 μ m, which is useful to get compression molded bonded magnets with a high density and thus a high remanence (Br). An epoxy-bonded magnet with (BH)max: 61.8 kJ/m^3 (7.76 MGOe), Br: 0.732 T (7.32 kG), HcJ: 427 kA/m (5.36 kOe), and density of 6.20 Mg/m^3 is obtained.

INTRODUCTION

As a next-generation hard magnets, nanocomposite alloys, that is so-called exchange spring magnets, have been interested since they were reported firstly [1]. Especially Fe_3B/NdFeB and α Fe/NdFeB alloys have been widely investigated. Recently, amorphous-phase remaining α Fe/NdFeB alloys were found to show a good hard magnetic properties by Inoue et al.[2-5]. The above works have been conducted almost using melt-spinning apparatuses having a single-roller. It have been suggested that, as an alternative method in the production of rapidly quenched materials, the two-step rapid solidification apparatus offers the significant advantage over the conventional melt-spinning apparatuses, that is, a large volume of quenched powders, not ribbons which need pulverization, can be directly produced in a single operation [6,7]. So far, however, the apparatus have been used to prepare non-magnetic or soft magnetic powders with the thickness below 10 μ m.

In the application for bonded hard magnets, however, the thickness of the powder must be about 40 μ m. The reason for this is that the optimum quench rate for higher HcJ should be attained in the present alloys and that the density of bonded magnets prepared by compression molding should be enough to have higher remanence Br, in the vicinity of 40 μ m as thickness.

In this paper, magnetic properties of rapidly quenched powders of nanocomposite α Fe/NdFeB alloys prepared using a two-step rapid solidification apparatus are reported.

EXPERIMENT

Alloys of α Fe/NdFeB with the composition of $Nd_xFe_{bal}Co_8Nb_yB_z$, where x: 8.0-8.5, y: 1.5-2.5, z: 6.0-6.5 in at%, were prepared by induction melting and casting. Using a two-step rapid solidification apparatus as shown in Fig. 1, about 500 to 1000 g of the alloy was melted in an Al_2O_3 crucible by induction under an Ar atmosphere. Then the melt was gas-atomized by high-pressure Ar gas, and atomized melt was immediately reach on a collision-and-solidificating rotor. Thus, rapidly-quenched and flat-shaped powders were directly obtained by this apparatus.

The quenching rate of the melt can be controlled by experimental conditions such as temperatures of the melt, the diameter of the melt injecting nozzle, the diameter of the gas

Mat. Res. Soc. Symp. Proc. Vol. 577 © 1999 Materials Research Society

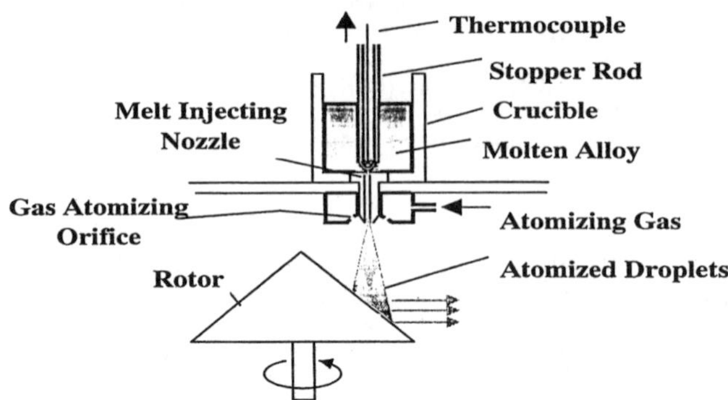

Fig. 1. Schematic illustration of the two-step rapid solidification apparatus incorporating high-pressure gas atomizing and collision-and-solidification systems.

atomizing orifice, pressures of atomizing Ar gas, rotating speeds of the rotor and the distance from the nozzle to the rotor. Some of these conditions were set up variously to get powders with optimally quenched state in order to realize higher HcJ. The quenched powders were sieved in air to be classified into three powder fractions of 50-150, 150-500 and more than 500 μ m.

For the sake of comparison the melt spun alloy ribbons were also prepared by using a single-roller melt spinning equipment having a copper roller of 200 mm in diameter and a quartz tube with an orifice diameter of about 0.5 mm. The surface velocity of the roller was determined to be in the range of 10-15 m/s for getting higher HcJ. Resulting ribbons were of 30-50 μ m in thickness and about 1 mm in width. Ribbons were then pulverized by ball milling and were sieved in air.

Annealing treatment of powders was performed with the vacuum-sealed quartz tubes in a furnace heated up to 700-780 ℃ for 180-300 s. Magnetic properties of heat-treated powders were measured with a vibrating sample magnetometer (VSM) in field up to 1.6 MA/m (20 kOe) at room temperature. No correction of the demagnetization field was made.

Bonded magnets were made from the classified powders in the range of 150-500 μ m. An epoxy resin of 1.5 wt% was mixed with powders and resulting compounds were compression molded at pressure of 980MPa. After curing treatment of the resin at 150 ℃ for 3.6 ks, bonded magnets with a size of 10 mm in diameter and 7 mm in height, that is nearly equal to a permeance coefficient Pc=2, were obtained. Magnetic properties of bonded magnets magnetized with a pulse current in field up to 4.8 MA/m (60 kOe) were measured with a B-H curve tracer in field up to 1.6 MA/m (20 kOe) at room temperature.

RESULTS

Table I shows some of the optimum parameters for the two-step rapid solidification and the maximum HcJ attained at composition of $Nd_{8\text{-}8.5}Fe_{bal}Co_8Nb_{2\text{-}2.5}B_{6\text{-}6.5}$, and the powder size fraction of 150-500 μ m. It is found that higher HcJ can be attained at the alloy compositions with Nb≒2-2.5 at%. In this case, as is seen in the table, when the diameter of the melt injecting nozzle orifice is set smaller, the optimum pressure of atomizing gas should be lower and the optimum diameter of gas atomizing orifice should be larger. On the gas atomized melt droplets, the geometric mean diameter D_{av} is given by the following equation [8].

$$D_{av} = K\left[\frac{\nu_m}{\nu_g} \bullet \frac{1}{W}\left(1 + \frac{M_m}{A_g}\right)\right]^{1/2} \bullet D_m \qquad (1)$$

Table Ⅰ. Optimum parameters for the two-step rapid solidification and HcJ at composition: $Nd_{8-8.5}Fe_{bal}Co_{8}Nb_{2-2.5}B_{6-6.5}$, (powder fraction: 150-500 μ m)

Diameter of melt injecting nozzle (mm)	**1.0**	**1.0**	**3.0**
Atomizing gas pressure (MPa)	**0.50**	**0.50**	**1.25**
diameter of atomizig gas orifice (mm ϕ)	**1.5**	**1.2**	**0.8**
Maximum HcJ (kA/m)	**449**	**442**	**449**

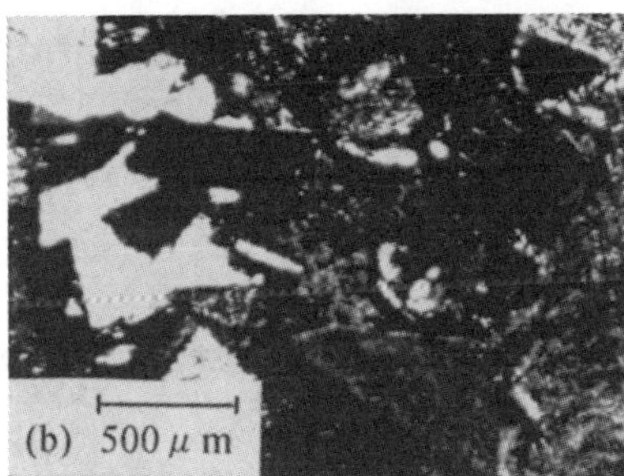

Fig. 2. Optical microscopy of powders prepared by the two-step rapid solidification method (a), and by the single-roller melt spinning method (b). (powder fraction: 150-500 μ m)

Where, K: constant, D_m: diameter of melt stream through the nozzle, ν m: Kinematic viscosity of melt, ν g: Kinematic viscosity of gas, M_m: mass flow rate of melt, A_g: mass flow rate of gas, and W: the Weber number ($W=V^2 \cdot \rho_m \cdot d_m / \gamma_m$, where V: velocity of gas at impact with melt stream, ρ_m: density of liquid, γ_m: surface tension of liquids)

In this study some of the parameters were fixed, that is, the rotating speed of the roter:5000 rpm, the conical angle of rotor:120° , and the distance from the nozzle to the rotor:200 mm. As described above, it is assumed that the diameter of the melt injecting nozzle is nearly equal to the D_m and that the product of the pressure of atomizing gas multiplied by the diameter of gas atomizing orifice is proportional to A_g. In Table Ⅰ, we can understand that when the product of the pressure of atomizing gas by the diameter of gas atomizing orifice is nearly constant, that means A_g is nearly constant, the maximum HcJ are realized. According to the above equation, this may be attributed to the optimum D_{av} which is directly related to the quenching rate of the melt droplet.

By the microscopic observation of the powder prepared by the two-step rapid solidification but not sieved yet, a mixture of the morphology of flaky, thorny and spherical was found. Powders of different shape may receive different quenching rate and hence may result in to have different HcJ that causes lowering of average HcJ of the non-sieved powder. Fig. 2 (a) shows a photograph of the powder in the sieved fraction of 150-500 μ m prepared by the two-step rapid solidification method in the conditions; the diameter of the melt injecting nozzle is 1.00mm ϕ, the pressure of atomizing gas is 1.0MPa, and the diameter of gas atomizing orifice is 1.20mm ϕ. In this sieved case, particles have a flat shaped structure and the thickness of 40～50 μ m. For comparison, a photograph of the powder prepared by the single-roller melt spinning method is shown in Fig. 2 (b). The shape of the powder is similar to that shown in Fig. 2 (b) excepting the micro-texture of the surface.

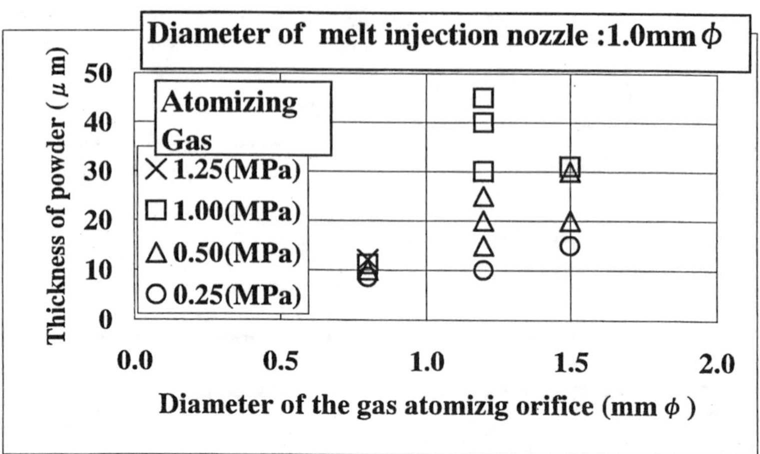

Fig. 3. Thickness of powders as a function of the diameter of gas atomizing orifice.

Table II. Comparison of magnetic properties of powders and bonded magnets of the $Nd_8Fe_{75.5}Co_8Nb_2B_{6.5}$ alloys prepared by different methods.

Method	Two-step rapid solidification	Single-roller melt spinning
Powder		
HcJ: kA/m	432	517
Bonded magnet		
Br: T	0.732	0.758
HcJ: kA/m	427	467
(BH)max: kJ/m^3	61.8	71.1

Fig. 3. shows the change in thickness of powders as a function of the diameter of gas atomizing orifice. It is seen that the thickness becomes thicker as increasing the diameter of gas atomizing orifice or the atomizing gas pressure. Although we can not elucidate the above tendency of the thickness obviously at present, it was found that thickness of the flat powder is controllable up to 70 μ m by this study which have not been reported so far regarding the two-step rapid solidification method. As mentioned in the early portion of this paper, controllability of the thickness of powder is much important to realize the higher performance of compression molded bonded magnets. The best HcJ measured in each sieved powder fraction is 525 kA/m(6.60 kOe) for a size more than 500 μ m, 449 kA/m(5.64 kOe) for 150-500 μ m, and 391 kA/m(4.91 kOe) for 50-150 μ m. These values are not so high as those measured for pulverized ribbons prepared using the single-roller apparatus as shown in Table II. This should be attributed to the powder size distribution even in the sieved powder fraction that causes fluctuation of the quenching rate.

Table II shows the comparison of magnetic properties of powders prepared using the two-step rapid solidification method and the single-roller melt spinning method and of bonded magnets made of the powders. It can be seen that all properties of the powders and bonded magnets prepared by the two-step rapid solidification method are lower than those by the single-roller melt spinning method. Further investigations to improve these magnetic properties are now in progress.

CONCLUSIONS

Using the two-step rapid solidification apparatus, powders of nanocomposite α Fe/NdFeB alloys with the composition of $Nd_xFe_{bal}Co_8Nb_yB_z$, where x: 8.0-8.5, y: 1.5-2.5, z: 6.0-6.5 in at%, have been directly prepared. The coercivity (HcJ) of flat-shaped powders is found to depend on powder sizes according to the quenching rate received during the gas atomizing process. The best HcJ in each sieved powder is 525 kA/m (6.60 kOe) for a size more than 500 μm, 449 kA/m (5.64 kOe) for 150-500 μm, and 391 kA/m (4.91 kOe) for 50-150 μm. These values are not so high comparing to the HcJ of pulverized ribbons of the same composition prepared using a single-roller melt spinning apparatus. Thickness of the flat powder is found to be controllable up to 70 μm by this study, which is useful to get compression molded bonded magnets with a high density and thus a high remanence Br. An epoxy-bonded magnet with (BH)max: 61.8 kJ/m^3 (7.76 MGOe), Br: 0.732 T (7.32 kG), HcJ: 427 kA/m (5.36 kOe), and density of 6.20 Mg/m^3 is obtained.

REFERENCES

[1] E.F. Kneller and R. Hawig: IEEE Trans. Magn., **27-4**, 3588 (1991).

[2] A. Inoue, A. Takeuchi, A. Makino and T. Masumoto: IEEE Trans. Magn., **31-6**, 3626 (1995).

[3] A. Inoue, A. Takeuchi, A. Makino and T. Masumoto: Mater.Trans., JIM, **36-7**, 962 (1995).

[4] A. Inoue, A. Takeuchi, A. Makino and T. Masumoto: Sci. Rep. RITU, A**42-1**, 143 (1996).

[5] A. Takeuchi, A. Inoue and A. Makino: Mater. Sci.and Engin., A**226-228**, 458 (1997).

[6] A. Inoue, T. Komura, J. Saida, M. Oguchi, H.M. Kimura and T. Masumoto: Int. J. Rapid Solidification, **4**, 181 (1989).

[7] M. Oguchi, A. Inoue and T. Masumoto: Mater. Sci. Eng., A**133**, 688 (1991).

[8] H. Lubanska: J. Met., **22**, 45 (1970).

MAGNETIC DOMAIN OBSERVATION ON MELT-SPUN Nd-Fe-B RIBBONS USING MAGNETIC FORCE MICROSCOPY

A. GAVRIN*, C. SELLERS[†], S.H. LIOU[§]
*Dept. of Physics, Indiana University Purdue University Indianapolis, Indianapolis, IN 46202, agavrin@iupui.edu
[†]Magnequench International, Inc., 6435 Scatterfield Rd., Anderson, IN 46013
[§]Behlen Laboratory of Physics, University of Nebraska, Lincoln, NE 68588

ABSTRACT

We have used Magnetic Force Microscopy (MFM) to study the magnetic domain structures of melt-spun Nd-Fe-B ribbons. The ribbons are commercial products (Magnequench International, Inc. MQP-B and MQP-B+) with a thickness of approximately 20 microns. These materials have identical composition, $Nd_{12.18}B_{5.36}Fe_{76.99}Co_{5.46}$, but differ in quenching conditions. In order to study the distribution of domain sizes through the ribbon thickness, we have prepared cross-sectional samples in epoxy mounts. In order to avoid artifacts due to tip-sample interactions, we have used high coercivity CoPt coated MFM tips. Our studies show domain sizes typically ranging from 50-200 nm in diameter. This is in agreement with studies of similar materials in which domains were investigated in the plane of the ribbon. We also find that these products differ substantially in mean domain size and in the uniformity of the domain sizes as measured across the ribbon. While the B+ material shows nearly uniform domain sizes throughout the cross section, the B material shows considerably larger domains on one surface, followed by a region in which the domains are smaller than average. This structure is presumably due to the differing quench conditions. The region of coarse domains varies in thickness, disappearing in some areas, and reaching a maximum thickness of 2.75 μm in others. We also describe bulk magnetic measurements, and suggest that the uniformity of the B+ domain structure is responsible for its superior magnetic performance.

INTRODUCTION

Since their discovery in 1984 [1, 2], a great deal has been learned about the properties of NdFeB-type magnetic materials; however, a wide gap remains between the best commercial products and the theoretical promise of these materials. As an example, it is expected that the $Nd_2Fe_{14}B$ phase is capable of producing energy products up to about 64 MGOe [3], however, commercial magnets are only available up to about 48 MGOe. Some of this gap is doubtless a result of our incomplete understanding of the interplay between composition and processing conditions on the one hand, and magnetic properties on the other. As an example of this interplay, consider the properties of NdFeB magnetic materials produced by melt-quenching. The magnetic properties of the as-quenched materials vary rapidly with the wheel speed used in the quenching process—too slow a speed results in too coarse a grain structure, whereas too high a speed results in large quantities of glassy material. However, the speed that results in optimum properties varies significantly in response to small changes in composition. To complicate matters further, most commercial materials are annealed after the quenching process, and the quenching conditions that produced optimal materials before annealing may not produce the best annealed material [1].

To help close the gap between theory and reality, it is necessary to gain a greater understanding of the interplay among composition, processing conditions, microstructure, and mag-

Mat. Res. Soc. Symp. Proc. Vol. 577 © 1999 Materials Research Society

netic properties. To accomplish this, one important step will be a detailed understanding of the magnetic microstructures formed in these materials, and the relationship of this structure to the other variables. The first such step was made by Lorentz imaging studies carried out by Chapman, et al., [4]. These studies showed a magnetic microstructure with a domain size several times larger than the crystalline microstructure. These domains, often termed interaction domains, were subsequently observed by MFM [5, 6].

In this work, we use Magnetic Force Microscopy (MFM) to study the magnetic structure of several NdFeB ribbons in the as-quenched state. In order to elucidate the effects of the quenching process on the magnetic microstructure, we have prepared cross sectional samples of the ribbons. Thus, we are able to discern differences in the magnetic microstructure caused by the different cooling rates across the thickness of the ribbon.

EXPERIMENT

Sample Preparation

We will compare two sample materials, designated MQP-B and MQP-B+. Both materials are commercial products from Magnequench International, Inc. MQP-B is a standard Magnequench product sold both as a powder and in finished magnets. The most important characteristics of MQP-B powder are its high remanence, coercivity, and energy product. MQP-B+ is a new variant of the original product. The new material will represent a significant improvement over its predecessor, particularly in an improved coercivity and energy product. The key properties of these materials (in the post-anneal state) are summarized in Table I. It is important to note that these materials have identical composition, $Nd_{12.18}B_{5.36}Fe_{76.99}Co_{5.46}$, but differ in quenching conditions (for MQP-B+, the processing conditions have been adjusted to improve the consistency of the heat transfer).

For our MFM studies, both materials are in the form of as-quenched ribbons set in epoxy; both samples are approximately 20 μm thick. Several pieces of ribbon share each epoxy mount so that typical regions may be identified and studied. In order to study the effects of differential cooling rates across the thickness of the ribbons, the sample pieces were placed between two thin sheets of stainless steel and mounted edge on. After setting the epoxy, the samples were polished using common metallographic preparation methods.

Table I Comparisons of the magnetic properties of the two materials after annealing.

Material	Residual Induction B_r (kG)	Intrinsic Coercivity H_{ci}(kOe)	Energy Product BH_{max} (MGOe)
MQP-B	8.0-8.4	8-10	12
MQP-B+	>8.4	8-10	13.8-15.0

Measurements

Magnetic Force Microscopy was performed on each of the samples using a Digital Instruments Dimension 3000 system operating in Tapping/Lift mode. In many studies of high remanence materials, MFM measurements are complicated by the strong sample-tip interaction. In many cases, the tip is remagnetized by the sample fields, resulting in artifacts such as streaks leading away from domain walls in the scan direction[6, 7]. In order to minimize these artifacts, we used standard silicon cantilevers coated at the University of Nebraska with a high-coercivity CoPt film. The coercivity of the tips is estimated to be 1.5 T [8]. This allows us to study the

NdFeB materials without tip reversal artifacts, even at very low lift heights. The measurements we will describe here were made at lift heights of 10-30 nm. The tips were magnetized in the plane of the sample using a 5 T field.

Bulk magnetic data were obtained using a Lakeshore model 7300 VSM. Annealed ribbons were ground to a powder, then approximately 0.67 g of each material were compacted to avoid particle rotations and free vibrations during the measurement. Each sample was pulsed magnetized in a field of approximately 5T. Samples were exposed to a 0.5 T field in the VSM then swept through the second quadrant.

RESULTS

MFM Measurements

Our studies show domain sizes ranging from 50 nm to approximately 400 nm in diameter. This is in agreement with studies of similar material, in which the domain structure was imaged in the plane of the ribbon [3-6]. However, we find that there are substantial differences between the domain structures of these two materials. Fig. 1 shows micrographs of each of the two ribbons. It is clear from these images that the domain structure is significantly smaller in the B+ ribbon. Furthermore, the domain structure of the B+ ribbon is quite uniform, whereas the MQP-B ribbon shows a region of coarse grains near the surface. This region varies in thickness, disappearing in some areas and reaching a maximum thickness of approximately 2.75 μm in others.

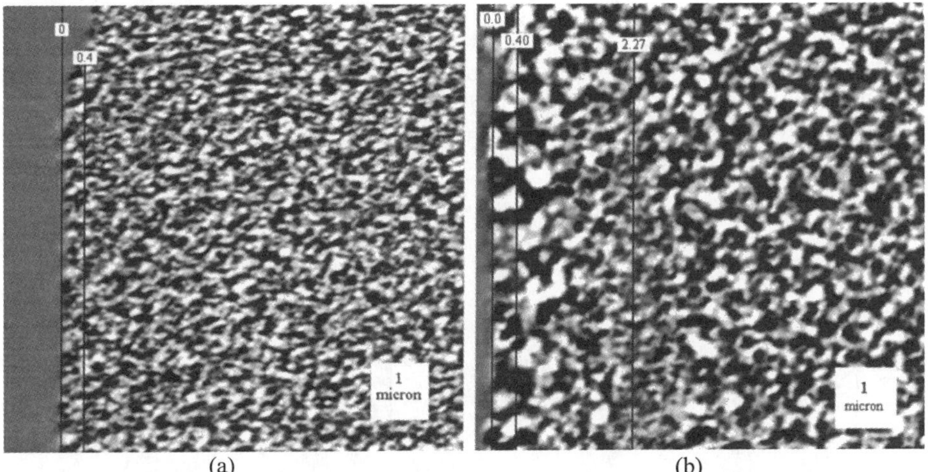

Figure 1. Cross-sectional MFM micrographs of (a) a Magnequench MQP-B+ ribbon and (b) a Magnequench MQP-B ribbon. The vertical lines marked 0.0 represent the edge of the ribbon, and the lines marked 0.4 represent the first line scan along which domain sizes were examined, a distance of 0.40 microns from the edge. The line marked 2.27 in (b) corresponds to the minimum of domain size shown in Fig. 2 (b).

Because the individual domains are difficult to resolve, and because the contrast in these measurements is a complex function of domain wall structure and domain orientation, we have not yet developed analytical methods that yield domain sizes directly. In order to characterize the domain structures we have calculated an "effective" domain size as follows. Beginning with a line profile through the image data, we take the spatial derivative of the data along the profile, then smooth the result using a 3-point boxcar average. Finally, we search for zero-crossings in the smoothed derivative, which indicate extrema in the image data. We define the effective domain size to be the mean distance between zero-crossings (hence, the mean distance between successive extrema).

Although we would prefer a direct measurement of the mean domain size, calculating the effective domain size does allow us to make quantitative comparisons among our samples. This method makes two significant errors: the effective domain size tends to underestimate the actual domain size because the profiles do not pass through the center of each domain, and this method will miss some domains, because the MFM contrast mechanism does not distinguish between domains oriented antiparallel to one another but perpendicular to the polarization direction of the tip. These errors will remain consistent among profiles so long as the domain shapes do not vary. Thus, we can use this method to compare domain sizes among similar samples and among regions within a sample. It should also be noted that the boxcar average and numerical differentiation result in a minimum domain size of three pixels. However, the images discussed here are all 512× 512 pixel images covering an 8×8 μm area. This corresponds to a pixel size of 15.63 nm, and a minimum domain size of 46.88 nm. In examining our data, we have found no evidence of magnetic structure below this length scale.

We have calculated the effective domain size as a function of distance from the ribbon edge for each of the images shown in Fig. 1. In each case, profile were spaced at ten pixel (0.156 μm) intervals, beginning 25 pixels from the edge of the ribbon. The results of this analysis, shown in Fig. 2, are quite striking. The analysis of the MQP-B ribbon (Fig. 2b) shows that in addition to coarse domains near the surface there is also a minimum in the domain size approximately 2.27 μm from the surface. In contrast, Fig. 2a shows that the domain size in the MQP-B+ ribbon is quite uniform. Furthermore, comparison of these two images shows that the domains in the MQP-B+ ribbon are approximately 25% smaller than those in the MQP-B ribbon.

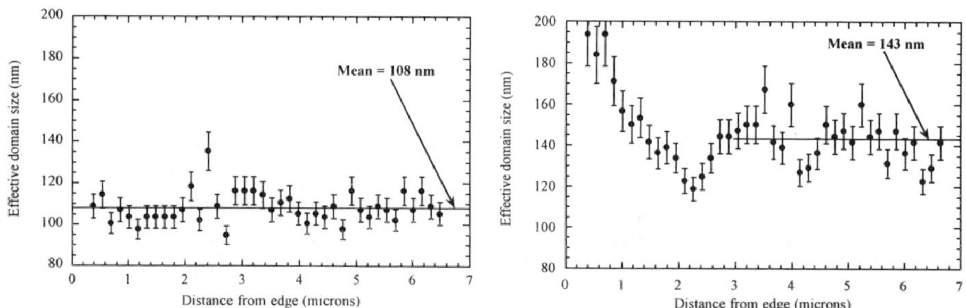

Figure 2. The effective domain size (as defined in the text) calculated for profiles parallel to the edges of (a) MQP-B+ and (b)MQP-B ribbons. These data were derived from profiles taken at 10 pixel intervals from the images shown in Fig. 1.

These results are consistent with our overall expectations of the domains sizes in quenched NdFeB. These materials are expected to have domain sizes that are comparable to the crystallographic grains size [9], which is in the range 50-75 nm for these ribbons [10]. The effective domain sizes that we calculate here are somewhat larger, perhaps due to the presence of interaction domains. Domains somewhat larger than the grain size have also been observed in similar materials by other groups [11].

Magnetization Measurements

The results of bulk magnetization measurements are shown in Fig. 3. As indicated in Table I, the MQP-B+ powder shows significantly better magnetic properties. The residual induction and shape of the curve are both improved, leading to a substantial increase in the maximum energy product. The data shown here yield a maximum energy product of 14.3 MGOe for the B+ powder compared to 12.7 MGOe for the B powder, an improvement of 12.6%.

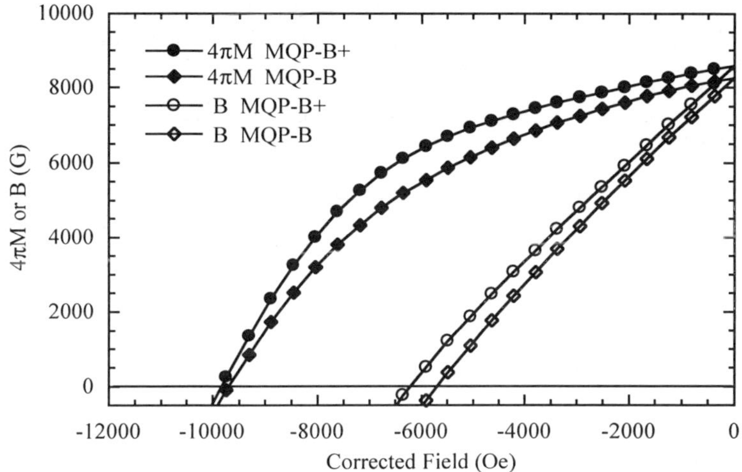

Figure 3. Magnetization and induction field plotted in vs. applied field in the second quadrant. The solid lines connect the data markers as a guide to the eye.

CONCLUSIONS

We have investigated two NdFeB-type materials in the form of as-quenched ribbons. The materials are commercial products of Magnequench International, Inc., and are designated MQP-B and MQP-B+. We investigated cross sections of these ribbons using magnetic force microscopy (MFM). We have also measured the bulk magnetic properties of annealed samples of these ribbons. Although the ribbons have identical compositions, the different quenching conditions lead to substantially different magnetic microstructures and magnetic properties. In particular, we have shown that the B+ material has a smaller and more uniform domain size distribution. As a result of these differences, the B+ material has substantially improved magnetic properties.

ACKNOWLEDGEMENTS

Research at the University of Nebraska was supported by the Army Research Office, grant DAAG55-98-1-0014. Research at IUPUI was supported by Magnequench International, Inc.

REFERENCES

1. J.J. Croat, J.F. Herbst, R.W. Lee, and F.E. Pinkerton, Appl. Phys. Lett. **4 4**, 148 (1984); J. Appl. Phys. **55**, 2078 (1984).

2. M. Saqa, S. Fujimori, N. Togawa, H. Hamamota and Y. Matsuura, J. Appl. Phys. **55**, 2083 (1984).

3. J. J. Croat, J. Mater. Eng. **10**, 7 (1988); J.F. Herbst, J.J. Croat, F. E. Pinkerton, and W.B. Yelon, Phys Rev. B **29**, 4176 (1984).

4. J.N. Chapman, S. Young, H.A. Davies, P.Z. Zhang, A. Manaf, and R.A. Buckley, in *Proc. 13th Int. Workshop on Rare Earth Magnets* edited by C.A.F. Manning, D. Jones, A. Williams, and I. Harris (University of Birmingham, 1994) p. 95.

5. M.R.J. Gibbs, M.A. Al-Khafaji, W.M. Rainforth, H.A. Davies, K. Babcock, J.N. Chapman, and L.J. Heyderman, IEEE Trans. Magn. **31**, 3349 (1995).

6. L. Folks, R.C. Woodward, K.L. Babcock, D.L. Bradbury, K. Humphrey, and R. Street, in *Magnetic Anisotropy and Coercivity in Rare-Earth Transition Metal Alloys* Vol. 2 edited by F.P. Missell, V. Villas-Boas, H.R. Rechenberg. And F.J.G. Landgraf (World Scientific, New Jersey, 1996) pp. 49–58.

7. K. Babcock, V. Elings, M. Dugas, and S. Loper, IEEE Trans. Magn. **30**, 4503 (1994).

8. S.H. Liou, and Y.D. Yao, J. Magn. Magn. Mater. **190**, 130 (1998).

9. P. Campbell, *Permanent Magnets and Their Applications*, Cambridge Univ. Press, 1994, p. 32.

10. V. Panchanathan, S. Koyama, and T. Nishio, private communication.

11. M.A. Al-Khafaji, W.M. Rainforth, M.R.J. Gibbs, H.A. Davies, and J.E.L. Bishop, J. Magn. Magn. Mater. **188**, 109 (1998).

Nd RICH Nd-Fe-B TAILORED FOR MAXIMUM COERCIVITY

Er. Girt*, Kannan M. Krishnan*, G. Thomas*, C. J. Echer* and Z. Altounian**
*Material Science Division, Lawrence Berkeley National Laboratory, University of California, Berkeley, California 94720, EGirt@lbl.gov
**Centre for the Physics of Materials, Department of Physics, McGill University, Montreal, Quebec, Canada

ABSTRACT

The coercive field, H_c, in rapidly quenched Nd-Fe-B was found to increase for samples with excess of Nd. In addition, H_c strongly depends on the Fe/B ratio, increasing from 1.67 T for Fe/B = 1.07 to 2.05 T for Fe/B = 14.6 in $Nd_{30}(Fe,B)_{70}$. The increase in H_c seems to correlate with an increase of the γ-Nd phase in the sample. The initial magnetization curves of $Nd_{30}(Fe,B)_{70}$ show that the domain wall pinning plays a more important role as the Fe/B ratio increases. Thus, γ-Nd may play an important role in promoting a pinning mechanism. The excess of Nd in rapidly quenched Nd-Fe-B samples was also found to promote growth of $Nd_2Fe_{14}B$ grains elongated in shape.

INTRODUCTION

The magnetic energy stored in permanent magnets is often measured by the maximum product of the magnetic induction, B, and the magnetic field, H, in the second quadrant of the B(H) loop. Therefore, permanent magnets with large (BH)max should exhibit both a large saturation magnetization, M_s, and a large coercive field, H_c. M_s is an intrinsic property of magnetic materials and H_c depends on both the intrinsic magnetic properties, such as magnetocrystalline anisotropy, H_A, and the microstructure of the magnetic material. $Nd_2Fe_{14}B$ is attractive, as a hard magnetic material, because of its large M_s and H_A at room temperature. The optimal microstructure of $Nd_2Fe_{14}B$ for obtaining large H_c was achieved by powder metallurgy [1] or melt-spinning [2]. However, H_c obtained using these processes in pure $Nd_2Fe_{14}B$ is at least four times smaller than the theoretically predicted value. In order to improve H_c, Nd and Fe were substituted by various elements. It was also found that an excess of Nd increases H_c as the magnetic $Nd_2Fe_{14}B$ grains are separated with non-magnetic Nd phase. In this paper we extend this work by studying the effect of the Fe/B ratio in Nd-rich Nd-Fe-B samples on H_c. The optimal microstructure of $Nd_2Fe_{14}B$ grains required to obtain the highest H_c using the melt-spin ribbons was also investigated in detail.

EXPERIMENTAL TECHNIQUE

Nd-Fe-B ingots, of composition $Nd_2Fe_{14}B$, $Nd_{30}Fe_{64}B_6$, $Nd_{30}Fe_{65}B_5$, and $Nd_{30}Fe_{65.5}B_{4.5}$, were prepared by arc-melting. Each ingot was melted at least five times to achieve a homogenous alloy. These alloys were subsequently melt-spun in an argon atmosphere using a Cu wheel with a tangential speed of 30 m/s. The as-quenched ribbons were annealed in vacuum at 873 K. Structural data on the Nd-Fe-B ribbons were obtained by X-ray diffraction

Mat. Res. Soc. Symp. Proc. Vol. 577 © 1999 Materials Research Society

using CuKα radiation. Magnetic measurements were done using both a thermo-gravitometric analyzer (TGA), and SQUID magnetometer. SQUID measurements were done at 290 K and in a magnetic field up to 5 T. The TGA measurements show that all annealed ribbons contain only one magnetic phase with a Curie temperature corresponding to the Curie temperature of the pure $Nd_2Fe_{14}B$ phase. Quantitative analysis of the Fe/Nd ratio in the $Nd_{30}Fe_{65.5}B_{4.5}$ sample was determined by EDX (energy dispersive X-ray microanalysis) using a 10 nm probe in an analytical electron microscope (AEM).

RESULTS AND DISCUSSION

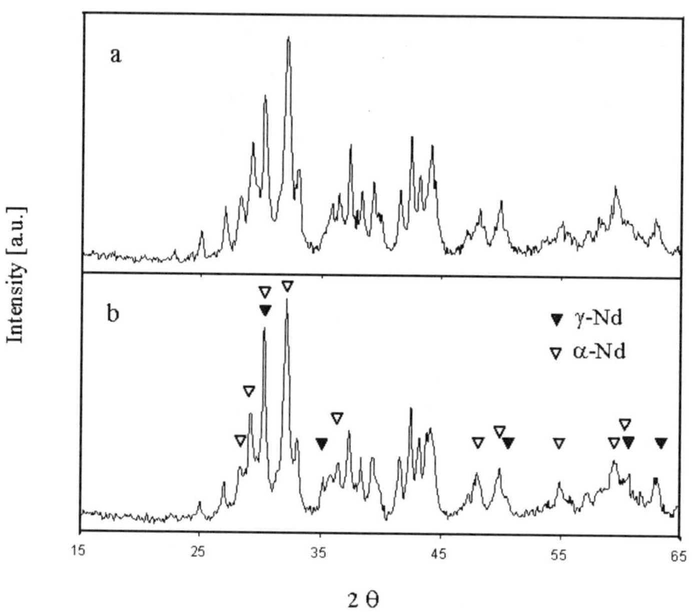

Fig.1 X-ray diffraction data of a) $Nd_{30}Fe_{64}B_6$ and b) $Nd_{30}Fe_{65.5}B_{4.5}$ annealed at 873 K for 4 min..

X-ray diffraction analyses showed that the as-quenched samples were partially amorphous with small α-Nd, γ-Nd, and $Nd_2Fe_{14}B$ crystals present in the ribbons. γ-Nd was also found to be present in all rapidly quenched $Fe_{100-x}Nd_x$ samples and it is stable up to 1000 K [3]. After annealing the ribbons were fully crystallized and α-Nd, γ-Nd, and $Nd_2Fe_{14}B$ were the only phases detected by X-ray diffraction. For a similar Nd-Fe-B composition, Buschow at el. [4] also observed a mixture of $Nd_2Fe_{14}B$ and Nd phases; however, the distinction between α-Nd, and γ-Nd was not specified. Fig.1 shows the X-ray diffraction patterns of $Nd_{30}Fe_{64}B_6$ and $Nd_{30}Fe_{65.5}B_{4.5}$ annealed at 4 min. at 873 K. The filled and empty triangles indicate the positions of the γ-Nd and α-Nd phases, respectively. All the other peaks correspond to $Nd_2Fe_{14}B$. Fig. 1 shows that both X-ray spectra are similar with a slight increase in γ-Nd observed for $Nd_{30}Fe_{65.5}B_{4.5}$. Mishra [5] found an amorphous Nd-rich film surrounding $Nd_2Fe_{14}B$ grains in the $Nd_{13.3}Fe_{81}B_{5.5}$ ribbons quenched with wheel speed 21 m/s. This Nd-rich phase was shown to be the main contributor to the domain wall pinning. After annealing at 870 K the amorphous Nd-rich phase crystallizes into the γ-Nd phase as was observed in $Nd_{13.75}Fe_{80.25}B_6$ samples [6]. Thus, the γ-Nd phase may also play an important role in promoting pinning mechanism in these materials.

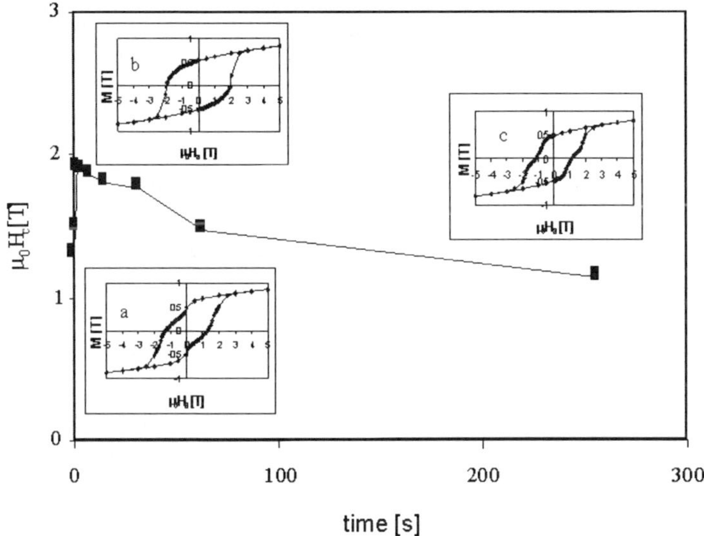

Fig. 2 Coercive field H_c as a function of the annealing time for $Nd_{30}Fe_{65}B_5$ annealed at 873 K. M-H loops for a) as quenched, b) annealed for 2 min., and c) annealed four hours, $Nd_{30}Fe_{65}B_5$ ribbons.

Fig. 2 shows the dependence of the coercive field H_c on the annealing time for $Nd_{30}Fe_{65}B_5$. As quenched samples contain a small amount of amorphous phase, which exhibits soft magnetic behavior, as can be seen from Fig.2a. The largest H_c was obtained after annealing for 2 min. at 873 K. The M-H loop is presented in Fig.2b. For $Nd_{30}Fe_{65}B_5$ samples annealed for 2 min. the grain size is close to the single magnetic domain size which is the optimal microstructure for obtaining a large H_c. The largest H_c in $Nd_{30}Fe_{64}B_6$, and $Nd_{30}Fe_{65.5}B_{4.5}$was also obtained after annealing for 2 min. at 873 K. Further annealing decreases H_c due to the formation of grains larger than that for single magnetic domains. For example, for samples annealed for four hours H_c decreases by about 40 %, Fig.2c. In $Nd_2Fe_{14}B$ the optimal annealing time for obtaining the largest H_c was found to be 8 min. at 873 K. The M-H loops for $Nd_2Fe_{14}B$ annealed for 8 min. and $Nd_{30}Fe_{64}B_6$, $Nd_{30}Fe_{65}B_5$, and $Nd_{30}Fe_{65.5}B_{4.5}$ annealed for 2 min. at 873 K are shown in Fig.3. It is clear that the samples with 30 at. % of Nd have larger H_c than $Nd_2Fe_{14}B$. Excess of Nd reduces the magnetic interaction between $Nd_2Fe_{14}B$ grains, thereby, increasing H_c. H_c also increases with an increase in Fe/B ratio from 1.67 T for $Nd_{30}Fe_{64}B_6$ (Fe/B = 10.7) to 2.06 T for $Nd_{30}Fe_{65.5}B_{4.5}$ (Fe/B = 14.6). The initial magnetization curve also changes with the Fe/B ratio indicating that the Fe/B ratio affects the hardening mechanism indicating that the domain wall pinning in $Nd_{30}Fe_{65.5}B_{4.5}$ plays a more important role than in $Nd_{30}Fe_{64}B_6$. This suggests that the γ-Nd phase (which was found to increases with an increase of the Fe/B ratio) may play the important role in promoting the pinning mechanism, in agreement with [5].

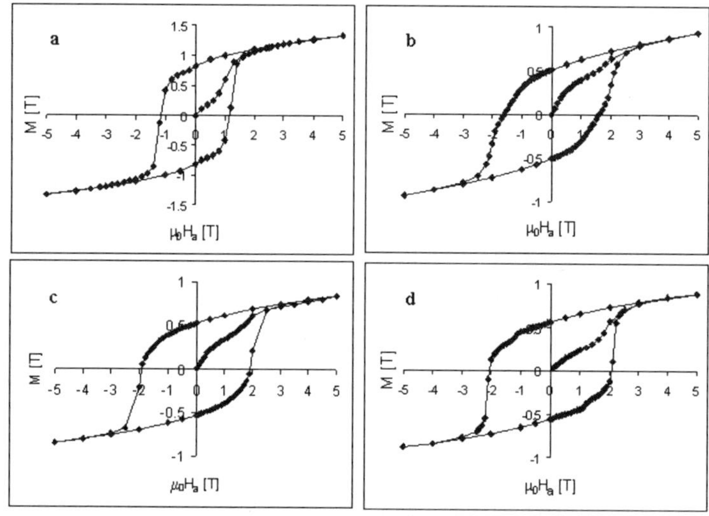

Fig. 3 M-H loops for a) $Nd_{12.7}Fe_{81.11}B_{6.19}$ annealed at 873 K for 8 min., b) $Nd_{30}Fe_{64}B_6$, c) $Nd_{30}Fe_{65}B_5$, d) $Nd_{30}Fe_{65.5}B_{4.5}$ annealed at 873 for 2 min.

The microstructure of $Nd_{30}Fe_{65.5}B_{4.5}$ annealed at 873 K for 4 min. is presented in Fig. 4a together with the selected area diffraction pattern Fig.4b, and micro-micro diffraction pattern of grain A Fig.4c. The results show that $Nd_2Fe_{14}B$ grains are elongated in shape.

Grain A, reveals the typical shape of the $Nd_2Fe_{14}B$ grains. Using micro-micro diffraction, Fig.4c, we found that the <110> direction of the A grain is parallel to the long axis of the grain. The <001> direction is perpendicular to <110>, as indicated in Fig.4a. Hadipanayis at el. [7] also observed elongated $Nd_2Fe_{14}B$ grains in addition to equiaxed grains in $Nd_{16}Fe_{76}B_9$. However, it seems that an increase in the Nd concentration promotes the growth of grains with elongated shape. The selected area diffraction pattern, Fig.4c shows a random distribution of the grains in all directions. This is an important information since any alignment between $Nd_2Fe_{14}B$ will affect the value of coercivity in these samples. Quantitative spectroscopic analysis show that the Fe/Nd ratio across grain A is 83/17 consistent with the composition of $Nd_2Fe_{14}B$. We also found that the Fe/Nd ratio in the sample is 69/31 close to the $Nd_{30}Fe_{65.5}B_{4.5}$ composition.

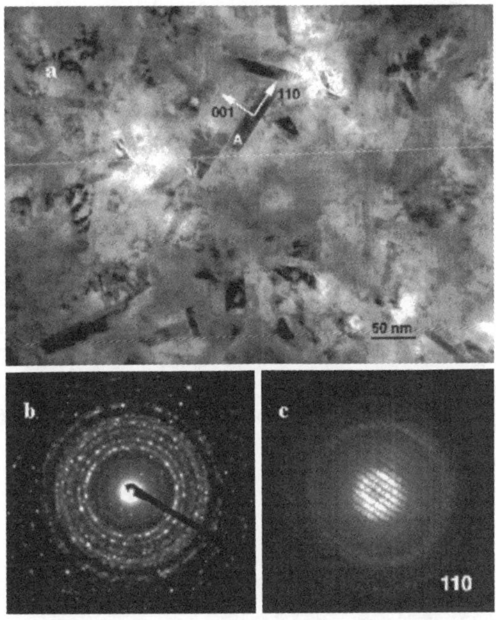

Fig. 4 a) Bright field image, b) selected area diffraction and c) micro-micro diffraction of $Nd_{30}Fe_{65.5}B_{4.5}$. The <001> and <110> directions of the grain A, figure a), are determined from micro-micro diffraction, figure c).

CONCLUSION

Magnetic measurements show that the coercive field in $Nd_{30}(Fe,B)_{70}$ can be increased with an increase in the Fe/B ratio. For Fe/B = 14.6 the coercive field is 2.06 T. The increase in Fe/B ratio also promotes a pinning mechanism. Structural measurements show an increase in γ-Nd content with an increase in the Fe/B ratio. This suggests that the γ-Nd phase may play an important role in improving the coercivity by forming thin layers between the $Nd_2Fe_{14}B$ grains. These intergranular layers would act as pinning centers for magnetic domains. Microstructural analysis shows the $Nd_2Fe_{14}B$ grains are elongated in shape and randomly oriented in the sample. The Fe/B ratio across the elongated particles was confirmed to correspond to $Nd_2Fe_{14}B$.

ACKNOWLEDGEMENT

This work was supported by the Director, Office of Energy Research, Office of Basic Energy Sciences, Materials Sciences Division of the U.S. Department of Energy, under Contract No. DE-AC03-76SF00098.

REFERENCES

1. M. Sagawa, S. Fujimura, M. Togawa, H. Jamamoto and Y. Matsuura, J. Appl. Phys. **55**, 2083 (1984).
2. J. J. Croat, J. F. Herbst, R. W. Lee and F. E. Pinkerton, J. Appl. Phys. **55**, 2078 (1984).
3. L. X. Liao, Z. Altounian and D. H. Ryan, J. Appl. Phys. **67**, 4821 (1990).
4. K. H. J. Buschow, D. B. de Mooij and H. M. Van Noort, J. Less Common Met., **115**, 357 (1986).
5. R. K. Mishra, J. Magn. Magn. Mat. **54-57**, 450 (1986).
6. Er. Girt, M. Koknaeva and Z. Altounian, J. Appl. Phys. **79**, 4833 (1996).
7. G. C. Hdjipanayis, R. C. Dickenson and K. R. Lawless, J. Magn. Magn. Mat. **54-57**, 557 (1986).

Part VI

Permanent Magnet Applications

NOVEL PERMANENT MAGNETS AND THEIR USES

S. CONSTANTINIDES
The Arnold Engineering Company, 300 N. West Street, Marengo, IL 60152

ABSTRACT

Two changes have gone hand in hand during the 20th century. First, applications were conceived that required new or greatly improved materials. Need being the mother of invention, materials were then developed to satisfy the application needs. Permanent magnets are today used in devices that could only have been dreamed about 25 years ago. Who could have foreseen, for example, the advances in data storage that have occurred over the last 15 years? Hard disk drives of under 10 megabytes in size and with random access times of over 65 milliseconds were the norm in 1984. Today, the standard drive is over 8 gigabytes, is smaller in size and costs less than a fourth of the old drives, with an access time of 10 milliseconds and more than 33 mbps burst throughput. Novel has two interpretations: first it may refer to the magnetic material or it may refer to the application of a material. Several examples of each are presented.

INTRODUCTION

Until the late 1930's, permanent magnets were predominantly steel compositions with low energy product and coercivity. Loudspeakers used an interaction of electromagnetic fields. Motors were of the induction type. The invention of alnico allowed product size reduction through the use of permanent magnets in place of induction coils.

Approximately every 12 years thereafter, a new magnetic material was discovered. Figure 1 shows how the maximum energy product has increased. It also illustrates that

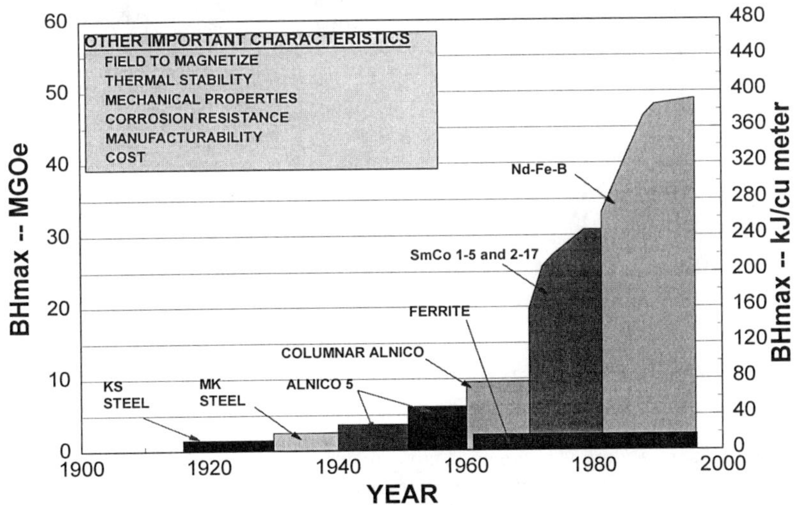

Figure I
Development of Permanent Magnet Materials

Mat. Res. Soc. Symp. Proc. Vol. 577 © 1999 Materials Research Society

materials with lower energy product, specifically ferrite, can be commercially successful. Introduced in 1961, ferrite remains the largest selling permanent magnet material on a weight basis primarily because of its relatively low price.

New materials have not obviated older ones: each has advantages and disadvantages. Alnico, though magnetically weaker than rare earth magnets, is much more temperature stable. Applications requiring stability over wide temperature ranges still rely on alnico. But the newer materials (ferrite, samarium cobalt, neodymium-iron-boron) all have a very important characteristic, a "square" second quadrant intrinsic curve, which allows use in applications which were not possible before.

NOVEL MATERIALS

The first question we must ask is: "What do we mean by novel?" It is proposed here that novel refers to a new magnetic material, a new application or design or a significant variation of an older design. To be considered, the material or application must also be commercially successful. Novel might also refer to the lack of general knowledge about a material or an application.

Table I is a summary of commercially available materials and processing methods. Prior to the announcement of SmFeN for use in bonded magnets, the newest material in the table was NdFeB, which is 15 years old. Old materials are not necessarily stagnant. Improvements in composition and processing of NdFeB powders for bonded magnets, for example, have raised the maximum recommended use temperature from 110°C to 180°C.

Table I
Commercially Available Permanent Magnet Materials

					BONDED		
MATERIAL	CAST	EXTRUDED OR ROLLED	SINTERED FULLY DENSE	INJECTION MOLDED	COMPRESSION BONDED	FLEXIBLE	RIGID EXTRUDED
ALNICO	Y		Y	Y			
IRON-CHROME-COBALT		Y					
CuNiFe		Y					
SmCo			Y	Y	Y		
SmFe(N,C)				Y			
NdFeB			Y	Y	Y	Y	Y
FERRITE			Y	Y		Y	
HYBRIDS				Y	Y	Y	

Neodymium-iron-boron is no longer a novel material in that it was commercially introduced in November 1984. However, the melt-spun isotropic powder for bonded magnets was not readily available until the late 1980's. Since it represented a new family of materials with new magnetic properties and limitations, applications had to be designed from scratch, a process requiring two to four years. The first products to take advantage of this new material were microelectronics produced in the Far East. Bonded neodymium-iron-boron magnets have become widely used in the United States only in approximately the last five years.

Bonded magnets represent a diverse set of capabilities and properties. Figure 2 shows the combinations of materials, binders and forming processes. The low processing temperatures allow mixing of heterogeneous materials within the binding matrix. Several combinations have been proposed or tried since the early 1970's including ferrite with samarium cobalt or ferrite with neodymium-iron-boron. Different grades of a single material family can also be mixed to achieve new properties, though with only an averaging effect. The advantage of ferrite and either of the rare earth compositions is that the ferrite has a positive temperature coefficient of coercivity while that of the rare earths is negative. There is a synergy in performance and in price.

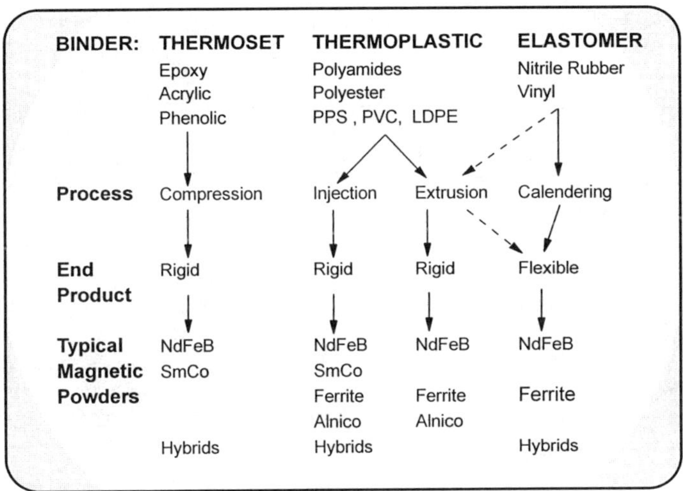

Figure 2
Diversity of Bonded Magnets

Another material is being prepared for commercial introduction: samarium-iron-nitride. Asahi Corporation received U.S. patents in 1987 for the material and processing method. Siemens, Hitachi and Sumitomo have all done extensive research on manufacturing SmFeN. Making the material has been challenging: when the nitride gas is forced into the crystal lattice of the samarium-iron base alloy, the alloy tends to decompose into samarium-nitride and alpha-iron. The nitrogen is interstitial, so it may be ejected from its position in the lattice causing decomposition. This is hastened when the temperature rises. Above about 450°C, decomposition is rapid. The claimed advantages include improved corrosion resistance over neodymium-iron-boron and improved temperature stability.

Developments have also taken place in rolled alloys. These are malleable alloy compositions rolled in continuous strips of thin foils. Recently introduced materials are "semi-hard" with coercivity ranging from 20 to 100 oersteds. The primary use for these alloys is in EAS (electronic article surveillance or anti-theft tags).

NOVEL APPLICATIONS

With so little new in materials, we will focus on applications. Hard disk drives are sold by the tens of millions each year. Who here does not have at least one? What is novel, I propose, is the extent to which the technology has progressed: magnets have increased in energy output, the magnet voice coil assemblies have become smaller in size and the drives have gone from 8" (width form factor) to 3.5 inches. Portable computers use 2.5" drives. IBM has introduced an even smaller drive (smaller than a pack of cigarettes). Several years ago, Ted Davis, President of Ted Davis Manufacturing (now part of Vacuumschmelze), complimented the industry on advances in the permanent magnets used in drives and the miniaturization that new materials allowed. He went on to challenge the industry to improve the flux carrying ability of the soft magnetic components so their size could also be reduced. Figure 4 shows example components. Notice the relative thickness of the magnet and the steel to which it is mounted.

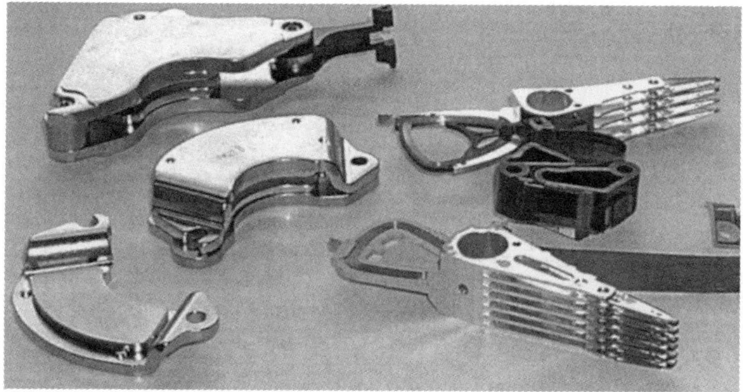

Figure 4
Hard Disk Drive Components
Items on the left are nickel plated NdFeB magnets and steel return paths; items on the right are the read/write head assemblies. All items are courtesy of Western Digital Corporation.

Voice coil motors in hard drives are either rotary, such as in Figure 4, or linear. The concept is a variation on a loudspeaker assembly wherein an electromagnetic field interacts with the field from a permanent magnet to produce motion. Another device using this principle is the "air core" gauge. These usually consist of two coils of wire, perpendicular to each other, with a disk shaped magnet inside. The field produced by the coils of wire is dependent upon the relative current in each coil. The internally located magnet, which is diametrally magnetized, rotates to align with the electromagnetic field. Two such gauges are shown in Figure 5 with magnets used in the gauge shown in front. The staff protruding from the gauge will have a pointer pressed onto the end. These gauges are used in cars, boats, trains, trucks, and more. The annual consumption of gauges in the western world is in excess of 100 million. Additional magnet examples are shown in Figure 6 and include oriented and unoriented injection molded ferrite, injection molded neodymium-iron-boron, sintered neodymium-iron-boron, sintered alnico and iron-chrome-cobalt.

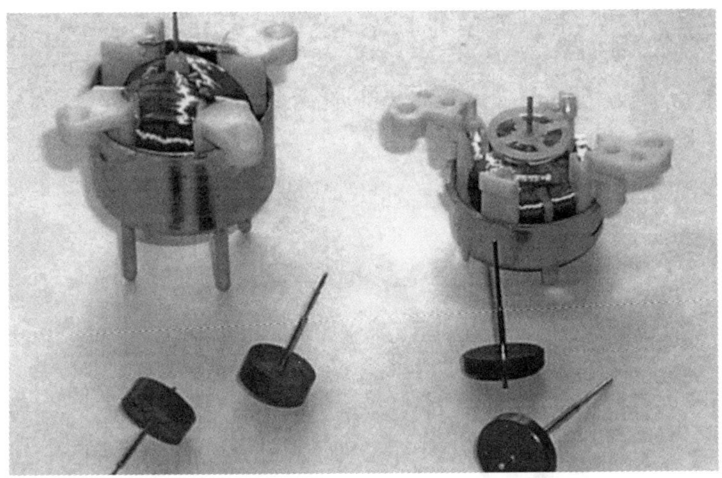

Figure 5
Air Core Gauges
Magnets used in these assemblies are shown in front. The ones on the left are injection molded ferrite; the ones on the right are injection molded NdFeB

Figure 6
Magnets Used in Air Core Gauges
Materials include injection molded ferrite, injection molded NdFeB, sintered alnico, sintered NdFeB and iron-chrome-cobalt.

What is unique about this gauge construction is that the coil is stationary and the magnet moves. Early gauge design used a stationary magnet and a moving coil, just as in the hard disk drive example given above. The can around the gauge is to provide magnetic shielding both from without and from within. Coincidentally, it also imparts some hysteresis to movement of the magnet, reducing vibration of the gauge pointer.

Hysteresis coupling can be used advantageously such as in drive devices with limited travel. Examples are electric window motors. When the window reaches end-of-travel, the

Figure 7
Hysteresis Rings and Drive Magnets
The rings are made of iron-chrome-cobalt or similar materials with coercivities of a few hundred oersteds. The Drive magnets are typically an injection molded ferrite.

hysteresis coupling allows the motor to spin with the output shaft stationary. Another major use of hysteresis coupled drives is in air-handling equipment in HVAC systems. Sample parts are shown in Figure 7. In this application the couple is typically made between a bonded ferrite magnet and a ring of iron-chrome-cobalt alloy. The ferrite magnet, usually magnetized with multiple poles on the OD, produces adequate field strength to saturate the iron-chrome-cobalt

Figure 8
Torque Coupled Pump
The inside magnet set is connected to the pump housing; the outside drive set is fastened to the drive shaft of the motor. A can seal separates the magnets. Photo courtesy of March Manufacturing, Inc.

ring. As the magnet rotates, the magnetized ring follows. When resistance to movement of the output shaft exceeds the breakaway torque, the magnet spins inside the ring, driving each part of the ring through a complete hysteresis loop. In effect, it is a non-contact clutch, operating without mechanical wear or vibration. When the secondary material in a hysteresis coupled device has adequate coercivity to resist demagnetization, increased torque output is possible. This is useful in pumps as a method to eliminate rotating seals. A cutaway view of a torque coupled pump is shown in Figure 8, courtesy of March Manufacturing. Typical magnets used in these applications include sintered ferrite, injection molded ferrite, injection molded hybrid magnets and sintered NdFeB. The advantage of having no rotating seal is obvious, especially for toxic, hot, chemically reactive or otherwise dangerous materials.

Another interesting application using a rotating magnet is a linear positioner used in automotive applications. The magnet assembly is shown in Figure 9. The core is an injection molded plastic with internal threads which has been molded onto a brass piece. The cylindrical magnet is then molded around this. The unit is magnetized on the OD and secured in the final device so that it becomes a rotor in a motor. The shaft on which it is mounted is threaded. As the rotor turns, the shaft is moved back and forth producing linear motion.

Figure 9

Rotating to Linear Motion Converter

These rotor assemblies consist of a plastic core with internal threads, a brass hub and an injection molded NdFeB outer cylindrical magnet.

The automotive industry has utilized magnets in many interesting ways. In the transmission of almost every car produced in the U.S. is a magnet, located so as to pull from suspension the fine wear particles. This magnet may be a flat, square sheet of flexible ferrite, a slug fit into the end of a bolt, or one of several other shapes. The example in Figure 10 is an injection molded ferrite magnet which is pressed into a cylindrical hole in a transmission fitting. The material must resist hydraulic fluid and be corrosion resistant.

Cars, trucks, buses and construction vehicles have become more sophisticated. Computers use sensors to monitor temperatures, air and fuel flow, oil pressure, transmission speed, wheel speed (each wheel!), etc. Using these inputs, the computer controls fuel/air ratio,

Figure 10
Transmission Chip Collector Magnet

prevents wheels from locking during panic stops, adjusts heating or cooling of the passenger compartment, and much more. Imagine the damage that could occur if one were to shift a car into reverse while it is speeding forward. In the off-road construction industry, the large (and expensive) vehicles must change gear under only suitable circumstances. To prevent improper operation, a sensor is placed on transmission components indicating which gearing is active. The computer will not allow shifting to take place if the conditions are incorrect. One

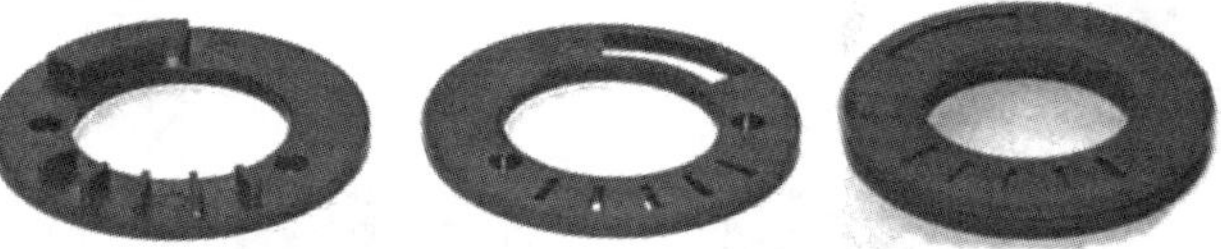

Figure 11
Four-Wheel Drive Transmission Position Sensor

such device utilizes an injection molded ferrite disk with selected points magnetized to signal Hall sensors. The example in Figure 11 is an automotive application which prevents improper operation of a FWD transmission. Two injection molded ferrite magnets are formed and magnetized. The two halves are mounted in opposition. The tips of the "fins" of one magnet show through to the back of the other producing very clearly defined poles and precise operation.

Transported goods receive bumps and jarring due to much more than improperly operated transmissions. Not all our highways are smooth nor our truckers infallible. Expensive devices are often shipped in air-ride trucks. Even so, it is necessary to monitor for excessive shock. One device for doing so is a shock sensor. The one shown in Figure 12 is meant to be attached to a crate. At least three would be used: one in each orthogonal direction. The small

Figure 12
Shock Sensor
Small round magnet at lower left has been dislodged from its normal position due to shock.

disk shown off-center in the lower left of the device is normally held in position with another magnet structure inside the case. Its inertia will cause it to break away from the other magnet when acceleration is too high. These devices may be calibrated and adjusted providing information on jarring that sensitive equipment received during transport.

CONCLUSION

There are many more applications, materials and manufacturing techniques that we have not talked about today. Some of these are: GMR, thin films (sputtered or condensed), roll-forming of malleable magnetic alloys, linear encoder applications, and rotary (non-contact) position sensors. Permanent magnets have become ubiquitous. We are aware of some in our day-to-day lives; many more reside below the level of our consciousness, tucked away in the devices we use: the coffee grinder, the cooling fan in the microwave, focusing magnets on CRT's (television tubes), and the magnet activated reed switch on flip phones.

Have we forgotten the most important permanent magnet of all: Earth itself. Perhaps next time.

APPLICATIONS OF RARE-EARTH PERMANENT MAGNET STRUCTURES TO ELECTRICAL MACHINERY

H.A. LEUPOLD*, A.S. TILAK*
*Army Research Laboratory
R.J. MARCHAND**, D.A. LAGRAFFE**
**U.S. Military Academy
All at U.S. Military Academy, Bartlett Hall, West Point, NY 10996, hh3416@exmail.usma.edu

ABSTRACT

A variety of promising permanent magnet structures based upon the magic cylinder and magic mangle concepts are investigated with regard to their suitability for application to electrical generators and motors. Such structures feature high magnetic flux densities in large volumes relative to their masses and bulks as well as minimal stray fields and net magnetic moments. Magic mangles also afford small moments of inertia of their rotating elements thereby resulting in improved rotational responsiveness. One computer calculation showed that a small conventional electric generator can be replaced by a magic mangle with 70% more flux amplitude but with only one half the mass and bulk.

INTRODUCTION

Magic cylinders or Halbach structures[1,2] are cylindrical shells of permanent magnet material that enclose cylindrical cavities in which uniform, completely confined transverse, fields are generated. Figure 1 shows sections of such arrays in which the direction, γ, of magnetization in the shells with respect to the polar plane is given by

$$\gamma = 2\theta \tag{1}$$

where θ is the azimuthal coordinate of the point at which the magnetization is viewed.[1,2]

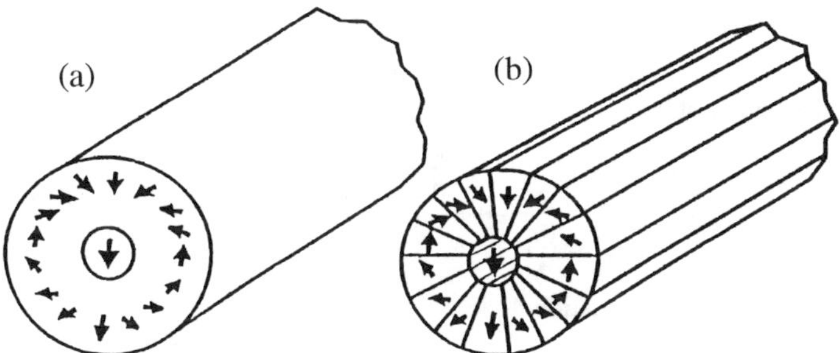

Figure 1: (a) Ideal magic cylinder. (b) Segmented approximation to (a).

The field in the central cylindrical cavity is given by

$$B = B_r \ln (r_o/r_i) \tag{2}$$

Mat. Res. Soc. Symp. Proc. Vol. 577 © 1999 Materials Research Society

Where B_r is the remanence of the magnetic material and r_i and r_o are the inner and outer radius respectively. Because of their cylindrical shape and because (2) shows that, in principle, there is no limit to the field that can be generated in their central cavities, it is tempting to consider such structures for use in rotating electrical machinery. For example a radius ratio of 4.2 would yield a working field of 20 kOe with a B_r of 14 kG. This is possible only with the highest energy product materials with nearly square hysteresis loops.

SIMPLE MAGIC CYLINDER GENERATOR

If one packs a flat, layered copper coil or a conducting bar split along a diametric plane into the cavity, the configuration of Figure 2 results. If the ends of the coil or semicylinder are

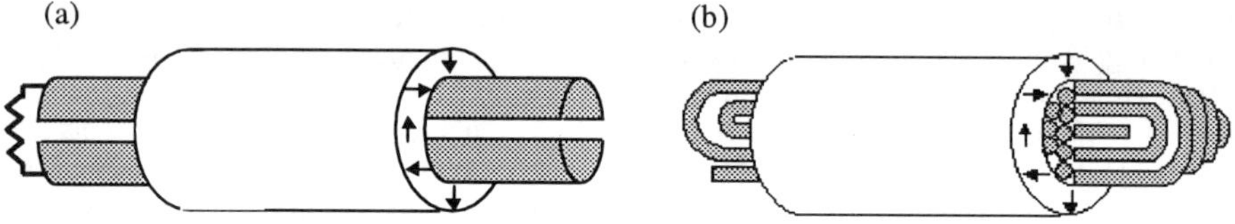

Figure 2: (a) Split copper bar threaded by the magnetic field of the cylindrical shell which if altered induces a current in the semicylinders electrically connected at their ends. (b) Stacked flat electrical coils connected in series and inserted into the cavity of a rotating magic cylinder.

bridged by loads or shorts and the magnetic shell is spun on its axis an A.C. generator results. The question then arises that if one is willing to devote only a limited value to this device's outer radius, what would be the optimal value of the inner radius. A smaller inner radius implies a larger field but also less area for magnetic flux to thread. Effective magnetic flux per unit axial length of the device is

$$\phi = BA = \pi r_i^2\, B = \pi\, r_i^2\, B_r \ln(r_o/r_i) \tag{3}$$

where B and A are the transverse field in the cavity and the circular cross-sectional area of the cavity respectively. Since the deliverable power of a generator is proportional to its flux amplitude squared we wish to maximize flux so (2) and (3) yield for maximum flux:

$$\frac{d\phi}{dr_i} = \frac{d\left[r_i^2 \ln\left(r_o / r_i\right)\right]}{dr_i} = 0 \tag{4}$$

for maximum tolerable bulk. Equation 4 yields

$$\frac{r_o}{r_1} = \sqrt{e} = 1.64 \tag{5}$$

COUNTER ROTATING NESTLED CYLINDERS

References [2] and [3] describe an adjustable magnetic field source wherein two nested magic cylinders produce equal field strengths as in Figure 3. For their field strengths to be equal the outer radius of the inner cylinder must be the geometric mean between the cavity radius and the outer radius of the outer cylinder. If the cylinders are rotated with respect to each other, their fields add vectorically with the resultant flux varying sinusoidally with half the angle between their polar planes. This flux can be used to generate an A.C. current in conductors placed in the cavity as in the preceding section. The generated currents are equivalent to those of the single rotating cylinder but because here the internal field magnitude is sinusoidal rather than constant, one would expect less loss due to eddy currents.

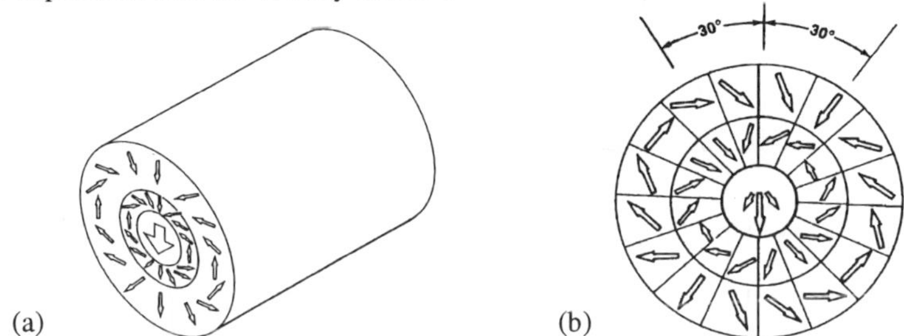

Figure 3: (a) Nested rings which are configured to generate equal fields in the cavity.
(b) Vector addition of fields of rings with a relative rotation of sixty degrees.

THE MANGLE APPROXIMATION TO A MAGIC CYLINDER

An approximation to a magic cylinder is obtained when its trapezoidal segments are replaced by inscribed cylindrical rods as in Figure 4a.[2,4,5] This is possible only for particular ratios of r_o/r_i if the component rods are to be tangential to each other in a circular array. Thus both the generated fields and the degree of approximation to a magic cylinder depend on the number of rods forming the circular array which is called a magic mangle. Unfortunately the relationship between a good approximation and field strength is inverse (Table I) but as can be seen from Table 1 and Figure 5, eight rods seems a good choice for many purposes. Fewer rods result in a rapid decline in efficiency as evidenced by declining field ratio with regard to

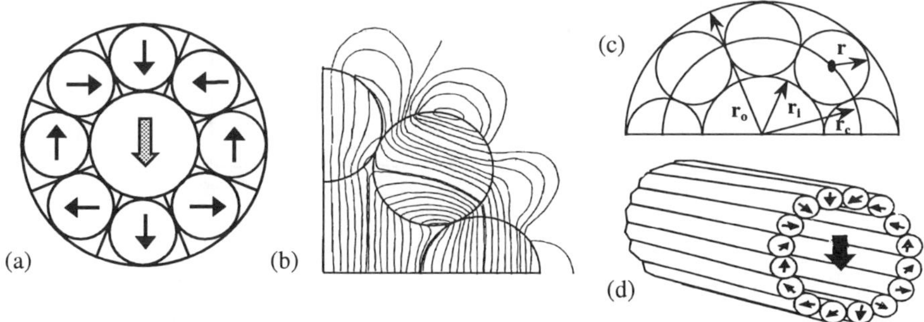

Figure 4 (a) A magic cylinder approximated with magnetized rods. (b) Mangle flux plot.
(c) Notational definitions. (d) An isometric view of a sixteen rod magic mangle.

Table I
A comparison of ideal magic cylinders with their equivalent mangles.

N	r (cm)	r_c (cm)	r_o (cm)	Mangle Field (kG)	Ideal Field (kG)	Field Ratio
4	2.41	3.41	5.83	14.00	24.68	0.57
6	1.00	2.00	3.00	10.50	15.38	0.68
8	0.62	1.62	2.24	8.24	11.29	0.73
10	0.45	1.45	1.89	6.71	8.95	0.75
12	0.35	1.35	1.70	5.64	7.42	0.76
14	0.29	1.29	1.57	4.88	6.34	0.77
16	0.24	1.24	1.49	4.26	5.53	0.77

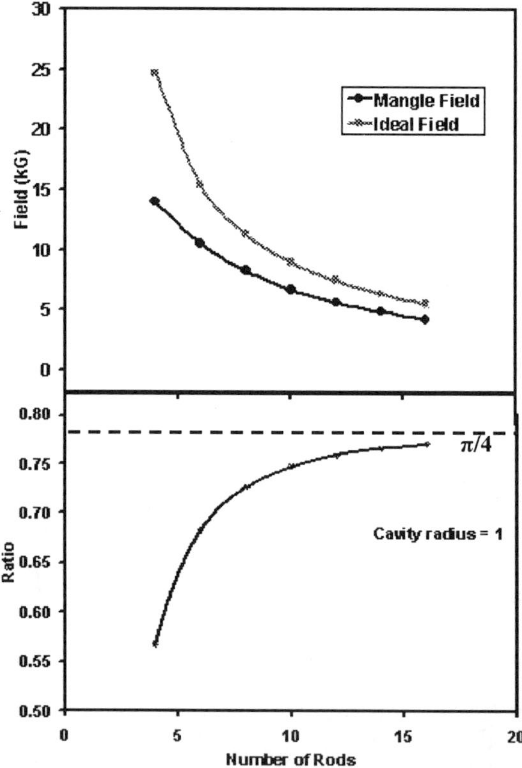

Figure 5: (a) Magnetic fields in mangles and corresponding cylinders. (b) Ratios of magic mangle fields to those of the cylinders they approximate.

equivalent magic cylinders. More rods result in a rapid decline in absolute field. Eight rods also maintain good field uniformity with only slight scalloping near the edges as in Figure 4b. Figure 5 illustrates the ratio of the mangle field to that of its magic cylinder counterpart. As the number of rods increases, that fraction approaches the circle-to-square packing fraction of $\pi/4$ as expected.

While mangles produce lower fields than their equivalent cylinders, they have advantages. One is that of simplicity of fabrication. Only one type of component is needed in its construction; identical transversely magnetized rods that are assembled within a circular ring with appropriately oriented magnetization. In a generator these rods would be rotated individually with the same angular velocity. Then by the two dimensional rotation theorem[2] the magnetic field would remain constant in magnitude while rotating in the sense opposite to that of the rods. This arrangement would be the mangle equivalent of that of the simple cylinder discussed in a previous section. See Figure 6a. If adjacent rods are rotated in opposite directions the arrangement would be equivalent to that of the counter rotating nested cylinders of a previous section Figure 6b. Of course the gaps between rods in each of the two virtual counter-rotating mangles may cause additional field scalloping but to a tolerable degree, especially when a large number of rods is used.

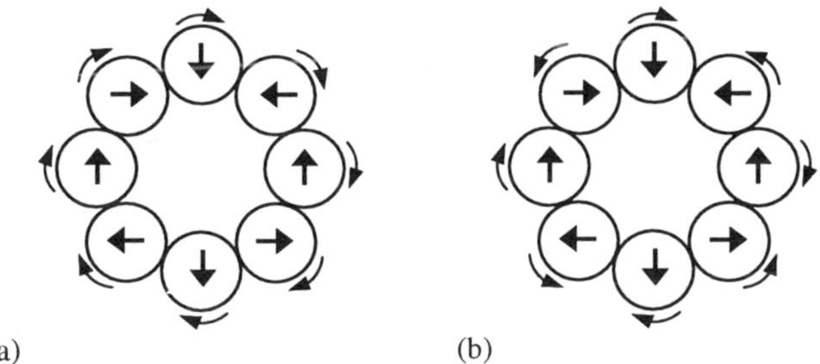

Figure 6: (a) Mangle equivalent of simple rotating magic cylinder. (b) Mangle equivalent of counter-rotating nested cylinders.

A twelve-rod mangle yields a r_o/r_i ratio of 1.69, close to the optimal value of 1.64 computed from equation 5. Computer analysis shows that a 12 rod mangle of half of the bulk of a conventional hand cranked generator that generates 2729 kMx of flux charge per rotor cycle, will produce 4760 kMx. This represents a bulk decrease of 50% and a flux increase of 75%. The mangle also has the advantage of a smaller total moment of inertia of the rotating parts, making for a more nimble response if used as a motor and less effort when used as a small generator and hand-cranked to the desired angular speed.

LAYERED MAGIC MANGLES

Although magic mangles are limited to discrete values of r_o/r_i, greater and intermediate fields can be attained by layered rod structures as in Figure 7a. The additional layer of rods shown in that figure would increase the field from 5.6 to 11.3 kG but increase the bulk to a value slightly more than twice that of the conventional hand-cranked generator. It would also increase the flux of the mangle from 1.75 times the value of the conventional structure to 3.5 times that value.

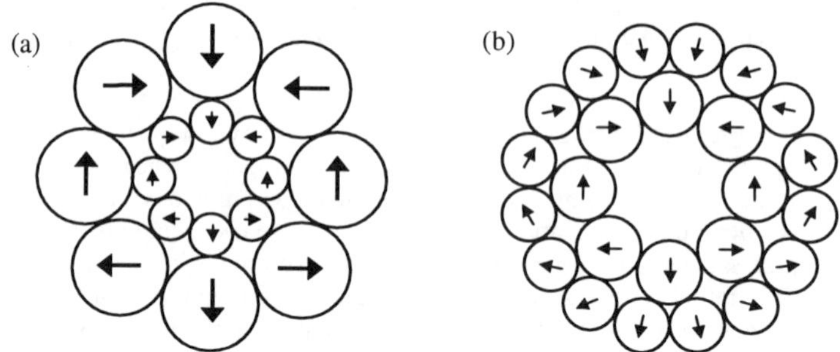

Figure 7: (a) Double-layered magic mangle. (b) Double layered mangle with different numbers of rods in the layers.

The second layer need not contain the same number (N) of rods as the first. If, for example, a large increase in total bulk is not desirable but a slight field increase is needed, a thinner layer with an even multiple of the original N rods can be used as in Figure 7b. The rods of the outer layer may also be rotated in the opposite sense to those of the inner to produce an alternating current in analogy to the structure of Figure 3. Indeed the relationship between the radius of the outermost rod envelope, the inner most envelope radius and that of the layer boundary is the same as for the analogous radii of Figure 3, namely the later is the geometric mean of the other two. If field equality of the inner and outer layers of different N cannot be achieved geometrically, the strength of magnetization in either or both may be adjusted.

CONCLUSION

Magic mangles show promise as flux sources in electric generators and motors where they may be used to produce considerable flux increases with concomitant decreases in bulk, mass, rotational moment of inertia and complexity of design and manufacturing.

REFERENCES

1. K. Halbach, *Nucl. Instr. and Meth.* **187**, p. 109 (1981).

2. H.A. Leupold in *Rare-Earth Permanent Magnets*, edited by J.M.D. Coey (Oxford Science Publications, Oxford University Press, Inc., New York, NY, 1996), pp. 381-429.

3. H.A. Leupold and E. Potenziani, II, and M.G. Abele, *J. Appl. Phys.*, **64**, p. 5994 (1988).

4. J.M.D. Coey and O. Cugat, *13th Int. Workshop on Rare-Earth Permanent Magnets and their Applications*, Birmingham, UK, 11-14 September (1994), p. 41.

5. Martin Gröuefeld, *Rare-Earth Magnets and their Applications, Vol 2.*, Edited by L. Schultz and K.H. Müller (Werkstoff-Information Gesellschaft, Frankfort, Germany) (1998), p. 807

SMALL BRUSHLESS DC-MOTOR WITH SENSORLESS CONTROL - A HOLLOW SHAFT PROTOTYPE

R. HANITSCH, B. FRENZEL
University of Technology Berlin, Institute of Electrical Energy Technology, D-10587 Berlin, Einsteinufer 11, Germany

ABSTRACT

The features of the prototype with NdFeB-magnets will be presented and also the sensorless control strategy will be outlined. Measured and calculated data show good agreement. Focus will be on the magnetic circuit and the thermal behaviour and not on the control aspects of the motor. Specific parameters demonstrate the good quality of the drive system. A thermal analysis was also done in order to meet the requirements given by the medical specialists.

INTRODUCTION

In recent years a number of different permanent magnets with a high energy product have been developed. Flux densities in the airgap well above 0.6 T can be realized with these novel rare-earth permanent magnets. The trend in machine design for a number of applications is to small packages, low power level, high efficiency and low noise [1,2]. For a special heart assist device a motor should drive a pump that is incorporated into a hollow shaft. Also the volume of the motor is restricted: outer diameter below 25 mm and active length below 50 mm.

MOTOR DESIGN

Due to the envisaged pump characteristic the nominal speed of the motor should be around 19,000 rpm. The medical experts also requested that the temperature increase in relation to the normal temperature of the human body should only be in the range of 2° K. Construction details of the small rotor are shown in Fig. 1. Premagnetized magnet blocks were glued on the rotor surface.

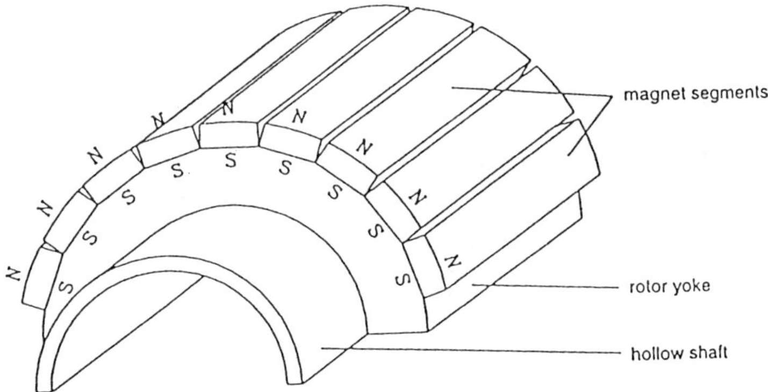

Fig. 1: Schematic of rotor construction

Mat. Res. Soc. Symp. Proc. Vol. 577 © 1999 Materials Research Society

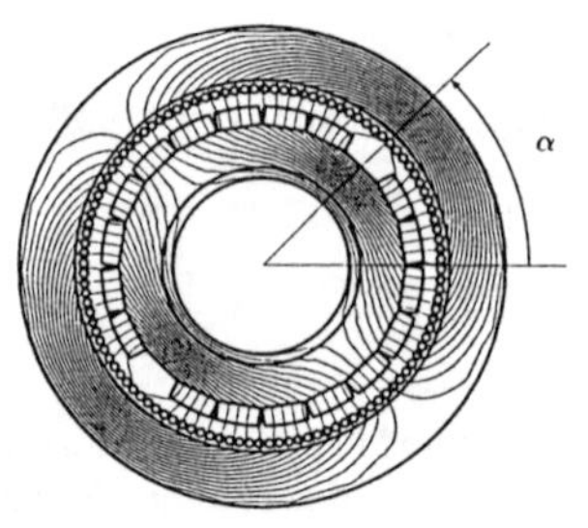

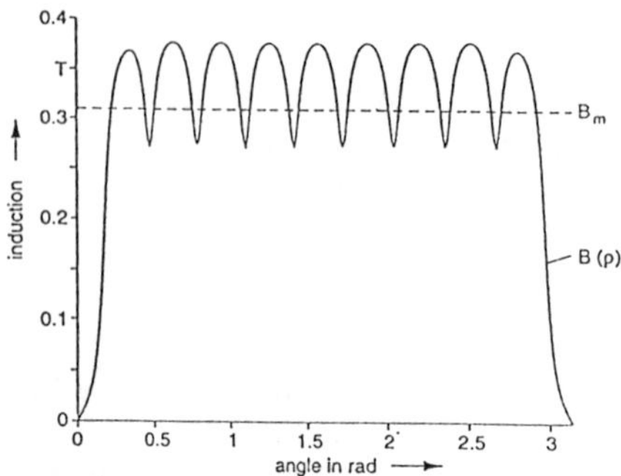

Fig. 2: Flux plot for rated current of 2 A

Fig. 3: Flux density in the airgap

The solution for this drive problem was a brushless dc motor with an airgap winding and permanent magnets on a hollow rotor. In order to reduce the dimensions of the end windings, a two-phase arrangement was selected. To make manufacturing easy each phase was divided into two coils. Because of the restricted space a single layer winding arrangement had to be used. In order to keep complexity on a normal level a sensorless approach was used to operate the brushless dc motor.

A hollow shaft, prepared for the incorporation of the pump, is surrounded by a softmagnetic ring on which the premagnetized magnet blocks are glued.

Remanence of the selected material was 1.0 T. Initial thickness of the magnets was 1.0 mm. In the final design 0.8 mm magnet thickness created an average airgap flux density of 317 mT. A typical flux plot for this type of motor is shown in Fig. 2, while the distribution of the induction is shown in Fig. 3. The outer diameter of the rotor was 14.6 mm. Because of the total airgap of 1.2 mm a single layer arrangement for the winding had to be used [3].

Due to the restricted space the flux density in the stator yoke is quite high. Fig. 4 shows the flux plot for the rated current of 2 A. The magnetic induction along the line A-B is $\hat{B} = 1.6$ T.

The two phase winding consists of four coils. Each coil has 7 turns and is arranged as shown in Fig. 5.

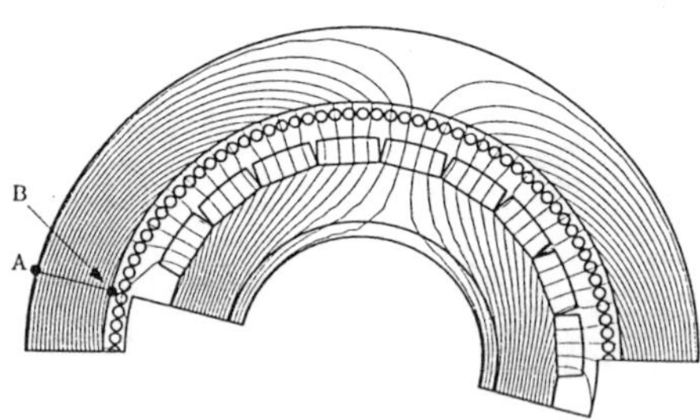

Fig. 4: Flux plot details

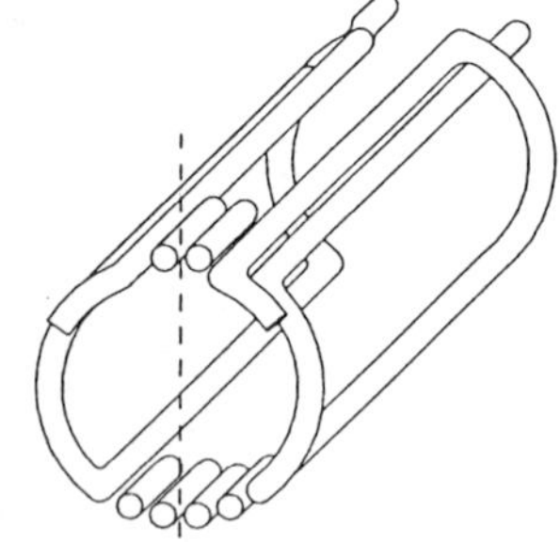

Fig. 5: Schematic of coil arrangement

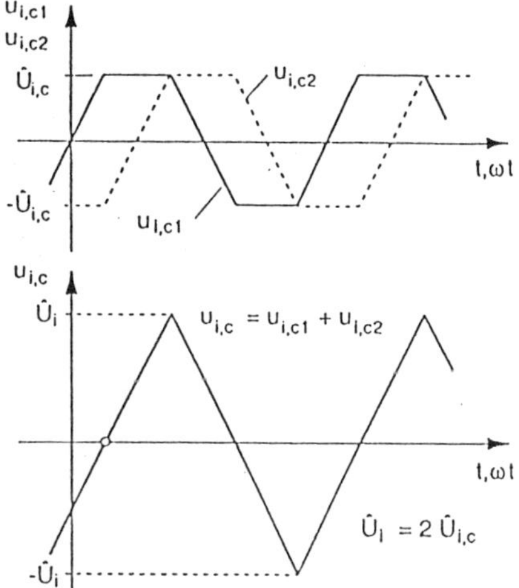

Fig. 6: Induced voltage

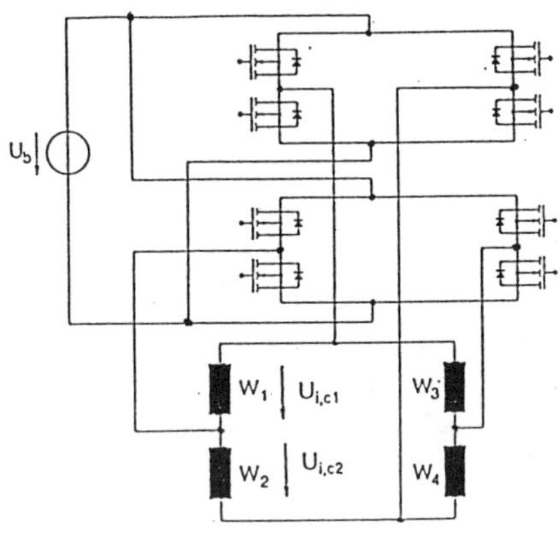

Fig. 7: Winding arrangement

The four coils were glued inside the stator bore. Due to the flux distribution the induced voltage of each coil is of trapezoidal shape. Because of the position of the coils we have a shift of 90° between two voltages as shown in Fig. 6. The resulting induced voltage therefore is of triangle-shape and can be used to identify the rotor position [7]. A short description of the control system is presented below [3, 9].

CONTROL

To get information about the rotor position in a brushless dc (BLDC) machine, usually sensors like Hall sensors or opto-electronic sensors are required. For small BLDC motors a sensorless approach is often required because of reduction of motor volume and weight. Therefore, the back EMF of one winding that is not excited will be used. For a two phase machine this means, that only 50 % of the copper can be used for torque generation. To avoid this disadvantage the windings can be connected in the form of a Wheatstone bridge as shown in Figure 7, where each winding will be divided into two parts.

In this circuit the difference of the back EMFs of, for example, windings W_1 and W_3 can be used to determine the rotor position and the switching point of the inverter, even though all motor windings are excited simultaneously. With a radial magnetization of the magnets the back EMF in one of the windings is trapezoidal, so that the back EMF of one phase is triangular with respect to the time delay between the back EMFs of the windings W_1 and W_3.

Due to the high speed the iron losses are significant. For the final design soft-magnetic material with a low loss-figure will be selected in order to improve the efficiency. The iron-plus friction-losses, the so called no-load losses, versus speed are given in Fig. 8.

The natural characteristic of the brushless dc motor for a reduced voltage of 4.1 V is shown in Fig. 9. Again there is good agreement between calculated and measured data [3,8].

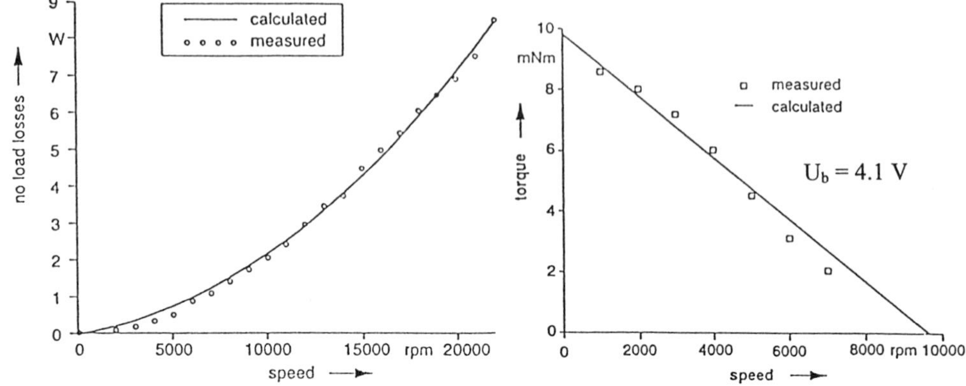

Fig. 8: No-load losses

Fig. 9: Torque-speed characteristic

Because of the end windings, the total length of the motor was 70 mm. The housing of the stator increased the diameter to 25 mm. The specific torque for this motor was calculated to 0.13 mNm/cm³.

THERMAL BEHAVIOUR

The circuit which was developed for the investigation of the thermal behaviour is shown in Fig. 10. Copper- and iron losses heat up the motor. Iron and friction losses were calculated and measured. Thermal resistance R_{14} and R_{34} represent the heat flow from the rotor to the environment and from the stator to the environment. The heat transfer from the winding to rotor and stator is given by the elements R_{12} and R_{23}. The heat capacity of rotor, winding and stator is represented by C_{14}, C_{24} and C_{34}. Details of the thermal investigation can be found in references [3,5].

For the final application "water cooling" is envisaged with a typical temperature of 37°C. The result of a thermal calculation is presented in Fig. 11. Overtemperature is only 2 K for the stator - this finding was essential for the medical experts.

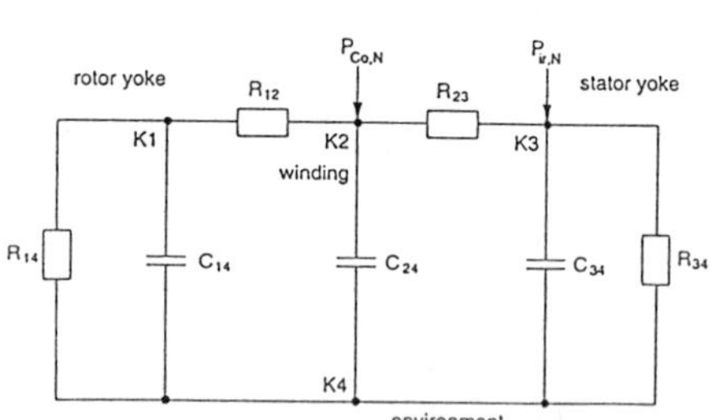

Fig. 10: Circuit for thermal simulation

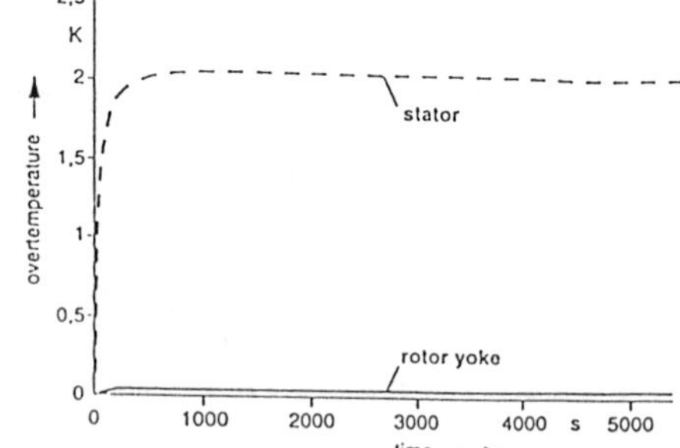

Fig. 11: Temperature increase in the case of liquid cooling

DISCUSSION

Based on a detailed analysis, all the different losses were determined and the efficiency was calculated. Efficiency versus torque is shown in Fig. 12. For the working point of 19,000 rpm and a torque of 4.5 mNm an efficiency of 50 % could be achieved. The characteristic data of the prototype are listed in Table I.

The specific torque was calculated as 0.13 mNm/cm³. A motor of similar dimensions, already available on the market, has 0.087 mNm/cm³.

Table I: Motor parameters

nominal power	9 W
nominal torque	4.5 mNm
nominal current	2 A
nominal speed	19000 rpm
nominal voltage	8.9 V
outer diameter (active material)	22 mm
active length	40 mm
inertia	14.5 gcm²
effective airgap	1.2 mm
motor constant	4.07 mVs
inductance	15 μH
phase resistance	0.4 Ω

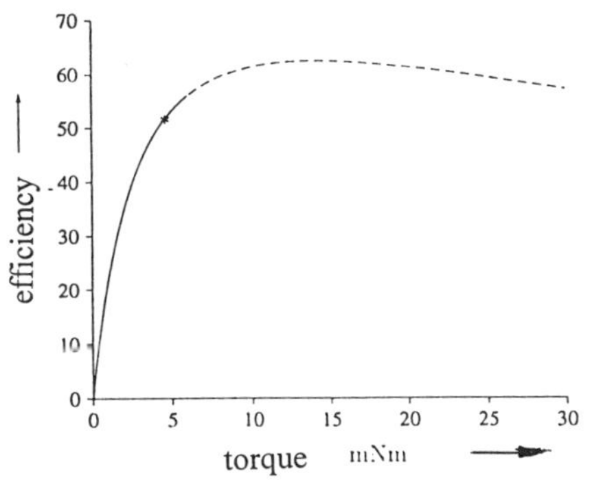

Fig. 12: Efficiency versus torque

CONCLUSION

A low noise, high speed mini motor has been designed with an overtemperature of only 2°K at full load. Premanufactured magnets were glued on the rotor to allow rapid prototyping. Because of the implementation of an airgap winding the low noise requirements are fulfilled.

REFERENCES

[1] J.M.D. Coey, Rare-earth iron permanent magnets, Clarendon Press (1996)

[2] R. Hanitsch, JMMM 80 (1989), pp. 119-130

[3] B. Frenzel, Entwurf und Aufbau einer bürstenlosen Gleichstrommaschine für ein Herzunterstützungssystem, Diss. TU Berlin, Verlag Dr. Köster (1998)

[4] R. Welch, PCIM (1995), Proc. pp. 58-66

[5] K. Atallah, D. Howe, JMMM (1994), pp. 578-582

[6] J.Y. Hung, Z. Ding, IEE Proc.-B, Vol. 140, No. 4, pp. 260-266

[7] L. Antognini, J. Paratte, Self synchronisation of PM step and brushless motors, a new sensorless approach, IMCSD (1990)

[8] B. Frenzel, R. Hanitsch, R. Stephan, IEE Power Electronics and Variable Speed Drives, No. 429 (1996), pp. 99-102

[9] B. Frenzel, R. Hanitsch, Optimal commutation scheme for sensorless brushless DC motors, ICEM '98, Istanbul, Proc. Vol. 2/3, pp. 1437-1440

New series of Sm_2TM_{17} magnet materials for applications at temperatures up to 550°C

Christina H. Chen, Marlin S. Walmer, and Michael H. Walmer
Electron Energy Corp., 924 Links Ave., Landisville, PA 17538, USA

Sam Liu and G. Edward Kuhl
University of Dayton, 300 College Park Ave., Dayton, OH 45469, USA

Gerard K. Simon
Air Force Research Laboratory, WPAFB, OH 45433, USA

ABSTRACT

A new series of $Sm(Co_wFe_vCu_xZr_y)_Z$ magnet products with ${}_iH_c$ up to 13 kOe at 400°C has been developed. These magnets have low temperature coefficients of ${}_iH_c$ and a straight line B vs. H (extrinsic) demagnetization curve up to 550°C. A straight line extrinsic curve provides greater design flexibility and facilitates reduced size and weight of magnetic circuits. The straight line extrinsic demagnetization B curves are required for magnets used in dynamic applications. The maximum use temperature, defined here as T_M, can be determined by finding the maximum temperature at which the B curve is a straight line. The results of this work show that increasing the Co content in the magnet enhances high temperature performance. A higher Co content provides a lower temperature coefficient β of ${}_iH_c$. Magnets with Co content w = 0.65 have β = 0.26%/°C from 25°C to 300°C. Magnets with w = 0.82 have β = 0.116%/°C from 25°C to 300°C. The lower β results in a higher $(BH)_{max}$ at high temperature, and a higher T_M. The T_M is 240°C for w = 0.65, 330°C for w = 0.68, 400°C for w = 0.76, 490°C for w = 0.78, and 550°C for w = 0.82. By increasing the Co content and adjusting the content of other elements, along with appropriate changes of processing parameters, a new series of magnets has been developed for use at temperatures from 400°C to 550°C. Magnetic properties vs. temperature for the new series of magnets, in comparison to those of the best commercial magnets, are shown in this paper to illustrate the improvements which have been made. Magnetic properties at different temperatures vs. Co content for the series are also reported in this paper.

INTRODUCTION

As described in our previous paper [1-2], permanent magnets for use at high temperatures (≥400°C) are in demand, especially for advanced power systems. Having the highest Curie temperature of the rare earth magnet alloys, and having moderately high saturation magnetization and magnetocrystalline anisotropy, the Sm_2TM_{17} compound is the best candidate of all known materials for developing improved magnets for high temperature applications [3-4]. Fig.1a and Fig.1b give a summary of $(BH)_{max}$ and ${}_iH_c$ vs. temperature for most commercial

Mat. Res. Soc. Symp. Proc. Vol. 577

permanent magnetic materials. All the data in the figures were obtained at EEC using a KJS hysteresigraph. $SrFe_{12}O_{19}$ is the only magnet that has a positive temperature coefficient of $_iH_c$. The $_iH_c$ of $SrFe_{12}O_{19}$ is 3.4 kOe at 25°C and 5.7 kOe at 250°C. However, the $(BH)_{max}$ of $SrFe_{12}O_{19}$ is 3.5 MGOe at 25°C and 1.12 MGOe at 250°C, which are too low for many applications. Alnico has the highest T_C (=840°C) of all materials. Alnico's $_iH_c$ is almost a constant, ~0.75 kOe, from −160°C to 300°C. Such low $_iH_c$ is not suitable for dynamic applications. PtCo has $(BH)_{max}$ = 8.8 MGOe at 25°C and 3.6 MGOe at 250°C, T_c = 475°C, and wonderful mechanical property. PtCo is not good for use at T > 300°C. The information about the T_c and other magnetic properties for all these materials in Figs.1a and 1b can be found in our previous paper [5].

Nd-Fe-B compounds have the highest $(BH)_{max}$ at temperatures up to 130°C. For applications at temperatures higher than 150°C, where concern for thermal magnetic losses exists, the Sm_2TM_{17} system is the choice. At temperatures above 150°C, Sm_2TM_{17} has the highest $(BH)_{max}$ of all materials, and it has a low temperature coefficient of B_r and of $_iH_c$. Nd-Fe-B, with 45.3 MGOe at 25°C, drops to 23.4 MGOe at 150°C, 16 MGOe at 175°C, and 9.6 MGOe at 200°C. The $(BH)_{max}$ of 45 MGOe Nd-Fe-B drops almost 50% when temperature reaches 150°C. Even the high coercive force Nd-Dy-Fe-Co-B, with $(BH)_{max}$=30MGOe and $_iH_c$ >25 kOe at 25°C, the $(BH)_{max}$ drops to 23 MGOe at 150°C. By comparison, Sm_2TM_{17} with 31.5 MGOe at 25°C only drops to 28.3 MGOe at 150°C, 27.5 MGOe at 175°C, and 26.7 MGOe at 200°C. Commercial Sm_2TM_{17} magnets maintain acceptable performance up to 330°C.

When the temperature goes above 330°C, the $_iH_c$ drops rapidly for most materials including the Sm_2TM_{17}. Thus, there is concern for both the coefficient of B_r and the coefficient of $_iH_c$, since a low $_iH_c$ limits the $(BH)_{max}$. For more than two decades, researchers have made efforts to improve the high temperature performance of RE-Co alloys. Temperature compensated TC-$RECo_5$ and TC-RE_2TM_{17} made by substituting heavy rare earth (HRE) elements for Sm [2,6-12] provides very low temperature coefficients of B_r and $(BH)_{max}$ up to 300°C. Since 1973, EEC has been making the TC magnets with various temperature coefficients of B_r, ranging from +0.08%/°C through zero to −0.04%/°C, which work very well for applications up to 300°C[2]. At T > 300°C, the $_iH_c$ of TC magnets also drops. Work is continuing, at EEC, to improve the $_iH_c$ of HRE substituted RE-Co materials.

Studies have shown improvement of the high temperature performance of $Sm(Co_wFe_vCu_xZr_y)_Z$ by adjusting w, v, x, y, and z [1,13-22]. Our work in 1997 [1] on $Sm_2(Co_w Fe_vCu_{.06} Zr_{.02})_{17}$ (w = 0.69–0.92) reported that Sm_2TM_{17} with an increased Co content had a straight extrinsic demagnetization B curve up to 400°C. A straight line B curve provides greater design flexibility and facilitates reduced size and weight of magnetic circuits. Straight line B curves are required for magnets used in dynamic systems. The maximum use temperature, defined here as T_M, can be determined by finding the maximum temperature at which the extrinsic B vs. H curve remains a straight line. In this paper, our results show that by increasing Co, adjusting other elements, and making process changes, a new series of magnets with increased T_M has been developed for use at temperatures from 400°C to 550°C.

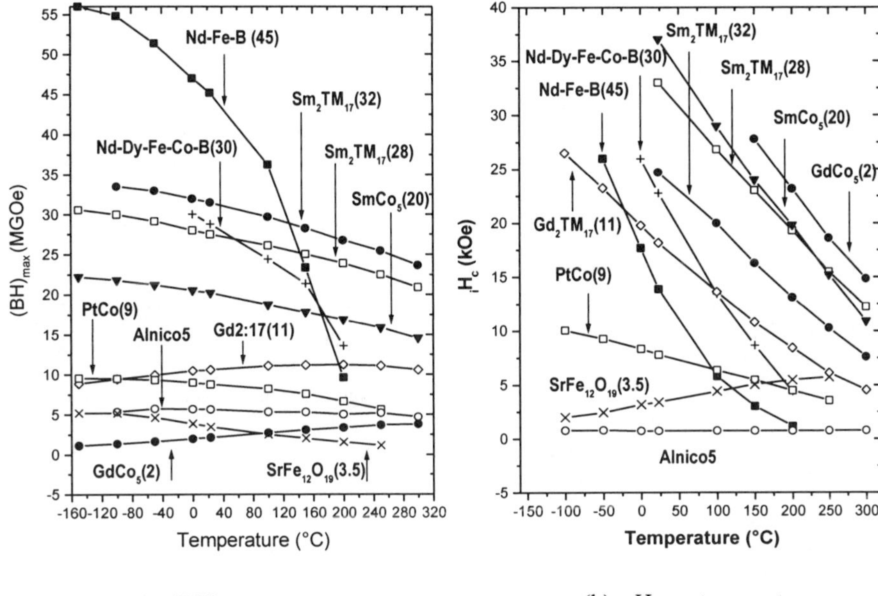

(a). $(BH)_{max}$ vs. temperature (b). $_iH_c$ vs. temperature

Fig. 1 Summary of magnetic properties for most commercial permanent magnetic materials at –160°C to 300°C. The numbers in the parentheses are the $(BH)_{max}$ at room temperature. All the magnet samples were tested at EEC using a KJS hysteresigraph.

Experimental Methods

Magnets of Sm_2TM_{17} type with compositions of $Sm(Co_wFe_vCu_xZr_y)_z$ (w = 0.6–0.92, v = 0–0.3, x = .04–0.12, y = 0.02–0.03, and z = 6.5–9.0) were prepared by both induction melting and arc button melting under Ar atmosphere followed by conventional powder metallurgy processing. Magnetic characterizations from room temperature up to 300°C were performed using a KJS hysteresigraph. At temperatures up to 600°C a vibrating sample magnetometer (VSM) was used. Because VSM characterization is an open-circuit measurement, careful calibration was carried out to ensure that the results obtained with the VSM were equivalent to the results of a closed-circuit hysteresigraph measurement.

RESULTS

Table I lists the magnetic properties of the new series magnetic materials in comparison to the commercial magnet materials. Table II shows the average temperature coefficient β of ${}_iH_c$ for each magnet in this series. Table III lists the magnetic properties at T_M. The formula for all the magnets is $Sm(Co_wFe_vCu_xZr_y)_z$ with w = 0.65–0.82, v = 0.27–0.06, x = 0.05–0.09, y = 0.02–0.03, z = 7.4 –7.7. The magnetic properties vs. temperature are shown in Fig.2a through Fig.2e. Figs.3a and 3b show the demagnetization curves for a magnet with w = 0.78. Figs. 4a and 4b show a comparison of the magnetic properties for the entire series. The magnetic properties vs. Co content are plotted in Fig.5a through Fig.5d.

1. Magnets for use up to 240°C and 330°C

EEC's commercial 2:17 magnets are EEC32-T240C for use up to 240°C, and EEC27-T330C for use up to 330°C. EEC32-T240C, with Co content w = 0.65, has a $(BH)_{max}$ = 31.5 MGOe and ${}_iH_c$ = 25 kOe at 25°C. The extrinsic demagnetization curves are straight line until the temperature exceeds 240°C. This enables the magnet to be dynamically stable and to perform well at temperatures up to 240°C. At 240°C, this magnet has ${}_iH_c$= 10.9 kOe, B_r= 10.7 kG, and $(BH)_{max}$ = 25.7 MGOe. Here ${}_iH_c/B_r$ = 1.02 Oe/G. Having ${}_iH_c/B_r \geq 1$ is important to ensure a straight line B curve. The magnetic properties vs. temperature are shown in Fig. 2a.

EEC27-T330C, with Co content w = 0.68, has $(BH)_{max}$ = 27.4 MGOe and ${}_iH_C$ = 34 kOe at 25°C. The B curves are straight line until temperature exceeds 330°C. At 330°C, ${}_iH_c$ is 9.8 kOe, Br is 9.4 kG, ${}_iH_c/B_r$ >1, and $(BH)_{max}$=20.1MGOe. The magnetic properties vs. temperature are shown in Fig. 2b. EEC27-T330C has served many applications up to 330°C since 1985 with no reported problems.

2. Magnets for use up to 400°C

One of the new magnets is EEC24-T400C, which has $(BH)_{max}$ = 24.5 MGOe and ${}_iH_c$ = 34 kOe at 25°C. By increasing Co from w = 0.68 to 0.76, a straight B curve at 400°C is obtained. At 400°C, the ${}_iH_c$ is 8.8 kOe, B_r is 8.6 kG, ${}_iH_c/B_r$ >1, and $(BH)_{max}$=16.5 MGOe. The magnetic properties vs temperature are shown in Fig.2c and Table I. Comparing EEC24-T400C with EEC-T240C, $(BH)_{max}$ of T400 is ~7 MGOe lower than the T240C at 25°C. T400C is 2.7 MGOe higher than T240C at 400°C. More important, T400C has a straight B curve at 400°C, which makes it to be dynamically stable at 400°C.

3. Magnets for use up to 490°C

Another new is EEC21-T490C, which has $(BH)_{max}$ = 20.8 MGOe, ${}_iH_c$ = 29 kOe, and H_K > 21 kOe at 25°C. By further increasing Co content to w = 0.78, the B curves are straight line until the temperature is increased to 500°C. The dependence of magnetic properties on temperature is shown in Fig. 2d. Fig.3a exhibits both intrinsic and extrinsic demagnetization curves at temperatures from 25°C to 300°C. Fig.3b shows the curves at temperatures from 400°C to 500°C. At 490°C, the magnet has $(BH)_{max}$ = 10.7 MGOe, ${}_iH_c$ = 7.7 kOe, B_r = 6.95 kG, and ${}_iH_c/B_r$ = 1.11 Oe/G. At 25°C, T490C is 10.6 MGOe lower than T240C, and at 400°C, T490C is 4.5 MGOe higher than T240C (See Fig.5b).

4. Magnets for use over 550°C

The highest use temperature in the new series is EEC16-T550C with has $(BH)_{max}$ = 16.4 MGOe and ${}_iH_c$ = 25 kOe at 25°C. By further increasing Co content to w = 0.82, straight line B curves can be obtained up to 550°C. Fig.2e shows the relationship between magnetic properties and temperature from 25°C to 600°C. At 550°C, T550C has $(BH)_{max}$ = 6.3 MGOe, ${}_iH_c$ = 6.4 kOe, B_r = 5.5 kG, And the ${}_iH_c/B_r$ is 1.16 Oe/G.

Table I. Magnetic properties at various temperatures for magnets $Sm(Co_wFe_vCu_xZr_y)_z$

Magnet	25°C		300°C		400°C		500°C		600°C	
	${}_iH_c$ [kOe]	$(BH)_{max}$ [MGOe]	${}_iH_c$ [kOe]	$(BH)_{max}$ [MGOe]	${}_iH_c$ [kOe]	$(BH)_{max}$ [MGOe]	${}_iH_c$ [kOe]	$(BH)_{max}$ [MGOe]	${}_iH_c$ [kOe]	$(BH)_{max}$ [MGOe]
EEC32-T240C	25	31.5	7.1	23.7	3.4	13.8	1.5	5.9	0.7	2.3
EEC28-T330C	33	27.5	11.6	21.0	5.4	16.7	2.9	8.7	1.2	3.5
EEC24-T400C	34	24.6	14.6	18.8	8.8	16.5	4.7	12.5	2.1	5.6
EEC21-T490C	29	20.8	16.7	15.6	12.4	13.2	7.3	10.4	3.6	6.8
EEC16-T550C	25	16.4	17.0	12.0	13.2	9.9	8.8	7.6	4.7	5.2

Table II

Co content, T_M and the temperature coefficient β of ${}_iH_c$ for the series $Sm(Co_w Fe_vCu_xZr_y)_z$

Magnet ID	Co content (w)	T_M * (°C)	β ** for ${}_iH_c$ (25 – 300°C) (%/°C)	β ** for ${}_iH_c$ (25 – 400°C) (%/°C)	β ** for ${}_iH_c$ (25 – 500°C) (%/°C)
EEC32-T240C	0.65	240	0.260	0.230	0.198
EEC27-T330C	0.68	330	0.236	0.223	0.192
EEC24-T400C	0.77	400	0.207	0.198	0.181
EEC20-T490C	0.78	490	0.154	0.153	0.157
EEC16-T550C	0.82	550	0.116	0.128	0.136

* T_M is the maximum use temperature for dynamic application.

** ${}_iH_c$ decreases as temperature increases. β is an average at the temperature range.

Table III. Magnetic properties at T_M for the series $Sm(Co_wFe_vCu_xZr_y)_z$

Magnet	T_M (°C)	Properties at maximum use temperature T_M			
		B_r (kG)	${}_iH_c$ [kOe]	$(BH)_{max}$ [MGOe]	${}_iH_c$ / B_r (Oe/G)
EEC32-T240C	240	10.7	10.9	25.7	1.02
EEC28-T330C	330	9.4	9.8	21.0	1.04
EEC24-T400C	400	8.6	8.8	16.5	1.02
EEC21-T490C	490	6.95	7.7	10.7	1.11
EEC16-T550C	550	5.5	6.4	6.3	1.16

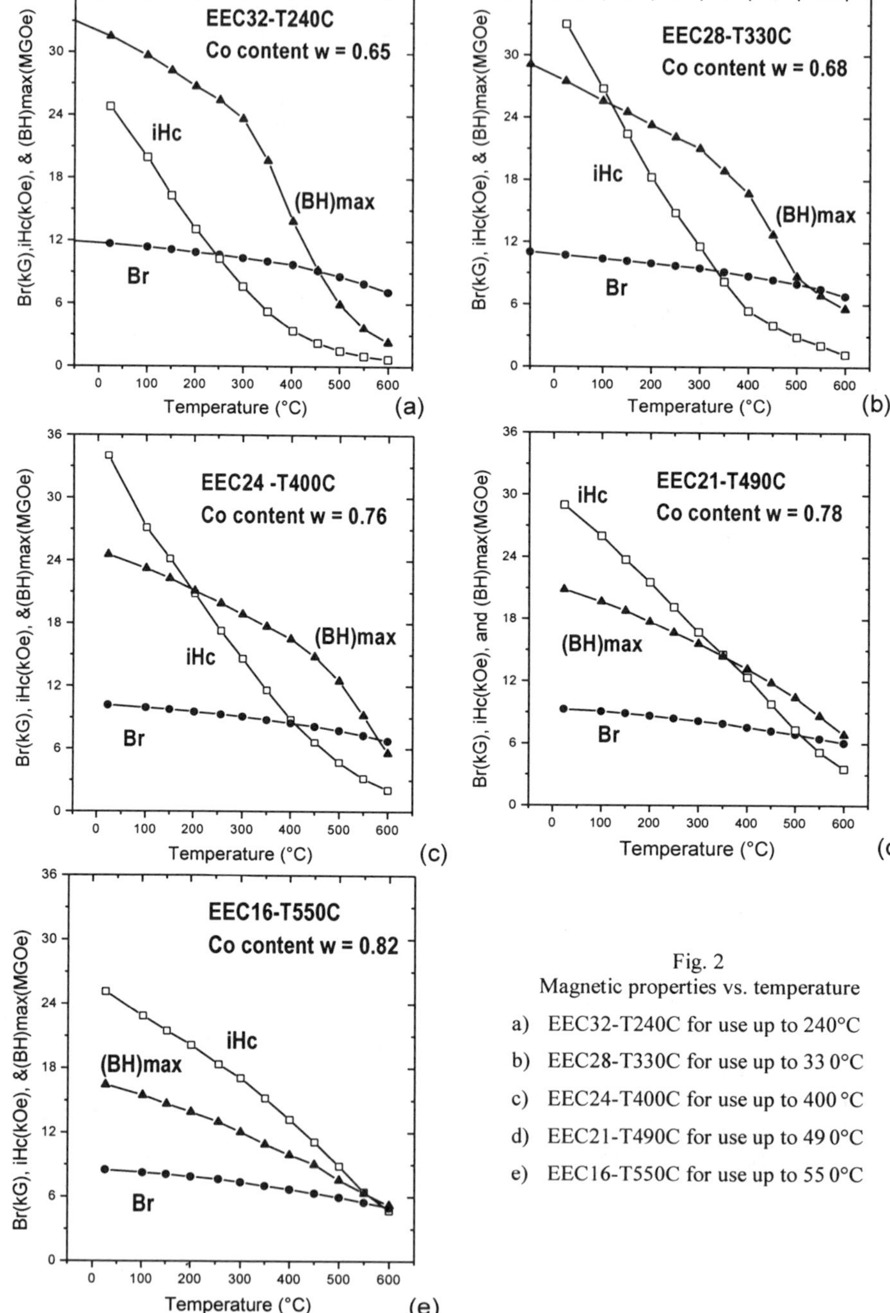

Fig. 2
Magnetic properties vs. temperature

a) EEC32-T240C for use up to 240°C
b) EEC28-T330C for use up to 33 0°C
c) EEC24-T400C for use up to 400 °C
d) EEC21-T490C for use up to 49 0°C
e) EEC16-T550C for use up to 55 0°C

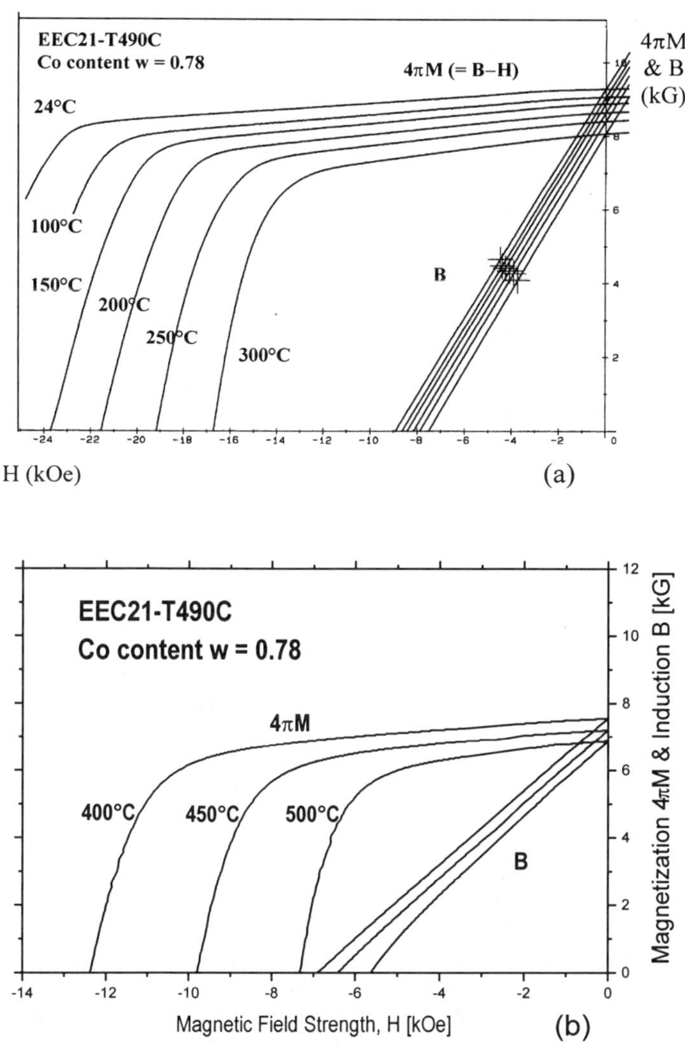

Fig. 3 Demagnetization curves for EEC21-T490
(a). Demagnetization curves tested with KJS hysteresigraph at 24°C to 300°C.
(b). Demagnetization curves tested with VSM at 400°C to 500°C.

5. Maximum use temperature and Co content

Comparing all the magnets described above and shown in Figures 2a through 2e, it is clear that the Co content plays an important role in high temperature performance. Our previous work [1] showed that higher Co (lower Fe) resulted in a higher Curie temperature and a lower thermal irreversible loss. In this paper, the results show that high Co content results in a lower temperature coefficient of ${}_iH_c$, which limits the reduction of $(BH)_{max}$ at high temperatures. The values of $(BH)_{max}$ vs temperature for all these magnets are plotted in Fig. 4a. from 230°C to 600°C. The values of ${}_iH_c$ vs temperature for all these magnets are plotted in Fig. 4b. from 100°C to 600°C The formula $Sm(Co_wFe_vCu_xZr_y)_z$ is used to describe the composition of this series. With the lowest Co content (w = 0.65) in this series, EEC32-T240C has 31.5 MGOe, the highest $(BH)_{max}$, at room temperature, but the lowest ${}_iH_c$ and $(BH)_{max}$ at any temperature above 450°C. T_M for EEC32-T240C is 240°C, its maximum use temperature in a dynamic system. With the highest Co content in this series, EEC16-T550C has 16.4 MGOe, the lowest $(BH)_{max}$ at 25°C, but the highest ${}_iH_c$ at any temperature above 300°C, and the highest T_M, 550°C, which determines its maximum use temperature 550°C in a dynamic system. As Fig.4b shows, at T >300°C, the ${}_iH_c$ and T_M increase as Co increases. Table II shows the Co content, T_M and the temperature coefficient of ${}_iH_c$ for the series of $Sm(Co_wFe_vCu_xZr_y)_z$ magnet materials.

The magnetic properties vs. Co content for the series are plotted in Fig.5a through Fig.5d. Fig.5a exhibits the dependence of temperature coefficient of ${}_iH_c$ on Co content w. The average temperature coefficient β (from 25°C to 300°C) of ${}_iH_c$ is decreased from β = 0.260%/°C for w = 0.65 to β = 0.116%/°C for w = 0.82. β is decreased by 55% when w is increased by 26%. Fig.5b shows the ${}_iH_c$ vs. Co content w at 300°C, 400°C, 500°C, and 600°C. Regardless what ${}_iH_c$ is at room temperature, the ${}_iH_c$ is dependent on Co content w at temperatures above 300°C: the higher the Co content, the higher the ${}_iH_c$ at T >300°C. Fig. 5c shows the $(BH)_{max}$ vs. the Co content at different temperatures. At room temperature, $(BH)_{max}$ is decreased as the Co content is increased. This is due to the fact that Co has lower saturation magnetization M_s than that for Fe. When temperature is increased, magnets with higher Co have lower β, higher T_c, and lower thermal losses. At 25°C, the highest $(BH)_{max}$ is obtained with w = 0.65. At 500°C, the highest $(BH)_{max}$ is obtained with w = 0.76. At 600°C, the highest $(BH)_{max}$ is obtained with w =0.78. Clearly, only those magnets with w ≥ 0.82 will survive at even higher temperatures. Fig.5d shows T_M vs. Co content for the five magnet alloys in the series. Fig.5d also shows the $(BH)_{max}$ at T_M. Table III lists the magnetic properties at T_M. Practically, when T_M is higher than 400°C, the $(BH)_{max}$ could not be expected to be very high due to the thermal agitation in the atomic lattice of the alloys. This will be discussed more in a future paper. The value of $(BH)_{max}$ is 25.7 MGOe for w = 0.65 at 240°C, 20.1 MGOe for w = 0.68 at 330°C, 16.5 MGOe for w = 0.76 at 400°C, 10.7 MGOe for w = 0.78 at 490°C, and 6.3 MGOe for w = 0.82 at 550°C. Increased T_M comes at the expense of $(BH)_{max}$. The significance is that the magnets with higher T_M have a straight line B curve at T_M, which makes the magnets to be stable in dynamic applications.

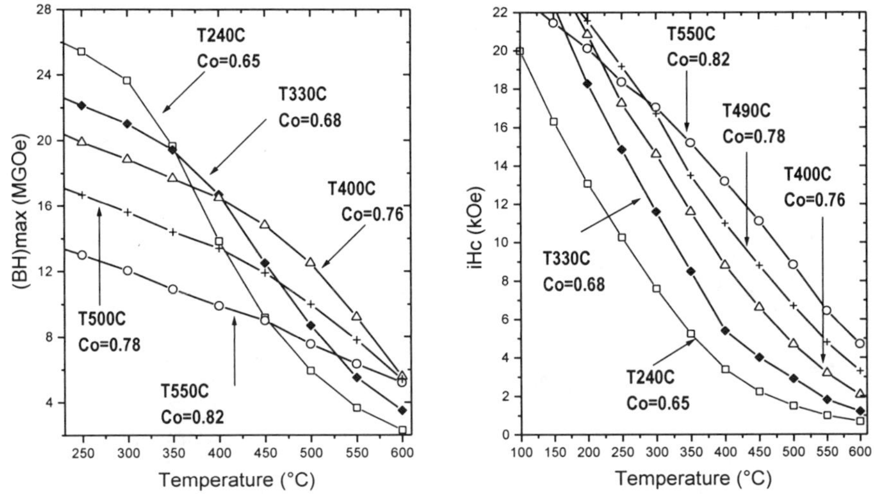

Fig. 4 Comparison for the series Sm-Co based magnetic materials.
(a). $(BH)_{max}$ vs. temperature (230 - 600°C)
(b). ${}_{i}H_{c}$ vs. temperature (100 – 600°C)

CONCLUSION

By increasing Co content and adjusting the contents of other elements, along with changing the process, temperature coefficient of ${}_{i}H_{c}$ is decreased, the ${}_{i}H_{c}$ is increased at high temperatures, and the maximum use temperature T_M is increased. A discrete Co content is required for each operating temperature. Also there is a discrete energy product for each operating temperature.

ACKNOWLEDGMENT

Authors wish to thank Dr. Jinfang Liu of EEC for helpful discussions. This work is supported by US Air Force under contract STTR # F33615-97-5017 and by US Navy under contract AMPS #.N0014-98-1-0636.

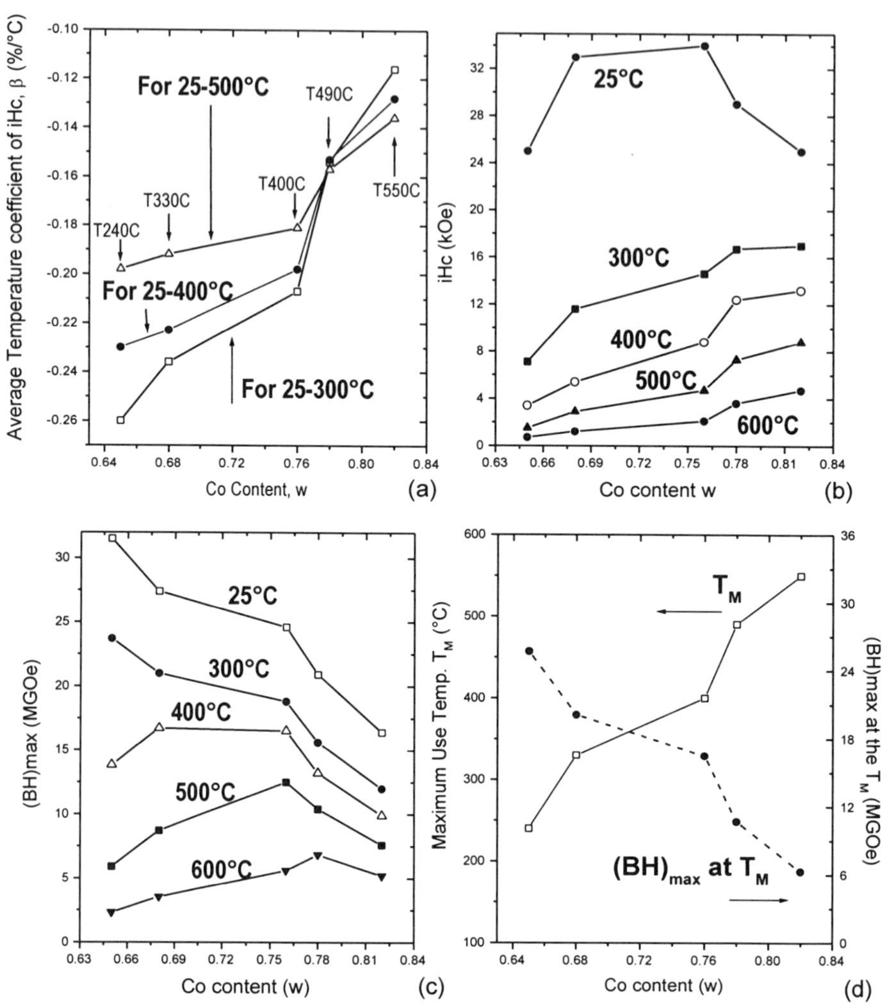

Fig. 5 Properties vs. Co content w

(a). Average temperature coefficient β of ${}_iH_c$ vs. Co content w.

(b) ${}_iH_c$ vs. Co content w at different temperature.

(c). $(BH)_{max}$ vs. Co content at different temperature.

(d). Maximum use temperature T_M and the $(BH)_{max}$ at T_M vs. Co content w.

REFERENCES:

[1] C.H. Chen, M.S. Walmer, M.H. Walmer, S. Liu, E. Kuhl, G. Simon, "$Sm_2(Co,Fe,Cu,Zr)_{17}$ magnets for use at temperature ≥ 400°C", 1998, J. Appl Phys. vol 83 (11), p.6706.

[2] M.S. Walmer, C.H. Chen, M.H. Walmer, S. Liu, G.E. Kuhl, G.K. Simon, "Use of heavy rare earth element Gd in $RECo_5$ and RE_2TM_{17} magnets for high temperature applications", 1998, Proc. **15th** Int. Workshop on Rare Earth Permanent Magnets and Their Applications, p. 689

[3]. A.E. Ray and K.J. Strnat, "Easy directions of magnetism in ternary $R_2(Co, Fe)_{17}$ phase", IEEE Trans. Magn., Vol 8, 1972, p.516.

[4] A.E. Ray and S. Liu, "recent progress in 2:17-type permanent magnets", J. Materials Engineering and Performance, vol. 1, 1992, pp183-191.

[5] Christina H. Chen, Marlin S. Walmer, Michael H. Walmer, Wei Gong, and Bao-Min Ma, "The relationship of thermal expansion to magnetocrystalline anisotropy, spontaneous magnetization and T_C for permanent magnets", Appl. Phys. April, 1999, to be published

[6] M.G. Benz, R.P. Laforce, and D.L. Martin, "AIP Conf. Proc. 1974, No.**18/2**, eds C.D.Graham and J.J. Rhyne (American Institute of Physics, New York) p. 1173

[7] F.G. Jones and M. Tokunaga, IEEE Trans. Magn., **MAG-12**, 1976, p. 968

[8] R.J.J. Martis, N. Gupta, S.G. Sankar and V.U.S. Rao, 1978, J. Appl. Phys., **49**, p. 2070

[9] K.S.V.L. Narasimhan, 1981, J. Appl. Phys., **52**, p.2512

[10] D. Li, E. Xu, J. Liu, and Y. Du, 1980, IEEE Trans. Magn., **MAG-16**, p. 988

[11] D. Li, J. Liu, S. Zhou, X. Jin, and E. Xu, 1983, Proc. 7th Int. Workshop on Rare Earth Permanent Magnets and Their Applications, Beijing (China Academic Publishers), p. 495

[12] H.F. Mildrum, J.B. Krupar and A.E. Ray, 1983, J. Less-Common Met. **93**, p. 261

[13]. S.Liu, H.F. Mildrum, and K.J. Strnat, "Demagnetization curves of four rare-earth-cobalt magnet types at temperatures 300-1000 K," J. Appl. Phys. Vol.53, 1982, p. 2383.

[14] S. Liu and E. P. Hoffman, "Application-oriented characterization of $Sm_2(Co,Fe,Cu,Zr)_{17}$ permanent magnets," IEEE Trans. Magn., Vol. 32, 1996, p. 5091.

[15] B. M. Ma, Y. L. Liang, J. Patel, D. Scott, and C. O. Bounds, "The effect of Fe content on the temperature dependent magnetic properties of Sm(Co,Fe,Cu,Zr)z and SmCo5 sintered magnets at 450°C," IEEE Trans. Magn., Vol. 32.1996, p. 4377.

[16] S. Liu, G. P. Hoffman, and J. R. Brown, "Long-term aging of $Sm_2(Co,Fe,Cu,Zr)_{17}$ permanent magnets at 300 and 400°C," IEEE Trans. Magn., Vol. 33, 1997, p. 3859.

[17] A. Kim, "High temperature stability of SmTM magnets," J. Appl. Phys., Vol. 83, 1998, p.6715.

[18] J.F.Liu,, Y. Ding, and G.C. Hadjipanayis, "Effect of iron on the high temperature magnetic properties and microstructure of $Sm_2(Co,Fe,Cu,Zr)_{17}$ permanent magnets", J. Appl. Phys., Vol. 85(3), 1999, p.1670

[19] J.F. Liu, T. Chui, D. Dimitrov and G.C. Hadjipanayis, "Abnormal temperature dependence of intrinsic coercivity in $Sm(Co,Fe,Cu,Zr)_z$ powder materials", Appl. Phys. Lett. Vol.73, 1998, p.3007

[20] J.F.Liu,, Y. Ding, D. Dimitrov and G.C. Hadjipanayis, "Microstructure and high temperature magnetic properties of $Sm_2(Co,Fe,Cu,Zr)_{17}$ (z = 6.7-9.1) permanent magnets", Appl. Phys. Vol.85(5), 1999, p.2800

[21] J.F.Liu,, Y. Ding, D. Dimitrov and G.C. Hadjipanayis, "New rare earth permanent magnets with an intrinsic coercivity of 10 kOe at 773 K", .

[22] Sam Liu, Jie Yang, Gerorge Doyle, G.Edward Kuhl, Christina Chen, Marlin Walmer, Michael Walmer, and Gerard Simon, " New sintered high temperature Sm-Co based permanent magnet materials", Inter-Mag, 1999, to be published.

Part VII

Microstructure and Micromagnetics

HIGH PERFORMANCE MAGNETS - MICROSTRUCTURE AND COERCIVITY

J. FIDLER, S. SASAKI AND E. ESTEVEZ-RAMS
Vienna University of Technology, Institute of Applied and Technical Physics,
Wiedner Hauptstr. 8-10, A-1040 Wien, Austria, http://atp6000.tuwien.ac.at/MAGNET/

ABSTRACT

The importance of newly developed permanent magnetic materials in many electro-, magnetomechanical and electronic applications can be attributed to the drastic improvement of the magnetic energy density product and coercive field. A systematic study has been undertaken in order to determine the influence of oxygen content on microstructure and coercivity of high remanence $Nd_2Fe_{14}B$ based sintered magnets. The energy density product >400 kJ/m^3 (50 MGOe) and the coercive field of 800 kA/m were obtained after a combination of rubber isostatic and transverse die pressing methods. Magnets of the composition $Nd_{15.1-x}Fe_{78+x}B_6Cu_{0.03}Al_{0.7}$ [x= 0-2.5] were prepared using strip cast materials. The high oxygen content of the magnets was gradually decreased from values of 4000-6000 ppm to a value < 1000 ppm. Abnormal grain growth (AGG) of the 2:14:1 grains occurred preferentially in magnets with low oxygen content, thus the squareness of the demagnetisation curve drastically decreased. The oxygen content strongly affects the AGG and the magnets with higher oxygen content have the higher critical temperatures at which the AGG occurs. On the other hand, isotropic magnets tend to have the lower critical temperatures than anisotropic magnets by 10-20°C. In the second part of the paper examples of the influence of intergranular phases, grain size and grain shape observed by TEM on coercivity of sintered and melt-spun Nd-Fe-B magnets are compared.

INTRODUCTION

The increasing impact of magnetic materials on many modern industries will continue well into the next century. Besides recording materials and soft magnetic devices, also hard magnetic materials are key components in transportation and information technologies, machines and many other systems. The improvement of the *magnetic energy density product* and *coercive field* of the new hard magnetic materials leads to new devices in many electro-, magnetomechanical and electronic applications. The rare earth intermetallic phases $SmCo_5$, Sm_2Co_{17} and $Nd_2Fe_{14}B$ are the basis for these high performance magnets [1-6]. Rare earth-Co magnets exhibit the highest coercive fields and (Nd,Dy)-(Fe,Co)-B:(M1,M2) magnets show the highest value of energy density product, obtained so far.

Hard magnetic materials are divided into the group of conventional metallic and oxide magnets and the group of modern magnets based on intermetallic compounds of rare earth elements with cobalt and/or iron. High performance $SmCo_5$/Sm_2Co_{17}- and $Nd_2Fe_{14}B$-based permanent magnets exhibit a complex, multiphase microstructure. The grain size of the magnets and the alignment of the grains strongly depend on the processing parameters. The formation and distribution of the phases is determined by the composition of the magnets and the annealing treatment.

For the better understanding and the development of high performance permanent magnets and for the search of new phases a detailed understanding of the hysteresis and coercivity mechanisms is necessary. This also includes the exact knowledge of the crystallography, phase

Mat. Res. Soc. Symp. Proc. Vol. 577 © 1999 Materials Research Society

diagram and phase relation of the magnet system. The hysteresis properties are governed by a combination of the intrinsic properties of the material, such as saturation polarisation, magnetic exchange and magnetocrystalline anisotropy [7-10]. The other important factors are the microstructural parameters, such as grain size, the orientation of the easy axes of the grains and the distribution of phases. The coercivity is determined by the long range dipolar interaction and short range exchange coupling between neighbouring grains. The doping of elements changes the phase relation and favours the formation of new phases. Additional secondary non-magnetic intergranular phases decrease the remanence and interrupt the magnetic interactions between the grains, thereby improving the coercivity of large grained sintered magnets. The role of intergranular structure between the grains plays a significant role determining the magnetic properties, if the grain diameter is in the nanometer scale.

The influence of oxygen on the hard magnetic properties is more complex. Special emphasis will be laid on the study of the role of oxygen, grain alignment and abnormal grain growth on the coercive field and the energy product of advanced, high remanence sintered Nd-Fe-B magnets.

EXPERIMENT

Highest energy density Nd-Fe-B permanent magnets (>*50 MGOe*) exhibiting a low oxygen content have been prepared by the sintering route according to the following experimental procedure [11,12]. The $(Nd,Pr)_{12.6-15.1}Fe_{72.8-80.5}B_{5.8-6.1}$ alloys were strip cast in an argon atmosphere from Nd-Fe, Fe-B, Fe and (Cu,Al) pre-alloys. The crushed flakes exhibit a homogeneous microstructure of fine dispersed Nd-rich phase within a $Nd_2Fe_{14}B$ matrix phase [13]. Using the hydrogen decrepitation (HD) process coarse powders of less than 500 μm were prepared. After jet milling under nitrogen-gas atmpsphere fine powders of about 2.2-3.3 μm in mean FSSS particle size (Fisher Sub Sieve Size) were obtained. The oxygen content of the milling gas was in the order of 10 ppm O_2. The distribution of the size of fine powders was measured with a scanning electron microscope.

The fine powders were pre-pressed and pre-aligned with a pulsed field in the glove box under argon atmosphere to a rectangular compact. The encapsled magnets were pressed in a transverse magnetic field press (combination of RIP and TDP) under a magnetic field of 1200 kA/m. This new pressing technique showed an improved alignment of the hard grains and therefore a high value of remanence. These compacts were sintered at 1233-1373 K for 10.8 ks in vacuum and were subsequently annealed at about 1073 K for 900 s and 783-843 K for 3.6 ks in order to obtain optimum density and coercive field.

Magnetic properties were measured by a B-H hysteresis tracer after being fully magnetised in a pulsed field up to 7 T. After each step of the preparation route the oxygen content of the alloy, power and the magnet, respectively was measured with a LECO oxygen analyser. The microstructure (grain size, phases and degree of alignment of grains) of the magnets was investigated by Kerr microscopy, electron probe microanalysis, scanning and transmission electron microscopy.

RESULTS

High remanent sintered Magnets

Magnets of the composition $Nd_{15.1-x}Fe_{78+x}B_6Cu_{0.03}Al_{0.7}$ [x= 0-2.5] were prepared by the powdermetallurgical sintering route [14]. The energy density product >400 kJ/m^3 and the coercive field of 800 kA/m were obtained after a combination of rubber isostatic and transverse die pressing methods under optimised sintering conditions. The misalignment of the hard magnetic grains with a diameter of 2~5 μm was calculated in the best case in the order < 14°. The high oxygen content of the magnets was gradually decreased from values of 4000 - 6000 ppm to a value < 1000 ppm. This high oxygen content was one limiting factor to decrease the Nd-content in order to improve the volume fraction of the hard magnetic phase.

The demagnetisation curve of an optimised magnet with the composition $Nd_{13.5}Fe_{bal}B_{5.95}Cu_{0.03}Al_{0.7}$ is shown in Fig.1. The magnets produced were sintered between 960°C and 1100°C. The sintering temperature was varied to get optimum density (7.5~7.6 g/cm^3) and $(B.H)_{max}$. The density of the samples and the remanence increased with increasing sintering temperature keeping the sintering time constant (3 hours), while the squareness of the demagnetisation curve only partly increased and drastically decreased as abnormal grain growth of the 2:14:1 grains occurred [15]. A single or a two step (800°C plus quenching) annealing treatment at 500~600°C for 60 min increased the coercive field by about 200 kA/m or about 20%. Besides the density and coercive field also the squareness of the demagnetising curve has to be optimised in order to get highest values of energy density products.

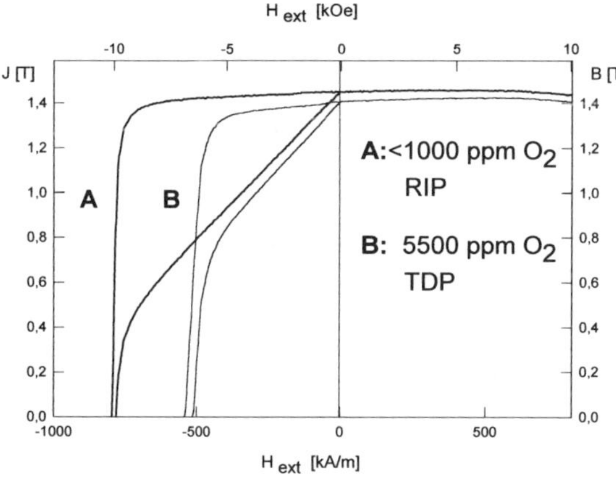

Fig.1:
Comparison of demagnetisation curves of $Nd_{13.5}Fe_{bal}B_{5.95}Cu_{0.03}Al_{0.7}$ sintered magnets manufactured by rubber isostatic die pressing (RIP) and transverse die pressing on air (TDP).
(A) J_r=1.45 T, $(B.H)_{max}$= 50.2 MGOe
(B) J_r=1.41 T, $(B.H)_{max}$= 41.9 MGOe

There are several factors that affect the Abnormal Grain Growth(AGG). In this study, we focused on the effect of the oxygen content and the alignment on AGG in the magnets with 13.7 at.% of Nd. Table 1 shows how much AGG occurred at each sintering temperature for each oxygen content. This also shows the difference between the isotropic magnets and the anisotropic magnets. For all conditions, the AGG occurs at higher temperatures. On the one hand, the oxygen content strongly affects the AGG and the magnets with higher oxygen content have the higher critical temperatures at which the AGG occurs. On the other hand, isotropic magnets tend to have the lower critical temperatures than anisotropic magnets by 10-20°C.

Even though too high oxygen content deteriorates magnetic properties, appropriate oxygen content works very effectively to prevent the AGG. We studied how much oxidation occurs at each step of the manufacture of sintered Nd-Fe-B magnets. Table 2 shows the change of oxygen content for two different ways. In high oxygen containing magnets controlled oxidation occurs during jet milling and compaction on air and insufficient sintering atmosphere. No significant oxidation occurs after the jet milling process in low oxygen magnets.

Table 1: Number of abnormally grown grains (AGG) in isotropic and anisotropic magnets (Nd=13.7 at.%) in dependence on sintering temperature and oxygen content.

oxygen content	magnet	AGG at each sintering temperature/°C					
		980	1000	1020	1030	1040	1050
0.10	isotropic	A	C	D			
0,20	isotropic		B	B		D	
	anisotropic		A	B		D	
0,40	isotropic		A	A		D	
	anisotropic		A	A		B	
0.55	isotropic			A	A	B	D
	anisotropic			A	A	A	C

amount of big grains: A<B<C<D, A: no big grain, D: mainly AGG´s

Table 2: Oxidation during the different steps of preparation of $Nd_{15.1-x}Fe_{78+x}B_6Cu_{0.03}Al_{0.7}$ [x= 0-2.5] magnets

oxygen content after each process wt.%		
	high oxygen magnet	low oxygen magnet
cast	0.01-0.03	
jet mill	0.25-0.35	0.08-0.15
compression	0.45-0.55	0.08-0.15
sintering	0.50-0.65	0.08-0.15

Remanence and energy product increase with decreasing Nd-content, whereas the coercive field shows its highest value at a high Nd-content. In the case when the oxygen content of the magnets was determined to be in the order of 4000 to 6000 ppm, a large part of the Nd was bound in the stable phase Nd_2O_3 phase. This is the why below 14 at.% Nd the density and the hard magnetic properties of the magnets drastically deteriorated in magnets with a high oxygen content. On the contrary, low oxygen content can cause the AGG and lower the magnetic properties.

Fig.2: SEM micrograph showing the jet-milled powder used for manufacturing of the low oxygen $Nd_{13.5}Fe_{bal}B_{5.95}Cu_{0.03}Al_{0.7}$ magnet with $(B.H)_{max}$= 50.2 MGOe

The best degree of alignment was estimated to be in the order of 14°. A measured density and a theoretical spontaneous magnetisation give an estimated remanence, which also depends on the degree of alignment. The measured densities and remanences and we can calculate the spontaneous magnetisation from the composition. In the case of $Nd_{13.5}Fe_{bal}B_{5.95}$, the volume fraction of the hard magnetic phase is estimated at 94.4vol.%, which corresponds to 1.51 T of the spontaneous magnetisation at the full density. The remaining 5.6 vol.% consist of the $Nd_{1+\varepsilon}Fe_4B_4$ and other boron-rich phases, the Nd-rich intergranular phases, Nd_2O_3, pores and other additional phases. However, a certain amount of Nd-rich phase is necessary to get the appropriate $_JH_c$.

Figure 2 shows scanning electron micrographs of fine powders of about 2.2 - 3.3 μm in mean diameter. The scanning electron micrograph (Fig.3) of a fractured surface of the sintered magnet of a low oxygen content shows besides hard magnetic grains with diameters up to several microns also large abnormally grow grains. The TEM investigation of high remanence and low Nd-content sintered magnets obtains Nd-rich phases only occasionally at grain boundary junctions (Fig.4a and b). There is again no Nd-rich phase between the 2:14:1- grains. The lattice plane image corresponds to the (001) basal planes. The three grains are nearly parallel to each other and the diffraction pattern of the grains only changed slightly in the distribution of the intensity among the diffracting spots (Fig.4a). Correspondingly a sluggish contrast of the same (001) planes is seen in the upper right grain and a still fainter basal plane lattice contrast is seen in the lower grain. The misalignment between the three grains is calculated to be less than 5°. The microstructure of the magnet did not changed significantly when observed with the c-axis perpendicular to the sample plane. The fibre-texture along [001] is observed and misalignments of the <100> directions in neighbouring grains was measured to be in the order of 20-25°. From the microstructural study it is clear that a well alignment of the grains is achieved. The TEM observations of the sample with the c-axis contained in the sample plane showed adjacent grains with nearly parallel c-axis, while the observations made in sample with the c-axis perpendicular to the sample surface showed that the formed crystallographic preferred orientation is a fibre texture (Fig.4b). Therefore our TEM observation agrees with the x-ray global texture measurements.

TEM investigation of magnets with a high oxygen content (> 4000 ppm) exhibit extensive occurrence of precipitation inside the hard magnetic grains. Figure 5 is a TEM micrograph of such precipitation with a size around tenth of nanometers.

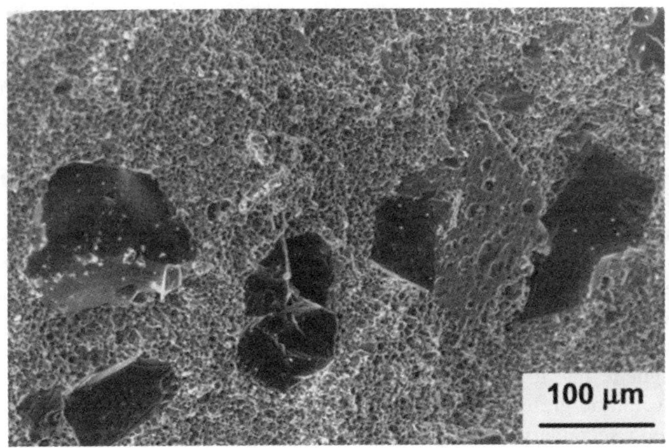

Fig.3:
Scanning electron micrograph of a fractured surface showing the abnormally grown grains in the isotropic $Nd_{13.7}Fe_{bal}B_{5.95}Cu_{0.03}Al_{0.7}$ magnet sintered at 1000°C / 3 hours.

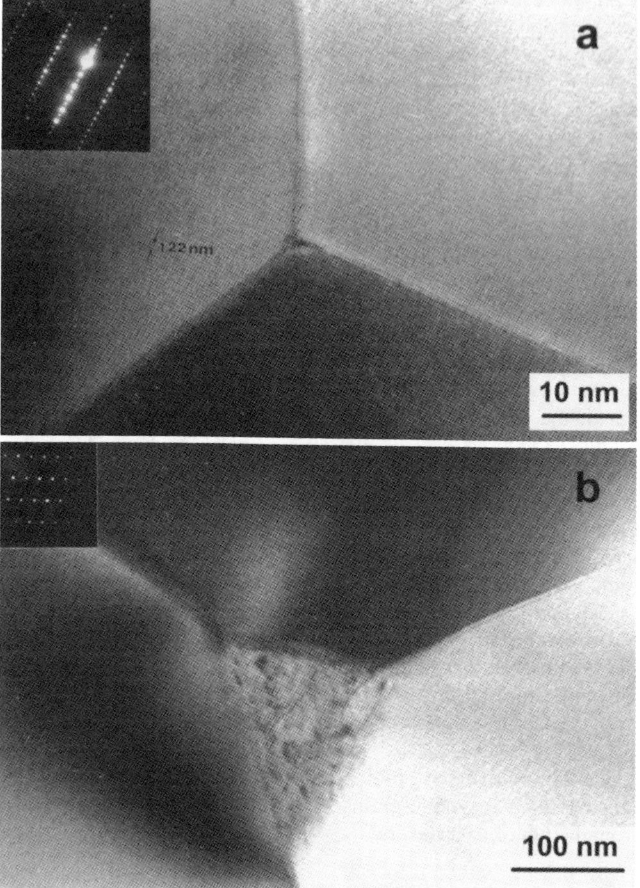

Fig.4:
The TEM investigation of grain boundary junctions of a $Nd_{13.7}Fe_{bal}B_{5.95}Cu_{0.03}Al_{0.7}$ sintered magnet with $(B.H)_{max}$=50.2 MGOe. C-axis of grains is parallel (a) and is perpendicular (b) to the image plane.

In order to improve coercive field, remanence and energy density, it is necessary to increase the volume fraction of the hard magnetic phase by decreasing the amount of oxygen, Nd and pores, and to improve the degree of alignment of the $Nd_2Fe_{14}B$ grains. Endoh and Shindo [16] reported about an energy density product of 410 kJ/m3 of a low oxygen (< 2000 ppm) $Nd_{12.7}Fe_{81.3}B_6$ sintered magnet.

In magnets where a high amount of dopant elements was added, it was found that the dopant elements Al, Ga and Cu were partly dissolved in the hard magnetic grains and were partly detected in the Nd-rich intergranular phase. A more detailed TEM analysis revealed various additional binary and ternary Ga- or Cu- or Al-containing Nd-phases in the intergranular region between hard magnetic grains [17,18]. The coercive field is determined by the long range dipolar interaction and short range exchange coupling between neighbouring hard magnetic grains. The doping of elements changes the phase relation and favours the formation of new phases. Additional secondary non-magnetic intergranular phases decrease the remanence and interrupt the magnetic interactions between the grains, thereby improving the coercivity of large grained sintered magnets. Soft magnetic secondary phases, such as α-Fe, destroy coercivity in large grained magnets. Numerical, micromagnetic calculations have shown that the critical diameter for the curling process of the polarisation vector inside the α-Fe phase, embedded between hard magnetic grains, is about 90 nm [19]. This is the reason why the formation of large soft magnetic grains considerably reduces the coercive field. For magnets with higher Nd-concentrations microstructural parameters, such as grain size, misorientation and distribution of grains and intergranular phases control the coercive field. The dipolar interactions considerably reduce the coercive field of ideally oriented particles by about 20% with respect to $_JH_c$ of an isolated particle. Exchange coupling between misaligned grains drastically reduces the coercive field to about 30% of the ideal nucleation field. The higher the Nd-content of the magnet and therefore the volume fraction of the Nd-rich intergranular phase is, the more reduced is the contribution of the exchange and also dipolar coupling between the grains.

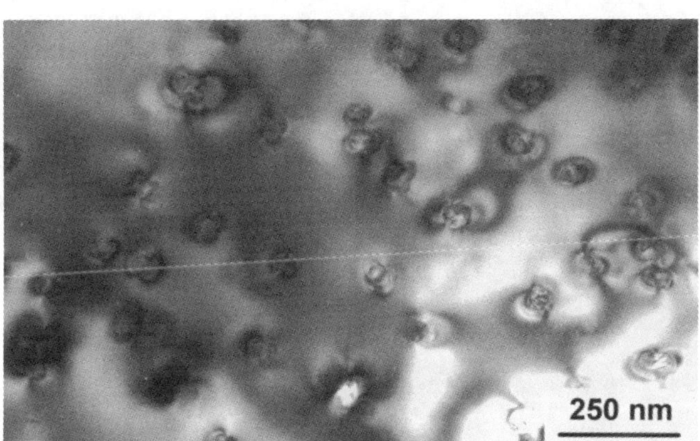

Fig.5:
The TEM investigation of Nd_2O_3 precipitation within the hard magnetic grains of a high oxygen (5500 ppm) containing magnet.

Microstructure of aligned melt-spun Magnets

In comparison with those of aligned sintered magnets, the microstructure of hot pressed and die-upset melt-spun magnets consists of platelet like grains with diameters less than 1 micron. These magnets are fully dense, and energy density product as large as 45 MGOe have been attained [20]. The degree of alignment, size and shape of the grains and the intergranular regions within the ribbons control the macroscopic magnetic properties. Die-upsetting modifies the spheroidal grains [21] after hot-pressing to platelets as shown in the TEM micrographs of Figs.6. Two magnets of $Nd_{14}Fe_{72}Co_7B_6Ga_1$, but with different coercive fields, were investigated. The magnet with the higher remanence (J_r=1.32 T) shows platelet shaped grains perpendicular to the pressing direction. Misaligned grains, which are clearly visible in Fig.6a, deteriorate the remanence. The c-axis for each grain runs perpendicular to the straight elongated edge. Nd-rich phase is found among the platelet shaped grains as a fine layer between the straight edges or as pockets at the end of the platelets or between the misaligned and aligned grains. On the other hand the magnet with the lower remanence (J_r=1.32 T) shows a microstructure with more equiaxed grains, i.e. the ratio between length and with of the grains is less than in the previous sample. The micrograph of Fig. 6b shows deformed grains, but not fully platelet shaped. In both magnets regions with abnormally grown, large grains were found. Some of these grains were fully developed, platelet shaped grains.

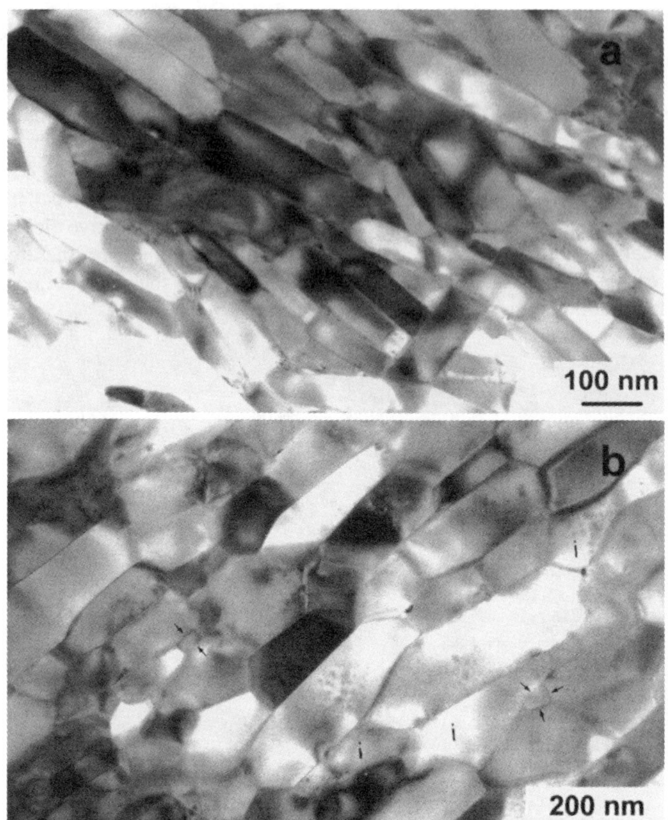

Fig.6: TEM micrograph of a hot pressed and die-upset $Nd_{14}Fe_{72}Co_7B_6Ga_1$ magnet with (a) J_r=1.32 T, $_JH_c$=1241 kA/m and (b) J_r=1.25 T, $_JH_c$=1265 kA/m. (perpendicular to c-axis view).

DISCUSSION

(Nd,S1)-(Fe,S2)-B:(M1,M2) magnets

High performance $Nd_2Fe_{14}B$-based permanent magnets are produced with different composition and various processing techniques. The composition and the processing route influence the complex, multiphase microstructure of the magnets, such as grain size, alignment and distribution of phases:

- grain sizes in the range between 10 nm and 500 nm are obtained by melt-spinning, mechanical alloying and the HDDR (hydrogenation-disproportionation-desorption-recombination) process
- sintered and hot worked magnets exhibit grain sizes above 1 μm

The coercive field is determined by the high uniaxial magnetocrystalline anisotropy as well as the magnetostatic and exchange interactions between neighbouring hard magnetic grains. The dipolar interactions between misaligned grains are more pronounced in large-grained magnets, whereas exchange coupling reduces the coercive field in small-grained magnets. Substituent and dopant elements influence the microstructure, coercivity and corrosion resistance of advanced (Nd,S1)-(Fe,S2)-B:(M1,M2) magnets. The replacement of the Nd-rich intergranular phase by secondary phases formed after doping by M1 and M2 type elements improves the corrosion resistance, especially in large-grained magnets.

The multicomponent composition of the magnets leads to the formation of non-magnetic and soft magnetic phases. Generally, two types of substituent elements, which replace the rare earth element or the transition element sites in the hard magnetic phase, and two types of dopant elements are distinguished [5]. Substituent elements mainly change the intrinsic properties, such as spontaneous magnetic polarisation, Curie temperature and magnetocrystalline anisotropy. Depending on the type, the dopant elements, which show a low solubility within the hard magnetic phase, form additional intergranular rare-earth-containing or boride phases. These phases change the coupling behaviour between the hard magnetic grains. Non-magnetic intergranular phases eliminate the direct exchange interaction and also reduce the long-range magnetostatic coupling between the hard magnetic grains; both effects lead to an increase of the coercive field. On the other hand, the decrease of the volume fraction of the hard magnetic phase within the magnet decreases the remanence. Insufficient temperature stability and poor corrosion resistance are the main factors limiting applications of $Nd_2Fe_{14}B$-based magnets. Secondary non magnetic phases, which replace the Nd-rich intergranular phase, considerably improve the corrosion resistance and are of great technological interest.

Nd-Fe-B based permanent magnets with a composition close to $Nd_{15}Fe_{77}B_8$ exhibit a complex multiphase microstructure. According to the ternary phase diagram at least three equilibrium phases occur:

- the hard magnetic $Nd_2Fe_{14}B$ phase
- the boride phase $Nd_{1+\varepsilon}Fe_4B_4$
- the low melting Nd-rich phase.

Other phases, such as Fe-rich and Nd-oxides, and pores are found depending on the composition and processing parameters. Selected substituent elements replace the Nd-atoms (S1=Dy,Tb) and the Fe-atoms (S2=Co,Ni,Cr) , respectively, in the hard magnetic ϕ-phase and considerably change intrinsic properties, such as the spontaneous polarisation, the Curie temperature and the magnetocrystalline anisotropy. The formation of intermetallic, soft magnetic Nd-(Fe,S2) phases, such as the Laves type $Nd(Fe,S2)_2$-phase, deteriorate the coercivity of the magnets. If dopant elements M1 or M2 are added to Nd-Fe-B, in some cases the coercivity is increased and the corrosion resistance is improved. This is the case, if the Nd-rich intergranular phase is replaced by other phases, such as $AlNd_6Fe_{13}$ and Nd_3Co. Our previous, systematic TEM-studies performed on sintered, melt-spun, mechanically alloyed and hot worked magnets have shown that two different types of dopants can be distinguished independently of the processing route. Both types influence the microstructure in a different way [11,12]:

- *Type 1 dopants* (M1 = Al, Cu, Zn, Ga, Ge, Sn)
  ⇨ form binary M1-Nd or ternary M1-Fe-Nd phases
- *Type 2 dopants* (M2 = Ti, Zr; V, Mo; Nb, W)
  ⇨ form binary M2-B or ternary M2-Fe-B phases

The processing route of the magnet strongly influences the grain size and grain size distribution. The coercive field in sintered magnets strongly depends on the sintering parameters, such as temperature and time. Nanocrystalline and submicron magnets are obtained by the melt-spinning route, or by mechanically alloying, or by the HDDR process [22-24]. Hot pressing and die upsetting of Nd-Fe-B ribbon materials reveals a densely packed, anisotropic magnetic material. Platelet-shaped grains are observed by TEM-investigations. The degree of orientation of the platelets, which are stacked transverse to the press direction with the easy c-axis perpendicular to the face of each grain, determines the remanence and coercive field of the magnet.

Highest energy density magnets

Nd-Fe-B sintered magnets possessing outstanding magnetic properties have developed into a major permanent magnet material in the 15 years since their invention. The drastic increase of the energy density product of newly developed $Nd_2Fe_{14}B$ based magnets enabled the invention of many new applications of permanent magnets. Applications of Nd-Fe-B magnets are expanding; voice coil motors (VCM) for hard disk drives, magnetic resonance imaging (MRI), small sized motors, electrical devices and so on. These magnets are produced by a conventional powder metallurgical process which is essentially based on alloy-making, coarse-milling, pulverising, pressing in a magnetic field, sintering, heat-treatment, grinding and surface coating.

In this process, it is very important to keep the processing atmosphere either in a vacuum or in an inert gas because rare earth elements, such as Nd, Pr, and Dy, which are essential for fabrication of Nd-Fe-B magnets are easily oxidizable.

The theoretical maximum value of the energy density product is in the case of a perfectly squared demagnetisation curve given by:

$$(B \cdot H)_{\max}^{theor.} = \frac{1}{4 \cdot \mu_0} \cdot J_r^2 \qquad \text{if} \qquad \left|{}_J H_c\right| \geq \frac{1}{2 \cdot \mu_0} \cdot J_r \qquad \text{(Eq. 1)}$$

The residual flux density $B_r = J_r$ in this case is expressed as the following equation:

$$J_r = J_s \cdot \frac{\rho}{\rho_0} \cdot V_{hm} \cdot F_{hm} = J_s \cdot \frac{\rho}{\rho_0} \cdot V_{hm} \cdot \cos\varphi \qquad \text{(Eq. 2)}$$

where, J_s is the saturation magnetisation of the hard magnetic phase (1.61 T), V_{hm} and F_{hm} are the volume fraction and the degree of alignment of the hard magnetic grains, respectively. In order to enhance J_r and therefore the energy density product, it is necessary to avoid pores and to densify the magnets up to the theoretical value ρ_0, increase the volume fraction V_{hm} and achieve a high degree of alignment F_{hm}.

The theoretical value of the maximum energy product of $Nd_2Fe_{14}B$ based magnets is calculated to be 512 kJ/m³ (64 MG0e) assuming 100% perfect alignment and 100% volume fraction of the hard phase. The origin of this magnetic property lies in the $Nd_2Fe_{14}B$ ternary tetragonal compound as a main phase. In addition, according to the ternary Nd-Fe-B phase diagram this magnet contains also a certain amount of $Nd_{1.1}Fe_4B_4$-phase and a Nd-rich phase. which is essential for sintering with liquid phase. In order to densify the magnets up to the theoretical density, it is very important to control the composition of magnets thus generating sufficient amount of liquid phase at sintering. Furthermore, controlling the volume fraction of the constituent phases is indispensable for enhancing residual flux density (B_r) and to keep intrinsic coercivity (${}_JH_c$) stable.

Several authors have reported to obtain Nd-Fe-B based magnets with 400 kJ/m^3 energy density product by

- keeping the oxygen content low [25]
- using the powder mixing technique [26]
- increasing the magnetising field and reducing the pressure during compaction [16]

The increasing demand for highest energy density magnets (> 400 kJ/m³), especially for voice coil motors in hard disc drives and for magnetic circuits for magnetic resonance imaging devices, needs an efficient manufacturing process for $Nd_2Fe_{14}B$ sintered magnets with improved energy density product. A new technology - Rubber isostatic pressing (RIP) - has been developed by Sagawa et al. [27,28] to improve the orientation of the particles in the green compact to obtain sintered magnets with perfect orientation. Rubber isostatic pressing (RIP) is one of the key technologies to approach for the theoretical limit, 64 MG0e at room temperature, of the magnets based on $Nd_2Fe_{14}B$. In RIP, magnet powder is subjected to such a strong pulsed field just before the compaction that the powder in the rubber mold is thoroughly oriented. Then the powder is compacted isostatically, while the orientation is completely held.

In the conventional die pressing which uses no rubber molds, the pressure applied to the powder is uniaxial. The uniaxial pressure tends to disturb the orientation of the particles during the pressing. To prevent this orientation disturbance, the powder has to be subjected to a strong magnetic field throughout the pressing. This is one of the reasons why a pulsed field cannot be adopted for the conventional die pressing. The high orientation of the magnet produced by RIP is attributed to:

- Application of a strong pulsed field which dissolves the agglomeration of the magnet powder particles, and then, impulsively orients the particles
- Isostatic pressing which holds the orientation high during the pressing

The influence of oxygen an the hard magnetic properties is more complex. Kim et al [29] reported that a controlled doping with oxygen improved grain alignment and resulted in an increase in remanence, coercivity and loop squareness. One possibility to improve the alignment factor is to optimise the alignment field and/or pressure during transverse pressing. Even the sintering process influences the degree of alignment of the grains [30]. On the other hand several authors [31,28] found that the highest alignment is obtained by isostatic die pressing (φ=11-14°), followed by the transverse field die pressing (18-20°) and the axial field die pressing (25-27°).

Acknowledgments

This work has been supported by the Austrian Science Foundation FWF through Grant No. P13433. The authors are grateful to M. Sagawa and Y. Hirose for their enthusiastic support and stimulating discussions.

References

[1] K. J Strnat, Hoffer, G., Oson , J. and Ostertag W., J. Appl. Phys. **38**, 1001 (1967).
[2] M. Sagawa, Fujimura, S., Togawa, N., Yamamoto, H. and Matsuura Y., J. Appl. Phys. **55**, 2083 (1984).
[3] J.J. Croat, Herbst, J.F., Lee, R.W. and Pinkerton F.E., J. Appl. Phys. **55**, 2078 (1984).
[4] J.F. Herbst and Croat, J.J., J.Magn.Magn.Mat., **100**, 57 (1991).
[5] J. Fidler and Schrefl, T., J. Appl. Phys., **79**, 5029 (1996).

[6] J Fidler, "Rare earth intermetallic magnets", Inst.Phys.Conf.Ser.No.152, Section G: Magnetic Materials, Paper presented at the 11th Int. Conf. on Ternary & Multinary Compounds, ICTMC-11, Salford, UK, September 1997, (1998), 805-813.
[7] J.F. Herbst, R2Fe14B materials: instrinsic properties and technological aspects, Rev. Modern Phys., **63,** 819-898 (1991).
[8] K.H.J Buschow,. (1988) Permanent magnet materials based on 3d-rich ternary compounds, in Ferromagnetic materials Vol.4. edited by E.P. Wohlfarth and K.H.J. Buschow, Elsevier Science Publishers, Amsterdam, 1-129.
[9] J.S. Cook and P.L. Rossiter, Rare earth iron boron supermagnets, CRC Critical Reviews in Solid State and Materials Sciences **15**, 509-550 (1989).
[10] K.H.J Buschow,. (1997) Magnetism and processing of permanent magnet materials, in Handbook of Magnetic Materials Vol.10. edited by K.H.J. Buschow, Elsevier Science Publishers, Amsterdam, 463-593.
[11] J. Bernardi, Fidler, J. and Födermayr, F., IEEE Trans. Magn. **28**, 2127 (1992).
[12] J Bernardi and Fidler, J., J. Appl. Phys. **76**, 6241 (1994).
[13] Y. Hirose, H. Hasegawa, S. Sasaki and M. Sagawa,
Proc. XV. Workshop on RareEarth Magnets and Applications, Dresden, 1998, Vol1., p.77-96.
[14] S. Sasaki, PhD Thesis, Vienna University of Technology, Austria, 1999.
[15] W. Rodewald, B. Wall and W. Fernengel, IEEE Trans. Magn. **33,** 3841 (1997).
[16] M. Endoh and M. Shindo, Proc. 13th Int. Workshop on rare earth magnets and their applications, Birmingham, ed. by Manwaring et al, Vol.1, (1994) 397.
[17] J. Bernardi, J. Fidler, M. Seeger and H. Kronmüller, IEEE Trans. Magn. **29,** 2773 (1993).
[18] J. Bernardi and J. Fidler, J. Appl. Phys. **76,** 6241 (1994).
[19] T. Schrefl and J. Fidler, J.Magn.Magn.Mat. **177-181,** 970-975 (1998).
[20] J.J. Croat, J. Less-Common Met., **148**, 7 (1989).
[21] R.K. Mishra, J. Appl. Phys., **62**, 967 (1987).
[22] I. R Harris, 1992, Proceedings of the XII. International Workshop on Rare Earth Magnets and their Appl., ed. by B.Street, The University of Western Australia, 347.
[23] R. Nakayama, Takeshita, T., Itakura, M., Kuwano, N. and Oki, K., JAP. **76**, 412 (1994).
[24] M. Uehara, Tomizawa, H., Hirosawa, S., Tomida, T. and Maehara, Y., IEEE Trans. Magn. **29**, 2770 (1993).
[25] M. Sagawa, S. Hirosawa, H. Yamamoto, S. Fujimura and Y. Matsuura, Jpn. J. Appl. Phys., **26** 785 (1987).
[26] E. Otsuki, T. Otsuka and T. Imai, Proc. 11th Int. Workshop on rare earth magnets and their applications, Pittsburgh, ed. by Shankar, Vol.1, (1990) 328.
[27] M. Sagawa and H. Nagata, IEEE Trans Magn. **29,** 2747 (1993).
[28] M. Sagawa, H. Nagata O. Itatani and W. Watanabe, Proc. 13th Int. Workshop on rare earth magnets and their applications, Birmingham, ed. by Manwaring et al, Suppl., (1994) 13.
[29] A.S. Kim, F.E. Camp and H.H. Stadelmaier, J. Appl. Phys. **76,** 6265 (1994).
[30] T.S. Chin, M.P. Hung, D.S. Tsai, K.F. Wu and W.C. Chang, J. Appl. Phys. **64,** 5531 (1988).
[31] W. Fernengel, A. Lehnert, M. Katter, W. Rodewald and B. Wall, J.Magn.Magn.Mat., **157/158,** 19 (1996).

HYSTERESIS LOOPS AND COERCIVITY MECHANISMS IN SINTERED AND NANOCRYSTALLINE PERMANENT MAGNETS

H. KRONMÜLLER, D. GOLL, I. KLEINSCHROTH, and A. ZERN
Max-Planck-Institut für Metallforschung, Heisenbergstr. 1, D-70569 Stuttgart, Germany

ABSTRACT

The hysteresis loops of nanocrystalline (nc) permanent magnets (pms) produced by the melt-spin technique have been investigated for compositions based on the intermetallic compounds $R_2Fe_{14}B$ (R = Nd, Pr) and the carbides $Sm_2Fe_{17-x}Ga_xC_y$. The following three types of pms have been studied: 1) High-coercivity pms with exchange decoupled grains. 2) High-remanence exchange-spring pms. 3) High-coercive–high-remanence composite pms with exchange coupled soft and hard magnetic grains. The temperature dependence of the coercive field $\mu_0 H_C$ for all three types of pms obeys a relation for a modified nucleation field, $H_C = (2K_1/J_S)\alpha - N_{eff} M_S$ (K_1 = first anisotropy constant, M_S = spontaneous magnetization). For an analysis of the characteristic differences between the microstructural parameters α and N_{eff} as obtained for the three types of pms, computational micromagnetism on the basis of the Finite Element Technique is applied. This powerful method allows a quantitative analysis of the role of grain size, grain boundaries (gbs), texture of easy directions and of soft magnetic phases in composite materials. In order to obtain satisfactory results, a self-adapting algorithm has been developed where the mesh size is adapted to the gradients of the direction cosines of the spontaneous magnetization. It turns out that excellent magnetic properties of composite pms can only be obtained if the gbs are as ideal as possible. Remanence and coercive field are found to decrease linearly with a corresponding reduction of both, the crystal anisotropy and the exchange constant within the gbs. In composite pms the diameters of the soft magnetic grains should be smaller than twice the domain wall width, δ_B^{hard}, of the hard magnetic phase in order to obtain a remarkable remanence enhancement. From these model calculations general rules for the development of optimized nc pms with large remanences and large coercivities are derived.

INTRODUCTION

High-performance magnetic materials are based on outstanding intrinsic magnetic properties, large anisotropy constants $K_1 > 10^6$ J/m³, large spontaneous polarization $J_S > 1.2$ T and large Curie temperatures $T_C > 300°$ C. In addition optimized microstructures as well as suitable alloy compositions are further pre-requisites for high-quality magnets. The interaction between microstructures and properties of the hysteresis loop in general corresponds to a complex micromagnetic problem which has been treated so far for special models of magnetic inhomogenieties. A fundamental property of permanent magnets (pms)is the so-called maximum energy product defined as

$$(BH)_{\max} = \frac{1}{4}\mu_0 M_R^2 \quad (1)$$

Mat. Res. Soc. Symp. Proc. Vol. 577 © 1999 Materials Research Society

1. High-Coercivity, Two-Phase Magnets

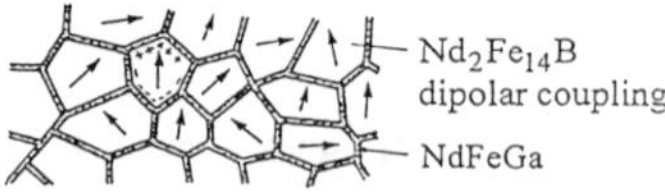

2. High-Remanence, Single-Phase Magnets

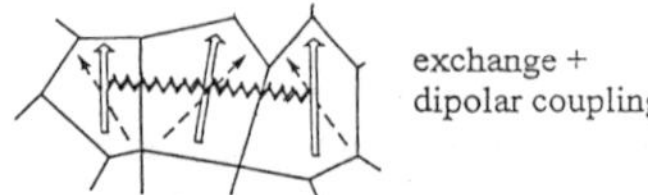

3. High-Remanence, Composite, Two-Phase Magnets

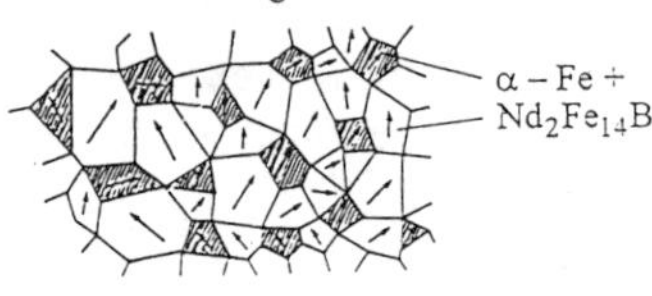

Fig. 1. Schematic microstructures of nc melt-spun pms. 1. High-coercivity two phase pm with exchange-decoupled grains. 2. High-remanence exchange-coupled single phase pm. 3. Exchange-coupled composite high-remanence two phase pm.

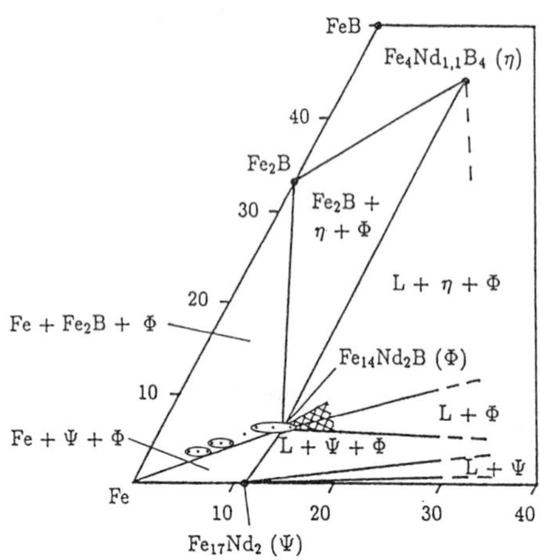

Fig. 2. Phase diagram of NdFeB with indications of the compositions used for the three types of nc pms shown in Fig. 1.

This quantity defines the magnetic energy stored by a pm outside of its volume. In eq. (1) M_R denotes the remanence and μ_0 corresponds to the vacuum permeability. It should be noted, however, that this value of the energy product only can be realized if the coercive field H_C, obeys the condition $\mu_0 H_C > (1/2) J_R = (1/2) \mu_0 M_R$ (J_R = remanent polarization, M_R = spontaneous magnetization), and if the hysteresis loop is rectangular and fully reversible up to an inverse field of $H_{ext} = -H_C$. Since this latter condition usually is not fulfilled the coercive field in general should be somewhat larger than 0.5 J_R.

In sintered magnets composed of magnetically decoupled ideal well-aligned particles of ellipsoidal shape and uniaxial anisotropy the coercive field would be given by [1,2]

$$H_C = \frac{2K_1}{J_S} - (N_{\parallel} - N_{\perp})M_S \quad , \tag{2}$$

where K_1 denotes the anisotropy constant and $N_{\parallel}$ and $N_{\perp}$ denote the demagnetization factors for magnetization parallel or perpendicular to the easy axis. As is well known so far the prediction of eq. (1) could not be realized neither for sintered nor for nc materials. Usually the coercive fields are a factor 4 – 5 smaller than predicted by eq. (2). This reduction has to be attributed to deteriorating effects of reduced anisotropies at grain surfaces or at gbs as well as to strong demagnetization fields at edges and corners of polyhedral grains. In many cases the coercive field of real pms is described by the following modification of eq. (2) [3,4,5]

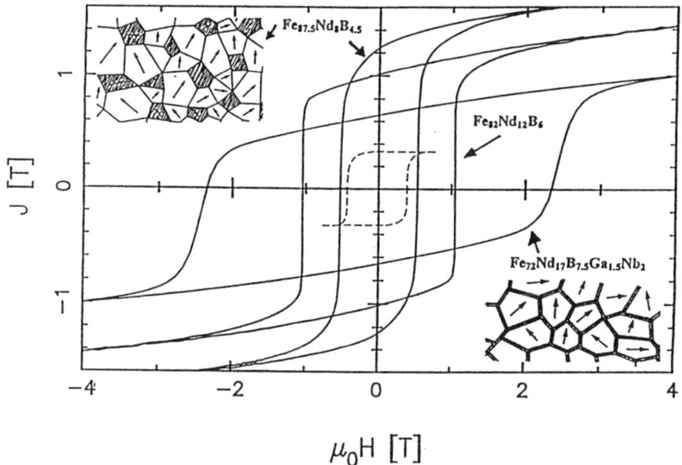

Fig. 3. Hysteresis loops of nc pms with the microstructures of Fig. 1. The broken curve corresponds to the conventional (Ba,Sr) based ferrite. The intermediate curve results from the stoichiometric $Nd_2Fe_{14}B$.

$$H_C = \frac{2K_1}{J_s}\alpha - N_{eff} M_s \tag{3}$$

Here α and N_{eff} denote microstructural parameters where α describes the reduction of the crystal field by defects and atomic disorder at the surface of grains as well as by misaligned grains. N_{eff} corresponds to an effective demagnetization factor at distinguished sites of the grains. The validity of eq. (3) has been tested for many types of pms and just recently also for nc isotropic pms. Three types of nanocrystalline pms with different microstructures can be prepared by means of the melt-spin method (see Fig. 1 – Fig. 3).

1) High-coercivity materials where hard magnetic nc grains are more or less magnetically decoupled by a paramagnetic intergranular film. In the case of $R_2Fe_{14}B$ based pms this is realized by an overstoichiometry of R and suitable additives as Ga and Nb [6,7,8]. There exist still some long-range dipolar interactions between the nc grains, however, N_{eff} of eq. (3) turns out to be rather small.
2) High-remanence–high coercivity pms are obtained if the nc grains are magnetically coupled by exchange interactions [9,10]. This type of pms is realized by highly stoichiometric alloys.
3) Composite high-remanence pms are obtained using alloys with overstoichiometric α-Fe. In this case the remanence is enhanced by exchange coupling between the grains but also by the large magnetization of the α-Fe grains [11,12].

The energy product $(BH)_{max}$ for these pms can only be used without destruction of the remanence if the demagnetization curve in the second quadrant behaves reversible up to an inverse field of $(1/2)\,J_R$. It will be the aim of this paper to discuss the effect of grain size, grain boundary properties, misaligned grains and soft magnetic grains on the characteristic properties of the hysteresis loop. Furthermore, the temperature dependence of H_C will be reviewed for a number of nc pms.

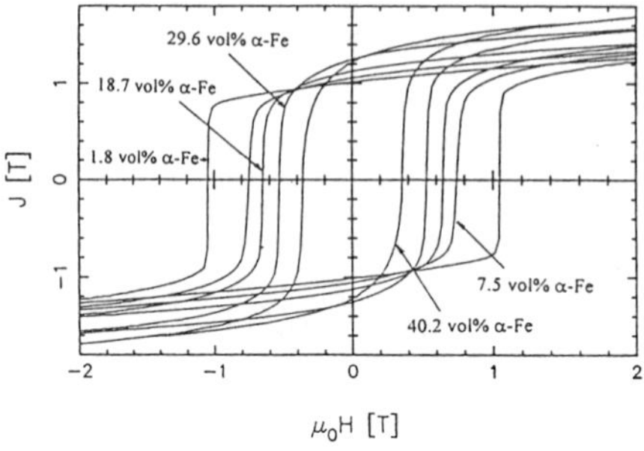

Fig. 4. Room-temperature hysteresis loops of melt-spun composite $Nd_2Fe_{14}B$ + α-Fe pms.

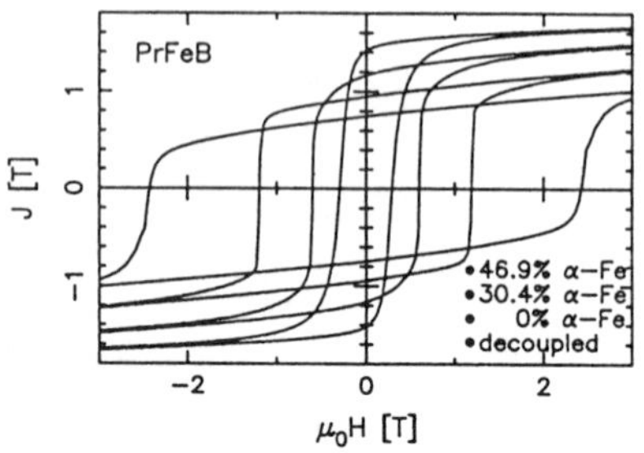

Fig. 5. Room-temperature hysteresis loops of melt-spun composite $Pr_2Fe_{14}B$ + α-Fe pms.

EXPERIMENTAL

The preparation of the three types of nc pms according to the concept outlined in the introduction is based on the phase diagrams of the ternary RFeB systems. Fig. 2 presents an isothermal section of the NdFeB system for 1000°C [13] in the iron-rich corner. On the right and left regions neighboring to the stoichiometric ϕ-phase, $Nd_2Fe_{14}B$, we obtain the decoupled high coercivity pms in the region L + ϕ and the composite iron-rich pms in the region Φ + Fe + Fe_2B. In order to avoid Fe_2B grains the compositions are chosen near to the phase boundary Fe → Φ. The well-known Coehoorn alloy [14] with a large boron content is obtained in the upper part of the Φ + Fe + Fe_2B region, but will not be considered here. Concerning the PrFeB phase diagram the situation is similar to that of NdFeB [13].

Hysteresis loops of the three types of magnets are shown in Fig. 3. Here the nearly stoichiometric composition NdFeB shows average but still large values of H_C and J_R, whereas the iron-rich composite pm reveals the expected large J_R value and the Nd-rich alloy shows the large coercive field. In Fig. 3 the broken lines correspond to the hysteresis loop of the conventional hard ferrite material, thus showing the considerable progress in remanent and coercive field in the case of rare earth-transition metal intermetallics.

Fig. 4 shows a series of hysteresis loops of melt-spun NdFeB alloys [15] containing over-stoichiometric iron. Similarly Fig. 5 presents corresponding hysteresis loops of PrFeB alloys [16]. In both cases similar features can be observed. Largest coercive fields and lowest remanences are observed for the alloys near to the stoichiometric composition. With increasing iron content J_R increases whereas H_C decreases.

Fig. 6 and Fig. 7 summarize the dependence of J_R, H_C and $(BH)_{max}$ as a function of the iron content. The remanences of the nearly stoichiometric alloys are 1.02 T for NdFeB and 0.95 T for PrFeB. Both values are considerably larger than the isotropic value of $J_R \sim 0.8$ T. Accordingly the remanence enhancement in stoichiometric alloys is of the order of 20% of the theoretical value. A further increase of J_R is obtained for the composite alloys with iron excess. Parallel to the increase of J_R the coercive field decreases monotonically. Due to the increase of

J_R an optimal maximum energy product in both cases is obtained for an excess iron content of 30%. Values for $(BH)_{max}$ of 185 kJ/m³ for NdFeB and 181 kJ/m³ for PrFeB are obtained. The H_C values for this composition are larger than $(1/2)J_R$ and of the order of 0.6 T. Here it should be noted that similar results have been obtained by Harland et al. [17].

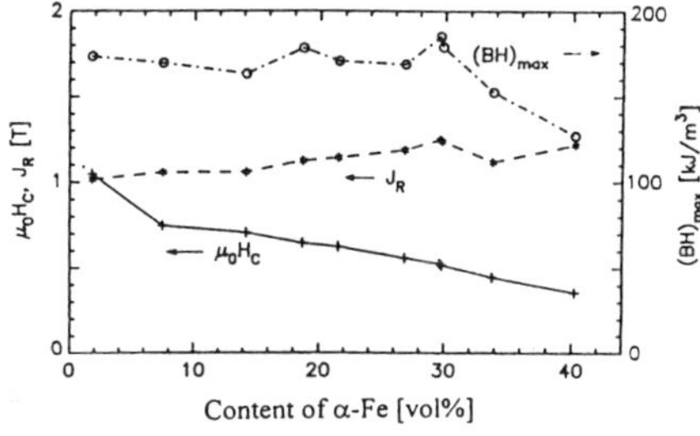

Fig. 6 $\mu_0 H_C$, J_R and $(BH)_{max}$ as a function of the α-Fe content for NdFeB composite pms.

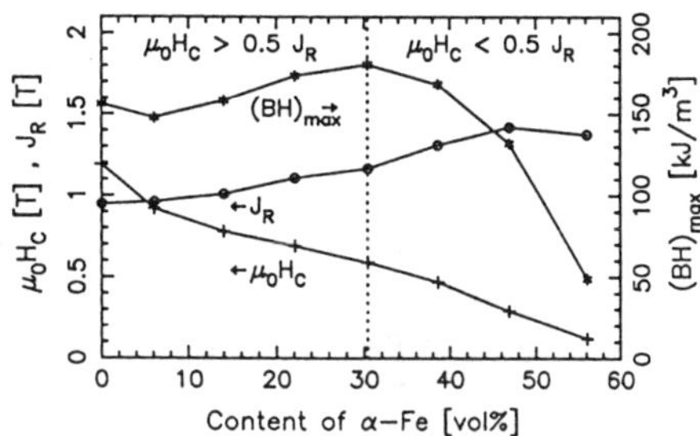

Fig. 7. $\mu_0 H_C$, J_R and $(BH)_{max}$ as a function of the α-Fe content for PrFeB composite pms.

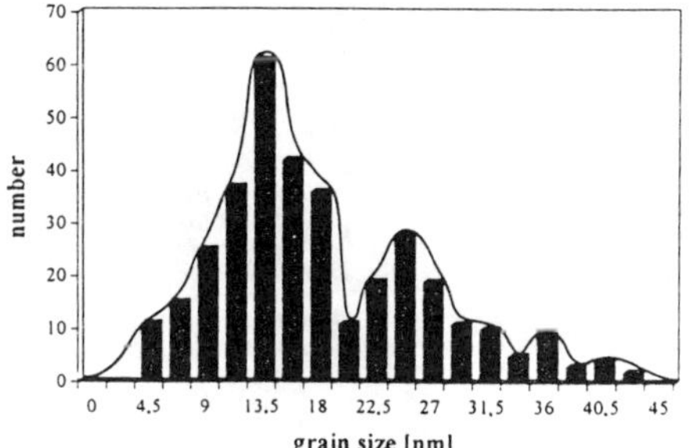

Fig. 8. Grain size distribution of melt-spun $Nd_{10}Fe_{84}B_6$ (14.2 vol% α-Fe).

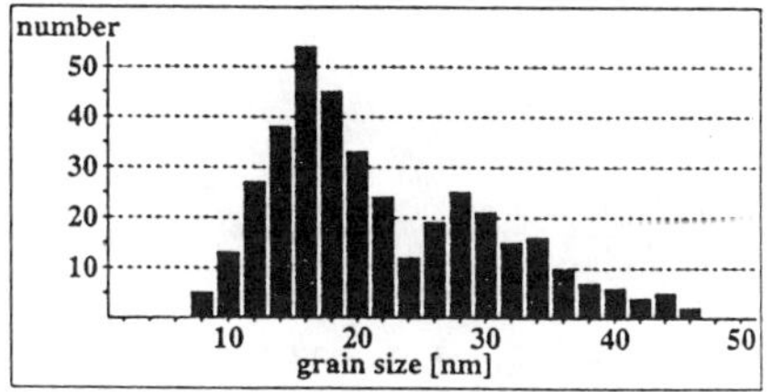

Fig. 9. Grain size distribution of melt-spun $Pr_{10}Fe_{84}B_6$ (14.0 vol% α-Fe).

The grain size distributions of the rapidly quenched alloys have been investigated by transmission electron microscopy (TEM). Fig. 8 and Fig. 9 show the grain size distributions for NdFeB and PrFeB for alloys with an excess iron content of ~ 14%. In both cases the distribution reveals two characteristic maxima. By determining the lattice spacing using HRTEM [15] it turned out that the smaller grains correspond to α-Fe and the larger ones to the hard phases $R_2Fe_{14}B$. The average grain sizes are 15 nm for α-Fe and 25 nm for the hard magnetic grains. It should be noted that these nc structures have been prepared by direct quenching into the nc state. This preparation technique leads to ideally exchange coupled grain structures without intergranular phases as shown in Fig. 10 and Fig. 11 where the gbs have a width of the order of the lattice parameter [8]. In the case of alloys with overstoichiometric Nd or Pr an intergranular phase exists. Fig. 12 shows the intergranular phase as a bright region between the grains. Chemical EDX microanalysis clearly shows the enrichment of the rare earth metal within the intergranular phase leading to a reduction of the strength of the exchange coupling or even to a complete decoupling between the grains. Fig. 13 presents a HRTEM micrograph of the intergranular phase of a Pr rich nc PrFeB pm. The intergranular phase reveals a disordered structure where three quasi-planes show a corrugated amorphous structure. These results clearly prove the concept of three characteristic types of pms as pointed out in the introduction.

Analysis of Hardening Mechanism

As is well known the experimental values of H_C are a factor 3 – 5 below the theoretical predictions for the nucleation fields, H_N, of small particles [3,4]. This discrepancy – known as Brown's paradoxon [1] – has been the topic of many articles. Two main sources of the low H_C-values in general are discussed:

1. The reduction of the nucleation fields by microstructures which may be the origin for reduced anisotropy constants near grain surfaces and gbs [18].
2. The reduction of the nucleation field by small reversed domains being produced by thermal activation. The critical field for the spontaneous expansion of the nucleus than determines H_C [19].

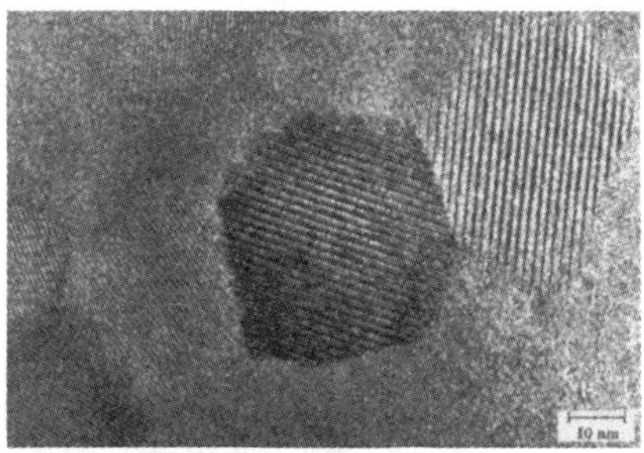

Fig. 10. High-resolution TEM micrograph of a single-phase nc $Nd_2Fe_{14}B$ pm.

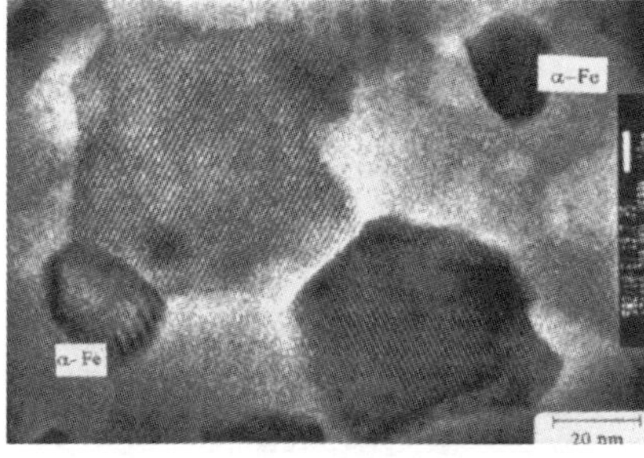

Fig. 11. High-resolution TEM micrograph of a composite nc pm of composition $Nd_2Fe_{14}B$ + 15 vol% α-Fe.

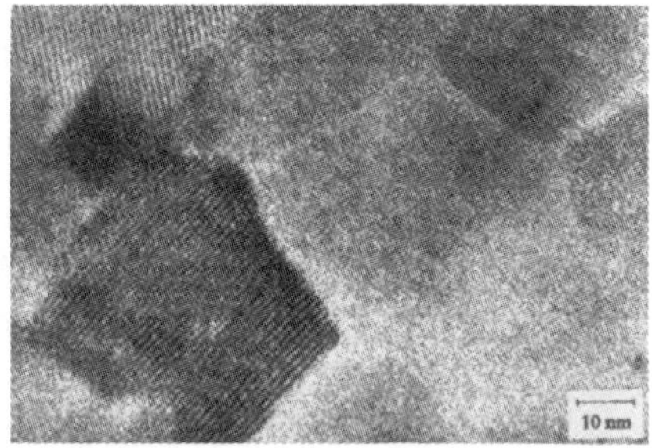

Fig. 12. High-resolution TEM micrograph of a two-phase nc pm of composition $Nd_{17}Fe_{72}B_{7.5}Ga_{1.5}Nb_2$. The bright intergranular phase shows increased contents of Nd, Ga, Nb.

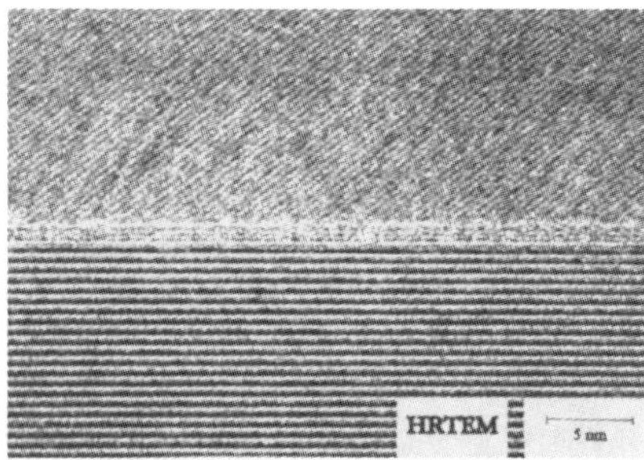

Fig. 13. High-resolution TEM of a Pr-rich nc PrFeB pm showing a corrugated intergranular phase.

The analysis of the hardening mechanism in general starts from the temperature dependence of H_C as given by eq. (2). We may assume that the grains with smallest H_C-values determine the actual H_C. According to the Stoner-Wohlfarth theory the nucleation field depends on the angle ψ_0 of the applied field with respect to the easy axis [2] and the smallest nucleation field have particles with a misalignment of 45° (Fig. 14). At larger misalignment angles the nucleation field again increases whereas the coercive field decreases and vanishes at ψ_0 = 90°. Up to misalignment angles of 45° the nucleation field coincides with the coercive field. For fields $H_{ext} < (K_1 + K_2)/J_S$ within the framework of the Stoner-Wohlfarth theory the change of magnetization in the case of isotropically distributed easy axes occurs exclusively by reversible rotations. For larger fields irreversible nucleation processes take place. The coercive field of an ensemble of isotropically distributed grains in a first approximation is given by

$$H_C \cong \frac{K_1 + K_2}{J_S} = H_N^{\min} , \tag{4}$$

which just corresponds to the minimum nucleation field $H_N^{\min}$. A rigorous calculation of $H_N^{\min}$ including K_2 has been given by Martinek et al. [20]. Assuming that the coercive field is determined by $H_N^{\min}$ eq. (3) may be rewritten as

$$\mu_0 H_C = \alpha_K \mu_0 H_N^{\min} - N_{eff} J_S . \tag{5}$$

For a test of eq. (5) it is usual to plot for different temperatures

$$\mu_0 H_C(T)/J_S(T)\big|_{\exp} \text{ vs. } \mu_0 H_N^{\min}(T)/J_S(T)\big|_{\text{theor}} , \tag{6}$$

where the left side presents the experimental and the right side the theoretical values. In case where a linear plot is found the microstructural parameter α_K and N_{eff} are obtained as the slope and the ordinate intersection of the linear plot, respectively. For ideal spherical particles $\alpha = 1$ and $N_{eff} = 0$ holds which in the case of $Nd_2Fe_{14}B$ gives a coercive field of $\mu_0 H_C = \mu_0 H_N^{\min}$ = 3.9 T ($K_1 = 4.33 \cdot 10^6$ J/m³, $K_2 = 0.65 \cdot 10^6$ J/m³, $J_S = 1.61$ T). In the case of the nc pms the coercive fields are at least a factor of 2 smaller which has to be attributed to the deteriorating effects of the microstructure.

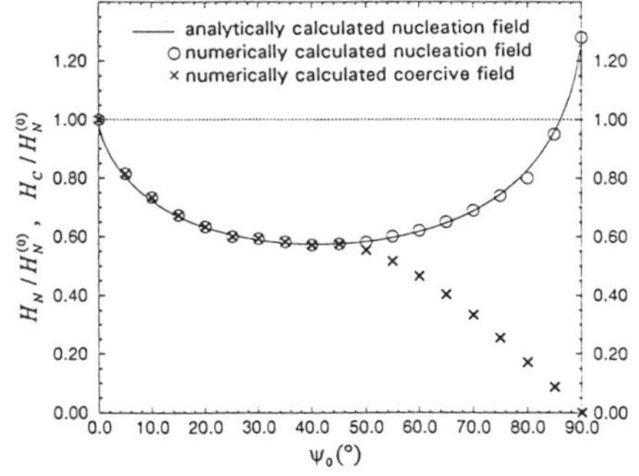

Fig. 14. Angular dependence of the nucleation and of the coercive field of a $Nd_2Fe_{14}B$ sphere. ψ_0 = angle between c-axis and the applied magnetic field.

The microstructural parameters α and N_{eff} have been determined for a large number of rare earth metal based magnets. Fig. 15 shows the $H_C(T)/J_S$-plots for nanocrystalline $Nd_2Fe_{14}B$ + α-Fe pms in the temperature range between 180 K – 560 K [15]. Similarly, Fig. 16 shows the same plot for $Pr_2Fe_{14}B$ between 20 K and 560 K [16]. Fig. 17 presents the $H_C(T)/J_S$-plots for carbides $Sm_{2+\delta}Fe_{14}Ga_2C_2$ for overstoichiometric Sm with δ= 0.00, 0.05 and 0.14 [21]. Over a wide temperature range down to 80 K a linear relationship is found. Fig. 18 shows the $H_C(T)/J_S$-plots for the nitride $Sm_{10.7}Fe_{89.3}N_x$ in the temperature range between 100 K and 575 K [22]. At temperatures T< 100 K deviations from the linear plot are observed, probably because the determination of $H_N^{\min}$ suffers from a large error of K_2 at low temperatures. The results presented so far clearly show that linear plots of eq. (6) are obtained for all nc pms discussed here, which has to be considered as a strong indication for the validity of the nucleation model. The α-values determined for nc pms in general are considerably smaller than those obtained for sintered aligned pms. For decoupled PrFeB pms α_K-values of 0.8 are found whereas the stoichiometric exchange coupled and the composite pms show α-values ranging from- 0.32 to 0.06. Also the N_{eff}-values of the decoupled magnets are large of the order of 1, whereas the values for the nc pms range between 0 and 0.2.

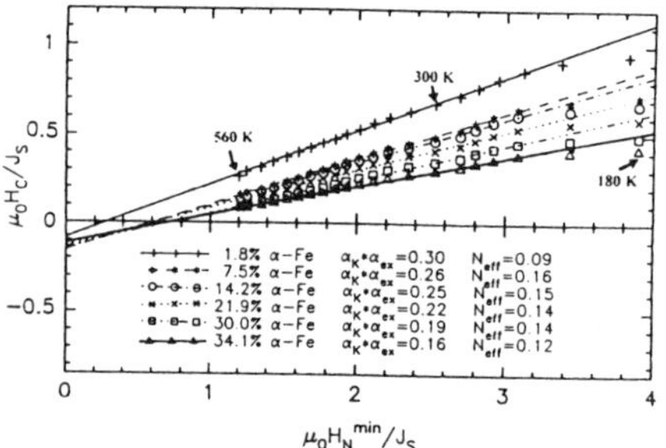

Fig. 15. $\mu_0 H_C / J_S$ vs. $\mu_0 H_N^{\min} / J_S$ plots to determine the microstructural parameters α_K and N_{eff} for different compositions of NdFeB.

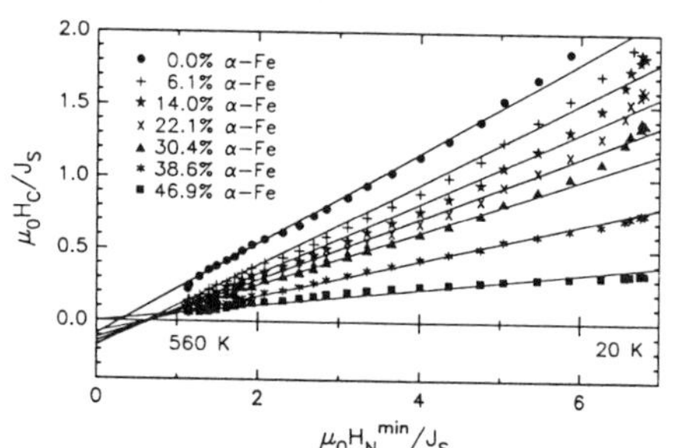

Fig. 16. $\mu_0 H_C / J_S$ vs. $\mu_0 H_N^{\min} / J_S$ plots for different compositions of PrFeB.

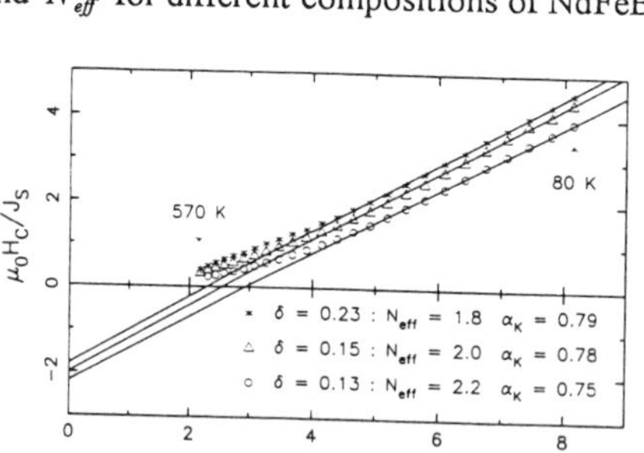

Fig. 17. $\mu_0 H_C / J_S$ vs. $\mu_0 H_N^{\min} / J_S$ plots for different compositions of the carbides $Sm_{2+\delta}Fe_{14}Ga_2C_2$ with δ = 0.0, 0.05, and 0.14.

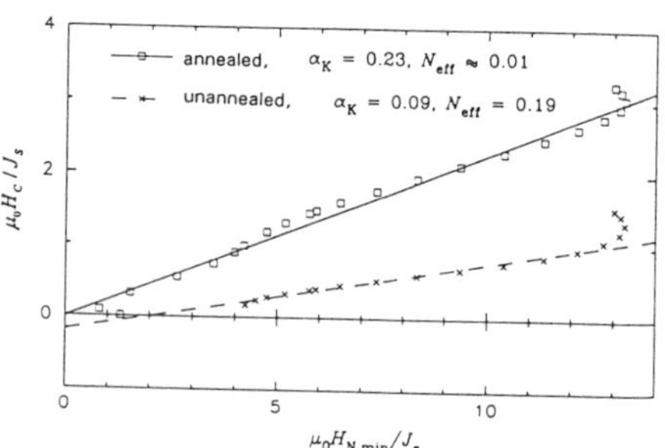

Fig. 18. $\mu_0 H_C / J_S$ vs. $\mu_0 H_N^{\min} / J_S$ plots for the nitride $Sm_{10.7}Fe_{89.3}N_x$ (–□–) annealed at T_a = 850°C, t_a = 17 min.

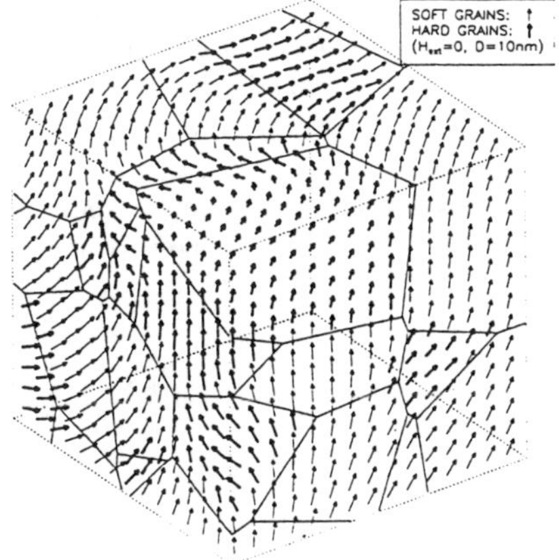

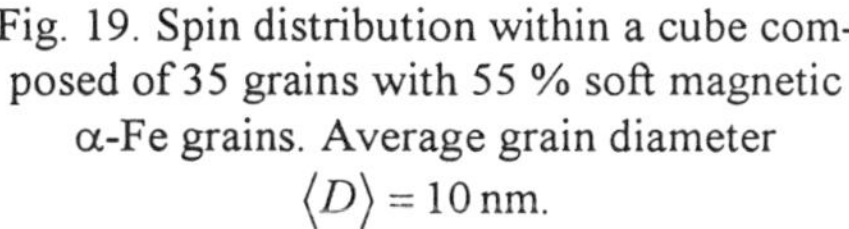

Fig. 19. Spin distribution within a cube composed of 35 grains with 55 % soft magnetic α-Fe grains. Average grain diameter $\langle D \rangle = 10\,\text{nm}$.

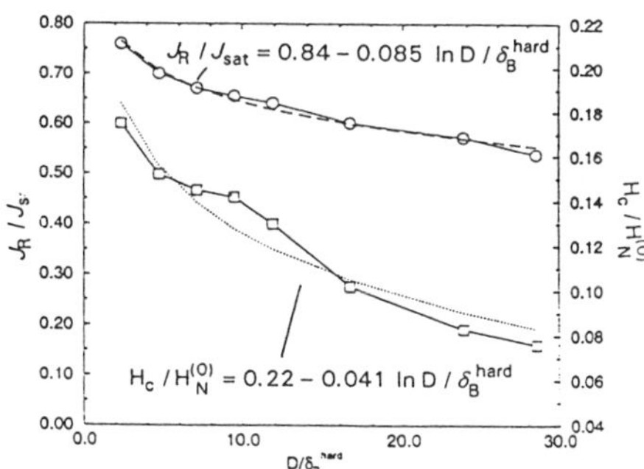

Fig. 20. Grain size dependence of H_C and J_R of a composite pm containing 55 % of α-Fe as a function of D / δ_B ($\delta_B = \pi \sqrt{A / K_1}$ = wall width of $Nd_2Fe_{14}B$).

Computational Micromagnetism and Microstructure

The experimental results for the coercive field known so far show a discrepancy by a factor of 4 – 5 with respect to the theoretical predictions. It is generally accepted that this discrepancy has to be attributed to thc deteriorating effects of the microstructure as, e.g., the isotropic distribution of the easy axes and the reduced magnetocrystalline anisotropy within the gbs. In order to treat the effect of such microstructures quantitatively the computational micromagnetism on the bais of the finite element method (FEM) has been proven to be a rather powerful tool [10,23]. Simulations of magnetization processes by this method originally have been performed for two-dimensional systems [10,24] and later on have been expanded to three-dimensional arrangements of polyhedral grains [23,25]. As the basis of numerical calculations usually a cubic particle composed of polyhedral irregular or regular grains is considered. This model allows the variation of the average grain size, the modification of the magnetic material parameters within the gbs and the formation of composite pms by the introduction of soft magnetic grains. As an example Fig. 19 shows the distribution of the spontaneous magnetization within a cube composed of 35 grains with an average grain diameter of 10 nm. Under the assumption of a perfect exchange coupling between the grains of a composite magnet with 55 % of α-Fe the coercive field and the remanence for an isotropic distribution of easy directions have been determined as a function of grain size by the FEM. The results shown in Fig. 20 reveal a decrease of J_R and of H_C with increasing grain size D. The numerical results have been fitted by ln (D / δ_B)-laws. It turns out that for grain sizes of 20 nm, i.e. $\sim 5\,\delta_B$, the coercive field is of the order of 15 % of the ideal coercive field. Fig. 20 also reveals the remanence enhancement far above the isotropic value of $J_R / J_S = 0.5$. Fig. 21 presents the numerical results for the three characteristic properties of the hysteresis loop as a function of the α-Fe content. The numcerical results are compared with experimental results as obtained for composite pms of the types NdFeB [15,26] and PrFeB [16]. Whereas the experimental results for the remanence show minor deviations from the numerical results in the case of the coercive field and the maximum energy product the deviations are appreciable. In particular the experimental coercive field is much smaller than the numerical one. It is also of interest to note that a regular grain structure in general results in larger values for J_R, H_C and $(BH)_{max}$. Whereas in the case of our experimental results for the composite pms the condition $H_C > 0.5\,J_R$ is valid for α-Fe concentrations smaller than 30 % the numerical results fulfill this condition up to 60 % of α-Fe.

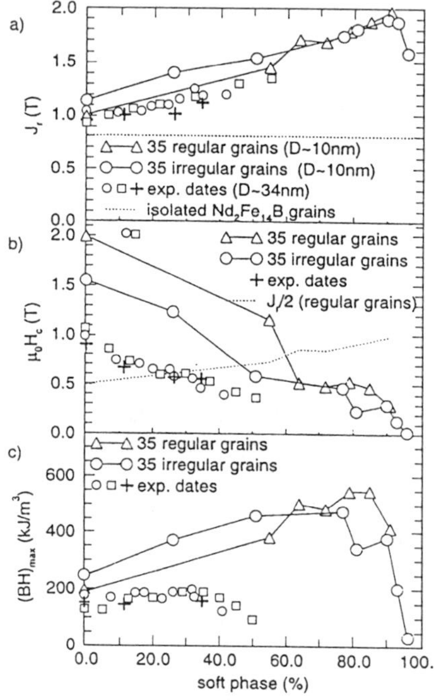

Fig. 21. Numerical results for J_R, $\mu_0 H_C$ and $(BH)_{max}$ of composite nc pms for regular and irregular distributions of grain sizes as a function of the amount of the soft magnetic phase. Average grain size 10 nm. Experimental results: (+) [26], o [15], □ [16].

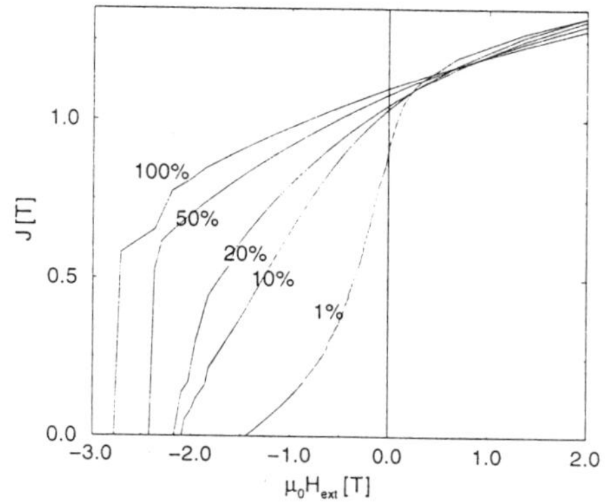

Fig. 22. Demagnetization curves for an ensemble of 64 grains of an average grain diameter of 20 nm and an intergranular phase of 3 nm thickness. The percentages refer to the relative reduction of the exchange and the anisotropy constant within the intergranular phase.

The observed discrepancies between experimental and numerical results have been attributed to reduced anisotropy and exchange constants within gbs [27]. In a review paper it has been shown that variation of the spontaneous magnetization near gbs do not fit the experimental results. Therefore only modifications of the exchange and anisotropy constants are considered. Fig. 22 shows demagnetization curves for reduced material contents within the gbs. These results were obtained for a cube of 64 grains of average diameter 20 nm. The gbs were assumed to have an intergranular width of 3 nm. According to Fig. 22 with decreasing exchange constant and crystal anisotropy within the gbs both properties, H_C and J_R, decrease smoothly. If we introduce the parameter f defined as the ratio $f = A^{gb} / A = K^{gb} / K_1$, where A^{gb} and K^{gb} denote the material parameters within the gb, the coercive field and the remanence [28]write

$$\begin{aligned} H_C / H_N &= 0.304 + 0.098 f \ , \\ J_R / J_{sat} &= 0.646 + 0.036 f \ . \end{aligned} \qquad (7)$$

Here H_N denotes the ideal nucleation field $2K_1 / J_S$ and J_{sat} the saturation polarization.

CONCLUSIONS

In nc pms the characteristic properties of the hysteresis loops depend predominantly on the following microstructural properties: 1. The grain size dependence for $\langle D \rangle > 2\delta_B$ may be described by a $\ln D / \delta_B$-law. 2. Exchange coupling between grains leads with increasing content of α-Fe to a linear increase of J_R and a linear decrease of $\mu_0 H_C$. 3. A maximum energy product of ~ 180 kJ/m³ is obtained for ~ 30 vol% of α-Fe in NdFeB and PrFeB. 4. Reduced exchange and anisotropy constants within the gbs of width 3 nm reduce the coercive field by ~ 50 % and the remanence by ~ 20 %.

REFERENCES

1. W.F. Brown Jr., Rev. Mod. Phys. **17**, p. 15 (1945).

2. E.C. Stoner and E.P. Wohlfarth, Philos. Trans. Roy. Soc. **240**, p. 599 (1948).

3. H. Kronmüller, K.-D. Durst and M. Sagawa, J. Magn. Magn. Mater. **74**, p. 291 (1988).

4. H. Kronmüller in *Supermagnets, Hard Magnetic Materials*, edited by G.J. Long and F. Grandjean (Kluwer, Dordrecht 1991), p. 461.

5. M. Sagawa and S. Hirosawa, J. Physique **49**, p. C8-617 (1988).

6. M. Tokunaga, Y. Nozawa, K. Iwasaki, S. Tamigawa and H. Harada, IEEE Trans. Magn. **25**, p. 3561 (1989).

7. J. Fidler, 7th *Int. Symp. on Magnetic Anisotropy and Coercivity in RE-TM Alloys*, edited by B. Street (The University of Western Australia, Perth 1992), p. 11.

8. H. Kronmüller, R. Fischer, M. Seeger and A. Zern, J. Phys. D: Appl. Phys. **29**, p. 2274 (1996).

9. E.F. Kneller and R. Hawig, IEEE Trans. Magn. **27**, p. 3588 (1991).

10. T. Schrefl, J. Fidler and H. Kronmüller, Phys. Rev. B **49**, p. 6100 (1994).

11. J. Ding, P.G. McCormick and R. Street, J. Magn. Magn. Mater. **124**, p. L1 (1993).

12. J. Ding, Y. Liu, R. Street and P.G. McCormick, J. Appl. Phys. **75**, p. 1032 (1994).

13. E.-Th. Henig and B. Grieb in *Supermagnets, Hard Magnetic Materials*, edited by G.J. Long and F. Grandjean (Kluwer, Dordrecht 1991), p. 171.

14. R. Coehoorn, D.B. de Mooij, J.P.W.B. Duchateau and K.H.J. Buschow, J. Physique **49**, p. C8-669 (1988).

15. J. Bauer, M. Seeger, A. Zern and H. Kronmüller, J. Appl. Phys. **80**, p. 1667 (1996).

16. D. Goll, M. Seeger and H. Kronmüller, J. Magn. Magn. Mater. **185**, p. 49 (1998).

17. C.I. Harland, H.A. Davies, B.E. Watts and F. Leccabue in *Rare Earth Magnets and Their Applications*, edited by L. Schultz and K.-H. Müller (Werkstoff-Informationsgesellschaft, Frankfurt 1998), p. 263.

18. H. Kronmüller, phys. stat. sol. (b) 144, p. 385 (1987).

19. D. Givord, P. Tenaud and T. Viadieu, IEEE Trans. Magn. MAG-**24**, p. 1921 (1988).

20. G. Martinek and H. Kronmüller, J. Magn. Magn. Mater. **86**, p. 177 (1990).

21. J. van Lier, M. Seeger and H. Kronmüller, J. Magn. Magn. Mater. **167**, p. 43 (1997).

22. J. Kleinschroth, Doctor Thesis, University Stuttgart 1997.

23. R. Fischer, T. Schrefl, H. Kronmüller and J. Fidler, J. Magn. Magn. Mater. **153**, p. 35 (1996).

24. T. Schrefl, H. Kronmüller and J. Fidler, J. Magn. Magn. Mater. **127**, p. L273 (1993).

25. T. Schrefl, J. Fidler and H. Kronmüller, J. Magn. Magn. Mater. **138**, p. 15 (1994).

26. M. Willcox, J. Williams, M. Leonowicz and A. Manaf in 8th *Int. Symp. on Magnetic Anisotropy and Coercivity in Rare Earth-Transition Metal Alloys*, edited by C. Manwaring, D. Jones, A. Williams and I. Harris (University of Birmingham, UK 1994), p. 443.

27. R. Fischer and H. Kronmüller, Phys. Rev. B **54**, p. 7284 (1996).

28. R. Fischer and H. Kronmüller, J. Magn. Magn. Mater. **184**, p. 166 (1998).

MICROSTRUCTURAL INVESTIGATION OF $RE_3(Fe.V)_{29}$ (RE = Nd, Tb) MAGNETIC MATERIALS

J. BERNARDI *, M. NONER *, J. FIDLER *, X.F. HAN **, F.M. YANG **
* Institut für Angewandte und Technische Physik, Vienna University of Technology, Wiedner Hauptstr. 8-10, A-1040 Wien, Austria, bernardi@email.tuwien.ac.at
** Institute of Physics, Chinese Academy of Science, P.O.Box 603, Beijing 100080, P.R.China

ABSTRACT

The microstructure of V stabilized $RE_3(Fe,V)_{29}$ (RE=Nd,Tb) has been investigated by transmission electron microscopy (TEM). The investigated samples were prepared by arc melting and subsequent annealing above 910°C. X-ray diffraction confirms that the samples can be indexed based on a monoclinic $Nd_3(Fe,Ti)_{29}$-type structure (3:29) with A2/m space group. Our TEM investigation confirms that $Nd_3(Fe,V)_{29}$ contains usually grains with 3:29 structure and A 2/m space group. In addition grains with rhombohedral Th_2Zn_{17} structure are observed regularly. The $Tb_3(Fe,V)_{29}$ alloy consists also of grains with monoclinic $Nd_3(Fe,Ti)_{29}$-type structure and contains a high density of planer defects like crystallographic twins or antiphase boundaries. Twinning occurs preferably on $(40\bar{2})$ of the monoclinic 3:29 structure. No tetragonal $RE(Fe,V)_{12}$ phase or Fe is found.

INTRODUCTION

The discovery of $Nd_2Fe_{14}B$ type permanent magnets [1] led to the search for new intermetallic compounds with attractive magnetic properties and a Curie temperature exceeding the relatively low one of $Nd_2Fe_{14}B$-magnets. The search for new rare-earth (RE) transition metal (TM) alloys led recently to the discovery of a novel phase that was initially reported as $Nd_2(Fe,Ti)_{19}$ [2] and a similar compound reported as $RE_2(Fe_{.91}V_{.09})_{17}$ [3]. The new phase was identified by x-ray powder diffraction (XRD) and by neutron powder diffraction as monoclinic $Nd_3(Fe,Ti)_{29}$ structure (3:29) with P2/c space group [4,5] and later more accurately with A2/m space group [6]. The intermetallic iron-rich $RE_3(Fe,M)_{29}$ alloys with M=Ti, V, Cr, Mn and Mo, gained much attention because they can absorb C or N and so improve the Curie temperature, saturation magnetization and anisotropy field significantly upon interstitial modification with nitrogen or carbon. The $Nd_3(Fe,Ti)_{29}$ structure and other pseudo-binary structures like the rhombohedral Th_2Zn_{17} (2:17) or tetragonal $ThMn_{12}$ (1:12) are all related to the hexagonal $CaCu_5$ structure in a way that a certain amount of RE atoms in hexagonal $RETM_5$ were replaced by transition metal dumbbells. The 3:29 structure can be described as an intermediate structure between the Th_2Zn_{17} and $ThMn_{12}$ structures consisting of a combination of 2:17 and 1:12 units in a ratio of 1:1.

The influence of the V-content on the phase formation has been recently investigated by x-ray diffraction (XRD) in the system $Tb_3(Fe_{1-x}V_x)_{29}$ [7]. It was shown that some 2:17 phase was detected for $x < 0.033$ and some 1:12 phase for $x > 0.061$. In this work we investigated the microstructure of two $RE_3(Fe,V)_{29}$ alloys by transmission electron microscopy and compared the results with data obtained by XRD studies.

Mat. Res. Soc. Symp. Proc. Vol. 577 © 1999 Materials Research Society

EXPERIMENT

Samples with the composition $Nd_3(Fe_{0.93}V_{0.07})_{29}$ and $Tb_3(Fe_{0.96}V_{0.04})_{29}$ were prepared from high-purity elements (at least 99.9%) by arc-melting in an argon atmosphere. The ingots were remelted at least four times to ensure mixing of the components. Excess rare-earth compared to the stoichiometric composition was used to compensate the loss of rare-earth during melting and annealing. The alloys were sealed in quartz tubes and subsequently annealed under protective argon atmosphere at 1000°C - 1050°C for up to 72 hrs. The phase formation was checked by x-ray diffraction using Cu-Kα radiation and the lattice parameters were calculated from the XRD patterns. The microstructure of the samples was investigated by means of transmission electron microscopy using a JEOL 200 CX analytical electron microscope. The electron transparent samples were prepared by conventional grinding and ion beam thinning.

RESULTS

The magnetic properties of the two investigated samples are described in detail in reference [7] and [8]. Table I summarizes the Curie temperature, the saturation magnetization and the anisotropy field of the two alloys. After annealing, the alloys were characterized by XRD. Figure 1 compares the strongest peaks from the x-ray diffraction pattern of $Nd_3(Fe_{.93}V_{.07})_{29}$ with a simulated XRD pattern for a monoclinic $Nd_3(Fe,Ti)_{29}$ structure in the A2/m space group. The sample appears nearly single phase. The slight difference of the intensities of some experimental peaks compared to the simulated peaks is caused probably by the existence of small amounts of impurity phases.

Table I: Curie temperature T_C, saturation magnetization Ms (μ_B/formula units) and anisotropy field H_A at room temperature (RT) of the investigated magnetic materials.

$RE_3(Fe,V)_{29}$	T_C [K]	Ms (RT) [μ_B/f.u.]	Ha (RT) [T]
$Nd_3(Fe_{.93}V_{.07})_{29}$	455	46.8	2.4
$Tb_3(Fe_{.96}V_{.04})_{29}$	474	25.8	4.5

Similar results and an even better agreement between experimental pattern and simulated x-ray diffraction pattern were obtained for $Tb_3(Fe_{.96}V_{.04})_{29}$. The XRD patterns were used to calculate the exact lattice parameters of the monoclinic $RE_3(Fe,V)_{29}$ phases (Table II).

Table II: Lattice parameters of the monoclinic $RE_3(Fe,V)_{29}$ phase (RE = Nd, Tb) of the investigated materials calculated from the analysis of XRD data for A2/m space group.

$RE_3(Fe,V)_{29}$	a [nm]	b [nm]	c [nm]	β [°]
$Nd_3(Fe_{.93}V_{.07})_{29}$	1.065	0.857	0.974	96,85
$Tb_3(Fe_{.96}V_{.04})_{29}$	1.057	0.849	0.967	96,90

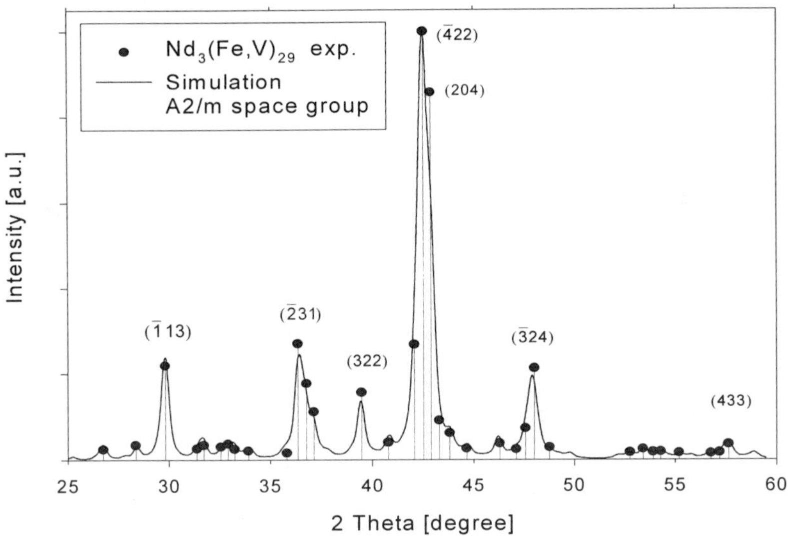

Fig. 1. Comparison of the strongest x-ray diffraction peaks of $Nd_3(Fe_{.93}V_{.07})_{29}$ with the simulated intensities for a monoclinic $Nd_3(Fe,Ti)_{29}$ structure in the A2/m space group.

The analysis of the x-ray diffraction pattern indicates that $Nd_3(Fe_{.93}V_{.07})_{29}$ consists mainly of monoclinic 3:29 phase. TEM investigation confirms that the $Nd_3(Fe_{.93}V_{.07})_{29}$ sample contains mostly 3:29 grains with diameters up to several microns (see Fig.2). The monoclinic grains show planar defects that are often aligned nearly parallel within a grain. In addition sev-

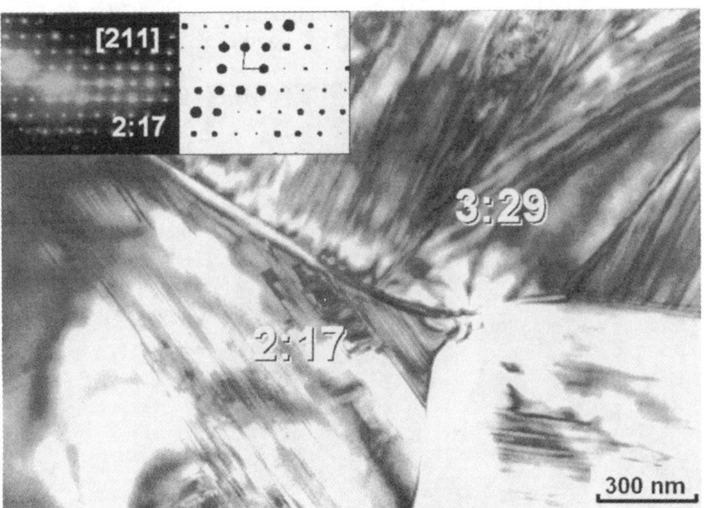

Fig. 2: Electron micrograph of $Nd_3(Fe_{.93}V_{.07})_{29}$ showing two monoclinic 3:29 grains next to a rhombohedral 2:17 grain. The corresponding electron diffraction pattern shown together with a simulated diffraction pattern is taken from the 2:17 grain close to [211] orientation.

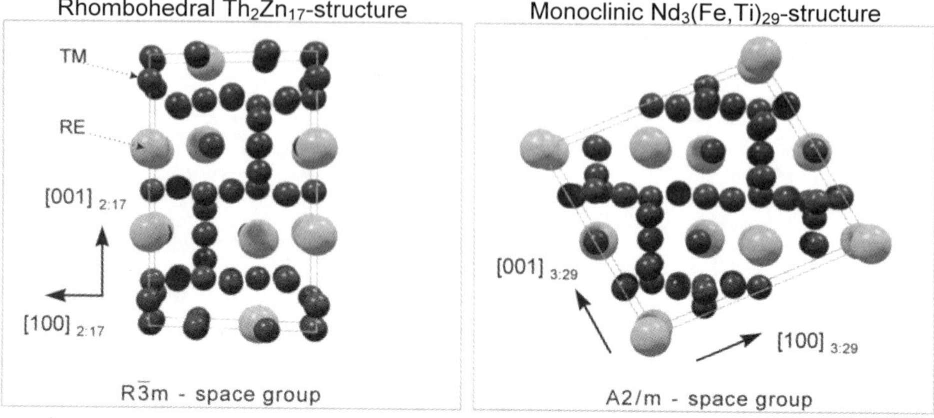

Fig. 3: Crystal structures of rhombohedral 2:17 and monoclinic 3:29 in [010] direction. The large spheres correspond to the rare-earth atoms (RE), the small spheres show the places of the transition metals (TM). The $(20\bar{1})_{3:29}$ plane is parallel to the basal plane $(001)_{2:17}$ of the 2:17 structure.

eral rhombohedral 2:17 grains throughout the sample are found with diameters of several microns as well as a few Nd-oxide grains. The close relationship between the rhombohedral 2:17 and the monoclinic 3:29 phase makes it usually difficult to distinguish between this two phases by means of electron diffraction unless several diffraction patterns are compared. Figure 3 demonstrates the similarities between the Th_2Zn_{17} and $Nd_3(Fe,Ti)_{29}$-structure.

The $Tb_3(Fe_{.96}V_{.04})_{29}$ alloy shows a microstructure with a grain size comparable to $Nd_3(Fe_{.93}V_{.07})_{29}$. In this sample with low V content we found only monoclinic 3:29 grains. Typically for $Tb_3(Fe_{.96}V_{.04})_{29}$ the formation of a high density of planar defects like twins and antiphase boundaries occurs in the 3:29 grains (Fig. 4 and Fig.5).

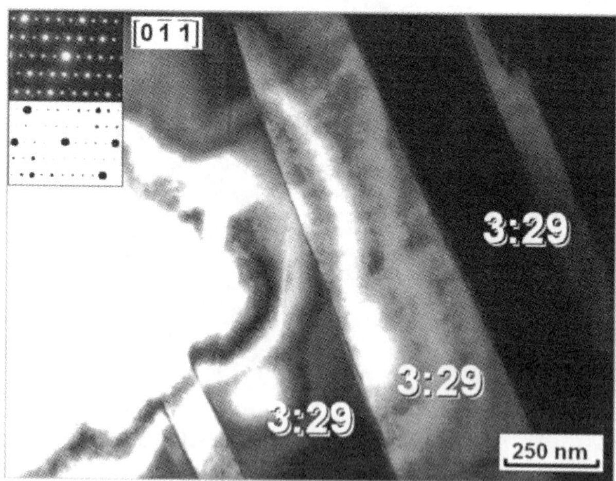

Fig. 4: Micrograph and corresponding electron diffraction pattern of $Tb_3(Fe_{.96}V_{.04})_{29}$ viewed close to $[0\bar{1}\,\bar{1}]$ showing the formation of crystallographic twins within a monoclinic 3:29 grain. The simulated diffraction pattern shows monoclinic 3:29 with A2/m space group in $[0\bar{1}\,\bar{1}]$ orientation.

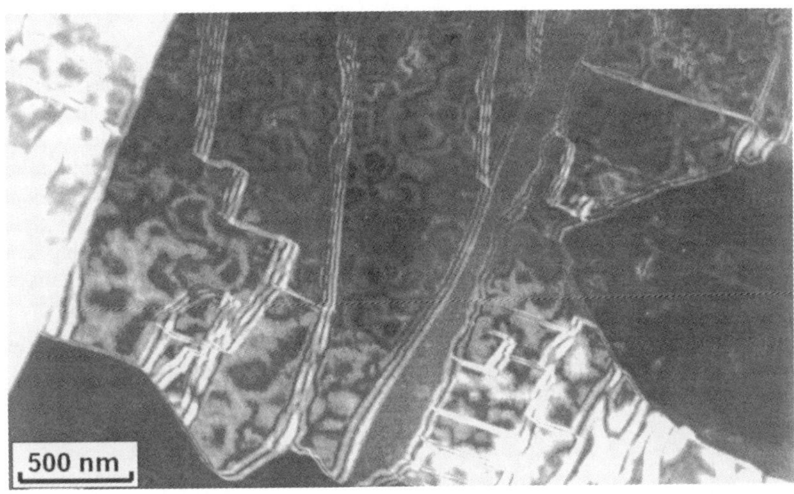

Fig. 5: Dark field micrograph of $Tb_3(Fe_{.96}V_{.04})_{29}$ showing a high density of stacking faults and antiphase boundaries within a monoclinic grain.

An analysis of several electron diffraction patterns of $Tb_3(Fe_{.96}V_{.04})_{29}$ showed that the preferable twin plane observed in this material is $(40\bar{2})_{3:29}$ (Fig. 6). This plane corresponds to the basal plane of the rhombohedral 2:17 structure.

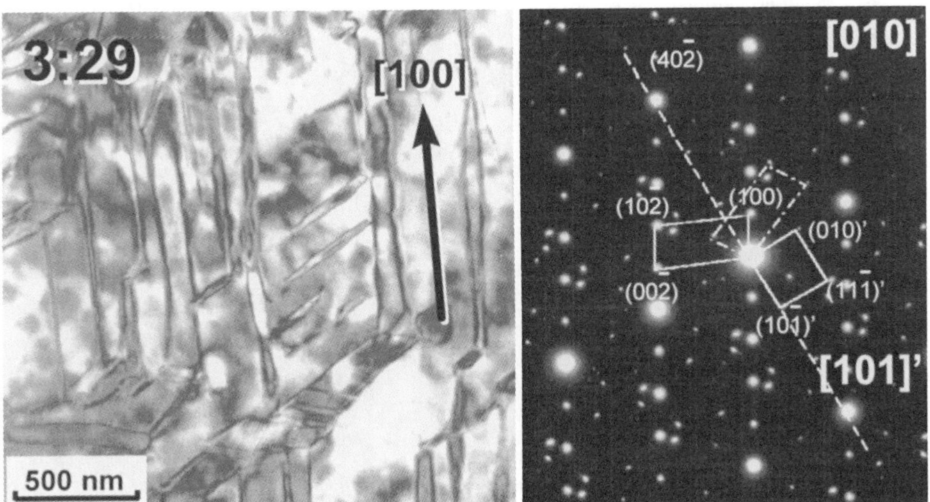

Fig. 6: Bright field micrograph of $Tb_3(Fe_{.96}V_{.04})_{29}$ and corresponding electron diffraction pattern showing a monoclinic 3:29 grain with [010] zone axis. One set of additional spots arise due to twinning at $(40\bar{2})$. There are also spots resulting from a grain in [101] orientation.

CONCLUSION

The microstructure of V-stabilized $RE_3(Fe,V)_{29}$ magnetic materials has been investigated by transmission electron microscopy. The grain size of the investigated alloys is up to several microns. In $Tb_3(Fe_{.96}V_{.04})_{29}$ only monoclinic 3:29 grains are found which was also confirmed by x-ray diffraction. $Nd_3(Fe_{.93}V_{.07})_{29}$ contains in addition to 3:29 grains some rhombohedral 2:17 grains and few Nd-oxide grains. All electron diffraction patterns of the 3:29 phase can be indexed using the monoclinic 3:29 structure with A2/m space group.

In all investigated 3:29 grains planar defects are found. In the Tb containing sample with the low V-content these defects occurred with a high density as crystallographic twins and antiphase boundaries. We could show that a preferable twin plane of the 3:29 phase in $Tb_3(Fe_{.96}V_{.04})_{29}$ seems to be $(40\bar{2})$, corresponding to the basal plane of the rhombohedral 2:17 structure. In both investigated $RE_3(Fe,V)_{29}$ alloys with (RE=Nd, Tb) no tetragonal 1:12 phase and no Fe-precipitates were found.

REFERENCES

1. M. Sagawa, S. Fujimura, M Tokunaga, and Y. Matuura, J. Appl. Phys. 55, 2083, (1984).
2. S.J. Collocott, R.K. Day, J.B. Dunlop, and R.L. Davis, in Proceedings 7th International Symposium on Magnetic Anisotropy and Coercivity in Rare Earth-Transition Metal Alloys, Canberra, 1992, ed. Hi-Perm Laboratory, Reserarch Center for Advanced Minerals and Material Procession, University of Western Australia, 437, (1992).
3. Y.V. Shcherbakova, G.V. Ivanova, A.S. Yermolenko, and V.S. Gaviko, J.Alloys Compounds 182, 199, (1992).
4. J.M. Cadogan, H.S. Li, A. Margarian, J.B. Dunlop, D.H. Ryan, S.J. Collocott, and R.L Davis, J. Appl. Phys. 76 (10), 6138, (1994).
5. Z. Hu and W.B. Yelon, J. Appl. Phys. 76 (10), 6147, (1994).
6. O. Kalogirou, V.Psycharis, L Saettas, and D. Niarchos, J. Magn. Magn. Mater. 146, 335, (1995).
7. X.F. Han; F.M. Yang, J.J.Zhu, H.G. Pan,Y.G. Wang, J.L. Wang, N. Tang, and R.W. Zhao, J. Appl. Phys. 81 (7), 3248, (1997).
8. X.F. Han; F.M. Yang, H.G. Pan,Y.G. Wang, J.L. Wang, H.L. Liu, N. Tang, R.W. Zhao, and H.S, Li, J. Appl. Phys. 81 (11), 7450, (1997).

MAGNETIZATION REVERSAL IN MELT-QUENCHED NdFeB

D.C. CREW*, L.H. LEWIS*, P.G. McCORMICK**, R. STREET**,
V. PANCHANATHAN***
* Materials Science Division, Bldg 480, Department of Applied Science, Brookhaven National Laboratory, Upton, NY, U.S.A. 11973-5000, dcrew@bnl.gov
** Special Research Center for Advanced Mineral and Materials Processing, The University of Western Australia, Nedlands, WA, Australia 6907
*** Magnequench International, Inc., Anderson, Indiana, U.S.A.

ABSTRACT

Melt-quenched NdFeB is an important modern permanent magnet material. However there still remains doubt as to the magnetization reversal mechanism which controls coercivity in material prepared by this processing route. To investigate this problem a new technique based on measurements of reversible magnetization along recoil curves has been used. The technique identifies the presence of free domain walls during magnetic reversal. For this study samples of isotropic (MQI), hot pressed (MQII) and die upset (MQIII) melt-quenched NdFeB were examined. The results indicate that in MQI free domain walls are not present during reversal and the reversal mechanism is most likely incoherent rotation of some form. Free domain walls are also not present during reversal in the majority of grains of MQII, even though initial magnetization measurements indicate that the grain size is large enough to support them. In MQIII free domain walls are present during reversal. These results are attributed to the reduced domain wall nucleation field in MQIII compared with MQII and the increased dipolar interactions in MQIII.

INTRODUCTION

Melt-quenched NdFeB is an important permanent magnet material which has many technological uses. The magnetization reversal mechanism of material prepared by this processing route is still the subject of doubt, however. Studies of the time dependence of magnetization and measurements of the irreversible susceptibility [1] have concluded that the reversal mechanism is consistent with nucleation. This conclusion is supported by other theoretical and experimental studies of the coercivity [2,3]. At the same time transmission electron microscopy studies performed on thin foils [4] have shown pinning of domain walls at grain boundaries and this supports a domain wall pinning model of magnetization reversal [5].

In this work we use a new technique, developed by Cammarano, McCormick and Street [6] based on measurements of recoil curves from the major hysteresis loop, to investigate magnetization reversal in melt-quenched NdFeB. This technique is based upon the idea that the reversible magnetization (M_{rev}) is not only a function of the field H but also of the irreversible magnetization (M_{irr}), an indicator of the magnetic state of the system. Thus plots of M_{rev} vs. M_{irr} at constant values of field can give insight into the reversal mechanism. These curves will be referred to as M_{rev} curves.

To experimentally determine the M_{rev} curves, a series of recoil curves are measured from the major hysteresis loop to zero internal field as shown in Fig. 1. A value of internal field (H) is chosen and the values of the reversible magnetization (M_{rev}) and the irreversible magnetization (M_{irr}) are calculated for each recoil curve measured. This procedure then defines one M_{rev} curve at a particular field H. By choosing different values of the internal field H a family of such curves may be obtained.

Models of magnetization reversal have shown [7,8] that if domain walls are free to move or bow under the influence of an applied field during reversal of magnetization then the M_{rev} curves exhibit a minimum. This result has been confirmed by experiment on a sample of a confirmed pinning controlled Sm_2M_{17} type magnet [9]. A minimum in the M_{rev} curves only occurs if the domain walls are able to reversibly bow under the action of an applied field. If this is not the case and the domain walls are unable to bow, or if domain walls are not present within the material

Mat. Res. Soc. Symp. Proc. Vol. 577 © 1999 Materials Research Society

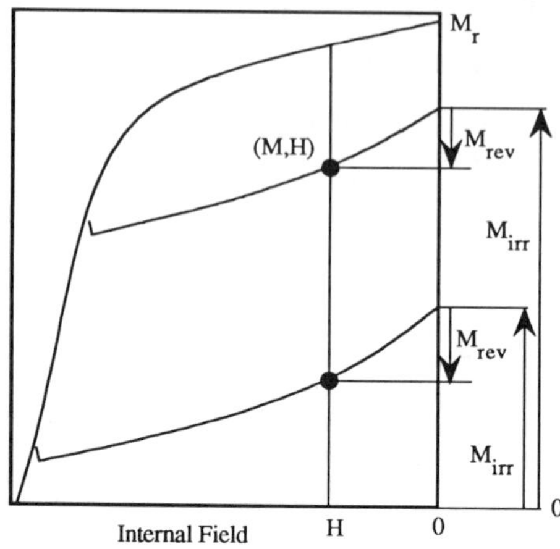

FIG. 1: The definitions of M_{rev} and M_{irr} used in this work plus how the reversible magnetization curves at any field were calculated.

during reversal, then the reversible magnetization behavior observed may only arise from rotation of the magnetization vectors away from the easy axis. In this case the M_{rev} curves are found to be straight lines with slopes that increase with increasing internal field H.

The difference in the M_{rev} curves measured from samples that reverse via free domain wall motion versus those that reverse by rotation of the magnetization permits the use of this technique to discriminate between different reversal mechanisms.

EXPERIMENT

In this study three samples of melt-quenched NdFeB, subject to different post-solidification processing steps, were used. The first sample, MQI, was optimally quenched ribbon which had been compacted with a resin binder to form a material with an isotropic distribution of easy axis orientations. The second sample, MQII, consisted of ribbons compacted with hot pressing, producing a very small amount of grain alignment. The last sample, MQIII, was comprised of quenched ribbons which had been hot pressed and then die upset to produce a material which was highly anisotropic and well aligned. All samples were supplied by Magnequench.

MQI and MQIII were ground and polished into spheres using the sphere polisher described by Folks et al. [10]. The final polishing medium was 0.25 μm diamond paste. Corrections to the magnetic measurements were made to account for the non-zero demagnetization factor of these samples. The MQII sample was cut into the shape of a square-based prism with a height-to-width ratio greater than 4. The surface finish was 300 grit fine sandpaper and the demagnetization factor was small enough that it could be neglected in this study.

Each of the samples was measured from the virgin magnetic state, i.e. after thermal demagnetization. Both the initial magnetization curve was recorded as well as the isothermal remanent magnetization (IRM) curve [11] which is an indication of irreversible processes which occur along the initial magnetization curve. After saturation in a 50 kOe positive field, recoil curves from the major hysteresis loop were measured on each sample. The recoil curves were measured in the second quadrant of the hysteresis loop and so all measurement fields were negative. The instrument used was either a vibrating sample magnetometer (MQI and MQIII) or a SQUID magnetometer (MQII). All measurements were performed at room temperature.

The data was analyzed using the method described above to obtain a series of M_{rev} curves at different values of field H for each sample.

RESULTS

The M_{rev} curves for MQI are shown in Fig. 2. These data consist of approximately straight lines which increase in slope as the internal field H at which they are measured increases. There is no evidence that a minimum occurs anywhere in this data.

The curves for MQII, shown in Fig. 3, are similar to those for MQI for most of the range of M_{irr} but exhibit a small, sharp minimum in M_{rev} near positive remanence for low values of field. These curves are similar to those measured previously in a Sm_2M_{17} pinning type magnet [9]. The magnitude of the reversible magnetization present in MQII is larger than that measured in MQI and over the majority of the range of M_{irr} there is less curvature as compared with MQI. The

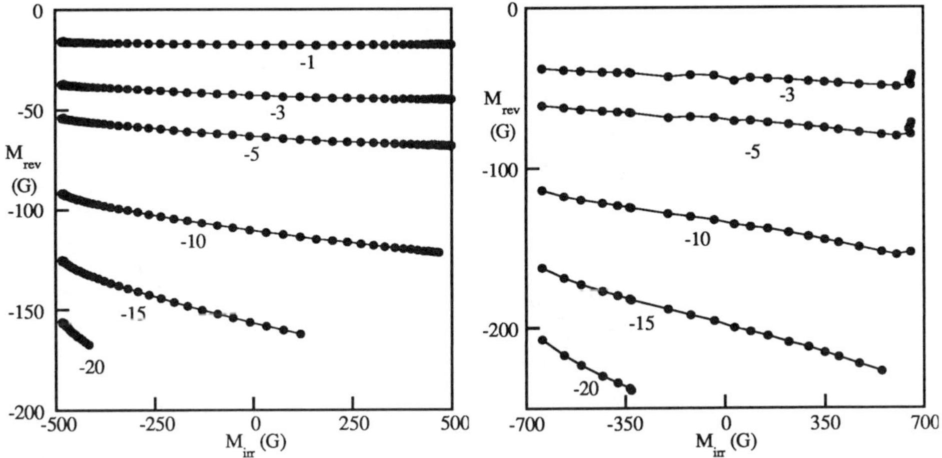

FIG. 2: Reversible magnetization curves for MQI. The fields at which the curves were calculated are shown in the figure in kOe.

FIG. 3: Reversible magnetization curves for MQII. The fields at which the curves were calculated are shown in the figure in kOe.

data of Fig. 3 were collected using a SQUID magnetometer and are thus more sparse than those of Fig. 2. Nevertheless the trend of the curves in Fig. 3 is clear.

The curves for MQIII, Fig. 4, exhibit a pronounced minimum in the vicinity of $M_{irr} = 0$. This is not a sharp minimum like that seen in MQII. The magnitude of M_{rev} is much smaller than that seen in either MQI or MQII.

The initial magnetization curves, as manifest in the IRM data, are shown in Fig. 5. In low fields MQI evidences little magnetization change, requiring a field near 12 kOe before substantial magnetization is induced. MQII shows two steps in the curve, each of approximately the same magnitude, the first finishing at approximately 10 kOe, the other starting at fields of 17 kOe. Upon careful inspection MQIII shows two steps as well [1], similar to MQII, but the steps are much closer together.

DISCUSSION

MQI

The straight lines seen in the reversible magnetization data for MQI in Fig. 2 suggest strongly that there are no free domain walls present during reversal and that rotation of magnetization vectors away from the easy axis is responsible for the reversible magnetization seen in this material. This conclusion is consistent with the measured initial magnetization curve which shows a long flat region at low field. These results from MQI indicate that domain walls, if present in the thermally-demagnetized state, are not free to move. Additionally, if domain walls are nucleated in a saturated specimen under the action of a reverse field they will be pinned and unable to bow reversibly. The estimation by Folks, Street and Woodward [1] that only 6% of grains in MQI are multi-domain is consistent with this conclusion.

The grain structure of MQI consists of very fine grains (30 nm diameter) which are equiaxed and randomly oriented [12]. This grain size is below the theoretical limit for single domains, calculated for an isolated non-interacting particle to be 300 nm [1]. As the grains in isolation cannot support a domain wall within them, one possible mechanism for reversal in MQI is incoherent rotation such as curling. Such a mechanism would explain both the lack of free domain walls seen in the reversible magnetization measurements and the initial flat region in low fields of the IRM data. It is also possible, however, that magnetostatic and exchange interactions among

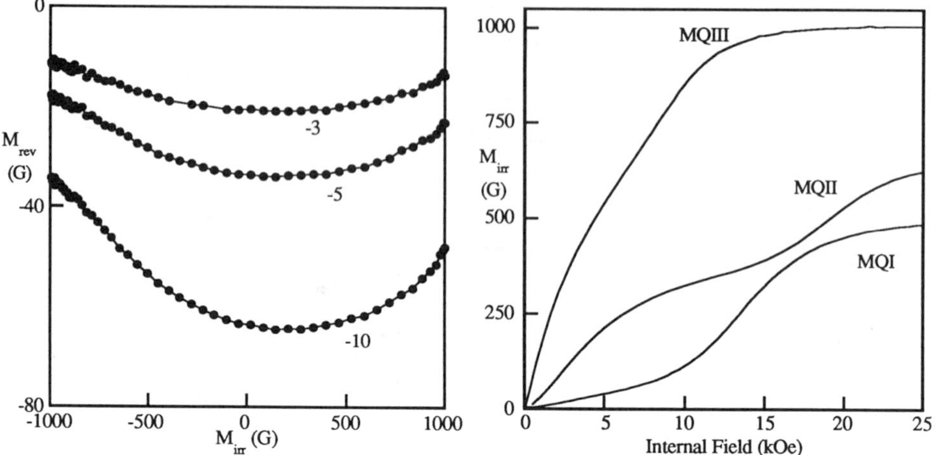

FIG. 4: Reversible magnetization curves for MQIII. The fields at which the curves were calculated are shown in the figure in kOe.

FIG. 5: Isothermal Remanent Magnetization (IRM) curves for MQI, MQII and MQIII showing the steps present in the curves for MQII and MQIII compared with MQI.

neighboring grains in MQI creates interaction domains and thus allows reversal by the movement of domain walls, a reversal route that many not be energetically favorable for an isolated single-domain size grain. If domain wall movement is operative in this material, however, the measured reversible magnetization data provide no evidence of domain wall bowing and thus requires the walls to be trapped at the grain boundaries by pinning sites that are too closely spaced to allow such reversible bowing to occur. This latter mechanism would resemble nucleation in form and indeed these two mechanisms of reversal are difficult to distinguish [13].

MQII

The first step in the IRM curve in Fig. 5 for MQII, in which a rapid increase in magnetization occurs for low fields, suggests that there is a significantly larger proportion of multi-domain grains in this material than were present in MQI. Folks, Street and Woodward [1] estimated that 44% of grains in a similar material were multi-domain in the thermally demagnetized condition. The reversible magnetization data however is only partly consistent with a reversal mechanism which relies upon movement of domain walls. Over the majority of the range of the M_{irr} data, the reversible magnetization curves are straight with a negative slope. This form for the M_{rev} curves is what would be expected if rotation was the dominant mechanism responsible for reversible magnetization. It is likely, therefore, that the reversal mechanism in this material is similar to that of MQI, i.e. incoherent rotation of some form or domain wall movement by domain walls trapped at grain boundaries (and hence unable to bow reversibly). Again the latter reversal mechanism is difficult to distinguish from a nucleation controlled process.

It has been observed that there is an intergranular phase present in this material [14] which may provide exchange coupling between the grains. If the grains are exchange coupled then it is expected that large multi-grain interaction domains will form. This in turn would mean that the most likely reversal mechanism will be domain wall movement or nucleation at the peripheries of reversed regions.

The small minimum in M_{rev} observed near positive remanence in MQII implies that there are, during this part of the reversal, free domain walls able to bow reversibly upon application of an external field. The form of the curves in Fig. 3 are similar to those observed in an Sm_2M_{17} type magnet although in MQII there is no evidence of a coherent intragranular network of pinning sites. It is logical therefore to attribute the small minimum in M_{rev} observed to the estimated 15% of grains which have been observed by transmission electron microscopy to be large and probably uncoupled [14] to the smaller-sized matrix grains. These large grains form at the positions of former ribbon boundaries during the hot-pressing process [14]. It is feasible therefore that these grains are large enough so that a nucleated domain wall is energetically stable and may bow reversibly upon application of a field. The fact that the minimum in the M_{rev} curve is so sharp indicates that the nucleation field of these grains is low but that little irreversible change occurs until the domain walls are driven from these grains by a higher field. The reason that a higher field is required to remove domain walls from these large grains is due to localized dipolar interactions which hinder the complete annihilation of reverse areas of magnetization. In this way these large grains in MQII behave in a similar manner to large grains of sintered NdFeB and PrFeB on the initial magnetization curve [15].

MQIII

In MQIII the minimum observed in the M_{rev} curves is much broader than that seen in MQII. Folks, Street and Woodward [1] estimate that 49% of the grains in MQIII are multi-domain, a similar proportion to that found in MQII. The occurrence of a notably broad minimum in the M_{rev} curves implies that there are free domain walls present in MQIII during much of the reversal process. The fact that free domain walls are present during reversal over a large range of M_{irr} in MQIII while domain walls are only present in a small proportion of grains in MQII is most likely due to the differences in alignment of these materials. Alignment differences can create changes in the magnitude of the relative dipolar interactions which will alter the reverse domain nucleation and unpinning fields. Additionally, the die-upsetting process can produce subtle chemical and topographical differences in the grain boundary environment.

In MQII the isotropic orientation of the grains means that the dipolar interactions are relatively low when compared with the nucleation or unpinning field, which is influenced by the grain boundary environment. Thus when a domain wall nucleates on the demagnetization curve, or equivalently is unpinned from a grain boundary, the external field is already high enough to overcome dipolar interactions and the domain wall is free to sweep across the grain and to become pinned in the grain boundary on the other side or be annihilated altogether. This process guarantees that there is never a free domain wall and the reversible magnetization data displays characteristics most reminiscent of rotational processes. In MQIII the enhanced grain alignment increases the strength of dipolar interactions and decreases the nucleation field, as evidenced by the lower coercivity of MQIII (12.3 kOe) compared with MQII (18.7 kOe). Additional support for this hypothesis is found in the decreasing trend in coercivity observed in a series of samples as the degree of die-upsetting, and hence alignment, increases [16]. Thus it is reasonable to suppose that when a domain wall is nucleated or unpinned in MQIII, the external applied field is not large enough to overcome dipolar interactions and the domain wall remains within the grain, free to move and bow reversibly. It is possible to observe the effects of the domain walls in MQIII through reversible magnetization measurements. The dipolar interactions in MQIII retard the pinning and annihilation of a domain walls in a similar manner to the mechanism proposed by Taylor et al. [15] to explain the correlation observed between coercivity and initial magnetizing field in sintered NdFeB and PrFeB magnets .

The change in the nature of the dipolar interactions between MQII and MQIII type magnets has been investigated by Gavigan and Givord [17] using measurements of the temperature dependence of coercivity. In their work they concluded that dipolar interactions in MQII type magnets reinforce coercivity, because domain walls are pinned in regions which minimize the divergence of magnetization. In MQIII type magnets they found that the dipolar interactions result in a decrease in the coercivity. It is not inconceivable therefore that this change in the nature of the dipolar interactions between MQII and MQIII could affect the manifestation of free domain walls in these materials.

CONCLUSION

It has been shown that measurements of the reversible magnetization can determine the existence of free domain walls within a material during magnetization reversal. This information then puts constraints upon the possible reversal mechanism.

In MQI, the reversible magnetization measurements point to a reversal mechanism which does not involve free domain walls during reversal. The small grain size of this material and its isotropic orientation favors incoherent rotation or nucleation, although domain wall movement from grain boundary to grain boundary through well coupled grains cannot be discounted.

In MQII a similar process must occur to that operative in MQI. An intergranular phase has been observed in this material which may provide an avenue for exchange coupling of the grains which makes domain wall motion more likely. The initial magnetization curve indicates that the grain size in this material is large enough to support domain walls although only a small minority of grains are found to be multi-domain during demagnetization, which gives rise to a sharp minimum in the reversible magnetization data.

In MQIII the minimum in the reversible magnetization data is much broader than that of MQII which indicates that domain walls are free to move in a large majority of grains during reversal. This result is attributed to well-coupled, aligned grains that may support a domain wall together with dipolar interactions which are large and able to prevent a domain wall from being driven to a grain boundary and annihilated once nucleated.

For both MQII and MQIII it is possible that either nucleation of a reverse domain, most likely at the periphery of already reversed regions, or domain wall unpinning and motion is responsible for the demagnetization process. The differences between the materials is that free domain walls are present in a large number of grains in MQIII while in the majority of grains in MQII free domain walls are not present. These differences are most likely due to differences in both the nucleation fields and the dipolar interactions within grains produced upon deformation processing.

ACKNOWLEDGEMENTS

Research performed in part under the auspices of the U.S. D.O.E., Division of Materials Sciences, Office of Basic Energy Sciences under contract number DE-AC02-98CH10886.

We wish to thank R. Woodward and L. Folks for valuable discussions and for kindly supplying their measurement data on MQI and MQIII.

REFERENCES

[1] L. Folks, R. Street and R. Woodward, J. Appl. Phys. **75**, p 6271 (1994)
[2] K.-D. Durst and H. Kronmüller, J. Magn. Magn. Mater. **68**, p 63 (1987)
[3] M. Grönefeld and H. Kronmüller, J. Magn. Magn. Mater. **88**, L267 (1990)
[4] R.K. Mishra, J. Magn. Magn. Mater. **54-57**, p 450 (1986)
[5] E. E. Pinkerton and C. D. Fuerst, J. Appl. Phys. **69**, p 5817 (1991)
[6] R. Cammarano, R. Street and P.G. McCormick, J. Phys. D **29**, p 2327 (1996)
[7] D.C. Crew, PhD Thesis, The University of Western Australia (1998)
[8] D.C. Crew, P.G. McCormick and R. Street, unpublished
[9] D.C. Crew, P.G. McCormick and R. Street, J. Appl. Phys. In press.
[10] L. Folks, R. Street, G. Warburton and R. C. Woodward, Meas. Sci. Technol. **5**, p 779 (1994)
[11] M. El Hilo, K. O'Grady and R. W. Chantrell, IEEE Trans. Magn. **27**, p 4666 (1991)
[12] J.F. Herbst, Rev. Mod. Phys. **63**, p 819 (1991)
[13] J.D. Livingston, IEEE Trans. Magn. **MAG-23**, p 2109 (1987)
[14] R.K. Mishra, J. Appl. Phys. **62**, p 967 (1987)
[15] D.W. Taylor, V. Villas-Boas, Q. Lu, M.F. Rossignol, F.P. Missell, D. Givord and S. Hirosawa, J. Magn. Magn. Mater. **130**, p 225 (1994)
[16] L.H. Lewis, T.R. Thurston, V. Panchanathan, U. Wildgruber and D.O. Welch, J. Appl. Phys. **82**, p3430 (1997)
[17] J.P. Gavigan and D. Givord, J. Magn. Magn. Mater. **84**, p 288 (1990)

CHEMICAL ORDERING AND MICROSTRUCTURAL AFFECTS ON MAGNETIC PROPERTIES OF Sm_2Fe_{17} AND $Sm_2Fe_{17}N_x$

B.E. MEACHAM*, K.W. DENNIS**, R.W. McCALLUM** AND J.E. SHIELD*
*Department of Materials Science and Engineering, University of Utah, Salt Lake City, UT 84112
**Ames Laboratory, USDOE, Ames, IA 50011

ABSTRACT

The effect of chemical order and defect structure on the magnetization process in Sm_2Fe_{17} and $Sm_2Fe_{17}N_x$ has been investigated. In Sm_2Fe_{17}, chemical disorder results in the development of random anisotropy and a small degree of magnetic hardness. The anisotropy is reduced as long-range order increases. The motion of domain walls in $Sm_2Fe_{17}N_x$ is more dependent on the antiphase domain structure than on the amount of long-range order.

INTRODUCTION

The effect of microstructural features on the magnetization process has been of tremendous interest for many years. Characteristics of the grain, such as size and morphology, significantly influence the magnetic properties, as does grain boundary structure. Intragranular planar defects such as stacking faults, twins and antiphase boundaries also influence magnetic behavior by acting as pinning sites for domain walls. For example, Fe-Pd alloys, with proper processing, possess significant densities of twins, stacking faults and antiphase boundaries, which can effectively pin domain walls [1,2]. Additionally, cell boundaries between $SmCo_5$ and Sm_2Co_{17} in two-phase Sm-Co permanent magnet alloys act to pin domain walls [3]. Effective pinning of domain walls leads to improved magnetic properties, especially at elevated temperatures.

The Sm_2Fe_{17} compound attracted initial interest as a possible Fe-rich permanent magnet material. This compound, however, is characterized by planar anisotropy and a low Curie temperature [4]. Dramatic changes in the magnetic properties were induced by modifying the structure with interstitial alloying elements [5-7]. The modified compounds have uniaxial anisotropy and, in the case of $Sm_2Fe_{17}N_x$ (where x is close to 3), the Curie temperature is increased by 350°C to 470°C [6]. The improved magnetic properties make this an attractive material for permanent magnet applications. The defect structure can be modified through processing, making this an ideal system to gain a better understanding of planar defects and chemical ordering phenomena on magnetic properties.

The defect structure in Sm_2Fe_{17} arises from the presence of an order-disorder transformation between the disordered $SmFe_7$ structure and the long-range-ordered Sm_2Fe_{17} structure, which results in varied order parameters and antiphase boundary densities and configurations. The Sm_2Fe_{17} is a derivative of the $CaCu_5$-type structure in that one-third of the Ca (Sm) sites are occupied by an Fe-Fe dumbbell pair. In the Sm_2Fe_{17} structure, the replacement of the Sm atoms by the Fe-Fe dumbbells is done in a regular fashion. However, in the $SmFe_7$ ($TbCu_7$-type) structure, the placement of the Fe-Fe dumbbells on Sm sites lacks long-range order. While the Sm_2Fe_{17} is the thermodynamically-preferred structure, rapid solidification suppresses the formation of the long-range order [8-11]. The long-range order can be developed through appropriate heat treatment, allowing the order parameter and defect structure to be altered. The variation in microstructure and amount of chemical order has important ramifications on the magnetic properties. In this paper, the effects of chemical order and antiphase boundary structure on the magnetic properties of Sm-Fe and its nitrided counterpart are reported.

Mat. Res. Soc. Symp. Proc. Vol. 577

EXPERIMENT

Samples of nominal composition $Sm_{11}Fe_{89}$ were alloyed by arc melting appropriate quantities of Sm of 99.5% purity with Fe of 99.9% purity. The arc-melted ingots were then remelted by RF induction heating and rapidly solidified with a single roller melt spinning apparatus at a tangential wheel velocity of 30 m/s. Samples were annealed by sealing lightly ground powder and unground ribbons wrapped in Ta foil in quartz capsules under high purity Ar. Annealing temperatures were 600 and 800°C. All anneals were for 15 minutes, and samples were quenched in water from the annealing temperature. Samples were prepared for nitriding by further grinding so that powder passed through a 45 μm sieve. The powder was then annealed under flowing N_2 at 475°C for 4 hours.

The structural characteristics of the as-solidified, annealed and nitrided samples were examined by x-ray diffraction utilizing a Philips powder diffractometer, with the powder mounted on a zero-background sample holder. Transmission electron microscopy samples were prepared by mounting ribbon specimens on a copper oval and ion milling to perforation on a Gatan Duomill at liquid nitrogen temperatures. A JEOL 2000FXII transmission electron microscope was used. The room temperature initial magnetization curves were measured from 0 to 2000 G on a Quantum Design SQUID magnetometer.

RESULTS AND DISCUSSION

X-ray diffraction was utilized to evaluate the phase purity of the as-solidified and annealed samples, and to monitor the nitrogenation of the samples. The as-solidified material was essentially single-phase, with peaks indexing to the $SmFe_7$ structure, meaning no long-range order was detected. X-ray diffraction of the nitrided samples also confirmed that the Sm-Fe-N interstitial phase was formed.

The presence and qualitative amount of long-range order was determined by electron diffraction, which is more sensitive to superstructure than x-ray diffraction. The electron diffraction patterns of the as-solidified $Sm_{11}Fe_{89}$ revealed that some degree of order does exist upon solidification, as indicated by the superlattice peaks marked "A" and "B" (Figure 1(a)). These correspond to the $(10\bar{1}1)$ and $(20\bar{2}2)$ reflections for the Sm_2Fe_{17} structure. Qualitatively, the intensity of the superlattice reflections relative to the fundamental $(30\bar{3}3)$ reflection increases with annealing temperature, indicating that the long range order increases upon annealing at 600 and 800°C (Figure 1(b) and (c)). It is interesting that the superlattice reflections are streaks rather than discrete spots for the sample annealed at 600°C, and to a lesser extent the sample annealed at 800°C. The streaking arises from the presence of planar defects normal to the direction of the streak.

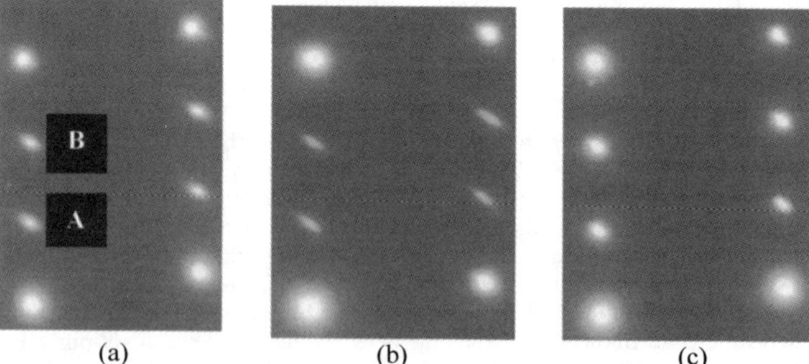

Figure 1: Selected area electron diffraction patterns of the same zone axis showing the superlattice reflections for (a) as-solidified and samples annealed at (b) 600°C and (c) 800°C.

The annealing treatments also had significant affects on the microstructure of the $Sm_{11}Fe_{89}$ samples. The microstructure of the as-solidified $Sm_{11}Fe_{89}$ sample consisted of more or less equiaxed grains on the order of 500 nm (Figure 2(a)). The internal microstructure lacked a significant number of defects, with only isolated features resembling dislocations. No features resembling antiphase boundaries were observed. This indicates that there is no significant development of long-range-order upon solidification. Upon annealing, however, a defect structure consisting of antiphase domains develops (Figure 2(b) and (c)). In the sample annealed at 600°C, the defect density is rather high. The APBs are more or less parallel and perpendicular to the c-axis, which is the direction of ordering. Further long-range order develops upon annealing at 800°C, leading to a decrease in APB density. In this case, however, the APB configuration is different than it is in the sample annealed at 600°C. APBs, while still normal to the c-axis, meander in other directions as well. The decrease in the number of APBs in the 800°C sample explains the less extensive streaking observed in the electron diffraction patterns. The effect of the chemical order and APB structure on the magnetic properties was investigated by measuring the initial magnetization curves. The samples were in a state of thermal demagnetization. The grains in the nitrided samples in all cases were in a multidomain configuration since the grain size exceeds the single domain limit. The single domain limit, estimated using the Kittell formula, is approximately 320 nm. The unnitrided sample is expected to have a rather large single domain limit and a relatively thick wall due to the planar anisotropy.

The initial magnetization curves revealed that the chemical order and associated APB defects affected the motion of domain walls in the $Sm_{11}Fe_{89}$ sample. The magnetization process became less difficult with increasing long-range order, as indicated by the increased slopes and initial susceptibility in the M/M_s vs. H curves (Figure 3). The coercivity also decreases with annealing, from 290 Oe for the as-solidified, to 190 Oe for the sample annealed at 600°C, to 145 Oe for the sample annealed at 800°C.

Because of the basal plane easy magnetization, in principle no in-plane anisotropy exists and the magnetization process should proceed almost without hindrance. This also implies virtually no coercivity. That the initial magnetization and coercivity vary with the chemical order, then, is unexpected. It appears that the presence of chemical disorder induces an in-plane random anisotropy, making magnetization more difficult than in the ordered state.

The nitrided $Sm_{11}Fe_{89}$ displayed much different behavior. The initial magnetization curves are shown in Figure 4. The magnetization M is normalized to the magnetization at 2 T. The overall initial susceptibilities are much lower than the $Sm_{11}Fe_{89}$ due to the increased anisotropy of the nitrided compound. Also, since the applied field is low compared to the anisotropy field, the magnetization process is considered to be in the regime dominated by domain wall motion. In this case, the motion of domain walls is most difficult in the sample annealed at 800°C, and the as-solidified and 600°C-annealed samples display similar behavior. While the sample annealed at 800°C possesses the most long-range order and thus fewest APBs, the configuration of the APBs hinders domain wall motion. In the case of the sample annealed at 600°C, the APBs are predominantly planar and normal to the c-axis, as seen in Figure 2(b). The domain walls are perpendicular to these defects, resulting in less-effective pinning because of the limited cross-sectional interaction. The APBs in the sample annealed at 800°C, on the other hand, are not confined to planes perpendicular to the c-axis, as seen in Figure 2(c). The portions of the APBs not orthogonal to the domain wall provide more effective pinning, making domain wall motion more difficult. The sample annealed at 800°C also has a higher coercivity, 2200 Oe as compared to 1540 Oe for the sample annealed at 600°C. It should be noted that this coercivity is the coercivity of a minor loop, as the maximum field was 2 T.

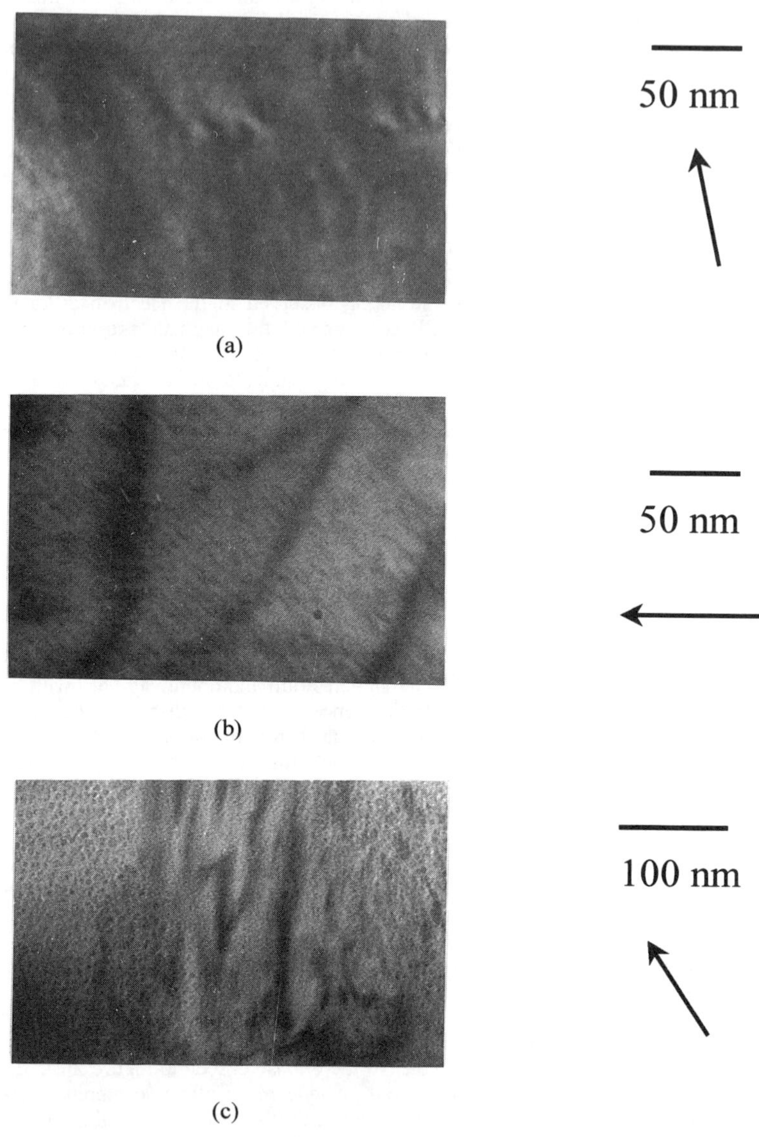

Figure 2: Bright field transmission electron micrographs displaying the APB structure of (a) as-solidified and samples annealed at (b) 600°C and (c) 800°C. All micrographs were taken at **g**=[30$\bar{3}$3], with the direction indicated by the arrow.

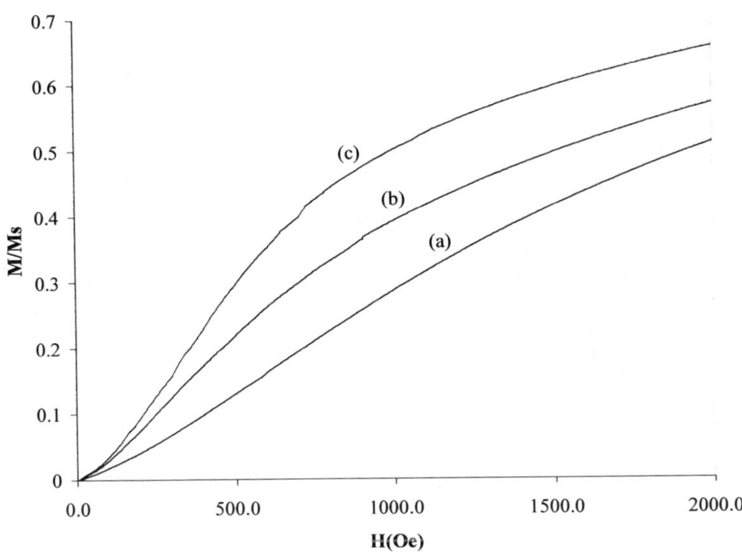

Figure 3: Initial magnetization curves for $Sm_{11}Fe_{89}$ (a) as-solidified and annealed at (b) 600°C and (c) 800°C.

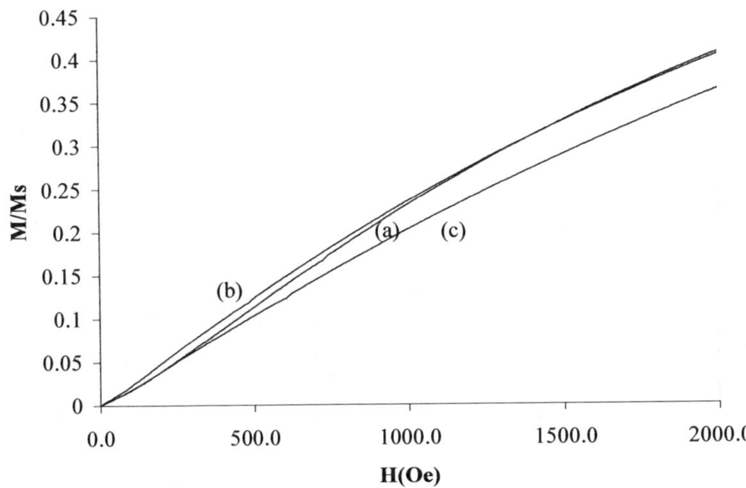

Figure 4: Initial magnetization curves of nitrided $Sm_{11}Fe_{89}$ (a) as-solidified and annealed at (b) 600°C and (c) 800°C.

CONCLUSIONS

The amount of long-range order and the antiphase boundary structure affect the magnetization process in Sm_2Fe_{17} and $Sm_2Fe_{17}N_x$. The amount of long-range order and antiphase domain structure were modified by annealing treatments of initially disordered melt spun $Sm_{11}Fe_{89}$ samples. The as-solidified material lacked significant long-range order and thus did not possess an APB structure. Annealing at 600°C developed long-range order and an APB structure of defects perpendicular to the c-axis. Annealing at 800°C further developed long-range order. The APB density decreased but became less confined to planes perpendicular to the c-axis.

The initial magnetization process was related to the amount of long-range order and APB structure in both $Sm_{11}Fe_{89}$ and its nitrided counterparts. In $Sm_{11}Fe_{89}$, the significant amount of chemical disorder produced random anisotropy on the basal plane, which was reduced significantly by inducing long-range order. In the nitrided samples, the APB structure of the sample annealed at 800°C was found to be more effective at pinning domain walls than the APB structure of the sample annealed at 600°C.

ACKNOWLEDGMENTS

The authors are grateful to the National Science Foundation's Division of Materials Research for support of this work (Grant No. 9714946). Work at Ames Laboratory was supported by the Director of Energy Research, Office of Basic Sciences, U.S. Department of Energy under contract no. W-7405-ENG-82.

REFERENCES

1. T. Klemmer, D. Hoydick, H. Okumura, B. Zhang and W. A. Soffe, Scripta Met. Mater. **33**, 1793 (1995).
2. B. Zhang and W.A. Soffa, Phys. Stat. Sol. (a) **131**, 707 (1992).
3. R. Kumar, J. Appl. Phys. **63**, R13 (1988).
4. W.E. Wallace, Prog. Solid State Chem. **16**, 127 (1985).
5. Zhong Xia-ping, R.J. Radwanski, F.R. de Boer, T.H. Jacobs and K.H.J. Buschow, J. Magn. Magn. Mat. **86**, 333 (1990).
6. J.M.D. Coey and Hong Sun, J. Magn. Magn. Mater. **87**, L251 (1991990).
7. Hong Sun, J.M.D. Coey, Y. Otani and D.P.F. Hurley, J. Phys. Condens. Matter **2**, 6465 (1990).
8. M. Katter, J. Wecker and L. Schultz, J. Appl. Phys. **70**, 3188 (1991).
9. F.E. Pinkerton and C.D. Fuerst, Appl. Phys. Lett. **60**, 2558 (1992).
10. H.T. Kim, Q.F. Xiao, Z.D. Ahang, D.Y. Geng, Y.B. Kim, T.K. Kim, and H.W. Kwon, J. Magn. Magn. Mater. **173**, 295 (1997).
11. J.E. Shield, C.P. Li and D.J. Branagan, J. Magn. Magn. Mater. **188**, 353 (1998).

Part VIII

Thin Film Permanent Magnets

PERMANENT MAGNETISM IN EXCHANGE-COUPLED NANOCOMPOSITES

R. SKOMSKI, J. P. LIU, D.J. SELLMYER

Department of Physics and Astronomy and Center for Materials Research and Analysis, University of Nebraska, Lincoln, NE 68588-0113, U.S.A.

ABSTRACT

The magnetic properties of permanent magnetic nanocomposites are investigated both experimentally and by model calculations. Rare-earth transition-metal films and nanocomposite thin films containing a $L1_0$ hard phase (PtFe) have been prepared by plasma sputtering and heat treatment. By choosing suitable heat-treatment processes a nearly ideal nanostructure and high energy products of up to about 50 MGOe have been achieved for the Fe-Pt system. The behavior of the nanocomposites is discussed in terms of a model applicable to nearly ideal magnets, where there are no pronounced shoulders in the hysteresis loops. Analytic expressions for M(H) are obtained for a number of cases, and it is discussed how the loops depend on the micromagnetic parameters of the involved phases. A novel aspect of magnetic nanostructures is that very small magnetic grains exhibit very large energy products, even if they are intrinsically soft-magnetic.

INTRODUCTION

Since the discovery of experimental evidence for remanence enhancement due to intergranular exchange between magnetically hard and soft phases [1], many experimental and theoretical investigations have been devoted to this fascinating area of magnetism [2-20]. Aside from potential applications in magnetic recording, interest in these nanocomposites has been sparked by the possibility of improving the energy product of hard-soft composites beyond that of the hard phase. For example, in modern rare-earth permanent magnets such as $Nd_2Fe_{14}B$ and Sm_2Co_{17}, the transition-metal sublattice is responsible for the saturation magnetization and ensures a high Curie temperature, whereas the rare-earth sublattice determines the magneto-crystalline anisotropy. However, the number of suitable single-phase compounds is limited, and there is a need for optimized nanocomposites.

Quantitative estimates for energy products achievable in two-phase nanocomposites were first reported by Skomski and Coey in 1993 [2]. As emphasized from the beginning [2, 21], the main aim of that calculation was to establish a reasonable upper limit for the energy product achievable in composites consisting of materials available at a given time. The point is that a further improvement of the intrinsic performance of hard-magnetic phases is by basic electronic-structure factors, such as the number of 3d and 4f electrons available per unit volume [2, 18]. The predictions made in [2] have therefore remained a realistic challenge during the last few years [3, 5, 8, 10-14, 16].

As clearly stated in [2], the predicted high energy products refer to ideally structured systems. In reality, there are always morphological inhomogenities which reduce the nucleation field and therefore the energy product. However, this problem is not restricted to two-phase nanocomposites. As pointed long ago by Aharoni [22], nanoscale inhomogenities reduce the

Mat. Res. Soc. Symp. Proc. Vol. 577 © 1999 Materials Research Society

coercivity of real materials and therefore explain Brown's paradox according to which the coercivity of ideally structured magnets is never achieved in practice. In 1987, Kronmüller [23] reported analytic calculations which may be considered as an early version of the concept of exchange-spring magnets established by Kneller and Hawig in 1991 [4]. Only recently, this problem has been reformulated in terms of a micromagnetic localization problem [24, 25]: disorder gives rise to a localization of the nucleation mode and thereby reduces the coercivity. In any case, recent experiments on thin-film magnets [11] confirm the predictions of [2] and yield energy products going beyond that expected for pure hard phases. Liu *et al.* [11] produced two-phase nanocomposites PtFe-$PtFe_3$ type nanocomposites with energy products of order 400 kJ/m^3 [50 MGOe]. In spite of the only moderate intrinsic performance of the hard-magnetic PtFe phase, this is close to the present world-record energy product in small-scale industrial Nd-Fe-B magnets (443 kJ/m^3).

Here we summarize some recent experimental results and present a semiquantitative theoretical explanation of the observed hysteresis loops. Emphasis is on a basic physical understanding of the magnets, requirements for advanced exchange-coupled permanent magnets, and conceivable future developments.

BASIC THEORETICAL CONCEPTS

A key problem in micromagnetism is to interprete and predict experimental hysteresis loops from the atomic composition and morphology of the magnet. The starting point of the consideration is the micromagnetic (free) energy functional. For simplicity, we restrict ourselves to the first uniaxial anisotropy constant K defined by the energy expression $E/V = K \sin^2\theta$, where θ is the angle between the crystal's c axis and the magnetization direction. In inhomogeneous magnets, the total magnetic energy is [2]

$$E = \int \left(A \left(\nabla \frac{\mathbf{M}}{M_o} \right)^2 - K \frac{(\mathbf{n}.\mathbf{M})^2}{M_o^2} - \mu_o \mathbf{M}.\mathbf{H} - \mu_o \mathbf{M}.\mathbf{H}_d(\mathbf{M}(\mathbf{r}))/2 \right) d\mathbf{r} \qquad (1)$$

Here A is the exchange stiffness, $\mathbf{M}$ denotes the magnetization, $\mathbf{H} = H\,\mathbf{e}_z$ is the external magnetic field, $\mathbf{H}_d$ denotes the magnetostatic self-interaction field, and $\mathbf{n}$ is the unit vector of the local easy-axis direction. The magnitude of the magnetization $|\mathbf{M}|$, that is the spontaneous magnetization, is denoted by M_o. The parameters A, M_o, K, and $\mathbf{n}$ exhibit a local variation denoted by the indices s (for soft) and h (for hard). The local and global minima of E as a functional of the 'spin configuration' $\mathbf{M}(\mathbf{r})$ determine the minor and major hysteresis loops.

Analytic and numerical approaches

Advanced computational methods are now able to reproduce and predict the hysteresis loops of particles whose size is of order 50 nm [26-28]. This size limitation largely originates from the necessity to divide the individual grains into subgrains, because the magnetization inside a nanocrystallite is in general nonuniform. This has to be contrasted to earlier model calculations [29, 30] where the magnetization in each grain is assumed to be uniform. Recently, Sabiryanov and Jaswal [7, 8] have calculated micromagnetic properties directly from the electronic structure of the composites, but this approach becomes extremely demanding when the volume of the considered regions is larger than 1 nm^3. However, this method is certainly of great interest for the study of interphase phenomena.

Permanent magnetism and equilibrium

The nonequilibrium character of permanent magnetism has a number of important consequences. This is seen most clearly by considering isotropic permanent magnets, produced by mechanical alloying or meltspinning [6, 15]. These magnets are random-anisotropy magnets, because they are composed of randomly oriented nanocrystallites. In a strict sense, isotropic materials are nonferromagnetic random-anisotropy spin glasses, because their ground state does not exhibit long-range ferromagnetic order. However, as emphasized in [31], the question of a random-field [32] or random-anisotropy [33] spin-glass type ground-state order is of minor importance in permanent magnetism. The virgin curve (initial curve) of a magnets reflects the ground-state order, because thermal demagnetization leaves the magnet in a low-lying energy minimum, but hysteresis means that the magnet is kept in an artificially maintained nonequilibrium state. A typical example is isotropic $Sm_2Fe_{17}N_3$, where coercivities in excess of 4 T [40 kOe] have been achieved [34].

The strict separation between equilibrium and nonequilibrium properties is of great importance for the interpretation of magnetic domains [18]. On the one hand, the absence of magnetic domains in a magnetic micrographs does not necessarily indicate a single-domain ground-state, because the magnet may be in a nonequilibrium spin configuration $\mathbf{M}(\mathbf{r})$. On the other hand, a single-domain particle does not necessarily reverse by coherent rotation, because the single-domain character of a particle does not exclude incoherent rotation, either delocalized (curling) or localized due to morphological defects. Figure 1 shows this relationship for a hard-magnetic particle.

For the ferromagnetic materials of interest, the transition from coherent rotation to curling occurs nearly invariably at particle radii of about 10 nm, independently of the magnetocrystalline anistropy. This striking but well-established effect originates from the fact that the coherence radius R_{coh}, below which the onset of magnetic reversal is coherent, is of order 5 $(A/\mu_o M_s^2)^{1/2}$ [18, 35]. For typical materials of interest here, A $\approx$ 10 pJ/m and $\mu_o M_s \approx$ 1.5 T, so that $R_{coh} \approx$ 10 nm. Above the coherence radius, the magnetostatic interaction dominates the exchange interaction, and the curling-type spin misalignment visible in Fig. 1(c) indicates a certain degree of magnetic flux closure. In a more general sense, the coherence

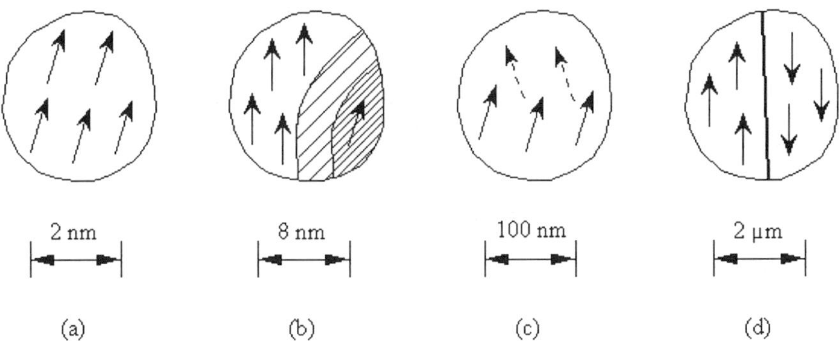

Fig. 1 Generic model of magnetic reversal: (a) coherently reversing small particle, (b) small particle where a defect gives rise to localized nucleation, (c) defect-free single-domain particle whose magnetic reversal starts with delocalized incoherent nucleation (curling), and (d) schematic ground state of a two-domain particle.

length reflects the competition between lowest-order relativistic interactions ('electrostatic' exchange) and higher-order relativistic (or magnetostatic) interactions. A 'relativistic' estimate of the coherence length is $l_o = a_o/\alpha$ [25], where $a_o = 0.5929$ Å is the Bohr radius and $\alpha = 1/137$ is Sommerfeld's fine-structure constant, which describes the strength of the electromagnetic interaction. Note that the numerical value $l_o = 7.25$ nm is very close to the exact coherence radii.

Magnetic nanostructures

The picture outlined in Fig. 1 becomes more complicated when the granular structure of the nanocomposites is taken into account. The proper distinction between equilibrium and nonequilibrium magnetization states is particularly important in the context of interaction domains [36-38], where the single-domain and nonequilibrium properties are often difficult to distinguish [24]. For example, for the Pr-Co structures considered in [39], the domain size, about 200 nm, is approximately ten times larger than the grain size (about 20 nm). However, this observation does not mean that the ground-state domain size is equal to 200 nm: there is a possibility that the true equilibrium domains are larger or smaller than 200 nm.

Figure 2 shows some frequently encountered loop shapes. The interesting point is that a superposition of hard and soft loops yields characteristic shoulders in the hysteresis loops. However, the shoulder smoothes on intergranular exchange [2, 24]. The limit of strong exchange, realized for example by very small soft inclusions in a hard matrix, corresponds to the 'virtual crystal' approximation in binary-alloy band structure theory. This limit, which has

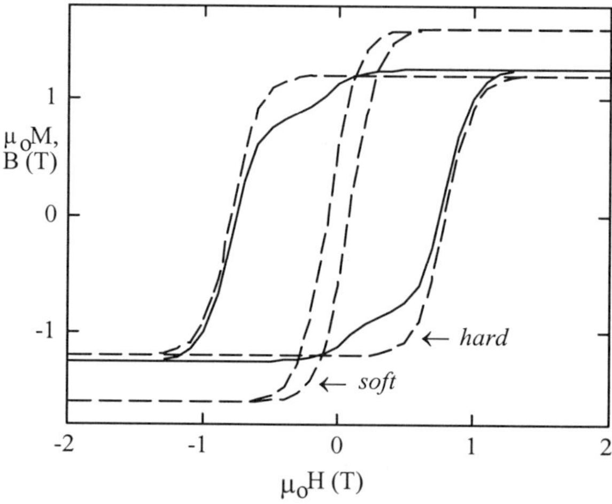

Fig. 2 Schematic inflection-type 'two-shoulder' hysteresis loop: (a) two-phase loop (solid line) as a superposition of two hard and soft loops (dashed lines), and (b) smoothing of of the original two-phase loop (dashed line) by exchange interactions (solid line).

been explored in [2], amounts to the use of volume averages such as $<M_o> = f_s M_s + f_h M_h$ and $<K> = f_s K_s + f_h K_h$ to calculate the energy product. Here we extend that work to misaligned grains, where n exhibits weak spatial fluctuations associated with the magnet's granular structure. This work is related to an earlier work by Chudnovsky *et al.* [40], who considered the ground state of a partially ordered magnet but did not attempt to calculate hyesteresis loops. It is also related to early investigations of the hysteretic properties of amorphous random-anisotropy magnets [41].

Weakly textured magnets

In this section we focus on an illustrative analytic treatment of 'nearly ideal' magnets, where the derivations from the ideal rectangular loop shape can be treated perturbatively. This approximation is reasonable for the advanced hard magnets in consideration but cannot be used when the coercivity is very low due to surface effects and bulk inhomogenities, that is if there is a pronounced shoulder in the loop. This amounts to two major assumptions. First, the magnet is only weakly textured, that is the hard nanocrystallites exhibit only weak deviations from the ideal c-axis alignment. Secondly, the influence of large soft grains is ignored. Furthermore, we restrict ourselves to the M-H loop in the second quadrant; explicit energy product calculations will be published elsewhere [18].

The main theoretical tool applied in this paper is an inverse operator expansion [42]. Furthermore, we make the common [40, 43] but rather crude approximation that the magnetostatic selfinteraction can be incorporated into the external field **H** [42]. In the case of nearly ideal hysteresis loops, we can restrict ourselves to small deviations from ideal spin alignment. The magnetization is written as $\mathbf{M} = M_o (1-\mathbf{m}^2)^{1/2} \mathbf{e}_z + M_o \mathbf{m}$, where M_o is the magnitude of the magnetization and $\mathbf{m}.\mathbf{e}_z = 0$. For small perpendicular spin projections **m** we obtain

$$\mathbf{M}(\mathbf{r}) = M_o(\mathbf{r}) \left(1 - \frac{\mathbf{m}(\mathbf{r})^2}{2}\right) \mathbf{e}_z + M_o(\mathbf{r})\, \mathbf{m}(\mathbf{r}) \qquad (2)$$

Similarly, we can write $\mathbf{n} = (1-\mathbf{a}^2/2)\, \mathbf{e}_z + \mathbf{a}$, where $\mathbf{a}(\mathbf{r})$ is the small local easy-axis misalignment. Note that the volume average $<\mathbf{a}^2>$ can be interpreted as a texture parameter: $<\mathbf{a}^2> = 0$ means a common c-axis throughout the magnet, whereas $<\mathbf{a}^2> > 0$ indicates some grain misalignment. The longitudinal magnetization component can then be written as $M_z(\mathbf{r}) = M_o(\mathbf{r})(1-\mathbf{m}(\mathbf{r})^2)$, so that

$$E = \int \left(A\, (\nabla \mathbf{m})^2 + K\, (\mathbf{m} - \mathbf{a})^2 + \frac{\mu_o}{2} M_o\, H\, \mathbf{m}^2 \right) d\mathbf{r} \qquad (3)$$

Minimizing this expression leads to the differential equation

$$- 2\, \nabla(A \nabla \mathbf{m}) + 2\, K\, \mathbf{m} + \mu_o\, H\, M_o\, \mathbf{m} = 2\, K\, \mathbf{a} \qquad (4)$$

where $A(\mathbf{r})$, $K(\mathbf{r})$, $M_o(\mathbf{r})$, and $\mathbf{a}(\mathbf{r})$ are random quantities.

In the simplest case, one considers a textured but chemically homogeneous *single-phase* materials where the only random quantity is **a**. This limit has been investigated for example in the context of random ferro- and antiferromagnetism [40, 43]. In particular, the equation $- \nabla^2 \phi$

$+ \kappa^2\phi = b$, where $<b> = 0$ and $<b(\mathbf{r})b(\mathbf{r}')> = b_o^2 \exp(-|\mathbf{r}-\mathbf{r}'|^2/R_o^2)$, has dimensionality dependent strong-coupling solutions [43]. Introducing the wall-width parameter $\delta = \sqrt{A/K}$ and the anisotropy field $H_o = 2K/\mu_o M_o$ we obtain the strong-coupling equations

$$<M_z> = M_o \left(1 - \frac{<\mathbf{a}^2> R_o^2}{8\,(1+H/H_o)\,\delta^2}\right) \qquad (d = 2) \qquad (5a)$$

$$<M_z> = M_o \left(1 - \frac{\sqrt{\pi}<\mathbf{a}^2> R_o^3}{16\sqrt{1+H/H_o}\,\delta^3}\right) \qquad (d = 3) \qquad (5b)$$

which break down when the reverse field – H approaches H_o. Equation (5a) describes magnetic films whose thickness is smaller than the 'grain radius' R_o. Figure 3 shows the strong-coupling predictions for typical permanent-magnet materials constants ($\delta = 2$ nm, $<a^2> = 0.4$, and $\mu_o H_o = 5$ T) and $R_o = 2$ nm. Note d = 4 is a marginal dimension below which the deviation from $M_z = M_o$ scales as $(1+H/H_o)^{(d-4)/2}$. We see that the anisotropy randomness, $<\mathbf{a}^2>$, reduces the magnetization whereas the intergranular exchange, A, tries to outweigh the magnetization deterioration due to the nonzero $<\mathbf{a}^2> > 0$. However, since the coercivity needs to be larger than about $M_o/2$, the validity of Eq. (5) is limited to very small grains where R_o is at most 3 nm.

To investigate the more realistic weak-coupling limit we obtain, after some calculation [42],

$$<\mathbf{m}^2> = \frac{4K^2}{(2K+\mu_o M_o H)^2} <\mathbf{a}^2> + \frac{8K^2 A}{(2K+\mu_o M_o H)^3} <\mathbf{a}.\nabla^2\mathbf{a}> + \ldots \qquad (6)$$

In the absence of exchange interactions the remanence is given by $M_r = M_o(1-<\mathbf{a}^2>/2)$. To evaluate the second average on the right-hand side of Eq. (6) we take into account that $\mathbf{a}.\nabla^2\mathbf{a} = \nabla(\mathbf{a}.\nabla\mathbf{a}) - (\nabla\mathbf{a})^2$. The volume average of the expression $2\nabla(\mathbf{a}.\nabla\mathbf{a}) = \nabla^2(\mathbf{a}^2)$ can be written as a surface integral $\int \nabla(\mathbf{a}^2).d\mathbf{S}$ and is equal to zero, because the magnet is macroscopically homogeneous. As a consequence, $<\mathbf{a}.\nabla^2\mathbf{a}> = -<(\nabla\mathbf{a})^2>$. This average scales as $<\mathbf{a}^2>/R^2$, where R is the grain radius. To obtain a numerical estimate we assume that the grains exhibit the texture profile $\mathbf{a}(\mathbf{r}) = \mathbf{a}_o\,(1 - r^2/R^2)$. Then, in d dimensions, $<(\nabla\mathbf{a})^2> = d(d+4)<\mathbf{a}^2>/2R^2$, and we obtain the remanence

$$M_r = M_o\left(1 - \frac{<\mathbf{a}^2>}{2}\left(1 - \frac{d\,(d+4)\,A}{2\,K\,R^2}\right)\right) \qquad (7)$$

In this limit, the power-law behavior of M(H) is independent of the dimension d.

The addition of some soft magnetic material having a high saturation magnetization may enhance the magnetization and improve the energy product. As outlined in [2], there is an ideal exchange coupling between hard and soft phases when the radius R_s of the soft regions is smaller than the Bloch-wall width $\delta_h = \pi_h\,(A_h/K_h)^{1/2}$ of the hard phase. When the size of the soft regions is comparable to that of the hard regions, then the use of average materials parameters overestimates the energy product, because the soft regions switch independently of the hard matrix.

Restricting ourselves to weak anisotropy inhomogenities [42] we write $K(\mathbf{r}) = <K> + \Delta K(\mathbf{r})$, expand M_z with respect to ΔK, and obtain after short calculation the remanence

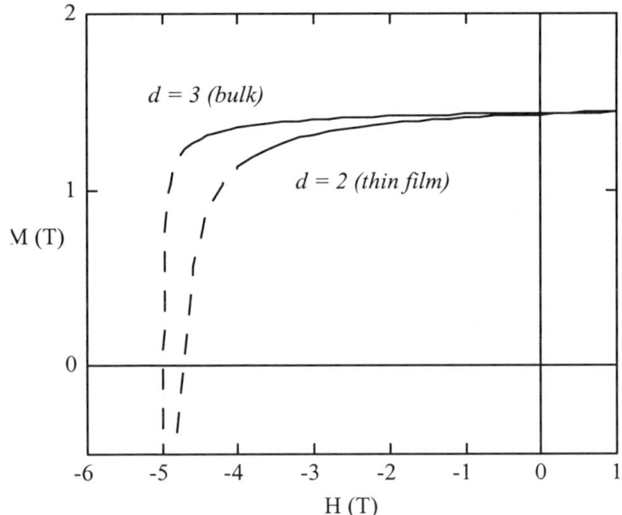

Fig. 3 Strong-coupling hysteresis loops for two and three-dimensional magnets.

$$M_r(H{=}0) = M_o\left(1 - \frac{<\mathbf{a}^2>}{2}\left(1 - \frac{d(d+4)A}{2<K>R^2}\left(1 + \frac{<\Delta K^2>}{<K>^2}\right)\right)\right) \quad (8)$$

Since $<\Delta K^2>/<K>^2 \approx f_s/(1 - f_s)$, Eq. (8) does not work for large volume fractions of the soft phase. Note that a moderate reduction of the anisotropy, for example associated with the temperature dependence of K, has actually a positive effect on the remanence and may *improve* the loop shape. Of course, Eq. (8) does not apply to very extended soft regions. Furthermore, we have neglected that the grain boundaries are sharp rather than smoothed over the whole grain. This effect modifies the effective intergranular exchange and amounts to the reintroduction of another length scale. The reason is that $\nabla\mathbf{a}$ divergenes at sharp grain boundaries, so that one has to introduce a cutoff parameter which turns out to be of order $\sqrt{A/K}$ [42].

EXPERIMENTAL EXAMPLES

In practice, the good overall hard-magnetic properties of $Nd_2Fe_{14}B$ have led to an emphasis on this material, and much attention has been paid to the microstructure of Nd-Fe-B magnets and their use in nanocomposites [1, 5, 6, 17, 19, 37, 44]. However, due to the complicated crystal structure of $Nd_2Fe_{14}B$ it is quite difficult to grow well-defined Nd-Fe-B based nanostructures such as $Nd_2Fe_{14}B$:Fe_3B and $Nd_2Fe_{14}B$:Fe. (Here we use notations such as PrCo:Co to represent the composites after annealing and the notations like PrCo/Co to represent the as-deposited multilayers.) Furthermore, neither the Curie temperature nor the magnetocrystalline anisotropy of $Nd_2Fe_{14}B$ reach the high values of e. g. $SmCo_5$ and $Sm_2Fe_{17}N_3$. This has stimulated work on the materials considered here, which have slightly inferior starting properties. Examples are $L1_0$-based composites such as FePt/Fe [10, 11] and 1:5 based composites such as $SmCo_{3+x}$:$Fe_{1-x}Co_x$ [3, 8, 14, 16] and $PrCo_3$:Co [12, 13].

For an effective intergranular exchange exchange coupling, the dimension of the soft phase should not be bigger than twice the domain-wall thickness of the hard phases [2]. By dimension here we mean either the grain size or the shortest distance to a neighboring hard region. Though theoretical calculations did not give a topological limitation to the hard phase, the dimension of the hard phase also should be small in order to increase the fraction of the soft phase to raise the remanent magnetization. Experimental attempts by mechanical alloying and rapid quenching have resulted in energy products only up to about 20 MGOe. This is because these methods have the technical difficulty in controlling the nanostructures, especially the grain size and the texture. This difficulty stimulated us to try to realize the ideal nanostructure and therefore better intergrain exchange coupling by preparing the nanocomposite materials by sputtering and subsequent heat treatment. The as-deposited films were then heat treated in various processes including rapid thermal annealing (RTA). In this process, heating and cooling rates up to 100 K/s were employed. Details of the preparation conditions can be found in our previous publications [10-13, 39].

The magnetization loops of the samples considered in this section were measured by an alternating gradient force magnetometer (AGFM) and a superconducting quantum interference device (SQUID) magnetometer, and the nanostructure of the films was studied by a transmission electronic microscope (TEM). Magnetic force microscopy (MFM) and atomic force microscopy (AFM) were used to observe the domain structures and the surface morphology. Finally, X-ray diffraction (XRD), transition electron microscopy (TEM) and energy-dispersive X-ray spectroscopy (EDX) have been used to identify the crystal structures of the phases in the nanocomposites studied.

The sputtering parameters in this investigation (including the argon pressure, the power rates of sputtering guns and the targets-to-substrate distance) were chosen to produce films with maximum coercivity. The thickness of each film was calculated by measuring the mass gain upon deposition and by low-angle X-ray diffraction.

Nanocomposites containing hard-magnetic $L1_0$ alloys

For the iron-platinum system, the multilayers were deposited by sputtering alternatively the Fe layers and the Pt layers onto the substrates at room temperature [11]. Subsequent heat treatment was necessary to form the magnetically hard phase, the equal-atomic $L1_0$ (fct) FePt phase.

In [11], energy products of about 420 kJ/m^3 (52.8 MGOe) have been obtained in fine-grained Pt-Fe thin films. In this structure, FePt and and an fcc phase close to the composition $PtFe_3$ act as hard and soft phases, respectively, and there is considerable exchange coupling in the material. Essentially, soft-magnetic grains having radii of about 5 nm are embedded in a matrix of hard grains whose radii are of order 50 nm. For large iron contents inflection-type hysteresis loops were obtained, in contrast to the nearly ideal hysteresis loops in the regime where the energy product was largest. Note, however, that the ratio $M_r/M_o \approx 0.8$ indicates only moderate texture ($\langle \mathbf{a}^2 \rangle \approx 0.4$).

Figure 4 shows the nanostructure of the sample with energy product 52.8 MGOe. A very small grain size of the fcc soft phase in the range from 5-8 nm was obtained after the RTA treatment. In the tetragonal FePt phase, the wall thickness is about 4 nm [45], which means

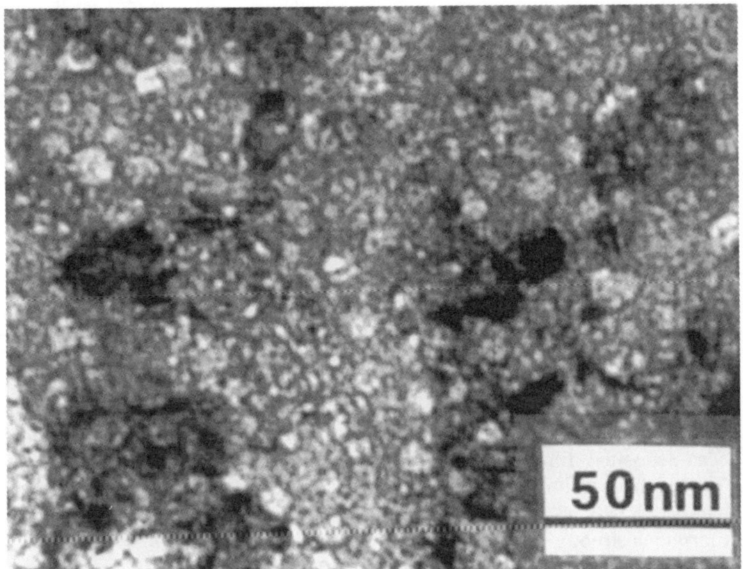

Fig. 4 TEM bright-field image of a heat-reated iron-platinum sample. The "white grains" having a size of about 10 nm are soft grains embedded in a hard matrix of FePt grains having a lateral grain size from 50 to 100 nm.

that the grain size of the soft phase in the exchange-coupled FePt system should be smaller than 8 nm. The fine soft-phase grains were embedded in the matrix of the fct hard phase whose grain size is in the range of 50-100 nm. In single phase FePt films, the as-deposited films had (111) texture. When the annealing temperature was higher than 500 C, the orientation became (001), that is perpendicular anisotropy. In case of the nanocomposites, the situation is more complicated, and the determination of the texture turned out to be more difficult. However, a similar change in anisotropy from in-plane to out-of-plane with increasing annealing temperature was observed. This shows that the out-of-plane anisotropy was not caused by the film growth preference but was developed after sufficient annealing.

Nanocomposites containing hard-magnetic 1:3 alloys

The production of the $PrCo_{3+\delta}$:Co and $SmCo_{3+\delta}$:Co films ($\delta \approx 0.5$) started with the deposition of multilayers in a multiple-gun DC and RF plasma sputtering system. The purities of all the target materials were higher than 99.9 at.%. Glass or silicon substrates were used in this investigation, and the substrates were attached to a rotating table that was cooled by water during sputtering. Furthermore, Cr underlayers with thicknesses from 50 to 70 nm and coverlayers having a thickness of 10 nm were deposited. The base pressure in the chamber was 2×10^{-7} Torr. Flowing high-purity argon gas was used for sputtering and the pressures were varied from 5 to 30 mTorr.

In the $PrCo_x$:Co and $SmCo_x$:Co systems, the hcp Co phase was detected in the heat-

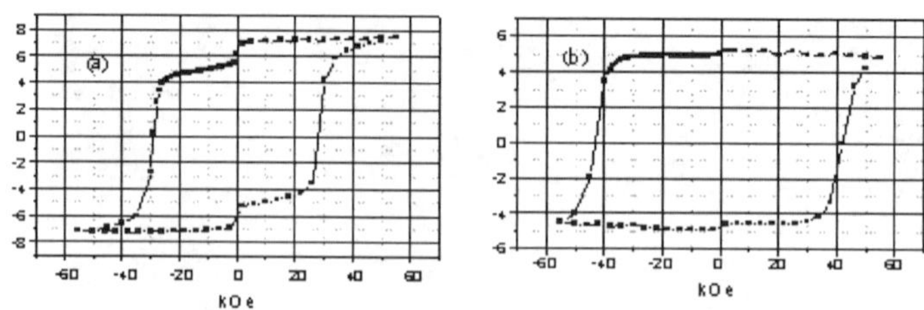

Fig. 5 The Sm-Co films treated with different processes: (a) 500°C RTA and (b) 500°C RTA plus 500°C 20 min annealing.

treated samples. The EDX spectra of a $PrCo_x$:Co sample measured in two adjacent grains [39] show that there are two phases, the hard $PrCo_x$ phase and soft Co phase existing in the composite. No third phase was found. In the $SmCo_{3.5}$:Co system, high-resolution TEM showed the hard phase to be a metastable $SmCo_3$ phase with the hexagonal $D0_{19}$ structure [12, 39], as opposed to the bulk equilibrium $SmCo_3$, which crystallizes in the rhombohedral $PuNi_3$ structure.The as-deposited films generally behaved like soft magnetic materials except for $SmCo_{3.5}$/Co in which coercivity up to 1 kOe was measured after the deposition. The soft magnetic behavior for $SmCo_{3.5}$/Co multilayers can be attributed to the fact that the as-deposited films have amorphous structure.

To optimize the loop shape, that is to reduce the size of the kinks (shoulders), a multi-step heat treatment have been used (Fig. 5). This is because if the as-deposited sample were treated with standard annealing directly, it could cause excessive grain growth. This would destroy the intergrain exchange coupling. However, sufficient annealing was necessary for high coercivity. To solve this dilemma, a special rapid thermal annealing process was first applied to the samples. In this process, annealing at above 500 °C can be done in 5 seconds; the heating rate can be about 200 K per second. Another effect on the exchange coupling is the grain boundary conditions. In view of metallurgical processes, more sufficient annealing leads to cleaner grain boundaries, which, as we have seen, helps the intergrain exchange coupling. In $SmCo_x$:Co systems, huge coercivities of more than 4 T [40 kOe] were obtained [39].

A NOVEL CONCEPT

The importance of magnetic nanocomposites is not restricted to the quantitative improvement of the performance of single-phase bulk materials. This is seen for example in the context of advanced recording media, where specifically designed nanocomposites play a role in the realization of ultrahigh recording densities [9].

Another, physically interesting example is the maximum energy product $(B.H)_{max}$ of small, slightly prolate nanoparticles of volume V (Fig. 6). Basically, there are two energy-product definitions [2], depending on whether $(B.H)_{max}$ refers to the useful magnetostatic energy stored outside the magnets, $E_a = (B.H)_{max}V/2$, or to the second quadrant of the B-H loop (see e.g. [18]). Here we show that these two definitions are not equivalent when very small particles are considered ($R < R_{coh}$). The small particle in Fig. 6 is a slightly prolate ellipsoid of revolution, so that the particle's shape anisotropy favors the magnetization alignment parallel to the axis of revolution. The corresponding magnetostatic energy stored outside the magnet is then $E_a = (1-D)D\mu_o M_s^2 V/2$. Since $D = 1/3 - \varepsilon$, this yields the energy $E_a = (1-3\varepsilon/2)\mu_o M_s^2 V/9$.

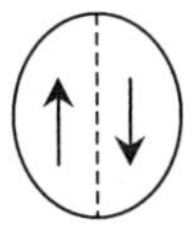

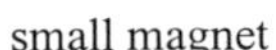

small magnet   macroscopic magnet

Fig. 6 Energy product of very small nanoparticles. In small, slighly prolate nanoparticles, the energy product is much bigger than predicted from the B-H loop.

For bcc iron and $Fe_{65}Co_{35}$, the corresponding energy products are about 800 and 1000 kJ/m^3 – huge values, realized in materials whose magnetocrystalline anisotropy is negligibly small. For macroscopic magnets this approach does not work, because the decay into domains causes the field outside the magnet to collapse.

It is impossible to extend this mechanism to compacted ensembles of small particles, because their mutual interaction is stronger than the small shape anisotropy. However, it is remarkable that it is possible to create a pronounced magnetostatic field in free space which is not directly related to the B-H loop: due to the smallness of the coercivity, the B-H loop area is negligibly small. This mechanism, which goes beyond a purely magnetostatic demagnetizing-field interpretation of shape anisotropy and B-H loops, indicates qualitatively new effects on very small length scales.

CONCLUDING REMARKS

In this paper, we have analyzed how and to what extent magnetic nanostructures can be used to improve the properties of single-phase bulk materials. Experiments on Fe-Pt, Sm-Co, and Pr-Co films confirm earlier theoretical predictions according to which it is possible to improve the permanent magnet energy product by adding magnetically soft material. The comparatively difficult production of well-performing hard-soft composites is ascribed to morphological inhomogenities. An analytic treatment of these effects has been developed for random-anisotropy disorder associated with the nanocrystalline structure of the composites. A particular aspect of nanomagnetism is that that small, softmagnetic grains may have huge energy products. This may be of interest for future applications in nanoelectronics.

ACKNOWLEDGEMENTS

This research was supported by several grants, including US DOE (DE-F6-02-86ER45262), AFOSR (F49620-98-1-0098), DARPA through ARO (DAAG-55-98-1-0268), and NSF (INT-9812082). The authors are grateful to Profs. Y. Liu and S.-H. Liou and to Mr. C. P. Luo for their experimental expertise and help in details.

REFERENCES

1. R. Coehoorn, D.B. de Mooij, and C. de Waard, J. Magn. Magn. Mater. **80** (1989) 101.
2. R. Skomski and J. M. D. Coey, Phys. Rev. B **48** (1993) 15812.
3. A. Al-Omari and D. J. Sellmyer, Phys. Rev. B **52** (1995) 3441.
4. E. F. Kneller, IEEE Trans. Magn. **27** (1991) 3588.
5. S. M. Parhofer, J. Wecker, Ch. Kuhrt, G. Gieres, and L. Schultz, IEEE Trans. Magn. **32** (1996) 4437.
6. J. M. D. Coey (Ed.), *Rare-earth Iron Permanent Magnets*, University Press, Oxford 1996.

7. R.F. Sabiryanov, S.S. Jaswal, J. Magn. Magn. Mater. **177-181** (1998) 989.
8. R. F. Sabiryanov and S. S. Jaswal, Phys. Rev. B **58** (1998) 12071.
9. D. J. Sellmyer and Z. S. Shan, in: *Magnetic hysteresis in novel magnetic materials* (Ed.: G. C. Hadjipanayis), Kluwer, Dordrecht 1997, pp. 419.
10. J.P. Liu, Y. Liu, C.P. Luo, Z.S. Shan, and D.J. Sellmyer, J. Appl. Phys. **81** (1997) 5644.
11. J. P. Liu, C. P. Luo, Y.Liu, and D. J. Sellmyer, Appl. Phys. Lett. **81** (1998) 5644.
12. J.P. Liu, Y. Liu, Z.S. Shan, and D.J. Sellmyer, IEEE Trans. Magn. **33** (1997) 3709.
13. J.P. Liu, Y.Liu, and D.J. Sellmyer, J. Appl. Phys. **83** (1998) 6608.
14. E. E. Fullerton, J. S. Jiang, C. H. Sowers, J. E. Pearson, and S. D. Bader, Appl. Phys. Lett. **72**, (1998) 380.
15. P. G. McCormick, W. F. Miao, P. A. I. Smith, J. Ding, and R. Street, J. Appl. Phys. **83** (1998) 6256.
16. R. J. Astalos and R. E. Camley, Phys. Rev. B **58** (1998) 8646.
17. D.H. Ping, K. Hono, and S. Hirosawa, J. Appl. Phys. **83** (1998) 7769.
18. R. Skomski and J. M. D. Coey, *Permanent Magnetism*, Institute of Physics, Bristol and Philadelphia 1999 (in press).
19. J. Jakubowicz, A. Szlaferek, M. Jurczyk, J. Alloys Compounds **283** (1999) 307.
20. Y.-H. Gao, J.-H. Zhu, Y.-Q. Weng, E. B. Park, and Ch. J. Yang, J. Magn. Magn. Mater. **191** (1999) 146.
21. R. Skomski, J. Appl. Phys. **76** (1994) 7059.
22. A. Aharoni, Rev. Mod. Phys. **34** (1962) 227.
23. H. Kronmüller, phys. stat. sol. (b) (1987) **144** 385.
24. R. Skomski, J. Appl. Phys. **83** (1998) 6503.
25. R. Skomski, H.-P. Oepen, and J. Kirschner, Phys. Rev. B **58** (1998) 3223.
26. T. Schrefl, J. Fidler, and H. Kronmüller, Phys. Rev. B **49** (1994) 6100.
27. R. Fischer and H. Kronmüller, J. Magn. Magn. Mater. **184** (1998) 166.
28. R. Fischer, T. Leineweber, and H. Kronmüller, Phys. Rev. B **57** (1998) 10723.
29. V. Christoph and K. Elk, J. Magn. Magn. Mater. **74** (1988) 143.
30. H. Fukunaga and H. Inoue, Jpn. J. Appl. Phys. **31** (1992) 1347.
31. R. Skomski, J. Magn. Magn. Mater. **157-158** (1996) 713.
32. Y. Imry and Sh.-K. Ma, Phys. Rev. Lett. **35** (1975) 1399.
33. D. J. Sellmyer and S. Nafis, Phys. Rev. Lett. **57** (1986) 1173.
34. C. Kuhrt, K. O'Donnell, M. Katter, J. Wecker, K. Schnitzke, and L. Schultz, Appl. Phys. Lett. **60** (1992) 3316.
35. A. Aharoni, *Introduction to the Theory of Ferromagnetism*, University Press, Oxford 1996.
36. D. J. Craik and E. D. Isaak, Proc. Phys. Soc. **76** (1960) 160.
37. R.K. Mishra and R.W. Lee, Appl. Phys. Lett. **48** (1986) 733.
38. W. Rave, D. Eckert, R. Schäfer, B. Gebel and K.-H. Müller, IEEE Trans. Magn. **32** (1996) 4362.
39. J.P. Liu, Y. Liu, R. Skomski and D.J. Sellmyer, to be published.
40. E. M. Chudnovsky, W. M. Saslow, and R. A. Serota, Phys. Rev. B **33** (1986) 251.
41. E. Callen, Y. J. Liu, J. R. Cullen, Phys. Rev. B **16** (1977) 263.
42. These calculations will be published elsewhere.
43. J. Richter and R. Skomski, phys. stat. sol (b) **153** (1989) 711.
44. G. C. Hadjipanayis and W. Gong, J. Magn. Magn. Mater. **66** (1987) 390.
45. T. Klemmer, D. Hoydick, H. Okumura, B. Zhang, and W. A. Soffa, Scripta Met. Mater. **33** (1995) 1793.

A STUDY ON HIGH COERCIVITY AND $L1_0$ ORDERED PHASE IN CoPt AND FePt THIN FILMS

R. A. Ristau, K. Barmak,
Department of Materials Science and Engineering, Lehigh University, Bethlehem, PA 18015

L. H. Lewis
Materials Science Division, Department of Applied Science, Brookhaven National Laboratory, Upton, NY 11973-5000

K. R. Coffey, J. K. Howard
IBM Storage Systems Division, 5600 Cottle Road, San Jose, CA 95193

ABSTRACT

This study relates the microstructure of equiatomic binary alloys of CoPt and FePt with their room-temperature hysteretic magnetic properties, particularly their high coercivity. A transformation from an atomically disordered, face-centered-cubic structure to the $L1_0$ ordered structure occurred during post-deposition annealing and was characterized using digital analysis of dark-field transmission electron microscopy (TEM) images. The transformation was observed to follow first-order nucleation and growth kinetics, and the ordered volume fraction transformed was quantified at numerous points during the transformation. The ordered volume fraction was then compared to the magnetic coercivity data obtained from a superconducting quantum interference device (SQUID) magnetometer. Although the relationship most commonly described in the literature is that the highest coercivity corresponds to a two phase ordered/disordered mixture, the maximum value for coercivity in this study was found to correspond to the fully ordered state. Furthermore, in samples that were less than fully ordered, a direct relationship between ordered volume fraction and coercivity was observed. The proposed mechanism for the high coercivity in these films is an increasing density of magnetic domain wall pinning sites concurrent with an increasing fraction of ordered phase.

INTRODUCTION

An extensive body of literature on CoPt and FePt alloys, spanning 50 years, exists because of their attractive magnetic properties, particularly high magnetic anisotropy, high coercive force, and good corrosion resistance [1, 2, 3, 4, 5, 6]. One of the issues that has been of great interest in these alloys is the source of the high coercivity. In the equiatomic alloy, it has long been known that the presence of the $L1_0$ ordered phase has the effect of greatly increasing the coercivity over that of the disordered alloy. It has been suggested that the presence of a finely dispersed mixture of the ordered and disordered phases is responsible for the high coercivity [1]. According to this proposal, the optimum state for the maximum coercivity is a mixture of approximately half ordered and half disordered phases. The purpose of this paper is to report on the results of experiments conducted to elucidate the relationship between the $L1_0$ ordered phase and the high coercivity in equiatomic alloy thin films of CoPt and FePt.

Mat. Res. Soc. Symp. Proc. Vol. 577

EXPERIMENTAL

Specimens for this study were fabricated by magnetron sputtering using an ultra high vacuum sputtering system. Targets of nominally equiatomic composition of CoPt, and elemental Fe and Pt were dc-magnetron sputtered onto room-temperature silicon substrates as a single 10 nm thick layer. After deposition, the samples were annealed at 700° C for a series of times between 2 and 210 minutes in a small tube furnace with a flowing gas atmosphere of Ar + 4% or 5% H_2. The tube furnace was modified so that the sample coupon could be inserted into, and withdrawn from the furnace rapidly while maintaining the flowing gas atmosphere to prevent contaminates from being introduced during annealing. The reported anneal time was measured from sample insertion to sample withdrawal, and included the time that the sample was heating up to the set temperature.

Measurements of magnetization versus applied field (M vs. H hysteresis loops) of selected samples were made on a Quantum Design 5 Tesla SQUID (Superconducting Quantum Interference Device) magnetometer. The measurements were not corrected for demagnetization effects, which in this case are expected to be small. A maximum applied field (H_{appl}) of ±50,000 gauss was utilized to obtain room-temperature hysteresis loops for selected samples. For most samples, however, only second-quadrant demagnetization curves were obtained.

The ordered volume fraction was measured from dark-field (DF) TEM images according to a method derived for this purpose. This was possible because the tetragonal $L1_0$ crystal structure (c/a = 0.970) [7] has lower symmetry than the disordered FCC structure, and therefore extra, or superlattice diffraction intensities, which do not appear in the FCC diffraction pattern, appear in the $L1_0$ diffraction pattern [Figure 1]. Note that some $L1_0$

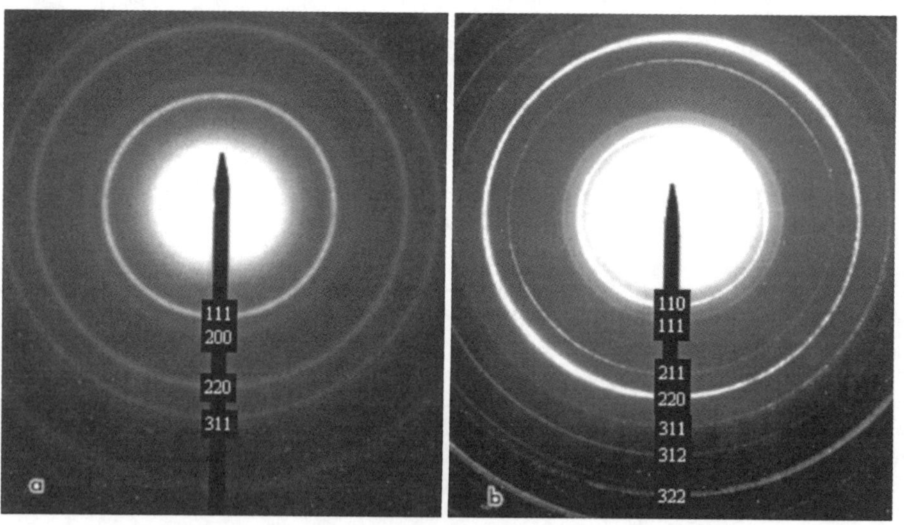

Figure 1. (a) Selected area diffraction pattern of CoPt 10 nm thin film as-deposited, showing only FCC reflections. (b) Diffraction pattern of CoPt annealed at 700° C showing many $L1_0$ reflections.

reflections are absent in Figure 1b (e.g. {200}) due to the orientation of the crystals relative to the electron beam. In a two-phase, ordered/disordered composite, the superlattice diffraction intensities will arise solely from ordered regions of the crystal, while both ordered and disordered regions of the crystal will contribute to the allowed FCC intensities. By using the superlattice diffraction intensities to form the DF image, only $L1_0$ ordered regions will appear illuminated in the image. It should be noted that the ordered phase in these alloys has a long-range-order parameter very near to 1, and that the transformation is thermodynamically first-order [8, 9, 10]. Therefore, it was assumed that the ordered regions in the sample are fully ordered and the disordered regions are fully disordered.

Based on these assumptions, it was possible to calculate the ordered-phase volume fraction by comparing DF images of ordered-phase regions with bright-field (BF) images of the same region. However, due to the type of crystal symmetry unique to the $L1_0$ ordered phase, in which the tetragonal c-axis is along one of the cube directions of the FCC lattice, there are three ordered variants which co-exist, each with the c-axis parallel to one of the cube edge directions (i.e. [001], [010], and [100]). While all three ordered variants can co-exist in a single grain, the electron beam will diffract in a slightly different manner when encountering each one. Therefore, in order to correctly measure the ordered fraction it was necessary to obtain three DF images of the grain to ensure that images of all three c-axis variants were included. This fact was overlooked in an earlier published work of ours [11], and as a result the reported values were much smaller than the true volume fraction as determined from subsequent measurements. This present paper should supercede those previously published results. A composite image from these three DF images and a BF image of the same grain were made. These images were digitally scanned, and the area of the illuminated regions in the composite DF image was measured, as was the total area of the grain from the BF image. The respective volumes were calculated assuming a constant sample thickness (10 nm), and the ordered-phase volume fraction was determined from the ratio of the ordered volume to the total volume of the grain.

RESULTS

The as-sputtered films were of the single-phase disordered FCC structure, as identified from the electron diffraction pattern [Figure 1a]. In the samples annealed at 700° C, the $L1_0$ ordered phase was detected by the presence of superlattice intensities in the diffraction pattern [Figure 1b]. Dark-field TEM images of the {110} superlattice reflections showed the ordered-phase regions as bright features in the DF images. In the films annealed for short times at 700° C, the ordered-phase regions appeared as small, precipitate-like regions within a disordered matrix. In samples annealed for longer times, the ordered regions appeared larger in size and fewer in number. When annealed for approximately 120 minutes or longer, the DF images of both CoPt and FePt samples showed only ordered-phase regions, when all three c-axis variants were properly imaged. It should be emphasized that the point at which the samples were single-phase ordered could only be determined by examining the DF images, as the information from the diffraction patterns was ambiguous. Because the disordered FCC reflections coincided with certain $L1_0$ reflections, it was not possible to detect the absence of the disordered phase from the diffraction patterns. Using the method described above, the ordered-phase volume fraction was calculated for several samples annealed for various times, the results of which appear in Figure 2.

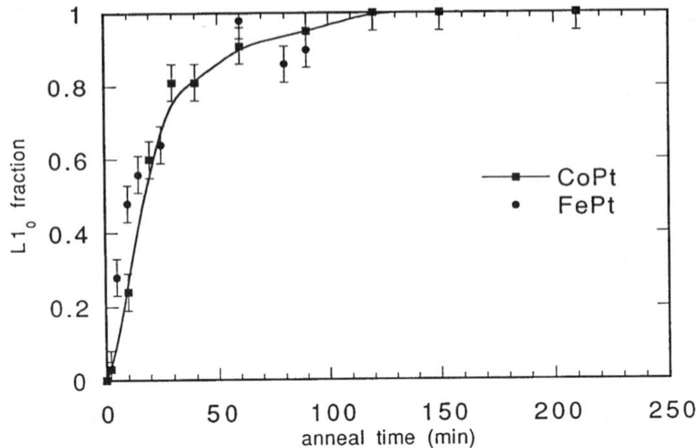

Figure 2. Plot of $L1_0$ ordered volume fraction versus anneal time for CoPt and FePt samples annealed at 700° C.

Room-temperature hysteresis measurements were also obtained using a SQUID magnetometer for as-sputtered samples and from numerous annealed samples of CoPt and FePt. The strength of the coercive force ranged from ~3 Oe and ~40 Oe in the as-sputtered FePt and CoPt samples to a maximum of 9100 Oe and 15,300 Oe in FePt and CoPt, respectively. The coercivity decreased in samples annealed for times longer than 120 minutes. Plots of the coercivity versus anneal time for both CoPt and FePt are shown in Figure 3. In Figure 4, a plot

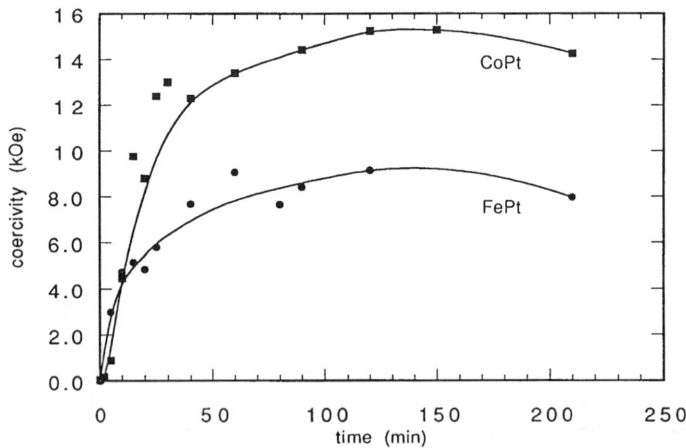

Figure 3. Plot of coercivity versus anneal time for CoPt and FePt samples annealed at 700° C.

of coercivity versus the ordered volume fraction is shown. As can be seen in this plot, there exists a direct relationship between the coercivity values and the ordered volume fraction in these films. This relationship was nearly linear until the samples reached an ordered fraction of 1, after which the coercivity began to decrease while the ordered fraction remained constant.

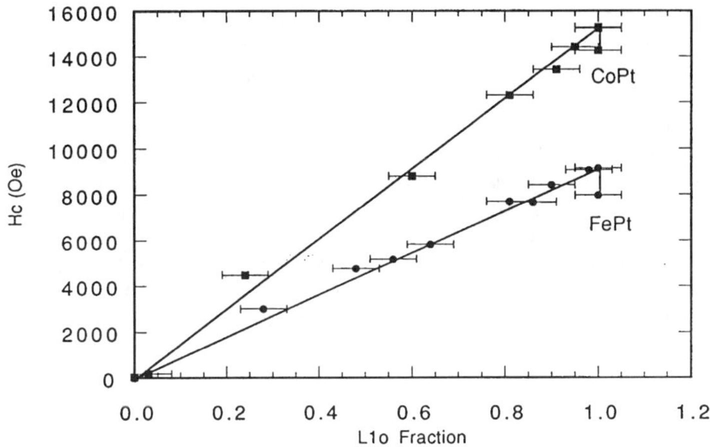

Figure 4. Plot of coercivity versus $L1_0$ ordered volume fraction for CoPt and FePt.

CONCLUSIONS

The results of this study clearly showed that a direct relationship exists between the magnetic coercivity and the $L1_0$ ordered volume fraction in CoPt and FePt thin films. Coercivity values were observed to increase in proportion to the ordered volume fraction until the samples were fully ordered at which point the maximum value for coercivity was recorded. After the sample was fully ordered, and only after this point, was a decrease in coercivity observed. Published results originating from the former Soviet Union describe very similar results for both CoPt [12] and FePt [13].

What is unique in the present study is that the coercivity exhibited a relationship that appeared to follow a simple rule of mixtures of the coercivities of the magnetically hard ordered phase and the magnetically soft disordered phase. Results of a concurrent study of grain growth in these systems found that there was little correspondence between grain size and coercivity. The rapid increase in coercivity with anneal time occurred only after grain growth attained a slow, monotonic growth regime. These results were surprising in that the coercivity was expected to be more dependant on changes to microstructure, rather than to ordered phase volume.

As a possible explanation for these observations it is suggested that the coercivity was not affected directly by the ordered volume fraction, but by a microstructural factor that is related to the ordered volume fraction and that scales similarly. Defects in magnetic materials, such as grain boundaries and phase boundaries, can form pinning sites that impede the movement of magnetic domain walls, leading to high coercivity [14]. In $L1_0$ ordered CoPt and FePt the magnetic easy axis corresponds to the tetragonal c-axis, and thus boundaries between adjacent c-axis variants are expected to be effective pinning sites because of the 90° change in the magnetic easy axis that occurs at this boundary. In addition, the boundaries between ordered and disordered regions might also represent domain wall pinning sites, although it is expected that these would be less effective than the c-axis variant boundaries, because in the former the boundary represents a change from the uniaxial anisotropic $L1_0$ phase to the FCC phase with multiple easy axes. Results obtained from initial magnetization curves indicate that these films exhibit behavior consistent with a pinning-controlled coercivity mechanism [15].

It is proposed, therefore, that the c-axis variant boundaries (antiphase boundaries) and the ordered/disordered phase boundaries formed the pinning sites for the magnetic domain walls, and therefore were the source of the high coercivity in these materials. Measurements from DF TEM images showed that the coercivity increase corresponded closely to the increase in ordered volume fraction. In actuality, as we are proposing, the increase in coercivity was due to the increase in magnetic domain wall pinning site density, which itself corresponded to the increase in ordered volume fraction. When the alloy was fully transformed to the ordered phase, a decrease in coercivity was noticed with increasing anneal time. This can probably be explained by a growth in ordered region size with further annealing. This would result in a decrease in the total boundary area, and thus represent a lower pinning site density for the magnetic domain walls, leading to a decrease in coercivity.

ACKNOWLEDGEMENTS

Research performed in part under the auspices of the U.S. Dept. of Energy, Division of Materials Science, Office of Basic Energy Sciences under contract no. DE-AC02-98CH10886. R. R. and K. B. acknowledge support under NSF DMR-9256332 and DMR-9458000 and thank R. Gambino for helpful discussion.

---

1. J. B. Newkirk, A. H. Geisler, D. L. Martin, R. Smoluchowski, Trans AIME **188**, 1249 (1950).
2. R. A. McCurrie, P. Gaunt, Phil. Mag. **13,** 567 (1966).
3. D. J. Craik, Platinum Metals Rev. **16**, 129 (1972).
4. J. A. Aboaf, S. R. Herd, E. Kolkholm, IEEE Trans. Mag. **19**, 1514 (1983).
5. M. Yanagisawa, N. Shiota, H. Yamaguchi, Y. Suganuma, IEEE Trans. Magn. **19**, 1638 (1983).
6. K. R. Coffey, M. A. Parker, J. K. Howard, IEEE Trans. Mag. **31**, 2737 (1995).
7. J. M. Pennison, A. Bourret, Ph. Eurin, Acta Metall. **19**, 1195 (1971).
8. D. R. de Fontaine, Met. Trans. **12A**, 559 (1981).
9. D. R. de Fontaine, M. Sluiter, P. Turchi, Phase Transformations '87, G. W. Lorimer, ed., (Institute of Metals, Cambridge) 8 (1988).
10. C. Leroux, A. Loiseau, D. Broddin, G. Van Tendeloo, Phil. Mag. B. **64**, 57 (1990).
11. R. A. Ristau, K. Barmak, K. R. Coffey, J. K. Howard, Mater. Res. Soc. Symp. Proc. **475**, 119 (1997).
12. Y. S. Shur, L. M. Magat, G. V. Ivanova, A. I. Mitsek, A. S. Yermolenko, O. A. Ivanov, Fiz. Metal. Metalloved. **26** (2), 241 (1968).
13. A. Y. Yermakov, V. V. Maykov, Fiz. Metal. Metaloved. **60** (4), 113 (1985).
14. J. D. Livingston, J. Appl. Phys. **52**, 2544 (1981).
15. J. Kim, Lehigh University, private correspondence.

MAGNETIC EXCHANGE-COUPLING IN CoPt/Co BILAYER THIN FILMS

J. Kim*, K. Barmak**, L. H. Lewis†, D. C. Crew† and D. O. Welch†
* Dept. of Materials Science and Eng., Lehigh University, Bethlehem, PA 18015
** Dept. of Materials Science and Eng., Carnegie Mellon University, Pittsburgh, PA 15213
† Materials and Chemical Sciences Division, Dept. of Applied Sciences, Brookhaven National Laboratory, Upton, NY 11973

ABSTRACT

Thin film CoPt/Co bilayers have been prepared as a model system to investigate the relationship between microstructure and exchange coupling in two-phase hard/soft composite magnets. CoPt films, with a thickness of 25 nm, were sputter-deposited from a nearly equiatomic alloy target onto oxidized Si wafers. The films were subsequently annealed at 700°C and fully transformed from the FCC phase to the magnetically hard, ordered $L1_0$ phase. The coercivity of the films increased rapidly with annealing time until it reached a plateau at approximately 9.5 kOe. Fully-ordered CoPt films were then used as substrates for deposition of Co layers, with thicknesses in the range of 2.8-225 nm, in order to produce the hard/soft composite bilayers. As predicted by theory, the magnetic coherency between the soft Co phase and the hard, ordered CoPt phase decreased as the thickness of the soft phase increased. This decrease in coupling was clearly seen in the magnetic hysteresis loops of the bilayers. At small thicknesses of Co (a few nanometers), the shape of the loop was one of a uniform material showing no indication of the presence of two phases with extremely different coercivities. At larger Co thicknesses, constricted loops, *i.e.*, ones showing the presence of a mixture of two ferromagnetic phases of different hardnesses, were obtained. The magnetic exchange present in the bilayer samples was qualitatively analyzed using magnetic recoil curves and the dependence of exchange coupling on the soft phase dimension in the bilayer hard/soft composite magnet films is discussed.

INTRODUCTION

The development of high-performance permanent magnet materials for energy-related applications is motivated by the need for increased maximum energy product, which describes the ability to store magnetostatic energy. The energy product increases with coercivity and remanence but can never exceed the intrinsic theoretical limit of $\mu_o M_s^2/4$. This limit corresponds to an ideal rectangular hysteresis loop, where the saturation magnetization (M_s) is equal to the remanence (M_r) and μ_o is the permeability in vacuum [1]. Thus, permanent magnet development has been focused on achieving this theoretical limit by increasing both coercivity and remanence. However, further increase of the energy product in single phase permanent magnets such as $SmCo_5$, and $Nd_2Fe_{14}B$ is intrinsically limited by their relatively low saturation magnetization.

The "exchange-spring magnets", which are nano-dispersed composites of hard and soft magnets, have been explored in attempts to approach the theoretical limit of the energy product [1,2]. It is expected that a significantly improved energy product may be obtained by exchange-coupling a soft phase with a high saturation magnetization to a hard phase with a high coercivity. However, appropriate control of the nano-scale microstructure is required to take full advantage of the attributes of both phases, since the exchange coupling depends strongly on microstructural factors.

To this end, fundamental studies of the role of microstructural attributes - grain size, spatial dimension of the phases, and structural coherency - on the interphase magnetic exchange

Mat. Res. Soc. Symp. Proc. Vol. 577 © 1999 Materials Research Society

coupling and nature of the domains and domain walls in the system must be carried out before a rational exploitation of these nanocomposite magnetic systems may be accomplished. In this work, bilayer thin films of CoPt/Co were prepared as a simplified one-dimensional exchange-coupled model system to investigate the fundamental relationships between microstructure and exchange-coupling.

EXPERIMENT

15-25 nm-thick CoPt films with near-equiatomic compositions (52-53^{at}% Co) were sputter-deposited on oxidized Si(100) substrates at room temperature. An arc-melted alloy of CoPt was used as the sputtering target cathode. Ar-4%H_2 was used as the sputtering process gas in order to ensure a reducing atmosphere during deposition, as suggested by K. Coffey [3]. The base pressure before deposition was below 5×10^{-8} torr and the sputtering gas pressure was 3×10^{-3} torr.

25 nm-thick CoPt films were deposited as underlayers for CoPt/Co bilayer films. 1.2-1.5 μm-thick freestanding films were prepared by a lift-off technique using Cu underlayers. These latter films were used for compositional analysis using wavelength dispersive spectrometry (WDS) and for determination of the order/disorder transition temperature using differential thermal analysis (DTA). The 25 nm-thick CoPt films were annealed in a tube furnace in Ar-4% H_2 at a flow rate of 100 ml/min at 700°C in the manner of Ristau [4] to produce magnetically hard, ordered $L1_0$ CoPt. For the preparation of thin film CoPt/Co bilayers, the 25nm-thick CoPt films were annealed for 120 minutes. These single-layer films were then reinserted into the UHV sputtering chamber for deposition of an additional layer of magnetically-soft Co. In this manner, bilayer CoPt/Co films with Co thicknesses in the range of 10 and 90% (Co thickness 2.8-225 nm) of the total bilayer thickness were prepared.

Plan-view transmission electron microscopy (TEM) samples of 25 nm-thick CoPt films subjected to various annealing times were prepared by chemical back-etching of the Si substrate. The microstructure was studied using a Philips EM400T operating at 120 kV. Magnetic sample coupons of 5 x 5 $(mm)^2$ films on Si substrate were prepared for SQUID magnetometry. Two oxidized Si wafer pieces of dimensions 5 mm x 60 mm were used to minimize the diamagnetic signals from the substrates during the magnetic measurements [5]. Magnetic characterization was performed at room temperature in the range $-5\ T \le H \le +5T$ with the field applied parallel to the film plane. In order to assess the exchange coupling qualitatively from our single-layer and bilayer samples, magnetic recoil curves were obtained for a number of recoil fields at 1000 Oe intervals along the demagnetization curve of a previously saturated sample. The mean slope of each recoil curve was then measured to analyze the change of reversible magnetic susceptibilities of the samples [6].

RESULTS AND DISCUSSION

The cross-sectional geometry of CoPt single-layer and CoPt/Co bilayer samples used in this study are shown in Fig. 1. The 25 nm-thick single-layer films were magnetically hardened through annealing as described above. The microstructural development of CoPt film as a function of annealing time is shown in Fig. 2. The films were transformed from a disordered fcc to an ordered $L1_0$ phase, evidenced by the appearance of the {110} and {011} superlattice reflections. Grain growth and <111> fiber texturing also occurred during annealing. The mean grain size of the film annealed for 120 minutes was 45 nm. Our results on ordering, grain growth and fiber texture development are in qualitative agreement with those of Ristau [4], who recently investigated the kinetics of these processes in 10 nm-thick, nominally equiatomic CoPt films.

(A)
CoPt (25 nm)
SiO_2(100 nm)
Si(100)(450 μm)

(B)
Co (2.8 - 225 nm)
CoPt (25 nm)
SiO_2(100 nm)
Si(100)(450 μm)

Fig. 1 - Schematic diagrams of the thin film samples used in this study.
(A) Single-layer CoPt
(B) Bilayer CoPt/Co

Fig. 2 - Transmission electron micrographs of 25 nm-thick CoPt films annealed for (A) 0 min (as-deposited), (B) 30 min, (C) 120 min, and (D) 900 min at 700°C.

The change of coercivity with annealing time for CoPt single-layer films of various thicknesses is shown in Fig. 3. The coercivity of 25 nm-thick CoPt films increased with annealing time and reached a plateau of 9.5 kOe after 120 min. This coercivity was maintained up to 900 min of annealing without agglomeration of the film. The lower coercivities in the present films compared with those of Ristau's are believed to be related to the Co-rich composition of our films. However, it is worth noting that, unlike Ristau's films [4], the samples used here can be annealed for considerably longer times without the films breaking up. Based on our coercivity vs. annealing time results, 120 min annealed CoPt films were used as hard phase underlayer for CoPt/Co bilayer composites. The average grain size of the 120 min-annealed CoPt films (45 nm) is about 6 times smaller than the calculated single magnetic domain size of 288 nm ($D_c \approx \gamma / M_s^2$, where γ (domain wall energy) ≈ 18.4 erg/cm^2, M_s ≈ 800 emu/cm^3[7,9]). It is postulated that the grains are exchange-coupled into multi-grain interaction domains which may or may not support distinct domain walls. The change of magnetization under influence of an external field may proceed by a rotational-type process, in which no clear domain walls exist but magnetization changes by an avalanche-type reversal sequence, or it may proceed by the more conventional route of domain wall motion. A combination of data obtained from diverse techniques is necessary to determine the presence and nature of domain walls in the system, and will be the subject of future work.

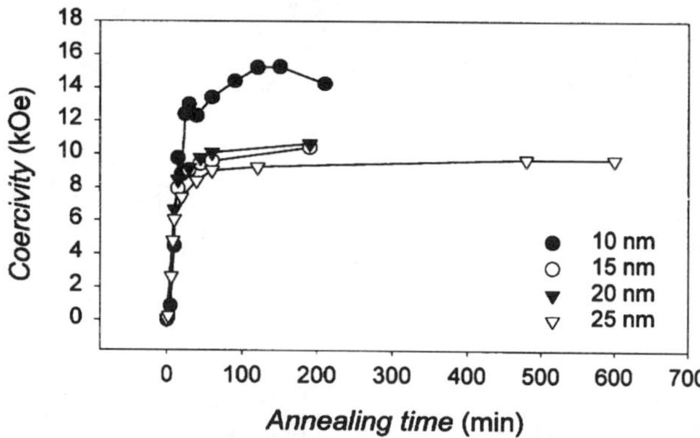

Fig. 3 - Coercivity as a function of annealing time at 700°C for CoPt films of differing thicknesses. Lines are drawn to guide the eye. The 10 nm-thick CoPt film data of Ristau [4] are included for comparison.

The fraction of $L1_0$ ordered, hard CoPt phase in the single-layer film increased with annealing time as the ordering proceeded [4]. Thus, this hard phase coexisted with the disordered fcc soft phase at all times during the microstructural development, resulting in a two-phase exchange-coupled nanocomposite. No discontinuities in the second-quadrant demagnetization curves were visible for any of the single-layer CoPt films. The mean slope of recoil curves to zero internal field, which, as the curves are nearly linear, is equivalent to the reversible susceptibility and is presented as a function of the recoil field in Fig. 4.

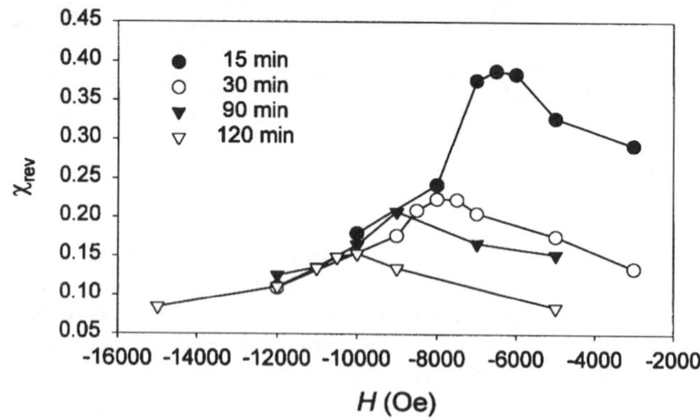

Fig. 4 - Mean slope of the linear recoil curves, as a function of recoil field, of 25 nm-thick CoPt single layer films subjected to different annealing times (χ_{rev} denotes reversible susceptibility).

All films show a maximum in the mean slope at the coercive field of the film. As the recoil field defines a particular remanent magnetization state for the film, and hence for a particular irreversible magnetization state, the plot in Fig. 4 shows equivalent information to a plot of reversible magnetization *vs.* irreversible magnetization. It has been shown by Crew, Woodward and Street [8] that such a plot with an extremum in the reversible magnetization indicates the presence of domain walls in the material which are able to bow reversibly under the action of a field. This conclusion is true even for the well-annealed film samples and indicates that although the grains are small enough to be single domain in isolation, exchange interactions group them into interaction domains which can support a domain wall. The observed increase in the coercivity with annealing time, and the decrease of the mean slope of the recoil curves with annealing time, is consistent with the increasing amount of hard phase. The increased ordered fraction was quantitatively measured for annealed 10 nm-thick CoPt films by Ristau [4].

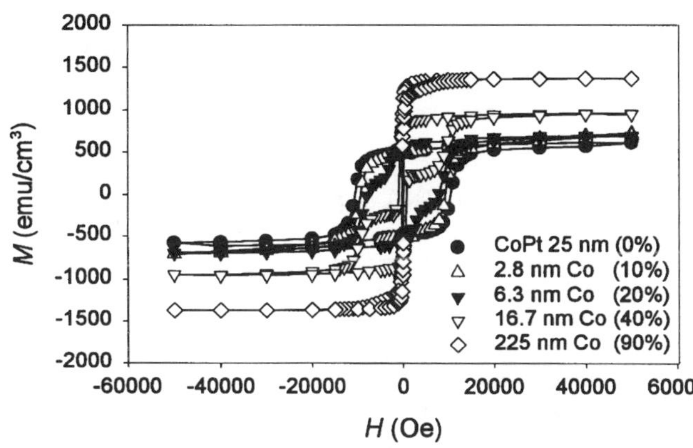

Fig. 5 - *M-H* loops for bilayer CoPt/Co films, with Co thicknesses ranging between 10% and 90% of the total film thickness.

The change of the shape of the magnetic hysteresis loop with Co (soft phase) thickness in the bilayer thin film systems is shown in Fig. 5. As the Co layer thickness increased, the coercivity decreased and the saturation magnetization increased. While the 2.8 nm Co layer (10% of the total thickness of the bilayer) was exchange-coupled to the CoPt layer for its entire thickness, the CoPt/Co bilayer with a Co layer thickness of 6.3 nm (20% of total thickness of the bilayer) began to decouple, as evidenced by the constricted hysteresis loop. Further increases in the Co layer thickness beyond 16.7 nm (40% of the total thickness) caused the coercivity to almost disappear. At these thicknesses, the magnetic behavior of the composite is dominated by that of the decoupled soft Co layer.

The theoretical domain wall width in ordered CoPt is calculated to be 4.6 nm (using the formula, $\delta \approx \sqrt{A/K}$, where A (exchange constant) $\approx$ 8.5 x 10^{-6} erg/cm^2 and K (magnetocrystalline anisotropy constant) $\approx$ 4 x 10^7 erg/cm^3 [7,9]). Thus, it is anticipated that complete exchange coupling will break down for Co overlayer thicknesses greater than approximately 5 nm, or 17% of the total bilayer thickness. The mean slope of the linear recoil curves as a function of recoil field is illustrated in Fig. 6 for the samples that are fully exchange-coupled (CoPt, 120 min; 10% Co/CoPt) and partially exchange coupled (20% Co/CoPt).

The mean slope of the linear recoil curves increased with thickness of the soft phase layer, indicating that a greater amount of material was contributing to the reversible remagnetization processes. As was demonstrated earlier with the single-phase CoPt films, there was a maximum value of the mean slope of the recoil curve which occurred at the coercive field of each sample. Thus it may be postulated that domain walls are also present in these bilayer system, and are free to move under the influence of a reverse field.

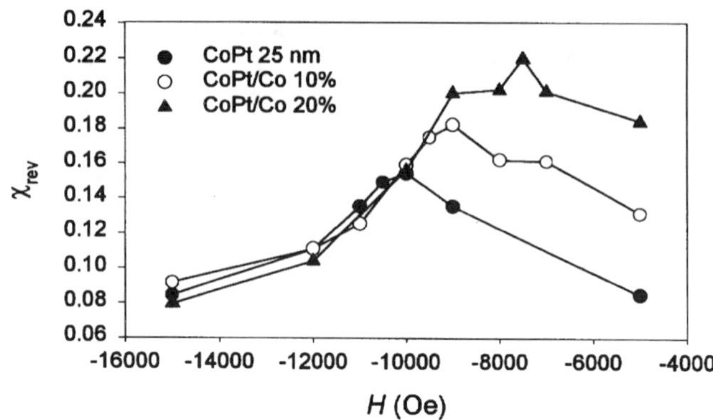

Fig. 6 - Mean slope of the linear recoil curves as a function of recoil field. (χ_{rev} denotes reversible susceptibility).

CONCLUSIONS

Interlayer exchange coupling was studied using CoPt/Co model bilayers to relate the exchange-coupling behavior of hard/soft nanocomposite magnets to fundamental microstructural factors. Exchange coupling is suggested to be present between the ordered $L1_0$ and disordered fcc phases in the single layer of CoPt. The interlayer exchange coupling between CoPt and Co increased with Co soft phase content below the critical dimension predicted from the domain wall width of the ordered CoPt phase, as evidenced by the second-quadrant demagnetization curves. The behavior of the mean slope of the linear recoil curves indicated that there were domain walls present in both the single-layer and the bilayer systems, and that these domain walls are free to move under the influence of a reverse field.

** This research was performed under the auspices of the U.S. Dept. of Energy, Division of Materials Science, Office of Basic Energy Sciences under contract DE-AC02-98CH10886. KB and JK gratefully acknowledge the support of the Horner Fellowship, Lehigh University, and NSF DMR-9458000 and DMR-9411146. Tencor and TA instruments are acknowledged for partial support on the purchase of a P2 long-scan profiler, and a TG-DTA 2960, respectively.

REFERENCES

1. R. Skomski, and J. M. D. Coey, Phys. Rev. B **48** (1993) 15812.
2. E. F. Kneller and R. Hawig, IEEE Trans. Mag. **27** (1991) 3588.
3. K. R. Coffey, M. A. Parker and J. K. Howard, IEEE Trans. Mag. **31** (1995) 2737.
4. R. A. Ristau, Ph.D. Thesis, Lehigh University, Bethlehem (1998).
5. L. H. Lewis and K. M. Bussmann, Rev. Sci. Instrum. **67** (1996) 3537.
6. E. H. Feutrill, L. Folks, P. G. McCormick, P. A. I. Smith and R. Street, Proc. of the 8th Int. Symp. on Mag. Anisotropy & Coercivity in RE-TM Alloys, Birmingham, UK (1994) 297.
7. A. Tsoukatos, H. Wan, G. C. Hadjipanayis, Y. J. Zhang, M. Waite, and S. I. Shah, J. Magn. Magn. Mater. **118** (1993) 387.
8. D. C. Crew, R. W. Woodward and R. Street, J. Appl. Phys. in press.
9. B. D. Cullity, Introduction to magnetic materials, Addison-Wesley Publishing Co., Menlo Park, California, USA (1972).

Epitaxial Co/NiAl Thin Film Growth on Si Substrates

Heng Gong, Wei Yang, David N. Lambeth and David E. Laughlin*
Department of Electrical and Computer Engineering, hgong@andrew.cmu.edu,
*Department of Materials Science and Engineering,
Data Storage Systems Center, Carnegie Mellon University, Pittsburgh, PA 15213-3890

ABSTRACT

Quad-crystal Co/NiAl/Ag thin films were epitaxially grown on hydrofluoric acid etched Si(111) single crystal substrates by sputter deposition. The orientation relationship was studied by x-ray θ/2θ diffraction and pole figure φ-scan, and it was determined to be Co($10\bar{1}1$) || NiAl(110) || Ag(111) || Si(111). The Ag layer contains two twin-related orientations of grains. NiAl grows with three variants of grains on the Ag(111) layer. Since Co($10\bar{1}1$) contains grains with four possible easy axis directions when grown on each NiAl variant, the quad-crystal media consist of grains with twelve easy axis directions and exhibit nearly isotropic in-plane magnetic properties. Meanwhile, uni-crystal Co($10\bar{1}0$)/NiAl(112) films were also epitaxially grown on Ag(110)/HF-Si(110) templates. The orientation relationships determined from these epitaxial films were very useful in understanding the texture evolution of the Co alloys on NiAl. It indicates that the Co ($10\bar{1}0$) texture evolution on the (110) texture of NiAl on amorphous substrates, such as NiP/AlMg or glass, is due to a difficult to observe NiAl (112) texture on the top of the (110) texture.

INTRODUCTION

Co-based alloys are currently the most popular thin film medium materials for magnetic recording. In order to achieve the optimal magnetic properties, control of the Co-alloy film texture is essential. Textures that induce in-plane Co easy axes are usually desired in order to obtain a high coercivity, and this is most often achieved by using Cr underlayers due to the lattice matching between Cr and Co. One of the most common lattice match relationships between Co and Cr for longitudinal recording is the quad-crystal structure, in which Co has four possible easy axis orientations on the Cr(110) lattice, and the Co easy axes are tilted 28° out of the film plane. Another lattice match relationship is the uni-crystal structure for which only one Co easy axis orientation is observed on the Cr(112) plane.

In order to investigate new underlayers while exploiting the unique properties of the Cr underlayer, NiAl intermetallic with B2 ordered crystal structure (space group $Pm\bar{3}m$, CsCl

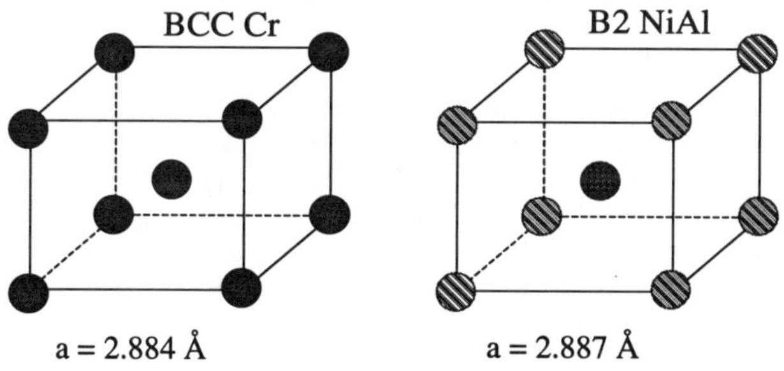

Fig. 1. Lattice structures of BCC Cr and B2 NiAl.

Mat. Res. Soc. Symp. Proc. Vol. 577 © 1999 Materials Research Society

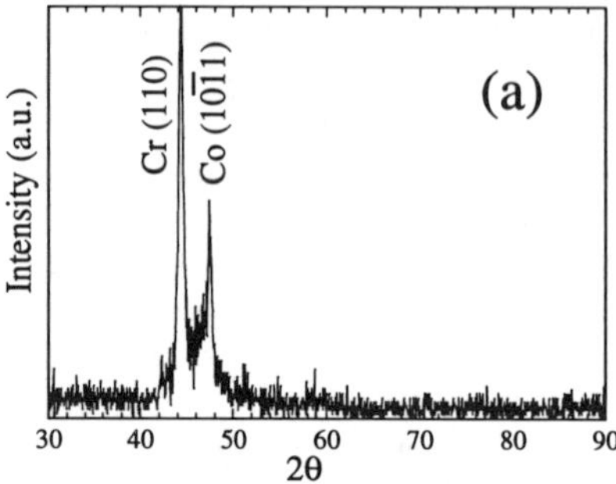

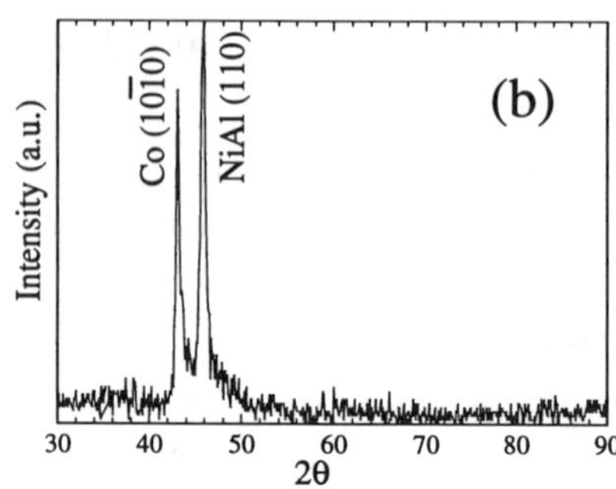

Fig. 2. X-ray diffraction spectra of (a) a Co(50nm)/Cr(100nm) film, and (b) a Co(50nm)/ NiAl(100nm) film, both samples were deposited at room temperature.

prototype) was successfully developed as an alternative underlayer material for low noise Co-alloy recording media [1]. Fig. 1 compares the very close lattice spacing values and similar unit cell structures of Cr and NiAl. The similarities would lead one to anticipate the same Co texture evolution on NiAl as on Cr. Moreover, compared with the widely used Cr underlayers, NiAl thin films exhibit smaller and more uniform grains, which in turn induces smaller grains in the Co alloy films resulting in an improved media signal to noise ratio [2]. It was reported that a NiAl underlayer was adopted in the media structure of IBM's 10 Gbit/in^2 magnetic recording demonstration [3].

Despite the similarity between Cr and NiAl, NiAl underlayers seem to show some distinct characteristics during the texture evolution. Fig. 2 shows the x-ray diffraction patterns of a Co(50nm)/Cr(100nm) film [Fig. 2(a)] and a Co(50nm)/NiAl(100nm) film [Fig. 2(b)]. Both of the samples were sputter deposited on glass substrates at room temperature and the x-ray diffraction scans indicated that the Cr and NiAl grew with the (110) textures. However, the textures of the Co films on the Cr and NiAl underlayers are different. The Co film on the (110) textured Cr film was observed to develop a $(10\bar{1}1)$ texture while that on the (110) textured NiAl developed a $(10\bar{1}0)$ texture. It has been suggested that a difficult to observe (112) NiAl texture evolves on top of the (110) texture [4]. In order to investigate this apparent growth contradiction, epitaxially grown NiAl films with highly pure textures were needed. In this study, NiAl(110) and NiAl(112) films were epitaxially grown by sputter deposition using Ag/HF-Si templates. Then Co films were epitaxially grown on these highly pure textured NiAl films and the epitaxial relationships between the Co and NiAl were studied.

EPITAXIAL GROWTH OF NiAl

To grow highly oriented NiAl films, HF-etched Si wafers were used as substrates, and epitaxially grown Ag films were used as templates to induce the desired orientations in the NiAl. In particular, NiAl(110) films were grown on Ag/HF-Si(111) to investigate the Co texture evolution on pure (110) textured NiAl underlayers. As we reported earlier, Ag grows epitaxially on Si(111) with a (111) orientation [5]. The NiAl layer then grows with three variants of grains on the Ag(111) layer, with the following orientation relationships:

$$\text{NiAl}(110)[001] \parallel \text{Ag}(111)[1\bar{1}0] \parallel \text{Si}(111)[1\bar{1}0],$$

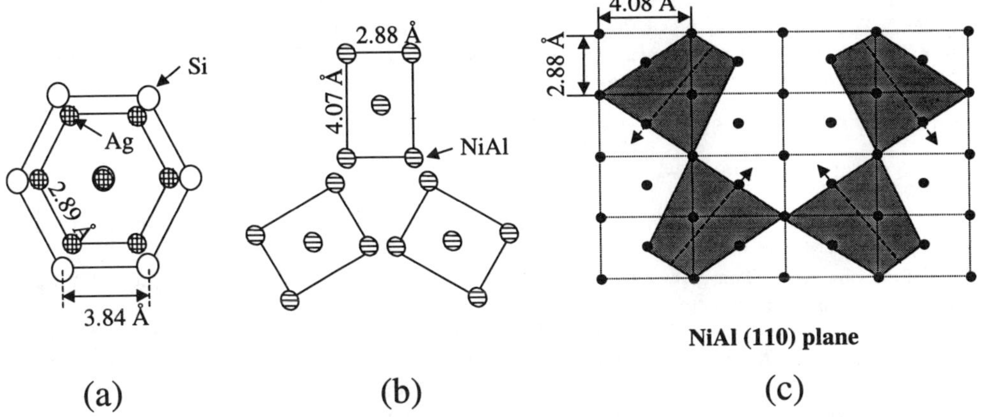

Fig. 3. Orientation and interatomic spacing relationships (a) between Si(111), Ag(111), and (b) NiAl (110) lattices. Note the three possible orientations of NiAl(110) on Ag(111). (c) Lattice match relationship between Co($10\bar{1}1$) and NiAl(110) planes.

NiAl(110)[001] ǁ Ag(111)[$10\bar{1}$] ǁ Si(111)[$10\bar{1}$],
NiAl(110)[001] ǁ Ag(111)[$01\bar{1}$] ǁ Si(111)[$01\bar{1}$].

The epitaxial orientation relationships for the quad-crystal Co/NiAl/Ag/HF-Si(111) films are shown in Fig. 3. The Co layer contains grains with four possible easy axis directions when grown on each NiAl variant. As a result, the grains in the quad-crystal media can take on one of twelve possible easy axis directions.

EXPERIMENTAL DETAILS

Si(111) wafers were first ultrasonically cleaned in organic solvents and rinsed in deionized water. The clean wafers were then immersed in HF to remove the native SiO_2 and obtain a hydrogen-terminated surface. The etched wafers were blown dry using N_2 gas, then placed into the sputtering system, and heated to about 200°C in vacuum before deposition. Ag, NiAl, and

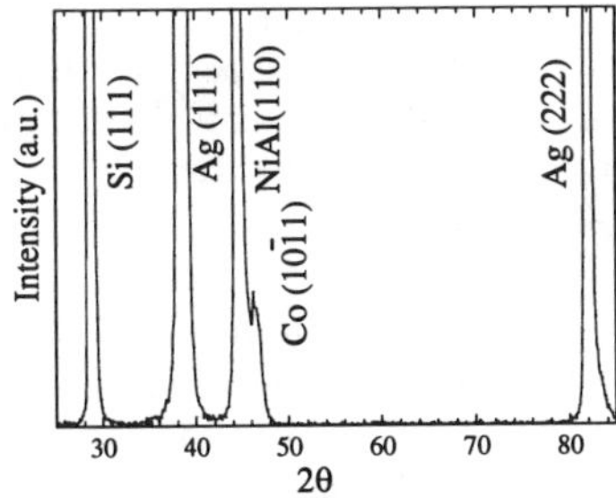

Fig. 4 X-ray diffraction patterns of a quad-crystal Co(30nm)/NiAl(50nm)/Ag(75nm) film on HF-Si(111).

Co thin films were then deposited by rf diode sputtering in a Leybold-Heraeus Z-400 sputtering system. The base pressure was below 5×10^{-7} Torr. The Ar sputtering gas pressure was fixed at 10 mTorr and the sputtering power density was about 2.3 W/cm^2. The epitaxial structure of the films was studied by x-ray $\theta/2\theta$ diffraction and pole figure ϕ-scan.

RESULTS AND DISCUSSION

Quad-Crystal Co/NiAl Films

The $\theta/2\theta$ x-ray diffraction spectrum for a Co(30nm)/NiAl(50nm)/Ag(75nm) film grown on HF-Si(111) is shown in Fig. 4. Only Ag(111), Ag(222), NiAl(110) and Co($10\bar{1}1$) diffraction peaks are observed. These very strong diffraction peaks, as compared to the background noise level, imply the epitaxial nature of the film.

The ϕ-scan spectra of the Co/NiAl/Ag/HF-Si(111) sample are shown in Fig. 5, along with Si (111), Ag (111), NiAl (110), and Co ($10\bar{1}1$) stereographic projections that are necessary for the interpretation of the spectra. As can be seen in Fig. 5(a), the Si{220}-pole scan spectrum of the single crystal Si(111) substrate shows three diffraction peaks with 120° separation. They correspond to the three {110} poles in the Si (111) stereographic projection [Fig. 5(a')], which are also 120° apart along the dotted circle showing the ϕ-scan path.

While three Ag{110} poles are expected for single crystal Ag(111), six peaks were observed in the Ag{220}-pole scan [Fig. 5(b)] of the Ag(111) film grown on Si(111). This suggests that there exist two twin-related orientations of Ag grains in the Ag(111) film. They may emerge at the Ag/Si epitaxial interface or result from Ag growth twinning. In fcc metals, the (111) plane is the most common twinning plane, and the twinning direction is [$11\bar{2}$]. The extra poles are indicated by the "×" symbol in Fig. 5(b'). Three of the peaks, the first, third and fifth, appear at the same ϕ positions as the three peaks in Si{220}-pole spectrum, confirming the parallel relationship between the Si[$1\bar{1}0$] and Ag[$1\bar{1}0$] directions. The twinning was also confirmed by electron diffraction patterns [5].

Similarly, while two NiAl{001} poles are anticipated for single crystal NiAl(110), the NiAl{002}-pole scan spectrum [Fig. 5(c)] shows six peaks that are 60° apart. They appear at the same locations as the six peaks in the Ag{220}-pole scan. This confirms not only the three possible growth variants of NiAl grains on Ag(111), but also that the three NiAl{001} directions are parallel to the three Ag{$1\bar{1}0$} directions. The two extra sets of poles are denoted by "×" and "+" in the streographic projection in Fig. 5(c').

Finally, with four possible orientations of Co grains on each NiAl variant, as indicated by the four ($10\bar{1}0$) poles (marked a_1 to a_4) or the four (0001) poles (marked c_1 to c_4, as these are also the easy axis directions), there will be altogether twelve ($10\bar{1}0$) poles. The two extra sets of poles are denoted by "×" and "+". As shown in Fig. 5(d'), the twelve poles can be divided into six groups, inside each group there is a separation of about 10° along the ϕ-scan path, while the neighboring two groups are 60° away from each other. The observed Co($10\bar{1}0$)-pole scan [Fig. 5(d)] agrees well with the projection. Also as expected, each two-peak group is centered about the six NiAl peaks in the NiAl{002}-pole scan spectrum.

Uni-Crystal Co/NiAl Films

Previously, Yang et al. [6] reported the epitaxial growth of the uni-crystal Co($10\bar{1}0$)/Cr(112) films on Ag(110)/HF-Si(110). Using the same technique, uni-crystal Co/NiAl films were grown on sputter deposited Ag(110)/HF-Si(110). Fig. 6 shows the x-ray diffraction pattern of a uni-crystal film, which can be compared to the quad-crystal pattern of Fig. 4. Only the NiAl(112) and Co($10\bar{1}0$) peaks are observed. With the x-ray pole figure ϕ-scan spectra, the epitaxial

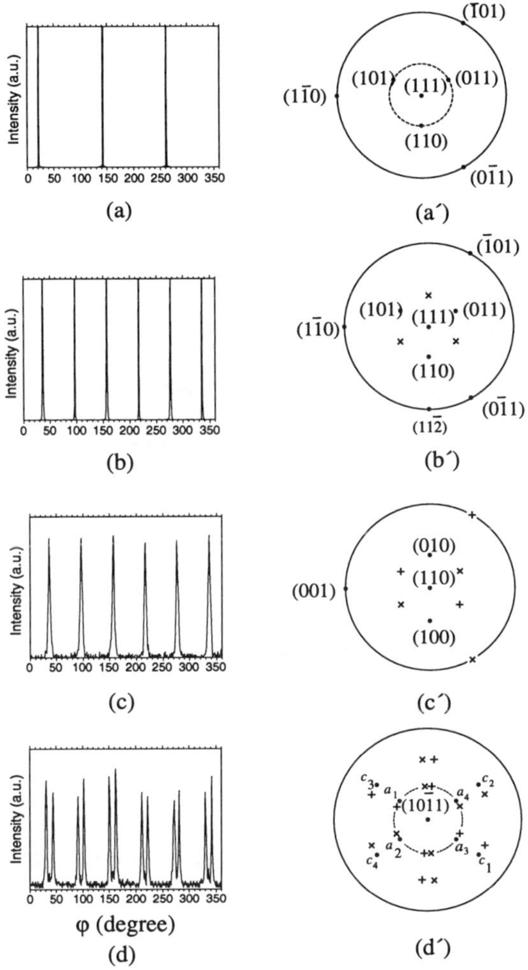

Fig. 5 (a) Si{220}-pole, (b) Ag{220}-pole, (c) NiAl{002}-pole, and (d) Co($10\bar{1}0$)-pole ϕ-scan spectra of a Co(30nm)/NiAl(50nm)/Ag(75nm) /HF-Si(111) film. The stereographic projections are for (a') Si(111), (b') Ag(111), (c') NiAl(110), and (d') Co ($10\bar{1}1$).

relationships were determined to be the same as those of Co($10\bar{1}0$)/Cr(112)/Ag(110)/HF-Si(110) [6].

Co Texture Evolution

With the epitaxial NiAl underlayers which show highly pure textures, it is determined that the NiAl(110) plane only induces Co($10\bar{1}1$), while the NiAl(112) only induces Co($10\bar{1}0$).

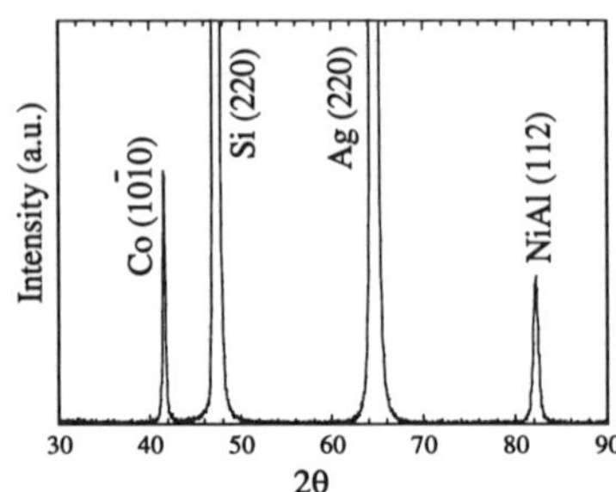

Fig. 6 X-ray diffraction patterns of a uni-crystal Co(30nm)/NiAl(50nm)/Ag(75nm) film on HF-Si(110).

Hence, the $(10\bar{1}0)$ Co texture on the (110) textured NiAl underlayers grown on amorphous substrates can not be a result of the epitaxial growth of Co$(10\bar{1}0)$ on the NiAl(110). The only possibility is that on top of the (110) textured NiAl films, a (112) texture has developed and induced the $(10\bar{1}0)$ Co texture. Since the (112) texture develops late, it was difficult for the x-ray diffraction to detect the relatively low (112) peak intensity at the high 2θ angle.

CONCLUSIONS

In this work, we have successfully fabricated quad-crystal Co thin-film media on NiAl (110)/Ag(111)/HF-Si(111) by sputter deposition. The epitaxial orientation relationship was determined by x-ray pole figure scan. Two twin-related orientations were found in the Ag(111) layer. NiAl grows with three possible variants on Ag(111) and this finally results in twelve easy axis directions in the Co$(10\bar{1}1)$ layer. Also the uni-crystal Co$(10\bar{1}0)$ films were grown on NiAl(112)/Ag(110)/HF-Si(110). Using these NiAl films with highly pure (110) and (112) orientations, the texture evolution of Co on NiAl underlayers was examined.

ACKNOWLEDGMENT

This work is supported by the National Science Foundation under grant No. ECD-8907068. The U.S. government has certain rights in this material.

REFERENCES

[1] L.-L Lee, D.E. Laughlin and D. N. Lambeth, IEEE Trans. Magn. **30**, 3951(1991).
[2] D.N. Lambeth, E.M.T. Velu, G.H. Bellesis, L.L. Lee and D.E. Laughlin, J. Appl. Phys. **79**, 4496(1996).
[3] M. Madison et al., IEEE Trans. Magn. **35**, 695(1999).
[4] L.-L Lee, Ph.D. Thesis, Carnegie Mellon University, Dept. of Materials Science and Engineering (1997).
[5] H. Gong, M. Rao, D.E. Laughlin, and D.N. Lambeth, J. Appl. Phys. **85**, 4699(1999).
[6] W. Yang, D.N. Lambeth, and D.E. Laughlin, J. Appl. Phys. **85**, 4723(1999).

THE INFLUENCE OF MICROSTRUCTURE ON THE MAGNETIC PROPERTIES IN NANOSTRUCTURED COPT $L1_0$ THIN FILMS

S. JEONG, M.E. MCHENRY, and D.E. LAUGHLIN

Data Storage Systems Center, MSE Department, Carnegie Mellon University, Pittsburgh, PA, 15213, sjeong@andrew.cmu.edu

ABSTRACT

We have applied the Neel's spin pair potential model to model the dependence of magnetocrystalline anisotropy on the long range order parameter. We have produced CoPt nanoscale thin films, with and without an immiscible second phase, by RF and DC magnetron sputtering. These thin films showed a shoulder in their magnetic hysteresis, indicative of a disordered soft magnetic phase for samples annealed above 600 °C while it was barely detectable in 500-600 °C annealed samples. A (111) texture is observed in as-deposited and annealed films is identified as one of the possible contributions to the lower coercivity squareness S* and perpendicular magnetic properties. Micromagnetic calculations confirmed a poor exchange interaction, identified rotational energy barriers and inferred that the ordered region was not fully ordered. The initial stage of ordering clearly showed a discontinuous type transformation.

INTRODUCTION

CoPt and FePt thin films have potential applications in extremely high density recording (EHDR)[1] where their large anisotropy energy density provides a barrier to thermally activated switching of the magnetization (superparamagnetism). It is well known that the Co-based HCP phase will exhibit significant thermal decay for sizes below 10 nm and superparamagnetic response below 7 nm[1,14]. Because of the high magnetocrystalline anisotropy, $L1_0$ materials is predicted to be thermally stable down to the 2-3 nm size range. However, increasing the anisotropy results in increasing the intrinsic anisotropy field, H_K, and consequently the coercive field, H_c, making it difficult to write magnetic information. It is important to determine the optimum anisotropy field and resulting coercivity, considering both the issues of writing and the thermal stability, simultaneously. If the coercivity can be effectively varied through control of the magnetocrystalline anisotropy, it would be applicable to the future media. In the case of $L1_0$ materials , when they are completely ordered, the coercivity can reach over 1 T.[18] However for recording media applications, we must adjust and control this intrinsic coercivity so as to attain a reasonable writing coercivity.

In the light of the above arguments, it is of interest to investigate $L1_0$ materials since their magnetic properties can be tailored by choice of annealing temperature, time and composition[3]. Prior studies have involved bulk $L1_0$ materials and correlation of the magnetic properties and their microstructures. However, many unsolved questions remain and the study of thin film $L1_0$ materials is in its infancy. Recent studies of CoPt $L1_0$ films (based on TEM measurements) showed (111) texture and heterogeneous transformation ordering kinetics[2,6]. These authors discussed exchange coupling between the magnetically hard $L1_0$ phase and remaining soft disordered phase. In bulk CoPt, it is known that the coercivity mechanism can be attributed to domain pinning due to anti-phase domains (APBs) and twins[4]. This might be different in the nanoscale polycrystalline thin films where we could assume the films to be a single domain assembly with neighbors coupled by relatively weak intergranular interactions. It is known that

Mat. Res. Soc. Symp. Proc. Vol. 577 © 1999 Materials Research Society

the exchange interaction is sensitive to atomic configurations, therefore, the grain boundary could have a weakened exchange interaction compared with bulk materials.
In this article, we have attempted to relate the magnetocrystalline anisotropy energy density to the long range order parameter, applying spin pair model. We've synthesized CoPt alloy thin films with and without immiscible second phases at the grain boundaries. A micromagnetic model was used to explain how the microstructure would influence the magnetic properties. Based on previous studies, our experimental and simulated results, we will discuss the possible relationship between the microstructure and magnetic properties in the nanoscale thin films.

EXPERIMENTAL PROCEDURE

CoPt alloy films were deposited on the silicon (100) wafers or smooth glass substrates by the RF or DC magnetron sputtering. The base pressure before sputtering was 5 x 10^{-7} torr. The sputtering pressure of Ar gas was 1-5mtorr. Sputtering was performed in the power range of 20 – 100 W. Alloy targets (CoPt, 50 at%) or chips were used to produce the thin films. The purity of metal sources was better than 99.9%. Post annealing process was performed by RTA (rapid thermal annealing) in an Ar atmosphere for 5-7 minutes. Magnetic properties were measured using VSM (Vibrating Sample Magnetometry), and AGFM (Alternating Gradient Force Magnetometry) in the fields up to 1.4 T or by SQUID magnetometry in fields up to 5 T. Structural analysis was made using X ray diffractometer (Cu-Kα, 35kV, 20mA). The chemical composition of the films was measured by X-ray fluorescence or EDS.

RESULTS AND DISCUSSION

A. Spin-pair Potential Model for $L1_0$ Ansiotropy

We have applied the Neel's spin pair potential model to model the strong uniaxial intrinsic anisotropy and explain the dependence of a long range order parameter on the intrinsic magnetocrystalline anisotropy in CoPt. This is used to explain the dependence of coercivity on thermal treatment. This is a phenomenological theory explains the surface anisotropy based on symmetry breaking ideas.[8,10]. The premise of this model is that anisotropy due to symmetry breaking and magnetostriction at surface or interfaces are connected quantitatively[7,8,10]. In this localized moments model, the elementary internal energy of the spin pair can be expanded in a series of Legendre polynomials, as

$$W(r,\psi) = g(r) + l(r)(\cos^2\psi - \frac{1}{3}) + q(r)(\cos^4\psi - \frac{6}{7}\cos^2\psi - \frac{3}{35}) + ... \quad \text{----- (1)}$$

Where r is the distance between atoms and ψ is the polar angle between the bond and spin direction **S**. The common direction of the spins is determined by the exchange interaction. The first term, g(r), which is spatially isotropic, corresponds to the exchange interaction. The second term is called the dipole-dipole interaction term. The magnitude of l(r) is much larger than q(r). therefore, in this study, we neglect from the third term. Magnetic anisotropy is calculated by summing over all spin pairs included in a unit volume of the crystal. Since the pair energy for distant pairs is small, we consider only the nearest neighbors. In an A-B binary system, considering AA, AB, BB atomic pairs and applying above assumptions, this internal energy per unit volume can be expressed as:

$$E_a = \frac{1}{2}N\sum_{j}^{NN=12} l_j(\cos^2\psi_j - \frac{1}{3}) = \frac{1}{2}\left[N_\alpha \sum_{j}^{NN=12} l_j(\cos^2\psi_j - \frac{1}{3}) + N_\beta \sum_{j}^{NN=12} l_j(\cos^2\psi_j - \frac{1}{3})\right] \text{-- (2)}$$

Here, $N_\alpha = N_\beta = N/2$, $l = l(r,\xi,C)$. N_α, N_β and N are the number of atoms per unit volume in α, β sitse and the total number of atoms per unit volume, respectively. ξ is the long range order parameter and C is the atomic fraction of A element. The anisotropy constant K_u can be expressed in terms of the long range order parameter ξ and the composition, as $E_a = K_u \sin^2 \alpha_3 + \text{const}$ ---- (3)

where $$Ku(\xi,c) = N/4\left[-\frac{4l_{\alpha-\beta}(\xi,c)}{1+\Delta^2}(2\Delta^2-1)+(l_{\alpha-\alpha}(\xi,c)+l_{\beta-\beta}(\xi,c))\right], \Delta=c/a$$

From this relationship, we can predict the variation of anisotropy constant with the degree of order and composition.

The $L1_0$ structure has alternating layer of Co and Pt. It is known that Pt can be spin polarized in CoPt alloys[16], though its moment is small. When completely disordered, the second order or dipole term makes no contribution to the anisotropy. However, in the ordered state or partially ordered state, the second term does not disappear and contributes to the uniaxial anisotropy. This is true for geometries in which we do not consider strain. In a strained state, $l(r)$ can be expanded in terms of the bond strain. The values of $l(r)$ for Co-Co, Pt-Pt, Co-Pt can be deduced from magnetostriction coefficients or from first principles calculated values by J. M. MacLaren[11,12]. Also a consideration is the fact that the lattice parameter of the CoPt alloy depends on the composition. Here then, $l(r)$ could have different values at each composition differing from the bulk values.

In the theory of phase transformations, $L1_0$ type ordering is classified as a first order. Therefore, above an instability line, it would exhibit a discontinuous transition involving nucleation processes. It has been reported that the transformation will occur on an ordering path with a lower energy barrier and a metastable structure based on $L1_0$ symmetry with (1) a non-equilibrium c/a ratio or (2) a non-equilibrium order parameter[4]. Also it has been proposed that ordered regions are not fully ordered in the initial stage of the transformation[13] and do not possess their maximum possible degree of tetragonality. The dependence of the anisotropy constant on annealing time and temperature can be attributed to these reasons and thus might be very sensitive to composition and degree of order due to the quadratic and linear dependences of K_u on ξ and composition, respectively. Our calculations can be useful in interpreting previous and present experimental results and the scatter of coercivity data.

B.Experimental and Simulation Results

Our preliminary results on CoPt alloy films synthesized by putting Co and Pt chips on a SiO_2 target showed degraded magnetic properties below 10 nm thickness. This can be attributed to several reasons. The composition of the sputtered thin films were 54 at% Pt by EDS and X-ray fluorescence analysis. However, a small thickness dependence of the composition was detected. Other reasons could include oxidation during sputtering or slow kinetics.

Further investigation of the magnetic properties was carried out for CoPt thin films produced using 50 at%Pt alloy target. The deposited alloy composition was 54 at% Pt and thickness ranged from 5 to 40 nm. X-ray analysis revealed random (111) texture with a tendency toward more random texture with increasing SiO_2. For the same thickness, the addition of SiO_2 contributes to the peak broadening probably due to non-uniform strain or smaller grains. The samples annealed below 600 °C by RTA for 7 minutes do not show any shoulder in their hysteresis curve suggesting highly exchange coupled grains (Figure 2.(b)). ΔM showed larger peaks with decreasing annealing temperature. This confirmed an increasing grain isolation during the grain growth process or oxidation of surface layer with increasing the annealing temperature. As the annealing temperature is increased, the hard phase comes out discontinuously, giving rise to a shoulder in the hysteresis curve. Both in-plane and perpendicular coercivities of films were almost the same, suggesting a spin rotation mechanism[19]. Ordering starts at 500 °C, and coercivities increase from below hundred Oe to the higher values depending on the thermal treatments. The coercivities abruptly increased from around 1000 Oe to 7000 Oe , between 600 °C and 650 °C. The saturation magnetization, M_s decreases with increasing annealing temperature (as deposited $M_S \cong 700$ emu/cc). This reduction might be due to tracing a minor loop or the dependence of M_S on order parameter.

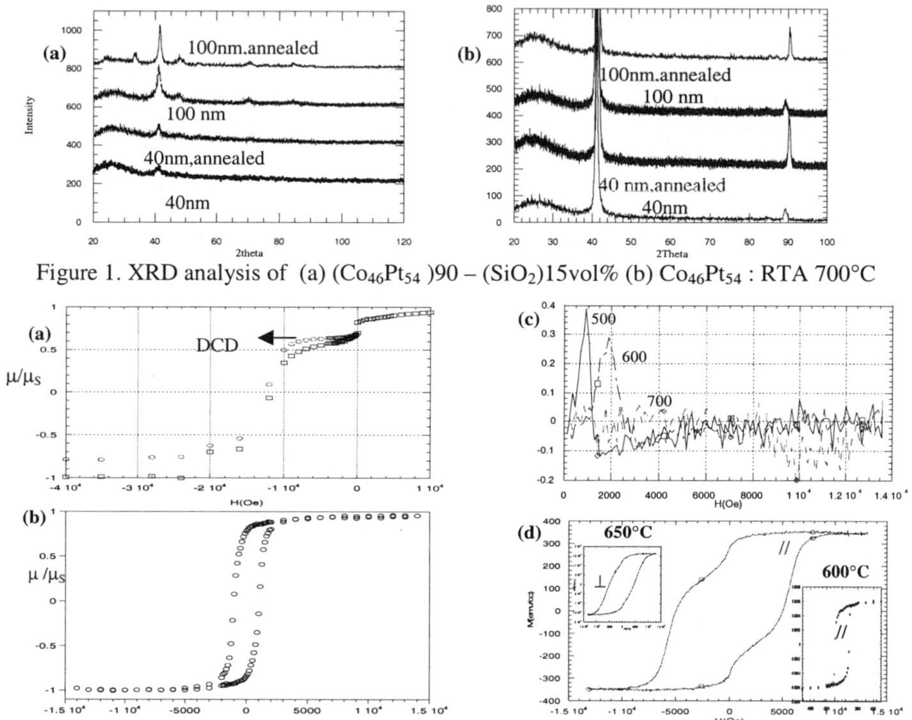

Figure 1. XRD analysis of (a) $(Co_{46}Pt_{54})90 - (SiO_2)15vol\%$ (b) $Co_{46}Pt_{54}$: RTA 700°C

Fig.2 (a) $Co_{46}Pt_{54}$, 20nm, 700°C by RTA , 7min, in-plane; (b) same as (a) but 550°C RTA,in-plane; (c)$\Delta M/Mr$,$Co_{46}Pt_{54}$ 550,600,700°C by RTA 7min,20nm; (d)CoPt- SiO2(30vol%), 10nm, RTA 600(Hc=905 Oe), 650 °C, 7min , // : in-plane, ⊥: perpendicular

It has been reported that the ordering at 500 - 600 °C from disordered phase is 'discontinuous' while the annealing over 600 °C promotes a 'continuous' transformation in bulk CoPt[13]. Our experiments for the nanoscale thin films showed a shoulder in the hysteresis loop for samples annealed by RTA to temperatures above 600 °C. The shoulder disappears with increasing annealing time. The samples annealed by RTA below 600 °C, do not show a shoulder at the initial stage and have much higher squareness, S and S*, values presumably due to enhanced exchange coupling. SiO_2 addition shows a nearly similar behavior (Figure 2(d)) except for a reduced exchange interaction inferred from the ΔM curve.

(111) texture results in magnetic easy axes, 36° out of plane has a nearly identical reversal process to the 3-D random distribution, based on the simple Storner-Wohlfarth model. Consequently, both textures result in the low squareness, S*, since most of the irreversible switching occurs after a large amount of rotation before the coercive field is reached (Figure 3(a)). Due to the intergranular coupling, texture effects may be lessened (higher S* and S), however, inappropriate texture could still be a possible reason for low S*. Introduction of magnetostatic interactions yield similar results, however, its effects will be lessened when there is the substantial increase of K_u (decrease of Ms/Hk) based on the previous micromagnetic studies. To identify the reversal process in more depth at the initial stage of ordering, a numerical simulation of the magnetization reversal process was performed using a 2-d nanostructured array model of the interacting single domain assembly. These were performed varying the ordered fraction and considering the intergranular interaction in terms of magnetostatic interaction and exchange coupling between the grains. A detailed description of micromagnetic calculation can be found in references 14 and 17. The previously observed grain sizes had a large variation, so we've used a size ranging from 10 to 60nm in our model. These values are all smaller than the domain wall width, δ_w (around 72 nm) of the soft disordered phase (for the ordered $L1_0$ phase: δ_w ~ 10nm)[15]. In each grain, the magnetocrystalline anisotropy is assumed to be uniaxial and the ordered tetragonal phase can be assumed to have non-equilibrium uniaxial anisotropy, as discussed above, which could range from ~10^5 to 10^7 erg/cc. Even though this model has limitations (due to the possibility of an incoherent reversal mechanism (exchange length l_{ex} << particle size), co-existence of two phases and structural variants within a grain and uniform particle size assumptions), we can easily visualize the mechanism of the reversal process qualitatively. Although there exsists a significant fraction of the soft phase, it is hardly detectable due to intergranular exchange coupling. Even if we remove the remaining soft phase by long time annealing, the resulting large anisotropy is not favorable in the recording media applications.

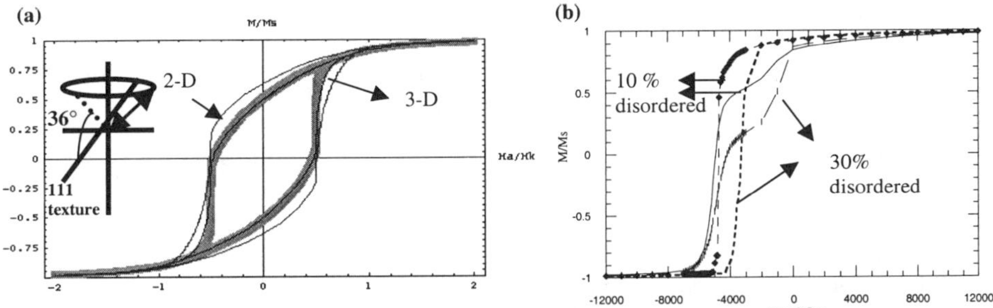

Fig.3 (a)Stoner-Wohlfarth model for 3-D random, 2-D random ,(111) texture of easy axis distributions(Cornical); (b)one of the micromagnetic simulation results containing soft phase 10%, 30% with and without effective exchange coupling $A^*=0.4 \times 10^{-11}$J/m ($h_{exchange}$ (normalized by H_k)$=A^*/KD^2=C^*=0.1$), D(particle size) =10nm, Ms common to both phases(=500 emu/cc), partially ordered phase assumed, $K_1 \approx 4 \times 10^6$ and $K_{1,disordered} \approx 6\times10^5$ erg/cc. Easy axis : 3-D random.

CONCLUSION

We have related the dependence of the magnetocrystalline ansiotropy to the long range order parameter in CoPt thin films. CoPt thin films deposited by RF or DC magnetron sputtering showed 'discontinuous' type transformation during the annealing procedure and consequently a disordered magnetically soft phase was retained. The (111) texture observed is not favorable for the longditudinal recording. A strong in-plane orientation must be induced using the underlayer to increase the S* which is related to 'transition parameter' in the longditudinal recording media.

APPENDIX

P (Probability of A atom occupying α site) : P_A^α , C: Atomic fraction of A atom.(C_A)

$N_A^\alpha + N_B^\alpha = 0.5N, N_A^\beta + N_B^\beta = 0.5N, N_A^\alpha + N_A^\beta = N_A, N_B^\alpha + N_B^\beta = N_B$

$USING, P = P_A^\alpha = N_A^\alpha / N_\alpha, P_B^\alpha = N_B^\alpha / N_\alpha, P_A^\beta = N_A^\beta / N_\beta, P_B^\beta = N_B^\beta / N_\beta, C_A = C, C_A + C_B = 1,$

$C_A / C_B \le 1 (= C_A \le 0.5), Then, P_B^\alpha = 0.5 - P, P_A^\beta = C - P, P_B^\beta = 0.5 - P_A^\beta = 0.5 - C + P, Therefore,$

$$l_{\alpha-\alpha} = P_A^\alpha P_A^\alpha l_{A-A} + 2 \bullet P_A^\alpha P_B^\alpha l_{A-B} + P_B^\alpha P_B^\alpha l_{B-B} = P^2 l_{A-A} + 2P(1-P) l_{A-B} + (1-P)^2 l_{B-B}$$

$$l_{\alpha-\beta} = P_A^\alpha P_A^\beta l_{A-A} + (P_A^\alpha P_B^\beta + P_B^\alpha P_A^\beta) l_{A-B} + P_B^\alpha P_B^\beta l_{B-B}$$
$$= P(2C-P) l_{A-A} + [P(P-2C+1) + (1-P)(2C-P) l_{A-B} + (1-P)(P-2C+1) l_{B-B}$$

$$l_{\beta-\beta} = P_A^\beta P_A^\beta l_{A-A} + 2 \bullet P_A^\beta P_B^\beta l_{A-B} + P_B^\beta P_B^\beta l_{B-B}$$
$$= (2C-P)^2 l_{A-A} + 2(2C-P)(P-2C+1) l_{A-B} + (P-2C+1)^2 l_{B-B}$$

ß laver

α laver

$$\xi = \frac{N_A^\alpha + N_B^\beta}{N} - \frac{N_A^\beta + N_B^\alpha}{N}$$

$\hat{S}$ ψ $\vec{r}_j$

$\hat{S}_i, \hat{S}_j = (\alpha_1, \alpha_2, \alpha_3)$

$\psi_j = \cos^{-1}(\hat{S} \bullet \vec{r}_j)$

Z α3 α2 Y X α1

ξ: Long range order parameter ξ=2(P-C)

P= P_A^α=ξ/2 + C => Plug this term into above equations, Then we get $l = l(\xi, C)$.

[1] D. J. Sellmyer, M. Yu and R. D. Kirby, Nano-98 June, 1998.
[2] Kevin R. Coffey, Michael A. Parker and J. Kent Howard, IEEE Trans., 2737(1995)
[3]G. Hadjipanayis and P. Gaunt, J. Appl. Phys. **50**(3), March 1979
[4] Bing Zhang, Univ. of Pittsburgh , Ph. D Thesis, 1991.
[5]J.Kent Howard ,invited talk MMM-Intermag conference in 98
[6]K.Barmak and R.A.Ritau, J. Appl. Phys. **79**(8), 5330 April ,1996
[7] Soshin Chikazumi, Physics of Ferromagnetism, Second edition.OXFORD,1997,pp.266
[8]Patrick Bruno,J.Phys. F: Met.Phys. **18**(1988) 1291-1298
[9]Bo-Bian, submitted to Nature.
[10]D.S.Chuang,C.A.Ballentine,R.C.O'Handley,Physical Review B,**49**,15084 (1994)
[11]J.M.MacLaren,R.H.Victora, J.Appl.Phys. **76**(10), 6069(1994)
[12]R.H.Victora, J.M.MacLaren, J.Appl.Phys. **73**(10), 6415(1993)
[13]H.N.Southworth and B.Ralph, Institute of Metals Monogragh No. **33**, 1969
[14]Pu-ling Lu, S,Charap, Ph.D Thesis 1995.
[15]Ferromagnetic materials, McCurrie, 1994, London, Academic.pp.230
[16]M. E. McHenry, J. M. MacLaren and D. P. Clougherty; J. Appl. Phys. **70**(10), 5902 (1991).
[17]Jian-Gang Zhu, Ph.D Thesis,1989
[18]S.H. Liou,Y.Liu,S.S.Malhotra,M.Yu and D. J. Sellmyer, J.Appl.Phys. **79**(8), 5060(1996)
[19]P.W.Jang,D.W.Kim,C.H.Park,J.G.Na and S.R.Lee, J. Appl. Phys. **83**(11), 6614 (1998)

Part IX

Fine Particle Magnets

CHEMICAL CHARACTERIZATION OF MAGNETIC MATERIALS AT HIGH SPATIAL RESOLUTION

JOHN HENRY J. SCOTT
Microanalysis Research Group, National Institute of Standards and Technology (NIST)

ABSTRACT

The fabrication of nanoscale magnetic materials and novel heterostructures has placed great demands on the analytical tools used to characterize the chemical and structural variations in these samples. Attempts to determine the composition, phase, crystallinity, and interface structure of nanoscale materials often involve tradeoffs between accuracy, sensitivity, and spatial resolution. Traditional techniques such as electron energy-loss spectroscopy (EELS), energy dispersive x-ray spectrometry (EDS), secondary ion mass spectrometry (SIMS), and scanning Auger microscopy (SAM) will be surveyed in this context. Powerful new tools for probing the chemical heterogeneity of materials at ultrafine length scales will also be discussed, including energy-filtered transmission electron microscopy (EFTEM) and spectrum imaging, as well as recent advances in new methods such as backscattered electron Kikuchi patterns (BEKP) and microcalorimetry.

INTRODUCTION

Advances in magnetic materials research rely increasingly on information about chemical distributions at near-atomic dimensions. Providing this information has placed great demands on characterization techniques. Several techniques that meet this challenge and provide detailed chemical information at high spatial resolution are compared here, including secondary ion mass spectrometry (SIMS), scanning Auger microscopy (SAM), electron probe microanalysis (EPMA), and analytical electron microscopy (AEM). Choosing among these techniques usually involves tradeoffs of spatial resolution, chemical sensitivity, ease of sample preparation, and ease of use. In many cases the instrument may also carry information about the sample's crystallinity or morphology to the analyst along with the chemical information, and this can also be a factor in choosing one technique over another. Indeed, these tradeoffs are often active within a given technique when the instrument may be operated in different modes. After a brief introduction to the problem and a discussion of analytical concerns common to all the techniques, each will be examined using recent data from magnetic systems to highlight the capabilities of the tool.

Characterization techniques used to probe the chemical distribution of magnetic materials can be classified according to their spatial resolution. Determination of chemical distributions at micrometer length scales has been performed for many years, and the field of microanalysis, exemplified by EPMA, is a mature science.[1,2] The limitations of this technique have been carefully examined and analyses are routinely performed with well-defined accuracy and precision by a large number of laboratories. Nanoanalysis techniques, such as AEM, are capable of determining chemical distributions at nanometer length scales. Often the limitations of these higher-resolution techniques are not understood fully, and some methods (e.g. quantitative EELS) are only beginning to gain wide acceptance as quantitative tools available for general use. Techniques which exhibit much better depth resolution than lateral resolution, such as SIMS and SAM, can be considered surface analysis, microanalysis, and nanoanalysis tools. When used to

Mat. Res. Soc. Symp. Proc. Vol. 577

Technique	Exciting Radiation	Analytical Radiation	Lateral Resolution	Depth Resolution	Trace Sensitivity	Accuracy (relative)
SIMS	ions	ions	50 nm-2 μm	2-30 nm	ppm-ppb	2-200%
Auger	e^-	e^-	50-100 nm	0.5-10 nm	1000 ppm	15-150%
SEM/EDS	e^-	x-rays	1 μm	1 μm	1000 ppm	5%
STEM/EDS	e^-	x-rays	2-10 nm	sample	1000 ppm	10%
SEM/WDS	e^-	x-rays	1 - 5 μm	1 μm	100 ppm	5%
SEM/Calorimeter	e^-	x-rays	1 μm	1 μm	100 ppm ??	5% ??
STEM/EELS	e^-	e^-	0.5-10 nm	sample	100 ppm	5-10%

Table 1. Capabilities of analytical techniques described in this work. All values are for typical samples and typical operating conditions. Most of the values listed are sample-dependent.

characterize a fully 3-dimensional microstructure, the limited lateral resolution of these methods classifies them as micro-techniques, but their high resolution depth profiling capabilities and surface sensitivity clearly qualify them as nano-techniques when applied to laterally homogeneous samples such as multilayers and thin films.

In contrast to grading the techniques according to their spatial resolving power, the task of classifying the methods according to their performance as chemical characterization tools is more complex. In some situations all that is required is the qualitative identification of major constituents (defined here as >10 wt.%), to distinguish a magnetic Co layer from an adjacent Cu layer for example. Other situations require the accurate determination of the relative concentrations of several minor constituents (≤10 wt.% but ≥1 wt.%), while a few applications demand sensitivity to trace concentrations (<1 wt.%) of analyte in the sample. In general, the needs of the magnetic materials community are less stringent than those of the semiconductor industry, where trace sensitivities on the order of 1 part in 10^9 are sometimes required. While the techniques mentioned here vary in their accuracy and sensitivity, all are suitable for the majority of analyses encountered in magnetic materials research. Further complicating the comparison, some techniques can go beyond identifying the elemental concentrations in the sample and provide chemical state or bonding information. This is true of Auger spectroscopy and EELS, where both the position of ionization edges (chemical shift) and their shapes (near-edge structure) can reveal features of the bonding state or valence interactions in the sample. A few auxiliary techniques such as backscattered electron Kikuchi pattern (BEKP) analysis and electron diffraction provide phase information which can yield chemical information through stoichiometry.

All of the characterization techniques described above excite the sample with a beam of incident radiation, often a focussed probe. After the probe interacts with the sample and scatters from it, analytical radiation emitted from the sample is collected and processed, yielding information about the sample composition and the scattering process. The radiation used for analysis is not necessarily the same as that used for excitation; in fact, it is usually advantageous (when possible) to collect multiple analytical signals from the sample simultaneously. Table 1 lists the exciting and analytical radiations employed by the primary techniques described in this work, along with estimates of their spatial resolution and ability to determine the chemistry of the sample.[3,4] The resolutions, sensitivities, and accuracies quoted here are typical of those achievable when analyzing magnetic materials, but it should be noted that the performance of a given technique is invariably sample-dependent. Particularly favorable cases can be found where an instrument will outperform these estimates, and in other cases factors will sometimes conspire to make analysis very difficult.

Because all of the techniques considered here rely on the collection of analytical radiation following excitation of the sample by an energetic probe, their ultimate sensitivities are governed by similar physics. To interrogate the sample efficiently and with high spatial resolution, it is usually necessary to focus the exciting radiation onto a small region of the specimen. The brightness of the primary excitation source, β, is a measure of the flux of incident radiation per solid angle that can be directed at the sample. Higher brightness means better excitation and better sensitivity. The success of field emission electron sources in the SEM, SAM, and AEM derives from the influence of this term on the ultimate sensitivity. Interaction of the exciting radiation with the sample, and the production of the analytical signal, can be described by a generation cross section, σ, a sample parameter usually expressed as an area. When processes with high cross section can be exploited for analysis, more sensitivity is expected since more signal is available for a given excitation of the sample. The collection and detection of the analytical radiation varies greatly from one technique to the next, but the overall efficiency of the detection chain can be characterized by a single number, ε, between 0 and 1. For some techniques, such as EELS, the collection efficiency approaches unity, while for other techniques ε can be 0.01 or less. Of course, the ability to detect small quantities of analyte is also dependent on spectral quality parameters such as the signal level of the characteristic radiation, P, and the background level, B, in the same region of the spectrum. Techniques with higher intrinsic signal-to-noise ratios are expected to be able to detect smaller quantities of analyte. Folding all these parameters together, the concentration limit of detection (expressed as a minimum mass fraction) can be expressed as,

$$C_{MMF} \cong (P/B)^{-1} (1/\sigma\varepsilon\beta\tau)^{1/2}, \quad (1)$$

where τ is the integration time used to collect signal, corrected for detector deadtime if appropriate.[5]

SIMS

SIMS is one of the most sensitive techniques available for the chemical characterization of magnetic materials. Detection limits for many elements are in the 10^{12}-10^{16} atoms/cm^2 range (i.e. ppb-ppm), but its lateral spatial resolution is poor compared to the other techniques discussed here. It has excellent depth resolution, however, as demonstrated by Figure 1, a SIMS depth profile of an artificial multilayer structure consisting of interleaved magnetic and nonmagnetic films upon a silicon nitride substrate layer. This data was acquired with a Cameca IMS-4F SIMS instrument[#] using a 3 nA, 2.5 keV Cs^+ primary ion beam incident at about 57° from the sample normal. The primary ions sputtered away successive layers of the sample at a rate of ~0.6 nm/min over a square area 250 μm on a side. In general, SIMS must consume (destructively) a significant mass of the sample to achieve its high sensitivity. All elements were detected as CsM^+ attachment species except Ta, which had to be detected as Ta^+ because of the mass-range limit of the spectrometer. This helps to reduce the sizable matrix effects that make accurate concentration measurements difficult with SIMS.

[#] Certain commercial equipment, instruments, or materials are identified in this paper in order to specify the experimental procedure adequately. Such identification is not intended to imply recommendation or endorsement by the National Institute of Standards and Technology, nor is it intended to imply that the materials or the equipment identified are necessarily the best available for the purpose.

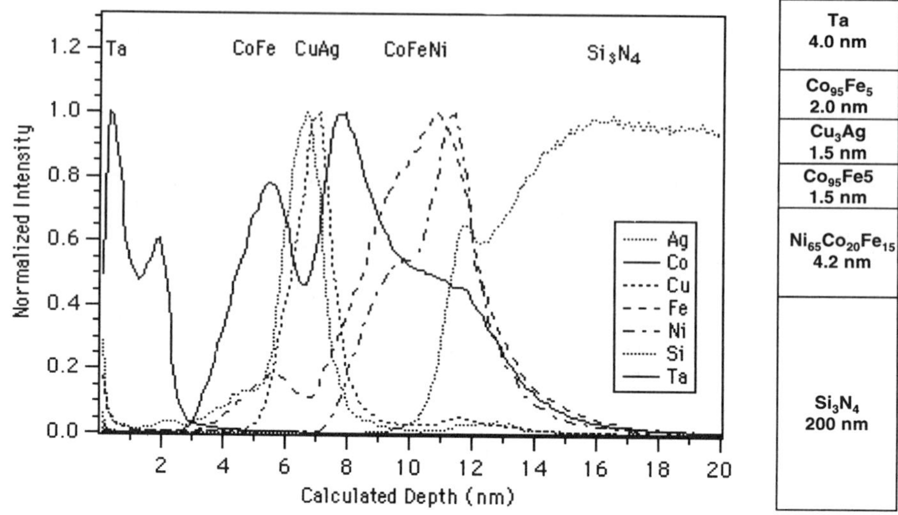

Figure 1. Depth profile of a multilayer structure acquired by secondary ion mass spectrometry (SIMS). The nominal structure is shown at right.

AUGER

Figure 2 shows an Auger depth profile of the same artificial multilayer structure appearing in Figure 1. Auger electrons were excited from the sample in a scanning Auger microscope by a 50 nA incident beam of 10 keV electrons. Simultaneously, the sample was sputtered by 0.5 keV Ar ions at a rate of ~0.6 nm/min. Auger electrons dispersed by an electron spectrometer were used to detect and quantify the elements present in the sample, and the depth profile was produced by plotting these intensities (normalized to unity) against sputtering time. As in the SIMS case, conversion of the abscissa from time to depth was accomplished by a simple linear scaling that assumed the CuAg layer was 6.8 nm below the surface. This model ignores the differences in sputtering rates between different materials, but is common in depth profiling because of its simplicity and the difficulty measuring the depth continuously and accurately. The Auger measurement shown here was performed after the SIMS measurement, and surface oxidation would explain the presence of the large oxygen surface concentration seen in Figure 2. Other differences between the profiles serve as an example of the uncertainty inherent in sputtering-based techniques.

EPMA

Electron probe microanalysis (EPMA) excites the sample using a focussed electron beam, usually from a scanning electron microscope (SEM) or a dedicated electron microprobe. The resulting x-rays emitted from the sample are collected and analyzed quantitatively to measure chemical concentrations with relatively high accuracy and spatial resolution. The two most common x-ray detection schemes are energy dispersive x-ray spectrometry (EDS) and

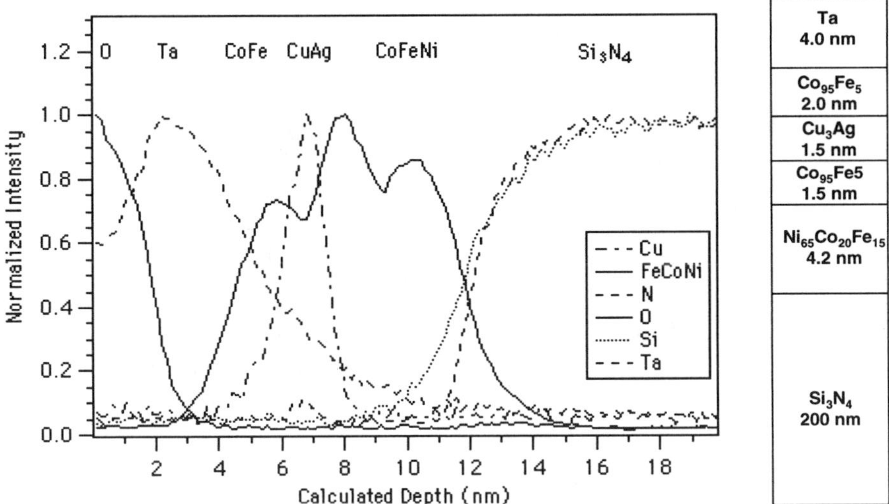

Figure 2. Depth profile of a multilayer acquired by scanning Auger microscopy (SAM). This is the same structure as seen in Figure 1.

wavelength dispersive x-ray spectroscopy (WDS). EDS has the advantage of parallel acquisition (all x-ray energies are collected simultaneously), but WDS has a much higher energy resolution and, hence, a much higher signal-to-background ratio. A typical energy resolution for WDS is 10 eV, while the best EDS detectors do no better than 130 eV. The $(P/B)^{-1}$ term in equation (1) explains how this added energy resolution translates into added chemical sensitivity. For quantitative measurements of elemental concentrations in flat, polished, bulk samples WDS is widely used despite the laborious need to scan the diffracting-crystal spectrometer sequentially over all the energies of interest. The spatial resolution of both WDS and EDS is limited by the finite excitation volume in the sample, not the electron probe diameter. Figure 3 shows two Monte Carlo simulations of electron trajectories in two materials. On the left, 15 keV incident electrons are seen to excite a much smaller volume in a ribbon of HiTPerm amorphous magnet than in a silicon nitride wafer, shown on the right. Unfortunately, the added interaction strength in materials with high average atomic number also results in partial absorption of the outgoing x-ray signal that carries the chemical information to the analyst. To counter this problem a set of numerical correction procedures (e.g. ZAF and phi-rho-Z corrections) have been developed to convert the observed x-ray intensities into elemental concentrations.[1,2,6]

The chemical information provided by EDS or WDS x-ray measurements can be used to augment the imaging modes of the SEM or microprobe. If an x-ray measurement is made at each pixel in the SEM scan raster, an elemental map can be generated that shows the distribution of the measured element over the field of view. Figure 4(a) is a conventional secondary electron image of mixed Fe and Co powders used as precursors in the synthesis of Fe-Co nanoparticles.[7] Figures 4(b,c) are x-ray maps of Fe and Co respectively, detailing the distribution of individual Fe and Co particles in the powder mixture. These maps show the level of heterogeneity remaining in the powder despite thorough mixing. The enormous amount of information conveyed by such maps compared to point spectra is obvious.

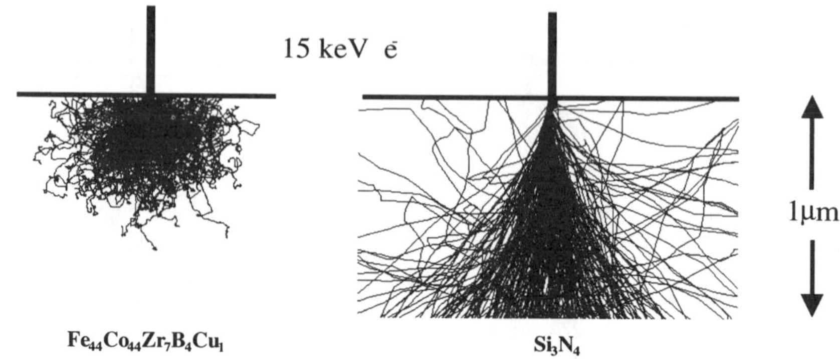

Figure 3. Monte Carlo simulation of 15 keV electrons incident on an amorphous ribbon of HiTPerm (left), and a silicon nitride substrate (right).

A recent advance in EDS technology promises to revolutionize x-ray spectrometry in EPMA. The NIST microcalorimeter EDS detector combines the best features of WDS and EDS in a single spectrometer, offering parallel detection of many energies simultaneously (like EDS) but with an energy resolution of 3 eV, better than many WDS spectrometers.[8,9] The microcalorimeter uses a superconducting transition-edge sensor cooled to 100 mK, and a superconducting quantum interference device (SQUID) current amplifier. Instead of measuring the number of electron-hole pairs created in a silicon wafer by incoming x-ray photons, as in conventional EDS detectors, the microcalorimeter measures the energy deposited in a small thermal mass kept at a constant temperature by a feedback loop. Figure 5 shows two simulated EDS spectra from an amorphous ribbon of HiTPerm,[10,11] one using a conventional lithium-drifted Si EDS detector with 138 eV energy resolution and the other using the NIST microcalorimeter EDS operating at an energy resolution of 3 eV. With the Si(Li) detector the Fe Kβ line interferes with the Co Kα lines due to severe peak overlap. The microcalorimeter not only resolves the Co Kα lines from the Fe Kβ line, it is also able to resolve the Co Kα doublet.

A second advance that promises to improve the chemical characterization capabilities of scanned electron beam instruments is backscattered electron Kikuchi pattern (BEKP) analysis, also known as electron backscattered pattern (EBSP) analysis.[12] In this technique a focussed electron probe is directed at the surface of a sample, usually in an SEM, and the spatial and angular distribution of backscattered primary electrons is captured by a scintillator mounted on a CCD camera. Figure 6(a) shows a conventional SEM image of NIST Standard Reference Material 485 (SRM485), 5% retained austenite in ferrite. Figure 6(b) shows a BEKP image from one of the austenite grains. The bright and dark Kikuchi lines result from diffraction of the backscattered electrons by crystallographic planes of atoms in the excited volume. Intersections of lines represent zone axes in the sample, and their position and distribution allow the phase and orientation of the excited region to be determined. Figure 6(c) shows a BEKP image from a ferrite grain on the same sample after matching the pattern to a crystallographic model. This tool, when coupled through software to a crystallographic structure database such as the JCPDS-ICDD powder diffraction file provides for phase identification in the SEM. At least for stoichiometric phases, this can lead to accurate chemical information as well.

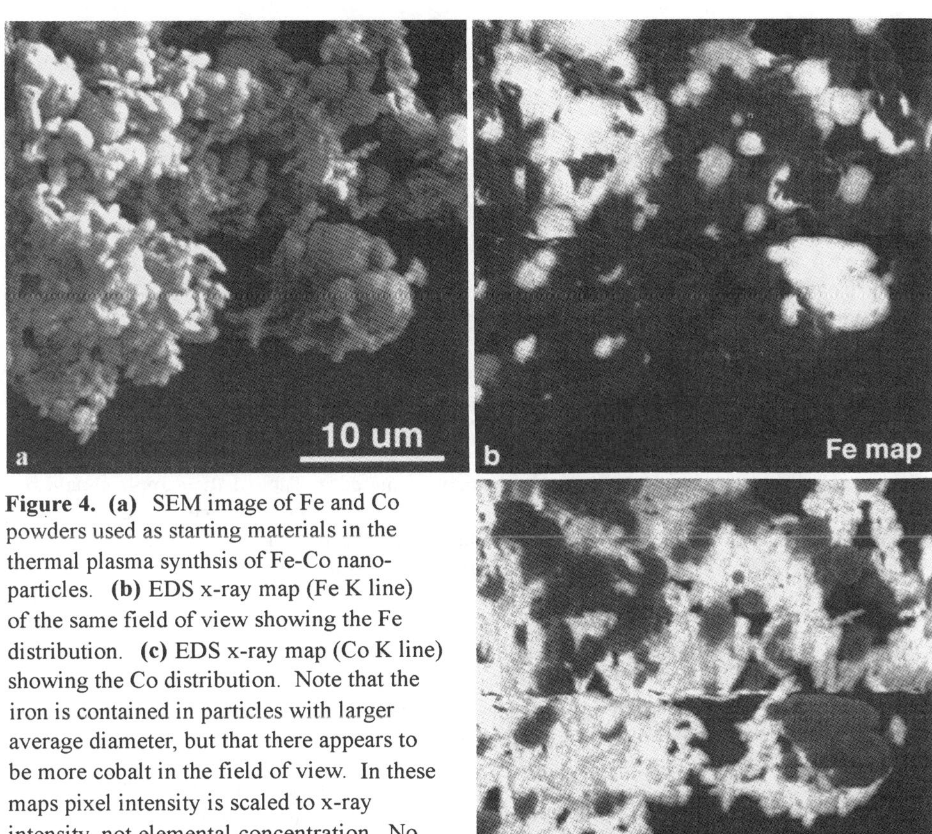

Figure 4. (a) SEM image of Fe and Co powders used as starting materials in the thermal plasma synthsis of Fe-Co nanoparticles. **(b)** EDS x-ray map (Fe K line) of the same field of view showing the Fe distribution. **(c)** EDS x-ray map (Co K line) showing the Co distribution. Note that the iron is contained in particles with larger average diameter, but that there appears to be more cobalt in the field of view. In these maps pixel intensity is scaled to x-ray intensity, not elemental concentration. No corrections have been applied.

AEM

The analytical electron microscope, typically consisting of an intermediate-voltage transmission electron microscope/scanning transmission electron microscope (TEM/STEM) equipped with EDS and EELS spectrometers, is capable of imaging and chemical analysis with very high spatial resolution.[13] In STEM mode, modern instruments can generate focussed probes 1 nm in diameter or less with sufficient current to excite analytical signals such as x-rays for EDS or inelastically-transmitted electrons for EELS. In TEM mode, the same microscope is often capable of acquiring high resolution lattice images with a point-to-point resolution of better than 0.2 nm. This high spatial resolution is realized, in part, because of the very thin samples demanded by the technique. To achieve the degree of electron transparency needed for analysis in transmission mode, the sample must be less than a few hundred nanometers thick, and (depending on the average atomic number and mass density of the material) sometimes much

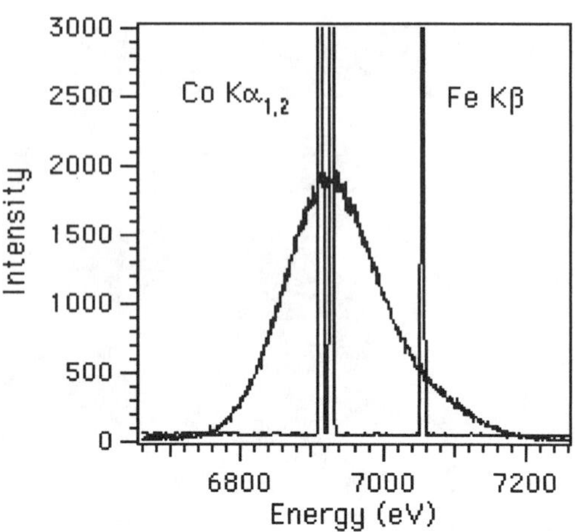

Figure 5. Simulated x-ray spectra from a bulk sample of HiTPerm ($Fe_{44}Co_{44}Zr_7B_4Cu_1$) amorphous ribbon, as detected by the NIST microcalorimeter (3 eV energy resolution, narrow peaks) and a conventional Si(Li) detector (138 eV resolution, broad peaks). While the Fe Kβ and Co Kα lines exhibit severe overlap in the Si(Li) spectrum, they are easily resolved by the microcalorimeter.

thinner. To yield the lateral resolutions for analysis quoted in Table 1 the sample should be no thicker than a few tens of nanometers. Meaningful high resolution lattice images are acquired from regions of the sample with thicknesses less than 10 nm. Figure 7 shows a Monte Carlo simulation of 300 keV electrons propagating through a 50 nm thick film of $Nd_2Fe_{14}B$. Clearly the lateral resolution depends on the sample thickness, as indicated by the widening cone of excitation with depth. Some samples, such as nanoparticle powders and free-standing ultrathin films, are intrinsically transparent and are ideally suited for study in the AEM. Most samples, however, must be artificially thinned from the bulk, a process that is difficult, time consuming, and prone to introduce structures or artefacts not present in the original material. These sample preparation issues are a serious drawback to analysis in the AEM.

Chemical information in the AEM is obtained through analytical techniques such as EDS and EELS. EDS in the TEM is similar to EDS in the SEM or electron microprobe. Because of the thin specimens, however, the complex partial-excitation, x-ray absorption, and fluorescence corrections needed in EPMA are much smaller in the AEM case, often negligible. The brehmsstrahlung background is typically smaller as well, leading to more accurate measurements of net peak counts. These factors simplify the calculation of chemical concentrations from raw x-ray intensities. Usually a simple relative sensitivity factor approach is used.[14] In all, EDS in the AEM is a trusted, robust technique.

Quantitative EELS in the AEM is less widely used, even though it offers several advantages over EDS, including better spatial resolution, better collection efficiency, better sensitivity to low-Z elements, better trace detection limits, better energy resolution, and sensitivity to chemical bonding effects.

Most of these benefits arise because EELS analyzes scattered primary electrons to characterize the sample, not secondary signals like characteristic x-ray photons. The scattering cross section of high energy (100-300 keV) primary electrons in solids is strongly peaked in the forward direction. For a target of intermediate atomic number (copper) and a relatively low primary energy (100 keV), the most probable scattering angle is 2°-3°.[15] This narrow solid angle of emergence allows most of the primaries to enter the entrance aperture of the EELS spectrometer, resulting in high collection efficiencies. In EDS the x-rays emitted by the de-exciting atoms are radiated isotropically over the full range of angles. Since Si(Li) detector chip areas are about 30 mm^2, and the detector must be located at least several millimeters away from

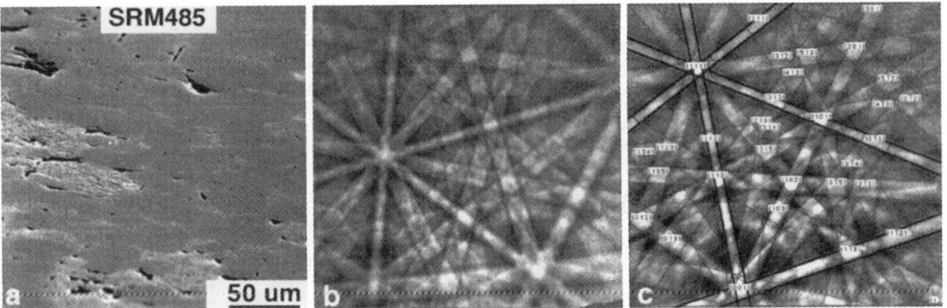

Figure 6. **(a)** SEM image of NIST Standard Reference Material 485, 5% retained austenite in ferrite. The surface has been metallographically polished and etched to produce optical contrast between the austenite and the ferrite grains. **(b)** Backscattered electron Kikuchi pattern (BEKP) of an austenite region acquired in the SEM. **(c)** BEKP after indexing the Kikuchi lines to identify the crystallographic phase and orientation.

the sample due to geometric restrictions and interference with the objective lens pole piece, an EDS spectrometer can only subtend a small fraction of the 4π steradian emission solid angle. This, combined with absorption losses in the detector window material, lead to overall EDS collection efficiencies of no more than a few percent, even for high-Z materials. The situation for low-Z atoms is more grim. The characteristic x-ray emission pathway competes with other mechanisms for atomic relaxation, notably Auger electron emission. The likelihood of photon emission, called the fluorescence yield (ω), increases rapidly with atomic number. While ω reaches about 90% at Z=54, it is only 2% for Z=13 (aluminum).[16] The soft x-rays emitted by low-Z atom relaxations are also more strongly attenuated by the EDS detector window. EDS thus suffers a triple penalty compared to EELS: fewer photons are emitted, fewer of those emitted travel towards the detector, and fewer that reach the detector are counted by the electronics.

Up to this point the discussion has focussed on acquisition of a single 1-dimensional EELS or EDS spectrum. Although acquisition of point spectra using static, focussed probes remains a popular and powerful tool for examining small volumes, much more information can be obtained by combining the AEM's imaging capabilities with the analytical modes. The EDS signal from a scanned probe can be used to produce x-ray maps that reveal elemental distributions across a 2-dimensional image. This is straightforward and is very similar to the equivalent process in the SEM, except for the much higher spatial resolution that can be achieved. As in the SEM, multiple elemental maps may be acquired simultaneously because of the parallel detection nature of EDS.

More exciting combinations are possible when 2-dimensional imaging is combined with EELS spectroscopy: electron spectroscopic imaging (ESI) and spectrum imaging. ESI is based on the acquisition of energy-selected images, or images formed from electrons whose energy loss falls within a narrow window.[17,18] These images are acquired in TEM mode, with a spread electron beam for illumination, not with a focussed probe in STEM mode. Since each energy-selected image only contains contributions from electrons of a single energy, if several such images are acquired at different energies they can be stacked to develop a data cube for the sample. Two dimensions of this cube are the familiar x and y spatial locations of the pixels on the sample. The third cube axis is energy, as shown schematically in Figure 8(a).

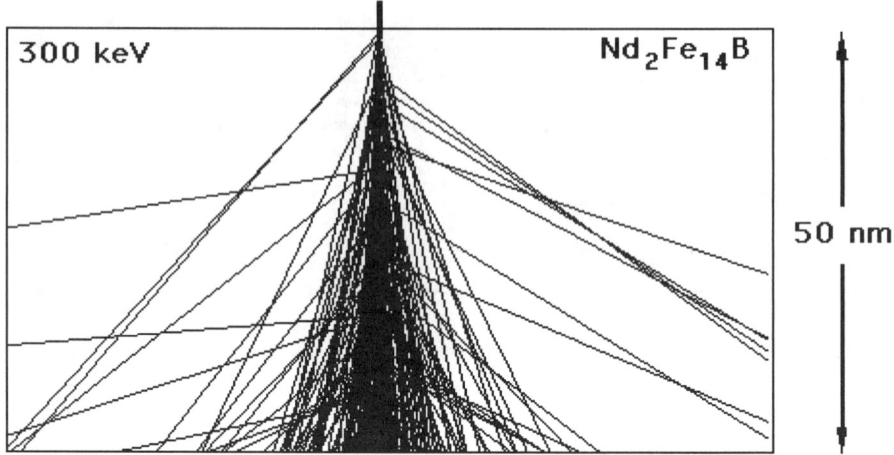

Figure 7. Monte Carlo simulation of 300 keV electrons propagating through a thin film of $Nd_2Fe_{14}B$, as might occur in a TEM/STEM. The beam broadening due to scattering within the film is evident, and results in decreased lateral resolution for thicker specimens. The sample is 50 nm thick.

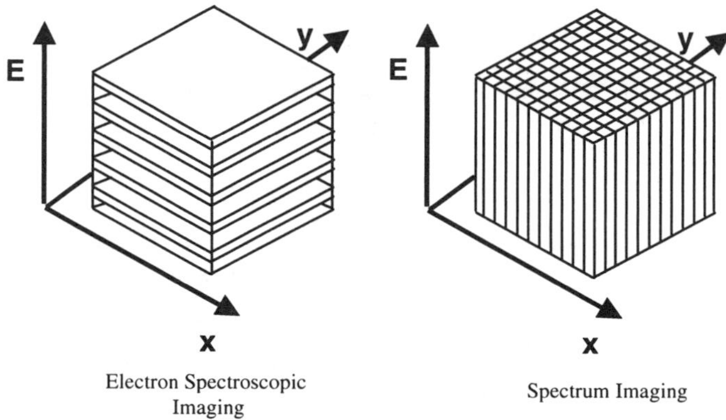

Figure 8. Schematic showing how ESI and STEM-based spectrum imaging build up an identical data cube using two different methods. In ESI the analyst acquires a series of 2-dimensional energy-selected images, each at a different energy. In spectrum imaging, the sample is interrogated by a focussed probe and an EELS spectrum is acquired at each pixel.

Spectrum imaging develops the same data cube for the sample, but in a different way.[19,20] In this technique a focussed STEM probe is used to acquire an entire EELS spectrum down the energy axis of the cube before the probe is rastered to the next pixel, as seen in Figure 8(b). Both techniques collect a large amount of information from the sample in a relatively short time and rely on cpmputer post-processing to generate quantitative chemical distributions from the dataset.

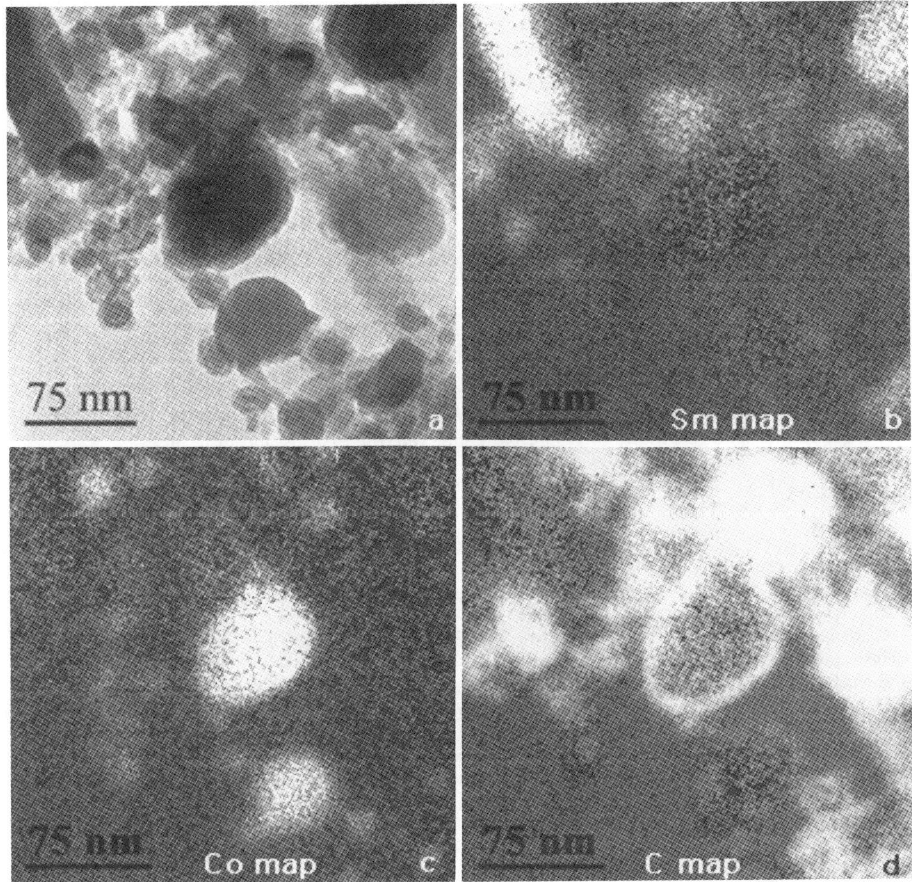

Figure 9. Electron sprectroscopic imaging (ESI) of a Sm-Co-C nanocomposite. **(a)** Bright field TEM image showing the morphology of the nanocomposite. A carbon-coated Co particle can be seen at center, and a Sm-containing rod is visible in the upper left. **(b)** Sm map of the same field of view, produced using three energy-selected images and a background subtraction model. **(c)** Co map. **(d)** C map. The carbon coating surrounding the central Co nanoparticle is clearly visible in this view.

Figure 9 shows an example of ESI used to map the spatial distribution of Sm, Co, and C in a nanocomposite sample produced by the carbon arc evaporation of graphite electrodes doped with samarium oxide and cobalt.[21] Figure 9(a) is a conventional TEM image of the sample, showing coated nanoparticles, rod-like structures, and carbonaceous debris. Figures 9(b,c,d) are Sm, Co, and C elemental maps produced by ESI using a Gatan imaging energy filter (GIF). For each of the three elemental maps, three energy-selected images were acquired. Two energy slices were collected below the element's ionization edge and an inverse power law background model was used to calculate, pixel by pixel, the background intensity at an energy loss above the ionization edge. This calculated image was subtracted digitally from the third observed energy slice, taken above the edge, to produce an elemental map. The goal of the synthesis experiment was to produce carbon-coated $SmCo_5$ nanoparticles. While the high spatial resolution of the EELS maps reveals a carbon coating around the central Co particle, the Sm remained segregated from the Co. This lack of mixing is not clear from the TEM micrograph in Figure 9(a).

ACKNOWLEDGMENTS

The author is indebted to David S. Simons and John T. Armstrong of the NIST Surface and Microanalysis Science Division for providing the data used in Figures 1 and 2, and to Joe Michael of Sandia National Laboratory for the BEKP images. He would also like to thank Sara A. Majetich, Elaine Kirkpatrick, Anit Giri, Krishna Chowdary, Zafer Turgut, Michael McHenry, and N. Tom Nuhfer of Carnegie Mellon University for help preparing the samples used in the EELS examples, for technical assistance, and for many useful discussions. The author gratefully acknowledges NIST and the NRC Postdoctoral Research Associateship program for support, and thanks Eric Steel, John Small, and Dale Newbury for suggestions and guidance.

REFERENCES

1. J.I. Goldstein, D.E. Newbury, P. Echlin, D.C. Joy, A.D. Romig, Jr., C.E. Lyman, C. Fiori, E. Lifshin, *Scanning Electron Microscopy and X-ray Microanalysis, 2nd Edition*, (Plenum Press, New York, 1992).
2. D.E. Newbury, D.C. Joy, P. Echlin, C. Fiori, J.I. Goldstein, *Adv. Scanning Electron Microscopy and X-ray Microanalysis*, (Plenum Press, New York, 1986).
3. C. Richard Brundle, C.A. Evans, Jr., S. Wilson, *Encyclopedia of Materials Characterization*, (Butterworth-Heinemann, Boston, 1992).
4. E. Fuchs, H. Oppolzer, H. Rehme, *Particle Beam Microanalysis*, (VCH, New York, 1990).
5. D.B. Wittry, 7th Eur. Cong. Electron Microscopy/9th Intl. Con. X-ray Optics and Microanalysis, eds. P. Breeders and V.E. Cosset (Leyden), **3**, 14 (1980).
6. K.F.J. Heinrich and D.E. Newbury, *Electron Probe Quantitation*, (Plenum Press, New York, 1991).
7. J.H. Scott, S.A. Majetich, Z. Turgut, M.E. McHenry, K. Chowdary, M. Boulos, *Thermal Plasma Synthesis of Magnetic Nanoparticles*, submitted to J. Mater. Res.
8. D.A. Wollman, K.D. Irwin, G.C. Hilton, L.L. Dulcie, D.E. Newbury, J.M. Martinis, J. Microscopy-Oxford **188**(3), 196 (1997).
9. K.D. Irwin, G.C. Hilton, D.A. Wollman, J.M. Martinis, Appl. Phys. Lett., **69**(13), 1945 (1996).
10. M.A. Willard, D.E. Laughlin, M.E. McHenry, D. Thoma, K. Sickafus, J.O. Cross, V.G. Harris, J. Appl. Phys. **84**, 6773 (1998).
11. M.A. Willard, M.-Q. Huang, D.E. Laughlin, M.E. McHenry, J.O. Cross, V.G. Harris, C. Franchtti, J. Appl. Phys **85**(8), 4421 (1999).
12. R.P. Goehner, J.R. Michael, J. Res. NIST **101**(3), 301 (1996).
13. R. D. Leapman and D.E. Newbury, Anal. Chem., **65**, 2409 (1993).
14. D.B. Williams and C.B. Carter, *Transmission Electron Microscopy,* (Plenum Press, New York, 1996), p. 599.
15. D.E. Newbury, in *Principles of Analytical Electron Microscopy*, edited by David C. Joy, Alton D. Romig, Jr., and Joseph Goldstein (Plenum Press, New York, 1986), p. 5.
16. M. Cardona and L. Ley, in *Photoemission in Solids I, General Principles*, edited by M. Cardona and L. Ley (Springer-Verlag, Berlin, 1978), p. 78.
17. R. F. Egerton, *Electron Energy-Loss Spectroscopy in the Electron Microscope*, (Plenum Press, New York, 1986), p. 312.
18. O.L. Krivanek, C.C. Ahn, R. B. Keeney, Ultramicroscopy **22**, 103 (1987).
19. C. Jeanguillaume and C. Colliex, Ultramicroscopy **28**, 252 (1989).
20. J.A. Hunt and D.B. Williams, Ultramicroscopy **38**, 47 (1991).
21. S.A. Majetich, J.H. Scott, E. Kirkpatrick, M.E. McHenry, in *Intl. Symp. on the Sci. & Technol. of Atomically Engineered Mater.*, eds. P. Jena, S. Khanna (World Scientific Pub. Co., 1996).

CONTROLLED ASSEMBLY OF MONODISPERSE ε-COBALT-BASED NANOCRYSTALS

S. SUN*, C. B. MURRAY AND H. DOYLE
IBM T. J. Watson Research Center, Yorktown Heights, NY 10598, ssun@us.ibm.com

ABSTRACT

General synthetic routes to monodisperse ε-cobalt (β-Mn type phase) nanocrystals (ε-Co) and controlled assembly of these nanocrystals are presented in this paper. The ε-Co particles are obtained by superhydride reduction of cobalt chloride (anhydrous or hexahydrate) in a high temperature solution phase (200°C) in the presence of a combination of long chain diol, oleic acid and trialkylphosphine. Monodisperse nanocrystals are isolated by size selective precipitation. As synthesized cobalt particles are each a single crystal with a complex cubic structure related to the beta phase of elemental manganese (β–Mn). Self-assembly of these uniform cobalt particles on solid substrates is induced by evaporation of the carrier solvent producing 2-D and 3-D magnetic superlattices. Annealing of assembled ε-Co nanocrystal arrays converts them to the hcp cobalt crystal arrays. The inter-particle distance can be adjusted by selected thermal treatments or by chemical ligand exchange. This control over particle dimensions, crystallinity and assembly offers a model system for the study of ultra-high density recording media.

INTRODUCTION

Nanometer size crystals, Nanocrystals (NCs), display many properties which are distinct from their respective bulk materials and discrete atomic or molecular species. These novel properties of NCs arise from the large fraction of atoms which reside on the surface of the particle and from the finite number of atoms in each crystalline core. Recent research has focused heavily on the synthesis of monodisperse nanoparticle materials and their assembly into nanocrystalline thin film with controlled engineering properties and technological functions. Thin granular films of ferromagnetic metal NCs are already the basis of conventional magnetic storage media. Progress in magnetic recording density is due in part to the development of media with finer and finer grains and controlled dispersity in size and shape of these grains [1-4]. Fabrication of granular magnetic materials in a dielectric or non-magnetic metal matrix will be of great interest due to the observation of magneto-resistive phenomena [5-11]. Due to their small size, small nanoparticles have large self charging energies, making it possible to build metallic nanoparticle based single electron devices [12-14].

A relatively new approach for constructing nanocrystalline thin film structures is to use monodisperse NCs as building blocks which assemble to form 2-D or 3-D structures (colloidal crystals). Sterically stabilized NCs can be organized into close-packed arrays simply by evaporating the solvent from a dispersion. This general approach to NC superlattices has been shown to apply to a variety of particle materials, including metal Co [15] Au [16, 17], Ag [18, 19], and semiconductor CdSe [20] NCs. Regular arrays of nanoparticles can also be fabricated from solution phase by either normal/reverse micelle method [21] or LB techniques [22].

Various synthetic approaches have been developed in make metal particles with a very narrow size distribution. Ideally, the growth of these particles can be terminated at a desired stage by confinement in a supporting matrix or passivation with organic capping molecules. Organic

Mat. Res. Soc. Symp. Proc. Vol. 577 © 1999 Materials Research Society

and inorganic matrices, such as monolayers, polymers, normal / inverse micelles, and zeolites, have been used to control particle size. Pyrolysis of organometallic reagents to initiate nucleation and subsequent growth via Ostwald ripening for size selective precipitation of nanocrystallites have proven to be an effective way to make semiconductor NCs with near atomic precision [23]. Significant progress has also been made in preparing monodisperse noble metal nanocrystals (Au, Ag, and Pd, Pt) [24-29]. Strong magnetic interactions in Ni, Fe, and Co particle systems make it difficult to form stable colloids. Uncontrolled agglomeration of the magnetic particles often makes it impossible to employ separation procedures which could narrow the distribution, and prevent the ready formation of the smooth, thin films required in practical applications.

We recently reported that reduction of anhydrous cobalt chloride by superhydride in the presence of oleic acid and trialkylphosphine yielded stable ε-Co NCs [15]. In this paper, we will describe the generalized procedures for monodisperse (σ <7% std. dev.) ε-Co NCs and demonstrate the formation of ordered nanocrystals arrays. The ε-Co NC assembly can be converted to the hcp-Co NC assembly by controlled annealing. We will also demonstrate our preliminary experimental results of controlling inter-particle spacing via thermal treatment and ligand exchange.

EXPERIMENT

The synthesis was carried out using standard airless procedures and commercially available reagents. Tetrahydrofuran (THF) solution of superhydride ($LiBEt_3H$) (1M), dioctylether (99%) ($C_8H_{17}OC_8H_{17}$), phenyl ether (99%) (Ph_2O), 1,2-dodecanediol (97%) [$1,2\text{-}CH_3\text{-}(CH_2)_9\text{-}CH(OH)\text{-}CH_2(OH)$], and oleic acid (90%) [$CH_3(CH_2)_7CH{=}CH(CH_2)_7COOH)$] were purchased from Aldrich Chemical Company. Cobalt chloride (anhydrous or hexahydrate) was from Strem Chemicals, Inc. Tributylphosphine {$(P[(CH_2)_3CH_3]_3$} and trioctylphosphine {$P[(CH_2)_7CH_3]_3$} were from Fluka Chemika. Dioctylether solutions of superhydride (2M concentration) were prepared by mixing the THF superhydride solution with dioctylether and evaporating THF under vacuum.

Synthesis of monodisperse ε-Co NCs:

Method A: Cobalt chloride (anhydrous) (1 mmol), oleic acid (1 mmol) and dioctylether (20 mL) were mixed under nitrogen and heated to 100°C. Tributylphosphine (3 mmol) was added and the mixture was then heated to 200°C. Dioctylether superhydride solution (1 mL, 2 mmol superhydride) was injected under vigorous stirring. A color change from dark blue to black was observed, indicating reduction of the blue Co^{2+} complex to form cobalt particles. The black solution was stirred at 200°C for 20 minutes and then cooled to room temperature. Co NCs were precipitated by the addition of ethanol to the dispersion. The supernatant was discarded either by decanting or by centrifugation. The waxy magnetic precipitate then redispersed in 10 mL hexane with 100 - 500 μl of oleic acid to ensure sufficient surface passivation to provide a stabile colloid. Size selective precipitation and centrifugation were employed to fractionate the colloids.

Method B: Cobalt chloride hexahydrate (4 mmol), oleic acid (4 mmol), and phenylether (40 mL) were mixed under nitrogen and heated to 200°C. The mixture was stirred at 200°C for 30 minutes. Tributylphosphine (4 mmol) was added and the mixture was then heated to 240°C. Phenyl ether solution of 1,2-dodecanediol (10 mmol in 6 mL Ph_2O) was added and the solution was stirred at 240°C for about 30 minutes, and cooled to 200°C. Dioctylether solution of superhydride solution (2x2 mL, 8 mmol) was injected stepwise during vigorous stirring. The black solution was stirred at 200°C for 15 minutes and cooled to room temperature. The following work-up procedure was the same as that described in **Method A**.

Method C: As in **Method B**, cobalt chloride hexahydrate, oleic acid, and phenylether, tributylphosphine and phenyl ether solution of 1,2-dodecanediol were mixed and heated. Dioctylether solution of superhydride solution (4 mL, 8 mmol) was added dropwise in a course of about 5 minutes during vigorous stirring. The black solution was stirred at 200°C for 15 minutes, followed by the treatment described in **Method A**.

NC superlattices (colloidal crystals) formed spontaneously when a thin layer of cobalt NC dispersion was spread on a solid substrate and allowed to slowly dry. Evaporation of a dilute dispersion (~2% particles) over a period of 20 min at temperatures between 20-80°C produced thin films containing close-packed monolayer and multilayer regions. Samples for transmission electron microscopy (TEM) analysis were prepared by drying an alkane dispersion of cobalt particles on amorphous carbon or silicon monoxide coated copper grids, while X-ray samples employed silicon (100) wafer or glass as substrates.

The annealing was performed either under vacuum (0.2 mmHg), N_2 or Ar + H_2(5%). During annealing, the protective organic shell was destroyed and subsequent exposure of unpassivated NCs to air can lead to the rapid formation of layers of cobalt oxide. Oxidation of the annealed samples was prevented by treating the annealed samples with acetone before exposing them to air.

Cobalt and phosphorus elemental analyses of the cobalt particles were performed on an ICP-Atomic Emission Spectrometer at Galbreith Lab, Tennessee. TEM imaging was used to determine particle size and size distribution. Commercial 30 nm Au particles (BBInternational, UK) were used as an internal standard for low resolution TEM measurements on a Philips EM 420 (120 KV) or EM 430 (300 KV). High resolution TEM images showing the lattice structure of the individual particles were obtained on a JEOL 4000FX (400 KV). X-ray powder diffraction was performed on an Inel CPS 590 diffractometer with Co Kα radiation (λ = 1.788965 Å).

Magnetic studies were carried out using MPMS2 Quantum Design SQUID magnetometer. Measurements of thin film were carried out on self-assembled and annealed thin films. The ordered thin films were prepared by evaporation of particle dispersion on silicon oxide substrates and loaded with the substrate parallel to the magnetic field.

RESULTS AND DISCUSSION

Solution phase metal salt reduction is a well known technique for the preparation of cobalt particle materials. The reducing agents employed to reduce cobalt salts include alkaline metal [30], borohydride (including superhydride) [31, 32], and polyol [33]. Electrochemical reduction has also been used to reduce cobalt salt to cobalt particles [34, 35]. A variety of stabilizing ligands, such as surfactants and polymers, have been used to control particle growth, stabilize particle dispersions and limit oxidation of the particles. The colloids are stable when a repulsive force of sufficient strength and range exists to counteract the combined attractive forces. Cobalt NCs experience strong van der Waals attractions which, combined with the magnetic dipole interactions, make stabilizing these systems quite challenging. We have employed a combination of surfactants - trialkylphosphine and oleic acid to control particle growth, stabilize the particles, and prevent them from oxidation. Trialkylphosphine reversibly coordinates neutral metal surface sites, slowing but not stopping NC growth. It can not prevent particles from eventually growing to undispersable aggregates at high temperature when used alone. Oleic acid is an excellent stabilizing agent for many particle systems including Fe and its oxide particle systems [36-38], but in our synthetic conditions, it does not permit controlled growth and adequately stabilize Co NCs. The combination of trialkylphosphine and oleic acid presents a necessary stabilizing shell protecting the Co NCs from aggregation and oxidation.

Co NCs are synthesized by rapid injection of superhydride solution in dioctylether solution (Method A). They can also be made by an alternate approach of stepwise injection of fraction volume of superhydride solution (Method B) or simply by dropping superhydride solution into the reaction mixture (Method C). The processes (Method B and C) permit large scale synthesis of cobalt particle materials. The quality of material is comparable to that reported for Method A and only two stages of size selection is required for the isolation of monodisperse particles. The step-wise injection or dropping should be finished in 5 minutes. Continued heating at 200°C for about 15 minutes allows steady growth of these clusters into nanometer sized, single crystals of cobalt. Prolonged heating results in excessive particle growth and the formation of unredispersed powdered cobalt. At desired stage, growth of the particles is quenched by cooling the dispersion. Average particle size can be coarsely controlled by selecting the type of alkylphosphine used in the combination with oleic acid during the growth. A bulky ligand such as $P(C_8H_{17})_3$ limits growth to produce smaller particles (3-7 nm) while $P(C_4H_9)_3$ leads to larger (6-11 nm) particles. Small particles can also be separated by adding superhydride to cobalt chloride solution at temperature lower than 200°C, e.g. down to 170°C. Fine tuning of the particle size can be done by fractionating the coarsely adjusted samples and selecting the desired size fraction. In this way a series of nearly monodisperse NC samples are produced. A typical X-ray diffraction pattern of 11 nm cobalt particles is shown in Fig.1. Shown in Fig.2 are TEM images of 8 nm and 11 nm particles. The measurement of X-ray line widths yields a crystal coherence length which matches with TEM particle size measurements, indicating that each particle is a single crystal, as can be seen in a HRTEM image of one 8 nm particle in the inset of Fig.2 (a).

The ε-Co NCs are very uniform in both size and shape, allowing them to self-organize readily into 2-D and 3-D superlattices during slow evaporation of a carrier solvent. TEM images of a monolayer region of two different sized particles are shown in Fig.2. The images show that each particle is separated from its neighbor, indicating each cobalt NC is sheathed in a robust organic ligand shell to limit oxidation and prevent irreversible aggregation. The use of solvent like dodecane or octane as the carrier solvent for the cobalt NCs allows slow evaporation of the solvent and facilitates particle diffusion to the lowest energy superlattice sites during evaporation, producing well-defined 3-D hcp superlattice structure. More rapid evaporation leads to the formation of randomly packed NC solids. Superlattices of 8 nm and 11 nm particles are shown in Fig 3.

Two stable forms (hcp and fcc) of bulk elemental cobalt are known. In practice, many bulk cobalt samples display mixed structures with hcp and fcc character. A metastable bcc structure of cobalt has been detected in MBE experiments [39]. Solution phase chemical synthesis is generally not thermodynamically controlled and thus can allow preparation of crystal phases which are metastable. Method A and B produced cobalt particles with the cubic symmetry of the β-phase of manganese structure [15, 40]. It was a soft magnetic material with a coercivity of approx. 500 Oe at 5K. The crystal structure of the NCs was confirmed by X-ray diffraction (Fig.1) and was also revealed by a high resolution TEM image (inset of Fig.2a). The HRTEM shows clear cobalt lattice planes with a spacing of 1.99 Å, close to d_{310} =1.997 Å spacing expected for ε-Co. Method C produced Co particles with bulk structures of ε-Co with some hcp stacking faults. Computer simulations of the X-ray patterns indicated that the samples contained about 70% of ε-Co and 30% of hcp Co. The metastable cubic ε-Co could be transformed to stable crystal phase by thermal annealing. Fig.4 displays a series of X-ray diffraction reflections from as synthesized 8 nm particles (Fig.4a) and annealed particle films (Fig.4b-d). It shows that

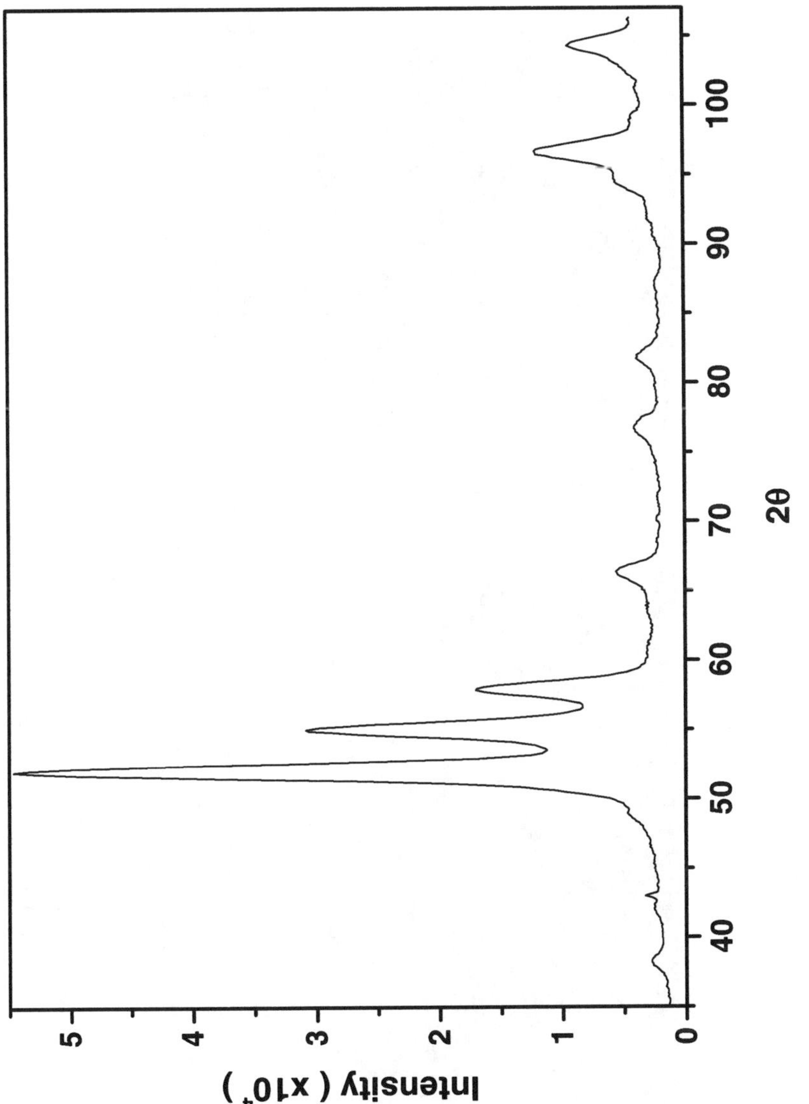

Fig.1. XRD pattern of 11 nm ε-Co nanocrystals on Si(100).

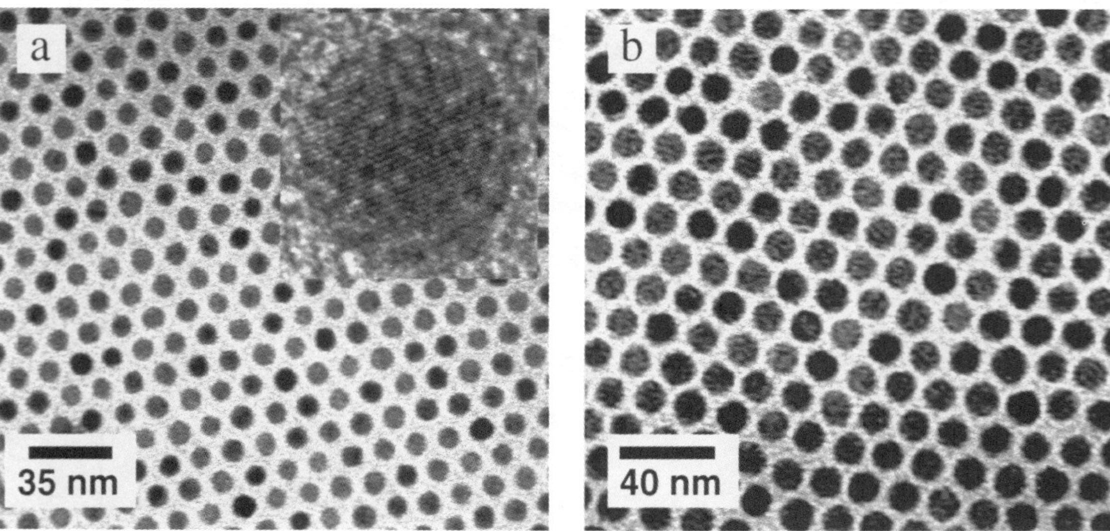

Fig.2. TEM micrographs of 2D arrays of ε-Co nanocrystals with sizes of (a) 8 nm and (b) 11 nm. Inset: HRTEM image of 8 nm ε-Co.

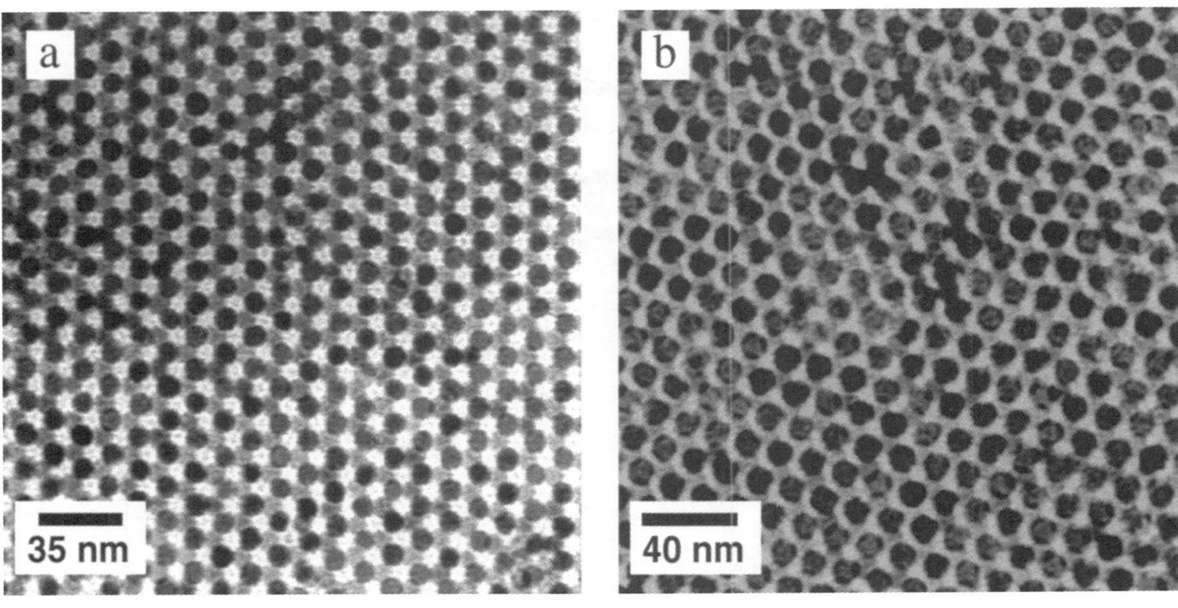

Fig.3. TEM micrographs of 3D arrays of ε-Co nanocrystals with sizes of (a) 8 nm and (b) 11 nm.

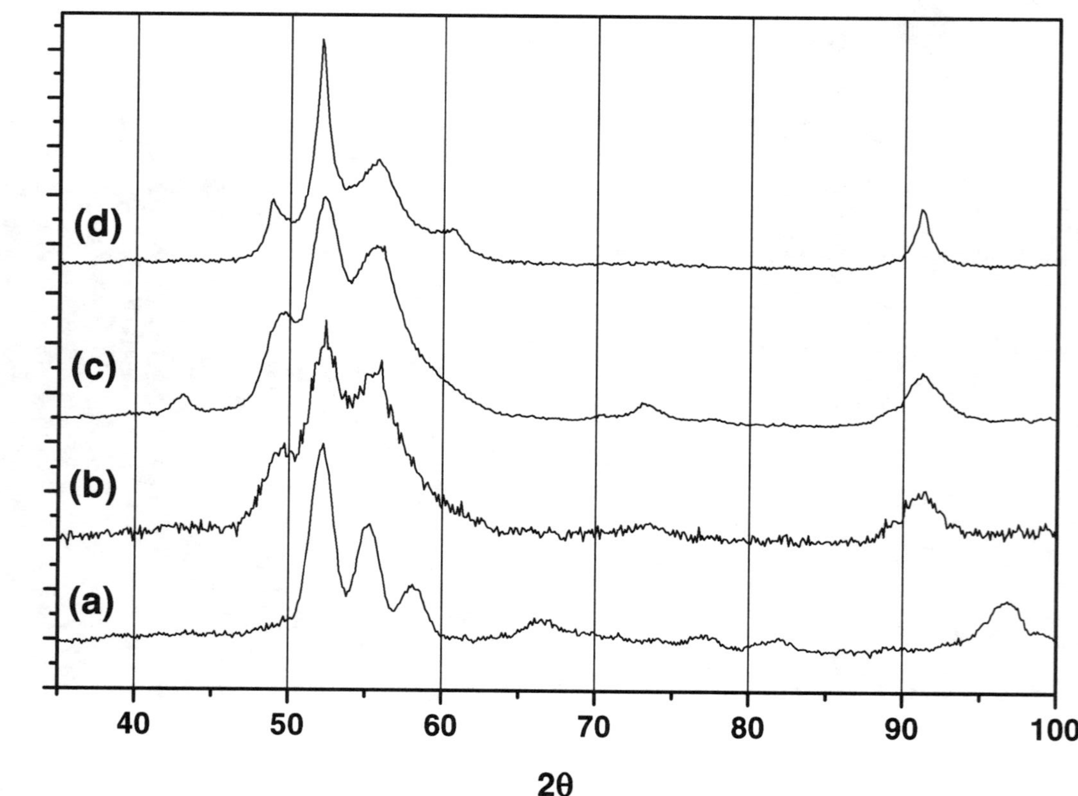

Fig.4. XRD patterns of as-synthesized 8 nm ε-Co nanocrystalline film (a), and annealed films at 300°C under vacuum (b), 350°C under vacuum (c), 450°C under Ar + H_2 (5%) (d).

upon annealing to 300°C, ε-Co NCs is converted to a predominantly hcp crystal phase (Fig.4b). Simulations revealed that the NCs contained 60% of hcp character and 40% of fcc character.

TEM images indicated that superlattice symmetry of the regions containing 2D arrays were unchanged by annealing at 300°C under vacuum or inert atmosphere, but the regions of 3D superlattices were converted to a cubic (Fig.5) rather than the original hexagonal superlattice structure (Fig.3). The driving force for this packing change might be due to the internal structure change of cobalt from cubic ε-Co to the uniaxial nature of the predominantly hcp structure. Further annealing of the samples at 500°C on particle film led to dramatic change not only in core crystal structure and the stacking of the particles but in particle size as well, with extensive fusing and sintering of the NCs and destruction of the superlattice. X-ray diffraction patterns and simulation showed that 500°C annealed particles film possessed NCs with predominantly fcc internal structure particles with only a small portion of hcp component (<15%). This structure change is consistent with the bulk cobalt behavior; the bulk hcp form is stable at temperatures below 425°C, while the bulk fcc form is the stable structure at higher temperature.

The facile conversion of ε-Co particles to more hcp-like nanoparticles at 300°C permits the magnetic properties of particle arrays to be modified after assembly. 5 K hysteresis measurements of as synthesized 8 nm ε-Co particle film on silicon oxide show a coercive field (H_c) of 500 Oe. Annealing at 300°C increased the coercivity to 1500 Oe. Further annealing of the particle film at 500°C for 3 h did not give higher H_c even though TEM and X-ray analysis show that grain growth occurred along with the conversion to a largely fcc structure. Instead, H_c decreased to 700 Oe. This coercivity change confirms that it is the internal phase transformation of the particles and not an increase in particle size that responsible for higher coercivity at 300°C.

No distinct peaks corresponding to CoO phase were detected in as synthesized particles and 300°C annealed particles. However particle thin film annealed at >350°C show sign of CoO. This can be easily seen in Fig.4c. Presumably, this CoO layer is formed during the decomposition of carboxylate groups on the particle surface. This oxide layer can be readily reduced to fcc cobalt layer by annealing CoO-coated film at 450°C under Ar + H_2 (5%), as shown in Fig.4d. It should be mentioned that annealed film was very stable towards further oxidation, probably due to presence of the oxy-carbide thin layer coating the particles. Complete oxidation of annealed thin film did not happen even if the film was treated at 300°C under air for 3 h. However as shown below, annealing under air did allow control of interparticle spacing in 2-D particle arrays.

Our goal is to prepare 2-D and 3-D superlattice with adjustable superlattice periods and composed of monodisperse particles with controlled composition and internal structure. Investigation of physical phenomena such as magnetostatic coupling, exchange coupling and electron tunneling between particles will require such precise control. Our annealing experiments on silicon oxide surface show that the inter-particle spacing of 2-D arrays did not change at 300°C . We have found that thermal treatment of annealed thin film under air at different temperatures resulted in the formation of 2-D particle arrays with different inter-particle distance. TEM measurements did not reveal obvious particle growth. The inter-particle distance, however, was decreased from 4 nm to ~2 nm (for 100°C air treated sample) and finally to ~1 nm (for 200°C air treated sample), as illustrated in Fig.6. Treatment of annealed sample under air at 300°C resulted in the formation of CoO layers and thus the increase of particle size. These transformations are readily followed by X-ray diffraction and TEM analysis. Thus a sequence of controlled thermal treatments of particle arrays on silicon oxide surfaces can be used to control interparticle spacing in the 2-D supperlattices.

A more established approach to control interparticle spacing, ligand exchange, is also applicable. Ligand exchange of oleic acid for shorter chains from dodecyl down to hexyl

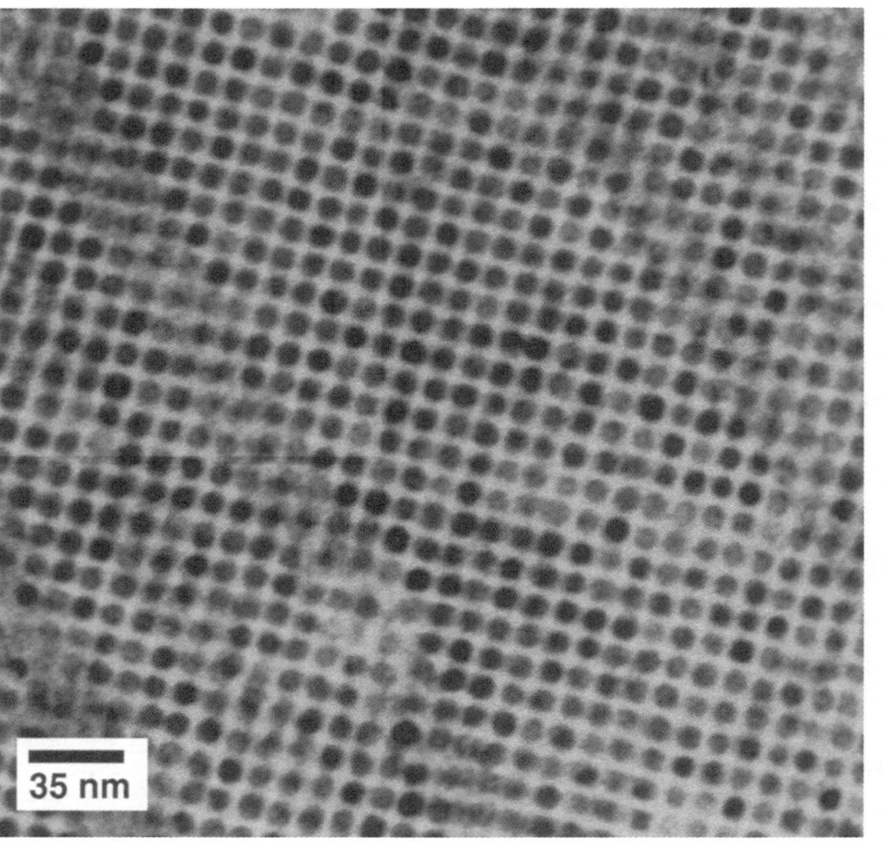

Fig.5. TEM micrograph of 3D particle arrays of 8 nm ε-Co nanocrystals after annealed at 300°C under vacuum.

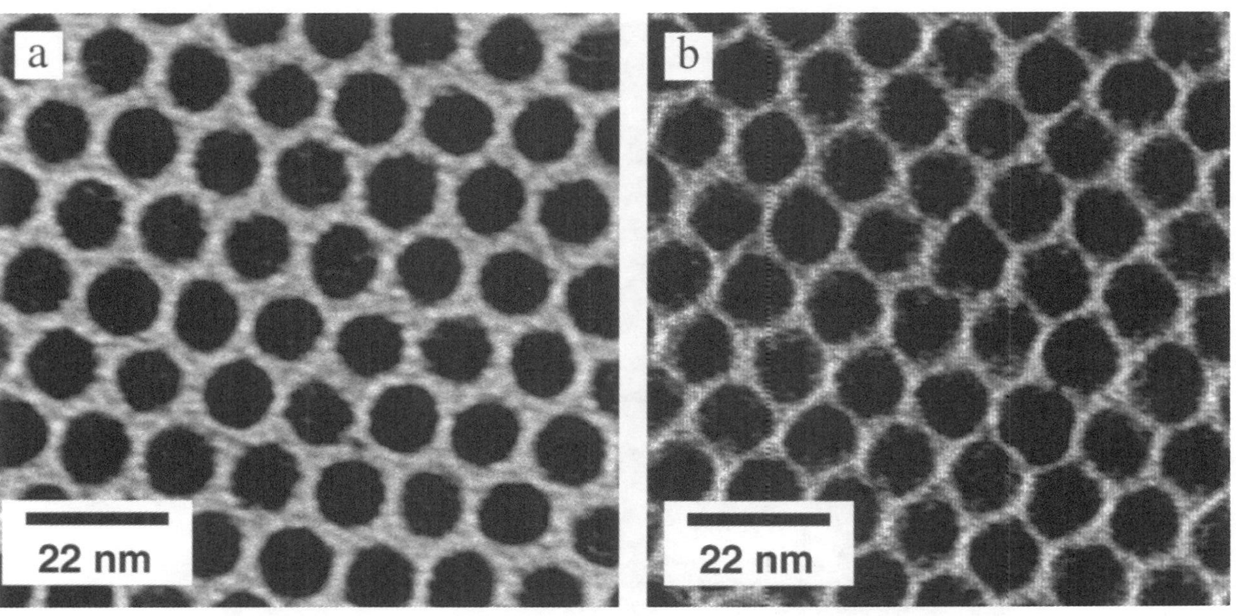

Fig.6. TEM micrographs of 300°C annealed 11 nm cobalt 2D particle arrays after (a) 100°C treatment under air and (b) 200°C treatment under air.

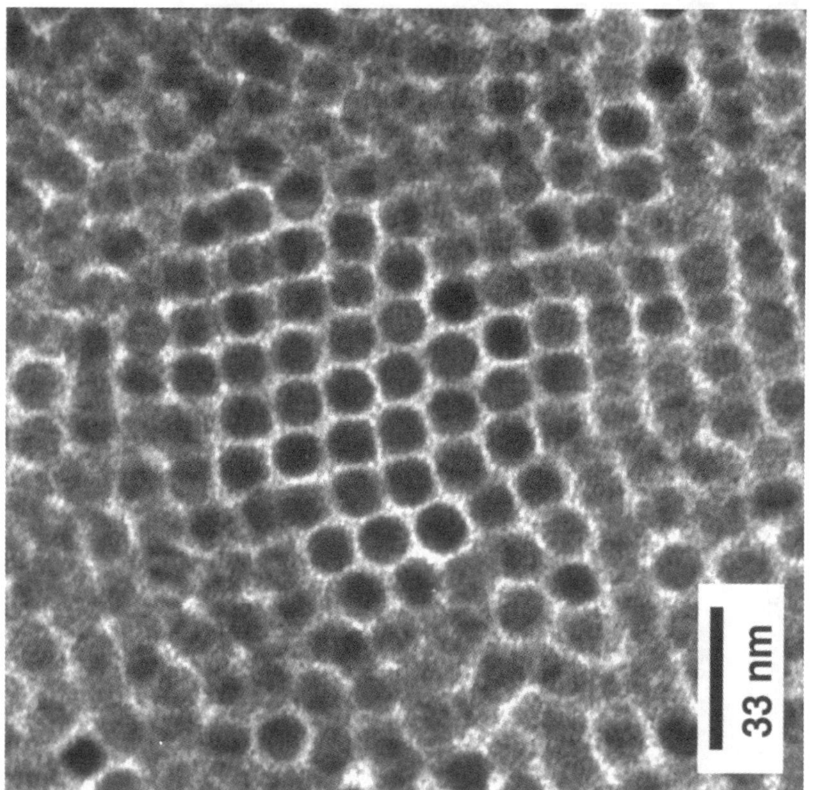

Fig.7. TEM micrograph of 2,4-hexadienoic acid protected 11 nm ε-Co nanocrystals.

hydrocarbon is possible at room temperature, allowing the interparticle distance to be decreased. Fig.7 is a result of 2,4-hexadienoic acid replacement of oleic acid on Co particles. Interparticle spacing was changed from originally 4 nm to around 1 nm due to the short chain length of 2,4-hexadienoin acid (<0.6 nm). Thus in addition to thermal treatment approach to control interparticle distance, engineering particle spacing of cobalt nanocrystal arrays can also be performed at room temperature with varied length and bulkiness of stabilizing ligand.

CONCLUSIONS

General synthetic routes to monodisperse Co NCs with an unusual cubic structure (ε-Co) with the symmetry of the beta phase of elemental Mn has been presented. The ε-Co particles were obtained by superhydride reduction of cobalt chloride in a high temperature solution phase (200°C). Controlled growth and steric stabilization of cobalt NCs was provided by the combination of oleic acid and trialkylphosphine. Monodisperse particle materials were isolated by size selective precipitation. Annealing the particles at 300°C resulted in the phase transformation of the ε-Co structure to the hcp structure. Procedures to induce the self-organization of these monodisperse particles into 2D and 3D regular arrays with control over inter-particle spacing were identified, permitting systematic investigations of magnetic properties and electronic transport phenomena in magnetic nanoparticle superlattices. This will provide new opportunities to engineer nanoscale magnetic systems.

REFERENCES

1. E. S. Murdock, R. F. Simmons, R. Davison, IEEE Trans. Mag., **28**, 3078 (1992).
2. T. Yogi, T. A. Nguyen, IEEE Trans. Mag., **29**, 307 (1993).
3. P.-L. Lu, S. H. Charap, IEEE Trans. Mag., **30**, 4230 (1994).
4. D. N. Lambeth, E. M. T. Velu, G. H. Bellesis, L. Lee, D. E. Laughlin, J. Appl. Phys., **79**, 4496 (1996).
5. C. L. Chien, J. Appl. Phys. **69**, 5267 (1991).
6. A. E. Berkowitz, J. R. Mitchell, M. J. Carey, A. P. Young, S. Zhang, F. E. Spada, F. T. Parker, A. Hutten, G. Thomas, Phys. Rev. Lett., **68**, 3745 (1992).
7. J. Q. Xiao, J. S. Jiang, C. L. Chien, Phys. Rev. Lett., **68**, 3749 (1992).
8. Y. G. Pogorelov, M. M. P. de Azevedo, J. B. Sousa, Phys. Rev. B, **58**, 425 (1998).
9. S. Takahashi, S. Maekawa, Phys. Rev. Lett., **80**, 1758 (1998).
10. S. Yamada, T. Kikutani, N. Aoki, H, Hori, Phys. Rev. Lett., **81**, 5422 (1998).
11. D. L. Peng, K. Sumiyama, S. Yamamuro, T. Hihara, T. J. Konno, Appl. Phys. Lett., **74**, 76 (1999).
12. R. P. Andres, T. Bein, M. Dorogi, S. Feng, J. I. Henderson, C. P. Kubiak, W. Mahoney, R. G. Osifchin, R. Reifenberger, Science, **272**, 1323 (1996).
13. A. Bezryadin, C. Dekker, G. Schmid, Appl. Phys. Lett., 71, 1273 (1997).
14. D. Davidovic, M. Tinkham, Appl. Phys. Lett., 73, 3959 (1998).
15. S. Sun, C. B. Murray, J. Appl. Phys., **85**, 0000 (1999).
16. R. P. Andres, J. D. Bielefeld, J. I. Henderson, D. B. Janes, V. R. Kolagunta, C. P. Kubiak, W. J. Mahoney, R. G. Osifchin, Science, **273**, 1690 (1996).
17. J. Fink, C. J. Kiely, D. Bethell, D. J. Schiffrin, Chem. Mater., **10**, 922 (1998).
18. S. A. Harfenist, Z. L. Wang, R. L. Whetten, I. Vezmar, M. M. Alvarez, Adv. Mater. **9**, 817 (1997).

19. B. A. Korgel, S. Fullam, S. Connolly, D. Fitzmaurice, J. Phys. Chem. B, **102**, 8379 (1998).
20. C. B. Murray, C. R. Kagan, M. G. Bawendi, Science, **270**, 1335 (1995).
21. M. P. Pileni, Langmuir, **13**, 3266 (1997).
22. C. P. Collier, R. J. Saykally, J. J. Shiang, S. E. Henrichs, J. R. Heath, Science, **277**, 1978 (1997).
23. C. B. Murray, D. J. Norris, M. G. Bawendi, J. Am. Chem. Soc., **115**, 8706 (1993).
24. G. Schmid, Chem. Rev., **92**, 1709 (1992).
25. M. T. Reetz, W. Helbig, J. Am. Chem. Soc., **116**, 7401 (1994).
26. M. T. Reetz, S. A. Quaiser, Angew. Chem., Int. Ed. Engl., **34**, 2240 (1995).
27. R. G. Osifchin, W. J. Mahoney, J. D. Bielefeld, R. P. Andres, J. I. Henderson, C. P. Kubiak, Superlattices and Microstructures, **18**, 283 (1995).
28. S. A. Harfenist, Z. L. Wang, M. M. Alvarez, I. Vezmar, R. L. Whetten, J. Phys. Chem., **100**, 13904 (1996).
29. T. Teranishi, M. Miyake, Chem. Mater., **10**, 594 (1998).
30. R. D. Rieke, Acc. Chem. Res., **10**, 301 (1977).
31. G. N. Glavee, K. J. Klabunde, C. M. Sorensen, G. C. Hadjapanayis, Langmuir, **8**, 771 (1992).
32. H. Bonnemann, W. Brijoux, T. Joussen, Angew. Chem. Int. Ed. Engl., **29**, 273 (1990).
33. F. Fievet, J. P. Lagier, M. Figlarz, MRS Bulletin, 29 (1989).
34. J. A. Becker, R. Schafer, R. Festag, W. Ruland, J. H. Wendorff, J. Pebler, S. A. Quaiser, W. Helbig, M. T. Reetz, J. Chem. Phys., **103**, 2520 (1995).
35. D. Hofmann, W. Schindler, J. Kirschner, Appl. Phys. Lett., **73**, 3279 (1998).
36. M. D. Bentzon, J. van Wonterghem, S. Morup, A. Tholen, C. J. W. Koch, Phil. Mag., **60**, 169 (1989).
37. K. S. Suslick, M. Fang, T. Hyeon, J. Am. Chem. Soc., **118**, 11960 (1996).
38. C.Y. Hong, C. J. Hsu, Y. D. Yao, H. C. Yang, J. Appl. Phys., **81**, 4257 (1997).
39. G. A. Prinz, Phys. Rev. Lett., **54**, 1051 (1985).
40. D. P. Dinega, M. G. Bawendi, Angew. Chem., to be published.

Thermal Plasma Synthesis of γ-FeN_x Nanoparticles as Precursors for the $Fe_{16}N_2$ Synthesis by Annealing

Z. Turgut*, D. E. Ferguson**, M. Q. Huang***, W. E. Wallace*** and M. E. McHenry*

*Carnegie Mellon University, Dept. of Materials Sci.& Eng.

*** and Carnegie Mellon Research Institute
Pittsburgh, PA 15213

**Virginia Polytechnic Institute and State University
Dept. of Physics, Blacksburg, VA 24061

Abstract

An ordered $Fe_{16}N_2$ phase has been reported with iron moments as high as 3.2 μ_B. It is precipitated from nitrogen martensite structures ideally containing 10.5 at.% nitrogen. Due to the highly distorted crystal structure and metastability of this phase non-equilibrium processing routes are sought to synthesize this phase. Here we report on radio frequency (RF) plasma torch synthesis which is used to produce FeN_x nanoparticles quenched into a body centered tetragonal (bct) structure as precursors for further annealing studies to form α"- $Fe_{16}N_2$ phase. We have employed a Tekna PL-50 type 50 kW, RF plasma torch. A plasma gas mixture containing 40 standard liters per minute (slpm) Ar and 8 slpm Hydrogen - 70 slpm Ar gas was used as a sheath gas. Iron powder (< 10 μm) was injected into the plasma stream using Ar flowing 15 slpm as a carrier gas. Nitrogen and Ammonia were used as a nitrogenization sources. Relatively low injection rates were used in order to achieve smaller particle sizes and thus faster quenching rates. We were able to produce particles containing up to 45 % of the quenched γ-phase. Observations based on x-ray diffraction (XRD) determination of lattice expansion and phase transition temperatures observed by differential thermal analysis (DTA) indicated that the quenched phase contains 6.5 atomic % nitrogen. Scherrer analysis of the fine particle broadening indicated that the average particle size for γ- phase is 27 nm, whereas this value is found to be 55 nm. for α-Fe. Nitrogen is well known for its grain size refinement in Fe thin films. Saturation magnetizations were found to be as low as 123 emu/g due to the presence of the nonmagnetic γ-FeN_x phase.

I. INTRODUCTION

α"- $Fe_{16}N_2$ was first discovered as a metastable phase[1] forming in the decomposition of nitrogen martensite to ferrite and Fe_4N. After this discovery most of the earlier work was focused on determining the crystal structure of this phase. In the early seventies *Takashi and Kim* [2] found out that this new phase had a so called " giant magnetic moment" observed in thin films. Since then a large amount of work has been done on Fe-N system motivated by the possibility of modifying the well known Slater-Pauling curve in which $Fe_{65}Co_{35}$ has remained as a highest moment alloy system known for over 70 years. The technical importance of this high moment iron nitride phase has been discussed for thin film recording head applications. Studies of the α"- $Fe_{16}N_2$ system have yielded puzzling and attractive magnetization values reported over the past two decades. Fe magnetic moments ranging from 3.2 μ_B (well above the average Fe moment for $Fe_{65}Co_{35}$) to 2.3 μ_B (slightly larger than pure iron) [3-6]. It has been proposed that variability in the reported magnetic moment in α"- $Fe_{16}N_2$ may be explained by different stress state, film thickness, or synthesis route, however a systematic variation with any of these parameters has not been reported [7,8]. Electronic structure calculations to date do not predict such a large moment on Fe. Band structure calculations performed by different workers[9-14] yielded average iron moments between 2.37-2.50 μ_B. In 1991 Sugita *et al.*[3,15] experimentally inferred average moments as high as 3.2 μ_B on epitaxially grown single crystals and TEM studies of single crystal films were also reported [16].

In terms of the crystal structure, α"- $Fe_{16}N_2$ phase is characterized by full and long range ordering of the interstitial nitrogen resulting in a distortion of the originally cubic lattice to a b.c.t structure with the c/a ranging from 1 to 1.10 [1] depending up on the nitrogen concentration. Three unequal iron sites Fe(4e), Fe(8h), Fe(4d) exist in the structure with different local symmetries and nearest and next nearest neighbor coordination by iron and nitrogen. It is suggested that the effect

Mat. Res. Soc. Symp. Proc. Vol. 577

of N atom on the electronic structure and magnetic properties is different for each iron site and it is well defined by both band structure calculations and Mossbauer studies[17-19].

In previous work[20,21] we employed the procedure first reported by Jack[1] and have reported average iron moments as high as 2.94 μ_B (approximately 34% higher than that of elemental Fe). In this paper we focus our attention to thermal plasma synthesis of nanocrystalline γ-FeN as a precursor for further annealing to form α"- $Fe_{16}N_2$.

II. EXPERIMENT

We have employed a radio frequency plasma system consisting of a TEKNA PL 50 type plasma torch head and a gas expansion-reaction vessel connected to a filtering unit. A 50 kW, 3 MHz power supply (Lepel) was used. Since cooling rates as high as 10^5 °C/s are achievable with this process we took advantage of this to quench iron nanoparticles (with nitrogen in solution) into austenite which then used for annealing studies. Iron powder <10 micron was injected axially into the plasma jet along with N_2 flowing 50 standard liters per minute (slpm) as a nitrogenizing agent. Argon gas flowing at 40 standard liters per minute (slpm) was used as the plasma gas and the sheath gas was consisted of 70 slpm Ar mixed with 8 slpm of hydrogen. During the synthesis, the chamber pressure was maintained at ~250 torr. At plasma temperatures, evaporation of iron powder is accompanied with the dissociation of molecular nitrogen resulting in a final product which is two phase consisting of γ-Fe and α-Fe with nitrogen in solution. It is believed that the amount of quenched phase in the final product is determined mostly by non-equilibrium synthesis conditions. While water cooled reactor walls act as splash surfaces, particles formed homogeneously through the reactor column have chance to complete γ-Fe --> α-Fe phase transformation and to grow larger, since the residence time of those particles in plasma jet is longer than that of particles quenched on the reactor walls. Both nitrogen and ammonia were used as nitrogen resource. No significant change in the amount of quenched phase was observed as nitrogen resource switched to ammonia.

Magnetic characterizations were carried on a vibrating sample magnetometer (VSM) equipped with a furnace which can operate at temperatures up to 1300 K. A Rikagu type 2-theta machine was used for X-ray measurements along with Cu-Kα radiation. A JEOL 120 CX transmission electron microscope (TEM) was employed for microstructural characterizations. Differential thermal analysis (DTA) measurements were performed on a Perkin-Elmer 1700 type differential thermal analyzer. Magnetic measurements were performed at 5 °C per minute while this value was 10 °C for DTA scans.

III. RESULTS AND DISCUSSION.

X-ray measurements performed on as produced powders revealed a two phase product. A Scherrer analysis of the peak broadening indicated an average particle diameter of d~ 55 nm for α-Fe while same average diameter for γ-Fe particles is ~27 nm. agreement with the TEM observations in Figures 1a and 1b respectively.

Following the heat treatment procedure as first introduced by Jack, annealing studies were performed at 125 °C for several minutes assuming faster diffusion values observed in nanoparticles may result in faster kinetics. Surprisingly it was not the case and annealing times up to 525 hours showed no structural change at this temperature. Further annealing studies were carried on at temperatures 150 °C and 200 °C and precipitation of α"- $Fe_{16}N_2$ was structurally observed (Figure 2, the calculated diffraction pattern is performed for 6 at.% nitrogen composition). As reported by Jack[1], not all of the calculated intensities were observed. Previously it was suggested [22-23] that , upon precipitation the α"- $Fe_{16}N_2$ phase might exhibit interstitial atom deficiency and structural vacancies on the nitrogen sublattice can occur leading to α"- $Fe_{16}N_{2-x}$. The observed peak shifts may be attributed to this deficiency effect which clearly effects position and integrated intensity of the α"- $Fe_{16}N_2$ reflections.

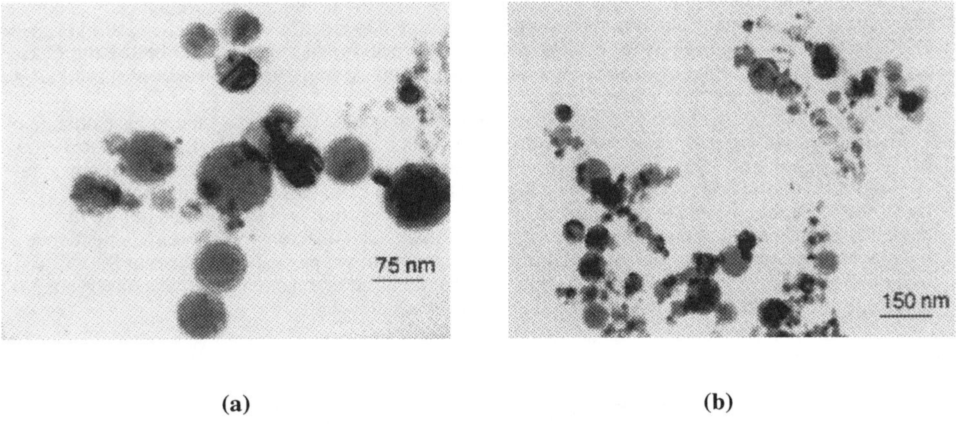

(a) (b)

Figure.1a and 1b. Bright field images of the as produced powders showing morphology and size distribution.

Thermomagnetic measurements were performed up to 800 °C on both as produced and annealed samples as illustrated in Figures 3a and 3b respectively. The various magnetic and structural transformations are also labelled and summarized in Table I. At temperatures near 200 °C the γ-Fe

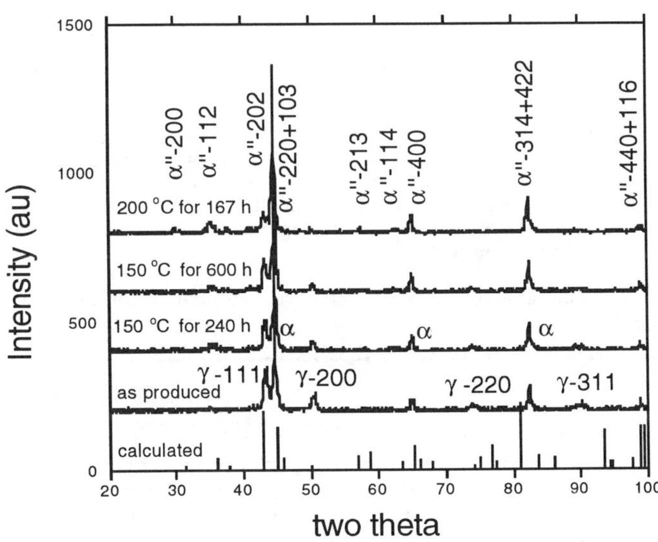

Figure.2. X-ray diffraction patterns of as produced and annealed samples.

begins to decompose into α-Fe and γ'-Fe_4N, after reaching the peak temperature the decrease in magnetization can be attributed to the strong temperature dependence of the magnetization of the γ'-Fe_4N phase. This is followed by the eutectic phase transformation. At temperatures below the Curie temperature, T_c, of alpha iron, an α-Fe + γ-Fe --> γ-Fe phase transformation was observed as is predicted in the equilibrium phase diagram [24].

On cooling the only observed transformation is paramagnetic to ferromagnetig phase change of γ'-Fe_4N (no eutectic transformation was observed). This may be argued as follows: by formation of α-Fe on heating nitrogen is rejected from formed alpha regions causing an increase in nitrogen content of the gamma phase. In regions where the nitrogen concentration is high enough, formation of stable γ'-Fe_4N is promoted resulting into segregation into two phases before the eutectic transformation temperature is reached. We observed a different temperature dependence for γ'-Fe_4N on heating and cooling. As an additional experiment we purged the system at 800 °C and looked at the change in magnetization again. Typical metallic Fe type response was observed. No change in the magnetization was observed after the heating cycle.

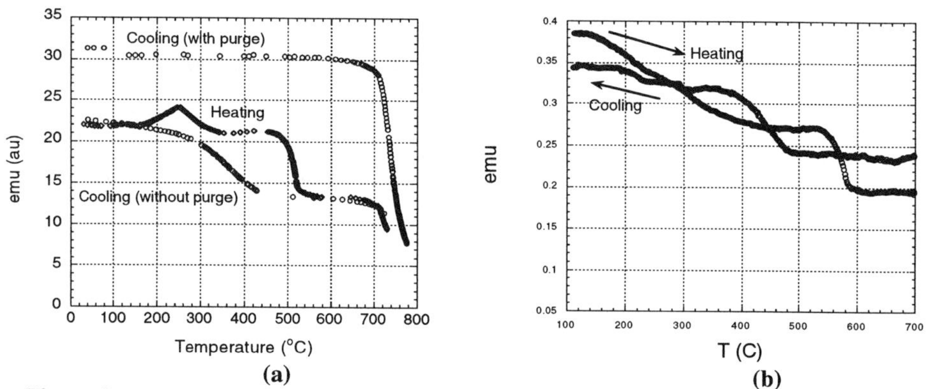

Figure.3. Temperature dependence of magnetization in as produced state **(a)** and after annealing at 200 °C for 167 h **(b)** taken at 500 Oe..

We also performed thermomagnetic, M(T) measurements on annealed samples and a magnetic response similar to that of unannealed samples was observed on both heating and cooling (Figure 3b). The only difference was in the magnetization after heating cycle. The difference is calculated as 11 % . A strong temperature dependence on the heating curve is believed due to transformation of α"- $Fe_{16}N_2$ to γ'-Fe_4N. Same type of temperature dependence was also observed by Kano *et al.* [25] and Peng *et al.* [26] in thin films.

Table.I. Observed structural and/or magnetic phase transformations on as produced powders with and without purge.

Temperature (°C)	Type of transformation
~ 250 (on heating)	γ-Fe --> γ'-Fe_4N + α-Fe
~ 575 (on heating)	(eutectic)
~ 765 (on heating)	α-Fe + γ --> γ-Fe
~ 300-450 (on cooling - without purge)	T_c of γ'-Fe_4N
~ 770 (on cooling - with purge)	T_c of Fe

DTA measurements indicated a similar pattern to that of the M(T) measurements performed on as produced samples (Figure 4). Table II. indicates observed structural phase transformations and temperatures. As it was concluded from the M(T) measurements on cooling only one peak was observed around 700 °C. This peak is believed due to trnsformation of γ-Fe --> γ-Fe + α-Fe and no eutectic transformation was observed. The hysteresis for γ-Fe --> γ-Fe + α-Fe transformation temperatures on heating and cooling can be again explained by argument made earlier. Slight temperature differences observed in DTA and magnetic measurements can be attributed to different heating rates used.

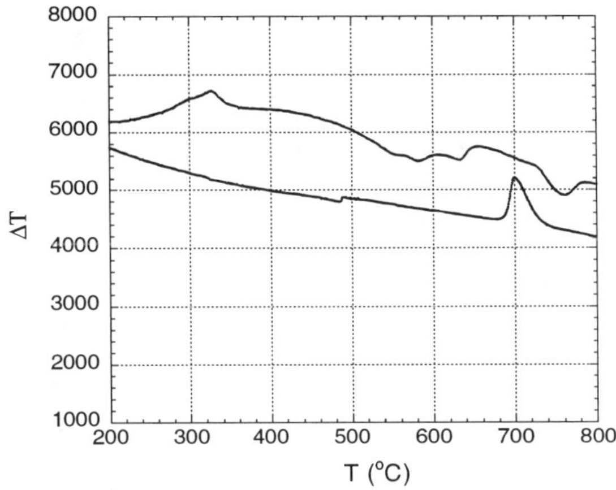

Figure.4. DTA data showing structural phase transformations.

Table.II. Structural phase transformations observed on DTA scan.

Temperature (°C)		**Type of transformation**
~ 325	(on heating)	γ-Fe --> γ'-Fe_4N + α-Fe
~ 551	(on heating)	unidentified
~ 581	(on heating)	eutectic
~ 630	(on heating)	α-Fe + γ --> γ-Fe
~ 760	(on heating	α-Fe --> α-Fe + γ-Fe
~ 700	(on cooling)	γ-Fe --> α-Fe + γ-Fe
~ 485	(on cooling)	unidentified

VI. CONCLUSIONS.

We have studied the effect of nitrogen on magnetization of iron in nanocrystalline powders. An increase in magnetization was observed after formation of α''- $Fe_{16}N_2$ phase. No attempt has yet been made to find out the average moment on iron since there is no precise way to measure volume fractions of the present phases in nanocrystalline powders.

V. ACKNOWLEDGMENTS

Effort sponsored by the Air Force Of Scientific Research, Air Force Material Command, USAF, under grant number F49620-96-1-0454. The U.S. Government is authorized to reproduce and distribute reprints for governmental purposes notwithstanding any copyright notation thereon.

VI. REFERENCES

1. K. H. Jack, J. Appl. Phys. **76**, 6620 (1994); Proc. R. Soc. London, Ser. **A 208**, 216 (1951).
2. T. K. Kim and M. Takahashi, Appl. Phys. Lett. **20** 492 (1972).
3. Y. Sugita, K. Mitsuoka, M. Komuro, H. Hoshiya, Y. Kozono and M. Hanazono, J. Appl. Phys. **70,** 5977 (1991).
4. P. Bezdicka, A. Klarikova, I. Paseka and K. Zaveta, J. Alloys Comps. **274** 10 (1998).
5. J. M. Coey, K. O'Donnell, Q. Qinian, E. Touchais and K. H. Jack, J. Phys.: Condens. Matter **6**, L23 (1994).
6. T. Weber, L. de Wit, F. W. Saris and P. Schaff, Thin solid Films **279**, 216 (1996).
7. R. M. Metzger, X. Bao and M. Carbucicchio, J. Appl. Phys. **76**, 6626 (1994).
8. K. H. Jack, J. Appl. Phys. **76**, 6620 (1994).
9. A. Sakuma, J. Magn. Magn. Mater. **102**, 127 (1991).
10. B. I. Min, Phys. Rev. B **46**, 8232 (1992).
11. R. Coehoorn, G. H. O. Daalderop and H. J. F. Jansen, Phys. Rev. B **48**, 3830 (1993).
12. K. Umino, H. Nakajima and K. Shiiki, J. Magn. Magn. Mater. **153**, 323 (1996).
13. H. Nakajima, Y. Ohashi and K. Shiiki, J. Magn. Magn. Mater. **167**, 259 (1997).
14. S. Ishida, K. Kitawatase, S, Fujii and S. Asano, J. Phys. Condens. Matter **4**, 765 (1992).
15. Y. Sugita, H. Takahashi, M. Komuro, K. Mitsuoka and A. Sakuma, J. Appl. Phys. **76**, 6637 (1994).
16. D. C. Sun, E. Y. Jiang and D. Sun, Thin solid Films **298**, 116 (1997).
17. K. Nakajima, S. Okamoto and T. Okada, J. Appl. Phys. **65,** 4357 (1989).
18. M. Takahashi, H. Shoji, H. Takahashi, T. Wakiyama, M. Dio and M. Matsiu, J. Appl. Phys. **76**, 6642 (1994).
19. J. M. D. Coey, J. Appl. Phys. **76**, 6632 (1994).
20. M. Q. Huang, W. E. Wallace, S. Simizu and S. G. Sankar, J. Magn. Magn. Mater. **135**, 226 (1994).
21. W. E. Wallace and M. Q. Huang, J. Appl. Phys. **76**, 6648 (1994).
22. M. J. van Genderen, A. Bottger, R. J. Cernik and E. J. Mittemeier, Metall. Trans. A **21A**, 1965 (1993).
23. M. J. Van Genderen, A. Bottger and E. J. Mittemeijer, Metall. Trans A **28A**, 63 (1997).
24. H. A. Wriedth, N. A. Gogcen and R. H. Nofzinger, Bull. Alloy Phase Diagrams **8,** 355 (1987)
25. A. Kano, N. S. Kazama and H. Fujimori, J. Appl. Phys. **53,** 8332 (1982).
26. D-L. Peng, K, Sumiyama and K Suzuki, , J. Alloys Comps. **259**,1 (1997).

A Novel Apparatus for the Synthesis of Graphite Encapsulated Metallic Nanocrystals

K.L. Klug *, D.L. Johnson *, V.P. Dravid *
*MSE Department, Northwestern University, Evanston, IL 60208, kevinklug@nwu.edu

ABSTRACT

Graphite encapsulated metal nanocrystals (GEM nanocrystals) are a promising new class of materials with both scientific and industrial significance. Several encapsulating graphite layers protect the nanoscale metallic core without compromising the inherent magnetic properties of the metal, thereby enhancing studies of small particle magnetism. The size and single domain nature of these particles immediately suggest a wide variety of potential applications ranging from increased density magnetic data storage to immunoassays. Encapsulated nanoparticles have traditionally been synthesized by evaporating a composite metal/graphite anode via an electric arc. While effective, that method also results in a great deal of unwanted amorphous carbon debris in the product. Based on past synthesis runs and observations, an improved apparatus for GEM production has been designed with the goals of increasing GEM yield and reducing soot production. The four segment system consists of 1) a chamber capable of arc or resistive evaporation of raw material(s), 2) a tube furnace for the chemical vapor deposition of carbon on raw material(s) via hydrocarbon dissociation, 3) an increased-efficiency powder collection unit, and 4) a final powder reservoir. Details of the new apparatus are presented here.

INTRODUCTION

Nanoscale materials have attracted a great amount of attention in the research community for their novel properties relative to the bulk. Magnetic, catalytic, optical and many other characteristics are enhanced by virtue of the materials' small size. However, the large amount of surface area per unit volume in nanoparticles renders them subject to violent oxidative and corrosive reactions. For that reason, methods to coat nanoparticles have also been of great scientific interest.

Many experimental methods have been used in an effort to synthesize both coated and bare nanoparticles of prescribed composition and shape. These techniques include arc evaporation[1], laser ablation[2], and sonochemical reactions[3], among others. The predominant method used at Northwestern University has been arc evaporation for the synthesis of GEM nanocrystals.[4, 5] In this process, an electric arc is struck between a tungsten cathode and a composite metal/graphite anode. Compared to experiments using a graphite rather than tungsten cathode[6, 7], this technique increases the percentage of encapsulated particles relative to amorphous carbon debris in the collected product. However, room for significant improvement still exists.

A new apparatus that utilizes hydrocarbons rather than solid graphite as the carbon precursor for GEM nanocrystal formation has been designed, patented[8], and built. The specifics of this machine are presented below.

Mat. Res. Soc. Symp. Proc. Vol. 577 © 1999 Materials Research Society

THEORY

Overall System

A major goal throughout the design and construction of the system described here has been to retain a great deal of versatility. The ability to monitor and control various parameters that are believed to govern the synthesis of both bare and GEM nanoparticles is considered paramount. A variety of experimental arrangements are also possible; the major components of the apparatus are independently supported by wheeled frames that can be moved relative to each other. This permits the removal of the reaction furnace if unencapsulated particles are desired. A schematic of the overall system can be seen in Figure 1.

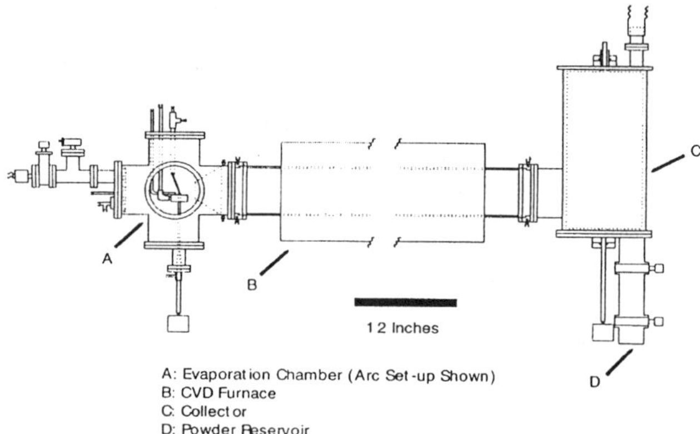

Figure 1: Overall schematic of novel nanoparticle synthesis apparatus

Chamber A is where unencapsulated nanoparticles are formed via the evaporation and subsequent quenching of raw materials. A coating may then be promoted on the nanoparticles in Chamber B, a tube furnace through which the particles pass and encounter reactive gases. Finally, the particles are collected Chamber C, where they are scraped and ultimately channeled into the powder reservoir (Chamber D). A more detailed explanation or each of the components is given below.

Material Evaporation (Chamber A)

The raw material for nanoparticle synthesis is evaporated in Chamber A. Arc evaporation and resistive heating configurations may be readily interchanged in this 6-way cross.

The arc evaporation design may be used when particles composed of single elements are desired. This geometry is shown in Figure 2. Electrons pass from the tungsten cathode to the metal of interest, evaporating the metal in the process. The electrons then travel up the two copper anode pipes, located behind the cathode. Modeled after the work of Araya et al.[9], this geometry provides "magnetic blowing forces" due to the direction of current flow. These

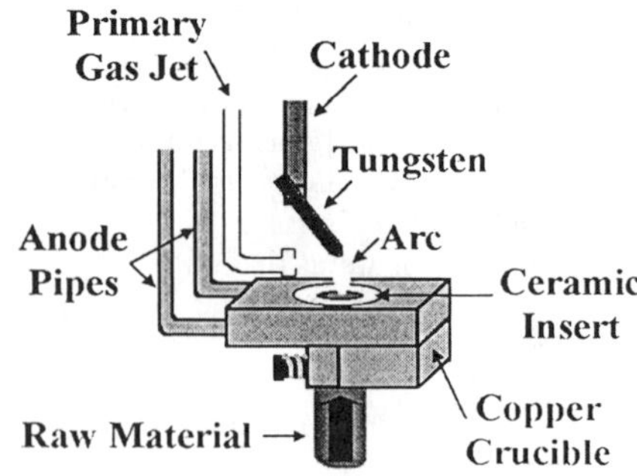

Figure 2: Arc evaporation geometry

blowing forces will be used to assist a high velocity helium jet in removing the metal vapor from the site of the arc, thereby increasing the rate of evaporation. Note that the raw material may be continuously fed through the ceramic insert that acts as a thermal and electrical buffer between the raw material and copper crucible.

When bimetallic or higher order particles are desired, an alternate geometry may be employed. This setup, shown in Figure 3, consists of an "inverted Y" composed of graphite. The metals of interest are independently fed into their respective ports at the bottom of the structure and resistively heated. A helium jet is introduced to the top of the inverted Y, quenching the metallic species as they are carried through an orifice in the wall of the structure. Depending on the vapor pressures of the respective metals, a wide variety of final particle compositions may be anticipated.

In both heating configurations, the particulate species are carried by helium into either the reaction furnace (if a coating is desired) or directly to the collection unit.

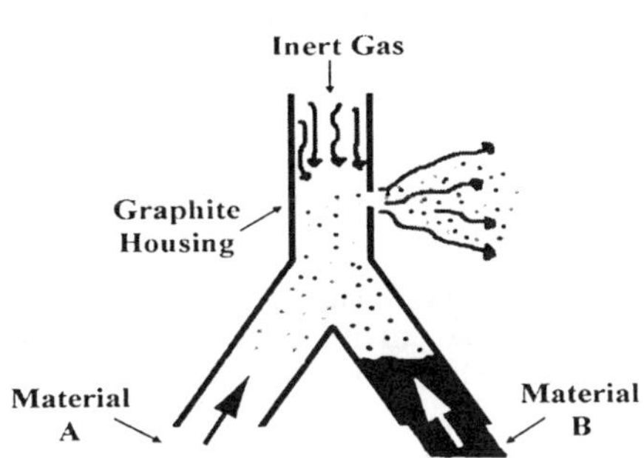

Figure 3: Resistive heating geometry

Tube Furnace Reactor

As eluded to earlier, the use of a solid graphite electrode in the arc evaporation process inherently generates excess amorphous carbon in the product. To avoid this, gaseous rather than solid carbon precursors may be introduced at the entrance to Chamber B. At elevated temperatures and under the catalytic action of the particles themselves, the gases will decompose, leaving behind surface species in the process. In the case of GEM nanocrystals, hydrocarbon gases may be used to deposit carbon on the particles' surfaces.

Through the use of mass flow controllers, it is believed that the amount of carbon supplied to the coating reaction may be much more accurately controlled than in the case of a metal/graphite hybrid electrode. This may significantly reduce the amount of carbon soot in the product. Additionally, the temperature at which the coating reaction occurs may be more stable in the furnace than in the case of the arc. It is hoped that this may lead to the coating of materials which have only formed carbides under the aggressive arc environment.

It is also important to note that the type of gas introduced into the furnace is not limited to hydrocarbons. If particles other than GEMs are desired, oxidative species may be supplied to form oxide coatings or completely ceramic particles. One can envision the use of other gases to form a variety of particle coatings.

Collection Unit and Powder Reservoir

After exiting either Chamber A (uncoated particles) or Chamber B (particles with a coating), the nanoparticles will be collected in the third main component. Chamber C consists of a vertically positioned stainless steel tube that houses a rotating cold finger filled with liquid nitrogen. As the particles flow into Chamber C, they are at an elevated temperature relative to the cold finger. This temperature gradient results in thermophoretic forces that drive the nanoparticles to the cold finger wall. They remain there until coming into contact with a stationary scraper located on the inside wall of the chamber housing, 180° away from the chamber entrance. The scraped particles are channeled to the powder reservoir (Chamber D), which may be sealed off from the rest of the system and safely removed for investigation following the experiment.

RESULTS

All major system components have been received and assembled. A trial run using nickel and the arc configuration coupled directly to the collection chamber is planned in the near future. Experiments using the reaction furnace to coat particles will be investigated shortly thereafter.

CONCLUSIONS

A novel apparatus for the synthesis of scientifically and industrially significant nanoparticles has been designed and constructed. The system has provisions for manufacturing both single element and bimetallic species, either of which can be left bare or coated via precursor gas decomposition.

ACKNOWLEDGMENTS

This work was supported through AFOSR grant # F49620-96-1-0461, and through the Institute for Environmental Catalysis at Northwestern University, which is funded by NSF.

The authors would also like to recognize Mark Seniw, Jim Hahn, and Ken Viani for their invaluable help in designing and constructing the synthesis equipment described herein.

REFERENCES

1. S.A. Majetich, J.H. Scott, E.M. Brunsman, and M.E. McHenry. *Formation of Nanoparticles in a Carbon Arc*. 1995: Materials Research Society.
2. R.F. Wood, J.N. Leboeuf, K.R. Chen, D.B. Geohegan, and A.A. Puretzky, Applied Surface Science **127-129,** p. 151-158 (1998).
3. K.S. Sislick, T. Hyeon, M. Fang, and A.A. Cichowlas, Materials Science and Engineering A **A204,** p. 186-192 (1995).
4. M.H. Teng, J.J. Host, J.-H. Hwang, B.R. Elliott, J.R. Weertman, T.O. Mason, V.P. Dravid, and D.L. Johnson, Journal of Materials Research **10 (2),** p. 233-236 (1995).
5. V.P. Dravid, J.J. Host, M.H. Teng, B. Elliott, J. Hwang, D.L. Johnson, T.O. Mason, and J.R. Weertman, Nature **374,** p. 602 (1995).
6. S. Bandow and Y. Saito, Japanese Journal of Applied Physics **32,** p. L1677-L1680 (1993).
7. J.P. Hare, W.K. Hsu, H.W. Kroto, A. Lappas, K. Prassides, M. Terrones, and D.R.M. Walton, Chem. Mater. **8,** p. 6-8 (1996).
8. D.L. Johnson, V.P. Dravid, M.-H. Teng, J.J. Host, J. Hwang, and B.R. Elliott, U.S. Patent No. 5,665,277 (September 9, 1997).
9. T. Araya, R. Okada, Y. Ibaraki, S. Hioki, M. Kanamaru, Y. Endo, and M. Haneda, U.S. Patent No. 4,610,718 (September 9, 1986).

THE KINETICS OF FORMATION OF NANOCRYSTALLINE γ-Ni-Fe DURING MECHANO-CHEMICAL SYNTHESIS

P. KNORR, B.S. KIM AND J.S. LEE
Dept. of Metallurgy and Materials Science, Hanyang University, Ansan, 425-791 Korea

ABSTRACT

An experimental study of the kinetics of the three partial processes involved in the mechano-chemical synthesis of nanocrystalline (nc) Ni-Fe is presented. These processes are the reduction of nc NiO, the reduction of nc Fe_2O_3, and the alloying of Ni and Fe to a γ-Ni-Fe solid solution. For the reaction rates of the reduction of nc NiO and nc Fe_2O_3 activation enthalpies of 82 kJ mol^{-1} and 47 kJ mol^{-1} were obtained. The progress of the overall synthesis was found to follow the sequence 'NiO red. → Fe_2O_3 red. → γ-Ni-Fe form.'.

INTRODUCTION

Among the number of approaches for fabricating nc materials the concept of the mechano-chemical synthesis appears to be one of the most promising for the production of nc materials on a commercial basis. Potential scale, process simplicity and cost of starting materials all appear favorable when compared to other production routes [1-4]. The method involves a two-step procedure consisting of high-energy ball-milling of metal oxide powders and their subsequent reduction with hydrogen (see fig. 1) [5,6]. The use of metal oxide powders as starting materials possesses some advantages over the employment of pure metals as in the case of mechanical alloying. Oxide particles are easier to fracture upon ball-milling due to their intrinsic brittleness. Moreover, while the introduction of internal strain and a potentially high density of lattice defects due to plastic deformation are inevitable in mechanical alloying, the use of metal oxide powders avoids this drawback. This is of particular concern for the fabrication of functional engineering materials such as low coercivity alloys. On the other hand, the advantage of mechanical alloying over the mechano-chemical synthesis is that the former is virtually applicable to all systems while the latter is confined to oxide mixtures which can be reduced with reasonable reaction rates. It is particularly the production of soft-magnetic Ni-Fe alloys that may hold the greatest potential for the mechano-chemical method, since (i) these alloys require the avoidance of defects and internal strain upon processing and (ii) the reduction of the involved metal oxide powders, Fe_2O_3 and NiO, meets favorable thermodynamic conditions [7].

The formation of nc Ni-Fe powders comprises two heterogeneous chemical reactions of the oxide phases, Fe_2O_3 and NiO, and the subsequent alloying of the reduced Fe and Ni phases by solid state diffusion:

$$Fe_2O_3\,(s) + 3\,H_2\,(g) \rightarrow 2\,Fe\,(s) + 3\,H_2O\,(g) \qquad (1)$$

$$NiO\,(s) + H_2\,(g) \rightarrow Ni\,(s) + H_2O\,(g) \qquad (2)$$

$$Ni\,(s) + Fe\,(s) \rightarrow \gamma\text{-Ni-Fe}\,(s) \qquad (3)$$

The reaction rates of reactions (1) and (2) will determine the onset of solid state diffusion in the reduced Ni and Fe phases. If the reaction rates differ remarkably from one another, the alloying of the Ni-Fe product phase (3) will be inhibited and particle growth will largely take place in either the pure Fe or the pure Ni phase. Moreover, reactions (1) and (2) may be accompanied by significant changes of the microstructure (pore structure, agglomeration) which again will have direct bearing on the nature of the subsequent alloying process (interdiffusion, pore diffusion). This paper reports about a systematic study on all three partial processes involved in the mechano-chemical synthesis of nc Ni-Fe with the objective to study their interdependence and to allow an optimization of the process conditions.

Mat. Res. Soc. Symp. Proc. Vol. 577 © 1999 Materials Research Society

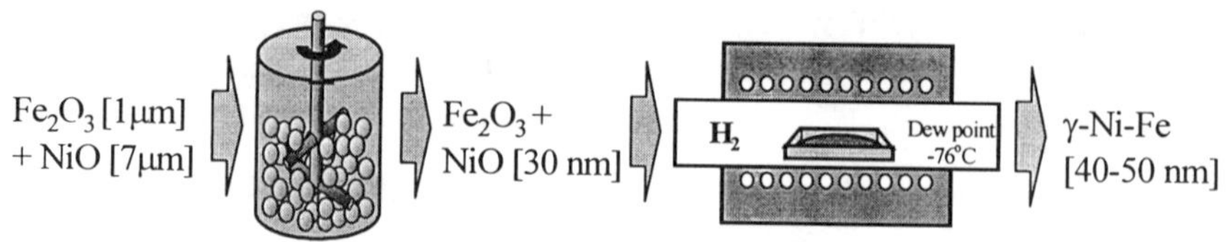

Fig. 1 Schematic diagram showing the mechano-chemical synthesis of nanocrystalline Ni-Fe powder.

EXPERIMENT

For the reduction experiments on nc hematite, commercial Fe_2O_3 powder with a specified average particle size of 1 µm was used as starting material. The powder was ball-milled for 10 hours in a stainless steel attritor at 350 r.p.m. Methyl alcohol was used as a milling agent. After ball-milling the powder was dried at 60 °C for several hours. The dried cake which formed was broken up and sieved down to a mesh size of 20 µm. X-ray diffraction analysis yielded an average grain size of 30 nm, while SEM observations revealed particle sizes of about 20-50 nm. For the preparation of nc NiO, commercial NiO powder with a nominal average particle size of 7 µm was employed. The method of preparation was identical to the one used for nc Fe_2O_3. Grain size and particle size of the nc NiO powder were found to be consistent with those obtained for nc Fe_2O_3.

The kinetics of reduction of the ball-milled metal powders were measured by means of the weight loss measurement technique. The thermogravimetric setup comprised a quartz reaction tube mounted vertically inside a tungsten lamp furnace. A small alumina crucible was used as a container which was mounted top-loading on an alumina pole extending from the microbalance. A shallow oxide powder bed of about 3 mm in height was spread uniformly in the container. A constant hydrogen flow rate of 0.3 l/min was maintained during the reduction. Previous experiments with different flow rates had shown that this rate is well above the level at which there could be any starvation of the reactant gas.

The alloying kinetics were followed by quantitative X-ray diffraction using the external standard method. For this purpose, a mixture of Fe_2O_3 and NiO with a composition of Ni-Fe(50 wt.%) was prepared by blending the respective nc powders obtained from the process described above. The powder blends were annealed isothermally at various temperatures for definite annealing times. XRD specimens were prepared by mounting thin powder layers of identical weight on glass plates. For a quantitative evaluation, the integral line intensity of the (111) diffraction line of γ-Ni-Fe was employed. The measurements were calibrated against standards which were prepared by using mixtures of γ-Ni-Fe, Ni and Fe of known composition with equal weight fractions for Ni and Fe. The grain size was calculated with the Scherrer formula. Appropriate allowance was made for the counting background, the $K_{\alpha 2}$-broadening and the instrumental line broadening. A careful analysis by the method of Hall and Williamson [8] had shown that virtually no residual lattice strain was present in the metallic nanocrystalline particles. This can be ascribed to the fact that the nc metallic Fe and Ni phases are formed by nucleation and growth from the ball-milled oxide phases and not by repeated fracturing as in the case of mechanical alloying.

RESULTS

The reduction curves for the reduction of nc Fe_2O_3 and NiO are depicted in figs. 2 and 3, respectively, as plots of the fractional reduction X vs. t. The fractional reduction X is defined as the ratio of the instant mass loss, Δm, to the theoretical total mass loss, Δm_0. Most X vs. t plots exhibit an incubation interval at the beginning of the reduction process which is followed by a steady-state region of undisturbed growth. In the case of the nc hematite, this steady-state region was chosen as the basis for determining the experimental rate constant which is given by the slope, $k = dX / dt$. When the experimental rate constant k is known, the reduction rate r_0 can be calculated according to the relation $r_0 = (k \times \Delta m_0 / 16)$ in units of g atom O $\cdot$ s^{-1}.

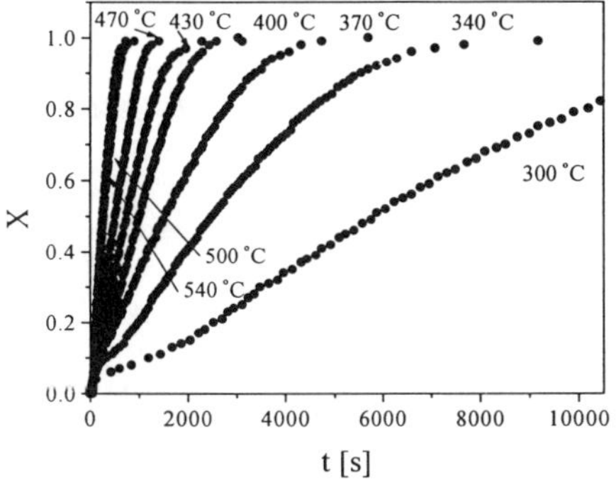

Fig. 2 X vs. t plots for hydrogen reduction of ball-milled NiO.

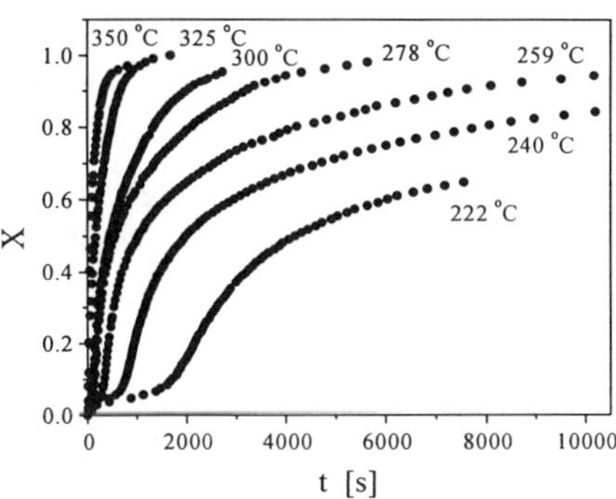

Fig. 3 X vs. t plots for hydrogen reduction of ball-milled Fe_2O_3.

For nc NiO, a different approach for the determination of the experimental rate constant k was chosen. Since the reduction of NiO proceeds without the formation of intermediate phases (which is not the case for the reduction of Fe_2O_3), the shape of the reaction curve contains information about the reduction mechanism. Detailed theoretical models have been developed for various reaction mechanisms (see e.g. [9]). Depending on whether the reaction is dominated by nucleation and growth, gas diffusion or the order of the reaction, different functional relationships $f(X)$ vs. t are predicted. The reduction curves of nc NiO were found to fit best to gas diffusion controlled kinetics. The experimental rate constant k was determined by plotting the corresponding $f(X)$ vs. t and calculating the slope of this plot. An Arrhenius plot of the reduction rates of nc Fe_2O_3 and NiO is shown in fig. 4. Regression analysis of the data provided the following expressions for the reduction rates: r_0 (NiO) = (6.1 ± 1.8) · exp(- (81984 ± 2660) J mol^{-1} / RT) g atom O · s^{-1} and r_0 (Fe_2O_3) = (2.1 ± 1.0) × 10^{-3} exp(- (47182 ± 2740) J mol^{-1} / RT) g atom O · s^{-1}.

Fig. 5 depicts a representative set of X-ray diffraction patterns measured for various annealing times at a temperature of 500 °C. The (111) diffraction peak of the emerging γ-Ni-Fe phase is located at 2θ = 55.75°, while the intensity maximum on the right hand side results from an overlap of the (111) diffraction peak of Ni at 2θ = 56.82° and the (110) peak of α-Fe at 2θ = 57.06°. The figure shows the increasing fractional conversion of Ni and Fe into γ-Ni-Fe with increasing annealing time.

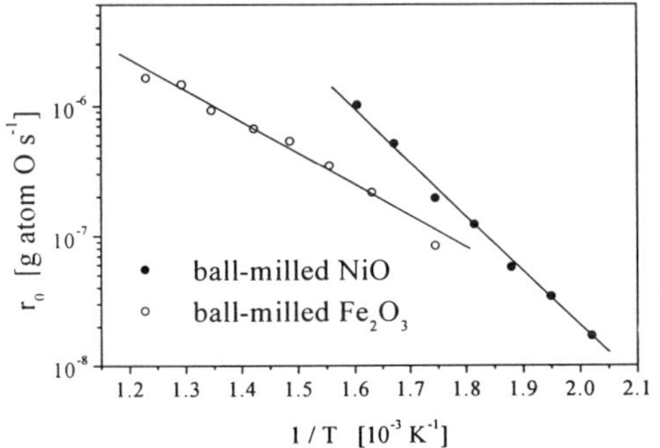

Fig. 4 Arrhenius presentation of the reduction rate constants.

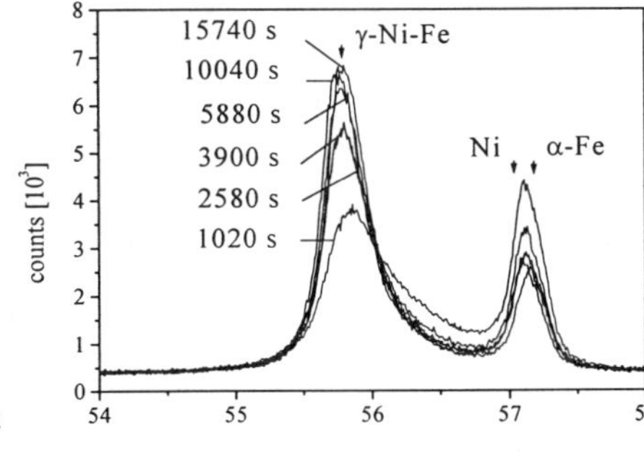

Fig. 5 X-ray diffraction pattern of specimens annealed at 500 °C for various annealing times.

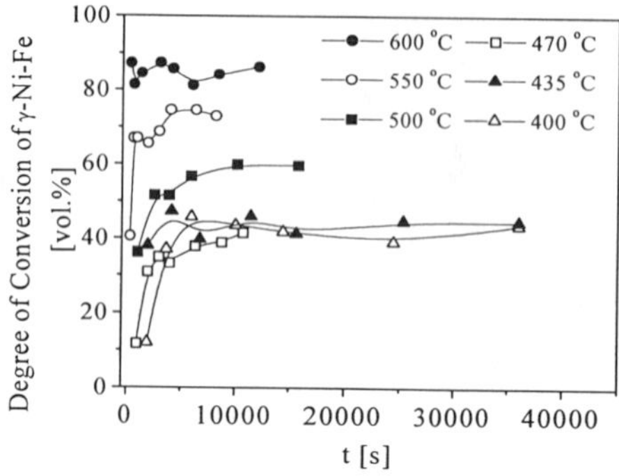

Fig. 6 X vs. t plot for the formation of the γ-Ni-Fe phase.

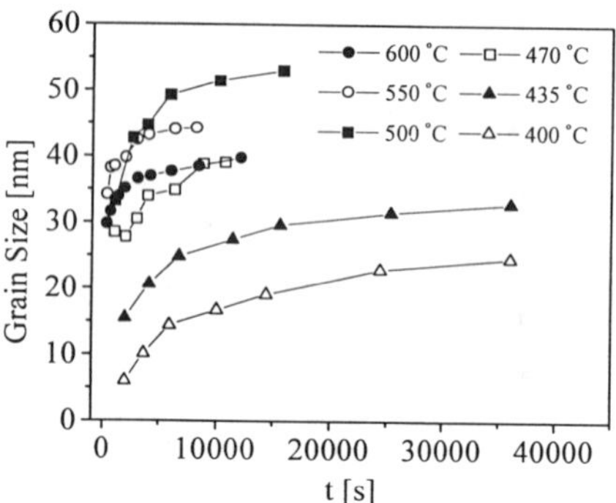

Fig. 7 Evolution of the grain size in the progress of the γ-Ni-Fe formation.

The time dependence of the formation of the γ-Ni-Fe phase is shown in fig. 6 for various temperatures. An inspection of the plot shows that for a given temperature the fractional conversion of the powder mixture into γ-Ni-Fe increases rapidly for short annealing times before it reaches an essentially constant degree of conversion. This apparent saturation level is about 40% for the lowest four annealing temperatures, and only for temperatures higher than 470 °C a discernible increase in the eventual degree of conversion is observed. It should be noted that the term 'saturation level' only refers to the obviously considerable slow-down in the rate of conversion for longer annealing times. Solid state diffusion and, hence, alloying will still continue to take place at such elevated temperatures. Fig. 6 suggests that even for the highest annealing temperature the fractional conversion was limited to about 90%. Apart from the fact that indeed not all oxide may have been reduced during the anneal, part of the apparently low degree of conversion can probably be ascribed to a certain reoxidation of the specimens *after* the process.

Fig. 7 depicts the variation of the grain size as a function of annealing time. For a given temperature, the grain size is monotone increasing with time, while the rate of grain growth decreases with increasing time. It is important to recognize that absolute values for the grain size given in fig. 7 are likely to underestimate the actual grain size. This is so, since the (111) diffraction peak associated with the formation of the γ-Ni-Fe(50 wt.%) alloy does not originate from a homogeneous alloy of this nominal composition but actually results from a system with a concentration gradient. This will give rise to an additional line broadening.

An interesting phenomenon is observed for the temperature dependence of the grain size for a constant but arbitrarily chosen annealing time. The grain size for e.g. an annealing time of 10000 s is found to increase rapidly from 15 nm at 400 °C to about 50 nm at 500 °C but then decreases to about 45 nm at 550 °C and finally to 40 nm at 600 °C. Obviously, grain growth is becoming successively impeded at higher annealing temperatures. The possible reason for this behavior will be discussed in the following paragraph.

CONCLUSIONS

Examination of the grain growth of the nc γ-Ni-Fe phase (fig. 7) showed that for an annealing time of 10000 s grain sizes of several 10 nm were observed. These values are far too large to be accounted for by assuming interdiffusion to be the dominant formation mechanism. Interdiffusion coefficients for the binary Ni-Fe system amount to $D = 1.24 \times 10^{-27}$ m^{2}s^{-1} at 400 °C and $D = 5.14 \times 10^{-22}$ m^{2}s^{-1} at 600 °C [10]. Taking $2\sqrt{Dt}$ as a mean diffusion distance, one arrives at values as low as 0.007 nm and 4.5 nm as typical interdiffusion distances at 400 °C and 600 °C, respectively. This is to be compared with grain sizes of about 15 nm and 37 nm for the

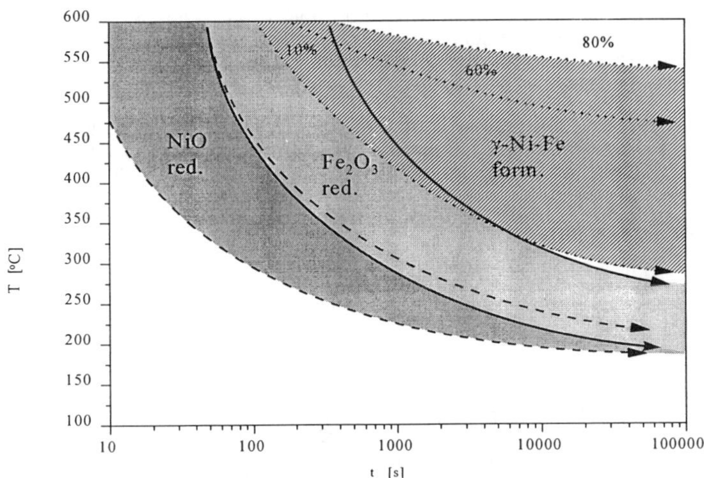

Fig. 8 Time-temperature-transformation (*TTT*-) diagram for the mechano-chemical synthesis of nanocrystalline γ-Ni-Fe. The dashed curves denote 10% and 90% reduction of NiO. The solid curves represent the same for Fe_2O_3. The dotted curves show various degrees of fractional conversion (10%, 60% and 80%) into γ-Ni-Fe.

respective temperatures. Lee et al. suggested that the reason for this rapid γ-Ni-Fe formation was a direct formation of the new phase occurring already during the oxide reduction [5]. The γ-Ni-Fe phase was believed to be formed 'in-situ' during the reduction process without the previous formation of a bulk Ni and Fe phase. This suggestion had been motivated by a series of TEM observations by Mori et al. which had revealed rapid 'spontaneous alloying' in nanosized clusters of Au-Cu [11] and Au-Al [12].

The present systematic experiments, however, clearly show that the idea of 'in-situ alloying' has to be dismissed for the Ni-Fe system. This is best visualized by a *TTT*-diagram comprising all three partial processes involved in the mechano-chemical synthesis of nc γ-Ni-Fe (see fig. 8). The *TTT*-diagram for the overall process can be constructed on the basis of the isothermal measurements of the reduction kinetics (figs. 2 and 3) and the alloying kinetics (fig. 6) by normalizing the fractional conversion data with respect to the total mass converted. In figure 8, the dashed curves enclosing the dark-shaded area designate 10% and 90% reduction of NiO, respectively. Correspondingly, the solid curves represent 10% and 90% reduction of Fe_2O_3. It is seen that the reduction of the nc NiO proceeds significantly faster than that of nc Fe_2O_3 for all temperatures. The fraction of Fe_2O_3 converted to Fe is only 10% when the reduction of NiO to Ni is already almost complete. However, both reduction processes occur largely *before* the alloying process sets in. This is most obvious for temperatures lower than 400 °C. Only at elevated temperatures the γ-Ni-Fe formation speeds up relative to the preceding reduction processes. And only at temperatures as high as 500 °C to 600 °C the conversion of Ni and Fe into the new γ-Ni-Fe phase begins to be discernible *before* the reduction of Fe_2O_3 is accomplished. It is at this point of the discussion that the observed impediment of grain growth at temperatures higher than 500 °C appears noteworthy. The observation that at elevated temperatures the alloying process begins to set in before the hematite reduction is complete suggests an inhibition of grain growth by unreduced Fe_2O_3 particles. These still unreduced hematite particles bordering the γ-Ni-Fe grains may inhibit further growth. This is not so for lower temperatures where both the NiO and the Fe_2O_3 reduction are essentially accomplished before the alloying process starts.

In summary, the experiments have shown that the mechano-chemical synthesis of nc Ni-Fe is essentially a stepwise process and not a quasi-simultaneous one as suggested previously. The progress of the overall synthesis follows the sequence 'NiO red. → Fe_2O_3 red. → Ni-Fe form.'.

The formation of the nc Ni-Fe proceeds very rapidly and cannot be accounted for by interdiffusion. A forthcoming TEM study aims to check whether rapid pore diffusion or diffusion induced grain boundary migration (DIGM) might come into play in the rapid alloying process. From a practical point of view, it is advisable to choose processing temperatures in the vicinity of 600 ^{o}C rather than lower temperatures for the following reasons: Firstly, only at such high temperatures the degree of conversion of the process reaches values close to 100%. Secondly, presumably because of a temporary inhibition of grain growth due to slower reducing hematite particles, the overall grain size turned out to be even smaller than at lower temperatures.

ACKNOWLEDGEMENTS

The authors wish to thank the Alexander-von-Humboldt foundation for support of this investigation through a fellowship granted to P. Knorr.

REFERENCES

1. C. Suryanarayana, Int. Mat. Reviews **40**, 41-64 (1995).
2. C. Cheung, F. Djuanda, U. Erb and G. Palumbo, Nanostr. Mat. **5**, 513-523 (1995).
3. S. Eroglu, S.C. Zhang and G.L. Messing, J. Mat. Res. **11**, 2131-2134 (1996).
4. F. Ge, L. Chen, W. Ku and J. Zhu, Nanostr. Mat. **8**, 703-709 (1997).
5. J.S. Lee, T.H. Kim, J.H. Yu and S.W. Chung, Nanostr. Mat. **9**, 153-156 (1997).
6. J.S. Lee, J.G. Nam and P. Knorr, J. of the Korean Institute for Metals and Materials, accepted for publication.
7. O. Kubaschewski and C.B. Alcock, *Metallurgical Thermochemistry*, edited by G.V. Raynor, (Pergamon Press, Oxford, 1979).
8. G.K. Williamson and W.H. Hall, Acta metall. **1**, 22-31 (1953).
9. J. Szekely, J.W. Evans, and H.Y. Sohn, *Gas-Solid Reactions*, (Academic Press, New York, NY, 1976).
10. J.I. Goldstein, R.E. Hanneman, and R.I. Ogilvie, Trans. Metall. Soc. AIME **233**, 812 (1965).
11. H. Mori, H. Yasuda, and T. Kamino, Phil. Mag. Lett. **69**, 279-283 (1994).
12. H. Mori and H. Yasuda, J. Microscopy **180**, 33-38 (1995).

GMR EFFECT AND PROPERTIES OF CoAg GRANULAR FILMS FORMED BY IMPLANTATION WITH A METAL VAPOR VACUUM ARC ION SOURCE

S.P. WONG*, M.F. CHIAH*, W.Y. CHEUNG*, N. KE*, J.B. XU*, X.X. ZHANG**
*The Chinese Univ. of Hong Kong, Dept of Electronic Engineering, Shatin, N.T., Hong Kong, China
**Hong Kong Univ. of Science and Technology, Physics Dept, Kowloon, Hong Kong, China

ABSTRACT

Cobalt-silver granular thin films were formed by Co implantation into Ag using a metal vapor vacuum arc (MEVVA) ion source. The magnetic field dependence and the temperature variation of the giant magnetoresistance (GMR) effect and their relation with the processing conditions were studied and discussed in conjunction with results of Rutherford backscattering spectrometry, atomic force microscopy, magnetic force microscopy (MFM), and SQUID measurements. Anomalous temperature dependence of the coercive field H_C in the perpendicular-to-film direction determined from GMR measurements was observed for some samples. For one sample, H_C shows a maximum value at around 240K and decreases with decreasing temperature from 240K to 20K. The temperature variation of the magnetization M of this sample exhibits a minimum. The maximum in the H_C-T curve corresponds well with the minimum in the M-T curve. The M-T curve suggests that there is more than one magnetic phase present in this sample. The domain structures of the implanted granular films as revealed by MFM images exhibit very different features compared with those of sputter deposited CoAg granular films.

INTRODUCTION

Giant magnetoresistance (GMR) effect was first observed in magnetic multilayer structures [1] in the late 80's. Since then, research interest in the study of magnetoresistance (MR) effects in various materials structures continues to grow. In particular, there are increasing reports on the study of MR effects in granular thin films [2-11]. Among these studies, there are a few recent attempts to synthesize magnetoresistive granular thin films by ion implantation [7, 9, 11]. One major difficulty of the implantation method for ion beam synthesis (IBS) applications is the high dose required. This problem is not an issue with the use of a metal vapor vacuum arc (MEVVA) ion source which was developed in the mid-80's [12] and has broad beam and high current capabilities. Recently, we have reported some preliminary results of our study on the IBS of CoAg granular thin films by MEVVA implantation [11]. In this work, we shall report results on the properties of these implanted CoAg granular films.

EXPERIMENTAL

Silver films of various thickness were deposited by thermal evaporation on thermally grown SiO_2 layers on Si substrates. The MEVVA implantation of Co ions to a nominal dose of 8×10^{16} cm^{-2} was performed at extraction voltages ranging from 35 kV to 65 kV. It should be noted that since there are three charge states, 1^+, 2^+, and 3^+, for the Co ions in the MEVVA source with corresponding particle flux fractions of 52%, 45%, and 3%, respectively [13], the implantation condition would be equivalent to a triple-energy implantation. The mean ion beam current density used was about 6 $\mu A/cm^2$ and the substrate temperature during implantation due to beam heating was estimated to be about 85°C. Thermal annealing was

Mat. Res. Soc. Symp. Proc. Vol. 577

performed in a vacuum chamber at a pressure of $2x10^{-5}$ torr for 2 hours at various temperatures. Rutherford backscattering spectrometry (RBS) was performed with a 2MV tandem accelerator using a 2 MeV $^4He^{2+}$ ion beam at a 170° backscattering geometry. Atomic force microscopy (AFM) and magnetic force microscopy (MFM) imaging were performed under ambient conditions using a Nanoscope III scanning probe microscope with a hard Co-alloy-coated silicon tip magnetized along the tip direction using tapping and lift modes. The GMR properties were measured in the temperature range from 20 to 300 K in the perpendicular (⊥) configuration where the applied magnetic field H is perpendicular to the film surface. The magnetization curves were measured with a SQUID magnetometer.

RESULTS AND DISCUSSION

For a fixed implant energy, because of the sputtering effect and the limited penetration depth of the implanted ions, one would expect that a correct combination of the implant dose and the thickness of the deposited Ag film is essential to maximize the GMR effect. To demonstrate such an effect, Ag films of various thickness were prepared and then implanted at the same extraction voltage of 35 kV to the same dose of $8x10^{16}$ cm^{-2}. Thermal annealing was performed at 450°C. The thickness and the composition of the implanted CoAg granular layer were then determined from the RBS spectra as shown in Fig. 1 where the initial Ag film thickness t_{Ag} before Co implantation is indicated. The RBS spectra reveal that for the sample with t_{Ag} = 90 nm, there is an equivalent CoAg layer of 15 nm thick with a composition of 20 at. % Co immediately on top of the SiO_2 layer in which some penetrating recoiled Ag atoms are present. For the other two samples with larger t_{Ag} values, there is effectively a shunting residual Ag layer remaining. Therefore, the sample with the smallest t_{Ag} is expected to show the largest GMR effect and vice versa.

The GMR effect is characterized by the ratio $\Delta R(H)/R_0 \equiv (R(H) - R_0) / R_0$ where R(H) is the resistance measured in the presence of a magnetic field strength H and R_0 is the resistance at zero field. Shown in Fig. 2(a) and (b) are the MR curves for the samples of Fig. 1 measured at 300K and 20K, respectively. The dependence of the magnitude of the GMR effect on t_{Ag} is shown in Fig. 3. It is seen that while the room temperature results indeed show the shunting effect of the residual Ag layer as expected, this shunting effect was not observed in the GMR results measured at 20K. Since the carrier mean free path at 20K is much longer than the thickness of the residual Ag layer, this is reasonable.

The MR curves for another set of samples prepared by implantation at various extraction voltage ranging from 35 to 65 kV are shown in Fig. 4. The variation of the magnitude of the GMR effect with the extraction voltage is shown in Fig. 5 and can be understood as follows. The low temperature results were found to be related to the composition of the CoAg films. The RBS results showed that while the Co content in both the samples implanted at 35 kV and 65 kV are about 20 at. %, in the other two samples it is about 33 at. %. It has been reported in the literature that for CoAg granular films, a composition of about 20 at. % Co is the optimum composition for the largest GMR effect [4]. This explains at least qualitatively the low temperature results. The sample prepared at 35 kV showed the smallest GMR effect at room temperature. This may also be attributed to the shunting effect since the penetration depth of the implanted ions is smaller at lower implant energy.

It is also noticed that in some of the MR curves measured at 300K shown in Fig. 2 and Fig. 4, hysteresis is clearly present. The fields at the two maximum points in the hysteresis loop correspond to the coercive field H_C of the sample [3]. However, a comparison of the MR curves measured at 300K and 20K reveals that H_C of the same sample is smaller at 20K than

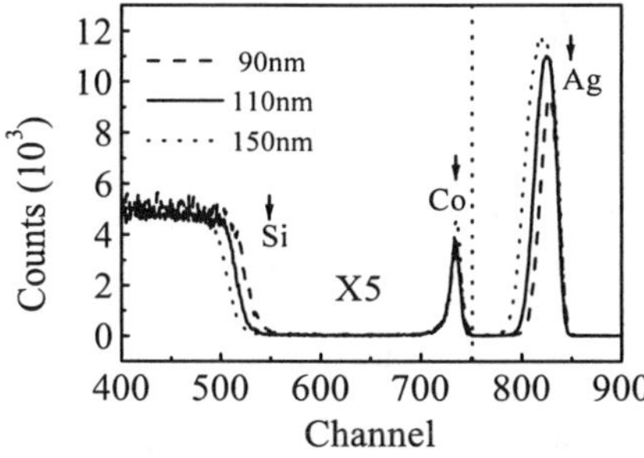

Fig. 1 RBS spectra for samples implanted at 35 kV to a dose of $8x10^{16}$ cm^{-2} and annealed in vacuum at 450°C for two hours. The thickness of the as-deposited Ag films before implantation are as indicated. Note the difference in scale at channel numbers smaller and larger than 750.

Fig. 3 Magnitude of the GMR effect versus the thickness of the as-deposited Ag films for the samples as in Fig. 1 and Fig. 2.

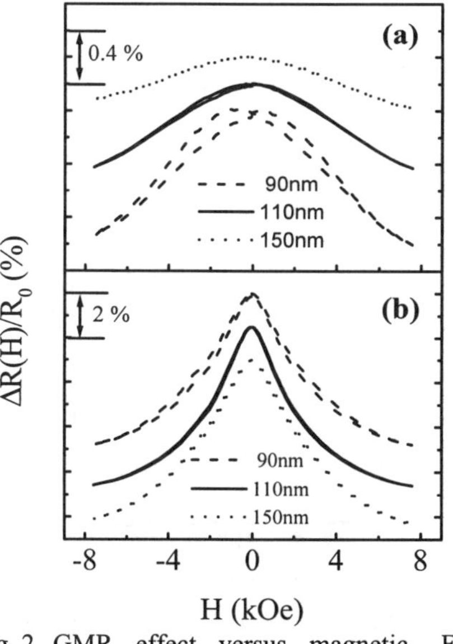

Fig. 2 GMR effect versus magnetic field strength measured at (a) 300K and (b) 20K for the samples as in Fig. 1. The MR curves were measured in the perpendicular configuration.

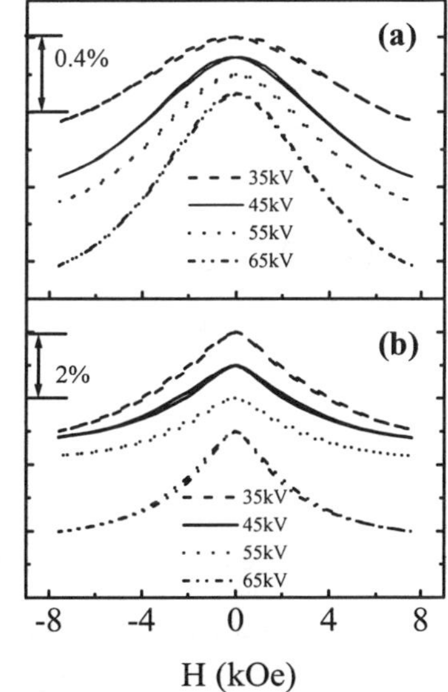

Fig. 4 Perpendicular GMR effect versus magnetic field strength measured at (a) 300K and (b) 20K for the samples implanted at various extraction voltages as indicated. The thickness of the Ag film before implantation is 85nm and the implant dose is $8x10^{16}$ cm^{-2}.

at 300K. This is highly unexpected and contradicts the opposite trend observed in CoAg granular films prepared by deposition methods [10]. Shown in Fig. 6(a) is the anomalous temperature variation of H_C determined from the MR curves for one of these samples in the temperature range from 20K to 300K. It is seen that H_C exhibits a maximum value at around 240K and decreases with decreasing temperature from 240K to 20K. An explanation for this observation is not available at the moment. Shown in Fig. 6(a) is also the temperature variation of the magnetization M of the sample measured at a magnetic field strength of 10 kOe using a SQUID magnetometer. It is seen that M first drops rapidly in the temperature range from 100 to 150K. After a slight further decrease, M reaches a minimum at about 240K and then begins to increase slowly with increasing temperature. The minimum in the M-T curve is seen to correspond well with the maximum of the H_C-T curve. The exact origin responsible for the M-T curve is also not clear. However, the M-T curve suggests that there is more than one magnetic phase present in the sample, and that they exhibit different coercive fields and different magnetic transition temperatures. Otherwise M should drop towards zero monotonically with increasing T. Therefore, the anomalous change of H_C with temperature is probably a manifestation of the combined effect of the different magnetic phases. Two typical magnetization curves measured at 100K and 200K up to a magnetic field strength of 30 kOe are shown in Fig. 7. The coercive field at 100K is clearly seen to be smaller than that at 200K. Typical variation of the GMR effect and the normalized resistance with temperature for one sample are shown in Fig. 6(b). The resistance of the films shows normal metallic behavior and the magnitude of the GMR increases as the temperature decreases as expected.

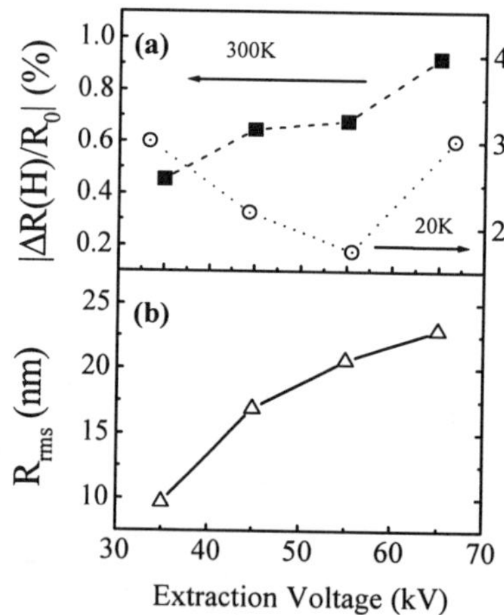

Fig. 5 (a) Magnitude of the GMR effect and (b) root-mean-square surface roughness versus the implant extraction voltage for the samples as in Fig. 4.

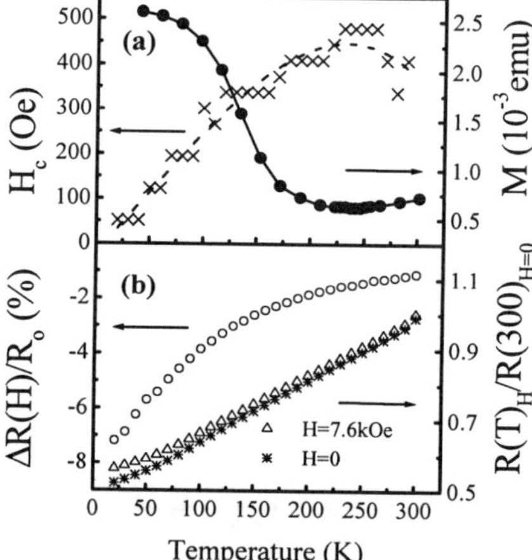

Fig. 6 Temperature dependence of (a) the coercive field H_C determined from perpendicular MR curves and the magnetization M by SQUID measurements; and (b) the GMR effect and the normalized resistance for the sample with an original Ag film thickness of 90nm and implanted at 35 kV to a dose of $8x10^{16}$ cm^{-2} after annealing at 400°C.

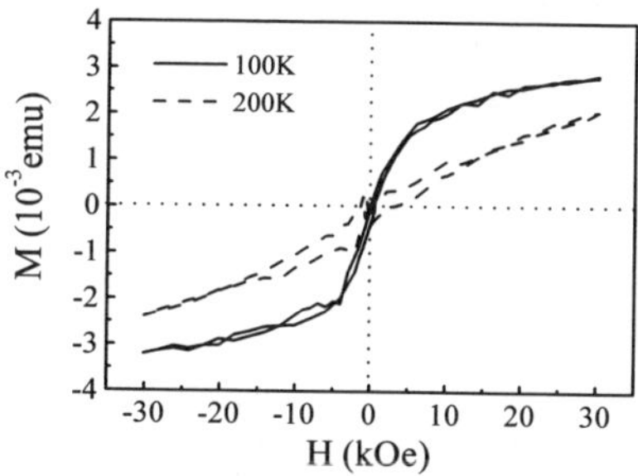

Fig. 7 Magnetization curves, measured at the temperatures as indicated, for the sample as in Fig. 6.

The surface morphology of the implanted CoAg granular films was observed by AFM. The AFM results showed that the root-mean-square roughness of the surface increases with the implant extraction voltage as shown in Fig. 5(b). A typical AFM micrograph for the sample of Fig. 6 is shown in Fig. 8(a) where a rough surface consisting of relatively large craters and small protrusions is seen to have resulted after implantation. In Fig. 8(b), exactly the same AFM image as in Fig. 8(a) but with the height data presented by the gray level is shown for convenience of comparison with the corresponding MFM image shown in Fig. 8(c).

The bright and dark regions in the MFM image represent magnetic domains with antiparallel perpendicular magnetization. However, the domain structure revealed in Fig. 8(c) for our implanted CoAg granular films shows very different features in comparison with those reported in the literature for sputter deposited CoAg granular films using MFM [8] or SEMPA [5] techniques, which exhibit structures consisting of elongated clusters or labyrinthine stripes with typical width of about 0.1 μm and length of 2 μm, and do not seem to have correlation with the surface morphology as revealed by AFM. The domain structure of the implanted films exhibits a much more chaotic feature in terms of shapes and size distribution of the domains and is also seen to have good correlation with the AFM image.

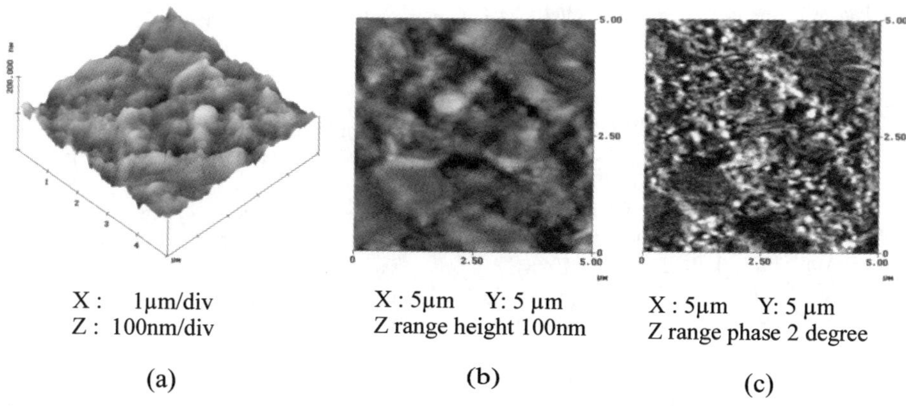

Fig. 8 Scanning probe micrographs of the sample as in Fig. 6 and Fig. 7. Shown in (a) is an AFM image in 3-D view. Shown in (b) is exactly the same AFM image as in (a) but with the height data presented by the gray level for convenience of comparison with the corresponding MFM image shown in (c).

The exact reason for this difference is not clear. As one possible explanation, it could be attributed to the surface roughness of the implanted films. In general, the magnetization vector favors lying along directions parallel to the surface to reduce demagnetization effect. Therefore, a rough surface could well disturb the formation of larger and more ordered domain structures and some of the features of the surface roughness will be reflected in the domain structure.

CONCLUSION

We have prepared CoAg granular films by the MEVVA implantation method and studied the MR properties of these films and their relation to the processing conditions. Anomalous temperature dependence of the coercive field H_C in the perpendicular-to-film direction determined from GMR measurements was observed for some samples. For one sample, H_C shows a maximum value at around 240K and decreases with decreasing temperature from 240K to 20K. The temperature variation of the magnetization M of this sample exhibits a minimum. The maximum in the H_C-T curve corresponds well with the minimum in the M-T curve. The M-T curve also suggests that there are more than one magnetic phases present in this sample. The domain structure of the implanted granular films as revealed by MFM images exhibit very different features as compared with those of sputter deposited CoAg granular films.

ACKNOWLEDGMENT

This work is supported in part by the Research Grants Council of Hong Kong (RGC ref. no.: CUHK374/96E).

REFERENCES

[1] M.N. Baibich, J.M. Broto, A. Fert, F. Nguyen Van Dau, F. Patroff, P. Etienne, G. Creuzet, A. Friederich, J. Chazelas, Phys. Rev. Lett. **61,** 2472 (1988).

[2] J.Q. Xiao, J.S. Jiang, C.L. Chien, Phys. Rev. Lett. **68,** 3749 (1992).

[3] A.E. Berkowitz, J.R. Mitchell, M.J. Carey, A.P. Young, S. Zhang, F.E. Spada, F.T. Parker, A. Hutten, G. Thomas, Phys. Rev. Lett. **68,** 3745 (1992).

[4] A. Tsoukatos, H. Wan, G.C. Hadjipanayis, K.M. Unruh, Z.G. Li, J. Appl. Phys. **73,** 5509 (1993).

[5] A. Gavrin, M.H. Kelley, J.Q. Xiao, C.L. Chien, Appl. Phys. Lett. **66,** 1683 (1995).

[6] H. Sang, G. Ni, S. Zhang,Y. Du, S.P. Wong, N. Ke, W.Y. Cheung, IEICE Trans. Electron E**80**-C, 1161 (1997).

[7] S.P. Wong and W.Y. Cheung, Mat. Res. Soc. Symp. Proc. **475,** 55 (1997).

[8] Y.J. Chen, W.Y. Cheung, I.H. Wilson, N. Ke, S.P. Wong, J.B. Xu, H. Sang, G. Ni, Appl. Phys. Lett. **72,** 2472 (1998).

[9] J.C. Soares, L.M. Redondo, C.M. de Jesus, J.G. Marques, M.F. da Silva, M.M. Pereira de Azevedo, J.A. Mendes, M.S. Rogalski, J.B. Sousa, J. Vac. Sci. Technol. A**16,** 1812 (1998).

[10] A. Butera, T.J. Klemmer, J.A. Barnard, J. Appl. Phys. **83,** 4885 (1998).

[11] S.P. Wong, M.F. Chiah, W.Y. Cheung, N. Ke, J.B. Xu, Nucl. Instr. and Meth. B**148,** 813 (1999).

[12] I.G. Brown, J.E. Gavin, R.A. MacGill, Appl. Phys. Lett. **47,** 358 (1985).

[13] I.B. Brown and X. Godechot, IEEE Trans. Plasma Sci. **19,** 716 (1991).

INTERFEROMETRY OF OPTICAL SECOND HARMONIC GENERATION FROM Gd-CONTAINING LANGMUIR-BLODGETT SUPERSTRUCTURES: MAGNETO-INDUCED EFFECTS

E.I. VISHNEVSKAYA, N.V. DIDENKO, A.A. FEDYANIN, G.B. KHOMUTOV, T.V. MURZINA, A.A. NIKULIN, O.A. AKTSIPETROV
Physics Department, Moscow State University, Moscow 119899, Russia, aktsip@astral.ilc.msu.su

ABSTRACT

Nonlinear magneto-optical Kerr effect in Gd-containing Langmuir-Blodgett superstructures is studied by second harmonic generation (SHG) interferometry. The intensity of the regular magneto-induced second harmonic (SH) is found to be by a factor of about 2 larger than the intensity of the regular SH component of a non-magnetic origin. The phase shift between the magneto-induced and non-magnetic contributions to the nonlinear polarization is about 46°. The intensity of the magnetization-independent hyper-Rayleigh scattering (HRS) is by two orders larger than that of magneto-induced HRS.

INTRODUCTION

Recently magnetic nanostructures has become a focus of attention stimulated by observation of new interesting magnetic effects, such as giant magnetoresistance in nanogranular films [1] and interlayer coupling through nonmagnetic spacer in layered magnetic/nonmagnetic metal nanostructures [2]. Magnetic properties of atomic monolayers of magnetic materials are of particular interest for fundamental studies (the problems of 2D magnetism) as well as for technological applications (fabrication of materials for high-density magnetic storage). In all these aspects metallo-organic Langmuir-Blodgett (LB) films containing monolayers of magnetic ions are challenging alternative to magnetic epilayers. In [3] the magnetic properties of such LB films containing monolayers of Gd^{3+} ions are studied by the EPR technique. The perspectives of using these materials as magneto-optical storage media for nonlinear-optical readout are discussed in [4].

Being highly sensitive to the structural, electronic, magnetic, etc. properties of the system in which it is induced, the effect of optical second harmonic generation (SHG) has become a base for a family of informative and versatile nonlinear-optical methods of study and characterization of thin films and nanostructures [5-7]. One of recent impressive examples of such techniques is SHG interferometry [8]. In magnetic systems, such as Gd-containing LB films, SHG interferometry implies simultaneous application of a dc magnetic field to induce the nonlinear magneto-optical Kerr effect (NOMOKE) that provides a much higher magnetic contrast than its linear prototype, the magneto-optical Kerr effect (MOKE) [9]. In this paper we study systematically the magnetization induced features of SHG interference patterns from superstructures containing monolayers of Gd ions.

EXPERIMENT

The preparation technique of the organo-metallic multilayers by the Langmuir-Blodgett method is shown schematically in Fig. 1. A monolayer of stearic acid molecules is formed by

Mat. Res. Soc. Symp. Proc. Vol. 577

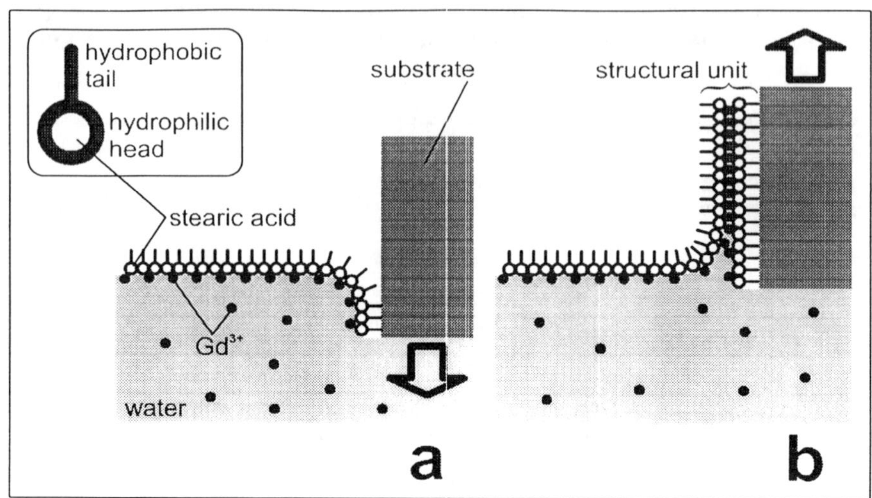

Fig. 1. Formation of the structural unit of the organo-metallic superstructure by immersing (*a*) and withdrawing (*b*) the substrate.

spreading a solution of stearic acid in chloroform on the surface of the water subphase. The monolayer adsorbs Gd^{3+} ions from the subphase being a $5 \cdot 10^{-4}$ M aqueous solution of gadolinium acetate. The stearic acid monolayers with adsorbed Gd^{3+} ions are subsequently deposited onto a fused quartz substrate with the use of the LB vertical lift method. The structural unit of the resulting superstructure is about 5 nm thick and consists of a Gd^{3+} monolayer "squeezed" in between two monolayers of stearic acid, with an area of approximately 0.1 nm^2 per Gd^{3+} ion.

In the nonlinear-optical experiments the output of a Q-switched YAG:Nd^{3+} laser at 1064 nm with a pulse duration 15 ns, repetition rate 12.5 Hz and pulse intensity of about 1 MW/cm^2 is used as a fundamental radiation that is focused onto a spot with a diameter of about 1 mm. The SH signal generated from LB film is selected by a set of bandpass filters and detected by a PMT and gated electronics.

The *s*-in, *s*-out polarization combination is used and two types of dependences are measured: the rotational (azimuthal) SHG anisotropy and interference patterns for magnetization induced second harmonic generation (MSHG). A thin SnO_2 film as a source of the reference SH wave is used for the MSHG interferometry. The phase shift between the SH waves generated by the sample and reference is varied by changing the distance between reference and sample [8].

A dc magnetic field of 1 kOe is applied parallel to both the film plane and that of incidence of the fundamental radiation (the longitudinal configuration of film magnetization).

RESULTS AND DISCUSSION

Figure 2 shows the azimuthal anisotropy of SHG in the absence of a dc magnetic field. Generation of the coherent isotropic SH is forbidden for the *s*-in, *s*-out geometry, therefore the presence of a large isotropic SH component indicates that coherent

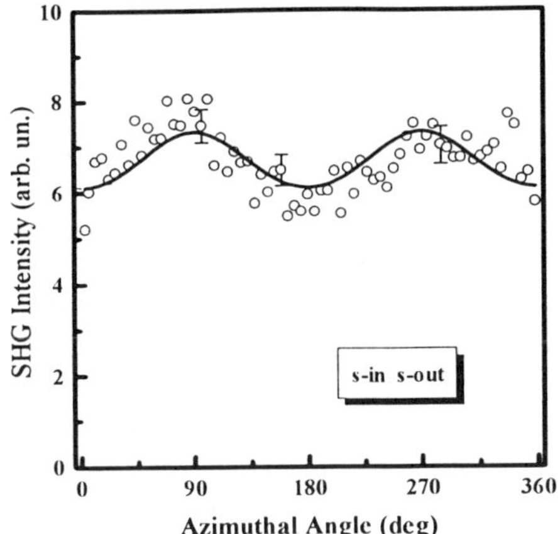

Fig. 2. The SH intensity as a function of the azimuthal angle ψ. Solid line: the dependence given by the expression $I_{2\omega}(\psi)=\tilde{I}+\left|A\sin\psi+B\sin^3\psi\right|^2$ with $\tilde{I}=6.10$, $A=0.99$, $B=0.12$, where $A\propto\chi_{yxx}+2\chi_{xyx}$, $B\propto\chi_{yyy}-\chi_{yxx}-2\chi_{xyx}$ and χ_{yyy}, χ_{yxx}, χ_{xyx} are components of the quadratic susceptibility tensor of the film in the crystallographic (rotating) frame with the x and y axes lying in the film plane.

is accompanied by hyper-Rayleigh scattering (incoherent SHG), which in turn reveals the existence of a partial disorder in the film structure. Apparently this is a disorder of an island-like 2D structure formed by Gd^{3+} ions in each Gd monolayer plane. Thus the total SH intensity is the sum of a anisotropic and incoherent (diffuse) isotopic components:

$$I_{2\omega}(\psi)=\bar{I}(\psi)+\tilde{I}, \tag{1}$$

where ψ is the azimuthal angle and the bar (tilde) over a quantity denotes a coherent (incoherent) component. The fact that the azimuthal dependence of the SH intensity has two maximums and two minimums indicates the existence of a chosen direction in the film plane.

Figure 3 shows MSHG interference patterns measured in the longitudinal configuration at two opposite directions of a dc magnetic field and at the azimuthal angle $\psi=12°$. An interference pattern is formed by the SH wave from the sample studied and that from the reference. Obviously the interference occurs only between the coherent component of the SH wave from the sample and the (coherent) SH wave from the reference. The phase shift between the two waves consists of the magnetization induced part and that resulting from the dispersion of the refraction index of air n, which yields the following dependence of the total SHG intensity $I_{2\omega}$ on the reference translation:

$$I_{2\omega}^{\alpha}(x_R)=I_R+\bar{I}^{\alpha}+\tilde{I}^{\alpha}+2\sqrt{I_R\,\bar{I}^{\alpha}}\cos\left[\Phi_0+\Phi^{\alpha}+2\omega\left(n(2\omega)-n(\omega)\right)\left(x_R-x\right)/c\right], \tag{2}$$

where I_R is the intensity of the SH wave from the sample, $\bar{I}^{\alpha}$ and $\tilde{I}^{\alpha}$ are the intensities of coherent and incoherent components of the SH radiation from the sample, respectively, Φ_0 is the phase shift between the two waves independent on the positions of the sample and reference as well as on the magnetic field, Φ^{α} is the magnetization induced part of the phase shift, x and x_R are the coordinates of the sample and the reference, respectively, measured in

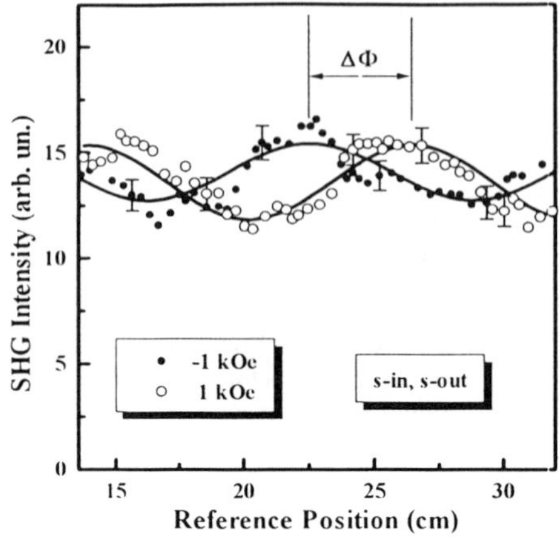

Fig. 3. The SH intensity as a function of the reference position measured at $H_{\parallel} = \pm 1\,\mathrm{kOe}$. Solid lines: the dependences given by Eq. (2) with $\bar{I}^{+} = 0.021\, I_{\mathrm{R}}$, $\bar{I}^{-} = 0.020\, I_{\mathrm{R}}$, $\tilde{I}^{+} = 0.979\, I_{\mathrm{R}}$, $\tilde{I}^{-} = 0.980\, I_{\mathrm{R}}$

the reference translation direction, ω is the fundamental radiation frequency, $\alpha = +, -$, superscripts + and – denote quantities pertaining to the opposite directions of the applied magnetic field. Thus the change of the direction of the applied magnetic field leads to a significant phase shift of the coherent part of the SH wave from the film: $\Delta\Phi = \Phi^{+} - \Phi^{-} = 115°$. To proceed with interpretation of the observed magnetization induced effects we introduce an in-plane component P of the nonlinear polarization being the source of the s-polarized SH radiation. This quantity is the sum of non-magnetic and magneto-induced parts. The latter is supposed to be linear in the longitudinal-configuration dc magnetic field $H_{\parallel}$:

$$P = p + q H_{\parallel}. \tag{3}$$

On the other hand, P is the sum of the regular and fluctuating parts:

$$P = \bar{P} + \tilde{P}, \qquad \bar{P} \equiv \langle P \rangle, \qquad \tilde{P} \equiv P - \langle P \rangle, \tag{4}$$

where brackets denote averaging over random realizations. Combining Eqs. (3) and (4) yields

$$\bar{P} = e^{i\varphi_0}\left(|\bar{p}| + e^{i\Delta\varphi}|\bar{q}|H_{\parallel}\right), \qquad \tilde{P} = \tilde{p} + \tilde{q}\, H_{\parallel}, \tag{5}$$

where in the regular term the phases φ_0 and $\varphi_0 + \Delta\varphi$ of the complex quantities $\bar{p}$ and $\bar{q}$ are introduced explicitly. For the coherent and incoherent SH components we have the following expressions:

$$\bar{I}(H_{\parallel}) \propto |\bar{p}|^2 \left[1 + 2\bar{\xi}(H_{\parallel})\cos\Delta\varphi + \bar{\xi}^2(H_{\parallel})\right], \quad \tilde{I}(H_{\parallel}) \propto \tilde{\Pi}\left[1 + 2\,\tilde{\eta}(H_{\parallel}) + \tilde{\zeta}(H_{\parallel})\right], \tag{6}$$

where $\bar{\xi}(H_\parallel) \equiv |\bar{q}| H_\parallel / |\bar{p}|$, $\tilde{\Pi} \equiv \int \langle \tilde{p}^*(\mathbf{r}) \tilde{p}(\mathbf{r}+\mathbf{r}') \rangle d\mathbf{r}'$, $\tilde{\eta}(H_\parallel) \equiv H_\parallel \operatorname{Re} \int \langle \tilde{p}^*(\mathrm{r}) \tilde{q}(\mathrm{r}+\mathrm{r}') \rangle d\mathrm{r}' / \tilde{\Pi}$, $\tilde{\zeta}(H_\parallel) \equiv H_\parallel^2 \int \langle \tilde{q}^*(\mathrm{r}) \tilde{q}(\mathrm{r}+\mathrm{r}') \rangle d\mathrm{r}' / \tilde{\Pi}$, $\mathbf{r}$ and $\mathbf{r}'$ are radius-vectors lying in the film plane. The parameters $\bar{\xi}$, $\Delta\varphi$, $\tilde{\eta}$ and $\tilde{\zeta}$ can be expressed through the quantities measured in experiment:

$$\bar{\xi}^2(H_\parallel = 1\,\mathrm{kOe}) = \left(\bar{I}^+ + \bar{I}^-\right) / \left(2\bar{I}(\psi = 12^\circ)\right) - 1, \qquad \Delta\varphi = \arcsin\left[\left(1 - \bar{\xi}^2\right) \tan\Delta\Phi / \left(2\bar{\xi}\right)\right], \quad (7)$$

$$\tilde{\zeta}(H_\parallel = 1\,\mathrm{kOe}) = \left(\tilde{I}^+ + \tilde{I}^-\right) / \left(2\tilde{I}\right) - 1, \qquad \tilde{\eta}(H_\parallel = 1\,\mathrm{kOe}) = \left(\tilde{I}^+ - \tilde{I}^-\right) / \left(4\tilde{I}\right). \quad (8)$$

From Eqs. (7) and (8) we obtain $\bar{\xi}^2 = 1.9$, $\Delta\varphi = 46^\circ$, $\tilde{\eta} = 3 \cdot 10^{-4}$, $\tilde{\zeta} = 7 \cdot 10^{-3}$. One can see that the intensity of the regular magneto-induced SH is by a factor of about 2 larger than the intensity of the regular SH component of non-magnetic origin, whereas for the incoherent (diffuse) SH the situation is quite different: the intensity of the magnetization-independent hyper-Rayleigh scattering is by two orders larger than that of its magneto-induced analog.

CONCLUSION

The method of nonlinear-optical interferometry has been used for determining the phase and amplitude characteristics of the nonlinear magneto-optical Kerr effect in Gd-containing multilayer Langmuir-Blodgett films. The relative magnitude of the magneto-induced hyper-Rayleigh scattering in these systems has also been found. These results prove the efficiency of nonlinear-optical methods in characterization of 2D magnetic superstructures with buried interfaces.

ACKNOWLEDGMENTS

This work was supported by INTAS-93 grant 370 (ext), special RFBR grants for Russian Leading Scientific Schools 96-15-96420 and 96-15-96476, RFBR grants 97-02-17919, 97-02-17923, 99-03-32218, INTAS Fellowship grants for Young Scientists YSF-8 and YSF-10, RFBR-DFG grant 98-02-04092, Federal Integration Program "Center of Fundamental Optics and Spectroscopy", Programs of Russian Ministry of Science "Physics of Solid State Nanostructures" and "Fundamental Metrology" and Program of Russian Ministry of Education "Universities of Russia"

REFERENCES

1. J.Q. Xiao, J.S. Jiang, and C.L. Chien, Phys. Rev. Lett. **68**, 3749 (1992).

2. S.S.P. Parkin, N. More, and K.P. Roche, Phys. Rev. Lett. **64**, 2304 (1990).

3. A.M. Tishin, Yu. A. Koksharov, J. Bohr, and G.B. Khomutov, Phys. Rev. B **55**, 11064 (1997).

4. O.A. Aktsipetrov, A.A. Fedyanin, A.V. Melnikov, E.D. Mishina, and T.V. Murzina, Jpn. J. Appl. Phys. **36**, 48 (1998); T.V. Murzina, N.V. Didenko, A.A. Fedyanin,

A.V. Melnikov, T.V. Misuryaev, O.A. Aktsipetrov, Mat. Res. Soc. Symp. Proc. **517**, 605 (1998).

5. T.F. Heinz, in *Nonlinear Surface Electromagnetic Phenomena*, edited by H.-E. Ponath and G.I. Stegeman (North Holland, 1991), p. 355.

6. J. Reif, C.Rau, and E. Matthias, Phys. Rev. Lett. **71,** 1931 (1993).

7. U. Pustogowa, T.A. Luce, W. Hübner, K.H. Bennemann, J. Appl. Phys. **79**, 6177 (1996).

8. R. Stolle, G. Marowsky, and G. Berkovic, Appl. Phys. B **63**, 491 (1996).

9. N.V. Didenko, A.A. Fedyanin, G.B. Khomutov, T.V. Murzina, O.A. Aktsipetrov, Mat. Res. Soc. Symp. Proc. **517**, 657 (1998).

A METHOD FOR CONTROLLED SYNTHESIS OF ANISOTROPIC NANOPARTICLES AND NANOSYSTEMS

G.B. KHOMUTOV*,***, S.P. GUBIN**, Yu.A. KOKSHAROV***, V.V. KHANIN***, A.Yu. OBIDENOV***, E.S. SOLDATOV***, A.S. TRIFONOV***
*Tetra Consult/Isle Nanoelectronics Ltd., M.V. Lomonosov Moscow State University, Moscow, 119899, Russia, GBK@bio383a.phys.msu.su
**Institute of General and Inorganic Chemistry, Russian Academy of Sciences, Moscow, 119899, Russia
***Faculty of Physics, M.V. Lomonosov Moscow State University, Moscow, 119899, Russia

ABSTRACT

In the present work it has been shown that the shape of ultrafine metallic magnetic particles is controlled by applied magnetic and/or electric fields and the state of reaction media during the nanoparticles formation. The ulraviolet (UV) decomposition of volatile organometallic compound iron pentacarbonyl was used to produce amorphous metallic nanoparticles. The two-dimensional photochemical synthesis of nanoparticles was carried out in a mixed Langmuir monolayer at the gas/water interface with stearic acid as surfactant matrix and also in a mixed film deposited onto graphite substrate. Scanning tunneling microscopy analysis of deposited nanoparticulate monolayers showed that the size and shape of nanoparticles formed were dependent on the initial mixed monolayer content, compression extent and exposure time to UV, and were changed dramatically from isotropic plate-like to aligned ellipsoidal and needle-like when external magnetic and/or electric fields were applied during the synthesis. Magnetic nanoparticles with diameter exceeding 5 nm formed chain structures alignable by external fields. The one-dimensional aligned metallic string-like nanostructures (nanowires) with cross-section size about 10÷20 nm were also formed at elevated temperatures under synergistically applied magnetic and electric fields. Magnetic properties of nanoparticulate multilayer Langmuir-Blodgett films were studied using electron paramagnetic resonance technique. The ferromagnetic resonance and superparamagnetic signals were observed in the material.

INTRODUCTION

Nanoparticulated composite materials are a subject of great interest due to properties important for fundamental studies and wide applications [1,2]. In particular, in magnetic recording applications materials are required with ultrasmall magnetic particles desirable for increased information storage densities [3,4]. Patterned magnetic structures may have also important technological applications for magnetic recording media to overcome the super-paramagnetic limit [5]. The preparation of nanoparticles and nanoparticulate condensed and liquid composite materials has become a major area of investigation. Search of ways to control the chemical content, phase morphology, size, shape and orientation of synthesized nanoparticles and to create patterned structures and ordered arrays of nanoparticles is of principal importance in this field. Langmuir-Blodgett (LB) technique allows to creating multicomponent organized ultrathin molecular films with incorporated or grown inside clusters and nanoparticles [6-9]. This approach allows for the formation of planar nanoparticulate structures of ultimately small thickness, corresponding to the dimensions of a

Mat. Res. Soc. Symp. Proc. Vol. 577 © 1999 Materials Research Society

single nanoparticulate monolayer, and to create corresponding multilayer nanoparticle-containing structures of the strictly controllable thickness, determined by a number of deposited nanoparticulate monolayers. Scanning tunneling microscopy (STM) and spectroscopy are an effective means of study the morphology of conducting nanostructures [10, 11]. The samples for investigations of single metallic nanoparticle morphology by this technique has to be monolayer on a conducting substrate.

In the present study, the amorphous iron nanoparticles were generated by ultraviolet (UV) decomposition of volatile organometallic compound iron pentacarbonyl directly in a surfactant monolayer (stearic acid) at the gas/water interface. Long-chain carboxylic acids are known as surfactant compounds to form insoluble monolayers at the air/water interface. Simultaneously, such compounds are widely used as stabilizing agents to prevent clumping of dipolar colloid particles in particulate liquids, for example, in magnetic fluids [12], due to short-range steric repulsion caused by surfactant layers bound to the surface of particles. In our approach the properties of fatty acids to form Langmuir monolayer and to hinder the total aggregation of dipolar particles are combined successfully. The microtopography of deposited nanoparticulate monolayers was studied using STM and indicated strong influence of applied magnetic and electric fields on the shape of the iron particles. Electron paramagnetic resonance (EPR) technique was used to characterize the magnetic properties of nanoparticulate multilayer LB films.

EXPERIMENT

Metallic magnetic nanoparticles were fabricated by the photo-decomposition of an iron pentacarbonyl $Fe(CO)_5$, a volatile compound, at ambient temperature (21 °C) in a mixed Langmuir monolayer formed at the gas/water interface with stearic acid (SA) as a surfactant matrix. Stearic acid (CH_3-$(CH_2)_{16}$-COOH) was obtained from Serva, iron pentacarbonyl $Fe(CO)_5$ was obtained from Alfa Inorganic and used without further purification. Water was purified by a Milli-Q system (Millipore Ltd.).

Surface pressure-monolayer area isotherm measurements and monolayer transfer to solid substrates were carried out on a full automatic conventional Teflon trough at 21 ^{0}C as described elsewhere [13]. Langmuir monolayers were formed by spreading a mixed chloroform solution of $Fe(CO)_5$ with stearic acid (2×10^{-4} M) on the surface of purified water (pH =5.6). Mixed $Fe(CO)_5$/SA monolayer ($Fe(CO)_5$/SA ratio varied from 50:1 to 1:1) was irradiated by UV for different time intervals using 300 mW conventional IVR UV source (λ = 300 nm). Monolayers were then compressed and deposited onto high oriented pyrolytic graphite (HOPG) substrate at a constant pressure (25 mN/m), temperature and dipping speed using the conventional horizontal lifting method with good transfer ratios to form monolayer nanoparticulate samples for STM investigations. Multilayer nanoparticulate LB films were formed by means of vertical lifting deposition onto polished silicon substrates with natural oxide layer at surface pressure value 25 mN/m. The microwave (MW) absorption signals were recorded with the use of Varian E-4 spectrometer at the X band at ambient temperature (21 °C).

The STM measurements of the monolayer films deposited on a HOPG surface by horizontal lifting method were performed using a modified Nanoscope-1 STM (Digital Instruments) device in the constant current mode. Topographic images were obtained by recording the STM tip height at constant tunnel current at an ambient temperature (21 °C). Tunnel current I = 0.3 nA, and a bias voltage V = 200 mV. The images were stable and reproducible.

RESULTS

STM topographic images of mixed UV-irradiated monolayers deposited onto the HOPG substrate revealed individual iron nanoparticles and aggregates formed. The particle size depended significantly on the mixed monolayer exposure time to the UV. Figures 1a) and 1b) show a typical STM image of a deposited nanoparticulate monolayer with quasi-isotropic plate-like nanoparticles of ~5 nm diameter formed in the absence of external fields. Figure 1c). shows a chain of circular nanoparticles also observed in such monolayer. It is known that particles with dipole moment aggregate into chains when dipole-dipole interaction energy exceeds k_BT (k_B - Boltzmann constant). Chains of magnetic ultrafine particles are usually observed in magnetic liquids [14] and it gives evidence that iron nanoparticles synthesized in our experiments have noticeable magnetic moments. The increase in the time of mixed monolayer exposure to UV or increase the $Fe(CO)_5$ content in the monolayer (above 30 for $Fe(CO)_5$/SA ratio) resulted in the formation of planar dense iron powder structures.

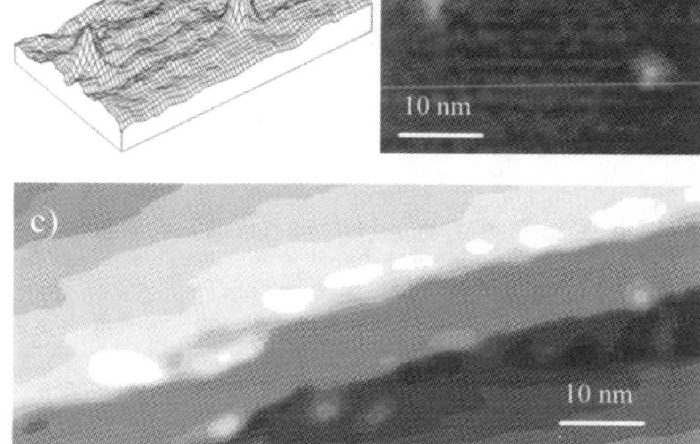

Figure 1. STM images of mixed nanoparticulate monolayers ($Fe(CO)_5$/stearic acid ratio 15:1, 5 min. exposure to UV), formed and deposited by Shaefer's method onto the surface of graphite substrate without external fields. (a): individual nanoparticles (3-D image). (b): individual nanoparticles (top view); (c): chain of magnetic metallic nanoparticles The scales of images are shown in the bottom of pictures. The transition from white to black corresponds to height difference 2 nm.

Figure 2a) displays STM image of anisotropic iron nanoparticles formed in a monolayer with initial content as in Figure 1a), and under the same conditions of compression and UV irradiation, but under applied external magnetic field 3×10^3 Oe. The shape of obtained nanoparticles is seen to be anisotropic ellipsoid-like with diameter ≅6 nm and axial ratio length/diameter ~ 4. Anisotropic particles were oriented in the field direction when synthesized and deposited in monolayer under external field. The size and shape of nanoparticles were reproducible, nanoparticles were arranged in monolayer individually and chain clusters of elongated nanoparticles were also observed.

The formation of nanoparticles under simultaneously applied magnetic and electric fields resulted in the substantial increase in an axial ratio length/diameter of nanoparticles formed (Figure 2b). Figure 2c) shows the part of one-dimensional string-like metallic nanostructure formed under 6.5×10^3 Oe external magnetic field and 10^3 V/cm electric field at 90 °C in deposited monolayer (in which nanoparticles were previously synthesized at room temperature). At this temperature surfactant matrix is molten and particle coalescence is promoted effectively.

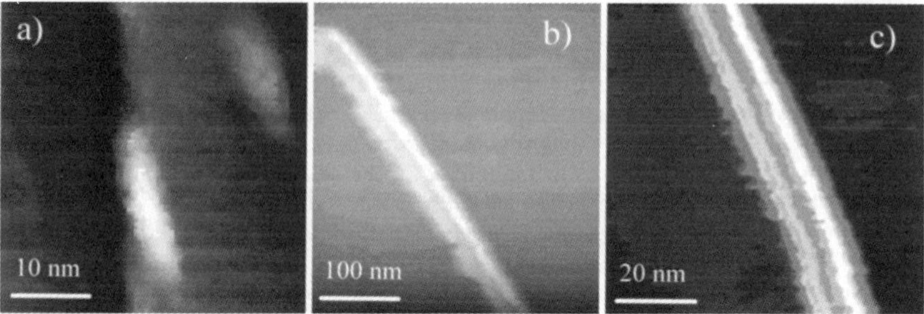

Figure 2. STM image (top view) of nanoparticulate monolayers.
(a): individual elongated nanoparticle synthesized in mixed Langmuir monolayer ($Fe(CO)_5$/stearic acid ratio 15:1, 5 min. exposure to UV) and deposited onto the surface of graphite substrate under applied magnetic field (2×10^3 Oe); (b): individual nanoparticle synthesized in mixed monolayer under applied magnetic (2×10^3 Oe) and electric (10^3 V/cm) fields and deposited by Shaefer's method onto the surface of graphite substrate; (c): One-dimensional metallic nanostructure synthesized in mixed nanoparticulated monolayer (T = 363 K) under applied magnetic (6×10^3 Oe) and electric (10^3 V/cm) fields onto the surface of graphite substrate. The scales of images are shown in the left bottom corner of pictures. The transition from white to black corresponds to height difference 3 nm.

X-ray diffraction study of particulated films formed showed no iron crystallite diffraction peaks, but a wide diffuse X-ray scattering typical for amorphous iron. A number of small peaks attributable to iron oxide, carbide or other iron-based phases were observed. This is consistent with known data that the decomposition of iron pentacarbonyl by ultrasound [15] and thermal treatment [16, 17] leads to the formation of dispersed amorphous iron with possible inclusion of other alloying elements.

In our experiments the UV-decomposition of iron pentacarbonyl into monolayer without external fields produced the plane-isotropic platelet iron particles. But the same reactions under external fields resulted in formation of stable highly anisotropic elongated metallic particles and extended nanostructures. Iron particles formed by decomposition of $Fe(CO)_5$ possess magnetic dipolar moment under external magnetic field and electric dipolar moment due to conducting particle polarization by external electric field. As a result the anisotropic interparticle dipole-dipole interactions and kinetic factors may play key role in the particle growth reactions under external fields.

In strong enough external field dipolar moments of the particles are aligned parallel to the applied field. When the dipole-dipole attraction energy is larger than the disordering thermal energy, the highly-anisotropic interaction between parallel dipoles of the colloid particles allows for rapid (milliseconds) aggregation of the particles into one-dimensional dense chains parallel to an applied field which themselves aggregate slowly into columns (a wider compact cylinder of particles) oriented in the field direction, the effect characteristic for electrorheological [18] and magnetic [19] fluids. Correspondingly, when dipole-dipole interaction energy exceeds k_BT the bi-molecular reaction rate of coalescence of particles with dipolar moments is substantially anisotropic and maximal when the line of centers of interacting particles is aligned with external field. Parallel applied magnetic and electrical fields synergistically increase anisotropy and maximal value of growth rate and correspondingly enhance the anisotropical growth of magnetic conducting particles.

The formation of chains and columns of dipolar particles oriented in the applied field directions and spanning the whole thickness of the particulated media [18, 19] can be related to the observed formation of string-like metallic nanostructures at elevated temperature under synergistically applied electric and magnetic fields when linearly chained particles were promoted to coalescence (Fig 2c.). The sticking together of couple such coalescenced chains can give double-line structure showed on this Figure.

To detect and investigate magnetic properties of nanoparticulate LB films formed we used the method of electron ferromagnetic resonance (FMR) [20]. Figure 3a shows the typical FMR spectra in the sample LB film at low field region (< 1500 Oe). The broad asymmetric line was detected in all samples investigated and characterized by high effective g-values (about 15-18) and sample-dependent linewidth (changing approximately from 500 Oe up to 800 Oe). Since paramagnetic Fe ions do not reveal such high effective g-value, probably, the low-field signal can be attributed to the ferromagnetic resonance. Numerical calculations for an assembly of isolated nanoparticles shown [21] that the ferromagnetic resonance appears at low fields only at low temperatures. It seems likely that the room temperature low-field signal results from collective interactions in the nanoparticle assembly. These interactions are determined by the spatial distribution of nanoparticles inside the film and can vary from sample to sample. It is typical for nanoparticulate systems [22] that besides FMR signal the resonance takes place due to isolated superparamagnetic particles. We managed to detect such signal in our films which is presented in Fig.3b.The line is characterized by $\Delta H \approx 900$ Oe and effective $g \approx 2.3$.

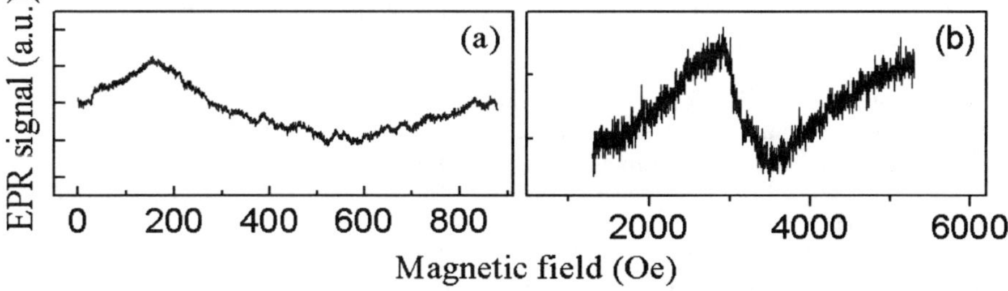

Fig.3. a) Low-field signal of microwave absorption in nanoparticulate LB film (45 layers) formed on the surface of Si substrate. Conditions of nanoparticles synthesis corresponds to those on Fig. 1a). T = 294 K. b) Middle-field signal of MW absorption in the same sample.

The simultaneous decomposition of a mixture of different volatile organometallic precursors can result in the formation of amorphous alloy powders with corresponding stoichiometric metal content [23]. The organic part of the material (stearic acid) can be removed by different treatments, for example, heating to temperatures of about 500 K in inert atmosphere. It is known also that amorphous metallic alloy powders and films can be crystallized by heat treating [15, 24]. As a result the approach presented can be useful for fabrication of phase-controlled (amorphous and/or nanocrystalline) nanocomposite materials including rare-earth transition metals alloys and ultrathin films.

CONCLUSIONS

UV decomposition of volatile organometallic compound iron pentacarbonyl in a mixed Langmuir monolayer at the gas/water interface with stearic acid as surfactant matrix resulted in the formation of amorphous iron alloy magnetic nanoparticles with plate-like shape without external fields. Our results demonstrate the controlled formation of elongated aligned amorphous metallic nanoparticles and one-dimensional nanostructures (chains and nanowires) under applied electric and/or magnetic fields what can be promising approach to nanophase engineering and obtaining bulk materials and ultrathin films with highly anisotropic electrical, optical, magnetic, thermal and mechanical properties.

ACKNOWLEDGMENTS

This work was supported by Russian Foundation for Basic Researches (Grants 99-03-32218 and 97-03-32199a).

REFERENCES

1. *Clusters and Colloids. From theory to Applications*, edited by G. Schmid (VCH, New York, 1994).
2. J. Shi, S. Gider, K. Babcock, D.D. Awschalom, Science **271**, pp.937-941 (1996).
3. B. Ao, L. Kummerl, D. Haarer, Adv. Mater. **7**, pp. 496-501 (1995)
4. J.C. Mallinson, *The foundation of Magnetic Recording*, Chapter 3, (Acad. Press, Berkeley, CA, 1987).
5. J.L. Simonds, Physics Today **48**, p.p. 26-32 (1995)
6. X.K. Zhao, S. Xu, and J.H. Fendler, J. Phys. Chem. **94**, pp. 2573-2581 (1990).
7. S. Ravaine, G.E. Fanucci, C.T. Seip, J.H. Adair and D.R. Talham, Langmuir **14**, pp. 708-713 (1998).
8. G.B. Khomutov, E.S. Soldatov, S.P. Gubin, S.A. Yakovenko, A.S. Trifonov, A.Yu. Obidenov, V.V. Khanin, Thin Solid Films **327-329**, pp. 550-553 (1998).
9. Y.S. Kang, D.K. Lee, and P. Stroeve, Thin Solid Films **327-329**, pp. 541-544 (1998)]
10. J.W.G. Wildoer, L.C. Venema, A.G. Rinzler, R.E. Smalley, and G. Dekker, Nature **391**, pp. 59-62 (1998).
11. S.A. Yakovenko, S.P. Gibin, E.S. Soldatov, E.S., A.S. Trifonov, V.V. Khanin and G.B. Khomutov, Thin Solid Films **284-285**, pp.873-875 (1996).
12. K.V.P.M. Shafi, S. Wizel, T. Prozorov, A. Gedanken, Thin Solid Films **318**, p.38 (1998).
13. G.B. Khomutov, S.A. Yakovenko, T.V. Yurova, V.V. Khanin, E.S. Soldatov, Supramolecular Science **4**, pp. 349-355 (1997).
14. *Magnetic Fluids and Applications - Handbook*, edited by B. Berkovsky and M. Krakov (Begel-House, New York, 1994).
15. K.S. Suslick, S.B. Choe, A.A. Cichowlas, M.W. Grinstaff, Nature **353**, p. 414 (1991).
16. J. van Wonterghem, S. Morup, S.W. Charles, S. Wells, J. Villadsen, Phys. Rev. Lett. **55**, pp. 410-413 (1985).
17. A.V. Kozinkin, V.G. Vlasenko, S.P. Gubin, A.T. Shuvaev, I.A. Dubovtsev, Inorg. Mater. **32**, pp. 422-428 (1996).
18. T.C. Halsey, Science **258**, pp. 761-766 (1992).
19. A.S. Silva, D. Wirtz, Langmuir **14**, pp. 578-581 (1998).
20. S.V. Vonsovsii, *Ferromagnetic resonance* (Pergamon, Oxford, 1966).
21. Yu.A.Raikher, V.I.Stepanov, Phys.Rev.B **50**, pp.6250-6259 (1992).
22. V.K. Sharma and F.Waldner J. Appl. Phys. **48**, pp. 4298-4301 (1977).
23. K.V.P.M. Shafi, A. Gedanken, R.B. Goldfarb, I. Felner, J. Appl. Phys **81**, pp. 6901-6905 (1997).
24. C. Prados, G.C. Hadjipanayis, Appl. Phys. Lett. **74**, pp. 430-432 (1999).

Part X

Nanocrystalline Antiferro- and Ferrimagnets

SYNTHESIS AND CHARACTERIZATION OF ANTIFERROMAGNETIC $KMNF_3$ NANOPARTICLES

C. SANGREGORIO, E. E. CARPENTER, C. J. O'CONNOR
Advanced Materials Research Institute University of New Orleans, New Orleans LA USA

ABSTRACT

The magnetic properties of nanosized antiferromagnetic particles of $KMnF_3$ are presented. The particles were synthesized using the microemulsion technique, i.e. by using the aqueous core of reverse micelles as constrained microreactors for the precipitation of the particles. The structural characterization of the samples, accomplished by TEM and XRD, reveal that the samples consist of cubic-shaped, crystalline $KMnF_3$ nanoparticles of uniform size. Control over the average size of the particles was achieved by changing the reaction time. Four different samples of average size in the range 13-35 nm were prepared. DC magnetic susceptibility measurements revealed superparamagnetic behavior of the particles. Hysteresis loops measured after field cooling the samples through T_N were shifted. The shift is ascribed to the exchange coupling between the antiferromagnetic core of the particles and the uncompensated spin shell surrounding it.

INTRODUCTION

The synthesis and characterization of nanometric-sized magnetic materials have been among the dominant research areas of the last decades. Due to the small size, magnetic nanoparticles possess unique properties. These properties are appealing from the theoretical point of view as well as for the wide range of important technological applications in which they can be employed, ranging from biological application [1,2] to magnetic data storage [3]. However, despite the large amount of work carried out on ferro- and ferrimagnetic nanosized materials, little research has been done on the magnetic properties of nanosized antiferromagnetic materials. Hence, their properties remain relatively unexplored.

The magnetic behavior of small-size grains of antiferromagnetic materials was first proposed by Néel in 1961 [4]. He suggested that small antiferromagnetic particles, due to the high surface/volume ratio, should possess a net magnetization below the ordering temperature. The origin of the magnetic moment lies in the uncompensation of the surface spins. Due to a non-zero magnetization, weak ferromagnetism and superparamagnetic behavior should be observed for antiferromagnetic nanoparticles, behavior analogous to that of ferro- and ferrimagnetic single domain nanoparticles. Néel also evaluated the magnetic moment of nanometer antiferromagnetic particles as being proportional to $\sqrt{(N/2)}$, where N is the number of magnetic atoms in the particle [5]. In the following years the superparamagnetic behavior was experimentally observed in different metal oxide small particles, such as α-Fe_2O_3, Cr_2O_3 and NiO [6,7].

The last decade has seen a renewed interest of many researchers in this field. For example, antiferromagnetic nanoparticles are constituent of some biological systems. The most investigated system is ferritin, an iron storage protein which consists of a spherical proteic shell (apoferritin) and an inorganic, antiferromagnetic-ordered spherical core of ferrihydrite [8]. The magnetic core can store up to 4500 Fe^{3+} ions and reaches size of ca. 80 Å. Another motivation is that these materials are promising candidates to observe low-temperature quantum tunneling of the magnetization [9].

In this paper we report on the synthesis and magnetic properties of nanosized antiferromagnetic particles of potassium fluoromanganate, $KMnF_3$. At room temperature,

Mat. Res. Soc. Symp. Proc. Vol. 577 © 1999 Materials Research Society

$KMnF_3$ is a paramagnet with the cubic perovskite structure. At 88.3 K it undergoes a transition to an antiferromagnetic ordered state [10]. This material presents some advantages that make it suitable for the synthesis in the nanophase state: it is stable and can be easily precipitated in crystalline form from a water solution.

The nanoparticles were synthesized using the microemulsion technique that utilizes the aqueous core of reverse micelles as a constrained microreactor for the precipitation of the particles. Reverse micelles are aggregates formed by dissolving amphiphilic molecules in organic solvents. Water is solubilized in the polar core of these aggregates, thus forming water droplets dispersed in the organic solvent. Inorganic salts can be dissolved inside the aqueous core of reverse micelles and then can be precipitated after chemical reaction as insoluble inorganic nanoparticles. This technique has been proven effective in synthesizing of highly homogeneous nanosized particles of different materials, including pure metals, oxide, and superconducting materials [11,12]. Furthermore, this method has the advantage of providing control over the average size of the particles in a relatively wide range (3-30 nm) by simply adjusting the water-to-surfactant molar ratio.

EXPERIMENTAL SECTION

The $KMnF_3$ nanoparticles were synthesized by using a microemulsion system with CTAB (cetyltrymethilammonium bromide) as surfactant, n-butanol as co-surfactant and n-octane as oil phase. All chemicals were purchased from Aldrich Chemical and used without further purification. Distilled and deionized water was used throughout the synthesis. Microemulsions were prepared by adding to 12g of CTAB, 12.3 ml of n-butanol and 62.6 ml of n-octane with a certain amount of an aqueous solution of the reactants, corresponding to the desired value of w = $[H_2O]/[CTAB]$. Different microemulsions, with w values in the range 5-20, were prepared. To precipitate the particles, two different microemulsions with the same composition but containing a 0.1 M aqueous solution of $MnCl_2{\cdot}4H_2O$ and a 0.4 M KF aqueous solution respectively, were prepared. The two microemulsions were then mixed together under constant stirring and allowed to react for different experimental times, ranging from few minutes to several hours. The micellar solution was then disrupted with the addition of a 1:1 methanol/chloroform mixture, resulting in a cloudy mixture. The nanoparticles were collected with a centrifuge, washed several times with methanol and dried in an oven at 50°C.

The structural characterization of the as prepared light pink colored samples, was carried out by TEM and X-Ray Powder Diffraction. The nanoparticles were observed in images taken with a JEOL 2010 transmission electron microscope. To prepare TEM samples, the powder was ultrasonically dispersed in methanol and a drop of the dispersion was deposited over a carbon coated copper grid. The average diameters of the particles were evaluated from the size distribution observed in different micrographs. XRD diffraction patterns were collected on a Philips X' PERT Diffractometer System equipped with a graphite monochromator and PMT detector using the Cu K_α wavelength. $KMnF_3$ patterns were simulated using a pseudo-Voigt profile with a Cauchy fraction of 50. No lattice contraction or expansion was included and the patterns were corrected for the $K\alpha_2$ wavelength.

Static magnetic susceptibilities were measured with a Quantum Design MPMS-5S SQUID magnetometer in the temperature range 2-300 K and with parallel field up to 50 kOe. The temperature dependence of the static susceptibility was investigated by cooling the sample in zero field, applying a measuring field, and then raising temperature (ZFC curve) or by cooling the sample in an external applied field and then raising temperature (FC curve). Both curves were collected with a measuring field of 100 Oe.

RESULTS AND DISCUSSION

A TEM image of a sample prepared with w=10 is shown in figure 1. The micrograph clearly shows the occurrence of cubic-shaped nanoparticles with highly uniform size. The average size of the edge of the cube, as evaluated from statistics over a large number of particles in different areas of the copper grid, is 35 nm. The XRD pattern of the sample is shown in the bottom part of figure 2. The pattern clearly displays all the peaks of the cubic perovskite structure of $KMnF_3$. The comparison of the experimental pattern with the simulated one (figure 2, top) shows that no other phases or traces of impurities are present. The average diameter obtained from the analysis of the peak broadening performed through the Scherrer equation is 22 nm, in good agreement with that observed in TEM micrographs.

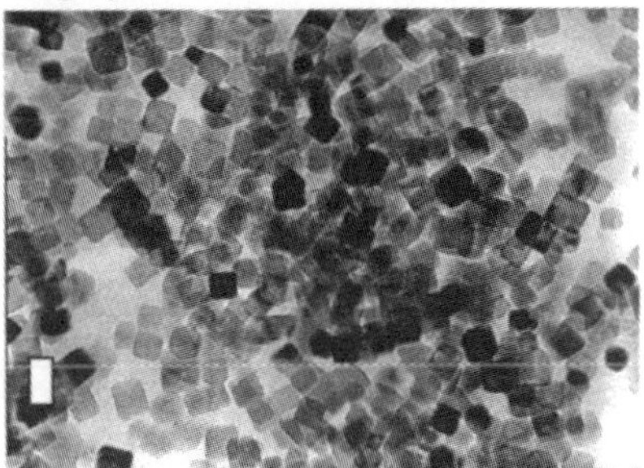

Figure 1. Transmission electron image of the 35 nm sample taken at 200,000X. The image clearly shows uniform cubic nanocrystals with an average edge of 22 nm. The white bar on the left side of the figure equals 50 nm.

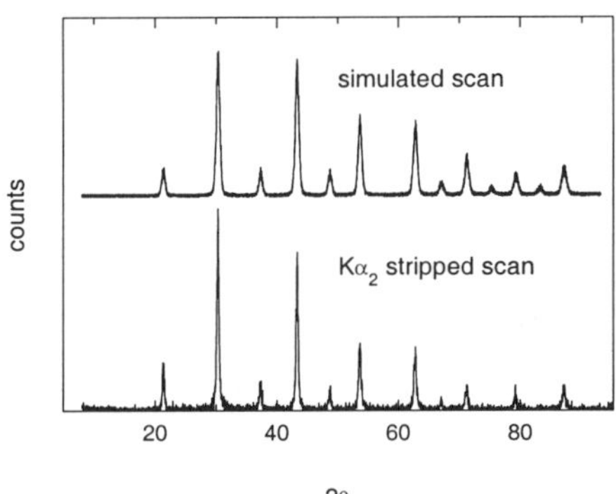

Figure 2. Experimental (bottom) and simulated (top) X-Ray diffraction patterns of the 22 nm sample. The good match indicates that no impurities are present. The experimental pattern was corrected for the $K\alpha_2$ wavelength.

Both electron micrographs and XRD patterns of other samples (prepared either by changing the *w* parameter or by keeping *w* fixed while the metal concentration was decreased) show that neither the morphology nor the average size of the particles is affected. On the contrary, we were able to tune the average size of the particles down to 13 nm by simply diminishing the time between the mixing of the two microemulsions and their disruption. This is shown in table 1, where the average size of the particles and the corresponding reaction times are reported. On reducing the reaction time the morphology of the particles evolves from a perfect cubic shape to a more spherical one with rounded edges. In addition the particles keep their full crystallinity, as evidenced by the good match between the average size evaluated by TEM and XRD.

Table 1. Structural and magnetic properties of a series of samples prepared with w=10 and with an initial Mn^{2+} and F^- concentration of 0.1 M and 0.4 M respectively.

Sample	*Reac.time*	*Av. Diam. TEM (nm)*	*Av. Diam. XRD (nm)*	T_B *(K)*	T_N *(K)*	H_c *(Oe) T=4.5K*	H_e *(Oe)*
A	2 h	35	33	36	89	200	80
B	10'	24	22	26	89	180	40
C	3'	19	18	15	89	/	/
D	1'	13	13	10	86	65	15

This result is quite unusual if compared to commonly observed products from the microemulsion technique. It is in fact well established that the water content and the ion concentration are the two most important parameters to control the particle's size. Roth and Hempelmann observed a similar behavior during the synthesis of NH_4MnF_3 nanoparticles in the water core of ammonia-exchanged AOT reverse micelles [13]. A possible explanation to this discrepancy is to admit that, once the particles are precipitated inside the micelle core, the surfactant is not effectively adsorbed on the micelle surface. The particles are then no longer protected against further growth and the only way to control their growth is to terminate the reaction.

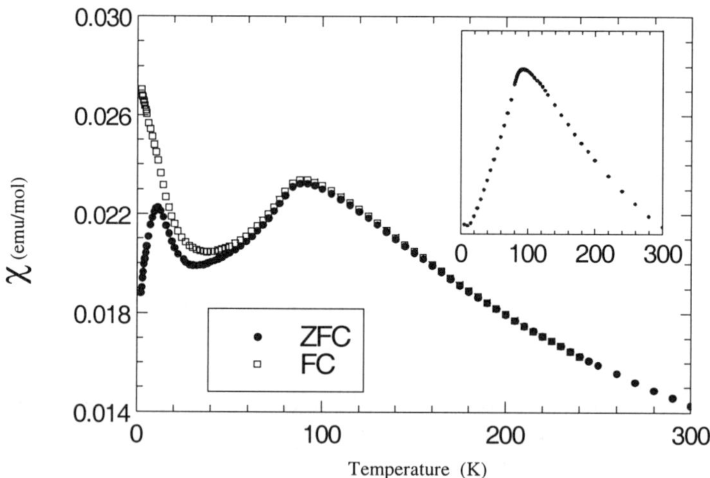

Figure 3. ZFC and FC magnetic susceptibilities of $KMnF_3$ nanoparticles with average size of 13 nm. Data were collected with an parallel external field of 100 Oe. The peak at high T (86 K) corresponds to the Néel temperature while the one at low T (10 K) corresponds to the blocking of the magnetic moment of the particles. In the inset is shown χ vs. T for a bulk powder of $KMnF_3$.

The magnetic properties of the four samples with average size of 13, 18, 22 and 35 nm respectively, were investigated. Figure 3 shows the temperature dependence of the magnetic susceptibility of the 13 nm sample measured after cooling the sample in zero field (ZFC) and after cooling it in the measuring field (FC). Above T_N the two susceptibilities coincide and obey the Curie-Weiss law with μ_{eff}= 5.86 μ_B and a Curie temperature θ of –193.0 K. On cooling the sample below T_N, after an initial decrease, the two curves separate. While the FC susceptibility data monotonically increase to the lowest temperature, the ZFC data display a cusp at 10 K. The low temperature ZFC maximum is absent in the magnetic susceptibility of a coarse grain powder of $KMnF_3$, shown in the inset of figure 3. The origin of this peak can be ascribed to the blocking of the magnetic moment of the particles arising from uncompensated surface spins. In other words, we can describe our particles according to the Néel model consisting of an ordered antiferromagnetic core (which causes the initial susceptibility decrease below T_N), and a thin disordered surface layer. The resulting magnetic moment will be blocked below T_B, while it will be free to reorient along the two directions of the easy axis above T_B, giving rise to the characteristic ZFC cusp. Due to this disordered surface layer, the overall shape has little affect on the coercivity of the particles.

The mechanism that gives rise to the magnetic moment of antiferromagnetic ultrafine particles is still a matter of investigation and different models have been proposed. For example, Kodama *et al.* [14] suggest that the reduced coordination of the surface spins might result in a stabilization of multi-sublattice configurations with respect the classical 2-sublattice one. However the observed magnetic behavior seems to indicate that our sample can be described in terms of the simple core-shell model.

On increasing the average size of the particles the blocking temperature increases from 10 K for the 13 nm sample to 36 K for the 35 one (see table 1), as predicted by the superparamagnetic model. The Néel temperature of an antiferromagnetic material is expected to be different from the bulk when the material is in the form of nanosized particles. Sako *et al.* [15] reported for MnO nanoparticles with average diameter between 3 and 6 nm, a decrease of T_N up to 50 K with respect to the bulk value. In our case, apart for the 13 nm sample, which displays an ordering temperature of 3 K lower than that of the bulk, no significant shifts of T_N were observed. This result indicates that only below *ca.* 15 nm does the finite size start to affect the macroscopic properties of the material.

Magnetic hysteresis loops were measured at 4.5 K cycling the field between ± 50 kOe. The data measured for the 22 nm sample are shown in figure 4. The sample is far from being saturated even at the highest field applied and the shape of the loop clearly resembles that of a bulk antiferromagnet. A non-zero coercivity was observed for all the samples investigated and the data are reported in table 1. The coercive field increases with the size of the particle and with an increase of their blocking temperature from 65 Oe for the smaller sample up to 200 Oe for the 35 nm one.

Hysteresis loops were also measured at the same temperature after cooling the sample through T_N with an external field of 10 kOe. The FC loop of the 22 nm sample is shown together with the ZFC one in the inset of figure 4. While the ZFC loop is symmetric with respect to the origin, the FC data are shifted along the cooling field direction. The exchange field H_e (*i.e.* the field shift measured) increases with the average size of the particles from 15 Oe for the smaller particles up to 80 Oe for the larger particles. Displacement of a FC hysteresis loop is a typical feature of exchange anisotropy. Exchange anisotropy was introduced by Meiklejohn [15] and is an additional term to magnetic anisotropy arising from the coupling of a ferromagnetic spin system to an antiferromagnetic spin system. In our case the observed shift can be ascribed to the exchange coupling between the ordered antiferromagnetic core and the uncompensated spin shell surrounding the particle.

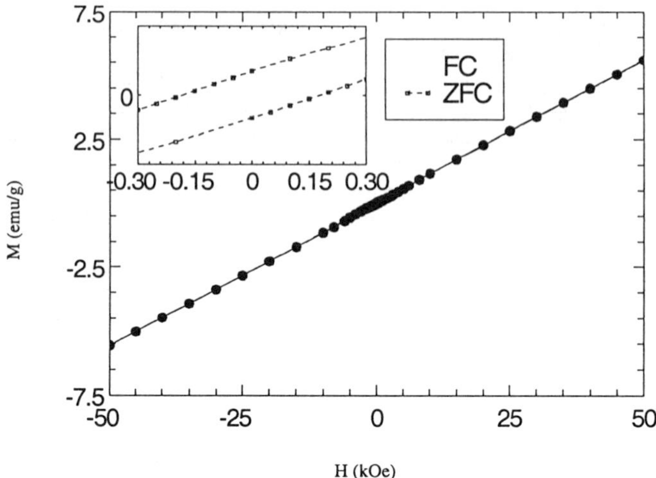

Figure 4. Hysteresis loop at 4.5 K of the 22 nm sample. In the inset is shown the enlargement between ±300 Oe of the ZFC and FC loops. The FC loop is shifted along the direction of the 10 kOe external cooling field of 40 Oe.

SUMMARY

$KMnF_3$ nanoparticles with different average size in the range 13-35 nm were synthesized with the microemulsion technique. Even though the control of the size could be achieved only by quenching the reaction, the particles obtained are characterized by high size uniformity and full crystallinity. Below T_N the particles display superparamagnetic behavior and a blocking temperature which increases with the average size of the particles. Below T_B coercivities were measured as well as a shift of the hysteresis loop when it was measured following a field cooling process. These results can be explained by describing our particles as containing an ordered antiferromagnetic core and a surface layer with uncompensated spins.

ACKNOWLEDGMENTS

The authors gratefully acknowledge Dr. T. Kodenkandath and J. Wiemann for support in XRD measurements. We are also in debt with Dr. W. Zhou for TEM images of our samples. This work was supported by the Advanced Materials Research Institute through DOD/DARPA Grant No. MDA972-97-1-0003.

REFERENCES

1. A. R. Schweneder, Chimica 46 69 (1992)
2. A. D. Gee "Cell Separation Methods and Applications" Eds. D. Recktenwald and A. Radbrush, pp. 175 Marcel Dekker, Inc., New York USA 1997.
3. Nanomaterials: Synthesis, Properties and Applications Eds. A. S. Edelstein and R. C. Cammarata; Institute of Physics Publishing, Bristol PA (1996).
4. L. Néel, *Compt. Rend. (Paris),* **252**, 4075 (1961).
5. L. Néel *Proc. Int. Conf. Magn. Crys. Tokyo, 1961, J. Phys. Soc. Jpn.* **17** 676 (1962)

6. J. Cohen; K. M. Creer; R. Pauthenet; K. Srivastava; *Proc. Int. Conf. Magn. Crys. Tokyo, 1961, J. Phys. Soc. Jpn.* **17** 685 (1962)
7. K. M. Creer; *Proc. Int. Conf. Magn. Crys. Tokyo, 1961, J. Phys. Soc. Jpn.* **17** 690 (1962)
8. S. Gider D. D. Awschalom, T. Douglas, S. Mann, M. Chapara Science 268 77 (1995)
9. B. Barbara; E. M. Chudnovsky; *Phys. Lett.* **145A** 205 (1990)
10. M. M Schieber "Experimental Magnetochemistry" Ed. E. P. Wohlfarth John Wiley & Sons. Inc. New York 1967
11. V. Pillai; P. Kumar; M. J. Hou; P. Ayyub; D. O. Shah; *Adv. Colloid Interface Sci.* **55** 241 (1995)
12. N. Moumen; M. P. Pileni *Chem. Mater.* **8** 1128 (1996)
13. M. Roth, R. Hempelmann, *Chem. Mater.*, **10**, 78 (1998).
14. R. H. Kodama, S. A. Makhlouf, A. E. Berkowitz, *Phys. Rev. Lett.,* **79**, 1393 (1997)
15. S. Sako; Y. Umemura; K. Oshima; M. Sakai; S. Bandow; *J. Phys. Soc. Jpn.* **65** 280 (1995)
16. W. H. Meiklejohn, C. P. Bean, Phys. Rev. 102 1413 (1956)

POLYMER COATED NANOPARTICULATE FERRITE POWDERS: PROPERTIES AND APPLICATION

D. VOLLATH, D. V. SZABÓ, J. FUCHS
Forschungszentrum Karlsruhe, Institut für Materialforschung III
P.O.Box 3640, D-76021 Karlsruhe, Germany, dieter.vollath@imf.fzk.de

ABSTRACT

Superparamagnetic behavior is characterized by a thermally fluctuating vector of magnetization, leading to magnetization curves free of hysteresis; it is a property of isolated ferrite particles with sizes below ca. 10 nm. These particles fulfil the condition $Kv < kT$ with K ... energy of unisotropy, v ... volume of the particle, kT has the usual meaning. To produce a superparamagnetic macroscopic part it is necessary to avoid the interaction of the particles. This can be achieved by coating the particles with a second non-magnetic phase. This special material can be synthesized using the microwave plasma process. Because of the specific interaction of charged particles with an oscillating electrical field, microwave plasmas excel in relatively low reaction temperatures. The low reaction temperature and the electrical charging of the particles in the plasma reduce the probability of agglomeration. Therefore, it is possible to pass the gas stream with the as produced particles through a second reaction zone, where the particles will be coated with a polymer. The thickness of the coating can be adjusted in the range from 1 to 5 nm. The composition of the ferrite kernel is selected either for maximum susceptibility or maximum frequency of application. They may consist either of maghemite, manganese-, manganese-zinc-, or magnesium iron spinell for maximum susceptibility or cobalt-iron spinelle type material for maximum frequency. Polymer coated ferrite nanoparticles can be consolidated by hot pressing with temperatures around 100 °C. Superparamagnetism is shown by static magnetic measurements and Moessbauer spectrometry. Dynamic measurements of the complex susceptibility show interesting properties up to frequencies in the gigahertz range.

INTRODUCTION

Nanoparticles of ferrites consisting of one domain excel in special magnetic properties. Very small magnetic nanoparticles are superparamagnetic. Superparamagnetic materials are characterized by a thermally fluctuating vector of magnetization. Provided the particles fulfil the condition

$$Kv < kT \qquad (1).$$

(K ... constant of magnetic unisotropy, v ... volume of the particle; Kv ... energy of magnetic unisotropy, kT ... thermal energy) the material is superparamagnetic [1,2]. This leads to a zero coercitivity. The thermal relaxation time τ, limiting the maximum frequency of application is according to Neél [1]

$$\tau = \tau_0 \exp(Kv/kT) \qquad (2).$$

As τ_0 is in the range between 10^{-9} and 10^{-13} sec, the frequency of thermal relaxation may be well beyond 1 GHz at room temperature provided that the composition of the ferrite is selected properly. Basis of this selection is the relationship for τ_0 given by Aharoni [3]:

$$\tau_0 \propto M/K = nm/K \qquad (3)$$

(M ... saturation magnetization, n ... number of particles, m ... magnetic moment of one particle) From (3) it is obvious that a small value of τ_0 is obtained only with materials showing a large magnetic unisotropy K because the variability in the magnetization is comparatively low. τ_0 increases with increasing magnetic moment m of the particles. On the other hand, the susceptibility μ of a superparamagnetic material is given by

$$\mu = nm^2/3kT \qquad (4).$$

μ increases quadratically with the moment of the particle. Equations (3) and (4) are contradictory. Therefore, superparamagnetic materials can be optimized either for high susceptibility or for high frequencies of application.

Superparamagnetism is a property of isolated non-interacting particles. In a macroscopic part, interaction via the dipole moments of the particles leads to magnetically large particles that are not longer superparamagnetic. This interaction can be reduced by embedding the nanoparticles in a second non-magnetic material. This leads to nanocomposites. The most progressed type of nanocomposite is obtained by the use of coated nanopowders. Nanocomposites – as they are described in this paper – are composite materials with at least one phase exhibiting the properties of a nanomaterial. Additionally, it is assumed that there is a random arrangement of the nanoparticles in the composite. As kernel and coating material are homogeneously distributed on a nanometer scale the magnetic properties of a densified solid can be changed gradually with the thickness of the coating. Depending on the requirements of the system in question, the coating material may be ceramic or polymer. By coating nanoparticles with a second layer, the following improvements are obtained:

- The kernels are kept on a well-defined distance. Therefore, the interaction of the particles can be adjusted.
- During sintering, the growth of the kernels is thwarted, provided that core and coating show no mutual solubility.
- The distribution of the two phases is homogenous on a nanometer scale.

Gas phase synthesis of coated nanoparticles is possible only by means of the microwave plasma process. This is because the particles leave the reaction zone with high electrical charges of the same sign avoiding agglomeration. Other gas phase processes like the inert gas condensation [4] or the conventional chemical vapor synthesis process [5] are not capable coat nanoparticles individually. Compared with chemical precipitation processes, the amount of ferrite in the polymer matrix obtained by the microwave plasma process is significantly higher.

SYNTHESIS AND MORPHOLOGY OF POLYMER COATED FERRITE PARTICLES

A process capable to synthesize coated superparamagnetic particles has to be designed in a way that avoids the formation of agglomerates. This goal can be achieved if the reaction temperatures are low and the particles leave the reaction zone with high electrical charges of the same sign. Both conditions are fulfilled with the microwave plasma process [6]. In a microwave plasma chemical reactions can be performed at low temperatures because the reactants are partly dissociated and ionized. A microwave plasma is a system not in equilibrium because the energy E transferred to a charged particle with the mass m in an oscillating electrical field with the frequency f is:

$$E \propto 1/mf^2 \qquad (5)$$

From formula (5), it is evident that there is more energy transferred to particles with small mass, like electrons as compared to heavier particles such as ions or radicals. Additionally, in a plasma environment the collision frequency z with the uncharged molecules and atoms in the gas phase has to be taken into account:

$$E \propto \frac{1}{m}\frac{z}{f^2 + z^2} \qquad (6)$$

As the collision frequency z is a function of the gas pressure, the energy transfer is also a function of the gas pressure exhibiting a maximum at $f = z$. Most stable operating conditions are found in the pressure range below this maximum.

For experimental or technical realization the standard industry frequencies 0.915 and 2.45 GHz are selected. From the formulae (5) and (6) it is obvious that 2.45 GHz is in those cases of advantage, where the temperatures have to be kept as low as possible. The main advantage of 0.915 GHz are found in the larger dimensions of the wave-guides, allowing the application of large sized reaction tubes allowing lager production rates.

In the technical realization of the microwave plasma process, a long reaction tube made of quartz glass passes a resonant TE_{01} or TE_{11} cavity. At the intersection of the reaction tube and the resonator, a microwave plasma is ignited. The precursors, chlorides, carbonyls or metal organic compounds, are introduced as vapors into the reaction tube directly in front of the plasma zone. For the synthesis of ferrites argon with the addition of 20 vol% oxygen is used as reaction gas. The temperature in the plasma zone, the reaction temperature, can be adjusted in the range from 100 to about 800 °C. The residence time in the reaction zone controlling the particle size of the product is in the range from 5 to 15 ms. Particles leaving the reaction zone are carrying electrical charges with the same sign. This, in combination with the low reaction temperature, reduces the probability for the formation of hard agglomerates and allows the coating of the individual particles with a polymer. Figure 1 shows a sketch of the arrangement to produce nano-

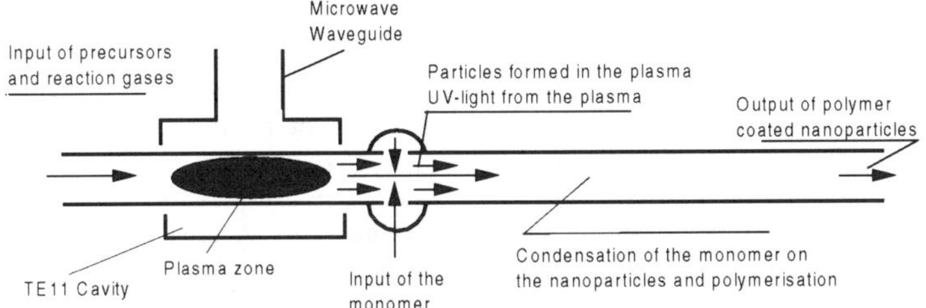

Figure 1: Experimental set-up for the synthesis of polymer coated ceramic nanoparticles.

particles coated with a polymer [7,8]. In a first step, the ceramic particles are synthesized in a microwave plasma zone. The operating conditions are selected to keep the temperatures as low as possible. After the plasma reaction zone, the monomer is introduced as a vapor. If the temperature is adjusted properly, the monomer condenses at the surface of the particles and polymerizes under the influence of the UV radiation. The monomers with the best performance in this respect are methacrylic acid (MA) and methyl methacrylate (MMA).

Figure 2 shows γ-Fe_2O_3 particles coated with PMA. The particles depicted in this micrograph show a uniform distribution of polymer at the surface. In the special example depicted the polymer coating is extremely thick to make the polymer coating visible. To obtain good magnetic properties, a polymer content of about 50 vol% is an optimum. The degree of polymerization was determined by gel permeation chromatography (GPC). The GPC plot of a typical Ferrite – PMMA specimen is depicted in figure 3. This graph shows maxima around 200 and 400 da, which are equivalent to about 2 respectively 4 masses of the monomer. Additionally, a broad range of larger mol masses up to 5000 da was found in low concentrations. This is an extremely low degree of polymerization. Evaluating these numbers, one has to take into account that the polymerization is limited by geometrical factors in a layer with a thickness of 1 or 2 nm on a 5 nm particle. Additionally, the chemistry in the gas phase behind the plasma zone is not favorable for the formation of long chains.

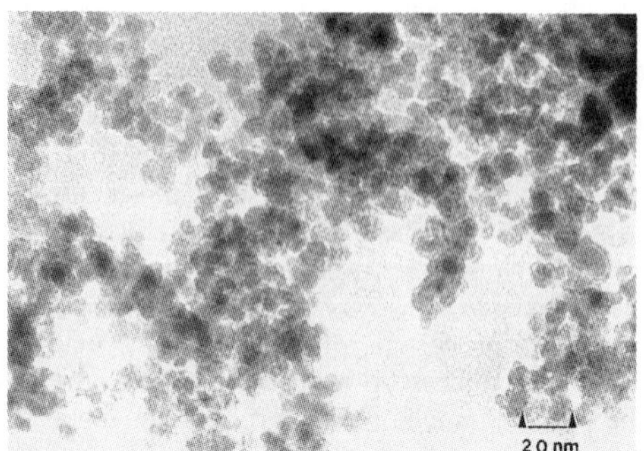

Figure 2: Maghemite (γ-Fe_2O_3) poly-methacrylic acid (PMA) composite powder

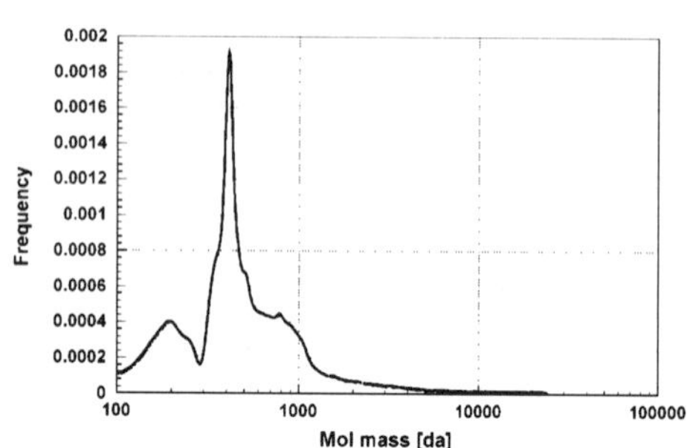

Figure 3: Distribution of mol masses in the PMMA coating of $MnFe_2O_4$ nanoparticles

MORPHOLOGY AND MAGNETIC PROPERTIES OF NANOCOMPOSITE CORES

Solids of polymer coated nanoparticles are produced by pressing the starting powder in the temperature range from 70 to 140 °C under a pressure in the range from 1 to 1.5 MPa. Figure 4 shows the morphology of a consolidated Fe_2O_3 – PMMA composite with 50 vol% polymer. Because of the low temperature during processing, any grain growth or reaction between the two phases of the composite is thwarted. The specimen for electron microscopy was prepared by cutting with an ultramicrotom. The thickness of the specimen was 20 nm. The layered structure of the material, visible in this micrograph, is an artifact of the specimen preparation. It is obvious from this figure that the distribution of particle sizes is in a very narrow range.
Figure 6 shows typical magnetization curves of small cylinders made of γ-Fe_2O_3 and $MnFe_2O_4$ – polymer composites. The magnetization curves depicted in this figure are obviously free of any hysteresis. Because of a significant polymer content the magnetization of this material is smaller than the one obtained with pure ceramic ferrite.

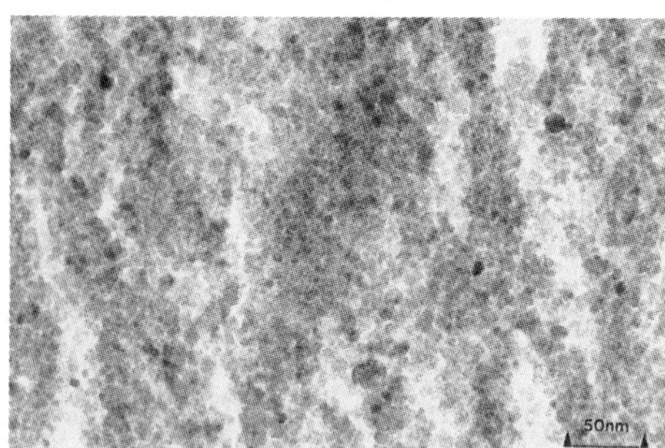

Figure 4: Morphology of a γ-Fe_2O_3 – PMMA nanocomposite. The visible layered structure is an artifact of the specimen preparation with an ultramicrotome.

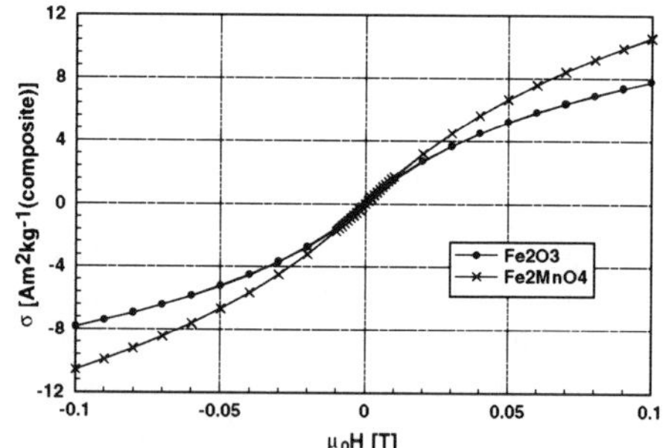

Figure 5: Magnetization curves up to 0.1 T of superparamagnetic γ-Fe_2O_3 and $MnFe_2O_4$-PMA composite cylinders (ca. 3 mm ∅, 4 mm high) at 300 K.

Figure 6 depicts a comparison between superparamagnetic Fe_2O_{3-x} - polymer composite and a commercial ferrite. For comparison the magnetization of the composite is referred to the pure ceramic content. This comparison shows that the magnetization at high fields (5 T) is quite similar whereas at low fields the susceptibility of the composite is higher.

The ultimate proof for superparamagnetism is obtained by Mößbauer spectroscopy. As in the superparamagnetic state the fluctuation frequency of the magnetization is higher than the Lamor frequency of the iron nuclei, these nuclei "do not realize" that there is a high magnetic crystal field. Therefore, the Mößbauer spectra of superparamagnets show the quadrupole doublet of non-magnetic material instead of the sextette normally observed with magnetic materials. This phenomenon can be used to determine the amount of non-superparamagnetic material. Figure 7 shows this for the γ-Fe_2O_3 - PMA composite specimen, whose magnetization curve is plotted in figure 5.

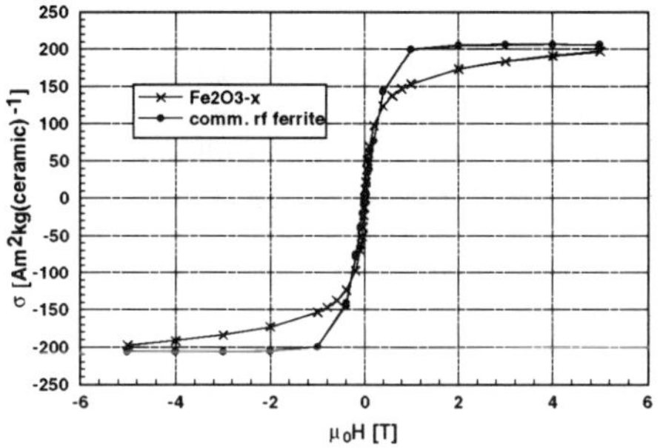

Figure 6: Magnetization curves up to 5 T of superparamagnetic Fe_2O_{3-x} PMA composite cylinders (ca. 3 mm ∅, 4 mm high) and a commercial ferrite at 300 K.

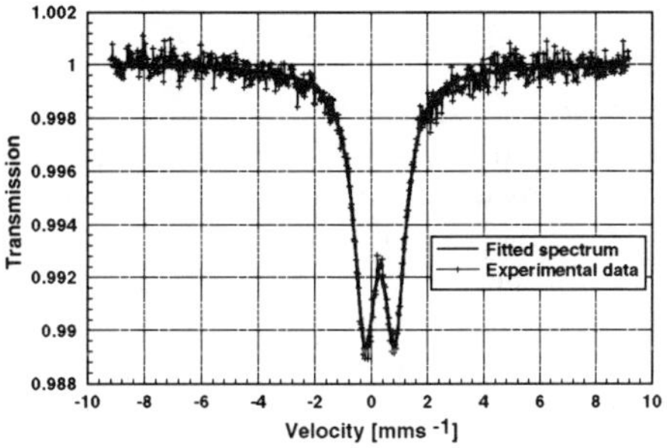

Figure 7: Mößbauer spectrum of a Fe_2O_3 - PMA composite specimen. The magnetization curve of this specimen is shown in figure 5.

The property, essential for technical application in the communication industry is the susceptibility. Figure 8 depicts the frequency dependence of the susceptibility of different ferrite based spinelle – PMMA composites in comparison to a commercial ferrite. The essential information from this figure is that for application the superparamagnetic material is comparable to today's commercial ferrite.

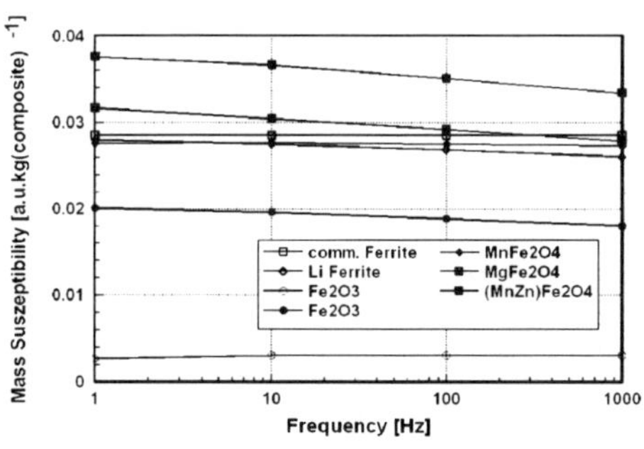

Figure 8: Susceptibility of different nanostructured ferrite – polymer composites in comparison to a commercial one.

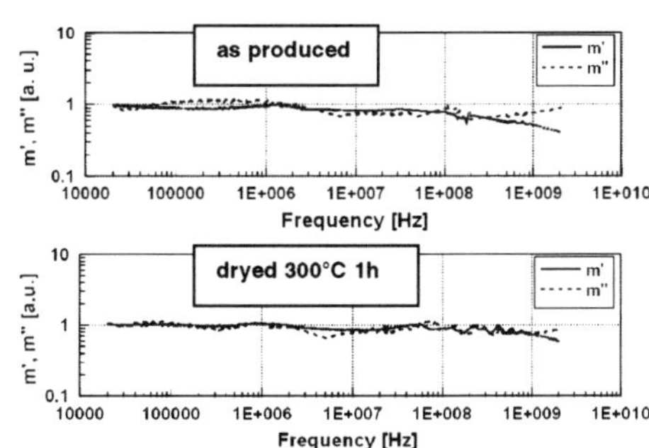

Figure 9: Influence of surface water on the real and imaginary suszeptibility for a $MnFe_2O_4$ containing nanocomposite.

Additionally, superparamagnetic ferrites have advantages over "classical" ferrites in lower losses and higher maximum frequencies for application. The resonance frequencies of these materials are well beyond 2 GHz. The data for the complex susceptibility depicted in figure 9 are slightly scattered. The real part m' decreases slightly with frequency. Additionally, a strong influence of the water adsorbed on the surface of the particles is observed in the frequency range beyond 0.1 GHz. Surface water reduces the real part m' and increases the imaginary part m" of the susceptibility. Careful drying of the nanocomposite is therefore a necessary prerequisite for any application. An influence of the spin resonance lines was not observed in the frequency range up to 2 GHz.

CONCLUSION

This paper introduces the concept of polymer coated ceramic nanoparticles as starting materials for improved ferrites. Synthesis of this special type of material is possible with the microwave plasma process. Nanocomposites made of coated nanosized ferrite particles have the advantage over similar nanocomposites made from mechanically blended powder that there are no kernels in direct contact. Therefore, it is possible to produce magnetic cores exhibiting properties characteristic for individual ferrite nanoparticles. The special properties of these materials were demonstrated in the systems γ-Fe_2O_3 – polymer with different doping of the ferrite. In this case, macroscopic bodies with superparamagnetic properties were obtained. These materials exhibit properties that are not possible with conventionally produced ferrite composites. Dynamic measurements up to the gigahertz range have shown that this material can be applied at least up to 2 GHz. The main problem for application is the water adsorbed on the surface of the powder particles before consolidation.

REFERENCES

1. L. Neél, Compt. Rend. **228**, 664 (1949)
2. D. Vollath, D. V. Szabó, R. D. Taylor, J. O. Willis, J. Mater. Res. **12**, 2175 (1997)
3. A. Aharoni, Phys. Rev. A **135**, 447 (1964)
4. H. Hahn, J. A. Eastman, R. W. Siegel, Ceram. Trans. **B1,** 115 (1988)
5. W. Chang, G. Skandan, B. H. Kaer, Nanostr. Materials **4**, 507 (1994)
6. D. Vollath, K. E. Sickafus, Nanostructured Materials **1**, 427 (1992)
7. D. Vollath, D. V. Szabó, B. Seith, German Patent 196 38 601.2-43 (1997)
8. D. Vollath, D. V. Szabó, J. Fuchs, Nanostructured Materials (1999) in the print

GIANT MAGNETORESISTANCE OF ELECTROSTATIC SELF-ASSEMBLED FE_3O_4 NANOCLUSTER AND POLYMER THIN-FILMS

YANJING LIU,* RICHARD O. CLAUS** and FAJIAN ZHANG***

*NanoSonic, Inc., P.O. Box 618, Christiansburg, VA 24068
**Department of Materials Science and Engineering, Virginia Tech, Blacksburg, VA 24060-0356
***Bradley Department of Electrical and Computer Engineering, Virginia Tech, Blacksburg, VA 24060

ABSTRACT

Giant magnetoresistance (GMR) as large as 25% at 25° C has been observed for multilayer ultrathin films of iron oxide (Fe_3O_4) nanoclusters and polyimide molecules alternately adsorbed onto single crystal silicon and quartz substrates using a novel self-assembly technique. This process involves the alternate dipping of a substrate into an aqueous solution of anionic polyimide precursor (polyamic salt, PAA) followed by dipping it into an aqueous solution of cationic polydiallyldimethylammonium chloride (PDDA)-coated Fe_3O_4 nanoparticles. The regular formation of alternating monolayers is verified by UV-vis spectroscopy and contact angle measurements. Vibrating sample magnetometry indicates the formation of ultrasoft films.

INTRODUCTION

Magnetic nanoparticles and ultrathin films of such particles have recently attracted attention from both theoretical and experimental points of view during the last few years [1], due to their potential advantageous uses in information storage media [2], magnetic imaging devices [3], ferrofluids [4] and magnetic separation systems [5]. Several different techniques for the formation of thin magnetic films have been reported, including sputtering [6], vapor trapping [7], carbon arc processes [7] and molecular beam epitaxy (MBE) [8]. However, there are some inherent limitations associated with these methods including area and thickness, the need to process at high temperature, and the complexity and time consuming nature of the formation methods.

EXPERIMENT

Here we report Fe_3O_4/polyimide thin-films with GMR behavior formed by a novel self-assembly process [9]. For this particular molecular system, several steps are involved. First, magnetite (Fe_3O_4) nanoclusters were synthesized using a process described elsewhere [10]. The agglomeration and precipitation of these nanoclusters was prevented by adsorption of cationic PDDA, and particle solutions were stable at ambient temperature and pressure for periods of several months. The average size of the PDDA-coated nanoclusters was less than 10 nm [11]. Second, a thermoplastic polyimide precursor polyamic acid (PAATEA) of ULTEM™, with molecular weight 30,000 g/mole after imidization [12], was used as a protective interlayer spacer material to mitigate surface oxidation of the Fe_3O_4 particles and an associated loss of magnetization [13]. Third, the fabrication of alternating monolayers of Fe_3O_4:PAATEA was carried out by electrostatic self-assembly (ESA) of the individual anionic PAATEA and cationic Fe_3O_4:PDDA complexes onto a cleaned and protonated substrate surface as described in detail elsewhere [11]. Immersion time in each solution was on the order of several minutes.

Mat. Res. Soc. Symp. Proc. Vol. 577

ANALYSIS

The regular self-assembly of consecutive monolayers of PAATEA and Fe_3O_4:PDDA was verified by several different methods, supporting the model of strong spontaneous ionic bond formation between the net surface-charged substrate and the oppositely charged molecules adsorbed during each immersion step. First, water contact angle measurements that indicate the relative hydropobic or hydrophillic nature of the outermost layer were used to qualitatively monitor the dipping process. Figure 1 shows the oscillatory variation in water contact angle as PAATEA and then Fe_3O_4 :PDDA monolayers are formed for the first few tens of layers.

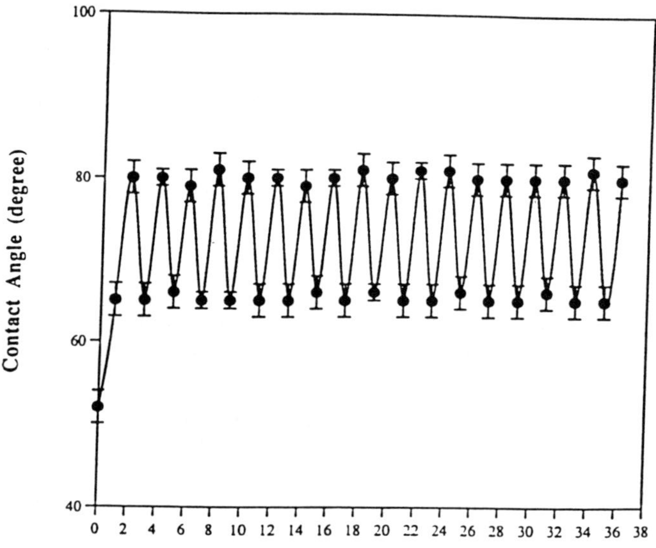

Fig. 1. Contact angle on the surface of (PDDA:Fe_3O_4)/PAATEA multilayer films as a function of the number of layers.

Similar repetitive behavior has been obtained by our group for thin-films with thousands of bilayer pairs [11]. Contact angle data was obtained at three locations on the surface and averaged for each data point shown, and the regularity of every other point demonstrates that the process is very uniform and repetitive. Second, UV-Vis absorption measurements, shown as a function of the number of bilayers on a quartz substrate in Figure 2, indicates that an approximately equal amount of material is formed on the surface with the addition of each subsequent bilayer.

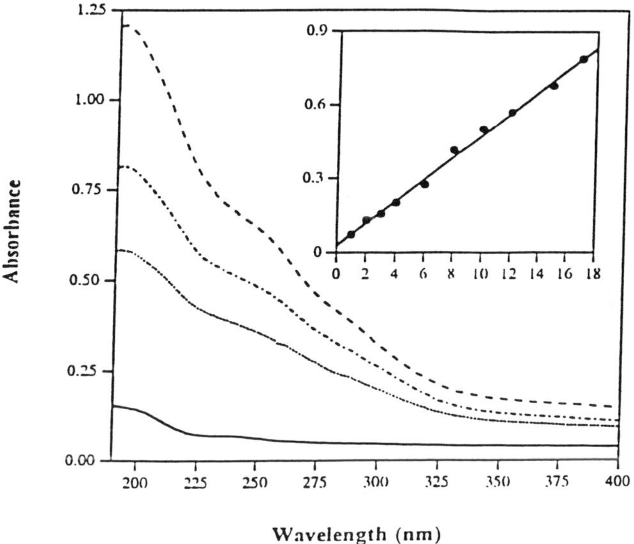

Fig. 2. Optical absorbance spectra of (PDDA:Fe_3O_4)/PAATEA multilayer films with inset showing the linear relationship between absorbance at 225 nm and the number of bilayers deposited.

Again, each absorption spectrum obtained was the average of three spectra taken at different positions on the substrate surface; maximum variations of several percent were observed among all measurements on the same layers. The linear increase in absorbance at 225 nm shown additionally suggests that nanoparticle size does not increase during the ESA process, and that vacancy and void defects do not occur.

MAGNETIC PROPERTY CHARACTERIZATION

Vibrating sample magnetometry (VSM, Lakeshore, Model 7307) was used to analyze the magnetic properties of the ESA-formed thin-films. Figure 3 shows the magnetic moment as a function of the amplitude of magnetic fields applied parallel and perpendicular to the plane of the substrate.

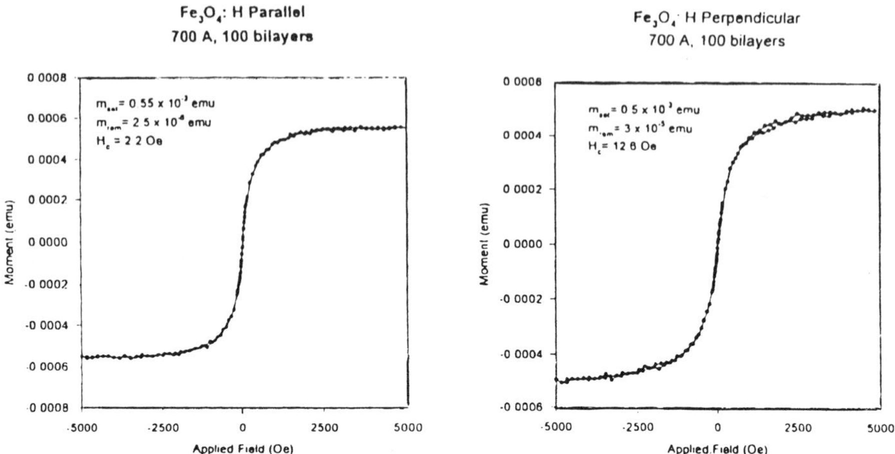

Fig. 3. Magnetic moment versus magnetic field applied parallel (left) and perpendicular (right) to substrate surface.

The very low coercivity and remanence and negligible hysteresis are consistent with a superparamagnetic description for nanoparticle behavior. The magnetoresistance ratio, $\Delta\rho/\rho$ was measured at room temperature. A maximum in $\Delta\rho/\rho$ of 25% is clearly observed.

DISCUSSION

Ultrasoft magnetic thin-films have been self-assembled from aqueous solutions of water-soluble polymers and polymer-coated magnetic nanoclusters. The adsorption process is very linear, suggesting good order in the films at the molecular level. The ultrasmall size and size uniformity of the clusters, and individual particle separation by both surface passivation coating layers and alternating polymer monolayers suggests that soft behavior should occur as confirmed experimentally. Modifications of the ESA process chemistry may be used to vary both the thicknesses of the particle coatings and the alternating polymer layers at the Angstrom level, as demonstrated for other particle-polymer combinations. This suggests an interesting possibility for control over the separation and spacing of magnetic domains, and over resulting magnetic properties.

ACKNOWLEDGMENTS

This work was supported in part by ARO Grant DAAG55-97-1-0101. The authors thank Dr. B. Dodrill of Lakeshore Inc. for assistance with magnetic measurements.

REFERENCES

1. Bonetti and D. Fiorani, Eds., *Nanophase Materials* (Trans Tech Publications, 1995), A. Hernando, Ed., *Nanomagnetism* (Kluwer Academic Publishers, 1993); R.F. Ziolo, E.P. Giannelis, B.A. Weinstein, M.P. O'Horo, B.N. Ganguly, V. Mehrotra, M.W. Russel and D.R. Huffman, *Science*, 257, 219 (1992).

2. R.G.L. Audran and A.P. Huguenard, U.S. Patent 4,302,523 (1981); H.N. Bertram, *Theory of Magnetic Recording* (Cambridge Univ. Press, 1994); M.H. Kryder, in *Applied Magnetism*, R. Gerber, C.D. Wright and G. Asti, Eds. (Kluwer Academic Publishers, 1993), vol. 253, chap. 39.

3. Q.A. Pankhurst and R.J. Pollard, *J. Phys.: Condens. Matter* **5**, 8487 (1993).

4. I. Anton *et al., J. Magn. Magn. Mater.* 111, 29 (1990).

5. R. Gerber, in Applied Magnetism, R. Gerber, C.D. Wright and G. Asti, Eds. (Kluwer Academic Publishers, 1993), vol. 253, pp. 165.

6. L.M. Falicov et al., J. *Mater. Res.* **5**, 1299 (1990).

7. G.N. Glavee, K. Eason, K.J. Klabunde, C.M. Sorensen and G.C. Hadjipanayis, Chem. Mater. 4, 1360 (1992).

8. J.J. Rhyne, R.W. Erwin, J.A. Borchers, M.B. Salamon, F. Tsui, R. Du and C.P. Flynn, in *Sci. and Tech. of Nanostructured Magnetic Materials*, G.C. Hadjipanayis and G.A. Prinz, Eds. (Plenum Press, New York, 1991), pp. 117.

9. G. Decher, J.D. Hong and J. Schmitt, *Thin Solid Films* **210/211**, 831 (1992); M. Ferreira and M.F. Rubner, *Macromolecules* **28**, 7101 (1995).

10. R. Massart, *IEEE Trans. Magn.* **17**, 1247 (1981).

11. Y. Liu, A. Wang and R. Claus, Appl. Phys. Lett. 71 (16), 2265 (1997).

12. S. Gardner, and R. Davis, in preparation.

13. G.C. Hadjipanayis, S. Gangopadhyay, C.M. Sorensen, and K.J. Klabunde, *IEEE Trans. Magn.* **29**, 2602 (1993).

Part XI

Ultrasoft Nanocrystalline and Amorphous Materials I

NANOCRYSTALLINE SOFT MAGNETIC Fe-M-B (M = Zr, Hf, Nb) "NANOPERM," Fe-M-O (M = Zr, Hf, RARE EARTH) ALLOYS AND THEIR APPLICATIONS

A. Makino *†, A. Inoue **, and T. Masumoto ***
* Central Research Laboratory, Alps Electric Co., Ltd., Nagaoka 940-8572, Japan
** Institute for Materials Research, Tohoku University, Sendai 980-8577, Japan
*** The Research Institute of Electric and Magnetic Materials, Sendai 982-0807, Japan

ABSTRACT

New soft magnetic nanocrystalline materials in Fe-M-B (M = Zr, Hf, Nb) and Fe-M-O (M = Zr, Hf, rare earth) systems have been fabricated by use of partial crystallization of melt-spun amorphous phase and sputtering-induced partial crystallization, respectively. These alloys have a mixed structure of nanoscale α-Fe grains and an amorphous phase containing large amount of M, B and M, O, respectively. This structural property should be a dominant factor for achieving good soft magnetic properties in both alloys. The $Fe_{85.5}Zr_2Nb_4B_{8.5}$ ribbon exhibits a high magnetic flux density (B_s) of 1.64 T and a high permeability of 60,000 at 1 kHz as well as a very low core loss of 0.09 W/kg at B = 1.4 T and 50 Hz. The soft magnetic Fe-M-O films with high O content of 10–35 at.% maintain high permeability up to the high frequency above 100 MHz because of higher electrical resistively of 6 to 23 μΩm than those of other conventional soft magnetic alloy films. The Fe-M-B ribbons "NANOPERM" should be more suitable for low frequency transformers. The Fe-M-O films are useful for high frequency applications such as thin-film inductors and transformers for micro switching converters for portable electric equipment.

INTRODUCTION

Since an amorphous phase in Pd-Si, Fe-P-C and (Fe, Co, Ni)-Si-B systems have been found to be useful as a precursor to prepare a nanocrystalline structure upon crystallization [1], a great effort has been devoted to the development of new type of high-functional or high-strength materials by utilizing the formation of crystallization-induced nanostructure. In a soft magnetic field, it has been found in 1988 that the crystallization of Fe-Si-B amorphous alloys containing Nb and Cu causes the formation of a nanoscale bcc structure and the bcc alloys exhibit good soft magnetic properties of 1.2–1.4 T for saturation magnetic flux density (B_s) and high effective permeability (μ_e) [2]. Although the good soft magnetic properties are obtained for the nanoscale bcc $Fe_{73.5}Si_{13.5}B_9Nb_3Cu_1$, the relatively low Fe concentration leads to the limitation of B_s less than 1.4 T and hence the development of a new soft magnetic alloy with high B_s above 1.5 T has strongly been desired because the simultaneous achievement of both properties enables the extension of application fields.

We tried to synthesize a nanocrystalline soft magnetic material with high B_s, and have developed nanocrystalline Fe-M-B (M = Zr, Hf or Nb) alloys with high B_s more than 1.5 T as well as good soft magnetic properties [3]. The formation of amorphous phases in Fe-metalloid systems such as Fe-Si-B and Fe-P-C alloys by the melt-spinning method usually occurs in a range of 70–84 at.% Fe [4]. On the other hand, amorphous phases have been obtained in the range of 88–91 at.% Fe for Fe-transition metal systems such as Fe-Zr [5] and Fe-Hf [6] alloys. In addition, it has been reported that the addition of a small amount of B to Fe-Zr and Fe-Hf alloys extends

† Present address: Department of Machine Intelligence and System Engineering, Faculty of System Science and Technology, Akita Prefectureal University, Honjo 015-0055, Japan.

Mat. Res. Soc. Symp. Proc. Vol. 577

the glass formation range [7, 8]. The high volume fraction of the α-Fe phase has been obtained by crystallizing the amorphous Fe-M-B alloys with the high Fe concentration.

And then we have found that dissolution of a large amount of oxygen into the remaining amorphous phase is effective for a drastic increase in electrical resistivity (ρ) of sputtered Fe-Hf-O, Fe-Zr-O and Fe-rare earth elements-O films, leading to the achievement of good high-frequency permeability in the high frequency range of 1–100 MHz [9]. This paper aims to review our recent results on two kinds of the nanocrystalline soft magnetic materials and their engineering applications.

MATERIALS PROCESSING

The Fe-M-B alloys "NANOPERM" with 15 mm in width and 15–20 μm in thickness was obtained by crystallizing an amorphous phase produced by single roller melt-spinning method. The crystallization was carried out by keeping the samples at 748–1073 K for 60–1800 s in an evacuated state. The Fe-M-O films with 2–3 μm in thickness were deposited onto indirectly water-cooled glass substrates by an rf reactive sputtering technique in a mixed atmosphere of pure argon and oxygen. The deposition was performed in no magnetic field or a static magnetic field to induce in-plane uniaxial magnetic anisotropy. The films were subjected to annealing for 10.8×10^3 s under a uniaxial magnetic field of 160 kA/m at lower temperatures than crystallization temperature of an amorphous phase in an evacuated atmosphere.

STRUCTURE AND PROPERTIES

Figure 1 shows typical examples of the structure for (a) crystallized F-M-B ribbon (NANOPERM) and (b) as-deposited Fe-M-O film. Both alloys have a similar structure composed of nanoscale α-Fe grains and an amorphous phase with high Curie temperatures (T_c) because of large amounts of M and B, or M and O elements, respectively [9, 10]. Good soft magnetic properties should be achieved mainly by the achievement of the decreased apparent anisotropy caused by the nanoscale α-Fe grains smaller than the domain wall size and easy magnetic coupling between the grains through the amorphous phase with high T_c [11]. The saturation magnetic flux density lager than 1.5 T results from the high volume fraction of α-Fe phase for the NANOPERM.

Figure 2 shows the changes in (a) B_s, (b) μ_e at 1 kHz, (c) α-Fe grain size (D), (d) T_c of the residual amorphous phase and (e) saturation magnetostriction (λ_s) as a function of annealing temperature (T_a) for 300 s for typical Fe-M-B alloys which have amorphous structure in an as-quenched state. The data of annealed structure are also shown. The saturation magnetic flux density of the Fe-M-B alloys increase rapidly with increasing T_a in the range from 748–823 K, where the structure changes from the amorphous phase with low T_c value resulting from an Inver effect [12] to α-Fe + residual amorphous phase with rather high T_c values. In the T_a range above 823 K, the change in B_s is getting small. The effective permeability of the alloys increase with increasing T_a and reaches the maximum value of 30,000–60,000, where small D and nearly zero-λ_s are achieved. The residual amorphous phase contains the large amount of M elements and B suppresses the grain growth of the α-Fe phase [10]. At higher T_a, microstructure of the alloys changes from α-Fe + residual amorphous phase to α-Fe + compounds and D rapidly increase. The soft magnetic properties are deteriorated due to a large magneto-crystalline anisotropy of the compounds with large grain size and increase of the apparent anisotropy resulting from the increase of D. The saturation magnetostriction shows the positive values of $0.4–1.8\times10^{-6}$ at T_a = 773 K and changes to the negative values of $0.7–3.9\times10^{-6}$ at T_a = 1023 K, passing through zero

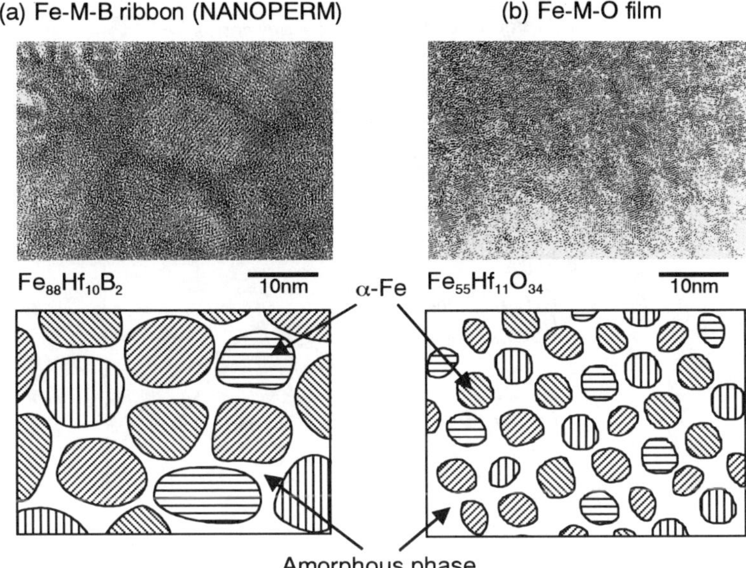

Figure 1. High-resolution transmission electron microscopy (TEM) images and schematic illustration of microstructure for nanocrystalline (a) Fe-M-B ribbon (NANOPERM) and (b) Fe-M-O film.

Table I. Saturation magnetic flux density (B_s), effective permeability (μ_e), coercivity (H_c), saturation magnetostriction (λ_s) and core loss (W) of nanocrystalline Fe-M-B ribbons (NANOPERM) and Fe-Si-B amorphous alloy.

	B_s (T)	μ_e*	H_c (A/m)	λ_s (10^{-6})	W** (W/kg)
$Fe_{90}Zr_7B_3$	1.70	30,000	5.8	-1.1	0.21
$Fe_{84}Nb_7B_9$	1.52	51,000	4.8	0.2	0.14
$Fe_{85.5}Zr_2Nb_4B_{8.5}$	1.64	60,000	3.0	-0.1	0.09
$Fe_{78}Si_9B_{13}$ (amor.)	1.56	10,000	3.5	27	0.28

* 1 kHz, 0.4 A/m

** 50 Hz, 1.4 T

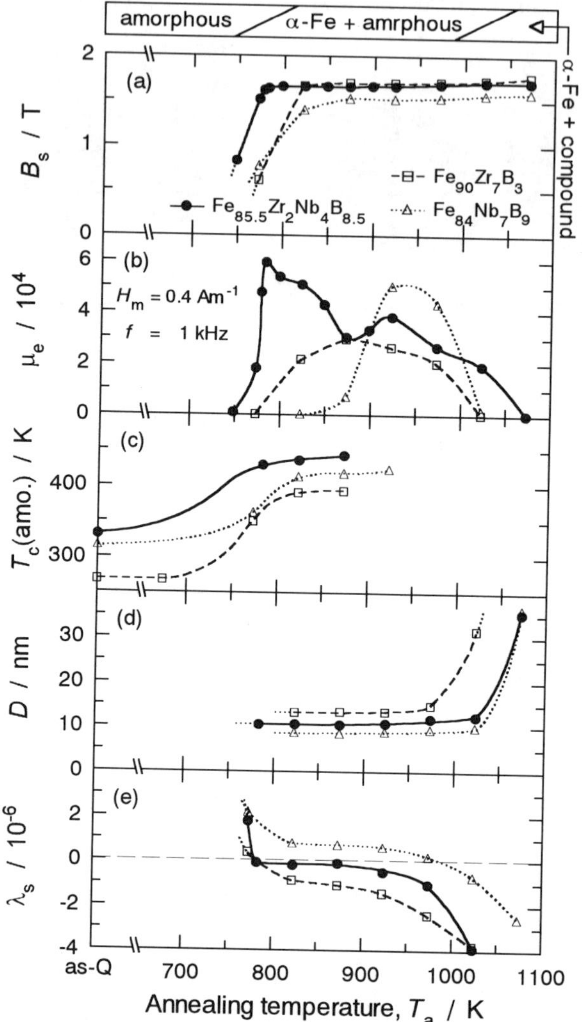

Figure 2. Changes in (a) saturation magnetic flux density (B_s), (b) effective permeability (μ_e), (c) Curie temperature (T_c) of amorphous phase, (d) mean grain size (D) and (e) saturation magnetostriction (λ_s) as a function of annealing temperature (T_a) for nanocrystalline $Fe_{90}Zr_7B_3$, $Fe_{84}Nb_7B_9$ and $Fe_{85.5}Zr_2Nb_4B_{8.25}$ alloys.

with increasing T_a. The change in λ_s with T_a is presumed to reflect the process in which λ_s approaches the value of polycrystalline α-Fe (-4.4×10^{-6} [13]). The change of λ_s with T_a is interpreted to originate from the decrease in the dissolution amount of the solute elements in the α-Fe phase [14].

Magnetic properties of the Fe-M-B ribbons "NANOPERM" are summarized in Table I. The NANOPERM with nearly zero-λ_s exhibits high B_s of 1.5–1.7 T, high μ_e at 1kHz of 30,000–60,000 and low core losses (W) at 1.4 T and 50 Hz of 0.09–0.21 W/kg. It is to be noted that these values are superior to those of $Fe_{78}Si_9B_{13}$ amorphous alloy. The relationship between B_s and μ_e at 1kHz for the NANOPERM and the conventional soft magnetic materials is shown figure 3. The NANOPERM is found to be situated in the most right and upper side among all the materials. Therefore, it can be said that the NANOPERM is best material for the achievement of high B_s and high μ_e.

On the other hand, oxygen in the Fe-M-O films preferentially combines with M element [15]. The chemical combination O with the M element, which are mainly contained in the amorphous phase, increases ρ of the amorphous phase. High electrical resistibility of the Fe-M-O films result from the existence of the amorphous phase with large amounts of M and O elements [9, 15]. Figure 4 shows the frequency dependence of the real part of permeability (μ') and the quality factor (Q = μ'/ μ", then μ" is the complex part of permeability) for the Fe-M-O films. As the ordinary metallic films such as Fe-Al-Si, Ni-Fe and Co based amorphous alloys have low electri-

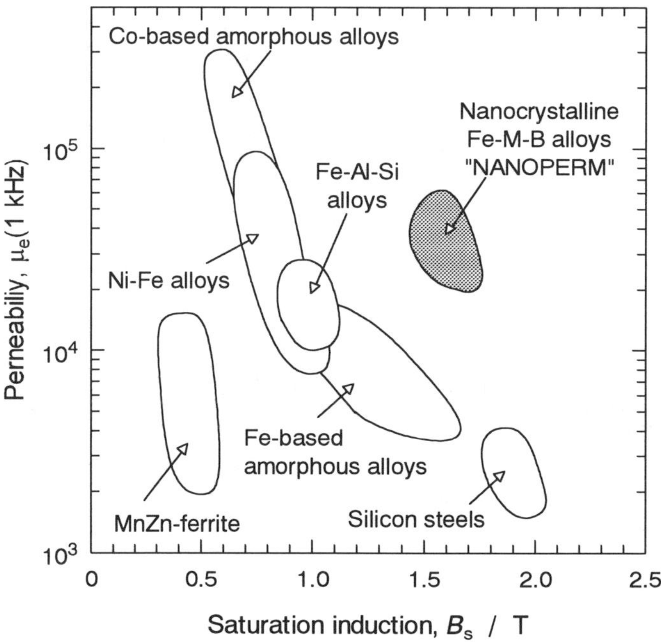

Figure 3. Relationship between saturation induction (B_s) and effective permeability (μ_e) at 1 kHz for nanocrystalline Fe-M-B ribbons (NANOPERM) and conventional soft magnetic materials.

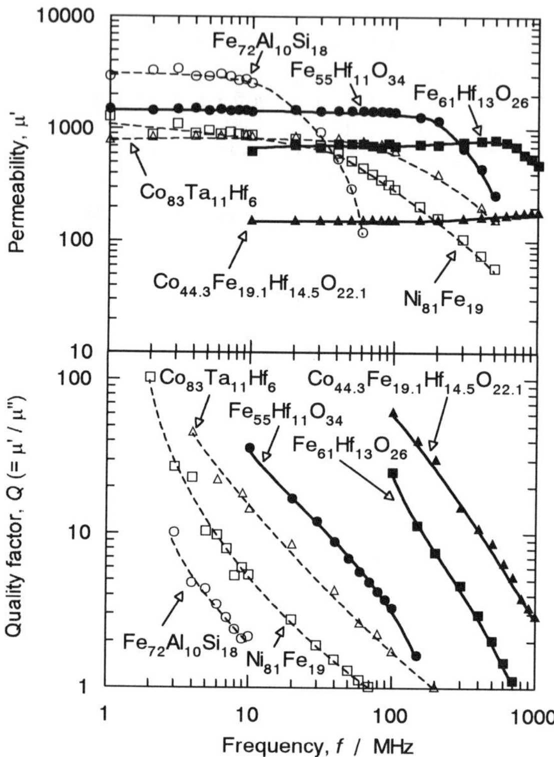

Figure 4. Frequency dependence of permeability (μ') and quality factor (Q) for nanocrystalline Fe-Hf-O and Co-Fe-Hf-O films compared with other soft magnetic films.

Table II. Saturation magnetic flux density (B_s), coercivity (H_c), electrical resistively (ρ) and structure of nanocrystalline Fe-M-O and Co-Fe-M-O films and other soft magnetic films.

	B_s (T)	H_c (A/m)	ρ (μΩm)	Structure
$Fe_{55}Hf_{11}O_{34}$	1.2	154	9.1	amor.+bcc
$Fe_{61}Hf_{13}O_{26}$	1.2	73	6.3	amor.+bcc
$Co_{44.3}Fe_{19.1}Hf_{14.5}O_{22.1}$	1.1	185	15.1	amor.+bcc
$Fe_{65}Zr_{9}O_{26}$	1.3	211	6.6	amor.+bcc
$Fe_{68}Y_{22}O_{10}$	0.9	207	22.6	amor.+bcc
$Fe_{67}Dy_{7}O_{26}$	1.3	322	9.3	amor.+bcc
$Fe_{72}Al_{10}Si_{18}$	1.1	40	1.3	bcc
$Ni_{81}Fe_{19}$	0.8	24	0.2	fcc
$Co_{83}Ta_{11}Hf_{6}$	0.8	33	1.7	amorphous

cal resistively smaller than 2 μΩm, μ' drastically decreases at the frequency below 100 MHz where an eddy current loss become relatively large. On the contrary, nanocrystalline Fe-M-O films with high ρ larger than 6 μΩm, as shown in Table II, exhibit superior μ frequency characteristics, and moreover, Q of the films is much higher than those of the ordinary ones.

APPLICATIONS

Figure 5 summarizes expected application fields for the NANOPERM together with the magnetic characteristics, which are required for their applications. As the application fields, one can list up power transformers, data communication interface components, electromagnetic interference (EMI) prevention components, magnetic heads, sensors, magnetic shielding and reactors. In this section, we describe the common mode choke coils for noise filters as an example of the application of the NANOPERM [16].

Recently, EMI considerations are increasingly important, as electronic devices become a part of our daily lives. This has rapidly increased the required performances for electromagnetic compatibility (EMC) components such as common mode choke coils, which provide protection against incoming and outgoing high frequency noise. They also protect electronic devices against high voltage pulse noise generated by spark discharge, etc. MnZn-ferrite cores have been widely used for common mode choke coils. However, the pulse attenuation characteristics of the choke coils made of MnZn-ferrite is inferior because of their low B_s (typically 0.5 T). The high μ_e and high B_s are required to obtain the good noise attenuation. If the magnetic flux density of the core is saturated by the magnetic field generated by the input pulse current (magnetic saturation), the choke coil cannot attenuate the input voltage because the permeability of the core material is extremely decreased. In order to prevent the magnetic saturation, high B_s is necessary.

In order to obtain the good noise attenuation, the high permeability for a pulse magnetic field is also important. Figure 6 shows the change in the pulse permeability (μ_p) as a function of the maximum flux density (B_m) when the pulse width is 1 μs. When B_m is small, μ_p of the NANOPERM is almost the same as that of the Mn-Zn ferrite. The pulse permeability gradually

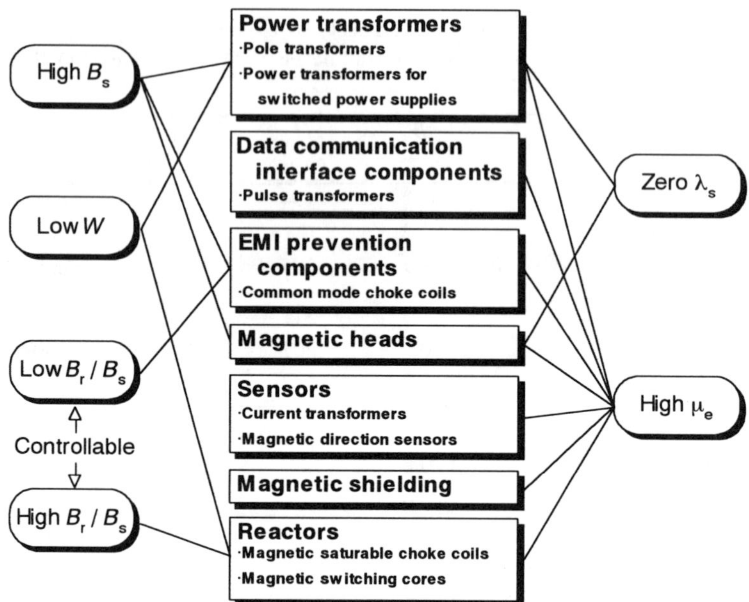

Figure 5. Magnetic characterizations and application fields for nanocrystalline Fe-M-B ribbons (NANOPERM).

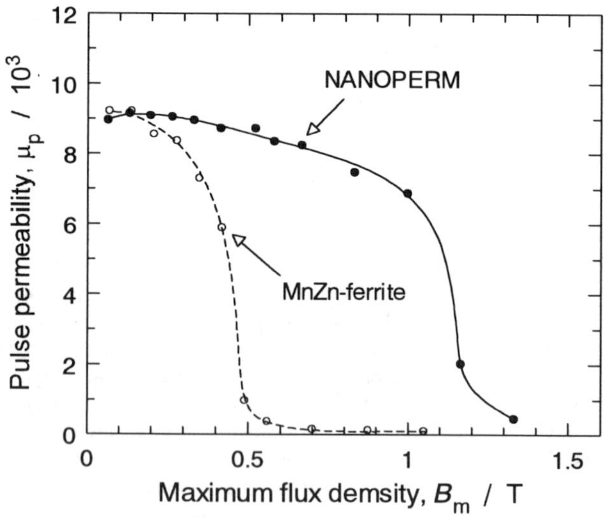

Figure 6. Change in pulse permeability (μ_p) of NANOPERM and MnZn-ferrite as a function of maximum flux density (B_m) when pulse width is 1 μs.

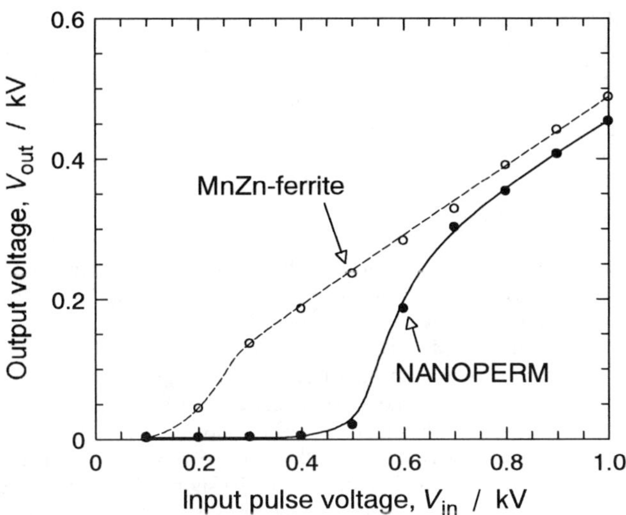

Figure 7. Output voltage of common mode choke coils made of NANOPERM and MnZn-ferrite as a function of input pulse voltage when pulse width is 1 μs.

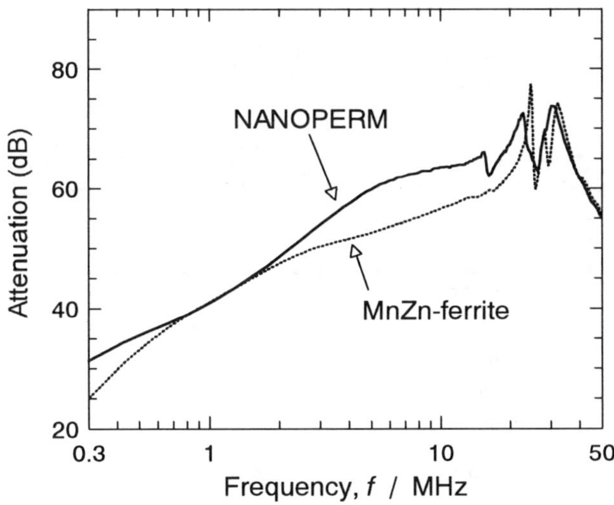

Figure 8. Frequency dependence of noise attenuation for common mode choke coils made of NANOPERM and MnZn-ferrite.

decreases with increasing B_m, and then falls abruptly due to the magnetic saturation. Figure 6 clearly shows that μ_p of the NANOPERM maintains high value up to higher B_m than that for the Mn-Zn ferrite. This means that the good attenuation characteristics against the high-voltage pulse noise can be obtained by using the NANOPERM as the core material. Figure 7 shows the output voltage of the choke coils as a function of the input pulse voltage when the pulse width is 1 μs. The NANOPERM choke coil shows superior attenuation.

Figure 8 shows the frequency dependence of the noise attenuation of the common mode choke coils made of the NANOPERM and MnZn-ferrite. The attenuation of the NANOPERM choke coil is higher than that of the MnZn-ferrite choke coil in the frequency range of 0–20 MHz, which includes the operating frequency range of normal switched power supplies. The NANOPERM choke coil is therefore useful for the attenuation of noise generated by switched power supplies. It can be concluded that the NANOPERM with high B_s and high permeability is suitable for the core material of the common mode choke coils.

As applications of the Fe-M-O films, micro switching converters, high density recording heads, noise filters and others used at high frequencies are now under investigation. For an example of the applications, figure 9 show a schematic illustration of a thin film inductor for the micro switching converter [17]. The inductor was made on a glass substrate by dry and wet process technique. The upper core is contact with the lower core at the outside and inside of the coil to make a completely closed magnetic loop. Figure 10 shows the frequency dependence quality factor (Q) of the thin film inductor. The quality factor of a thin film inductor using an amorphous Co-Zr-Ta film [18] is also shown. As can be seen fig. 10, Q exhibits a maximum value of 5.5 at 8 MHz for the inductor using the Fe-Hf-O film. On the other hand, the frequency that Q shows a maximum value of 4.3 is 0.7 MHz, which is only 1/10 that of the inductor using the Fe-Hf-O film. Therefore, the inductor using the nanocrystalline Fe-Hf-O films enables higher frequency operating and higher efficiency than the inductor using the amorphous Co-Zr-Ta film for the microswitching converters owing to the low loss characteristics of the magnetic film.

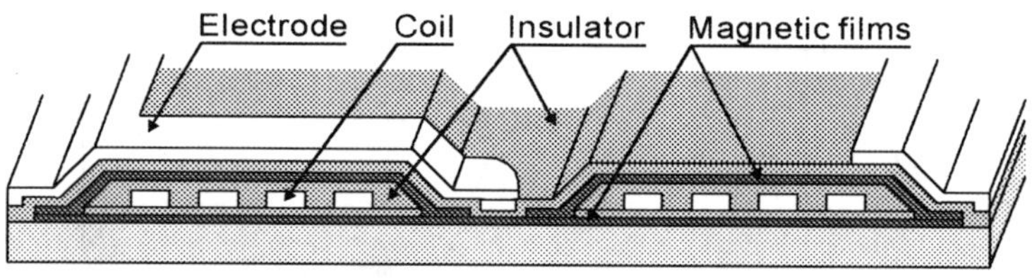

Coil type	Rectangular spiral
Coil size	3.5×3.5 mm^2
Number of turns	12
Conductor width	60 μm
Conductor thickness	20 μm
Magnetic film	$Fe_{55}Hf_{11}O_{34}$
Mgnetic film thickness	3 μm

Figure 9. Schematic illustration of thin film inductor using nanocrystalline Fe-Hf-O film.

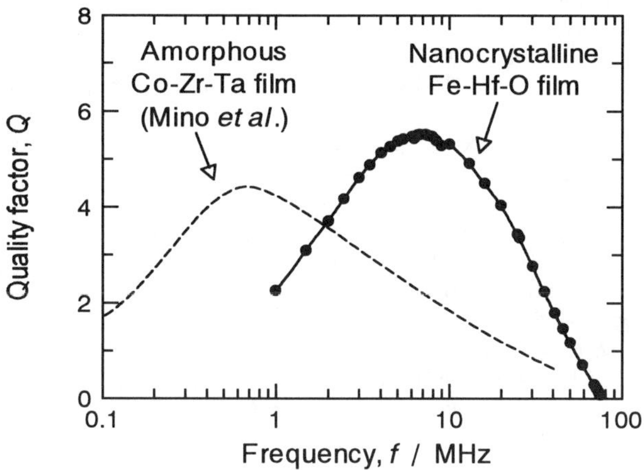

Figure 10. Frequency dependence of quality factor (Q) of thin film inductor using nanocrystalline Fe-M-O film and amorphous Co-Zr-Ta film [18].

CONCLUSIONS

We have developed the new type of nanocrystalline soft magnetic materials in the Fe-M-B (M = Zr, Hf, Nb) ribbons "NANOPERM" and the Fe-M-O (M = Zr, Hf, rare earth) films. The NANOPERM shows high B_s above 1.5 T as well as excellent soft magnetic properties. The Fe-M-O films with high ρ exhibits good high-frequency permeability in the high frequency range of 1–100 MHz. The excellent characteristics of a common mode choke coil made of NANOPERM and a thin film inductor using the Fe-M-O films were proved experimentally. The NANOPERM and the Fe-M-O films are expected to be used widely in these magnetic application fields.

REFERENCES

1. T. Masumoto, H. Kimura, A. Inoue, and Y. Waseda, Mater. Sci. Eng. **23**, p. 141 (1976).
2. Y. Yoshizawa, S. Oguma, and K. Yamauchi, J. Appl. Phys. **64**, p. 6,044 (1988).
3. A. Makino, K. Suzuki, A. Inoue, Y. Hirotsu, and T. Masumoto, J. Magn. Magn. Mater. **133**, p. 329 (1994).
4. T. Masumoto et al. (ed.), *Materials Science of Amorphous Metals*, Ohmu, Tokyo, 1982, p. 281.
5. M. Nose and T. Masumoto, Sci. Rep. Res. Inst. Tohoku Univ. **A28**, p. 232 (1980).
6. A. Inoue, K. Kobayashi, and T. Masumoto, in *Proceedings of Conference on Metallic Glasses, Science and Technology*, edited by C. Hargitai et al. (Cent. Res. Inst. Phys., Budapest, 1980), vol. II, p. 217.
7. A. Inoue, K. Kobayashi, M. Nose, and T. Masumoto, J. de Phys. **C8-41**, p. 31 (1980).
8. S. Ohnuma, M. Nose, K. Shirakawa, and T. Masumoto, Sci. Rep. Res. Inst. Tohoku Univ. **A29**, p. 254 (1981).
9. A. Makino and Y. Hayakawa, J. Jpn. Inst. Metals **57**, 1,301 (1993).

10. A. Makino, T. Hatanai, A. Inoue, and T. Masumoto, Mater. Sci. Eng. **A226-228**, p. 594 (1997).
11. G. Herzer, IEEE Trans. Magn. **25**, p. 3,327 (1989).
12. K. Shirakawa, S. Ohmuma, M. Nose, and T. Masumoto, IEEE Trans. Mag. **MAG-16**, p. 910 (1980).
13. E. W. Lee, Per. Prog. Phys. **18**, p. 184 (1955).
14. K. Suzuki, M. Kikuchi, A. Makino, A. Inoue, and T. Masumoto, Mtaer. Trans., JIM **32**, p. 961 (1991).
15. Y. Hayakawa, A. Makino, H. Fujimori, and A. Inoue, J. Appl. Phys. **81**, p. 3,747 (1997).
16. Y. Naitoh, T. Bitoh, T. Hatanai, A. Makino, A. Inoue, and T. Masumoto, Sci. Rep. Res. Inst. Tohoku Univ. **A43**, p. 161 (1997).
17. Y. Sasaki, S. Morita, T. Hatanai, A. Makino, T. Sato, and K. Yamasawa, in *Record of the 29th Annual IEEE Power Electronics Specialists Conference*, 1998), vol. 2, p. 1,555.
18. M. Mino, K. Tsukamoto, K. Yanagisawa, A. Tago, and T. Yachi, in *Proceedings of 11th Annual Applied Power Electronics Conference and Exposition*, 1996), vol. 1, p. 422.

MAGNETIC PROPERTIES OF HITPERM $(Fe,Co)_{88}Zr_7B_4Cu_1$ NANOCRYSTALLINE MAGNETS (INVITED)

M. A. WILLARD*, M. GINGRAS*, M. J. LEE*, V. G. HARRIS**, D. E. LAUGHLIN*, AND M. E. McHENRY*

* Materials Science and Engineering, Carnegie Mellon University, Pittsburgh, PA 15213-3890
** Naval Research Laboratory, Washington, D.C., 20375-5000

ABSTRACT

Alloys consisting of Fe-Co-M-B-Cu (with M = Zr, Hf, Nb), called HITPERM alloys, have been developed. Synchrotron X-radiation studies have been used to show that the ferromagnetic phase in an equiatomic FeCo-based alloy is the α'-FeCo phase. Since both the α'-FeCo phase and the FeCo-based amorphous phase of the nanocrystalline alloy have high Curie temperatures, a high magnetization persists up to the α -> γ structural phase transformation temperature of 980°C. Room temperature AC permeability measurements have shown that the alloys maintain a high permeability of ~2000 up to a frequency of 20 kHz. The room temperature core loss has also been shown to be competitive with commercial high temperature magnetic alloys with a value of 1 W/g at B_s = 10 kG and f = 10 kHz. Analysis of extended X-ray absorption fine structure (EXAFS) data is consistent with a two-phase mixture of nanocrystalline body centered cubic derivative FeCo structure and an amorphous Zr-rich phase. A differential scanning calorimetry study of the primary crystallization reaction shows an activation energy of 323.3 kJ/mol. As a preliminary study of phase and grain stability, broadening of X-ray diffraction peaks indicates little grain growth after annealing at 600 °C for 3072 hours.

INTRODUCTION

During the past decade, magnetic materials have been developed that demonstrate the favorable role that a nanocrystalline microstructure can have on extrinsic soft magnetic properties. The first of these alloys was based on the nanocrystallization of α-Fe_3Si from an amorphous precursor alloys with components Fe-Nb-Si-B-Cu (FINEMET)[1]. The next class of nanocrystalline alloys, relying on α-Fe nanocrystals, was developed by Suzuki et al. from an alloy of the composition Fe-Zr-B-Cu (NANOPERM)[2-3]. Most recently, magnets based on the α'-FeCo phase have been developed from alloys with composition Fe-Co-Zr-B-Cu.[5] These materials have been found to consist of a two-phase microstructure with a crystalline phase and an amorphous grain boundary phase. As seen in Table I, each alloy has its own distinct properties which makes it good for a certain set of soft magnetic applications.

Sample	Nanocrystal Phase	T_{x1} (K)	T_{x2} (K)	B_s (T)	μ_e	H_c (A/m)
FINEMET $Fe_{73.5}Si_{13.5}B_6Nb_3Cu_1$	α -FeSi	770-815	905-965	1.21 - 1.24	7 – 100k	0.5 – 1.4
NANOPERM $Fe_{88}B_4Zr_7Cu_1$	α -Fe	750-800	990-1000	1.52 – 1.64	3.5 – 48k	2.4 – 4.5
HITPERM $Fe_{44}Co_{44}B_4Zr_7Cu_1$	α' -FeCo	720-750	990-1000	1.6 - 2.0	1 - 2.5k	80-200

Table I: Comparison of various properties for three different classes of nanocrystalline soft magnetic materials.[4]

The focus of this paper is to describe the structure and magnetic properties of the latter alloy (HITPERM)[5-7]. One of the proposed applications for HITPERM is high temperature power generation in jet aircraft. For these applications the demands on bulk soft magnetic materials include (1) higher combined induction and permeability, (2) magnets capable of operating at higher

Mat. Res. Soc. Symp. Proc. Vol. 577 © 1999 Materials Research Society

temperatures, and (3) many non-magnetic issues such as mechanical properties, corrosion resistance, etc. The niche for these materials has been found due to their compromise between the good high frequency response of the residual amorphous phase and the high magnetization of the nanocrystalline phase. As discussed below, HITPERM combines the high permeability of nanocrystalline alloys of this genre with the attractive high induction and Curie temperature associated with FeCo alloys in general.

EXPERIMENT

Ingots with composition $Fe_{44}Co_{44}Zr_7B_4Cu_1$ were arc melted in an argon atmosphere starting with high purity Fe, Co, Zr, Fe_3B, and Cu components. Amorphous ribbons were then produced from the ingots using a standard single wheel melt spinning technique. The amorphous ribbons obtained by this process were about 1 mm in width and 20-50 μm in thickness. After obtaining the as-spun ribbons from the melt spinning procedure, they were isothermally annealed (510 °C < T_{ann} < 700 °C) in an argon atmosphere for 1 hour followed by water quenching. The annealing process resulted in a nanocrystalline microstructure.

Differential thermal analysis (DTA) was used to examine the crystallization temperatures of the as-cast ribbon. A scan rate of 10 °C/min was used between temperatures of 400 and 1200 °C. Synchrotron X-ray diffraction experiments were performed at the National Synchrotron Light Source. These were used to identify superlattice reflections which are signatures of the ordered α'-FeCo phase (B2). X-rays with energy of 7112 eV (or equivalently a wavelength of 1.748 Å) were used for the synchrotron X-ray experiments. This wavelength was chosen to take advantage of the anomalous scattering just above the Fe K-edge.

Kinetics studies were carried out using a Perkin Elmer DSC-7 differential scanning calorimeter (DSC). Temperature scan rates were varied between 5 and 40 °C/min over the temperature range 100 °C to 580 °C. Standard Kissinger analysis was used to find the activation energy of primary crystallization in the amorphous HITPERM alloy.

Long term annealing experiments were conducted at 600 °C for 10, 31, 100, 308, 1000, and 3072 hours. The crystal structure and grain sizes of these samples were examined by X-ray diffraction (XRD) using a Rigaku Θ/ Θ diffractometer, and Cu K_α radiation. The grain sizes were estimated by standard Scherrer analysis with background removal by linear fit.

Extended X-ray absorption fine structure (EXAFS) measurements were used to examine the two-phase microstructure of the HITPERM alloy. X-ray absorption spectra of the Fe, Co, Zr, and Cu K edges were collected from amorphous and nanocrystalline ribbons on Beamlines X23B (NRL) and X23A2 (NIST) at the National Synchrotron Light Source (Brookhaven National Laboratory). Data collection was performed on ribbon samples in transmission and fluorescence yield modes simultaneously. Artifacts were identified by comparison of the two data sets and multiple scans of each sample were performed for improved statistics.[8]

A Vibrating Sample Magnetometer (VSM) was used to measure the magnetization as a function of temperature of the as-quenched alloy from room temperature to 1000 °C, in a field of 500 Oe, and with a heating rate of 2 °C/min. A Walker AC permeameter was used to measure the room temperature permeability and coercivity of toroidal samples. The as-spun ribbons were prepared for permeability measurement by winding them into laminated toroids followed by the standard isothermal anneal for nanocrystallization prior to measurement of AC properties.

RESULTS

A. Microstructure Development

DTA data were collected to identify the region where a single ferromagnetic phase would be crystallized from the amorphous precursor. As seen in Figure 1, the primary crystallization peak (T_{x1}), corresponding to the crystallization of α'-FeCo, was found at 500 °C. Other crystalline phases began to form at temperatures above 700 °C, which allowed for a 200 °C range where nanocrystallization of only the desired ferromagnetic α'-FeCo formed.

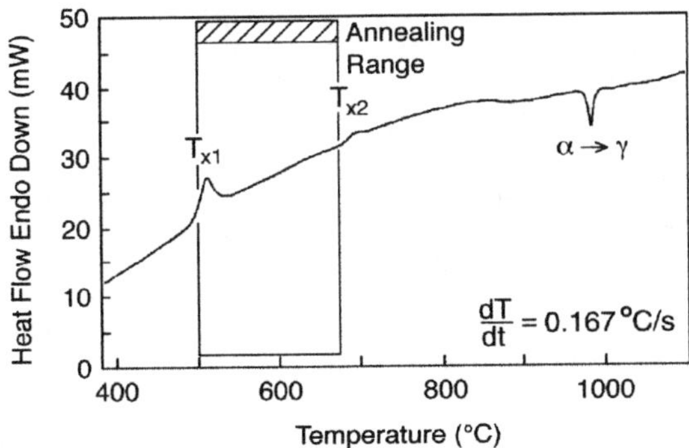

Figure 1: DTA data showing the range of annealing temperatures that allow only the ferromagnetic phase to crystallize (primary crystallization).

To confirm the crystallization of α'-FeCo, synchrotron X-ray diffraction was used. By conventional X-ray diffraction, it is difficult to differentiate between the disordered α-FeCo (bcc) and the ordered α'-FeCo (CsCl structure). The X-ray atomic scattering factors for Fe and Co are almost identical. Thus, the only difference between the α-FeCo and α'-FeCo diffraction patterns would be the faint superlattice reflections. The superlattice intensities are related to the square of the difference between the atomic scattering factors for Co and Fe, which remains very small. Therefore, using conventional X-ray diffraction would require long count times for the superlattice reflections to be resolved from the background radiation.

An alternative solution to the long count times necessary for conventional X-ray diffraction is to use synchrotron radiation. This has advantages due to the high intensity of the X-ray beam and the ability to choose a wavelength to take advantage of anomalous scattering. As seen in Figure 2, the (100) superlattice reflection verifies the α'-FeCo as the ferromagnetic phase.

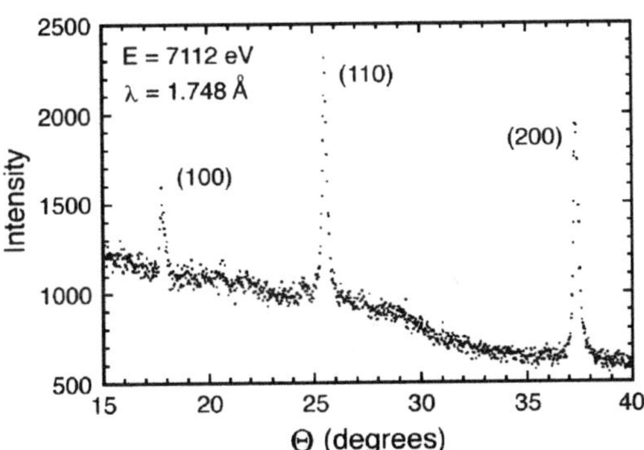

Figure 2: Synchrotron X-ray diffraction data showing the presence of α'-FeCo as the ferromagnetic phase from primary crystallization.

The formation of the desired ferromagnetic phase during primary solidification is critical for obtaining excellent soft magnetic properties since other phases might act as pinning sites for domain walls. Referring to Figure 1, the upper limit of the good annealing region is reached with the formation of undesired phases. Another important factor in the formation of the desired properties is proper nanocrystallization. According to the Herzer's theory, low coercivity is only obtained for grain sizes well below (or well above) the exchange correlation length (~ domain wall width).[9-10] Thus, optimization of grain size is a necessity.

To obtain the optimum grain size in nanocrystalline alloys, an understanding of the crystallization kinetics is important. As a preliminary analysis of crystallization kinetics, the Kissinger technique has been used to find the activation energy for primary crystallization.[11-12] Figure 3(a) shows crystallization peaks at various heating rates. As the heating rate is increased, the primary crystallization peak shifts to higher temperatures due to a thermal activation necessary for crystallization. The Kissinger plot in Figure 3(b) shows this activation energy to be 323 kJ/mol (3.35 eV/atom). These values are similar to those for Fe-based nanocrystalline soft magnetic alloys.[14-15] By controlling the composition, number and distribution of nucleation sites, and quench rate of the amorphous precursor, we will be able to control the kinetics in similar materials and ultimately optimize the extrinsic magnetic properties.

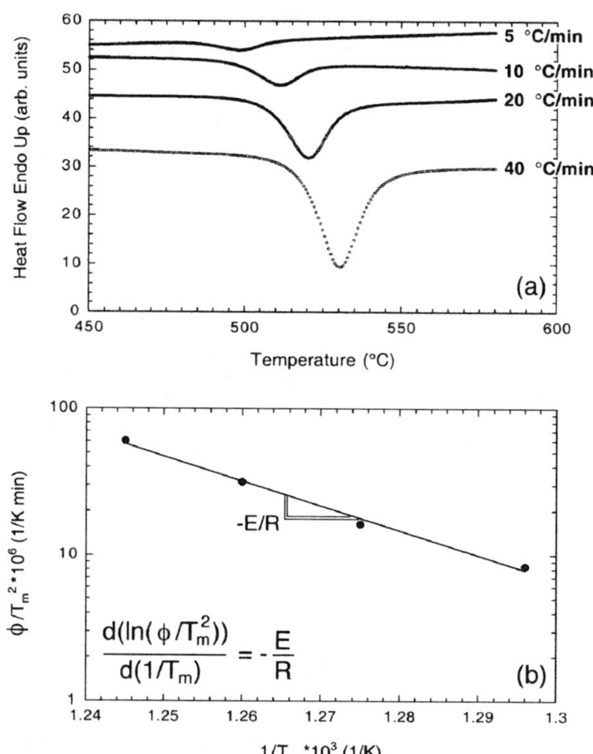

Figure 3: (a) DSC data showing the primary crystallization peak shift due to changes in heating rate (b) Kissinger plot used to determine the activation energy of primary crystallization

The stability of the nanocrystalline grains at high temperatures is essential for maintaining optimal magnetic properties. The chemical stability of the phases and minimization of grain growth of these alloys are both parameters which must controlled for use in high temperature applications. Both of these are concerns established by the Herzer and no pinning criteria already discussed.

To examine the stability of HITPERM, amorphous samples were annealed at 600 °C for up to 3072 hours. Each sample was examined by X-ray diffraction and the breadth of crystalline peaks were measured. The X-ray diffraction results shown in Figure 4, for the sample annealed for 3072 hours, showed that only the fundamental diffraction peaks for the α-FeCo were observed and that no secondary phase crystallization has occurred! This indicates that the stability of the nanocrystalline phase is maintained at high temperatures for long times and therefore these materials may be successfully used for high temperature applications.

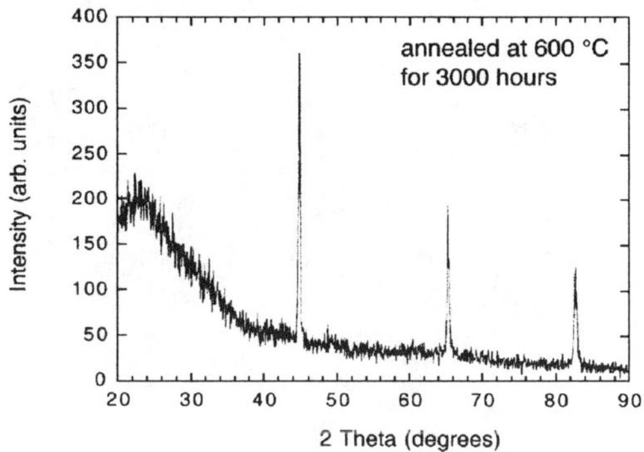

Figure 4: Conventional X ray diffraction showing nanocrystalline α-FeCo after annealing at 600 °C for 3072 hours

Nanocrystalline alloys, because of their small grain size exhibit significant fine particle broadening in the Bragg diffraction peaks. An empirical treatment of the fine particle broadening reveals that the particle size, t, can be determined from the breadth, B, of the x-ray peak as given by:

$$t = \frac{0.9\lambda}{B\cos\theta_B} \qquad [1]$$

that is known as the Scherrer formula.[13] It states that for a crystal of thickness, t, we should expect a particle broadening of B of the Bragg peak at $2\theta_B$. For 50 nm particles a broadening of ~0.20° is expected and is easily measured. Scherrer analysis has been used as a rough estimate of grain size for our annealed samples.

The Scherrer analysis results are summarized in Table II. The mean grain size seems to remain small over the whole range of annealing temperatures. Further support by transmission electron microscopy will be necessary to verify these preliminary results.

Annealing Time (hours)	D_{mean} (nm)	σ (nm)
10	49.39	17.12
31	36.52	10.33
100	70.89	12.97
308	57.97	32.96
1000	57.13	8.92
3072	61.18	14.03

Table II: Grain size by Scherrer analysis

In the similar FINEMET and NANOPERM alloys, a two phase microstructure has been observed as a ferromagnetic nanocrystalline phase and a ferromagnetic amorphous phase at the grain boundaries. The two-phase microstructure has been shown to aid the soft magnetic properties by: (a.) increasing the resistivity of the alloy thus reducing eddy current formation in

AC applications; (b.) decreasing the effects of magnetostriction since the amorphous and crystalline phases have magnetostrictive coefficients which effectively cancel out due to their sign differences, thus increasing the permeability. For these reasons, we would like to determine whether HITPERM has a similar two-phase microstructure.

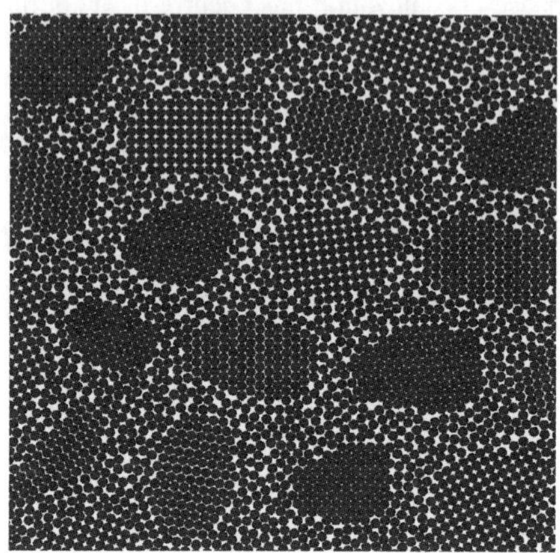

Figure 6: Schematic of a nanocrystalline alloy with amorphous grain boundary phase

Extended X-ray absorption fine structure (EXAFS) was used to examine the microstructure of the HITPERM alloy. As found in other nanocrystalline soft magnetic materials produced by the annealing of ribbons, the nanocrystalline phase is surrounded by a thin amorphous phase (see the schematic Figure 6). Since EXAFS is a technique that examines short range order, the very small amorphous phase region between nanocrystalline grains can be probed. Additionally, this technique is element sensitive allowing one to determine information as to the short-range order of the atoms whose absorption edge lies near the K edge of the element of interest. EXAFS studies described below have been used to determine the local environments of atoms in the nanocrystalline and amorphous intergranular phases. In particular, this study shows EXAFS data for ribbons annealed at 550, 650, and 800 °C at the Fe, Co, and Zr K edges.

Figures 7(a) and 8(a) show the Fourier transform (FT) Fe and Co EXAFS respectively. The data were collected from the as-spun sample and those annealed at 550, 650, and 800 °C. For comparison to a foil with a bcc structure, data from an Fe foil were also plotted. The ribbons are plotted using the left axis and the standard foil was plotted using the right axis.

The as-spun spectrum displays a well-defined near-neighbor (NN) peak with no signs of higher shell ordering. This is consistent with an amorphous structure. The sample annealed at 550 °C shows an increase in amplitude of the NN shell and a shift to higher radial distances. Additionally, higher order shells begin to appear. The peaks can be correlated directly to the bcc standard foil, but there is not enough r-space resolution to isolate each bond (i.e. $r_{1,2}$ and $r_{4,5}$). Comparison of the Fe or Co FT EXAFS with the Fe foil FT EXAFS shows a consistent bcc structure for our sample.

By plotting the NN peak amplitude and NN centroid as a function of the annealing temperature, insight into the structural evolution can be obtained. These results are plotted in Figures 7(b & c) and 8(b & c) respectively for Fe and Co. Both sets of data behave in a similar manner. As the annealing temperature increases, there is an abrupt increase in the NN average distance followed by a plateau (Figures 7(c) & 8(c)). This shows the transition of the NN peak from a single disordered atomic shell to that of a more ordered bimodal distribution of atoms. Increases in annealing temperature do not change this distance which is consistent with Fe and Co remaining in a bcc coordination.

The plots of the NN amplitude versus annealing temperature reveal a different trend. The amplitude increases without a plateau that indicates an increase in coordination and atomic ordering

(Figures 7(b) & 8(b)). This continuing increase in ordering and coordination indicates that even at annealing temperatures of 800 °C these features have not reached their equilibrium values.

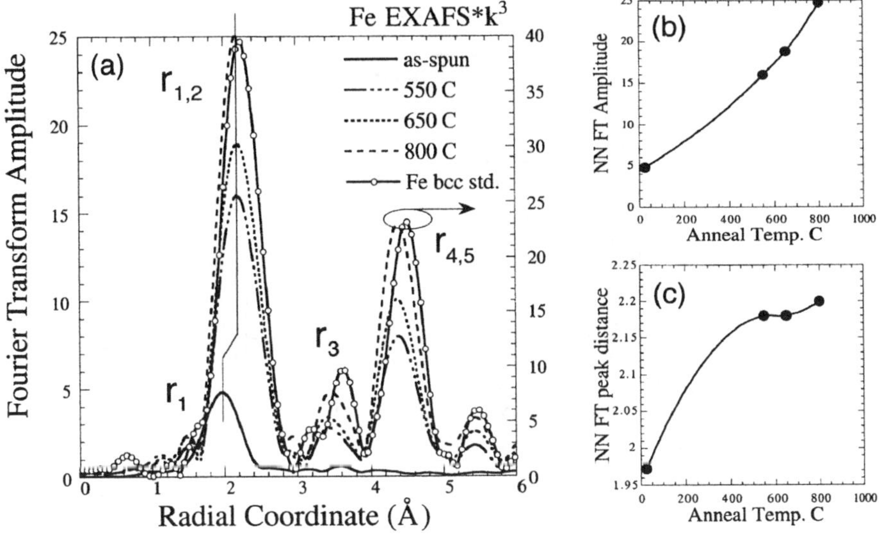

Figure 7: Fourier transform amplitude of Fe EXAFS showing BCC coordination of Fe

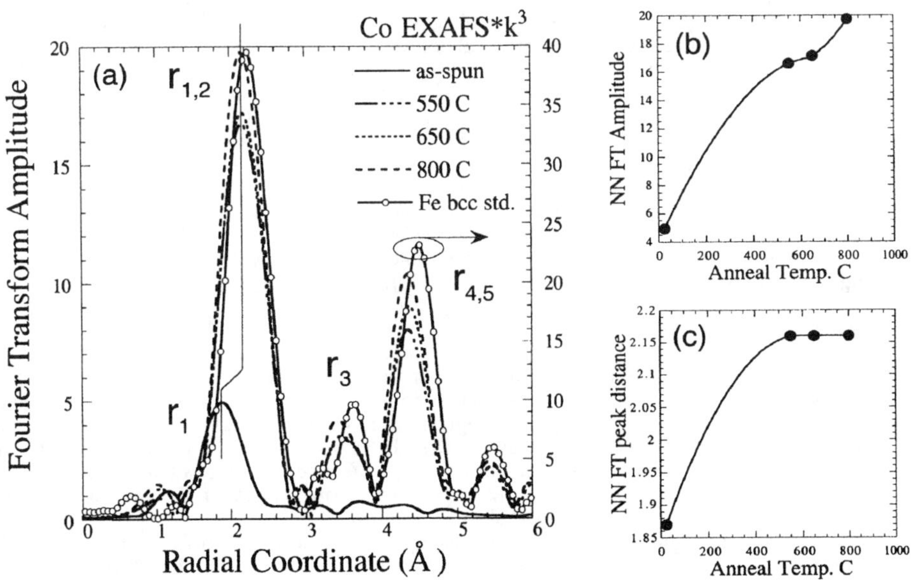

Figure 8: Fourier transform amplitude of Co EXAFS showing BCC coordination of Fe

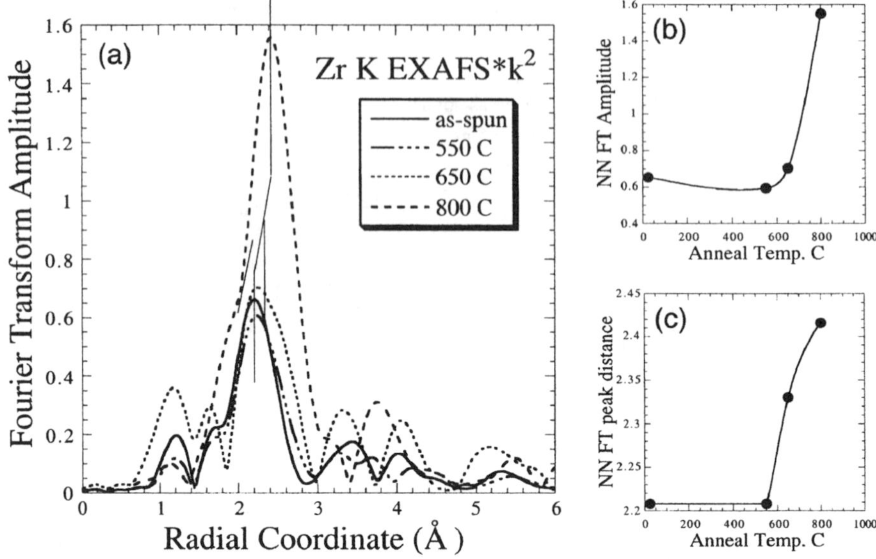

Figure 9: Fourier transform amplitude of Zr EXAFS showing coordination of Zr

The evolution of the local atomic environment of the Zr atoms is quite different from that of the Fe and Co atoms. The as-spun spectrum displays a well defined NN peak with two broad peaks having lower amplitude and at higher radial coordinates (Figure 9(a)). Although these peaks with higher radial distances may have structural origins, the atomic environment of Zr is clearly disordered and may be described as amorphous. After annealing at 550 °C, the NN coordination and distances have not changed much from the as-spun material. It is not until the sample is annealed at 650 °C that the peak distance begins to shift, but the amplitude still does not change much (Figure 9(b & c)). The sustained amplitude of the NN peak is evidence that the local environment of the Zr atoms remains amorphous. The change in the NN amplitude indicates that the local chemical order of the Zr atom is changing due to the formation of the FeCo crystallites. At 800 °C the amplitude of the NN peaks increases significantly and the distances continue to increase to higher radial values. This is a clear indication that the Zr atoms are significantly changing symmetry to a more ordered state (i.e. $(Fe,Co)_3Zr$).

B. Magnetic Properties

Curie temperature and magnetization are important intrinsic properties for high temperature applications. An amorphous precursor ribbon has been heated to observe the magnetization at high temperatures (Figure 10). Admittedly, the magnetization shown in Figure 10 is not the actual magnetization of the HITPERM alloy as it would be used. The magnetization in the nanocrystallized ribbon would be much higher since the effective annealing of the ribbon during this experiment is very short. However, this experiment shows some of the interesting phase transformations that occur in the HITPERM alloy.

As indicated in Figures 1, 10, and 11, heating of the amorphous ribbon with HITPERM composition shows primary and secondary crystallization, magnetic transformations, an order/disorder transformation, and a polymorphic phase transformation at the effective Curie temperature of the alloy.

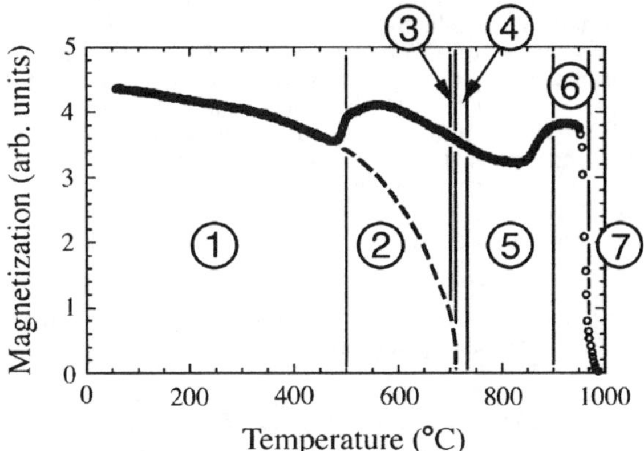

Figure 10: Thermomagnetic data taken from an amorphous sample with HITPERM composition.

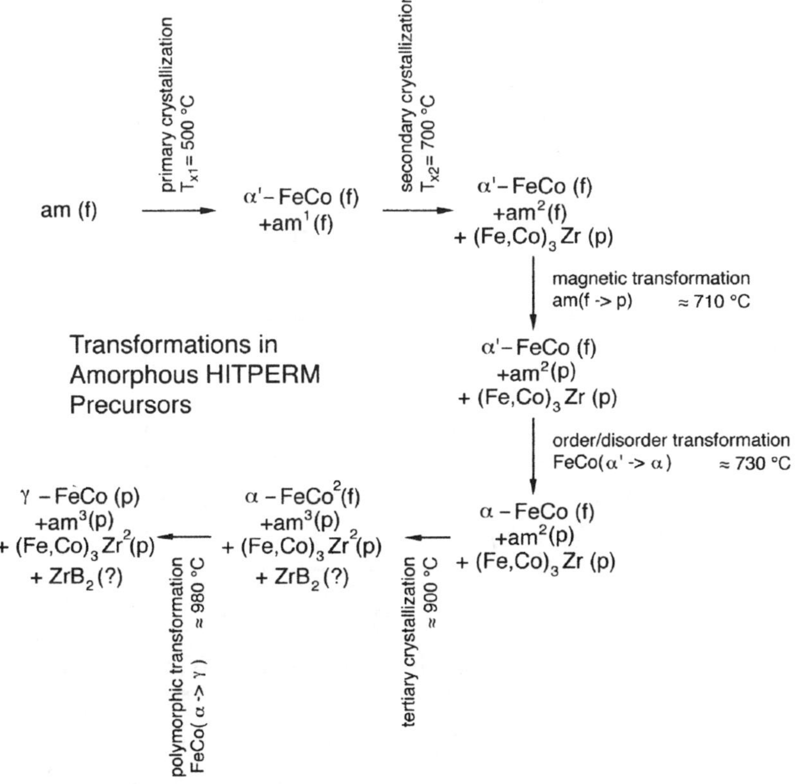

Figure 11: Phase transformations in HITPERM

The temperatures in Figure 11 are approximate for many of the transformations, and not all transformations are certain, but qualitatively the diagram shows many interesting aspects of the alloy. First of all, it is critical that the Curie temperature of the amorphous phase remain above the operation temperature of the alloy to avoid destroying the exchange coupling between grains and forming superparamagnetic particles of the nanocrystalline grains (or at least increasing the coercivity extensively). In all of the Fe-based nanocrystalline alloys, this is violated if the operating temperature is chosen as 500 °C. Therefore, operation of HITPERM should be limited to regions (1) and (2) of Figure 10, where the amorphous phase is ferromagnetic and no other crystalline phases have formed.

Another interesting aspect is the increase of the magnetization at high temperatures. This has been attributed to the formation of ZrB_2 or ZrO phases, yielding more α-FeCo. At 800 °C, the formation of α-FeCo is the only phase that will increase the magnetization, so formation of this ferromagnetic phase at the expense of paramagnetic $(Fe,Co)_3Zr$ seems likely.

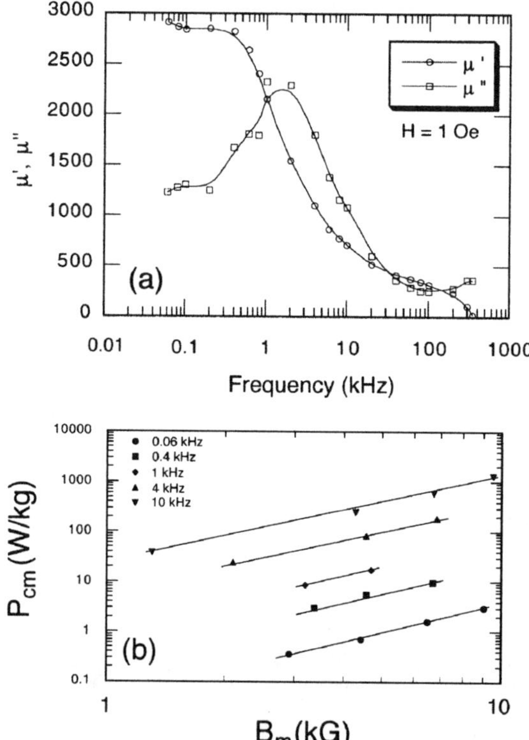

Figure 12: (a) Frequency dependent permeability for a field amplitude of 1 Oe (b) the core loss as a function of the maximum induction at various frequencies.

The extrinsic properties of interest include permeability and ultimately core loss in these alloys. In Figure 12(a), the frequency dependent permeability has been plotted for a field amplitude (H_m) of 1 Oe. The maximum permeability was determined to be 2900 and the peak in the loss portion of the permeability occurred at 2 kHz. However, an acceptable permeability of

1000 was maintained to 10 kHz. Figure 12(b) shows the core loss as a function of the maximum induction at various frequencies. Although the core loss seems quite high when compared to most Fe-based alloys, it is quite comparable to other FeCo-based alloys, which are used for high temperature applications

CONCLUSIONS

This work shows the formation of α'-FeCo phase as the ferromagnetic phase in a ribbon with a two phase microstructure. After annealing of nanocrystalline ribbons for up to 3000 hours at 600 °C, the ribbons show only small grain growth and no formation of secondary phases. Kinetics studies have shown an activation energy of 323 kJ/mol. Magnetization of the amorphous precursor ribbons show a high magnetization maintained to 980 °C. Permeability and core loss data show that HITPERM achieves the necessary magnetic properties for the intended application.

ACKNOWLEDGMENTS

Effort sponsored by the Air Force Office of Scientific Research, Air Force Materiel Command, USAF, under grant number F49620-96-1-0454. The U.S. Government is authorized to reproduce and distribute reprints for governmental purposes notwithstanding any copyright notation thereon. Work carried out in part at the National Synchrotron Light Source, Brookhaven National Laboratory, which is supported by the U.S. DOE, Divisions of Materials Sciences and of Chemical Sciences. Special thanks to Johnny Kirkland and Joseph Woicik for their help on Beamlines X23B and X23A2. Work carried out in part at Los Alamos National Laboratory, which is supported by the U.S. Department of Energy. DEL acknowledges partial support from NEDO project.

REFERENCES

1. Y. Yoshizawa, S. Oguma, and K. Yamauchi, *J. Appl. Phys.* **64**, 6044 (1988).
2. K. Suzuki, A. Makino, N. Kataoka, A. Inoue, and T. Masumoto, *Mat. Trans.* **JIM-32**, 93 (1991).
3. A. Makino, T. Hatanai, Y. Naitoh, T. Bitoh, A. Inoue, and T. Masumoto, *IEEE Trans. Magn.* **33**, 3793 (1997).
4. M. E. McHenry, M. A. Willard, and D. E. Laughlin. *Progress in Materials Science* (1999), in press.
5. M. A. Willard, D. E. Laughlin, M. E. McHenry, D. Thoma, K. Sickafus, J. O. Cross and V. G. Harris. *J. Appl. Phys.* **84**, 6773 (1998).
6. M. A. Willard, M.-Q. Huang, D. E. Laughlin, and M. E. McHenry, C. Franchetti, J. O. Cross, and V. G. Harris. *J. Appl. Phys.* **85** (1999) 4421.
7. H. Iwanabe, B. Lu, M. E. McHenry, and D. E. Laughlin. *J. Appl. Phys.* **85**, 4424 (1999).
8. J. D. Ayers, V. G. Harris, J. C. Sprague, and W. T. Elam, *Appl. Phys. Lett.* **64**, 974 (1994).
9. G. Herzer, *IEEE Trans. Magn.* **26**, 1397 (1990).
10. G. Herzer, *J. Magn. Mag. Mat.* **112**, 258 (1992).
11. H. E. Kissinger. *J. Res. Nat. Bur. Stand.* **57**, 217 (1956).
12. H. E. Kissinger. *Ana. Chem.* **29**, 1702 (1957).
13. B. D. Cullity, *Elements of X-ray Diffraction*, 2nd ed. (Addison-Wesley Publishing Company, Inc., Reading, MA, 1978), p. 102.
14. L. K. Varga. *Mat. Sci. and Eng.* **A179/A180**, 567 (1994).
15. C. F. Conde and A. Conde. *Mat. Lett.* **21**, 409 (1994).
16. K. Hono, A. Inoue, and T. Sakurai. *Appl. Phys. Lett.* **58** (1991) 2180
17. K. Hono, J. –L. Li, Y. Ueki, A. Inoue, and T. Sakurai. *Appl. Surf. Sci.* **67** (1993) 398
18. S. H. Kim, M Matsuura, N. Sakurai, and K. Suzuki. *Jpn. J. Appl. Phys.* **32** (1993) Supp. 32-2, 676
19. J. D. Ayers, V. G. Harris, J. C. Sprague, W. T. Elam, and H. N. Jones, *Acta Mater.* **46**, 1861 (1998).

AIR FORCE APPLICATION OF ADVANCED MAGNETIC MATERIALS

R.T. Fingers, C. S. Rubertus
Propulsion Directorate, Air Force Research Laboratory, Wright-Patterson Air Force Base, OH 45433-7251

ABSTRACT

A national initiative is underway to develop and test a more electric aircraft (MEA) and is being led by the U.S. Air Force Research Laboratory at Wright-Patterson Air Force Base, Ohio. The MEA concept is based on utilizing electric power to drive aircraft subsystems which are currently driven by a combination of hydraulic, pneumatic, electric and mechanical power transfer systems. A major objective of this effort is to increase military aircraft reliability, maintainability and supportability and to drastically reduce the need for ground support equipment. These improvements will be realized through the further advancement of key MEA technologies, including magnetic bearings, aircraft integrated power units (IPU), and starter/generators (IS/G) internal to an aircraft main propulsion engine. These advanced developments, as well as weapon and space power applications, are the driving force for the new emphasis on high temperature and high strength magnetic materials for power applications. In determining the best magnetic material for an application it is typically necessary to conduct an engineering trade-off analysis which takes into consideration mechanical behavior, electrical loss, and magnetic properties under the conditions of actual usage. New materials solutions are required to meet these challenges, as designers often find the magnetic material performance to be the technological limitation.

INTRODUCTION

Magnetic materials research is truly pervasive as it is crucial for the success of aircraft, space, and weapon power systems. The use of non-contact magnetic bearings for space applications is envisioned as a requirement for sustained performance in spacecraft power systems. Integrated power generation and thermal management systems, which require advancements to be made in the area of magnetics, are being considered for space applications as well. Also, a national initiative is underway to develop and test a more electric aircraft and is being led by the U.S. Air Force Research Laboratory at Wright-Patterson Air Force Base, Ohio.

Non-propulsive aircraft power is required for a number of on-aircraft functions. Some form of onboard or ground power is necessary to start the main propulsion engine. Once this engine has been spun through its light-off and is operating under its own motive force, the main engine gearbox and the airframe mounted gearbox are utilized to engage the power transfer mechanisms, which, in turn, provide the aircraft auxiliary power. Prior to the main engine starting, auxiliary power is required for the operation of many of the secondary subsystems, such as those which provide ground checks of specific equipment, avionics, and flight control actuation. Also, in-flight emergency power is required in the event of a generator or main engine failure. The emergency power unit is typically designed to provide the necessary power for a specific period of time allowing the aircraft to continue flying until the situation is corrected or until it can return to base safely.

Conventional aircraft secondary power systems, addressing the auxiliary, starting, and emergency power requirements are comprised of complex, high maintenance, hybrid subsystems. Mechanical, electrical, hydraulic, and pneumatic power systems are used to transfer power from the point of generation to the final utilization. The combined logistics to support

Mat. Res. Soc. Symp. Proc. Vol. 577 © 1999 Materials Research Society

these subsystems is substantial. Currently, sixteen C-141 cargo aircraft loads of supplies, equipment, and personnel are required for the deployment of a squadron of twenty-four F-16 fighter aircraft [1]. Based on rapidly evolving power electronics, fault tolerant electrical power distribution systems, and electric driven flight control actuator systems, increasing the use of electric power is seen as the direction of technological opportunity enabling a significant reduction in the required logistics support. Predicted aircraft improvements will be realized with the continued development of key power technologies, which include internal starter/generators, integrated power units, magnetic bearings, fault tolerant distribution networks, and electric actuators.

APPLICATIONS

As with many advanced developments, there are a variety of technical challenges that must be addressed to enable the successful transition of technology from the laboratory to the field. A particular design challenge involves the structural integrity of the IPU and the IS/G rotors, as well as the rotating components of magnetic thrust and radial bearings. This challenge is especially formidable at elevated operating temperatures. These units require a high power density in order to provide electrical power of sufficient quantity for all of the aircraft secondary power functions. Due to the mechanical stress, switched reluctance machines with robust rotors are preferred. As the rotor spins on the main shaft of the gas turbine engine, the rotor poles come into and out of alignment with the stator poles. As the pole pairs sequentially align and unalign, the magnetic flux changes direction through both the rotor and the stator. This reversal in direction causes eddy currents to be generated in the flux paths. As the rotor speed, and therefore frequency, is increased, the magnitude of the eddy currents also increases. To address these higher frequency losses, laminated rotor and stator structures are often used. By electrically insulating each laminate from the laminates above and below, the dimension over which the eddy currents flow will limit the corresponding losses. Eddy current losses increase as the square of frequency and the square of the laminate thickness, and linearly with the electrical conductivity. One particular IPU design incorporates a rotor stack comprised of 800 laminates. Each laminate is 0.015 cm thick and needs to be electrically insulated. A typical rotor and stator laminate and an actual IPU rotor stack are shown in Figure 1.

The number of rotor and stator poles can vary as a function of the size and desired application of the machine. The IS/G rotor and stator rings are significantly larger in diameter than those associated with an IPU design in order to fit around the shaft, which runs through the center of the main propulsion engine. The current carrying windings, for both electromechanical applications, are located on the stators rather than the rotors, thereby allowing a higher rotational speed and operating temperature for the rotor. Nevertheless, the resulting angular velocities of the rotor induce a substantial state of stress throughout the rotor at all operating temperatures. Also, the strength of assembled laminate stacks is critical since both the ISG and IPU locate the laminate stack between bearing supports. Therefore, a high strength magnetic material capable of good magnetic performance as well as sustained integrity in a high temperature environment is needed.

Integrated Power Unit (IPU)

The conceptual drawing illustrated in Figure 2 represents the basic configuration of the IPU. It is apparent that the IPU is a complete turbomachine with an air-cooled switched reluctance starter/generator mounted on the central shaft of the gas turbine engine [2]. The direct drive nature of the unit eliminates the need for speed control through a gearbox, thereby simplifying power transformation by providing only electrical power. This particular design places the switched reluctance starter generator aft of the compressor, which adds work to the airflow as it is compressed. Hence, the compressor discharge air, which is used for cooling the rotor, is hot. Some estimates of the available cooling air are as high as 300°C. True operating temperatures for the rotor materials would also be affected by heat generated from the spinning poled rotor due to the magnetic core losses as well as aerodynamic pumping losses.

Analytical investigations, as well as finite element analyses have been conducted to predict the resulting stress profiles and corresponding deformations of rotor laminates. The operating speed of the IPU is about 50,000 to 60,000 rpm. This high speed combined with the poled rotor geometry results in a tangentially oriented maximum stress, located at the edge of the central hole, of about 600 MPa. Also, due to the location of the switched reluctance motor in the IPU configuration, the mechanical bending resistance of the laminate stack is critical to acceptable rotordynamics. Effective stack stiffness is only about 10% - 20% that of a solid rotor, placing the first bending mode for this high-speed machine close to the desired operating speed. The operating temperature of the rotor is currently being limited to about 300°C, however, the MEA Generation II goals state an operating temperature between 500°C and 600°C by 2005. The relatively high stresses and temperatures create an operating environment in which creep deformation is likely to occur. The small air gap length between the rotor and stator poles requires a high priority to be placed on the mechanical behavior of the soft magnetic rotor material as well as the entire rotor stack.

Internal Starter/Generator (IS/G)

A cut away view of an F-110 main propulsion engine, as illustrated in Figure 3, details the incorporation of the rotor and stator components of the IS/G into the engine. The development of the IS/G represents one of the highest power-rated motor/generator systems in switched reluctance technology [3]-[5]. The incorporation of these advanced power technologies will enable the elimination of the troublesome conventional auxiliary components, such as: oil lubrication/ball bearing systems, instrumentation, accessory gearbox driven fuel and lubrication pumps, and hydromechanical fuel controls.

Designs for the GE F110-220 turbofan and other advanced engines have illustrated limitations in current magnetic material technology. Sheet material for high strength laminates is currently limited to approximately 25.4 cm width. Due to the lower rotating speed of large engines, optimized designs for the ISG require 28 cm or larger rotors for 375KW rated machines. If conventional bearings are present oil cooling of the rotor and stator can be employed to keep material temperatures under 200°C. A key benefit of magnetic bearings is elimination of liquid lubricants from the engine, and thus forces air cooling of the rotor. For advanced engines this design configuration could push material temperatures above 400°C with required strengths greater than 825 MPa. The strength limitations of the laminate stack are exacerbated by the long distances between support bearings. The shafts already operate above the first bending mode, but radial tolerances are extremely small and excessive flex results in catastrophic failure of the compressor and turbine in addition to the generator.

Magnetic Bearings

Magnetic bearings are intended to be an integral part of both, the IPU and IS/G applications. As the main shaft and mounted rotor are supported by magnetic forces, friction and the corresponding wear of the rotating components are essentially eliminated. In order to fully realize the overall aircraft system benefits of a more electric engine and power generation system, conventional bearings must be replaced with lubrication free and oil free components.

The functional requirements of magnetic radial and thrust bearings could be addressed using designs incorporating either soft or permanent magnets. The designs utilizing permanent magnets can be made smaller, somewhat lighter, and will utilize less power to operate than soft only designs. However, permanent magnet material is currently limited to a much lower operating temperature than is soft magnetic material. The temperature limitations of permanent magnet materials currently prevent high temperature applications from obtaining the design advantages that could otherwise be obtained. The critical applications for magnetic bearings for large turbofan engines are the thrust bearing, usually mounted in the front half of the engine with cooler ambient temperatures, and the rear radial bearing. The rotor support system in a large turbofan engine utilizing magnetic bearings is also expected to require nonmagnetic auxiliary, or touch down, bearings for the infrequent high loads such as during hard landings or high g-force maneuvering. The thrust bearing requires the highest mechanical stress because the diameter of the thrust disk is several centimeters larger than other rotating bearing components. The rear radial bearing requires the highest operating temperature, exceeding 550°C.

MAGNETIC MATERIAL REQUIREMENTS

An attempt has been made to avoid qualifying statements such as, "better…stronger…higher," when describing the needed magnetic material improvements. However, it is sometimes difficult to specify autonomous performance characteristics. Engineering trade-off analyses must be conducted to examine the complete system, or subsystem, and compromises are typically made between core loss, strength, thermal stability, and creep resistance to obtain an optimized system. With an overall improvement in mechanical, electrical, and magnetic behaviors in mind, specific quantities are presented as important characteristics to target.

The bottom-line needs for soft magnetic materials are strongly coupled to specific high temperature applications. The materials must maintain their structural and magnetic integrity for at least 5000 hours at nearly 600°C. Machine core loss is obviously a function of the operating parameters as well as the geometry and material properties. A core loss goal of 480 W/Kg at a frequency of 5KHz, an induction of 2 T, and an operating temperature of 500°C has been established. As resistivity is inversely proportional to the eddy current loss component of the total core loss, an increase from 40 to 60 micro-ohm-cm at room temperature has been stated to contribute to the overall machine loss reduction. A yield strength need of 700Mpa at 500°C has been established as well as a yield strength of 825 MPa at about 300°C. The limited capacity for radial growth of the rotor, due to the small air gap distances between machine rotor and stator poles, allows for minimal tolerance for time dependent creep deformation. Although creep characterization and modeling of various rotor configurations is continuing, the initial limit for steady-state creep deformation has been established as 2×10^{-6} hr^{-1} for a life of 5000 hours at 600Mpa and 550°C.

Permanent magnet needs are also focused on high temperature Air Force applications. Magnetic bearing systems could benefit from the lower weight and reduced feedback controls associated with permanent magnet designs if the high temperature capabilities were increased. The bottom line need for hard magnets is an energy product of 30 MGOe, at an operating temperature of 450°C, for a life of 5000 hours, without degradation to mechanical behavior.

Reproducible soft and permanent magnets with these specified performance characteristics would meet current Air Force magnetic material needs, while increasing

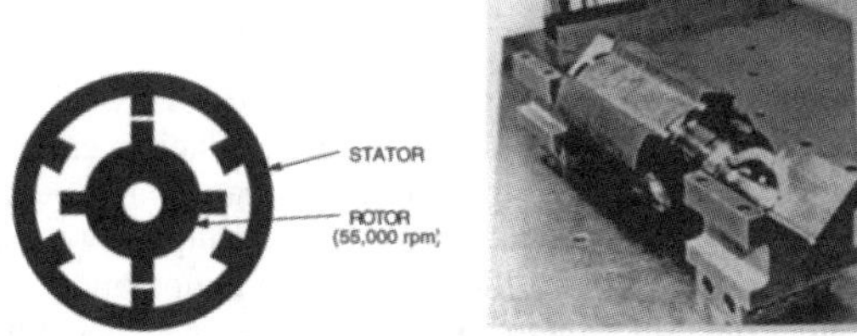

Figure 1 Cross Sectional View of Rotor and Stator Laminates and Actual IPU Laminated Rotor.

Figure 2 Conceptual Design of Integrated Power Unit (IPU).

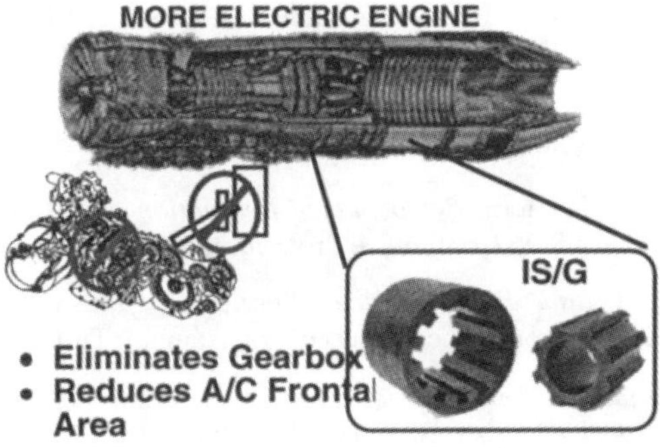

Figure 3 Internal Starter/Generator (IS/G) With an Expanded View of the Rotor and Stator Laminate Stacks.

significantly the number of potential applications throughout the Air Force as well as commercial markets.

SUMMARY

In the past emphasis has been on advancing the state of the art of magnetics by improving the magnetic characteristics. Today, the Air Force is placing emphasis on maintaining the magnetic properties while increasing maximum operation temperatures. With emerging microstructure control processes and nanoengineering techniques, it may be feasible to design an application dependent magnet. The best magnet at room temperature may not necessarily be the best magnet at the desired operating temperature. New magnets being developed for today's Air Force application must be characterized at elevated temperatures.

Researchers from the Propulsion and Materials Directorates have combined resources and expertise in recent years to establish a baseline for soft magnetic iron-cobalt alloys [6]-[7]. It is this combined effort that has enabled the focusing of related DOD, academia, and industrial efforts in magnetics. Air Force needs have been key in defining and promoting a Multi-Disciplinary University Research Investigation that is managed by the Air Force Office of Scientific Research (AFOSR). Multiple research organizations have focused their efforts toward the high temperature needs of the Air Force are being addressed. Also, the Advanced Magnets for Power Systems (AMPS) program has been established by the Defense Advanced Research Program Agency (DARPA). The AMPS program goals are based on the needs of the military, with the Air Force requirements being the most challenging.

Continued magnetics research is needed. As the Air Force continues to push the envelope by developing an advanced air and space force, power system requirements continue to rise. The trend toward higher speed and higher temperature electrical power generation, distribution, and utilization systems is obvious. New magnetic material systems to enable these advanced applications are crucial for continued dominance of air and space.

REFERENCES

1. R.E. Quigley, *More Electric Aircraft*, Conference Record, IEEE Applied Power Electronics Conference, March 7-11, 1993, p. 906-911.

2. P.G. Colegrove, *Integrated Power Unit for a More Electric Airplane*, AIAA Paper No. 93-1188, Aerospace Design Conference, February 16-19, 1993.

3. E. Richter, and C.A. Ferriera, *Performance Evaluation of a 250 kW Switched Reluctance Starter Generator*, IEEE-IAS 95 Conference Proceedings, Orlando FL, 1995.

4. E. Richter, *Design and Performance Simulation for a Gas Turbine Integrated Starter/Generator*, ICEM98 Conference Proceedings, Istanbul, Turkey, 3 August, 1998.

5. E. Richter, *Design and Performance Prediction for an Internal Starter/Generator for Application in the F110-129 Engine*, SAE TOPTEC Workshop Proceedings, Dayton, Ohio, 21-22 August, 1997.

6. R.T. Fingers, J.E. Coate, and N.E. Dowling, *Mechanical Properties of Iron-Cobalt Alloys for Power Applications*, IECEC Proceedings, Paper No. 97364, August, 1997.

7. R.T. Fingers, J.E. Coate, and N.E. Dowling, *Creep Deformation of a Soft Magnetic Iron-Cobalt Alloy*, J. Applied Phys. **85** (8) (1999).

INFLUENCE OF THE ALLOY COMPOSITION ON THE MAGNETIC PROPERTIES OF NANOCRYSTALLINE $Fe_{80}M_7B_{12}Cu_1$ (M: Ti, Ta, Nb, Mo)

M. KOPCEWICZ *, A. GRABIAS * and B. IDZIKOWSKI **
*Institute of Electronic Materials Technology, Wólczyńska 133, PL 01-919 Warsaw, Poland
**Institute of Molecular Physics, Polish Academy of Sciences,
M. Smoluchowskiego 17, PL 60-179 Poznań, Poland

ABSTRACT

Formation of the nanocrystalline bcc Fe phase due to thermal treatment of the amorphous $Fe_{80}M_7B_{12}Cu_1$ (M: Ti, Ta, Nb, Mo) precursors is studied by the Mössbauer and differential scanning calorimetry techniques. The dependence of the formation of the bcc Fe phase on the alloy composition is discussed. In order to determine the optimal soft magnetic properties of these nanocrystalline alloys the rf-Mössbauer technique is used in which rf collapse effect induced by a radio-frequency (rf) magnetic field is employed. It was found that anisotropy fields in the nanocrystalline phase were smaller in Nb- and Mo-containing alloys as compared to the alloys which contain Ti or Ta. Variations of the anisotropy field vs. alloy composition and annealing temperature are discussed in detail.

INTRODUCTION

Novel soft magnetic materials whose properties are governed by reduced dimensions at the nanometer length scale have been developed in recent years. The nanocrystalline alloys exhibit unique structural and magnetic properties which are different from those of the conventional coarse-grained polycrystalline materials. The soft magnetic properties can be obtained by the controlled crystallization of certain amorphous alloys. A new class of ternary iron-based NANOPERM-type alloys is of particular interest both for basic research and for practical application because of their superior magnetic properties (higher saturation magnetization and permeability) [1,2] as compared to more conventional FINEMET alloys [3]. Excellent soft magnetic properties of the nanocrystalline alloys are due to the refinement of the grain size and are well described in terms of the random anisotropy model [4]. Various experimental techniques were applied for studying the structure and magnetism of nanocrystalline alloys. One of the most powerful technique is the Mössbauer spectroscopy. Conventional Mössbauer measurements in transmission geometry allow the identification of phases formed due to annealing of the amorphous precursor and the evaluation of their relative contents. They, however, do not provide information regarding the anisotropy fields or magnetostriction. Therefore, in the study of the structure and magnetic properties of the nanocrystalline alloys we applied an unconventional rf-Mössbauer technique which combines the Mössbauer effect with the phenomenon induced by an external magnetic radio-frequency (rf) field (rf collapse effect) [5,6]. The collapse of the magnetic hyperfine splitting occurs due to fast magnetization reversal induced by the external rf field. If the frequency of the rf field is larger than the Larmor precession frequency and the rf field is sufficiently strong to overcome local magnetic anisotropy, then the magnetic hyperfine field is averaged to zero at the Mössbauer nuclei. The rf-collapsed spectra consist of a single line or a quadrupole doublet in the place of the magnetically split six-line pattern, though the sample remains in the ferromagnetic state. The rf collapse is very sensitive to small changes of local magnetic anisotropy, which allows the distinction of the soft

Mat. Res. Soc. Symp. Proc. Vol. 577

nanocrystalline phase from magnetically harder microcrystalline phases formed by annealing of the amorphous precursor. The rf collapse effect is sensitive enough to detect differences in magnetic anisotropy fields in the bcc Fe nanophase formed in various alloys.

Recently, we applied various Mössbauer techniques (conventional, rf-Mössbauer, conversion electron Mössbauer spectroscopy) for the study of iron-based nanocrystalline alloys [7-12]. Our present work concerns the influence of the alloy constituents, such as Ti, Ta, Nb and Mo, on the magnetic properties of NANOPERM-type alloys. These results are compared with our recent studies of FeZrBCu alloys [8,10]. The structure and magnetism of $Fe_{80}M_7B_{12}Cu_1$ nanocrystalline alloys are investigated by Mössbauer spectroscopy and are supplemented by differential scanning calorimetry (DSC) measurements.

EXPERIMENT

Amorphous $Fe_{80}M_7B_{12}Cu_1$ (M: Ti, Ta, Nb, Mo) alloys were prepared by a melt-spinning technique. The ribbons were 4 mm wide and about 25 μm thick. In order to obtain the nanocrystalline phase the amorphous alloys were annealed in an Ar atmosphere for 1 hour in the temperature range of 390-620°C. The DSC measurements at the heating rate of 20 K/min were performed to determine the first and second crystallization stages. Conventional Mössbauer measurements were performed for the as-quenched and annealed samples prior and after the rf exposure. The unconventional rf-Mössbauer measurements were performed during the exposure of the samples to the rf field of intensity ranging from 0 to 20 Oe at 61.2 MHz, which is about three times higher than the Larmor frequency. The rf field was applied in the plane of the samples which played the role of an absorber. Adequate cooling of the samples preventing excessive heating was applied [8,10-12].

RESULTS

DSC Measurements

The crystallization behavior of $Fe_{80}M_7B_{12}Cu_1$ alloys was studied by DSC. All alloys crystallized through two steps as revealed by two exothermic peaks in DSC curves. The temperatures of the onset of crystallization, T_{x1} and T_{x2}, were determined (Table I). The first stage with the crystallization temperature T_{x1} corresponds to the structural transformation from the amorphous to the nanoscale bcc phase. The second exothermic peak (T_{x2}) reflects the complete transformation of the remaining amorphous phase to a mixture of the coarsed bcc Fe grains and iron borides. The nanocrystalline bcc Fe phase can be formed in the temperature range between T_{x1} and T_{x2}.

Table I. Crystallization temperatures for $Fe_{80}M_7B_{12}Cu_1$ alloys determined from DSC curves.

M	First peak		Second peak	
	Onset temperature T_{x1} [°C]	Peak position [°C]	Onset temperature T_{x2} [°C]	Peak position [°C]
Ta	357	426	568	584
Mo	415	446	658	679
Nb	441	503	727	769
Ti	444	479	719	744

Conventional Mössbauer Measurements

The conventional Mössbauer measurements performed in transmission geometry at room temperature reveal the changes in the microstructure of the amorphous $Fe_{80}M_7B_{12}Cu_1$ alloys induced by annealing. As an example the spectra recorded for the $Fe_{80}Nb_7B_{12}Cu_1$ alloy are shown in Figs. 1a-1f.

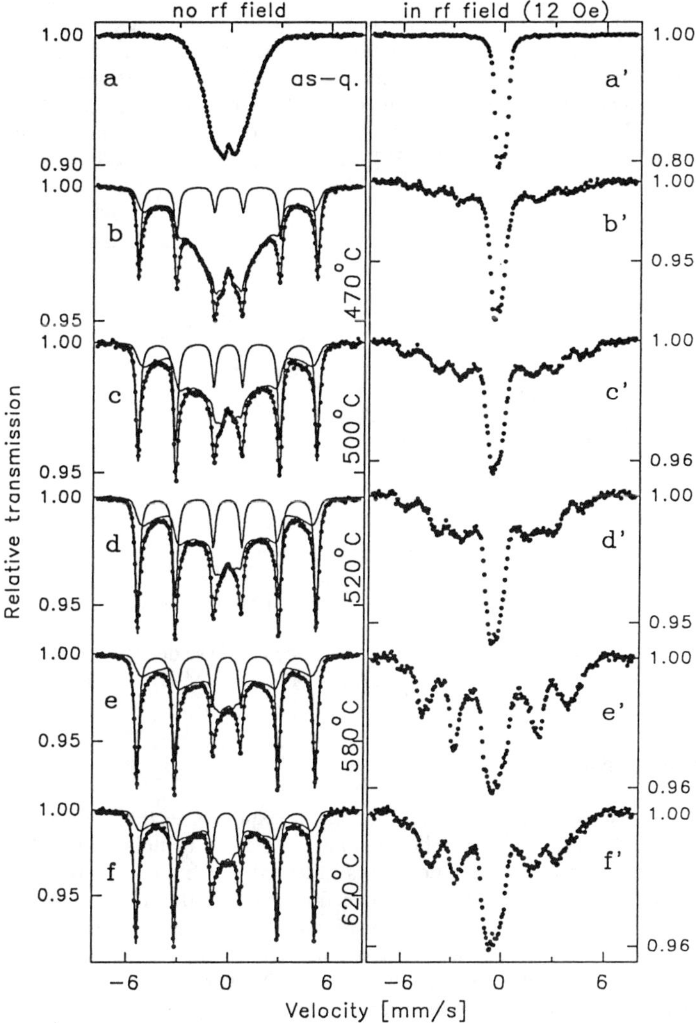

Fig. 1. Mössbauer spectra of the as-quenched and annealed $Fe_{80}Nb_7B_{12}Cu_1$ alloys recorded without (a-f), and with rf field of 12 Oe (a'-f').

The spectrum of the amorphous alloy (Fig.1a) was fitted with the hyperfine field distribution, P(H), method. The spectra of the annealed samples (Figs.1b-1f) consist of a sextet with sharp lines related to the bcc Fe phase, in addition to the broad sextet due to hyperfine field distribution corresponding to the retained amorphous phase. These spectra were fitted with the combination of P(H) and the sextet with a discrete value of the hyperfine field of 32.95T characteristic for the bcc Fe. As can be seen, the annealing at 470°C causes the formation of the bcc Fe phase (Fig.1b). The increase of the annealing temperature leads to the increase of the spectral contribution characteristic for the bcc Fe phase. The relative atomic fraction of Fe as bcc phase determined from the area of the spectral component with the hyperfine field of about 33T in the Mössbauer spectra (as in Figs.1b-1f) vs. annealing temperature is shown in Fig.2 for all alloys studied. For comparison the results obtained earlier for the $Fe_{79}Zr_7B_{12}Cu_2$ alloy [8,9] are included in Fig.2. As can be seen, the increase of annealing temperature, T_A, increases the relative fraction of the bcc Fe phase in all alloys. At higher T_A the relative fraction of bcc Fe begins to saturate at around 50%. The highest relative content of Fe (56%) was obtained for the $Fe_{80}Ti_7B_{12}Cu_1$ alloy annealed at 620°C.

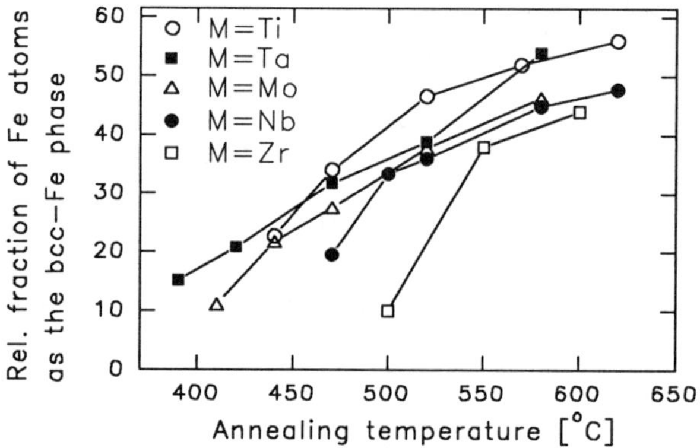

Fig.2. The relative abundance of the bcc Fe phase determined from the conventional Mössbauer spectra of the $Fe_{80}M_7B_{12}Cu_1$ alloys.

<u>The Rf-Mössbauer Results</u>

Typical Mössbauer spectra measured during the exposure to the rf field of 12 Oe at 61.2 MHz for the $Fe_{80}Nb_7B_{12}Cu_1$ alloy are shown in Figs.1a'-1f'. A complete collapse of the magnetic hyperfine structure to a quadrupole doublet is observed in the amorphous state (Fig.1a'). The formation of the nanocrystalline bcc Fe phase dramatically affects the shape of the collapsed spectra. The spectra recorded for the two-phase alloy (bcc Fe grains embedded in the retained amorphous matrix) consists of two components: (i) a fully collapsed central QS doublet originating from the retained amorphous phase, and (ii) a partially collapsed six-line component corresponding to the nanocrystalline bcc Fe phase (Figs.1b'-1d'). At higher annealing temperatures (580-620°C) the partially collapsed part becomes better resolved (Figs.1e',1f'). The partially collapsed six-line spectral component is related to the bcc Fe grains with magnetic anisotropy sufficiently large to limit the magnetization reversal. However, the hyperfine split pattern reveals an average hyperfine field markedly reduced as compared to α-Fe, suggesting that the magnetic anisotropy in nanograins is much smaller than in the bulk α-Fe, for which no rf-induced narrowing was observed. Thus, the partially collapsed component corresponds to the soft magnetic nanoscale bcc Fe grains whose magnetic anisotropy is, however, considerably larger than that of the amorphous phase, for which a complete rf collapse was observed.

The rf-Mössbauer measurements performed as a function of the rf field intensity allowed the determination of the anisotropy fields in each phase present in the nanocrystalline alloy. The rf-Mössbauer spectra recorded for all alloy compositions studied were compared. As an example the spectra of the samples annealed at similar temperatures and recorded during the exposure to the rf field of the same intensity (16 Oe) are shown in Fig.3. The spectra recorded in the absence of the rf field (Figs 3a-3e) reveal similar relative fractions of the bcc Fe phase in all alloys. However, the rf-Mössbauer spectra differ considerably. The most effective collapse of the magnetic hyperfine structure was observed in Figs. 3b' and 3c' for Nb- and Mo-containing alloys, respectively, and in Fig.3a' for FeZrBCu alloy.

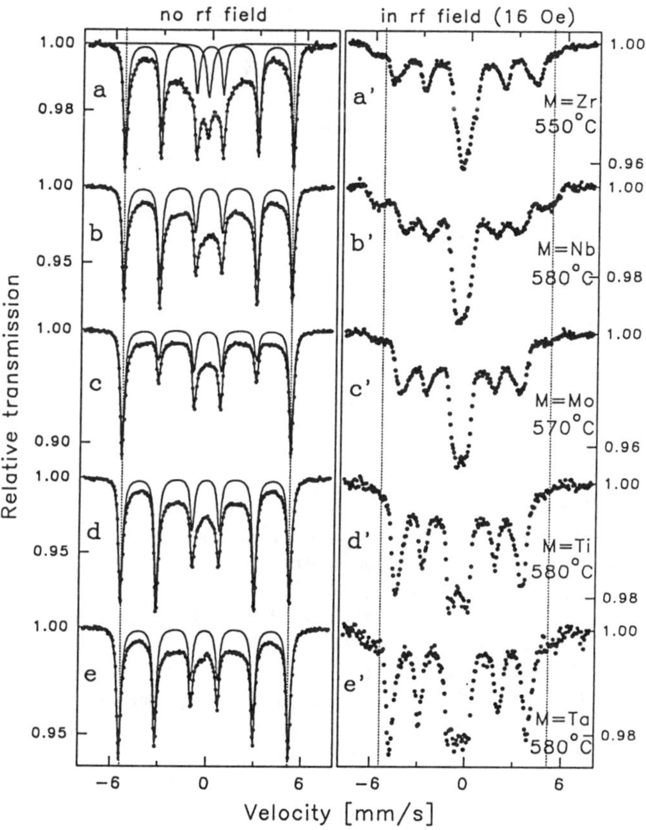

Fig.3. Mössbauer spectra of the annealed $Fe_{80}M_7B_{12}Cu_1$ alloys recorded without (a-e), and with rf field of 16 Oe (a'-e').

The fully collapsed central part dominates in these spectra and the partially collapsed six-line component is narrowed most. This suggests that the alloys containing Nb, Mo or Zr are magnetically softest, both as regards the retained amorphous phase and the nanocrystalline bcc Fe grains. The spectra of the Ti- and Ta-containing alloys (Figs.3d',3e') are much less narrowed (the splitting of the partly collapsed spectral component is larger than in Figs.3a'-3c'). The spectral contribution of the central collapsed part is much smaller than in the case of Nb- and Mo-containing alloys. This suggests that the magnetic anisotropy is considerably larger in Ti- and Ta-containing alloys most probably due to larger bcc Fe grains as compared with Nb-, Mo- and Zr-containing alloys. Similar characteristic features were observed for the samples annealed at lower temperatures.

CONCLUSIONS

Conventional Mössbauer measurements allow the identification of phases formed by annealing of the starting amorphous alloys but do not provide information concerning the magnetic anisotropy. The rf-Mössbauer experiments, in which the rf collapse effect is observed, permit us to distinguish the soft nanocrystalline bcc phase from magnetically harder microcrystalline α-Fe. The rf-Mössbauer results show that the nanocrystalline bcc Fe phase is magnetically soft (reveals much smaller anisotropy field that the bulk α-Fe), however, its anisotropy is markedly larger than that of the parent amorphous phase and of the retained amorphous matrix in the composite alloy. The complete rf collapse could be induced only in the fully amorphous as-quenched alloy and in the retained amorphous phase. The rf-Mössbauer measurements revealed that the magnetic anisotropy field depends strongly on alloy composition. The anisotropy fields in the nanocrystalline phase were smaller in Nb- and Mo-containing alloys as compared to the alloys which contain Ti or Ta.

ACKNOWLEDGEMENTS

The financial support from the Grant No. KBN 2 P03B 06815 from the Polish Committee of Scientific Research is gratefully acknowledged.

REFERENCES

1. K. Suzuki, A. Makino, A. Inoue, T. Masumoto, J. Appl. Phys. **74**, 3316 (1993).
2. A. Makino, T. Hanai, A. Inoue, T. Masumoto, Mater. Sci. Eng. **A226-228**, 594 (1997).
3. Y. Yoshizawa, S. Oguma, K. Yamauchi, J. Appl. Phys. **64**, 6044 (1988).
4. G. Herzer, Phys. Scr. **T 49**, 307 (1993).
5. L. Pfeiffer, in *Mössbauer Effect Methodology*, edited by I.J. Gruvermann, (Plenum, New York, 1972), Vol. 7, p. 263.
6. M. Kopcewicz, Structural Chem. **2**, 313 (1991).
7. M. Kopcewicz, J. Jagielski, T. Graf, M. Fricke, J. Hesse, Hyperfine Inter. **94**, 2223 (1994).
8. M. Kopcewicz, A. Grabias, P. Nowicki, D.L. Williamson, J. Appl. Phys. **79**, 993 (1996).
9. M. Kopcewicz and A. Grabias, J. Appl. Phys. **80**, 3422 (1996).
10. M. Kopcewicz, A. Grabias, D.L. Williamson, J. Appl. Phys. **82**, 1747 (1997).
11. M. Kopcewicz, A. Grabias, I. Škorvánek, J. Appl. Phys. **83**, 935 (1998).
12. M. Miglierini, M. Kopcewicz, B. Idzikowski, Z.E. Horváth, A. Grabias, I. Škorvánek, P. Dłużewski, Cs.S. Daróczi, J. Appl. Phys. **85**, 1014 (1999).

SOFT MAGNETIC PROPERTIES AND STRUCTURE OF NANOCRYSTALLINE ALLOYS BASED ON FINEMET

I.Todd[1], H.A.Davies[1], M.R.J.Gibbs[2], D.Kendall[3] * and R.V.Major[3]
[1] Department of Engineering Materials, University of Sheffield, Sheffield S1 3JD, UK
[2]Department of Physics and Astronomy, University of Sheffield, Sheffield S3 7RH, UK
[3]Telcon Ltd, Crawley RH10 2PY,UK
*Present address: Carpenter Technology (UK) Ltd, Crawley RH10 2PY, UK

ABSTRACT

The effects of partial substitution of Al for Fe in $Fe_{73.5-X}Al_XSi_{13.5}B_9Nb_3Cu_1$ nano composite alloys both on the a.c. magnetic properties such as permeability and core loss, saturation magnetostriction and core loss are presented. The observed improvements in material properties are discussed in relation to the nanocrystallite and residual glass component chemistry.

INTRODUCTION

The Finemet-type of nanocomposite, ultra-soft magnetic alloy, originally developed at Hitachi Metals in Japan [1] and having a composition of the type $Fe_{73.5}Si_{13.5}B_9Nb_3Cu_1$ (at%), is now well established commercially. It is prepared by melt-spinning to amorphous structured ribbon which is then partially devitrified. In the optimum magnetic state, the alloy exhibits a 3 phase nanocomposite structure consisting of ~70 vol.% of randomly oriented nanocrystallites, of composition Fe~20at% Si and average size 10-20nm, embedded in a matrix of ~30 vol.% uncrystallised glassy phase. A very small volume fraction (~1%) of nanometre scale Cu particles (~5nm) is also present. These act as nucleation sites for the Fe-Si crystallites during devitrification [2].

The excellent soft magnetic properties of the alloy have been explained by Herzer [3] on the basis of the random anisotropy model (RAM) first proposed for amorphous soft magnetic alloys by Alben *et al.* [4]. This model relates the exchange correlation length (or domain wall thickness) L , to the fluctuation length of the local easy axis orientation which, in the case of the nanocomposite alloys, approximates to the average grain diameter, d_g. When $L >> d_g$ the macroscopic magnetocrystalline anisotropy averages out to give an effective anisotropy $<K>$ which is approximately 3 orders of magnitude smaller. In addition, it is considered that the negative saturation magnetostriction (λ_s) of the high Si nanocrystalline phase is closely counterbalanced in volume terms by the larger positive λ_s for the glassy matrix so that the net λ_s is very small and positive and in the range +2 to +4 $\times 10^6$.

These two factors contribute to the low coercivity (<1A/m) and high permeability (~10^5 at 50Hz) reported for this alloy composition. The high electrical resistivity (~ 120 $\mu\Omega$.cm) and the small thickness dimension (~20μm) of the melt spun ribbon also contribute favourably to its low a.c. power loss characteristics.

Several attempts have been made to improve the properties of the $Fe_{73.5}Si_{13.5}B_9Nb_3Cu_1$ alloy composition by investigating the effects of partial substitution of additional alloying elements for Fe (e.g. Co [5,6] , Al [7,8]) as well as full or partial substitution of V, Cr , Mo , Ta etc. for Nb [e.g. 9,10]. In the case of partial substitution of Fe by Al in $Fe_{73.5-X}Al_XSi_{13.5}B_9Nb_3Cu_1$ Lim *et al.* [7] found that, in the range $0 \leq X \leq 1$, Al improved the d.c coercivity but caused a deterioration in both μ_i and λ_s . In a previous contribution we extended this study to cover the range $0 \leq X \leq 10$ [8] and reported that H_c showed a significant decrease upon the substitution of 2% Al for Fe. It was proposed that this reduction in H_c was the result of a decrease in the magneto-crystalline anisotropy caused by incorporation of Al into the nanocrystallites. In this paper we report further on the magnetostriction behaviour and, additionally, on the a.c. properties of $Fe_{73.5-X}Al_XSi_{13.5}B_9Nb_3Cu_1$ alloys over the range $0 \leq X \leq 6$.

EXPERIMENT

Four alloys in the series $Fe_{73.5-X}Al_XSi_{13.5}B_9Nb_3Cu_1$ with X= 0,2,4 and 6 were prepared from high purity elements by r.f. melting in an evacuable melting unit under flowing argon. Ribbons of cross-section typically 20 μm × 2 mm and roll contact surface roughness less than 1.5μm were produced by free jet melt spinning (FJMS) in an enclosed atmosphere unit in helium at 1/3 atmospheric pressure, whilst wider ribons, of cross section typically 18-20 μm × 8-10 mm and ~70-100 m continuous length and with roll contact roughness of 1.5-2 μm, were produced by planar flow casting (PFC) under 1 atm. He in a second enclosed atmosphere unit.

The narrow ribbons were cut into 100 mm lengths and thermal treatments were performed in a static Ar atmosphere in a conventional furnace for 1 hour after which the samples were allowed to air cool. The wide tapes were wound into toroidal cores and heat treated in a flowing hydrogen atmosphere for 1 hour before being pushed through to the cold zone of the furnace and allowed to cool under protective atmosphere.

The structure of the alloys, both in the as-cast state and after subsequent heat treatment, was confirmed by x-ray diffraction using Co $K\alpha$ radiation, and the mean crystallite size (d_g) for the Fe based nanocrystalline phase estimated from the half peak breadth of the (110) x-ray diffraction peak via the Scherrer equation. Transmission electron microscopy was also employed for selective studies of the nanostructure and grain size.

The saturation polarisation J_s, d.c. H_c and λ_s were measured using a vibrating sample magnetometer (VSM), hysteresis loop tracer [11] and the small angle magnetisation rotation (SAMR) method [12], respectively.

A.c. permeability measurements were made on toroidal cores at frequencies up to 100 kHz using a sinusoidal current power source, connected to a primary winding, to provide the magnetising field. The voltage across an open circuit primary winding was used to evaluate the value of flux density.

RESULTS AND DISCUSSION

Fig. 1 shows that d_g varies little from the nominal value of 10±1 nm either as a function of Al content or T_{ann} for those alloys and annealing conditions under investigation (1 h anneals).This was also confirmed by direct observation of the nanostructure via TEM [8].

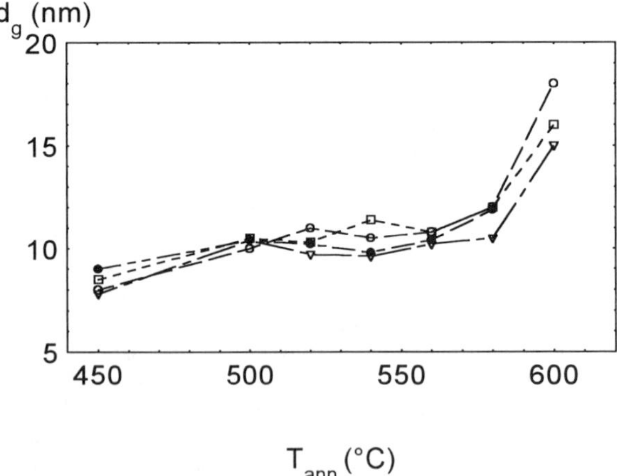

Fig.1: Variation in crystallite size d_g as a function of T_{ann} (1h anneals) and Al content X. ○ :0% Al, ∇:2% Al, □: 4% Al , ● :6% Al

Fig.2 shows the response of the saturation magnetostriction λ_s to heat treatment temperature T_{ann} for all alloy compositions studied. With the exception of the X=2 alloy, there is little response of λ_s to the presence of Al. In the case of the 2% Al alloy, however, a material with $\lambda_s \approx 0$ is produced after annealing at 550°C or above which is significantly smaller than the minimum value of 3ppm measured for the Al-free composition. An advantage of the Al containing alloys is that ultra -low coercivities are developed over a wider range of T_{ann} than is the case for the base alloy [13] so that optimisation of the heat treatment of the 2% Al alloy with respect to minimising magnetostriction does not have an unacceptably large effect on coercivity. The value of H_c, for this anneal condition , 0.5A/m is similar to H_c^{min} for Finemet.

The data for $\mu_{0.4}$ at 50Hz as a function of T_{ann} for all X are presented in Fig.3. ($\mu_{0.4}$ can be taken as μ_i as H_c at 50 Hz will be considerably larger than 0.4 Am^{-1}).The permeability of the X=2 at.% Al alloy shows an increase at all T_{ann} whilst alloys with higher X show no improvement (X=4) or a deterioration (X=6) compared with the $\mu_{0.4}$ of the base composition.

The values of $\mu_{0.4}$ obtained experimentally, combined with previously reported saturation polarisation measurements [8], allow the variation in effective anisotropy, <K>, to be estimated, both as a function of alloy composition and heat treatment condition, on the basis of the random anisotropy model (RAM) [3]. According to this model, for L>>d_g, <K> may be approximated by:

$$\langle K \rangle \approx \frac{K_1^4 d_g^6}{A^3} \qquad (1)$$

and μ_i and <K> are related thus:

$$\mu_i = p_\mu \times \frac{J_s^2}{\mu_0 \langle K \rangle} \qquad (2)$$

where p_μ is a constant and A is the exchange stiffness within the crystal with all other terms as previously defined.

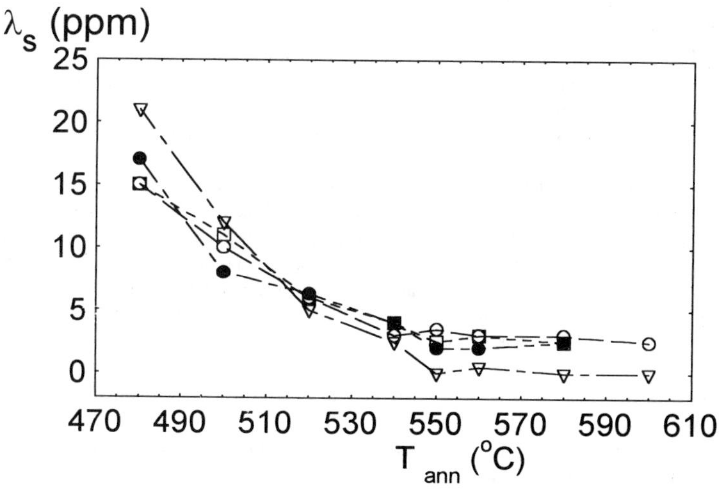

Fig.2: Variation in saturation magnetostriction λ_s as a function of T_{ann} (1h anneals) and Al content X. ❍ :0% Al, ∇:2% Al, □: 4% Al , ● :6% Al

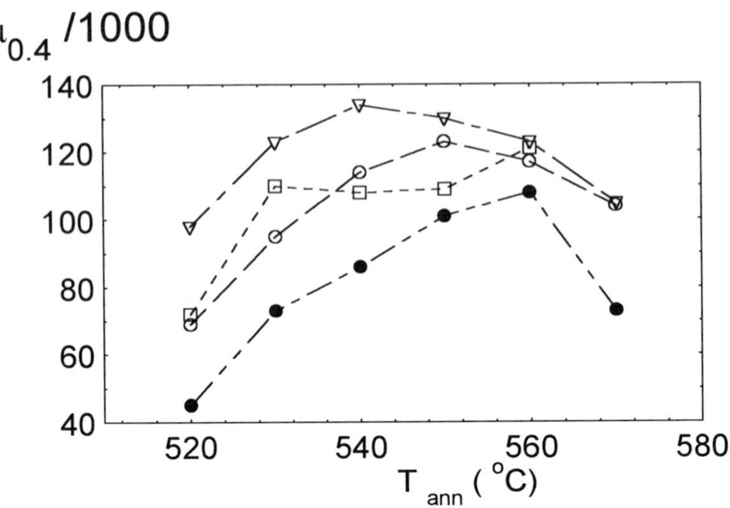

Fig.3: Variation in $\mu_{0.4}$ as a function of T_{ann} (1h anneals) and Al content X.
○ :0% Al, ∇:2% Al, □: 4% Al , ● :6% Al

Combining equations 1 and 2 the value of the constant p_μ may be determined as 0.26, assuming that the composition of the Fe rich nanocrystalline phase in the 0% Al alloy after heat treatment at 540°C is Fe~20% Si and that ,therefore, $A = 1\times 10^{-11}$ Jm^{-1} and $K_1 = 8 \times 10^3$ Jm^{-3} as employed previously and taking J_s=1.5T [8]. This value of p_μ is close to the 0.33 predicted by the RAM. Assuming this p_μ to be applicable over the whole range of annealing conditions and compositions and using the experimentally determined $\mu_{0.4}$ and the values of J_s determined by us previously [8] we can estimate <K> using Eq. 2. Values of <K> thus calculated for all X and T_{ann} are shown in Fig.4.

As there is no measurable change in d_g with either T_{ann} or X it is reasonable to assume that the decrease in <K> results from the incorporation of Al into the nanocrystallites as we previously postulated [8,10]. Alloy compositions in the Fe-rich corner of the Fe-Si-Al phase diagram [14] are known to exhibit lower magnetocrystalline anisotropies than for binary Fe-Si, with the Sendust composition (Fe 84.9 Si 9.62 Al 5.48 wt.%), in particular, exhibiting a value very close to zero. The presence of Al in the crystallites, therefore, would be expected to reduce K_1,and hence <K>. Recent studies of these Al containing alloys using a three dimensional atom probe (3-DAP) technique [15] have shown that the Al is present almost exclusively within the nanocrystallites, thereby lending support to this hypothesis.

The effect of Al at levels above 2 at.% on saturation polarisation is,however, detrimental to permeability, with J_s decreasing by ~0.06T / at% Al from 1.5T for X=0 to 1.14T for X=6 [8]. Equation 2 shows that whilst permeability is inversely proportional to <K>, it is directly proportional to J_s^2. Thus, whilst an initial, significant, decrease in the value of <K> more than compensates for the effect of decreasing J_s , leading to an overall increase in μ of 12,000, further reductions in J_s as X is increased lead to an overall reduction in initial permeability.

The optimum frequency response curves for $\mu_{0.4}$ as a function of X (Fig. 5)show that, in general, the permeability of the alloys containing Al is less stable with respect to increasing frequency than in the Al-free composition. In the case of the X=6 alloy , however, it is worthy of note that $\mu_{0.4}$, although initially smaller in magnitude than the other compositions, is less sensitive to operating frequency making it a suitable candidate material for components where a wide operating frequency range is desirable.

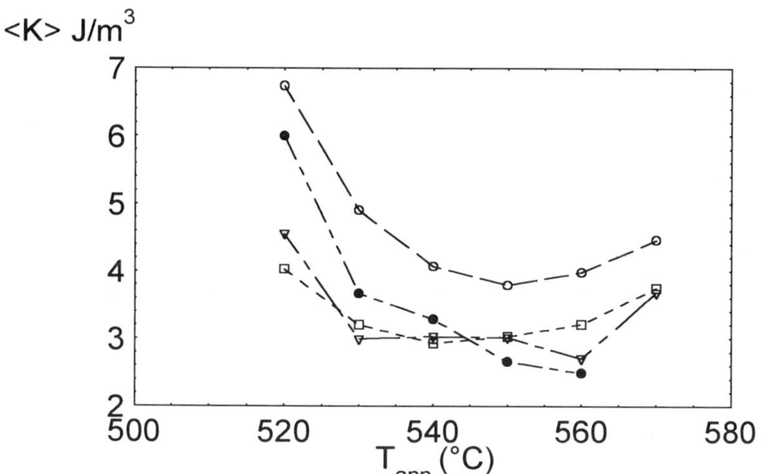

Fig.4: Effect of T_{ann} (1h anneals) and Al content X on effective anisotropy <K>.
O :0% Al, ∇:2% Al, □: 4% Al , ● :6% Al

The core loss characteristics of the Al-containing compositions show no improvement over the Finemet composition. It is likely that this is a result of a complex combination of factors , not least of which is resistivity which is known to be smaller in low Al, Fe-20%Si -Al crystals than in Fe-20% Si [16].

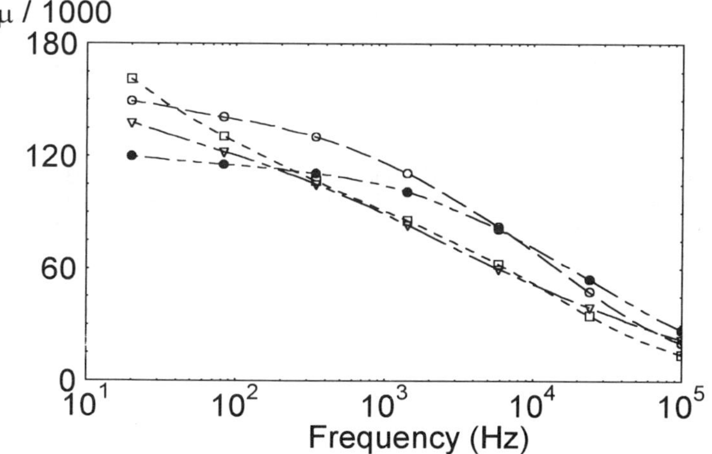

Fig.5: Frequency response of $\mu_{0.4}$ as a function of Al content X.
O :0% Al, ∇:2% Al, □: 4% Al , ● :6% Al

CONCLUSIONS

1) Partial substitution of Fe by Al in the Finemet composition leads to a reduction in the coecivity developed upon partial devitrification and also to an increase in the range of annealing temperatures over which a low coercivity may be developed.

2) Substitution of 2% Al for Fe produces an alloy which after after annealing at specific temepratures exhibits a lower coercivity and magnetostriction and a higher $\mu_{0.4}$ than that of the base composition.

3) Further substitution of Al for Fe leads to a deterioration of $\mu_{0.4}$ as a result of decreasing J_s.

ACKNOWLEDGEMENT

This research was supported by the Engineering and Physical Sciences Research Council through the Advanced Magnetics Programme.

REFERENCES

1. Y.Yoshizawa, S.Oguma and K.Yamauchi, J. Appl. Phys. **64**, 6044 (1988)
2. K.Hono, D.H.Ping, M.Ohnuma and H.Onodera, Acta Mater **47**, 997 (1999)
3. G.Herzer, IEEE Trans. Magn. **MAG-26**, 1397 (1990)
4. R.Alben, J.J.Becker and M.C.Chi, J. Appl. Phys. **49**, 1653 (1978)
5. S.C.Yu, K.S.Kim, Y.S.Cho and T.K.Kim, IEEE Trans. Magn. **28**, 2421 (1992)
6. G.Bordin, G.Buttino, A.Cecchetti and M.Poppi, J. Phys.D: Appl.Phys. **30** 2163 (1997)
7. S.H.Lim, W.K.Pi, T.H.Noh, H.J.Kim and I.K.Kang, J. Appl. Phys. **73**, 6591 (1993)
8. B.J.Tate, B.S.Parmar, I.Todd, H.A.Davies, M.R.J.Gibbs and R.V.Major, J. Appl. Phys. **83**, 6335 (1998)
9. Y.Yoshizawa and K.Yamauchi, Mater. Sci. Eng. **A133**, 176 (1991)
10 M.Frost, I.Todd, H.A.Davies and M.R.J.Gibbs, J. Magn. Magn. Mater., in the press
11. P.T.Squire, S.M.Sheard, C.H.Carter and M.R.J.Gibbs, J.Phys.E: Sci. Instrum. **21**, 1167 (1988)
12. K.Narita, J.Yamasaki and H.Fukunaga, IEEE Trans.Magn. **Mag-16**, 435 (1980)
13. I.Todd, B.J.Tate, H.A.Davies, M.R.J.Gibbs, D.Kendall and R.V.Major, To be submitted to J. Phys.D.
14. M.Takahashi, S.Nishimaki and T.Wakiyama, J. Magn. Magn. Mater. **66**, 55 (1987)
15. P.J.Warren, I.Todd, H.A.Davies, M.R.J.Gibbs, D.Kendall and R.V.Major, to be published
16. R.M.Bozorth., Ferromagnetism, New York: Van Nostrand (1959)

PROPERTIES OF GIANT MAGNETO-IMPEDANCE MATERIAL FOR A NOVEL INTEGRATING MAGNETIC SENSOR

J.G. GORE*, G.J. TOMKA*, J. MILNE*, M.G. MAYLIN*, P.T. SQUIRE**, D. ATKINSON**

*MSS, DERA FARNBOROUGH, FARNBOROUGH, GU14 0LX, UNITED KINGDOM
**DEPT. OF PHYSICS, UNIVERSITY OF BATH, BATH,BA2 7AY, UNITED KINGDOM

ABSTRACT

A novel magnetic field sensor, based on the giant magneto-impedance effect is described and analysed. The sensor utilises an amorphous wire, annealed under optimised conditions. The conditions required for the optimisation of the magnetic properties of the wire are shown. The wire can be up to 2m long, and integrates the field over the length of the wire. It operates in open-loop mode. With a bias field of ~500 A/m it achieves a sensitivity of 0.02 A/m and linearity of better than 5% for external fields between ±300 A/m.

INTRODUCTION

The giant magneto-impedance (GMI) effect, observed in amorphous wires, ribbons and fibres is a change in the complex impedance at high frequencies on the application of an axial magnetic field. The change in the impedance is a consequence of the dependence of the skin effect on the permeability [1]. An appropriate domain geometry in low anisotropy amorphous wires can make them extremely sensitive to axial magnetic fields. In the as-quenched state (AQ), residual stress in the wires can couple with the magnetostriction, hindering domain rotation, and consequently reducing the GMI effect.

An ideal domain geometry consists of circumferential easy-axis 'bamboo' domains magnetised alternately in clockwise and anticlockwise directions (only the outer shells are of interest as domain activity is limited by the skin effect at higher frequencies). An ac current in the wire produces an effective transverse field. The change in the ac transverse component of magnetisation (i.e. the transverse permeability, and consequently the impedance) rapidly increases as the moments rotate towards the axial direction on the application of an axial field. Simple models of this behaviour [2,3] are in good quantitative agreement with observed experimental behaviour. The form of the impedance response with field lends itself to closed and open loop configurations with high sensitivity and high dynamic range applications respectively.

The GMI effect has underpinned much recent work in the development of advanced magnetic sensors with high sensitivity and spatial resolution [4]. For example, it has been shown that a 1mm x 30μm FeCoSiB amorphous wire incorporated into a closed loop configuration can achieve a sensitivity better than 1nT and a bandwidth >100kHz [5]. For applications such as surveillance and control of the magnetic signature of ships, however, a more modest sensitivity integrated over an extended path is required. A novel magnetometric approach is to employ a single wire sensing element spanning the required spatial domain [6]. The mode of operation is inherently integrating over the length of the wire, provided the wire can be produced with uniform magnetic properties along its length. One way of releasing internal stresses which

Mat. Res. Soc. Symp. Proc. Vol. 577 © 1999 Materials Research Society

develop in the wires during their melt-spinning is to anneal them under carefully controlled conditions. To this end we have studied the properties of a Co-rich amorphous wire $(Co_{0.94}Fe_{0.06})_{72.5}Si_{12.5}B_{15}$ to find the optimum annealing environment. This paper shows changes in the properties of the material after being annealed under different conditions of furnace annealing, d.c. current annealing and pulsed current annealing. The performance of the optimally prepared wire incorporated into an integrating sensor is presented.

EXPERIMENTAL

The properties of a Co-rich Unitika Ltd AC20 amorphous wire $(Co_{0.94}Fe_{0.06})_{72.5}Si_{12.5}B_{15}$ were studied in an attempt to find the optimum annealing environment for the operation of a ~500A/m field biased integrating magnetic sensor. The amorphous wire, AC20, was chosen for its small magnetostriction (λ_s= -0.08ppm) rendering it relatively insensitive to applied stress. Measurements were undertaken on lengths of wire from 0.2-2m and the length of the wire used in the sensor demonstrator was 80cm.

The GMI properties of the material, after being annealed under different conditions of furnace annealing, d.c. current annealing and pulsed current annealing, were measured using a Hewlett-Packard 4194A impedance analyser. The magneto-impedance data were obtained at a frequency of 1 MHz, with a constant amplitude a.c. voltage of 0.4V. The performance of an optimally prepared wire incorporated into a demonstration ac bridge circuit, operating in open-loop mode at a fixed bias field of 500 A/m, was determined.

The demonstration sensor consisted of a straight sensing wire mounted under light tension inside a 1.5m solenoid to provide the bias field. The wire was connected via co-axial leads to a bridge circuit. The component values of the bridge are critical. The behaviour of the a.c. bridge is complicated by the fact that the impedance of the sensing wire depends on both the applied field and the measuring current. Source voltage and resistance, and stray capacitance are significant, and have been parameterised within a numerical model of the bridge. The bridge output was measured with a Stanford SR844 lock-in amplifier. Real and imaginary components were balanced at the chosen bias field and measured as a function of applied field. It was found that drift, arising as a result of source voltage and temperature changes could be largely compensated by modifying the real component of the output through a linear combination with the imaginary component.

RESULTS AND DISCUSSION

Measurement of the field dependence of the resistance for AC20 wire annealed under various conditions indicated that the greatest sensitivity for a closed loop system is obtained at an a.c. frequency of around 1 MHz. Figures 1-4 therefore show the behaviour of 20cm lengths of AC20 wires at 1 MHz. Figure 1 shows the effect of furnace annealing at 472°C for different lengths of time. Initially, annealing was found to remove internal stresses. Increasing the annealing times, however, seemed to lead to a weak anisotropy, with the easy axes along the wire axis. Although such annealing improved GMI sensitivity, it also led to brittleness in the wires.

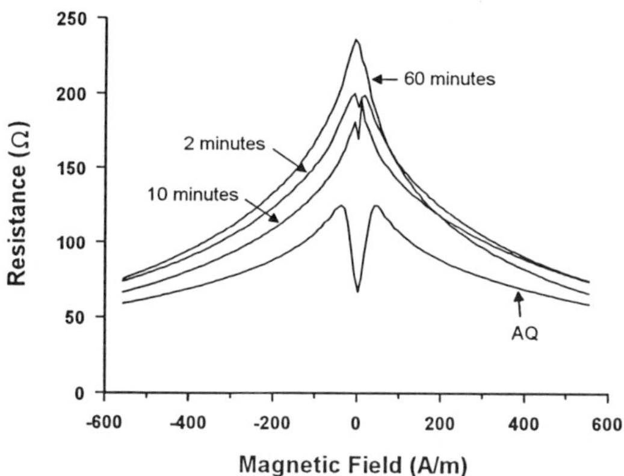

Figure 1 Influence of furnace annealing at 472°C on the field response of the resistive component of the impedance of a 20 cm as-quenched (AQ) amorphous AC20 wire.

Annealing through the application of a heating current along the wire length is complicated by the fact that in addition to the resistive heating effect, the current can also produce a circumferential magnetic field, which can induce anisotropy in the wire. On annealing AC20 wire with a 300mA current, it was found that a temperature was achieved at which such circumferential anisotropy was induced, as indicated by the increase in the field H_c at which the resistance R is maximum in figure 2. It is interesting that with a 500mA current there was a large increase in R and a decrease in H_c, as seen in figure 3. This is indicative that only stress relief was taking place. For currents of 700mA and above, it was found that the amorphous structure of the wire rapidly crystallised.

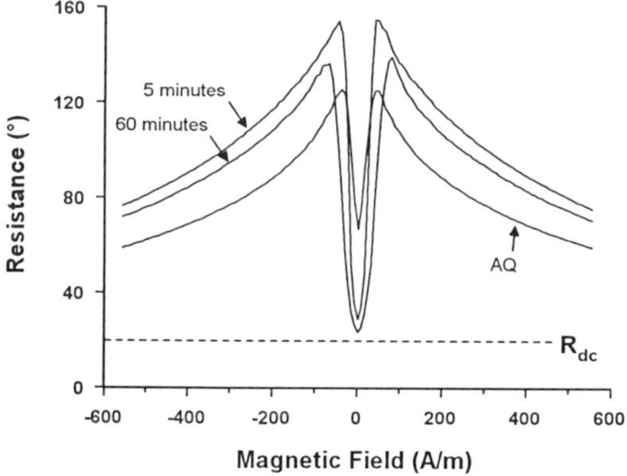

Figure 2 Influence of 300 mA annealing, for the durations shown, on the field response of the resistive component of the impedance of as-quenched (AQ) amorphous AC20 wire.

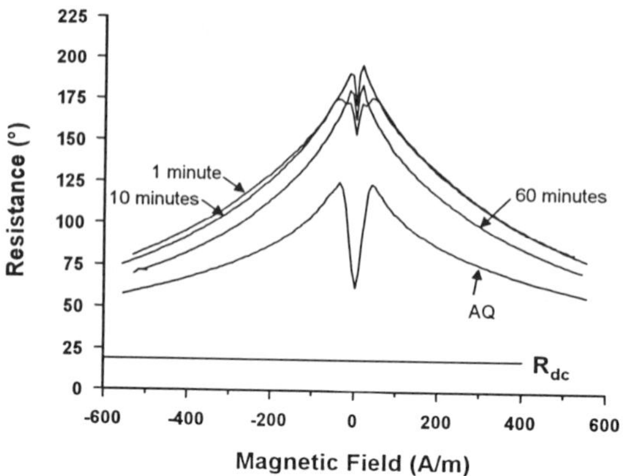

Figure 3 Influence of 500 mA annealing, for the durations shown, on the field response of the resistive component of the impedance of as-quenched (AQ) amorphous AC20 wire.

Annealing using short duration pulses allows large currents to be passed through the AC20 wire, which can then reach high temperatures without crystallisation occuring. This has been proved to be successful in achieving very soft magnetic properties, as shown in figure 4. The best response is obtained by pulse annealing at high currents, leading to a very large low-field GMI response.

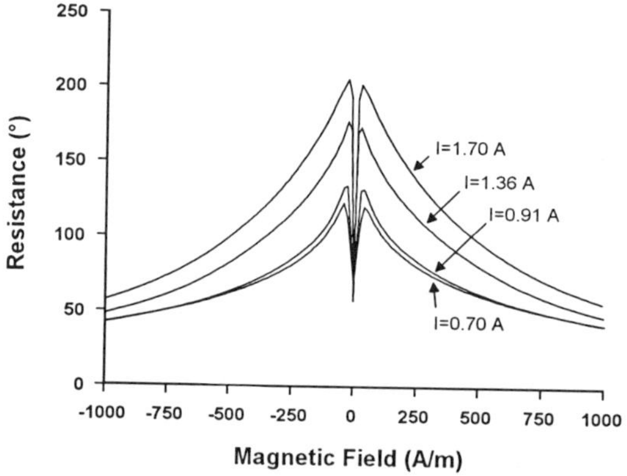

Figure 4 Influence of pulse current annealing, at the current levels shown, on the field response of the resistive component of the impedance of as-quenched (AQ) amorphous AC20 wire.

For long lengths of wire (>1 metre), direct current and pulse current annealing are prohibitive due to the high resistivity of the wire (R_{DC} ~ 1 Ω/cm). However, it has been demonstrated that carefully controlled furnace annealing can produce a homogeneous and enhanced GMI response for relatively long lengths (2 metres) of wire, as shown in figure 5.

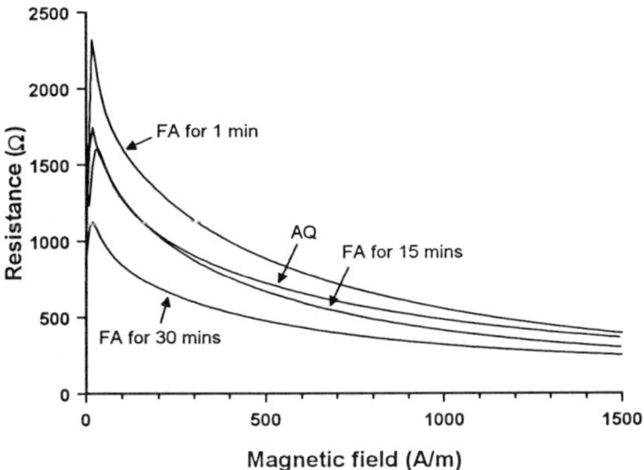

Figure 5 Influence of furnace annealing on the positive field response of the resistive component of the impedance of a 2 m length of as-quenched (AQ) amorphous AC20 wire.

The dynamic response of our bridge circuit sensor to a series of field steps of both signs was tested and it was established that the response is stable, linear and reproducible. The sensor incorporated an 80cm AC20 wire treated by direct current annealing at 400mA for 60s. The system was operated at a bias field of 500 A/m. It demonstrated a sensitivity of 0.02 A/m and a linearity of better than 5% for external fields between ±300 A/m at an optimum drive frequency of 1MHz. An example of the sensor response to homogeneous fields is shown in figure 6.

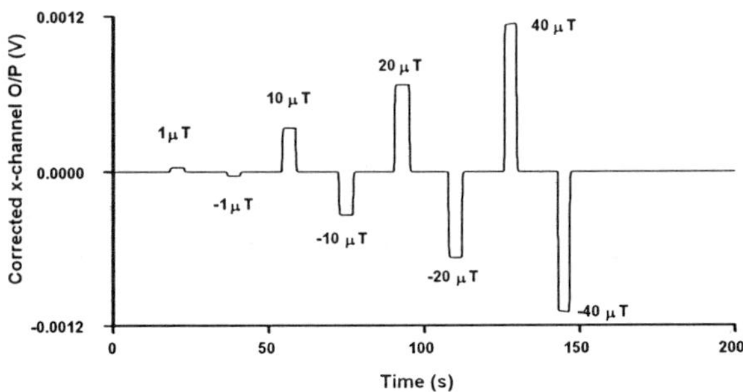

Figure 6 Response of the 80 cm GMI demonstration sensor system to homogeneous fields.

Results show that the sensor system is effective for integrating inhomogeneous fields. The response of our integrating sensor to inhomogeneous fields, emanating from a complex ferromagnetic body to which field is applied, is compared with averaged fluxgate readings in figure 7. In this case, peak-to-peak fields were of the order of 100 A/m.

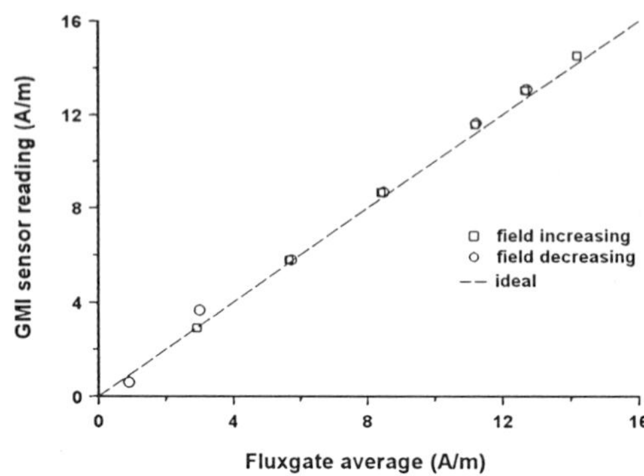

Figure 7 Response of the 80 cm GMI demonstration sensor system to inhomogeneous fields.

CONCLUSIONS

The giant magneto-impedance properties of AC20 $(Co_{0.94}Fe_{0.06})_{72.5}Si_{12.5}B_{15}$ amorphous wire improve on annealing. Pulse current annealing is effective for improving the low-field response of the impedance response and is therefore suitable for producing short lengths of amorphous wires for high sensitivity (closed-loop) magnetic sensor applications. Furnace annealing can be used to optimise the response of longer lengths of wire (>1m) for high-field (open-loop) sensor applications.

Direct current annealing at 400mA for 60s provided a wire for a demonstration a.c. bridge circuit sensor. This operates in open-loop mode (at a fixed bias field of 500 A/m) with a sensitivity of 0.02 A/m and a linearity of better than 5% for external fields between ±300 A/m at an optimum drive frequency of 1MHz.

ACKNOWLEDGEMENTS

This work was carried out as part of Technology Group 1 of the MoD corporate research programme. ©British Crown copyright 1999. Published with the permission of the Defence Evaluation and Research Agency on behalf of the Controller of HMSO.

REFERENCES

[1] L.V. Panina and K. Mohri, Applied Physics Letters, 65 (1994) 1189.
[2] P.T. Squire, J. Magn. Magn. Mater. 140-144 (1995) 10.
[3] D. Atkinson and P.T. Squire, J.App.Phys. 83 (1998) .
[4] L. Panina, K. Mohri, K. Bushida and M. Noda, J.App.Phys. 76 (1994) 6198.
[5] T. Kanno, K. Mohri, T. Yagi, T. Uchiyama & L.P. Chen, IEEE Trans.Magn. 33 (1997) 3358.
[6] D. Atkinson, P.T. Squire, M.G. Maylin and J. Gore, Sensors and Actuators (in press).

Part XII

Ultrasoft Nanocrystalline and Amorphous Materials II

EFFECT OF Cu ON MICROSTRUCTURAL EVOLUTION OF NANOCRYSTALLINE SOFT AND HARD MAGNETIC MATERIALS

K. HONO*, D. H. PING* AND S. HIROSAWA**
*National Research Institute for Metals, 1-2-1 Sengen, Tsukuba 305-0047, Japan
**Sumitomo Special Metals Co., Ltd., 2-15-17 Egawa, Osaka 618-0013, Japan

ABSTRACT

The nanocrystallization processes in Fe-Si-B-Nb-Cu and Fe-Nd-B(-Cu-Nb) amorphous alloys have been studied by transmission electron microscopy (TEM) and a three dimensional atom probe (3DAP). Cu additions are effective in refining the nanocrystalline microstructures of both alloys, because Cu atom clusters formed prior to the crystallization reaction serve as heterogeneous nucleation sites for the primary crystals. However, the clustering behaviors of Cu atoms in these two alloy systems are different, *i.e.*, Cu completely dissolves in the $Nd_2Fe_{14}B$ phase in the final microstructure of the $Nd_{4.5}Fe_{75.8}B_{18.5}Cu_{0.2}Nb_1$ alloy, whereas Cu clusters grow to fcc-Cu particles in the $Fe_{73.5}Si_{13.5}B_9Nb_3Cu_1$ alloy. The nanocrystallization processes in these two alloys clarified by the 3DAP results are compared.

INTRODUCTION

In 1988, Yoshizawa *et al.* [1] reported that excellent soft magnetic properties are obtained when the grain size is reduced to a nanometer scale by crystallizing an Fe-Si-B-Nb-Cu amorphous alloy (FINEMET). Nanocrystallization itself has been known for a long time, but this was the first report that nanocrystalline ferromagnetic materials exhibited soft magnetic properties. The coercivity decreases as a result of the averaging effect of the magnetocrystalline anisotropy of the randomly oriented nanograins. To make ferromagnetic alloys soft, both magnetocrystalline anisotropy and magnetostriction have to be reduced. In the conventional soft magnetic materials, controlling these properties was achieved by optimizing the alloy compositions of the ferromagnetic alloys, but it was impossible to obtain high saturation magnetization by increasing the alloying element. On the other hand, magnetic softening by nanocrystallization does not necessarily sacrifice the magnetization of the alloy, because the content of alloying elements can be kept low.

Yoshizawa *et al.* [2] reported that combined addition of Nb and Cu was required to obtain the nanocrystalline microstructure in Fe-Si-B based alloys, thus the role of Nb and Cu in the nanocrystallization effect has been of great interest. Using a conventional atom probe (AP), Hono *et al.* [4, 5] characterized the microstructures of the various stages of the crystallization process of an $Fe_{73.5}Si_{13.5}B_9Nb_3Cu_1$ amorphous alloys, and reported that Cu form clusters prior to the primary crystallization reaction of the α-Fe(Si) phase. However, due to the limitation of the technique, they did not succeed in clarifying the role of the Cu clusters at the nucleation stage of the α-Fe(Si) primary crystals. They also measured approximate compositions of the constituent phases in the final microstructure, and reported that Nb and B are partitioned in the remaining amorphous phase, while Si is partitioned in the α-Fe(Si) phase.

When a nanocomposite microstructure is composed of exchange coupled soft and hard magnetic phases, it behaves as a single phase magnet. Coehoorn *et al.* [6] reported that such microstructures can be achieved by crystallizing Fe-Nd-B based amorphous alloys. The nanocomposite magnet produced by this way exhibited very high remanence (M_r/M_s~0.8), and a relatively high energy product, $(BH)_{max}$~95 kJ/m^3, in spite of the fact that the fraction of the hard

Mat. Res. Soc. Symp. Proc. Vol. 577

magnetic $Nd_2Fe_{14}B$ phase was only 15% of the alloy. Depending on the alloy compositions, α-Fe, Fe_3B or $Fe_{23}B_6$ phases appear as the primary crystal and these soft magnetic phases have high saturation magnetization. On the other hand, the hard magnetic $Nd_2Fe_{14}B$ phase that appear in the subsequent crystallization stage has a very high magnetocrystalline anisotropy. By making a nanocomposite microstructure from these phases, the hard and soft magnetic phases are magnetically coupled and remanence higher than 1 T and energy product $(BH)_{max}$ higher than 100 kJ/m^3 are obtained. These nanocomposite magnets are regarded as promising candidate materials for bonded magnets with an adequate performance while keeping the content of expensive Nd to less than 50% of the single phase $Nd_2Fe_{14}B$ magnet. This type of magnet is called an *exchange-spring magnet* [7] or a *nanocomposite magnet*, and making the grain size smaller than 20 nm is required for obtaining a sufficiently high intergrain exchange interaction.

Many attempts were made to improve the magnetic properties of nanocomposite magnets by adding quaternary and quinternary elements in the Fe-Nd-B ternary alloys. Hirosawa *et al.* [8-10] reported that a combined addition of Co and Ga is very effective in reducing the grain size, thereby enhancing the hard magnetic properties. However, recent APFIM and TEM work [11] suggested that the magnetic property improvement in the Co and Ga bearing alloy was not due to the grain size reduction but due to the increase in the volume fraction of the $Nd_2Fe_{14}B$ phase. Ga atoms were found to partition only in the $Nd_2Fe_{14}B$ phase by APFIM, and this causes an increase in the volume fraction of the $Nd_2Fe_{14}B$ phase. On the other hand, in the case of $Nd_5Fe_{77}B_{18}Cr_3$ alloy, Sano *et al.* [12] found that Cr atoms are partitioned in the Fe_3B phase by employing APFIM. The effect of Cr was interpreted as a stabilizer of the Fe_3B phase, which modifies the tie lines in the ternary phase diagram, thereby suppressing the formation of a magnetically harmful $Nd_2Fe_{23}B_3$ phase. Thus, the effect on the microstructures and their mechanisms for improving the hard magnetic properties are different depending on the types of microalloyed elements.

For understanding the correlation between the microstructure and the hard magnetic properties better, varying the grain size without a major change in the constituent phases is necessary. For this purpose, we have chosen Cu and Nb as microalloying elements added in the $Nd_{4.5}Fe_{77}B_{18.5}$ alloy expecting that Cu atom clusters may stimulate the nucleation of primary particles as in the case of nanocrystalline soft magnetic $Fe_{73.5}Si_{13.5}B_9Nb_3Cu_1$ and $Fe_{89}Zr_7B_3Cu_1$ alloys.

EXPERIMENTAL

A melt-spun $Fe_{73.5}Si_{13.5}B_9Nb_3Cu_1$ amorphous ribbon was provided from Dr. Yoshizawa of Hitachi Metals Co. Ltd. Arc-melted alloys with nominal compositions of $Nd_{4.5}Fe_{77}B_{18.5}$, $Nd_{4.5}Fe_{76.8}B_{18.5}Cu_{0.2}$ and $Nd_{4.5}Fe_{75.8}B_{18.5}Nb_1Cu_{0.2}$ were melt-spun onto a Cu roll at a rolling speed of 20 m/s in an argon atmosphere. The thickness of the ribbon specimen was approximately 20 μm and the width was 2 - 3 mm. The as-melt-spun ribbons were all confirmed to be in the amorphous state by TEM and x-ray diffraction. Thin foil specimens for TEM observations were prepared by ion milling from both sides of the ribbon. Conventional TEM observations were made by Philips CM200. High resolution electron microscopy (HREM) observations were carried out by JEM-4000EX, whose point to point resolution is 0.17 nm. Specimens for atom probe analysis were prepared by mechanically grinding the ribbon to a square rod of approximately 20 μm × 20 μm × 6 mm followed by micro-electropolishing to a sharp needle shape specimen. These specimens were observed by field ion microscope (FIM) and then analyzed by a three dimensional atom probe (3DAP). The 3DAP was equipped with CAMECA tomographic atom probe (TAP) detection system [14] which was installed to a locally designed FIM chamber. The atom probe analyses were performed with a pulse fraction (V_p/V_{dc}) of 0.2 and a pulse repetition rate of 600 Hz at a specimen temperature of 35 K under an ultrahigh vacuum of ~$1x10^{-8}$ Pa. The data analysis and visualization of 3DAP data were carried out using KINDBRISK SDV and SSA 3DAP data analysis software under Advance Visualizaiton System (AVS).

NANOCRYSTALLINE SOFT MAGNET

The HREM image of an melt-spun $Fe_{73.5}Si_{13.5}B_9Nb_3Cu_1$ alloy, Fig. 1 (a), shows isotropic maze contrast typical for an amorphous structure. The selected area diffraction pattern (inset figure) exhibits typical halo ring suggesting that the as-melt-spun specimen is amorphous. Annealing at 400°C for 60 min does not cause any structural change as shown in Fig. 1(b). No recognizable difference is observed in the selected area diffraction pattern compared with Fig. 1 (a). Crystallization occurs by annealed the specimen at 450°C for 5 min as shown in Fig. 1 (c). Although 450°C is still lower than the crystallization temperature, ~517°C, determined by differential scanning calorimetry (DSC) with a heating rate of 10 °C/min [2], crystallization occurs by isothermal heat treatment as long as 5 min. On the other hand, no evidence for primary crystallization has been observed at 400°C, thus the specimens annealed at 400°C for 60 min has been examined by 3DAP for investigating pre-crystallization clustering of Cu atoms.

Figures 2 (a), (b) and (c) show 3DAP elemental mappings of Cu within analyzed volumes of ~10 × 10 × 40 nm in the as-melt-spun specimen, in the specimens annealed for 5 and 60 min at 400°C, respectively. In the as-melt-spun specimen, Cu distribution is uniform, confirming that a chemically homogeneous solid solution is obtained in the as-melt-spun state. In the specimen

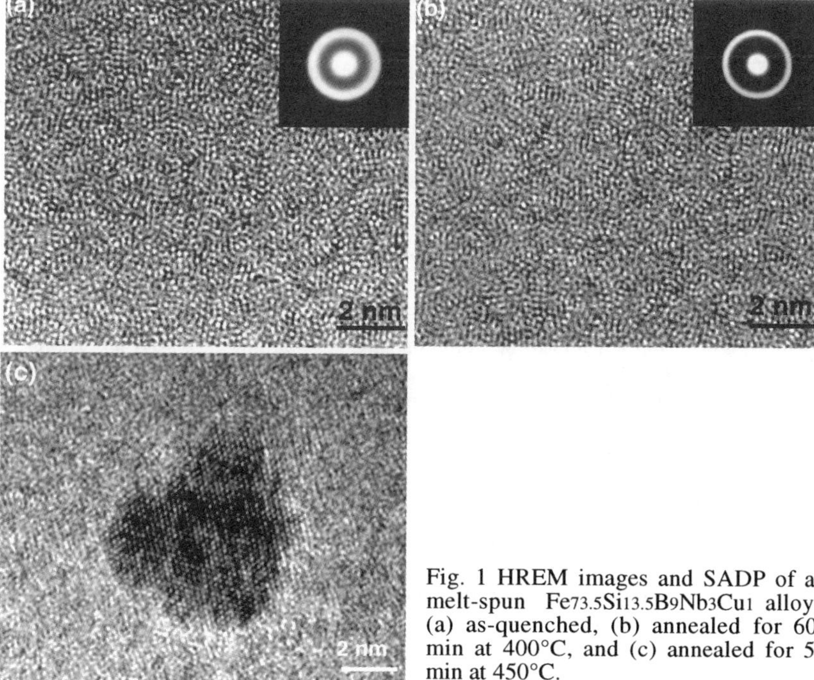

Fig. 1 HREM images and SADP of a melt-spun $Fe_{73.5}Si_{13.5}B_9Nb_3Cu_1$ alloy (a) as-quenched, (b) annealed for 60 min at 400°C, and (c) annealed for 5 min at 450°C.

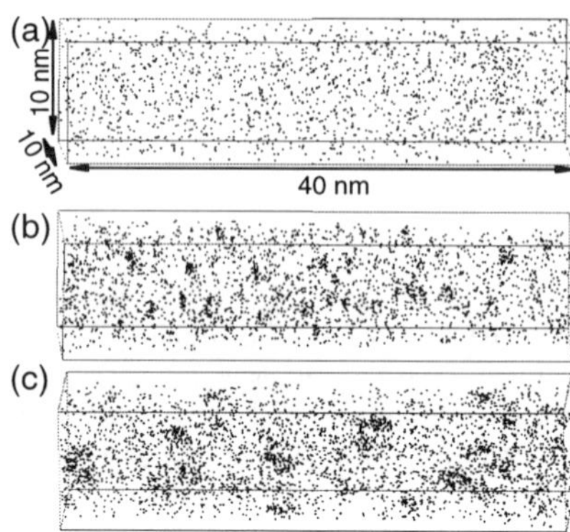

Fig. 2 3DAP elemental maps of Cu in a melt-spun $Fe_{73.5}Si_{13.5}B_9Nb_3Cu_1$ alloy (a) as-quenched, (b) annealed for 5 min at 400°C, and (c) annealed for 60 min at 400°C.

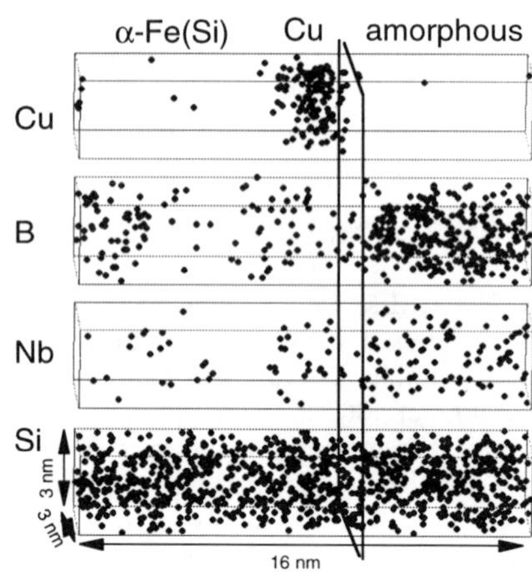

Fig. 3 3DAP elemental maps of Cu, B, Nb and Si in a melt-spun $Fe_{73.5}Si_{13.5}B_9Nb_3Cu_1$ alloy annealed for 10 min at 500°C.

annealed for 5 min at 400 °C, heterogeneous distribution of Cu atoms is apparent, indicating that clustering of Cu atoms occurs. After annealing for 60 min at 400°C, Cu atom clusters are observed more clearly. The HREM observation result shown in Fig. 1 (b) confirmed that no crystallization occurs up to 60 min at 400°C, thus it is concluded that Cu clustering occurs in the amorphous phase. The number of atoms in each cluster is in the range of 50 to 100 (taking the detection efficiency of the microchannel plate detector into consideration), and the size of the clusters is approximately 3 nm. The density of the Cu clusters estimated from the analyzed volume is in the order of 10^{24} m^{-3}. The composition of the Cu cluster has been estimated to be approximately 12 at.% Cu initially, but it increases as the clusters grow in size. The structure of these clusters was not identified by HREM, because the HREM image did not give any fringe contrast corresponding to any crystalline structure at these stages.

Based on the extended x-ray absorption fine structure (EXAFS) measurement results, Ayers *et al.* [14] reported that Cu atoms form clusters having near-fcc symmetry from the very early stage of the heat treatment. Thus, the Cu clusters observed in the 3DAP data are believed to have a fcc-like short range ordered structure, but they do not appear to be distinct fcc-Cu in the initial stage. Such Cu clustering appears to be common in Cu-containing Fe-based amorphous alloys. For example, Suzuki *et al.* [15] reported that the addition of 1 at.% of Cu in $Fe_{90}Zr_7B_3$ amorphous alloy causes a reduction of the grain size leading to better soft magnetic properties in the final nanocrystalline microstructure. Clustering of Cu in this system was also confirmed prior to the crystallization reaction by AP analysis work [16].

Fig. 3 shows 3DAP elemental maps near a Cu precipitate in the $Fe_{73.5}Si_{13.5}B_9Nb_3Cu_1$ alloy annealed at 550°C for 10 min. In this figure, the interface between α-Fe and amorphous phases is clearly seen from the drastic change in concentrations of B and Nb. The densities of Nb and B are higher in the remaining amorphous phase, while they are lower in the α-Fe. Cu precipitate is at the interface in direct contact with the α-Fe particle, which suggests that Cu clusters provide nucleation sites for the α-Fe primary crystals.

The microstructure of the $Fe_{73.5}Si_{13.5}B_9Nb_3Cu_1$ nanocrystalline alloy with the optimal soft

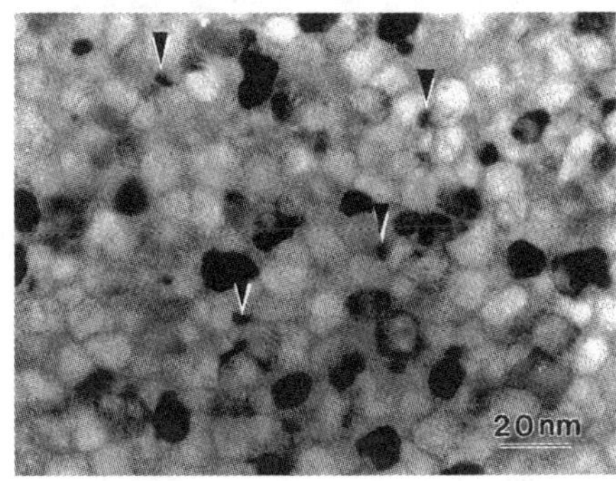

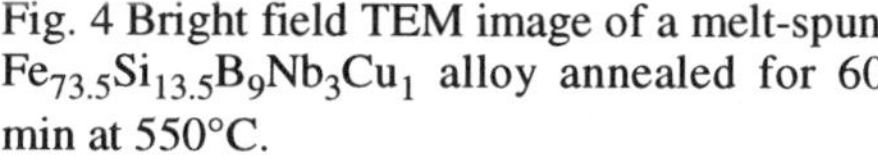

Fig. 4 Bright field TEM image of a melt-spun $Fe_{73.5}Si_{13.5}B_9Nb_3Cu_1$ alloy annealed for 60 min at 550°C.

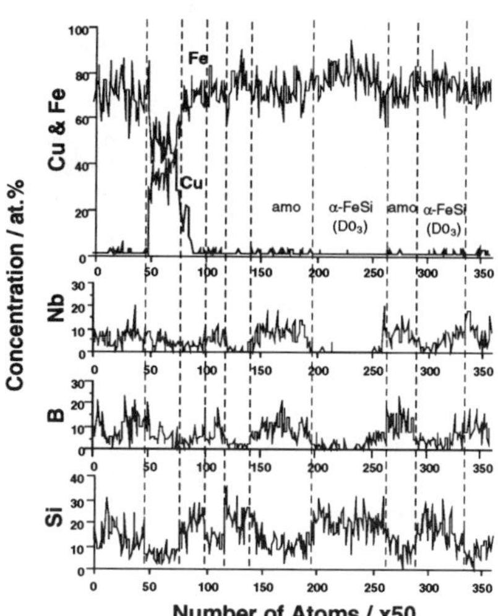

Fig. 5 Atom probe concentration depth profile of a melt-spun $Fe_{73.5}Si_{13.5}B_9Nb_3Cu_1$ alloy annealed for 60 min at 550°C.

magnetic properties which was produced by annealing the melt-spun amorphous ribbon for 60 min at 550°C is shown in Fig. 4. This consists of the nanocrystalline $D0_3$-ordered Fe_3Si phase with an average grain size of approximately 10 nm. The grains are randomly oriented without any preferred orientations. In addition to the grains of ~10 nm, smaller grains can be recognized as indicated by the arrowheads. These are the fcc-Cu particles which precipitate in the early stage of crystallization [5]. Atom probe concentration depth profiles obtained from the same microstructure are shown in Fig. 5 [4]. The chemical compositions of the three types of phases observed in Fig. 4 are quantitatively analyzed. The $D0_3$ phase contains ~20-25 at.%Si with little Nb and Cu. A few atomic percent of B remains in this phase. Boron dissolution in the α-Fe primary particles was commonly observed in other Fe-based nanocrystalline magnetic materials as well [17]. The remaining amorphous phase between the nanocrystalline particles is enriched in B and Nb containing a small amount of Si. In addition to these two major phases, a Cu-enriched particle is observed. This phase is strongly enriched in Cu (~60% or higher) but still contains appreciable amounts of the other elements. A separate nanobeam electron diffraction study revealed that the Cu-enriched particles were fcc-Cu [5]. This phase appears as a grain having a diameter of approximately 5 nm as indicated by the arrows in Fig. 4.

NANOCOMPOSITE HARD MAGNET

Figures 6 (a) - (c) show bright field TEM micrographs of $Nd_{4.5}Fe_{77}B_{18.5}$, $Nd_{4.5}Fe_{76.8}B_{18.5}Cu_{0.2}$ and $Nd_{4.5}Fe_{75.8}B_{18.5}Nb_1Cu_{0.2}$ alloys annealed at 660°C for 10 min, respectively [18,19]. The values of $(BH)_{max}$ obtained from these three specimens are also shown. The average grain size of the ternary alloy is approximately 30 nm, while those of the Cu containing quaternary and Cu and Nb containing quinternary alloy is approximately 17 nm and 12 nm, respectively. This indicates that adding only 0.2 atomic percent of Cu to the ternary alloy is effective to reduce the grain size

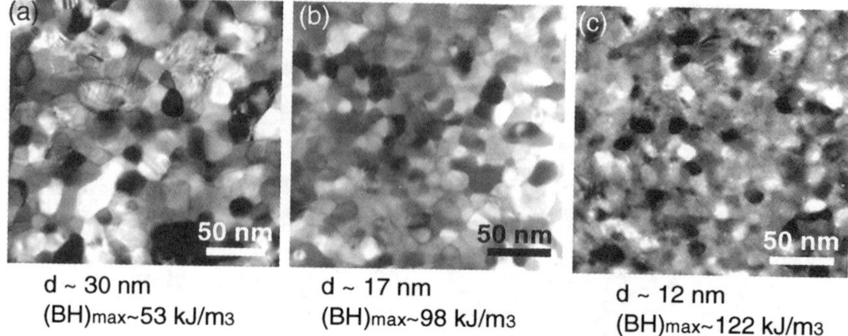

Fig. 6 TEM bright field micrographs of (a) $Nd_{4.5}Fe_{77}B_{18.5}$, (b) $Nd_{4.5}Fe_{76.8}B_{18.5}Cu_{0.2}$ and (c) $Nd_{4.5}Fe_{75.8}B_{18.5}Nb_1Cu_{0.2}$ alloys annealed at 660°C for 10 min. Average diameter of the grains and $(BH)_{max}$ are indicated.

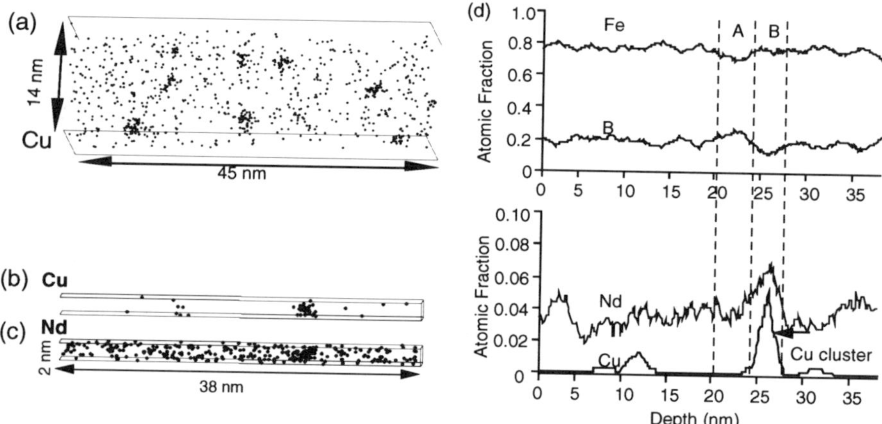

Fig. 7 (a) 3DAP elemental map of Cu in the $Nd_{4.5}Fe_{76.8}B_{18.5}Cu_{0.2}$ alloy annealed at 530°C for 10 min. 3DAP elemental maps of (b) Cu and (c) Nd atoms in a selected volume crossing a Cu cluster, and (d) the corresponding concentration depth profiles of Fe, B. Nd and Cu calculated from the selected volume.

of the optimum nanocomposite microstructure, resulting in improved $(BH)_{max}$. Combined addition of Cu and Nb is effective for reducing the grain size further, but addition of Cu alone has a very large effect. The microstructure is composed of Fe_3B and $Nd_2Fe_{14}B$ grains in the ternary and quaternary alloys, and there is no change in the constituent phases in these two alloys. Cu addition causes a lower crystallization temperature than that of the ternary alloy by 20 K, and this implies that Cu addition affects the kinetics of crystallization. However, the addition of Nb changed the constituent phases, and the $Fe_{23}B_6$ phase was formed in addition to Fe_3B, $Nd_2Fe_{14}B$ in the quinternary alloy, thus Nb addition affects the phase equilibrium. Combined addition of Nb and Cu further increase the value of $(BH)_{max}$, but this is not only because of the crystal grain refinement, but also because of the change in the constituent phases.

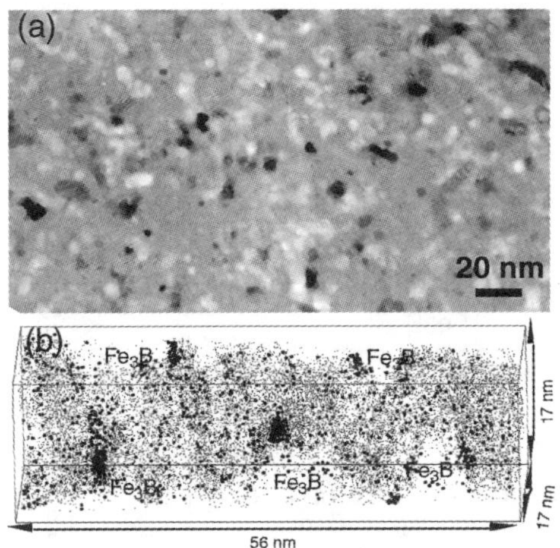

Fig.8 (a) TEM bright field micrograph showing a high density of Fe_3B particles precipitate in the amorphous matrix in the $Nd_{4.5}Fe_{76.8}B_{18.5}Cu_{0.2}$ alloy after annealing at 530°C for 30 min; (b) 3DAP elemental mapping of Cu and Nd. Cu and Nd atoms are represented by large and small black dots, respectively.

In order to understand the role of Cu atoms during the nucleation and growth stage of the primary Fe_3B crystals, the specimens crystallized at 530°C for 10 min has been studied by 3DAP analysis [18]. Figure 7 (a) shows a 3DAP elemental map of Cu within an analyzed volume of ~14 × 14 × 45 nm. It is apparent that Cu atoms form clusters at this stage with a density of about 10^{24} m^{-3}. Figures 2 (b) and (c) show Cu and Nd elemental maps in a selected volume of about ~2 × 2 × 38 nm cutting through one of the Cu clusters in Fig. 7 (a). The size of the Cu cluster is around 2 nm. Nd is also enriched in the Cu enriched region as can be clearly seen from the Nd elemental map in Fig. 7 (c). In order to evaluate the local concentration change near the cluster more quantitatively, compositional profiles determined from the selected volume are shown in Fig. 7 (d). The Cu concentration in the Cu cluster is estimated to be around 5.0 at.%. In the Cu-enriched regions, the Nd concentration is approximately 7.0 at.%, which is slightly higher than that in the amorphous phase (4.5 at.%). The concentration of Nd and Cu in these regions is much lower than that of any known Nd-Cu compounds, thus we believe that they do not correspond to any Nd-Cu compounds. The concentrations of Fe and B around the Cu clusters are slightly different from the other region. In Fig. 7 (d), the concentration of region A which is adjacent to the Cu enriched region is higher in B and lower in Fe than the average values. This is probably because Fe_3B phase is present in region A, suggesting that the Cu clusters act as heterogeneous nucleation sites for the Fe_3B particles.

Figure 8 (a) shows a TEM bright field micrograph of the alloy annealed at 530°C for 30 min. The crystalline particles observed in this micrograph are all Fe_3B (bct structure) embedded in the amorphous matrix, and the second stage crystallization of the $Nd_2Fe_{14}B$ phase does not occur yet in this stage. Figure 8 (b) shows a 3DAP elemental map of Cu and Nd in an analysis volume of approximately ~17 × 17 × 56 nm, where small and large black spots correspond to Nd and Cu atoms, respectively. Fe_3B particles can be easily recognized based on the atomic distribution of Nd because the Nd atoms are fully rejected from the Fe_3B phase. From Fig. 8 (b), it can be clearly seen that each Cu cluster is located at the interface between Fe_3B crystals and the amorphous matrix in direct contact with the Fe_3B particles. Cu atoms are found to be rejected

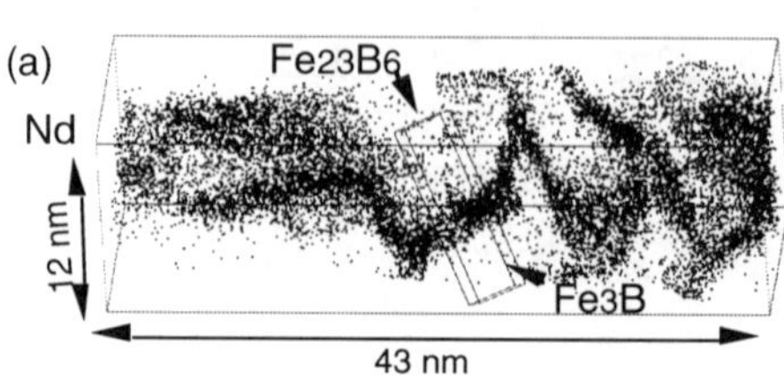

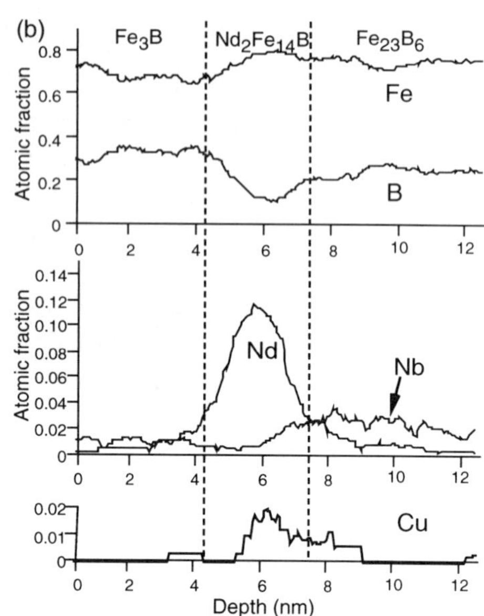

Fig. 9 (a) 3DAP elemental map of Nd in a melt-spun $Nd_{4.5}Fe_{75.8}B_{18.5}Nb_1Cu_{0.2}$ alloy annealed for 10 min at 660°C. (b) Concentration depth profiles form the selected volume in (a). Nb is partitioned in the $Fe_{23}B_6$ phase.

from the Fe_3B particles, so the Cu cluster is located at the amorphous phase side. This result suggests that Cu-enriched regions (or clusters) serve as heterogeneous nucleation sites for the Fe_3B primary crystals.

Several 3DAP data have demonstrated that one Cu cluster is present corresponding to each Fe_3B crystal. This strongly suggests that Fe_3B nanocrystals nucleate at the site of Cu clusters. The reason why Cu clusters serve as heterogeneous nucleation sites of Fe_3B would be explained as follows. From Fig. 8(d), Nd is also enriched in the Cu cluster region, but the total atomic fraction of Cu and Nd in the cluster is less than 15%. Thus, it is not likely that they are any Nd-Cu compounds, but they are believed to be still amorphous. So it is difficult to consider that the Cu clusters work as nucleation sites from the crystallographic point of view. Rather, the chemical factor would play an important role in enhancing the Fe_3B nucleation. As shown in the atom probe results of Fig. 7 (d), B atoms pile up at the Cu/amorphous interface (region A) when Cu clusters form. For the formation of Fe_3B crystals, Fe atoms do not have to diffuse too much because the matrix Fe composition (76.8 at.%) is very close to the concentration of Fe in the Fe_3B phase, but B atoms have to diffuse in by rejecting Nd atoms. Since Nd has strong affinity with Cu (large negative enthalpy of mixing), when Cu atoms aggregate, Nd atoms are attracted to the Cu atoms, resulting in co-segregation of Cu and Nd. As a result, enrichment of B and depletion of Nd occur at the Cu/amorphous interface, which benefit for the formation of Fe_3B phase adjacent to the Cu clusters.

Combined addition of Cu and Nb further reduce the grain size. 3DAP analysis results showed that Cu clusters formed in the early crystallization stage trigger the nucleation of the Fe_3B primary particles in the $Nd_{4.5}Fe_{75.8}B_{18.5}Nb_1Cu_{0.2}$ alloy as well. Nb is rejected from the Fe_3B primary particle and partitioned in the remaining amorphous phase. When the remaining amorphous phase crystallizes, $Nd_2Fe_{14}B$ and $Fe_{23}B_6$ phases are formed in the quinternary alloy, and Nb is exclusively partitioned in the $Fe_{23}B_6$ phase. The $Fe_{23}B_6$ phase is not observed in the ternary and quaternary alloy, hence it is concluded that Nb addition induces the formation of the $Fe_{23}B_6$ phase in the second stage of the crystallization. The 3DAP analysis result of the final microstructure of the $Nd_{4.5}Fe_{75.8}B_{18.5}Nb_1Cu_{0.2}$ alloy is shown in Fig 9. Fe_3B, $Fe_{23}B_6$ and $Nd_2Fe_{14}B$ are observed in the analysis volume. Cu is partitioned in the $Nd_2Fe_{14}B$ phase, whereas Nb is exclusively partitioned in the $Fe_{23}B_6$ phase.

NANOCRYSTALLIZATION PROCESSES

Based on the present 3DAP study, the microstructural evolution processes in Fe-Si-B-Nb-Cu and Fe-Nd-B-Cu alloys are summarized as shown in Fig. 10. In both cases, Cu clusters form in the early stage of annealing and these serve as heterogeneous nucleation sites for the primary crystals. In the case of Fe-Si-B-Nb-Cu alloy, the final microstructure is obtained after the primary crystallization reaction of the α-Fe(Si) phase. The remaining amorphous phase is stabilized by enrichment of Nb and B, thus the randomly oriented primary crystalline particles are embedded in the amorphous matrix. Each crystalline particle is magnetically coupled through a ferromagnetic amorphous phase, thus the crystal anisotropy is averaged out. In the case of the Fe-Nd-B-Cu alloy, the nanocomposite microstructures are obtained after the two step crystallization process. The primary crystals of Fe_3B nucleate heterogeneously at the site of Cu clusters by the primary crystallization reaction. Nd atoms are rejected from the Fe_3B phase and enriched in the remaining amorphous phase. In the second stage of the crystallization reaction, the remaining amorphous phase crystallize by the polymorphous crystallization, and the nanocomposite microstructure composed of Fe_3B and $Nd_2Fe_{14}B$ phases is formed. Since Cu has a

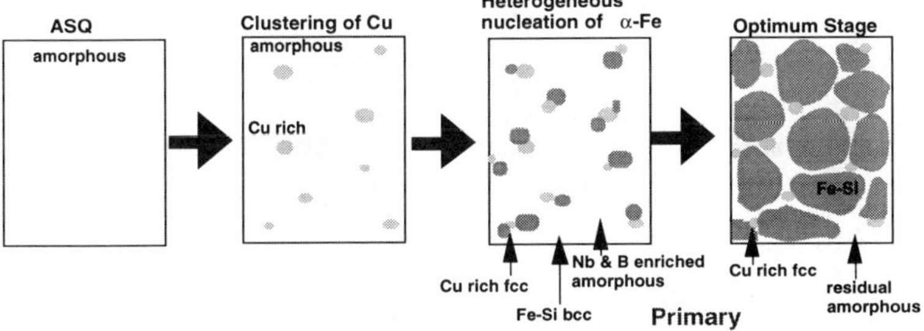

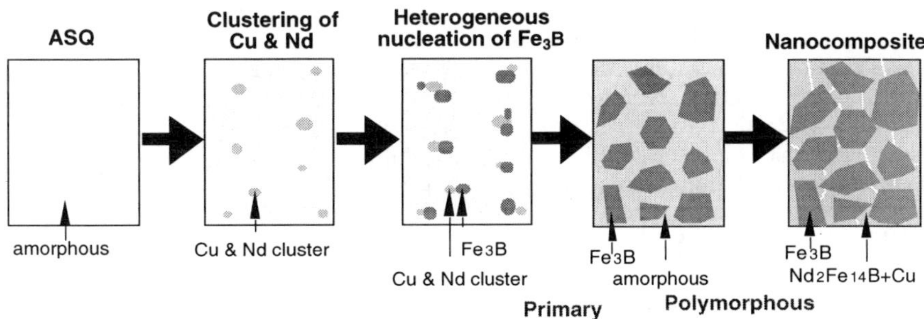

Fig. 10 Schematic illustration of nanocrystalline and nanocomposite microstructure evolution processes in $Fe_{73.5}Si_{13.5}B_9Nb_3Cu_1$ and $Nd_{4.5}Fe_{76.8}B_{18.5}Cu_{0.2}$ alloys.

large solubility in the $Nd_2Fe_{14}B$ phase, Cu is partitioned in this phase. Thus the Cu clusters that initially form are dissolved in the $Nd_2Fe_{14}B$ phase in the final microstructure. These particles are magnetically coupled, thus all these particles behave as a single phase magnet or an exchange-spring magnet. When Nb is added to the Cu containing alloy, $Nd_2Fe_{14}B$ and $Fe_{23}B_6$ phases are produced by the crystallization of the remaining amorphous phase. As this reaction cannot occur by the polymorphous reaction, the crystal grain size that was formed in the 2nd stage become smaller than the NdFeBCu ternary alloy. Since the $Fe_{23}B_6$ phase is not magnetically harmful (soft magnetic phase), its formation does not give any adverse effect to the magnetic properties in the final microstructure.

Recently, Hamano *et al.* [4] studied the microalloying effect of various additives in α-Fe/$Nd_2Fe_{14}B$ nanocomposite magnets and reported that the nanocomposite produced from a $Nd_8Fe_{76}Co_8Nb_2B_6$ melt-spun alloy exhibits good hard magnetic properties. In addition to Nb, they investigated the effect of Cu in the magnetic properties, but they did not find any beneficial effect in Cu additions. Although Cu atoms tend to form clusters by annealing when they are dissolved in Fe-based amorphous alloys [12, 16, 18], no evidence for Cu clustering was found in this type of alloys [21]. This is because the initial amorphous phase of this alloy composition can not be produced by the melt-spinning technique. Both α-Fe and $Nd_2Fe_{14}B$ crystalline particles are formed during solidification by melt-spinning, then Cu atoms are dissolved in the $Nd_2Fe_{14}B$ phase. Since there is no supersaturation of Cu in such a case, Cu atoms do not form clusters by the subsequent annealing. This is why Cu additions are not effective in refining the nanograin structures of the α-Fe/$Nd_2Fe_{14}B$ nanocomposite alloys. In order to refine the grain size by Cu addition, the starting material must be a fully amorphous alloy, and Cu clustering must occur prior to the nucleation process of the primary particles.

CONCLUSIONS

The nanocrystallization processes in Fe-Si-B-Nb-Cu and Nd-Fe-B amorphous alloys have been studied by the 3DAP technique and the results are compared. In both cases, Cu form a high density of clusters in the early stage of annealing. The number density of these clusters are in the order of 10^{24} m^{-3}, and these serve as heterogeneous nucleation sites for the primary crystals. However, the detailed mechanism how these clusters work as heterogeneous nucleation sites for the primary crystals are different in these two cases. Cu and Nd cosegregate in the early stage of annealing of $Nd_{4.5}Fe_{76.8}B_{18.5}Cu_{0.2}$ and $Nd_{4.5}Fe_{75.8}B_{18.5}Nb_1Cu_{0.2}$ alloys, and these serve as heterogeneous nucleation sites for the Fe_3B primary particles compositionally. When the remaining amorphous phase crystallizes to the $Nd_2Fe_{14}B$ phase, the Cu clusters completely dissolve in this phase, whereas Cu clusters grows to fcc-Cu particles in the final microstructure of the $Fe_{73.5}Si_{13.5}B_9Nb_3Cu_1$ nanocrystalline alloy. Nb atoms induce formation of $Fe_{23}B_6$ soft magnetic phase after the second stage crystallization of the $Nd_{4.5}Fe_{75.8}B_{18.5}Nb_1Cu_{0.2}$ alloy. Cu addition is not beneficial for reducing the grain size of the α-Fe/$Nd_2Fe_{14}B$ nanocomposite magnet, because the crystallization reaction occurs during solidification in the melt-spinning process, thus there is no opportunity for Cu atoms to form clusters to provide heterogeneous nucleation sites.

ACKNOWLEDGMENTS

This work was partly supported by the NEDO International Joint Research Grant.

REFERENCES

1. Y. Yoshizawa, S. Oguma, and K. Yamauchi, J. Appl. Phys. **64**, p. 6044 (1988).
2. Y. Yoshizawa and K. Yamauchi, Mater. Trans. JIM, **31**, p. 307 (1990).
3. G. Herzer, IEEE Trans. Magn. **25**, p. 3327 (1989).
4. K. Hono, K. Hiraga, Q. Wang, A. Inoue, and T. Sakurai, Acta metall. mater. **40**, p. 2137 (1992).
5. K. Hono, J-L. Li, Y. Ueki, A. Inoue, and T. Sakurai, Appl. Surf. Sci. **67**, p. 398 (1993).
6. R. Coehoorn, D. B. De Mooji, and C. De Waard, J. Magn. Magn. Mater. **80**, p. 101 (1989).
7. H. E. Kneller and R. Hawig, IEEE Trans. Magn. **MAG-27**, p. 3588 (1991).
8. S. Hirosawa, H. Kanekiyo, and M. Uehara, J. Appl. Phys. **73**, p. 6488 (1993).
9. H. Kanekiyo, M. Uehara, and S. Hirosawa , IEEE Trans. Magn. **29**, p. 2863 (1993).
10. S. Hirosawa and H. Kanekiyo, Mater. Sci. Eng. A **217**, p. 367 (1996).
11. D. H. Ping, K. Hono and S. Hirosawa, J. Appl. Phys. **83**, p. 7769 (1998).
12. K. Hono, D. H. Ping, M. Ohnuma and H. Onodera, Acta mater. **47**, p. 997 (1999).
13. D. Blavette, B. Deconihout, A. Bostel, J. M. Sarrau, M. Bouet, and A. Menand, Rev. Sci. Instrum. **64**, p. 2911 (1993).
14. J. D. Ayers, V. G. Harris, J. A. Sprague, W. T. Elam, and H. N. Jones, Acta mater. **46**, p. 1861 (1998).
15. K. Suzuki, N. Kataoka, A. Inoue, A. Makino and T. Masumoto, Mater. Trans. JIM, **31**, p. 743 (1990).
16. Y. Zhang, K. Hono, A. Inoue and T. Sakurai, Scripta metall. mater. **34**, p. 1705 (1996).
17. Y. Zhang, K. Hono, A. Inoue, A. Makino and T. Sakurai, Acta mater. **44**, p. 1497 (1996).
18. D. H. Ping, K. Hono, H. Kanekiyo and S. Hirosawa, J. Appl. Phys. **85**, p. 2448 (1999).
19. D. H. Ping, K. Hono, H. Kanekiyo and S. Hirosawa, IEEE Trans. Mag, in press.
20. M. Hamano, M. Yamasaki, H. Mizuguchi, H. Yamamoto and A. Inoue in Rare-Earth Magnets and their Applications, vol. 1, Proc. 5th Inter. Workshop on Rare-Earth Magnets and their applications, eds by L. Schultz and K. H. Müller, Werkstoff-Informationsgesellschaft, Germany, 1998, pp. 199.
21. Y. Q. Wu, D. H. Ping and K. Hono, IEEE Trans. Mag. in press.

LORENTZ MICROSCOPY AND ELECTRON HOLOGRAPHY OF NANOCRYSTALLINE MAGNETIC MATERIALS

MARC DE GRAEF
Department of Materials Science and Engineering, Carnegie Mellon University, Pittsburgh, PA 15213, degraef@cmu.edu

ABSTRACT

This paper employs an analytical model for the phase shifts imparted on a beam electron by a spherical particle with a uniform magnetization. Model image simulations of Fresnel and Foucault type Lorentz microscopy contrast are presented. Estimates of a lower particle size limit for which no magnetic information can be extracted from the images are given and compared to similar estimates for the electron holography observation method. It is concluded that conventional Lorentz methods have a lower particle size limit of around 30 nm (for cobalt particles), while electron holography can be used to even smaller sizes.

INTRODUCTION

The continuously decreasing length scale of magnetic recording media and other applications based on magnetic particles necessitates the use of advanced characterization methods, both for chemical and structural analysis and for the study of magnetic domains and magnetic interactions. In recent years transmission electron microscopes have become equipped with increasingly more sensitive detectors and cameras, and this in turn has dramatically improved the spatial resolution for measurement of chemical compositions, often at the unit cell length scale. Determination of the fine scale magnetic structure in these materials is essential for the prediction of, among other things, the bulk magnetostrictive strains. In addition, since virtually all engineering parameters of these materials are directly affected by microstructural features, the nature of these materials makes it necessary to closely combine the study of magnetic structure and microstructure. Lorentz microscopy, in particular the Fresnel and Foucault observation modes [1], has for several decades been the dominant observation technique for qualitative magnetic domain observations, and this paper deals with the application of Lorentz methods to the study of nanocrystalline magnetic materials.

The study of magnetic domain configurations by means of electron scattering has a long history; the early days of *Lorentz microscopy* have been summarized in [2]. The basic idea behind the Lorentz observation modes is that an electron traveling through a specimen with a non-zero magnetic induction experiences a force, the *Lorentz force*, normal to both its velocity vector and the local magnetic induction direction. This force causes a net deflection of the electron trajectory away from the path the electron would follow in the absence of a magnetic induction. The magnitude of the Lorentz deflection angle is usually in the range of a few tens of microradians, or about two orders of magnitude smaller than typical Bragg angles for electron scattering.

Lorentz observations are only meaningful if the sample can be observed in a field-free region; the usual immersion-type objective lens produces a field in the range of $1-2$ Tesla, which would saturate nearly all magnetic materials. One can use special low-field Lorentz pole pieces, mount the sample above the objective lens field, or use an objective minilens located below the sample

Mat. Res. Soc. Symp. Proc. Vol. 577 © 1999 Materials Research Society

(with the main objective lens switched off) to obtain a low-field sample environment. The consequent increase in the focal length results in a reduced final image magnification as compared to that of conventional transmission electron microscopy. This represents a serious limitation in the quantitative study of nanoscale magnetic structures. Furthermore, inelastic scattering in the sample contributes noise to the images, which further limits the attainable resolution of the standard Lorentz modes. We have reported a novel Lorentz microscopy setup, which combines a high resolution top-entry JEOL 4000EX TEM with a post-column Gatan imaging filter (GIF) [3]. With the main objective lens turned off, the sample is mounted in an essentially field-free region. The objective mini-lens, which sits below the plane of the specimen, provides sufficient field strength to focus the electrons into a diffraction pattern at the *selected area* (SA) aperture plane, thus providing Fresnel and Foucault observation modes with a maximum magnification of 3,000X. The energy filter provides an additional magnification of approximately 20X, and allows for the acquisition of zero-loss Lorentz images, increasing the signal-to-noise ratio by at least an order of magnitude [3]. The magnetic resolution of this system is better than 5 nm, with a maximum magnification of 60,000X at the CCD camera. beam coherence.

Magnetic induction mapping in the TEM was first introduced in [4] and adapted to this 4000EX + GIF combination in [5]. The mapping method consists of the acquisition of four series of Foucault images at increasingly larger beam tilts. Subsequent digital processing of these images can then be shown to result in a 2D map of the in-plane component (normal to the electron beam) of the magnetic induction multiplied by the local specimen thickness. While this technique has proven its usefulness for "bulk" thin foils, it is unclear at this point whether or not similar information could be obtained for nanoscale samples; in particular, it is unclear whether or not fringe fields (or demagnetization fields) can be analyzed using the magnetic induction mapping method.

Off-axis electron holography has successfully been used to obtain quantitative measurements of a variety of magnetic structures, e.g., the magnetic dipole and quadrupole moments in coated MFM tips [6] and the magnetization reversal behavior in patterned cobalt nanostructures [7]. Analysis of the phase distribution of reconstructed electron holograms allows one to obtain electrical and magnetic information from very localized regions or nano-particles, with good spatial resolution and significant phase resolution.

As standard Lorentz observations have been carried out on a variety of materials systems, the increasingly smaller length scales require that we answer the question: for how small a particle can we still expect to observe magnetic contributions to the image contrast? Or, in other words, what is the lower size limit for Lorentz microscopy observations. The present paper first provides an overview of the Lorentz image formation modes and electron holography as applied to the study of spherical nanoparticles with uniform magnetization. From an analytical expression for the electron wave function we then obtain simulated images for a range of different experimental conditions and particle sizes. We emphasize the resolution limits of both Fresnel and Foucault methods and point out the experimental conditions which need to be satisfied to obtain reliable images containing magnetic contrast. We conclude the paper with a section on electron holography, as applied to magnetic cobalt nanoparticles.

IMAGE FORMATION FOR MAGNETIC STRONG PHASE OBJECTS

Magnetic phase contrast images can readily be computed if the object phase function is known. In fact, magnetic image formation can be considered to be a special case of what is commonly known as *high resolution TEM*. For an in-depth description of HRTEM we refer the interested reader to [8]. For the purposes of this paper it will be sufficient to briefly outline the image formation theory. A monochromatic, parallel incident electron beam (plane wave) is modulated by a thin foil sample. Both phase and amplitude of the electron wave function $\psi(\mathbf{r})$ are modified. The objective lens of the microscope blurs the wave function, and the amount of blurring is described by the microscope *point spread function* $T(\mathbf{r})$. The intensity distribution observed on the viewing screen is then given by $I = |T(\mathbf{r}) \otimes \psi(\mathbf{r})|^2$, ignoring image rotation and magnification. The symbol $\otimes$ denotes the convolution product. The point spread function is usually expressed in the back focal plane of the objective lens, and is then known as the *microscope transfer function* $T(\mathbf{q})$, with $\mathbf{q}$ a vector in the back focal plane (reciprocal space). Since a magnetic lens can be regarded as an analog Fourier Transformer, the transfer function is related to the point spread function by a Fourier transform operation $T(\mathbf{q}) = \mathcal{F}[T(\mathbf{r})]$. The transfer function contains contributions from spherical aberration, defocus, defocus spread, beam divergence, back focal plane aperture, and other aberrations, as described on page 85 in [8].

In the case of magnetic phase contrast one can show [1] that the object wave function is a pure phase function $\psi(\mathbf{r}) = e^{i\phi(\mathbf{r})}$. The phase shifts involved can be quite large (several radians), which means that the so-called *weak-phase object approximation* commonly used for the simulation and interpretation of standard high resolution electron images should be used with caution in the case of magnetic phase contrast.

To facilitate the discussion of Lorentz imaging modes and a subsequent comparison with electron holography observations, we will restrict all image simulations to spherical particles with a uniform magnetization. Since an analytical expression for the vector potential of such a configuration is available [9], this allows for a detailed numerical analysis of the various Lorentz observation modes. It also permits a comparison with the more quantitative holography method, which allows for the complete retrieval of both phase and amplitude of the electron wave function. The phase function $\phi(\mathbf{r})$ consists of two components $\phi_m(\mathbf{r})$ and $\phi_e(\mathbf{r})$, the former magnetic and the latter electrostatic. The magnetic phase can be calculated from the Aharonov–Bohm expression [10], combined with the analytical vector potential for this configuration:

$$\phi_m(x,y) = M_0 \frac{e}{\hbar} \frac{4\pi a^3}{3\,(r^3 > a^3)} \int_{-\infty}^{\infty} \hat{\mathbf{z}} \cdot \{y(\hat{\mathbf{x}} \sin\theta + \hat{\mathbf{z}} \cos\theta) - \ell \hat{\mathbf{y}}\}\, d\ell. \tag{1}$$

The electron travels along the positive z-direction of a Cartesian reference frame defined by the unit vectors $\hat{\mathbf{x}}$, $\hat{\mathbf{y}}$, and $\hat{\mathbf{z}}$; ℓ parametrizes the distance of the electron to the plane $z = 0$. The particle with radius a is located at the origin, and its magnetization is given by $\mathbf{M} = M_0(\sin\theta\hat{\mathbf{x}} + \cos\theta\hat{\mathbf{z}})$ where θ is the angle between the electron trajectory and the magnetization vector. The distance r is given by $r = \sqrt{x^2 + y^2 + \ell^2}$ and the notation $(a > b)$ indicates that the larger of a and b should be used.

Introducing normalized coordinates $\bar{r} = \frac{r}{a}$, $\bar{x} = \frac{x}{a}$, ..., and using $M_0 = \frac{3}{8\pi}B$, with B the saturation induction, we find after a simple integration:

$$\bar{\phi}_m(x,y) = \frac{\phi_m(x,y)}{B_\perp a^2} = \frac{e}{\hbar} \frac{\bar{y}}{\bar{r}^2} \left\{1 - \beta(\bar{r}, \frac{3}{2})\right\}, \tag{2}$$

where $\bar{r}^2 = \bar{x}^2 + \bar{y}^2$, $B_\perp$ is the in-plane component of $\mathbf{B}$, and

$$\beta(r,q) = \left[1 - (r^2 < 1)\right]^q.$$

One can similarly compute an electrostatic phase shift for a spherical particle with a mean inner potential V_s; it is easy to show that this phase shift is given by

$$\phi_e(x, y) = 2\sigma \left[V_m t + a(V_s - V_m)\beta(\bar{r}, \frac{1}{2}) \right] \tag{3}$$

for a particle embedded in a matrix with mean inner potential V_m and thickness $2t$, and

$$\phi_e(x, y) = 2\sigma \left[V_m t + aV_s\beta(\bar{r}, \frac{1}{2}) \right] \tag{4}$$

for a particle lying on top of a support film with mean inner potential V_m and thickness $2t$. $\sigma = \frac{2\pi m|e|\lambda}{h^2}$ is the interaction constant, m the relativistic mass of the electron ($m = m_0\gamma$) and λ the relativistic electron wave length. In the absence of a support or embedding film — a situation which may arise when chains of particles are suspended across a hole in the support film — one can simply put V_m equal to zero.

An important component of standard Lorentz images of nanoparticles is the presence of background contrast caused by the support film or the embedding medium. Any realistic modeling of the image contrast must take this into account. There are three types of supports/embedding media:

- amorphous carbon films, typically with a thickness of around $1 - 10$ nm; the particles lie on top of the support film.
- thermoplastic matrix, consisting of long polymer chains which maintain their individual character in the hardened state, and with a thickness determined by the specimen preparation technique (typically around $5 - 20$ nm); the particles are embedded inside the polymer.
- a metallic embedding matrix with a thickness of a few times the particle diameter.

A mean inner potential V_m can be assigned to all three types. However, only in the third case is a *uniform* mean inner potential a realistic and accurate description of the embedding medium. The amorphous or polymeric nature of the other two media causes random fluctuations in the local electrostatic potential, and such fluctuations may well give rise to strong contrast features, in particular for the Foucault imaging mode to be discussed below. Both media have densities in the same range (around 2 g/cm^3). The fluctuations are truly random for the first medium, but have long range correlations for the polymeric medium, since the polymer chains retain their individual character.

The background contrast can be taken into account in image simulations in the following way:

- *amorphous carbon support*: for the density stated above there are about 100 atoms per nm^3, with a typical spacing of about 0.15 nm. Since only the projection of the atom positions along the incident beam direction counts, it is straightforward to randomly generate an array of carbon atoms, in a field of view of 512×512 nm^2. For a film thickness of 5 nm, this would imply the random generation of about 131 million carbon atom locations. Each atom can be represented by a gaussian disc, and the average potential for the film can be taken equal to the mean inner potential of graphite. An example of such a projected potential $V_m(\mathbf{r})$ is shown in Fig. 1a.

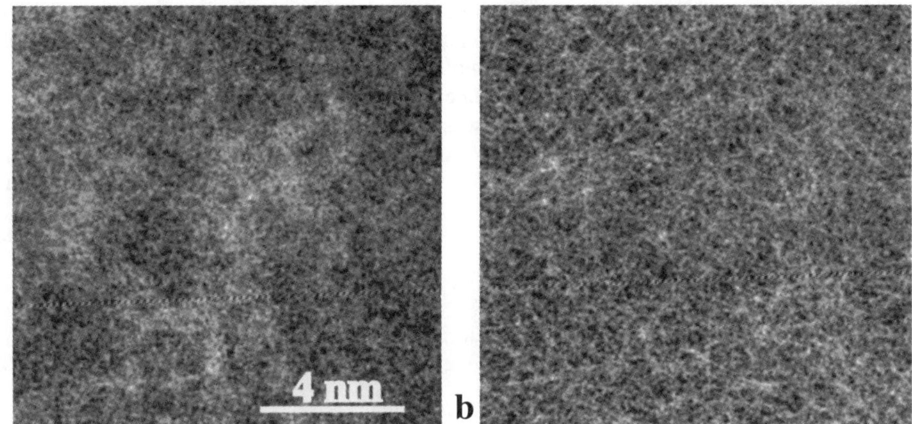

Figure 1: Projected potential for a) an amorphous carbon support film and b) a thermoplastic embedding medium. White corresponds to a high projected potential.

- *thermoplastic matrix*: The procedure to generate a model of such a matrix is nearly identical to that used for the amorphous carbon support film, except that atom positions are restricted to lie along chains. Using periodic boundary conditions it is straightforward to generate a projected potential for a single chain of 131 million carbon atoms; a random direction variation with an angular spread of $\pm 2^\circ$ is included in each bond, which gives rise to the structure shown in Fig. 1b. The average potential is then again rescaled to match the mean inner potential of graphite.

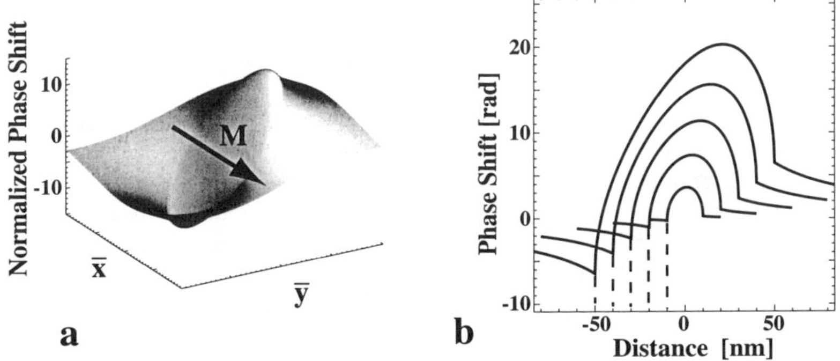

Figure 2: a) shaded surface representation of the magnetic contribution $\bar{\phi_m}(\bar{x},\bar{y})$ in rad/T·nm^2; b) total phase shift for an isolated cobalt particle as a function of particle radius at 400 kV.

The normalized functions $\bar{\phi}_m$ and $\bar{\phi}_e$ show that the electrostatic component of the total phase shift varies linearly with particle radius, whereas the magnetic component depends in a quadratic way on a. This means that with increasing acceleration voltage (and hence decreasing σ), the magnetic contribution becomes relatively stronger; however, the Lorentz deflection angle decreases with increasing accelerating voltage, indicating that standard Lorentz observations become more difficult in higher voltage microscopes. The normalized magnetic contribution to the total phase shift is shown in Fig. 2a; the function has mirror symmetry with respect to the plane $\bar{x} = 0$, and is antisymmetric along the line $(0, \bar{y})$. The linear and parabolic increases of the electrostatic and magnetic components, respectively, are illustrated in Fig. 2b, for a cobalt particle ($V_s = 29.46$ V, and $B = 1.7$ T) at 400 kV. The electrostatic component corresponds to the height of the central "bump", the magnetic component gives rise to the offset between the left and right shoulders of the curve.

Simulation of Magnetic Phase Contrast Images: Fresnel Mode

Lorentz image simulations can now be carried out, starting from the analytical results for the electrostatic and magnetic phase shifts. In the back focal plane of the objective lens the wave function is given by [8, p. 85]:

$$\psi(u, v) = \mathcal{F}_{u,v}[e^{i\left(B_{\perp}a^2\bar{\phi}_m(x,y)+\phi_e(x,y)\right)}]\left\{a(u, v)e^{-i\pi\lambda\Delta fq^2}e^{-\frac{(\pi\theta_c\Delta f)^2}{\ln 2}q^2}T(q^3, q^4, \ldots)\right\}, \quad (5)$$

with $\mathcal{F}_{u,v}$ the direct Fourier transform operator, u, v spatial frequency coordinates in the back focal plane, $a(u, v)$ the aperture function, θ_c the beam divergence angle, $q^2 = u^2 + v^2$, Δf the defocus, and λ the electron wavelength. Because of the small Lorentz deflection angles, typically in the range of tens of microradians, contributions of order 3 and higher in q (the function T in the equation above), including spherical aberration, are small and can be safely ignored when a small radius aperture is used. Furthermore, large defocus values are commonly used for Lorentz microscopy, again because of the small deflection angles.

The image intensity is as usual given by (ignoring image rotation and magnification)

$$I(x, y) = |\mathcal{F}^{-1}_{x,y}[\psi(u, v)]|^2.$$

The Fresnel or out-of-focus observation mode can readily be simulated by varying the defocus parameter Δf, for a given set of microscope parameters. The image simulations presented in the remainder of this paper are then carried out for two different microscopes, a 200 kV field emission TEM with a beam divergence θ_c of 10^{-5} radians, and a 400 kV LaB_6 thermal emission TEM with a beam divergence θ_c of 10^{-4} radians. The spherical aberration C_s for both machines is taken to be large, of the order of 100 mm.

In order to determine the particle size for which magnetic contributions to the image can still be observed, image simulations were performed for a configuration of five particles, with radii of $25, 20, 15, 10$ and 5 nm. All particles are pure cobalt, with an in-plane induction component of $B_{\perp} = 1.7$ T, and a mean inner potential of 29.46 V. The magnetic induction points from left to right in all simulated images. A diffraction aperture with radius 0.3 nm^{-1} is used in all simulations.

Fig. 3 shows simulated Fresnel images for particles without suspension medium (left column), particles suspended on a 20 nm thick amorphous carbon film corresponding to Fig. 1a (middle column), and particles embedded in a 40 nm thick polymer matrix corresponding to Fig. 1b (right-most column). The top row shows the total phase distribution for 200 kV electrons with a beam divergence of $\theta_c = 10^{-5}$. The remaining images correspond to defocus values ranging from -20 to $+20$ μm.

The vacuum particles show a clear fringe pattern with a bright feature slightly off-center. This is due to the magnetic contribution to the total phase shift; a purely electrostatic phase shift would

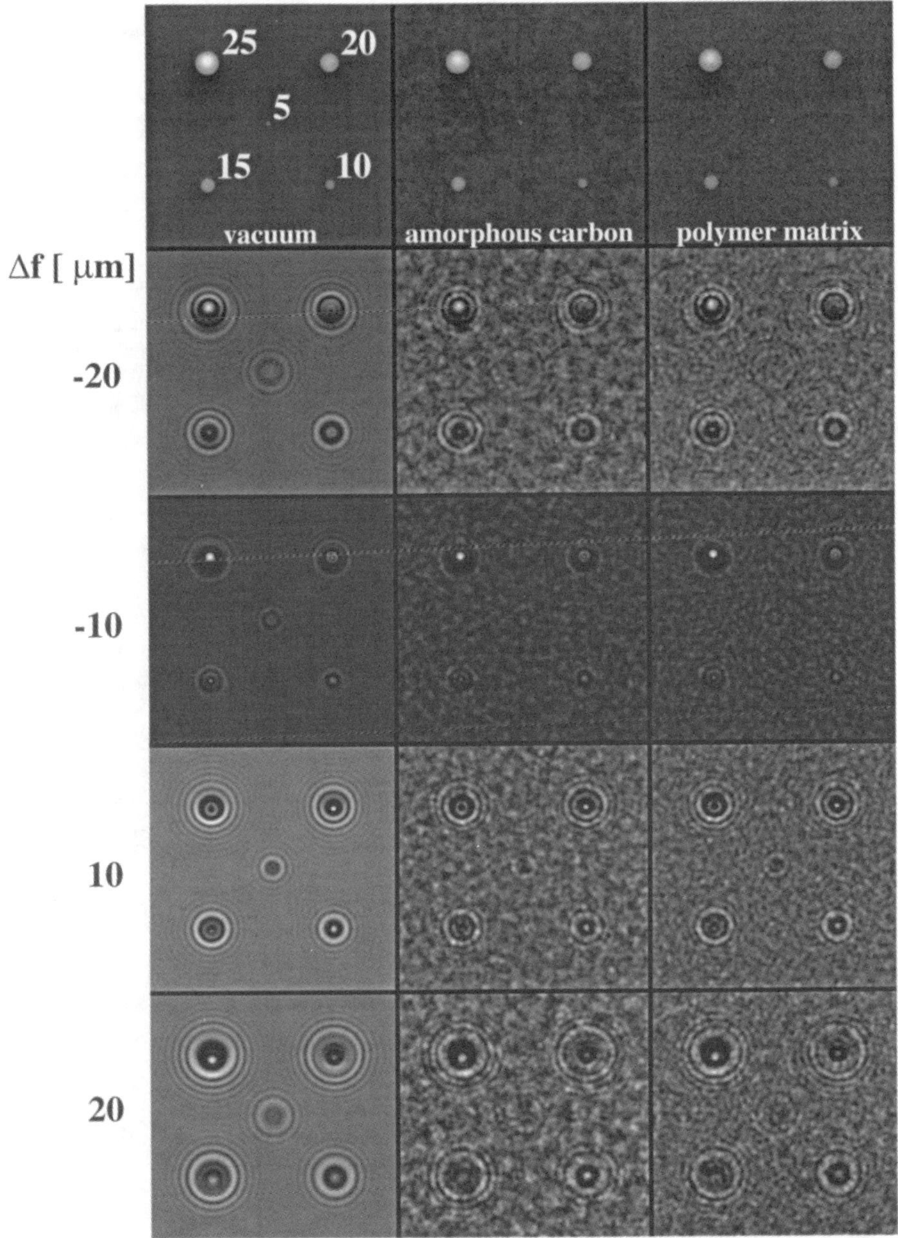

Figure 3: Through-focus Fresnel mode images for a 200 kV field-emission microscope for 5 spherical particles with radii of 25, 20, 15, 10 and 5 nm suspended in vacuum (left column), on a 20 nm thick amorphous carbon support film (center column), and embedded in a 40 nm thick polymer matrix (right column). Image simulation parameters are described in the text.

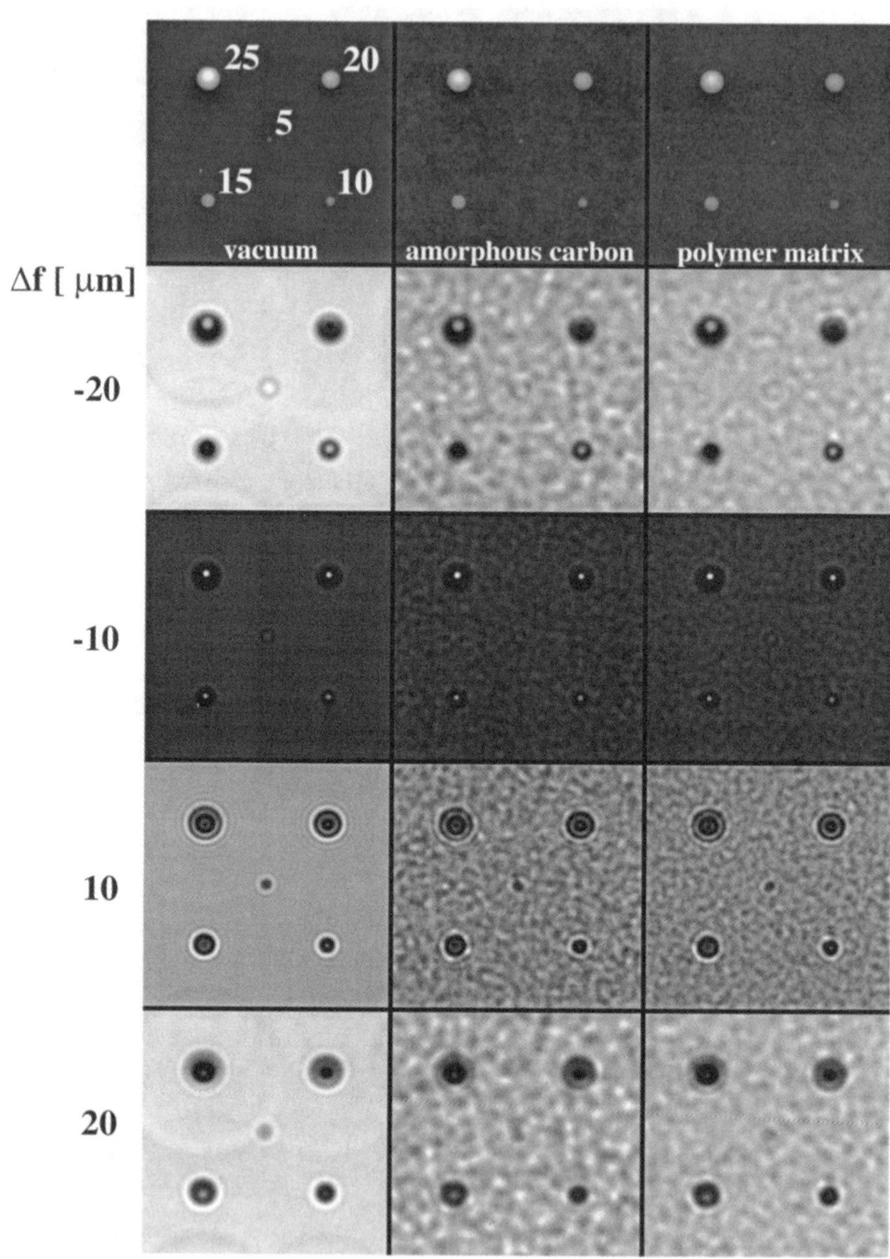

Figure 4: Same as Fig. 3 for a 400 kV conventional filament microscope.

give rise to concentric fringes without any shifts. The asymmetry between the top and bottom part of the fringe patterns is still visible for the 15 nm radius particle, but can no longer be discerned for the smaller particles. For larger defocus values one can obtain much larger shifts of the central feature, so that the magnetic contribution may also be discerned in the smaller particles; however, larger defocus values are also associated with large magnification changes, so that the use of the Fresnel mode becomes questionable for the smallest particle sizes.

In the case of particles suspended on or embedded in a support matrix, the contrast from the matrix itself overlaps with that of the particles, making it more difficult to separate out the magnetic component of the image contrast. The asymmetry is barely visible in the computed images of the 20 nm radius particle. Increasing the defocus will not be very useful in this case, because the contrast from the support matrix will also become coarser.

Fresnel image simulations for the same configuration of particles and support/embedding matrix and for a 400 kV microscope with lower beam coherence are shown in Fig. 4. The reduced beam coherence is reflected in the fact that fewer interference fringes are visible, and that the fringes are rapidly damped with increasing defocus. The support film also has a much coarser appearance than for the more coherent illumination conditions of Fig. 3. The increased accelerating voltage, while resulting in a relatively larger magnetic contribution to the total phase shift, also results in smaller Lorentz deflection angles, which means that for a given defocus value the asymmetry in the 400 kV image will be less than in the corresponding 200 kV image. This makes it more difficult to use the Fresnel mode at high accelerating voltages.

Simulation of Magnetic Phase Contrast Images: Foucault Mode

For the Foucault imaging mode the defocus Δf is set equal to zero, so that the image contrast is entirely determined by the radius and position of the diffraction aperture. For all image simulations the aperture radius was taken to be the same as for the Fresnel mode, i.e., $0.3\ \mathrm{nm}^{-1}$.

Fig. 5 shows Foucault images for three different situations; the leftmost column shows 5 images for unsupported particles. The arrows on the left indicate the direction of the aperture shift with respect to the magnetization direction (which is from left to right). The magnitude of the shift is such that the aperture nearly touches the central transmitted beam, similar to what one would do experimentally. The resulting image contrast consists of two components: inside the particle disks, there is a bright feature, which is off-center in the direction opposite to the aperture shift direction. This feature is mostly due to the electrostatic phase shift. The contrast outside the particle contains only magnetic information; one observes the typical double-lobe contrast features associated with magnetic image contrast. The bright and dark lobes switch positions when the aperture is shifted in the opposite direction, and if this is observed experimentally, this may be taken to be a direct signature of magnetic image contrast. Note that for the 10 and 5 nm radius unsupported particles, the magnetic component of the contrast is no longer as clear.

When the particles are embedded in a polymer matrix, the contrast outside the particles may no longer be clearly discernible, as is shown in the rightmost column. The parameters are again taken to be those of cobalt, and the images were computed for a 200 kV field emission microscope. The center column shows identical simulations *for particles with zero magnetic induction.* Comparison of the center and rightmost columns indicates that the magnetic component of the image contrast may be clearly defined for the larger particles, but becomes subtle or disappears completely for the smaller particles.

Since it is notoriously difficult experimentally to translate the diffraction aperture reproducibly

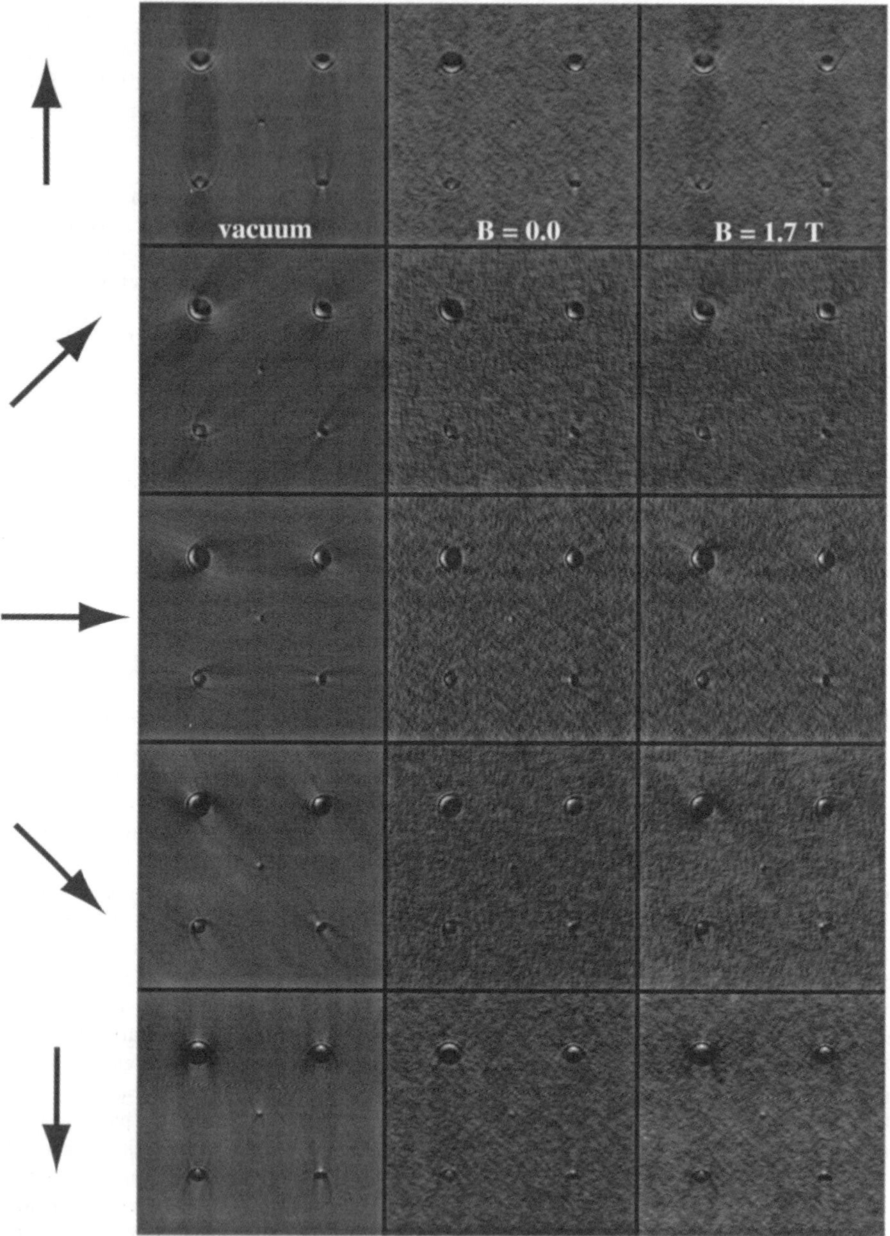

Figure 5: Foucault images for the same configuration of spherical particles as in Fig. 3. The arrow indicates the direction of the aperture shift. The left column shows the images for unsupported particles, the rightmost column for a 40 nm thick polymeric embedding matrix. The center column shows the corresponding images in the absence of magnetic induction ($B = 0$).

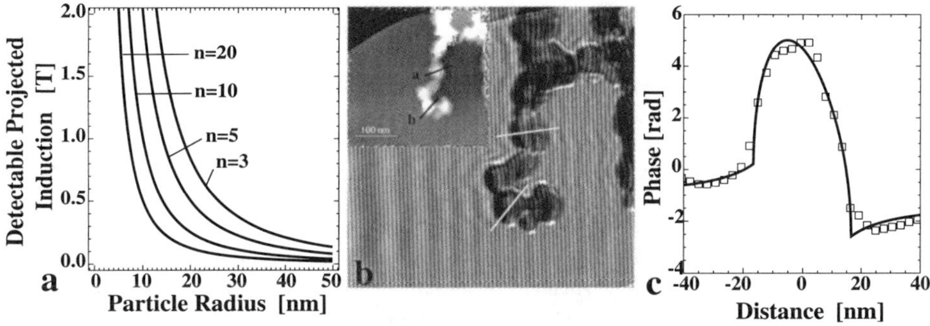

Figure 6: a) Detectable magnetic induction as a function of particle size and holographic phase resolution; b) experimental hologram of Co-nanoparticles with the reconstructed and unwrapped phase in the inset; c) non-linear least squares fit of the phase profile across particle **a** in b).

by a given amount, it becomes clear that the Foucault imaging mode must also fail to provide quantitative magnetic information for the smaller particle size range.

Simulation of Magnetic Phase Contrast Images: Holography Mode

In the off-axis holography method, a biprism is used to split the incident coherent electron wave into a reference wave, which passes through vacuum, and a wave which passes through the sample. The interference pattern between the two waves constitutes the electron hologram. By means of Fourier analysis one can then reconstruct the amplitude and phase of the electron wave, and simultaneously remove the effects of the microscope transfer function. If we assume that the smallest resolvable phase difference $\Delta\phi$ is written as $\frac{\pi}{n}$, with n an integer, then one can show [11] that the smallest detectable in-plane magnetic induction component for a spherical particle with radius a is given by:

$$B_{\perp}^{\mathrm{min}} = \frac{1011.74}{na^2} \text{ Tesla.} \tag{6}$$

Fig. 6a shows this relation for four different phase resolutions $n = 3, 5, 10$ and 20. Since a phase resolution of $\frac{\pi}{10}$ can routinely be obtained, the lower particle radius limit for cobalt nanoparticles is about 7.7 nm, or a diameter of about 15 nm. To detect the magnetic induction in smaller particles requires a much better phase resolution, and is thus much more difficult experimentally.

Fig. 6b shows an experimental hologram with the reconstructed phase in the inset for a string of cobalt nanoparticles suspended across a hole in a support film. A trace across the particle labeled **a** is shown in Fig. 6c, along with a least squares fit to the phase profile (continuous line). The parameters in the fit procedure were the particle radius (16.54 ± 0.58 nm), the in-plane magnetic induction $B_{\perp}$ (1.68 ± 0.34 T), the mean inner potential V_s (24.24 ± 2.13 V), and the particle center location and phase offset. We refer to [11] for more details about the fit procedure and the experimental holography procedure. The value for the magnetic induction is in good agreement with the bulk value of $B = 1.7$ T; the mean inner potential is somewhat lower than the calculated value of $V_s = 29.46$ V. This discrepancy could be caused by elastic scattering in the cobalt particle. The approximation used in this paper ignores such diffraction effects.

CONCLUSIONS

From Figs. 3 and 4 we find that there is a lower limit to the particle size for which the Fresnel imaging mode can produce useful magnetic image contrast. While it is difficult to state a particular value for this lower limit (the actual value depends somewhat on the microscope parameters and on the support or embedding medium), it is not unreasonable to state that the lower limit is probably in the range of $10 - 20$ nm particle radius for the cobalt particles used in the image simulations. For particles with a higher saturation induction, magnetic image contrast may be present down to smaller sizes, although the contrast from the support/embedding medium will always be present. One should be cautious in assigning magnetic origins to Fresnel type image contrast observed for particles in this size range or smaller. A similar remark holds for the Foucault observation mode, which is less reproducible than the Fresnel mode. Depending on the embedding medium and the magnitude of the in-plane component of the magnetic induction, one may be able to define a lower size limit. Most likely this limit will also be in the range of $10 - 20$ nm radius particles. Off-axis electron holography has been shown to produce accurate results for both the in-plane magnetic induction and the mean inner potential. The lower size limit for the detection of magnetic contributions to the electron wave phase can be expressed analytically, and for the cobalt particles used throughout this paper this limit is estimated to be around 8 nm radius. Contrary to Fresnel and Foucault observation modes, electron holography directly produces the phase information from which magnetic parameters may then be derived.

ACKNOWLEDGEMENTS

The author would like to acknowledge collaborations and/or conversations with N.T. Nuhfer, J. Dooley, M.E. McHenry, S. Majetich, J.H. Scott, M.R. McCartney, and S. Supapan. This work was carried out with partial financial support from the National Science Foundation through contracts DMR-9403621 under the guidance of Dr. L. Inglehart, and DMR-9501017 under the guidance of Dr. B. MacDonald.

References

[1] L. Reimer, *Transmission Electron Microscopy : Physics of Image Formation and Microanalysis*, Third Edition, Springer-Verlag, Berlin, 1993.

[2] P.J. Grundy, and R.S. Tebble, Adv. Phys., **17**, p. 153 (1968).

[3] J. Dooley, and M. De Graef, Micron, **28**, p. 371-380 (1997).

[4] A. Daykin, and A. Petford-Long, Ultramicroscopy, **58**, p. 365 (1995).

[5] J. Dooley, and M. De Graef, Ultramicroscopy, **67**, p. 113 (1997).

[6] D.G. Streblechenko, M.R. Scheinfein, M. Mankos, and K. Babcock, IEEE Trans. Magn., **32**, p. 4124 (1996).

[7] R.E. Dunin-Borkowski, M.R. McCartney, B. Kardynal, and D.J. Smith, J. Appl. Phys., **84**, p. 374 (1998).

[8] J.C.H. Spence, *Experimental High-Resolution Electron Microscopy*, 2nd Edition, Oxford University Press, Oxford, 1988.

[9] J.D. Jackson, *Classical Electrodynamics* (Second Edition), J. Wiley & Sons, New York, 1975.

[10] Y. Aharonov, and D. Bohm, Phys. Rev., **115**, p. 485 (1959).

[11] M. De Graef, N.T. Nuhfer, and M.R. McCartney, J. Microscopy, in press (1999).

MAGNETIC FORCE MICROSCOPY STUDY OF NEW NANOCRYSTALLINE SOFT MAGNETIC RIBBONS. M. E. Hawley , G. W. Brown, and D. J. Thoma, Los Alamos National Laboratory, Los Alamos, NM; M. A. Willard, D. E. Laughlin, and M. E. McHenry, Carnegie Mellon University, Department of Materials Science and Engineering, Pittsburgh, PA.

ABSTRACT

New nanocrystalline, multicomponent extremely soft magnetic materials with superior high temperature magnetic properties hold great promise in power applications. Fabricated in ribbon form by rapid solidification methods, the initial material is amorphous. By controlled annealing procedures, the amorphous material was transformed into a nanocrystalline form with the degree of crystallinity determined by the annealing temperature and time. The magnetic structures of ribbons, as-fabricated and annealed at temperatures from 550 to 750 °C were examined by magnetic force microscopy to determine the impact of residual stress and nanocrystallinity on the observed structure. A correlation was seen between the magnetic structures and surface microstructure. The wheel side of the as-processed ribbon was rougher than the top side of the ribbon and a complicated magnetic domain structure was present in the amorphous material. After annealing, nanocrystals formed, increasing in size with increasing temperature. The lowest temperature annealed sample had a bimodal grain size distribution and a combination of stripe and localized domains. After annealing little difference was seen between the two sides of the ribbons. Stripe domains were absent in the ribbons annealed at the highest temperatures.

INTRODUCTION

Both intrinsic and extrinsic effects play an important role in determining the ultimate properties of materials. While considerable work has focussed on chemically tuning properties, increasing attention is being directed toward understanding the role of process-induced effects such as stress and microstructure, i.e. the structure-property relationships. Although these process-induced effects can impact adversely on the functionality of the material, they can also be viewed as new degrees of freedom for producing unique tunable properties. Of particular interest here, is how they impact on the properties of magnetic alloy materials, e.g. stress-induced anisotropies in amorphous alloys [1]. In this work, a new alloy, HITPERM [2], with a composition of $(Fe_{0.5}Co_{0.5})_{88}Zr_7B_4Cu_1$, was produced in the form of amorphous melt-spun ribbons. Some ribbons were annealed at temperatures between 550 °C and 750 °C to produce nanocrystalline material. X-ray diffraction results combined with differential thermal analysis determined that the as-produced material was amorphous and that α'– FeCo crystallites form above 510 °C with secondary crystallization of small amounts of $(FeCo)_3Zr$ appearing above 700 °C. The addition of Zr and B is believed to aid glass forming properties while a small amount of Cu depresses the temperature at which α'– FeCo crystallites form.

Previously, atomic and magnetic force microscopies (AFM and MFM) were used to study amorphous $Fe_{80}B_{16}Si_4$ alloy ribbons [3]. In that study, stress was deliberately introduced in the ribbon by bending and the subsequent stress-induced microstructures and magnetic structures were studied. In that case, the MFM data faithfully reproduced the results obtained by more traditional techniques [4]. Although MFM, at this time, cannot give unambiguous quantitative

Mat. Res. Soc. Symp. Proc. Vol. 577 © 1999 Materials Research Society

magnetic information, that study demonstrated that it is powerful tool for revealing detailed relatively high resolution (10's of nanometers) magnetic structure information.

In the present study, AFM and MFM were used to study the micro- and magnetic structure of HITPERM ribbons resulting from the fabrication and annealing processes. This alloy is particularly attractive for use in high temperature applications because it retains a high magnetization to elevated temperatures (> 900 °C). The as-produced HITPERM ribbons produced in the melt-spinning process are amorphous. HITPERM has a non-zero magnetostrictive constant and is, therefore, subject to a number of factors during the fabrication and annealing processes that can lead to the formation of magnetic domains. Magnetoelastic effects during production due to a temperature gradient between the top and bottom sides of the ribbons as it is melt-spun, can result in residual stress creating a magnetic easy axis along a tensor direction. Annealing is often used to relieve this stress and eliminate magnetic domains and resulting ac losses. In this case, the annealing temperatures were deliberately chosen to create a nanocrystalline material to tailor magnetic properties by creating a two phase system composed of two magnetic phases, a nanocrystalline phase and an amorphous grain boundary phase. The current work is part of an ongoing study.

EXPERIMENTAL

Ingots of $Fe_{44}Co_{44}Zr_7B_4Cu_1$ were fabricated by arc melting electrolytic Fe(99.9% pure – low carbon), Co (99.9% pure), Fe_3B (90.75% metals basis), and Cu (99.9% pure) in an argon atmosphere. Amorphous ribbons were produced from the ingots using a single wheel melt spinning technique [5]. The melt spinning process involved remelting the arc-melted ingot in a boron nitride crucible in an argon atmosphere. A small positive pressure of argon was then used to quench the molten alloy onto the Cu-Be wheel rotating at 35 m/s. The ribbons produced by this procedure were approximately 1 mm wide and 20 to 50 μm thick. Following the melt spinning procedure, sections of the ribbon were isothermally annealed for 1 hour at 550 °C, 650 °C, or 750 °C in an inert (argon) atmosphere followed by a water quench. The annealing and quenching procedure produced nanocrystalline microstructures.

The ribbons were characterized using AFM and MFM. The microstructure, obtained in TappingMode™, and the force gradient magnetic structure, taken in constant-height lift mode, were acquired in a two-pass method that yields a one-to-one correspondence between microstructure and magnetic domain structure. Three types of commercially available magnetically coated Si tips were used for imaging. High coercivity (HC), moderately high coercivity CoCr, and superparamagnetic Fe/SiO_2 coated tips were used. Prior to imaging, the tips were polarized normal to the sample surface. To minimize the effect of static charges on the magnetic imaging, data was collected in an isolation box with moisture introduced by a humidifier (~45% relative humidity). The samples were grounded to the probe ground via a Cu strip.

Field-dependent MFM data were collected on the 550 °C ribbon. Preliminary field-dependent data were also collected on the 650 °C ribbon but will not be presented here. An external permanent magnet was used to apply an in-plane field of 34 to 250 Oe with MFM scans taken at a 25 to 50 Oe increment.

Very preliminary potentiometric measurements were made on some of the ribbons using the conductive magnetic tips. Typically 2 volts DC were supplied by a battery with a 22 MΩ current limiting resistor in series with the sample. 2 V AC was applied to the tip.

RESULTS

Figure 1 is a composite 15 μm x 15 μm images of the microstructures of the ribbons with their corresponding magnetic domain structures. The as-produced HITPERM ribbon appeared to be amorphous, i.e. no nanocrystalline structure was observed by AFM (Figure1a). The corresponding magnetic structure for this ribbon (Figure 1a - column 2) consisted of inclusion surrounded by a complicated array of parallel domain walls. Many of those walls appear to emanate from the inclusions. At the annealing temperatures selected in this study, nanocrystals were formed, at least on the surface of the treated ribbons, increasing in size with increasing temperature. As can be seen in Figure 1b, the ribbon annealed at 550 °C had a bimodal distribution of nanocrystal sizes (about 200 to 300 nm and 350 to 500 nm). Its magnetic structure (Figure 1b - column 2) consisted of stripe domains where the smaller crystallites existed and localized domains located in the larger crystallite regions. By 650 °C (Figure 1c), the surface RMS roughness of the ribbon had increased from about 40 nm for the 550 °C ribbon to around 60 nm. This increased roughness was a reflection of the wide distribution in crystallites sizes, ranging from 100 nm to over 700 nm for minority crystallites. The magnetic structure of the 650 °C ribbon (Figure 1b - column 2) consisted of a combination of meandering stripe walls/domains containing localized single domains located at the position of larger crystallites. The 750 °C ribbon surface crystallites were larger and more uniform in size, clustered between 450 nm and 650 nm with a surface RMS roughness of over 80 nm. The crystallites also appear more faceted, some showing the beginnings of island growth. The increased faceting suggests that these nanocrystals have improved crystal quality. One has to emphasize that all of the images are of the ribbon outer surface and do not necessarily reflect the structure of the interior.

A comparison between the differences in the microstructure and magnetic structure for the two sides of the as-processed ribbon and the ribbons annealed at 550 °C and 650 °C is presented in Figure 2a and 2b and Figure 2c and 2d, respectively. Although the wheel sides of these ribbons were rougher than their corresponding upper surface, the most dramatic difference existed for the as-prepared ribbon, Figure 2a and 2b column 1. The structure of the wheel side of the amorphous ribbon probably reflected the roughness of the surface of the wheel itself. The bottom sides of the ribbons annealed at 550 °C and 650 °C were only slightly rougher than their corresponding top sides (Figure 2a and b – column 2 and 3). The differences were not significant and might only reflect the particular selected area imaged. The annealing process led to the formation of nanocrystals on both sides of the ribbons, reducing the initial fabrication differences. The differences in structure between the upper side of the ribbons and their respective wheel side were also reflected in the magnetic structures, Figure 2c and 2d, and were again most notable for the as-fabricated ribbon. As mention above, the magnetic structure for the top side of that ribbon appears to consist of stress lines radiating out from inclusions. The character of the magnetic structure on the wheel side (Figure 2d – column1), although it had features running in the same

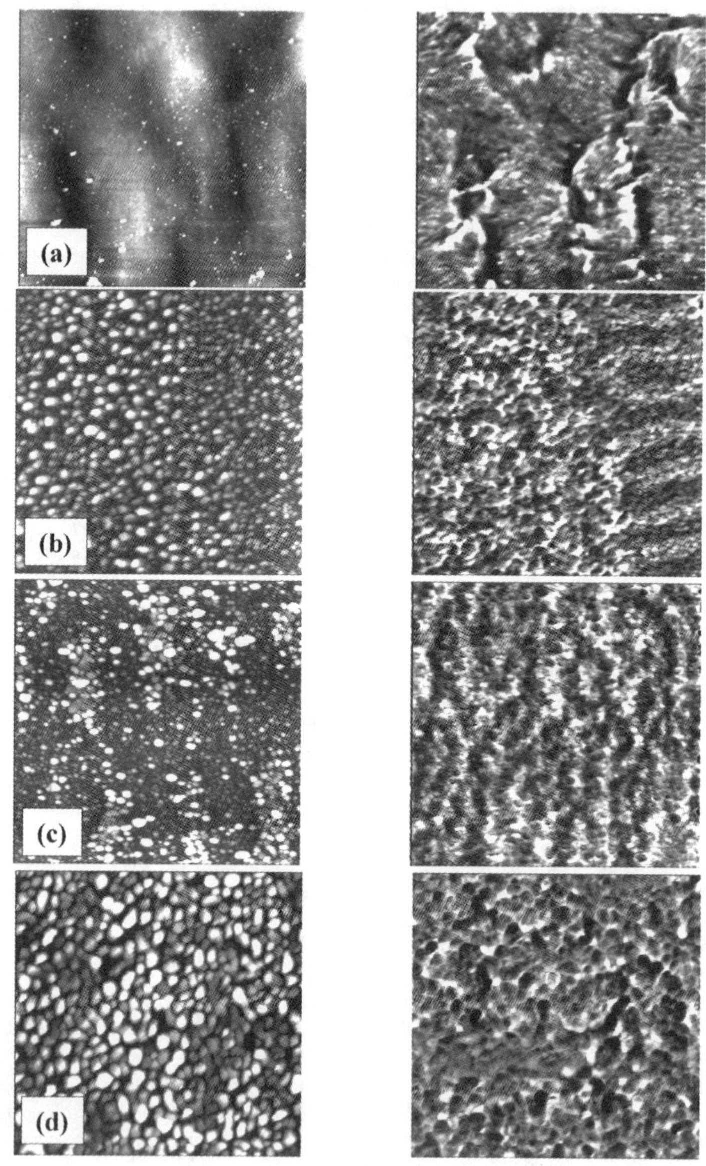

Figure 1. AFM (first column) and MFM (second column) 15 μm x15 μm images of the surfaces of $Fe_{44}Co_{44}Zr_7B_4Cu_1$ alloy ribbons: a) as-processed, b) 550 °C anneal, c) 650 °C anneal, and d) 750 °C anneal.

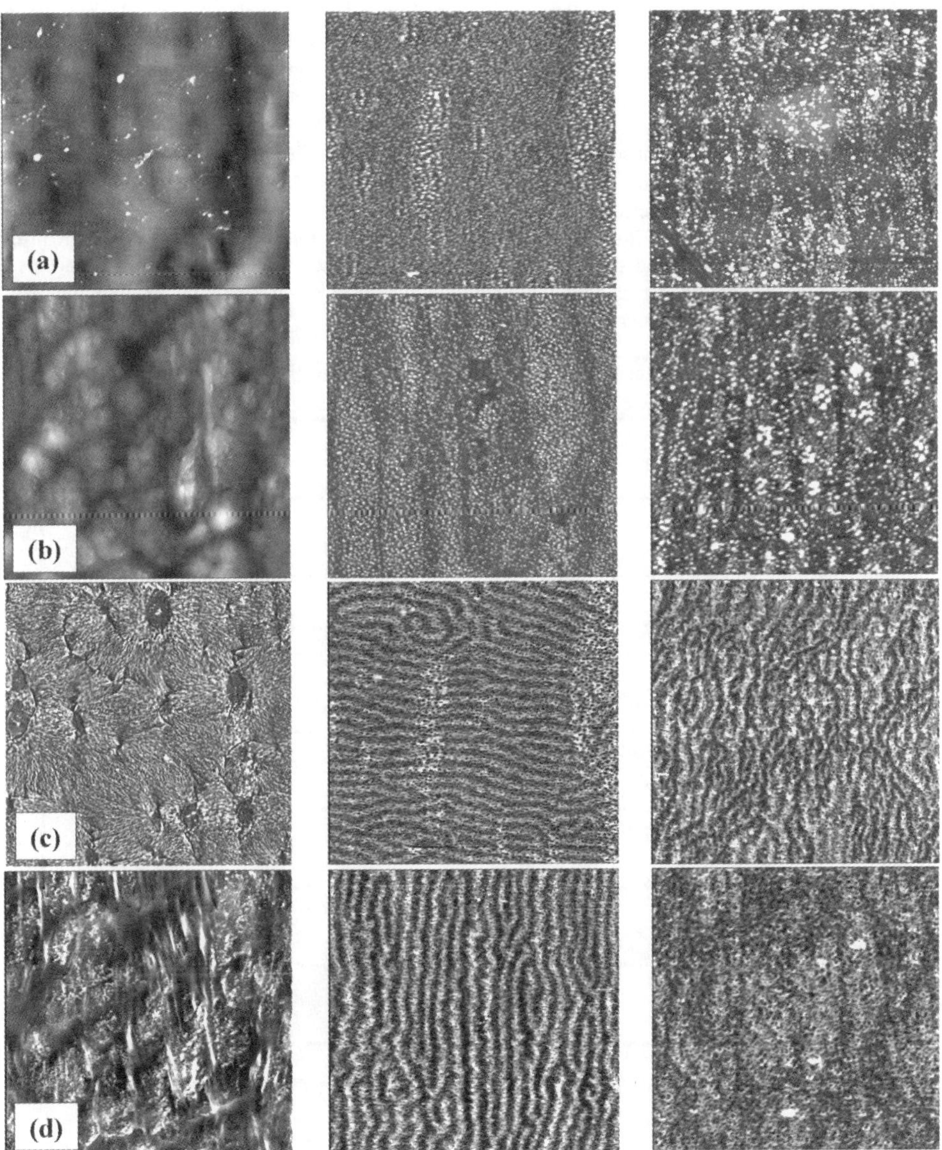

Figure 2. AFM, rows (a) and (b), and MFM, rows (c) and (d), 40 μm x 40 μm images of the surfaces of $Fe_{44}Co_{44}Zr_7B_4Cu_1$ alloy ribbons: Rows (a) and (c) are the top side of the ribbon, rows (b) and (d) are the wheel side. First column is the as-fabricated ribbon, the middle column was annealed at 550 °C, and the last column is the sample annealed at 650 ° C.

direction as the ribbon length, was too complicated to allow analysis. As noted above, the magnetic structure for the 550 °C ribbon - top and bottom (Figure 2c and d – column 2), and 650 °C ribbon - top (2c – column 3) had stripe or meandering stripe domains. Only the 550 °C ribbon had similar magnetic structures on both surfaces. The 650 °C ribbon – bottom appeared to have only localized domains. Again, these observations were limited to the small areas characterized. No extended domains were observed for the 750 °C ribbon. The magnetic structure for this ribbon, Figure 1, appeared to be localized to the individual crystallites. The correlation between the magnetic structure and microstructure is apparent from these images. The differences in the magnetic structures shown in Figure 1 reflect the evolution of the microstructure as the annealing temperature was increased and the average nanocrystallite sizes increased. The differences in Figure 2 reflect the thermal stress across the ribbons during fabrication and cool down and how the annealing temperature affects these differences.

Table I. Summary of crystallite sizes and surface RMS roughness.

Sample	Crystallite Size (nm)	RMS Roughness (nm)
Amorphous Top	N/A	7.7
Amorphous Wheel Side	*3 μm – 8 μm structural features	77
550 °C Top	200 – 300 & 350 - 500	40
550 °C Wheel Side	230 - 400	51
650 °C Top	150 - 350	50
650 °C Wheel Side	100 - 700	60
750 °C	450 - 650	80

Table I summarizes the microstructural data from Figure 1 and Figure 2. The spread in crystal size is shown in Table 1, not the average size.

Figure 3 is another topographic/magnetic 15 μm x 15 μm image set of the top surface of the as-processed ribbon, this time taken with a HC magnetic tip. These images clearly show a vortex-like the relationship between the inclusion and the surrounding magnetic structure. The magnetic image is free of streaks, a clue to the magnetic hardness of this stress-induces domain structure.

One concern with MFM imaging is the influence of the tip on the sample's magnetic structure and, therefore, the resulting MFM image. One counter strategy is to use tips with different magnetic hardness, in particular a very hard tip whose magnetic structure that isn't altered by the sample and a very soft tip whose magnetization is changes spontaneously under the influence of domains in the sample's. Figure 4 contains images taken with a paramagnetic tip of the amorphous (a) and 550 °C (b) and 650 °C (c) annealed ribbons. A comparison with the corresponding images in Figure 2 reveals very little differences, giving confidence that the field from the MFM tip has had little effect on domain structures of these ribbons.

In an attempt to identify whether the annealed ribbons had nonconductive phases, preliminary surface potential imaging was attempted. This technique could potentially eliminate some of the uncertainty in interpretation of some features in the magnetic images, i.e. whether structures in the MFM images are due to the presence of nonmagnetic material or whether they are due to the presence of in-plane magnetic domains. Briefly, this scanning probe two-pass technique using a conductive tip, the first pass in standard tapping mode to obtain the topography and the second pass in a nonvibrating mode but with oscillating voltage, $V_{ac}\text{Cos}\ \omega t$, applied to the tip and a DC voltage applied across the sample. This results in an oscillating electrostatic force $F = dC/dzV_{dc}V_{ac}$ on the cantilever at the ac frequency ω where dC/dz is the tip-sample capacitance gradient and V_{dc} is the voltage difference between the tip and the sample. The technique entails sensing variations in the voltage difference resulting from local variation in resistance within the sample. An equal and opposite voltage is applied to the tip to create a null voltage difference. Since the oscillating electrostatic force, F, is the product of the ac and dc components and $V_{dc} = 0$, the electrostatic force is also zero. The value of the applied field to the tip needed to accomplish the null condition becomes the intensity scale on the potentiometric image. Like the MFM technique, the technique is sensitive to changes within the sample not just the surface. Figure 5 is an image taken by this technique and reveals some areas over the 40 μm x 40 μm region where the resistivity is higher (brighter in the image). Although the impetus for using this technique was to identify nonconductive material in the ribbons, it is not possible at this time to speculate as to the actual origin of this variation in resistance and the resolution in this image is too low to identify individual insulating inclusions.

Finally, although field-dependent magnetic hysteretic data has not collected for any of these ribbon, MFM data was collected for the 550 °C ribbon at different values of in-plane external field applied via a permanent magnet. Data was collected for field values from 34 Oe to over 250 Oe. A selection of the images taken within this field range is presented in Figure 6 to show the evolution of the domain structure. By 34 Oe, the lowest field possible in the present experimental setup, the local domains had disappeared, leaving behind only the maze-like structure. As the field was increased to 100 Oe the spacing between domains decreased from about 2.4 μm to around 1.6 μm with a concomitant decrease in the number of branches resulting in a more stripe-like structure. The domain spacing was further reduced to about 1.3 μm at 125 Oe, 1.1 μm at 150 Oe, 0.9 μm at 200 Oe, and unmeasureable at 250 Oe. The wider stripes at the bottom of Figure 6f are probably due to electrostatic effects. The decrease in intensity of the MFM signal is primarily due to the rotation of the magnetization into the plane of the sample. The light and dark contrast in these images depend on whether the stray field out of the sample is aligned anti- or parallel to the tip magnetization direction. The only information about in-plane magnetic structure is indirect, that is the domain spacing observed at various fields between light and dark regions is

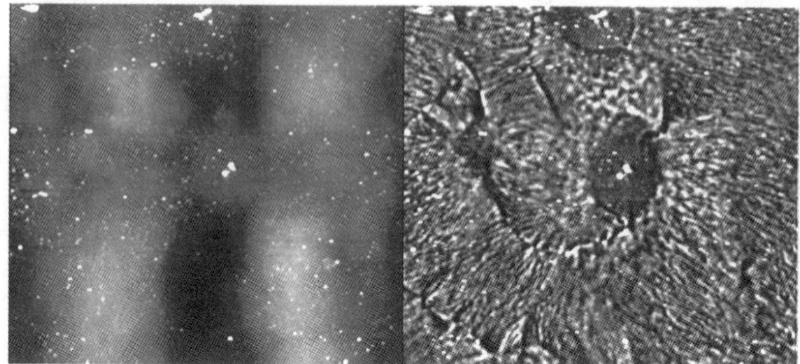

Figure 3. AFM and MFM 15 μm x 15 μm images taken on as-fabricated ribbon. The magnetic image (right) revealed an apparent inclusion not obvious in the topograph (left).

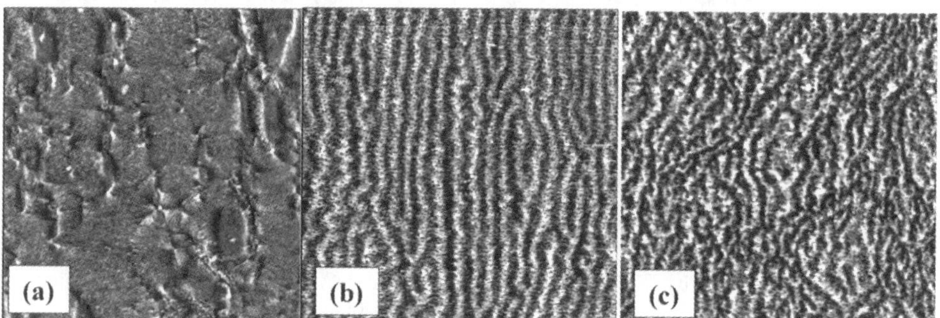

Figure 4. MFM 40 μm x 40 μm images taken on $Fe_{44}Co_{44}Zr_7B_4Cu_1$ alloy ribbons (a) as-fabricated, (b) annealed at 550 °C, and (c) annealed at 650 °C. These images were taken with superparamagnetic Fe/SiO_2 coated tips.

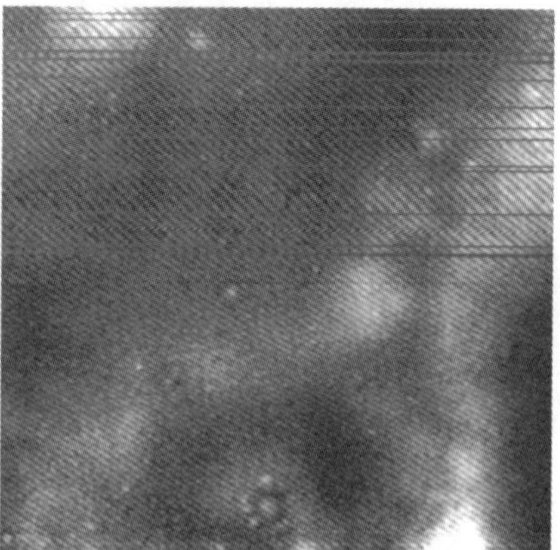

Figure 5. Preliminary 40 μm x 40 μm potentiometric image taken on a $Fe_{44}Co_{44}Zr_7B_4Cu_1$ alloy ribbons annealed at 550 °C with CoCr coated tips.

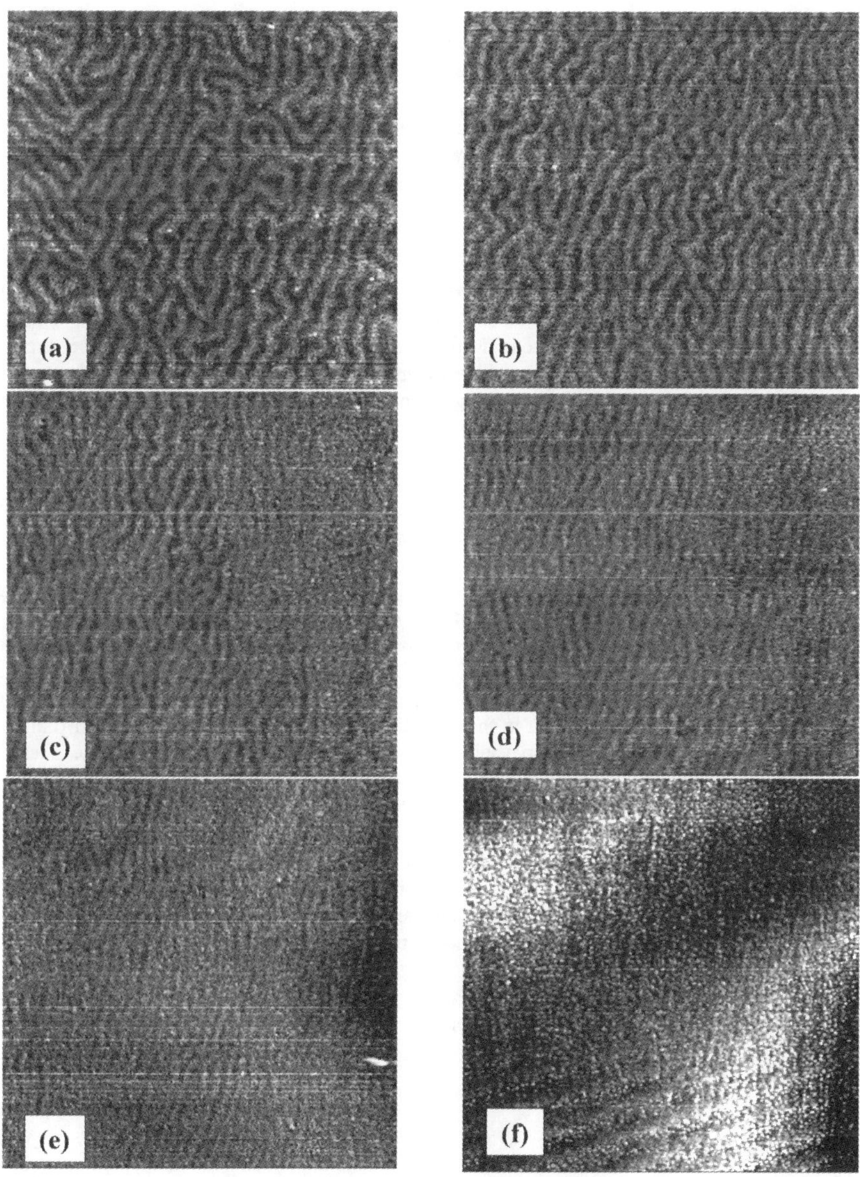

Figure 6. MFM 40 μm x 40 μm images taken on an alloy ribbon annealed at 550 °C with an in-plane magnetic field of (a) 34Oe, (b) 100Oe, (c) 125Oe, (d) 150Oe, (e) 200Oe, and (f) 250Oe.

uniform suggesting that any in-plane magnetization is oriented in approximately the same direction across the sample. Although not presented here, preliminary field-dependent MFM measurements on the 650 °C ribbon show the same trend. Note that this investigation examined the evolution of the domain structure for a field applied parallel to the ribbon thickness. Geometric (demagnetization) effects will cause the fields required to be much larger than if the field were applied along the ribbon length as would be the case in a toroidal core, for example.

DISCUSSION

The AFM data on the HITPERM ribbons revealed the evolution in the microstructure from the amorphous as-produced form to a nanocrystalline structure above 550 °C due with both crystal size and faceting increasing with increasing temperature. The low crystallization temperature is principally due to the addition of Cu. MFM measurements of the out-of-plane sample stray field reflect these microstructural changes evolving from magnetoelastically driven domains radiating from inclusions to maze or stripe domains due to both local exchange and longer range interactions. The magnetic structure of the larger nanocrystalline, higher temperature annealed ribbons appear monodomain and localized. One difficulty, however, remains in interpreting the MFM data, i.e. distinguishing between nonmagnetic areas from in-plane magnetization. This is particularly true when a large amount of complicated structure is present in the images. The complicated structures seen for the amorphous ribbon and those ribbons annealed at the higher temperature where phase separation is likely. In addition, some features in the magnetic images could be due to electrostatic effects or damping of the cantilever oscillation from other tip-sample interactions. The situation is complicated in the case of the field-dependent measurements when the in-plane applied fields approach the coercivity of the tip magnetic coating. The magnetization orientation of the tip in this case is expected to start rotating in response to the field in a direction cant to the sample surface normal as the external in-plane field approaches the coating coercivity. Nonetheless, the MFM measurements reveal some interesting trends in the evolution of the microstructure and corresponding magnetic structure as the as-fabricated sample was annealed at increasing temperatures.

The field-dependent MFM data does not reflect directly what is happening with the in-plane magnetization. As mentioned above it can give indirect information about whether in-plane domains are aligned anti parallel or not. It doesn't tell you how much of the magnetization is out-of-plane versus in-plane. The MFM data is not a substitute for vibrating sample magnetometry data but compliments that information. For instance, the MFM may shed light on why a hysteretic loop is not square due to out-of-plane magnetization due. The MFM-field data can also tell you about wall pinning and wall motion. The MFM is a powerful tool for revealing the presence and nature of domain structures as well as the presence of magnetoelastic effects to aid in interpretation of bulk type measurements.

SUMMARY

The magnetic images of the amorphous ribbon show that the magnetic structure is probably due to magnetoelastic effects from residual stresses present upon cooling. Inclusions, are more visible in the magnetic images, forming a vortex for magnetic domains. The complicated magnetic domain structure for the annealed samples reflects the presence of the nanocrystals. Stripe domains were present in the 550 °C ribbon and persisted somewhat in the 650 °C sample but were absent in the ribbon annealed at the highest temperatures. As the nanocrystals became larger the domains became more localized. The sample annealed at the highest temperature had localized domains, each nanocrystal having a single domain..

REFERENCES

1. S. Chikazumi, Physics of Ferromagnetism, 2nd Edn., (Oxford University Press, N.Y. 1997), p. 339.
2. M. A. Willard, D.E. Laughlin, and M.E. McHenry, *J. Appl. Phys*. **84 (12)**, 6773 (1998).
3. M.E. Hawley, G.W. Brown, D.J. Markiewicz, F. Spaepen, E.P. Barth, *J. Magn. & Magn. Mat.* **190**, 89 (1998).
4. V. Lakshmanan, J.C.M. Li, *Mat. Sci. Eng*. **98**, 483 (1988).
5. D. J. Thoma, E.M. Schwartz, S.R. Bingert, D.R. Korzekwa, R.D. Field, and L.A. Jacobson, "Microsegregation during Melt-Spinning of Dilute Palladium Alloys" in Melt Spinning, Strip Casting and Slab Casting, eds. E.F. Matthys and W.G. Truckner (TMS, Warrendale, PA) p. 173-184 (1995).

SURFACE ANALYSIS OF THE NANOCRYSTALLINE Fe-BASED ALLOYS BY CONVERSION ELECTRON MÖSSBAUER SPECTROSCOPY

A. GRABIAS *, M. KOPCEWICZ * and B. IDZIKOWSKI **
*Institute of Electronic Materials Technology,
Wólczyńska 133, PL 01-919 Warsaw, Poland
**Institute of Molecular Physics, Polish Academy of Sciences,
M. Smoluchowskiego 17, PL 60-179 Poznań, Poland

ABSTRACT

A combined conversion electron and γ-transmission Mössbauer spectroscopy is applied to investigate the differences in the surface and bulk crystallization of the amorphous Fe-M-B-Cu (M: Zr, Ti, Nb, Ta) alloys. It is shown that the surface crystallization occurs at an annealing temperature lower than in the bulk. The nanocrystalline bcc Fe phase is significantly more abundant at the surface than in the bulk in the entire annealing temperature range.

INTRODUCTION

Nanocrystalline soft magnetic Fe-M-B-Cu (M: Zr, Ti, Nb, Ta) alloys can be produced by first step crystallization of the amorphous precursors [1,2]. The nanocrystalline bcc Fe grains are formed due to annealing of the amorphous alloys at temperatures exceeding the crystallization onset temperatures, T_x, which were determined from differential scanning calorimetry measurements [2] ($T_x = 533°C$, 444°C, 441°C, 357°C for alloys with M = Zr, Ti, Nb or Ta, respectively). It is known that amorphous ribbons prepared by the melt-spinning technique have two distinct surfaces: dull (which was in contact with the quenching wheel) and shiny (free surface), whose atomic compositions may differ from that of the bulk. Therefore, the onset of crystallization in the annealed samples may be significantly affected by the local composition fluctuations and may be different at the surfaces and in the bulk of the ribbon. In order to compare the bulk and the surface crystallization induced by annealing we used a combined γ-transmission and conversion electron (CEMS) Mössbauer techniques, which make possible the simultaneous study of the bulk and of the surface layer (about 100 nm deep) of the sample, respectively. Both techniques allow qualitative and quantitative analysis of the crystallization process. Earlier Mössbauer effect studies of the crystallization process in amorphous FeBSiC [3], FeCoB [4] or FePC [5] alloys revealed distinct differences between the surface and bulk crystallization. Recently, the CEMS technique was applied for the study of the $Fe_{81}Zr_7B_{12}$ alloy in which the nanocrystalline bcc Fe phase was formed due to 1 hour [6] and short time [7] annealing.

In this paper, we report on the CEMS and transmission Mössbauer studies of Fe-M-B-Cu (M: Ti, Nb, Ta) alloys in the amorphous and nanocrystalline states. The aim of this study is to compare the crystallization behavior in the bulk and in the surface regions depending on the annealing temperature and alloy composition. For comparison the results obtained recently for the $Fe_{79}Zr_7B_{12}Cu_2$ alloy are included.

Mat. Res. Soc. Symp. Proc. Vol. 577 © 1999 Materials Research Society

EXPERIMENT

The $Fe_{80}M_7B_{12}Cu_1$ (M: Ti, Nb, Ta) and $Fe_{79}Zr_7B_{12}Cu_2$ amorphous ribbons were prepared by the melt-spinning technique in an Ar atmosphere. The ribbons were about 25 μm thick and 4 mm wide. In order to obtain the nanocrystalline structure the amorphous ribbons were annealed for 1 hour in the temperature range of 440-620°C. The CEMS and transmission Mössbauer measurements were performed for the as-quenched and annealed samples.

The CEMS spectra were measured at room temperature by using a gas flow electron counter with He + 6% CH_4 gas in which the investigated sample was placed. A proportional counter for measuring the transmitted Mössbauer γ-radiation was placed behind the electron counter which was transparent for the γ-radiation. Thus both the CEMS and transmission Mössbauer measurements were performed simultaneously for the same part of the sample. CEMS spectra were recorded for both the shiny and dull sides of the ribbons. A conventional constant acceleration Mössbauer spectrometer with ^{57}Co in Rh source of an activity of 25 mCi was used. The isomer shift data are given with respect to α-Fe standard.

The transmission and CEMS spectra were fitted by a combination of the hyperfine field distribution, P(H), related to the amorphous phase, and the subspectrum with discrete value of the hyperfine field (H_{hf}) and isomer shift (δ) corresponding to the bcc Fe phase (H_{hf} = 32.95 T, δ= 0.00 mm/s). The P(H) distribution was calculated by the histogram method [8,9] using the NORMOS program [10].

RESULTS

The surface and bulk crystallization of the $Fe_{80}Ti_7B_{12}Cu_1$, $Fe_{80}Nb_7B_{12}Cu_1$, $Fe_{80}Ta_7B_{12}Cu_1$ and $Fe_{79}Zr_7B_{12}Cu_2$ amorphous alloys induced by 1 hour anneals at temperatures ranging from 440°C to 620°C was studied by CEMS and standard transmission Mössbauer techniques. The CEMS results obtained for the shiny side of the samples are discussed here.

The Mössbauer γ-transmission and conversion electron spectra recorded for the as-quenched and annealed $Fe_{80}Ti_7B_{12}Cu_1$ and $Fe_{80}Nb_7B_{12}Cu_1$ alloys are shown in Figs.1 and 2, respectively. The spectra of the as-quenched ribbons are typical for ferromagnetic amorphous alloys (Figs.1a, 1a', 2a, 2a'). However, traces of a crystalline phase can be observed at the surface of the $Fe_{80}Ti_7B_{12}Cu_1$ as-quenched ribbon (less than 2% of the total spectral area, Fig.1a'). The spectra of the amorphous alloys were fitted with the hyperfine field distribution, P(H), method [8,9]. The transmission and CEMS spectra of all annealed samples reveal, in addition to the P(H) component related to the amorphous phase, a discrete spectral component – the sextet with the hyperfine field H_{hf} =32.95 T and isomer shift of 0.00 mm/s, corresponding to the bcc Fe phase (Figs.1b-1e, 1b'-1e', Figs.2b-2e, 2b'-2e').

Clear differences between the bulk and surface crystallization can be observed already for the $Fe_{80}Ti_7B_{12}Cu_1$ sample after annealing at 440°C (Figs.1b, 1b') and for $Fe_{80}Nb_7B_{12}Cu_1$ annealed at 470°C (Figs.2b, 2b'). The spectral contribution of the bcc Fe sextet is significantly larger in the CEMS spectra than in the corresponding transmission spectra. This behavior is preserved throughout the entire annealing temperature (T_A) range for both FeTiBCu and FeNbBCu alloys (Figs.1 and 2). The transmission and CEMS spectra of $Fe_{80}Ta_7B_{12}Cu_1$ closely resemble those recorded for the $Fe_{80}Ti_7B_{12}Cu_1$ alloy.

The evolution of the retained amorphous phase with annealing temperature is clearly seen both in CEMS and transmission spectra. The average hyperfine field characteristic of the amorphous phase increases with increasing T_A and clearly exceeds that of the as-quenched FeTiBCu and FeNbBCu alloys. This strongly suggests that the change of the composition of the retained amorphous phase (due to the formation of the bcc Fe grains) towards iron-poorer one leads to the increase of the hyperfine field at room temperature which is most probably related to the increase of the Curie temperature of the retained amorphous phase with the decrease of iron content in that phase. This behavior is particularly clear in the case of the $Fe_{80}Nb_7B_{12}Cu_1$ amorphous alloy which exhibits relatively low Curie temperature of about 60 °C [2].

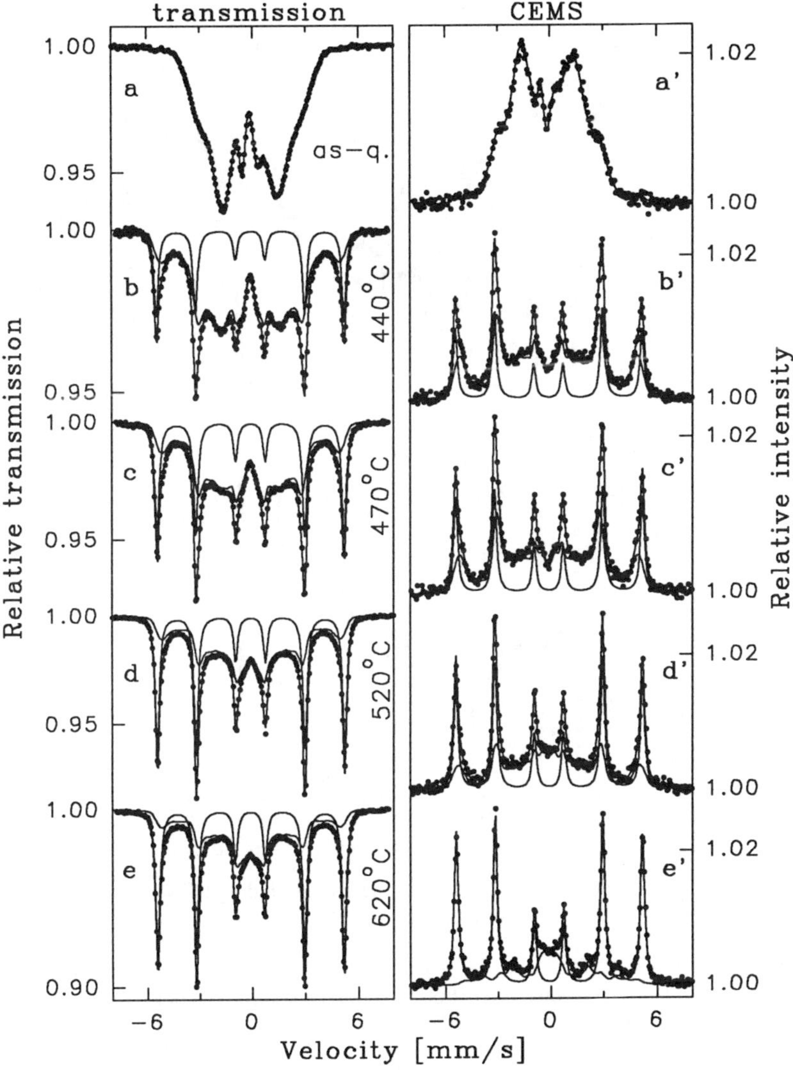

Fig. 1. Mössbauer transmission (a-e) and corresponding CEMS (a'-e') spectra recorded for the as-quenched and annealed $Fe_{80}Ti_7B_{12}Cu_1$ alloys.

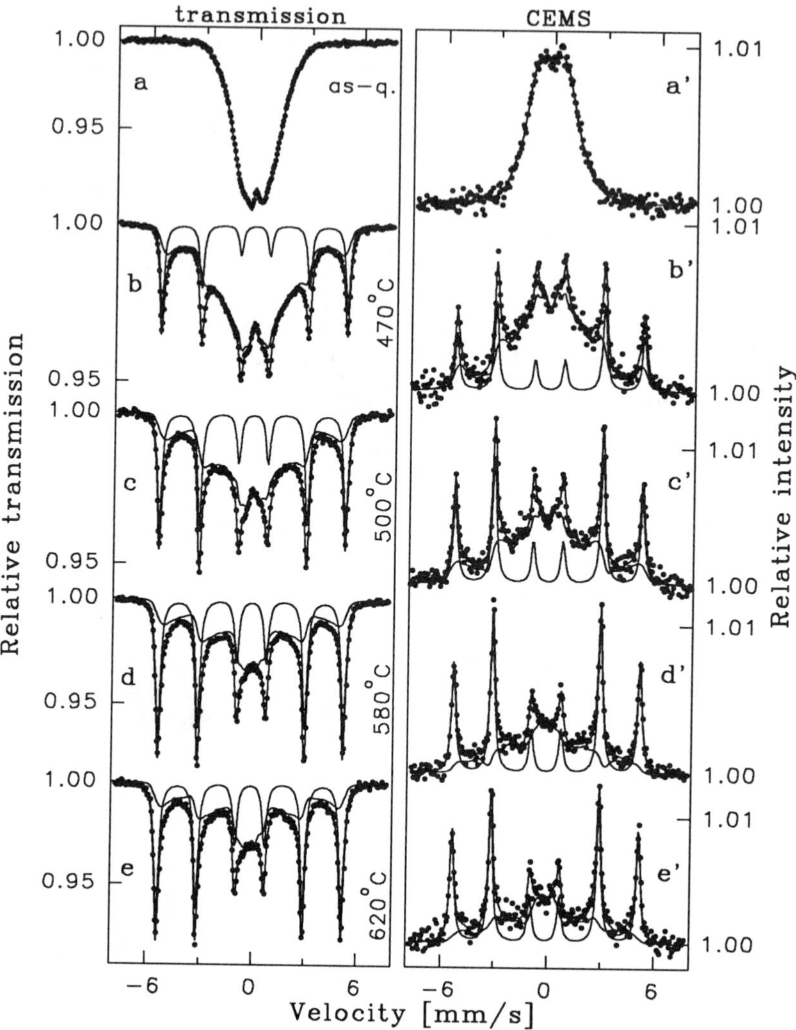

Fig.2. Mössbauer transmission (a-e) and corresponding CEMS (a'-e') spectra recorded for the as-quenched and annealed $Fe_{80}Nb_7B_{12}Cu_1$ alloys.

The results obtained for Fe-M-B-Cu alloys are compared with those obtained recently for $Fe_{79}Zr_7B_{12}Cu_2$ [11]. The as-quenched $Fe_{79}Zr_7B_{12}Cu_2$ alloy is completely amorphous (Fig.3a); there are no traces of crystallization at the surface (Fig.3a'). Annealing at 430°C, temperature much lower than T_x, causes onset of the surface crystallization; the sextet characteristic for the bcc Fe phase appears in the CEMS spectrum only (Fig.3b'). No traces of crystallization in the bulk were detected (Fig.3b). Annealing at 500°C induces already bulk crystallization resulting in a formation of the bcc Fe phase (Fig.3c). Surface crystallization is much more pronounced than the bulk one as shown by the CEMS spectrum in Fig.3c'.

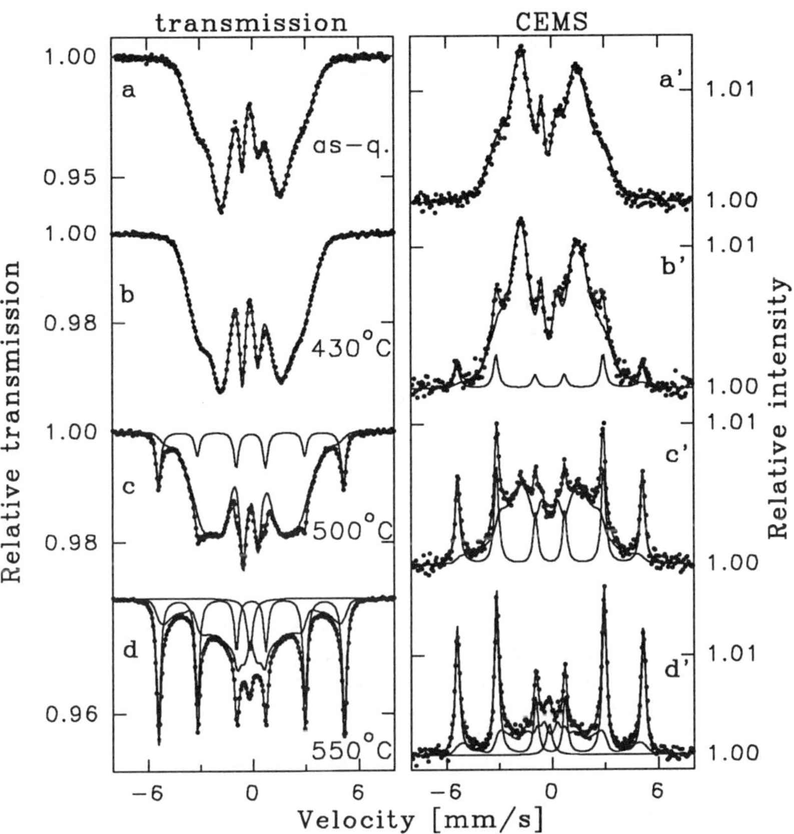

Fig.3. Mössbauer transmission (a-d) and corresponding CEMS (a'-d') spectra recorded for the as-quenched and annealed $Fe_{79}Zr_7B_{12}Cu_2$ alloys.

After annealing of the $Fe_{79}Zr_7B_{12}Cu_2$ alloy at 550°C, in addition to the bcc Fe phase, a crystalline paramagnetic phase is formed. The single-line spectral component with δ= -0.06 mm/s detected both in transmission (Fig.3d) and CEMS (Fig.3d') spectra allowed the identification of this phase as fcc Fe. This phase is seen also after annealing at 600°C. Annealing at higher temperature (T_A=780°C) causes complete crystallization of the FeZrBCu alloy. The transmission and CEMS spectra consist of the α-Fe sextet only. In our recent study of the $Fe_{81}Zr_7B_{12}$ alloy [6] different phases were detected at the surface of the sample annealed at 600°C than in the bulk. In distinction to the bulk crystallization, which leads to the bcc Fe phase only, the iron boride phase (Fe_2B) was clearly seen at the surface [6,11].

From the CEMS and transmission spectra the relative atomic fraction of Fe as bcc phase was determined from the spectral area of the component with the hyperfine field of about 33T characteristic of the bcc Fe phase. The results obtained for all alloys studied vs. annealing temperature are shown in Fig.4. The relative spectral contribution of the bcc Fe sextet increases with increasing annealing temperature at the expense of the P(H) component related to the retained amorphous phase, indicating the increase of the volume fraction of the

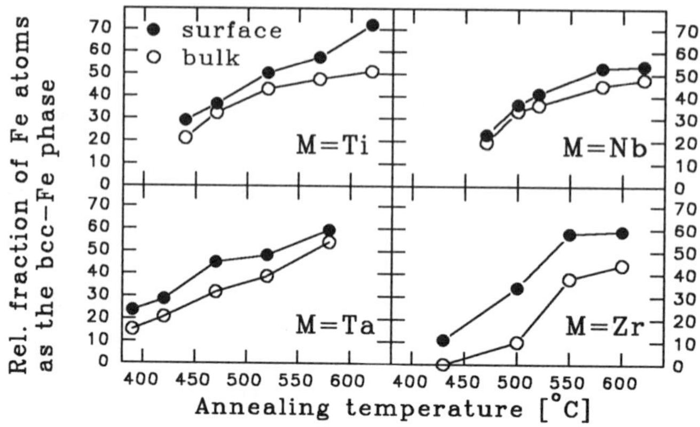

Fig.4. The relative abundance of the bcc Fe phase determined for the $Fe_{80}M_7B_{12}Cu_1$ alloys.

bcc Fe phase. Transmission spectra of the annealed samples (Figs.1b-1e, 2b-2e, 3c-3d) are similar to the corresponding CEMS spectra. However, by determining the relative abundance of each phase one can see the difference between the amount of the bcc Fe phase in the bulk and in the surface layer of the sample (Fig.4). In all cases the crystalline phase is more abundant at the surface than in the bulk of the sample at each annealing temperature. The more effective surface crystallization can be related to the boron depletion of the surface regions of the samples [3,4]. Most probably some boron atoms escaped from the surface region during melt spinning process or during annealing. The decrease of boron content in the amorphous Fe-M-B-Cu alloys decreases the crystallization temperature as revealed by differential scanning calorimetry and Mössbauer measurements for, e.g., $Fe_{93\text{-}x\text{-}y}Zr_7B_xCu_y$ alloys [12]. Hence, the crystalline phases, especially bcc Fe, appear at the surface at lower T_A than in the bulk.

ACKNOWLEDGEMENT

Financial support from the Grant No. KBN 2P03B 068 15 from the Polish Committee of Scientific Research is gratefully acknowledged.

REFERENCES

1. M. Kopcewicz, A. Grabias, B. Idzikowski, J. Magn. Magn. Mat. **177-181**, 1434 (1998).
2. M. Miglierini, M. Kopcewicz, B. Idzikowski, Z.E. Horváth, A. Grabias, I. Škorvánek, P. Dłużewski, Cs.S. Daróczi, J. Appl. Phys. **85**, 1014 (1999).
3. N. Saegusa and A.H. Morrish, Phys. Rev. B **26**, 6547 (1982).
4. T. Zemčik, K. Závěta, I. Škorvánek, J. Mater. Sci. **28**, 654 (1993).
5. U. Gonser, M. Ackermann, H.G. Wagner, J. Magn. Magn. Mat. **31-34**, 1605 (1983).
6. M. Kopcewicz and A. Grabias, J. Appl. Phys. **80**, 3422 (1996).
7. A. Grabias and M. Kopcewicz, Mater. Sci. Forum **269-272**, 725 (1998).
8. J. Hesse and A. Rübartsch, J. Phys. **E 7**, 526 (1974).
9. G. Le Caer and J.M. Dubois, J. Phys. **E 12**, 1083 (1979).
10. R.A. Brand, J. Lauer, D.M. Herlach, J. Phys. **F 13**, 675 (1983).
11. M. Kopcewicz, A. Grabias, P. Nowicki, Proc. Rapidly Quenched and Metastable Materials RQ9 (Bratislava, Slovakia, 1996), Mater. Sci. Eng. A Suppl., Elsevier, (1997), p. 95.
12. M. Kopcewicz, A. Grabias, D.L. Williamson, J. Appl. Phys. **82**, 1747 (1997).

Part XIII

Ultrasoft Nanocrystalline and Amorphous Materials III

CRYSTALLIZATION AND NANOCRYSTALLIZATION KINETICS OF Fe-BASED AMORPHOUS ALLOYS

A. HSIAO*, Z. TURGUT*, M.A. WILLARD*, E. SELINGER*, D.E. LAUGHLIN*, M.E. McHENRY*, R. HASEGAWA**
*Materials Science and Engineering, Carnegie Mellon University, Pittsburgh, PA
**Allied Signal Corporation, Morristown, NJ

ABSTRACT

In this work we describe crystallization kinetics as inferred from time-dependent magnetization studies and thermal analysis for an Allied Signal amorphous Fe-based METGLAS® 2605SA-1 alloy and a NANOPERM ($Fe_{88}Zr_7B_4Cu_1$) alloy. We illustrate and contrast several phenomena important to understanding crystallization kinetics in particular to the NANOPERM alloy system. In METGLAS® 2605SA-1 primary and secondary crystallization events are observed in differential scanning calorimetry data (DSC) at temperatures of 504 ^{0}C and 549 ^{0}C, respectively for data taken at a 10 ^{0}C/min scan rate. Both temperatures are greater than the Curie temperature of the amorphous alloy. For the NANOPERM alloy primary crystallization (as determined from differential thermal analysis (DTA)) occurs at 500 ^{0}C and secondary crystallization at 730 ^{0}C and M(t) at temperatures near the primary crystallization temperature is dominated (at short times < 1 hour) by the primary crystallization event. Using the Johnson-Mehl-Avrami equation for isothermal transformations and the Kissinger equation for constant heating transformations, we find corresponding models for the crystallization kinetics of the NANOPERM alloy.

INTRODUCTION

The understanding of the crystallization kinetics of magnetic amorphous alloys is of scientific interest as it represents a phase transformation occurring under extreme conditions far from equilibrium. The subject is also of technological interest because of the drastic deterioration of promising magnetic properties in amorphous alloys upon crystallization. Yet by controlling crystallization, amorphous alloys can be used as precursor starting materials for producing desirable fine microstructures. The mechanisms involved in the transformation from the metastable amorphous state to the stable crystalline state and their kinetics depend on various parameters, such as composition, concentration of nucleation sites, diffusion coefficients, and thermal history of the sample [1]. Crystallization kinetics is often determined from differential scanning calorimetry (DSC) where the onset of crystallization, the temperature of crystallization, and secondary crystallization events can all be revealed. In amorphous ferromagnets the magnetization can also change upon crystallization making it an appropriate probe of crystallization kinetics.

As an example of the manifestation of crystallization events in thermal analysis, Figure 1(a) illustrates a DSC scan and Figure 1(b) illustrates magnetization vs. time for a METGLAS® 2605SA-1 amorphous Fe-based sample. METGLAS® is a Fe-based magnetic alloy trademarked by Allied Signal. It has applications as sensors, high frequency cores, and transformers. Its characteristics include low core loss and the ability to be annealed for very high DC permeabilities. The DSC scan shows a primary crystallization and a secondary crystallization event at 504°C and 549^{0}C, respectively. Both of these crystallization events occur at temperatures above the Curie temperature of the amorphous alloy. As a result, the magnetization

Mat. Res. Soc. Symp. Proc. Vol. 577 © 1999 Materials Research Society

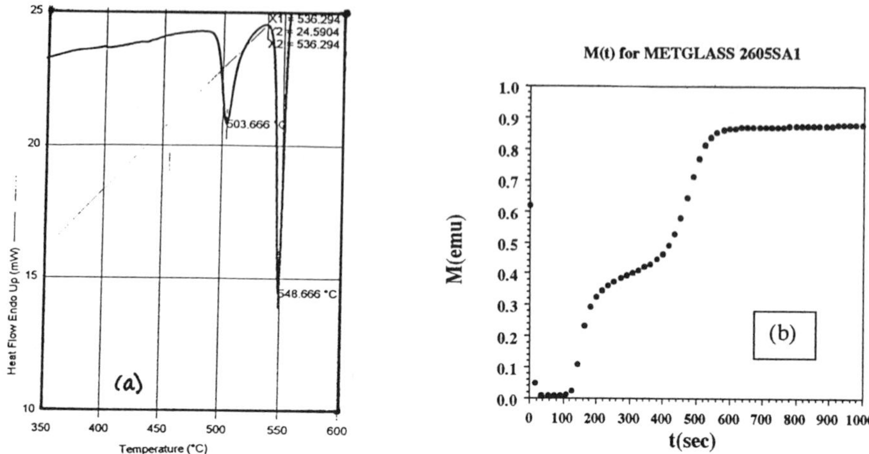

Figure 1: (a) DTA trace and (b) magnetization vs. time for Allied signal METGLAS alloy

is nearly zero below the crystallization temperature but increases upon primary and then again upon secondary crystallization.

NANOPERM is among one of the premiere soft magnetic materials of interest for AC magnetic and high temperature aircraft applications. It is a trademark for the amorphous alloy composition $Fe_{88}Zr_7B_4Cu_1$, which possesses nanocrystalline microstructure in its solid state. Its crystallization process occurs as:

$$\text{Amorphous I} \rightarrow \text{BCC } \alpha\text{-Fe} + \text{Amorphous II (777 K)} \rightarrow \alpha\text{-Fe} + Fe_3Zr \text{ (977 K)} \quad (1)$$

Target properties for soft magnetic materials for high temperature applications include a working temperature of 500 °C, induction of greater than 2 Tesla, thermal stability at 600 °C for 5000 hours, and a core loss of 480 W/kg at 5 kHz and 500 °C. In alloys of Fe-Zr-B, Fe-Hf-B, and Fe-Nb-B, permeabilities of up to 3.2×10^5, core losses of 0.066 W/kg at 1 Tesla and 50 Hz, inductions of up to 1.63 Tesla, and magnetostrictions as low as -1×10^{-6} were observed [2].

This paper presents two methods for observing the crystallization kinetics of NANOPERM. The DSC is a standard way of measuring the heat flow versus temperature or time of the crystallization process. The heat flow, whether endothermic or exothermic, is related to the volume of the solid transformed. A novel way of measuring the crystallization process is by Vibrating Sample Magnetometry (VSM). This method observes the magnetization versus temperature or time. Figure 2 illustrates a characteristic of the NANOPERM material, which makes the VSM an excellent method for modeling crystallization kinetics. Since the Curie temperature T_c of the amorphous precursor phase is below the crystallization temperature, T_p, of NANOPERM, the alloy initially has zero magnetization when heated to its crystallization temperature. As the alloy crystallizes, the magnetization is directly proportional to the volume fraction of BCC α-Fe that forms. Thus the time dependence of the magnetization is directly proportional to the volume fraction transformed in the isothermal case.

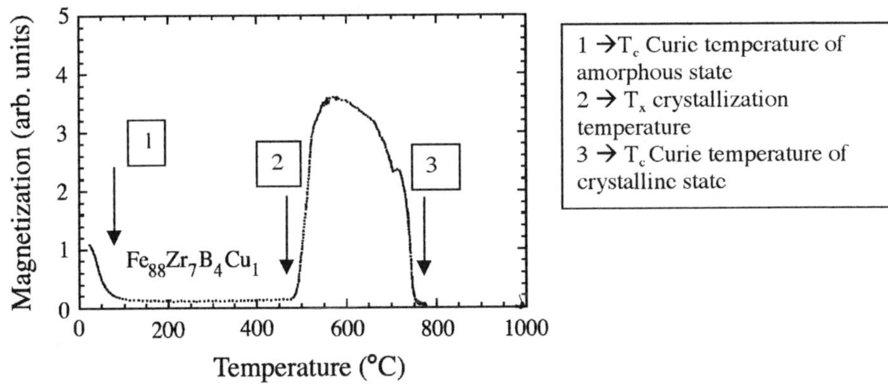

Figure 2: Magnetization vs. temperature for the NANOPERM alloy $Fe_{88}Zr_7B_4Cu_1$

EXPERIMENT

The Kissinger method is one way of quantifying the crystallization kinetics of a continuous heating process[8]. The Kissinger equation:

$$\ln (\alpha/T_p^2) = -Q/RT_p + \text{constant} \tag{2}$$

relates the natural log of the heating rate (α) divided by the square of the crystallization temperature (T_p) to the activation energy (Q) divided by the ideal gas constant (R) and T_p.

A Perkin-Elmer DSC was used to carry out crystallization occurring under continuous heating conditions. Melt-spun ribbons of NANOPERM were cut into pieces weighing approximately 10 mg, put into an aluminum dish, and heated from 50°C to 580°C at varying rates of 2°C/min, 5°C/min, 10°C/min, 20°C/min, 40°C/min and 60°C/min.

The Johnson-Mehl-Avrami (JMA) model is one common way of describing the crystallization kinetics of an isothermal process. JMA transformation kinetics are described by the equation:

$$X(t) = 1 - \exp(-kt^n) \tag{3}$$

Here the volume fraction transformed (X) as a function of time (t) is related by the exponential of the rate constant (k) and time (t) to a power of the morphology index (n). The value of the morphology index depends on the specific nucleation and growth mechanisms of the process. The rate constant can be written as:

$$k(T) = k_o \exp(-Q/RT) \tag{4}$$

where k_o is a rate constant coefficient and Q is the activation energy. The JMA equation can be rearranged into an Arrhenius-like equation, which is useful experimentally.

$$1/t_1 - 1/t_2 \sim -Q/R \exp(1/T_1 - 1/T_2) \tag{5}$$

A Lakeshore Cryogenics VSM was used to carry out crystallization occurring under isothermal conditions. Pieces from the same melt-spun ribbon of NANOPERM were placed into a quartz holder and subjected to a magnetic field of 500 Oersteds for at least one hour at varying temperatures of 490°C, 500°C, 520°C, 550°C, and 570°C.

Given the JMA model and the Kissinger model, it is of interest to compare important values such as activation energy (Q), which can be calculated from experimental data for each model. The morphology index (n) calculated from isothermal experiments on the VSM and the DSC can be compared also. These experimental values serve to describe the kinetics and provide new information to be added and compared to current literature published on magnetic amorphous alloys used for high temperature aircraft applications.

RESULTS AND DISCUSSION

The DSC for continuous heating showed one primary crystallization peak within the range of 50°C to 580°C. It is known that NANOPERM has a secondary crystallization peak at 704°C, but the DSC was limited to 600°C so it could not detect the second peak. At 2°C/min,

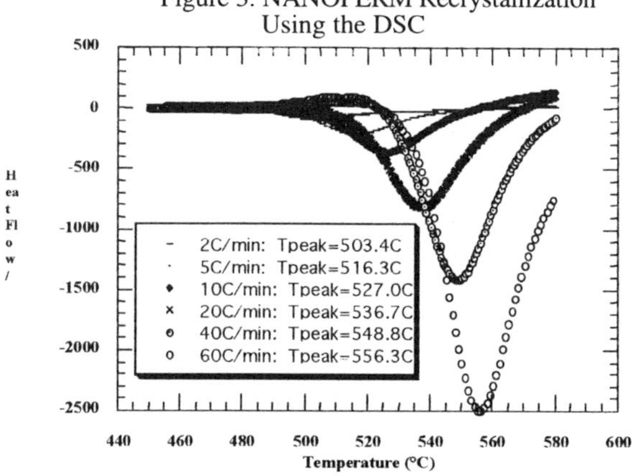

Figure 3: NANOPERM Recrystallization Using the DSC

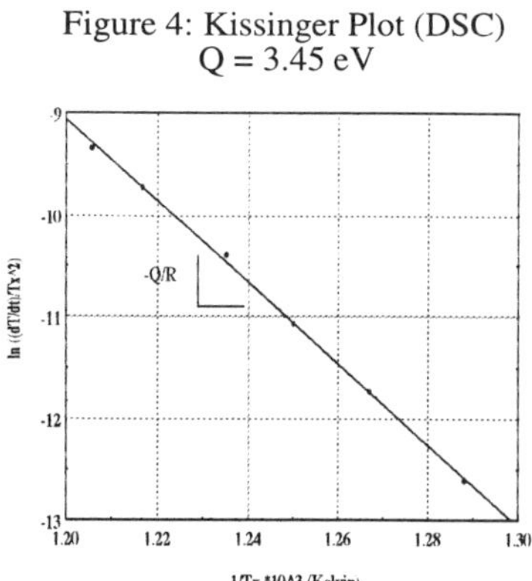

Figure 4: Kissinger Plot (DSC) Q = 3.45 eV

the experimental value for the crystallization peak temperature is 503 °C, which is comparable to the value found in literature. In Figure 3, T_p increases as the heating rate increases suggesting a dependence of the crystallization temperature upon the heating rate. This suggests that the Kissinger model of transformation can be used to relate the rate of heating to the activation energy and temperature. The activation energy from the Kissinger model was found to be 3.5 eV as shown in Figure 4. This is the fitted value for the slope of a Kissinger plot, where the regression coefficient was R^2=0.999.

The magnetization as determined by the volume fraction transformed (dX/dt) from isothermal heating on the VSM also indicates that the primary crystallization temperature is in the range of 500°C to 520°C. Figure 5(a) shows that at temperatures higher than the actual T_p, the magnetization reaches saturation within an hour. At temperatures below the actual T_p, saturation does not occur even after two hours. Figure 5(b) shows the inflection point as the maximum rate of volume transformed. The activation energy from isothermal heating was found to be 3.5 eV or 340 kJ/mole between 490-500°C. The activation energy values at higher temperatures are observed to increase. This suggests that a sample heating correction may be important where the kinetics are fast. The activation energy for Fe self-diffusion in pure BCC α-Fe is 239.7 kJ/mole [7] at atmospheric pressure. Fe-diffusion, as the majority constituent, is thought to have an important contribution to the kinetics of primary crystallization.

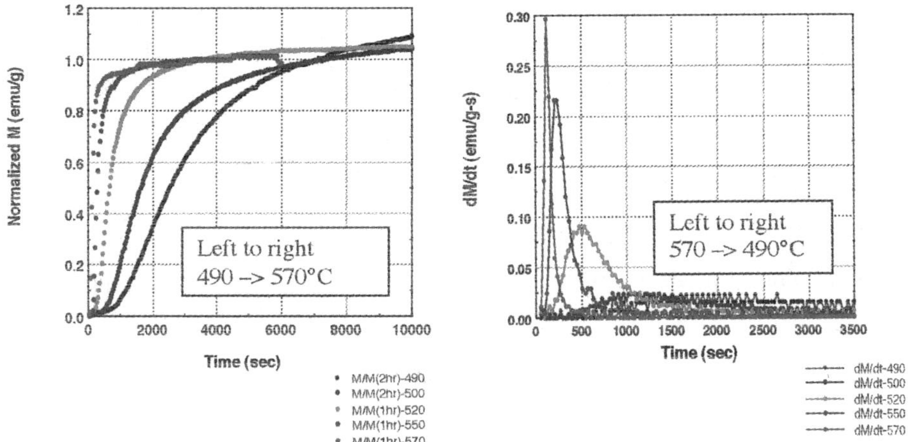

Figure 5: (a) Magnetization vs. Time (b) dM/dt during crystallization of a NANOPERM alloy

Table 1 shows the values of the morphology index calculated experimentally. When the morphology index, n=3/2, the reaction is three-dimensionally diffusion limited with immediate nucleation. This suggests that other mechanisms besides self-diffusion of α-Fe, perhaps contributed by the additional atoms, is occurring during the transformation. Values of Q and n are comparable to those published in literature on similar amorphous magnetic alloys. Table 2 indicates that kinetic analysis has been done either using the JMA model or the Kissinger model, but never both. However, for the similar alloy, FINEMET, experimental values of Q lie in the range of 3.5 - 4.0 eV as well, and the morphology index compares with n = 3/2.

	Volume Fraction Transformed	Morphology Index (n)
490°C	0.33 ± 0.15	1.95 ± 0.70
500°C	0.23 ± 0.08	1.37 ± 0.18
520°C	0.32 ± 0.07	1.66 ± 0.26
550°C	0.38 ± 0.10	2.19 ± 0.70
570°C	0.31 ± 0.17	1.86 ± 0.63

Table 1: Morphology Index of NANOPERM crystallization JMA kinetics

Alloy	Kinetic Analysis	Reference	Q (eV)	JMA index (n)	Experiment
FINEMET	JMA	L.K. Varga, Mat. Sci. and Eng. **A179/A180**, 567, (1994).	3.84-4.08		Vickers Hardness
FINEMET	Kissinger	Same as above	3.7-4.5		DSC
FINEMET	JMA	J. Bigot, J. Mag.Mag. Mat. **133**, 299, (1994).	4.3	n=1.5-2.0	Thermomagnetic gravimetry
FINEMET	Kissinger	C.F. Conde and A. Conde, Mat. Lett. **21**, 409, (1994)	3.8		DSC

Table 2: Activation Energies and Morphology Indices for FINEMET alloys

CONCLUSION

The time dependence of the magnetization is a direct and accurate probe for understanding NANOPERM crystallization kinetics. The DSC is a useful tool for characterizing crystallization temperatures of amorphous alloys. The Kissinger (constant heating) and Johnson-Mehl-Avrami (isothermal heating) models were compared, resulting in similar values for the activation energy. Values of activation energy and morphology index are comparable to those found in the literature for similar amorphous magnets.

ACKNOWLEDGMENTS

Effort sponsored by the Air Force Office of Scientific Research, Air Force Material Command, USAF, under Grant No. F49620-96-1-0454.

REFERENCES

1. Yoshihiro Takahara, *Mat. Trans.* **JIM-37** (9), 1996), 1453-1457 (1996).
2. A. Makino, K. Suzuki, A. Inoue, and T. Masumoto, *Mat. Sci. Eng.* **A179/A180**, 495, (1994a); 127, (1994b).
3. K. Suzuki, A. Makino, N. Kataoka, A. Inoue, and T. Masumoto;,*Mat. Trans.* **JIM-32**, 93, (1991).
4. A. Makino, T. Hatanai, Y. Naitoh, T. Bitoh, A. Inoue, and T. Masumoto, *IEEE Trans. Magn.* **33**, 3793, (1997).
5. Y. Yoshizawa, S. Oguma, and K. Yamauchi, *J. Appl. Phys.* **64**, 6044, (1988).
6. M. A. Willard, D. E. Laughlin, M. E. McHenry, D. Thoma, K. Sickafus, J. Cross and V. Harris, *J. Appl. Phys*. **84**, 6773-6777, (1998).
7. D.A. Porter and K.E. Easterling, Phase Transformations in Metals and Alloys, 2nd ed. (Chapman & Hall, London, 1992), 78.
8. L.V. Meisel, P.J. Cote, *Acta metall.* **31(7)**, 1053-1059, 1983.

Quantitative Crystallization And Nano-grain Size Distribution Studies of A FeCuNbSiB Nanocrystalline Alloy

M.S. Leu* and T.S. Chin**

* Department of Materials Sci. & Eng., Tsing Hua University, 300, Taiwan, ROC

** Materials Research Lab., Industrial Tech. Research Institute, 310, Taiwan, ROC

ABSTRACT

Superior soft magnetic properties of nanocrystalline two-phase FeCuNbSiB alloy can only be attainable from a suitably nano-crystallized structure in the amorphous matrix. In this study a thermal-cycling-annealing process was adopted to anneal the amorphous specimens. The crystallization fraction of $Fe_{73.5}Cu_1Nb_3Si_{13.5}B_9$ ribbons after different annealing was quantitatively measured by a DTA method through the integrated crystallization enthalpy from annealed samples versus that of as-cast amorphous specimen. The crystallization fraction was estimated to be 70 % as the specimen was annealed for 550°C×7.63 ks, the optimal annealing condition. Using this method, the crystallization quantity of the crystallized FeCuNbSiB amorphous alloy was accurately estimated and controlled during nano-crystallization processes.

The nanostructure of the annealed flake were investigated by high resolution TEM. The population of grain size is dominant in the range of 3-9 and 3-11 nm, corresponding to isothermal annealing at 520 and 550 °C, respectively, for 910 s. The average grain size associated with the above two annealing conditions is estimated to be 7 and 8.5 nm, respectively. The portion of nanocrystals with grain size smaller than 10 nm is about 80 % for the annealing at 520°C×910 s and 72 % for the annealing at 550°C×910 s, respectively. However, there is only one α-Fe(Si) phase evidenced in the TEM diffraction patterns.

INTRODUCTION

In recent years there has been an upsurge interest in the science and technology of nanocrystallized soft magnetic alloys, e.g., the $Fe_{73.5}Cu_1Nb_3Si_{13.5}B_9$ alloy, developed by Yoshizawa et al. in 1988 [1,2]. Much effort has been devoted to investigating the correlation between the microstructure and magnetic properties during nanocrystallization process. The superior soft magnetic properties can be attained for this $Fe_{73.5}Cu_1Nb_3Si_{13.5}B_9$ alloy by subjecting it to a suitable annealing process to induce a nanocrystallized structure in the amorphous matrix [3]. However, proper crystallization including proper amount of nanocrystalline phase and proper grain size, in the amorphous matrix is necessary in order to minimize the magnetostriction and improve the magnetic properties of the specimen.

Herzer, by a close analysis of the magnetization curves, estimated that α-Fe(Si) nanocrystals, being the major constituent phase, is with a volume fraction of about 70-80% when $Fe_{73.5}Cu_1Nb_3Si_{13.5}B_9$ alloy was annealed for 3.6 ks at 540 °C [4]. But there has been no direct proof to date to justify such a result. Therefore, the quantitative crystallization study as well as the grain size effect on magnetic properties of the amorphous alloys is fundamentally important. These are the major purposes of this research.

Mat. Res. Soc. Symp. Proc. Vol. 577 © 1999 Materials Research Society

Several coordinating parameters should be taken into consideration as the specimen is annealed to determine its optimal annealing conditions. Ribbon thickness, core mass, cooling rate during ribbon casting, annealing temperature and time are such parameter. This study adopted a simple and time-saving method to work out suitable annealing conditions by means of in-situ measurements on the temperature and time dependencies of initial permeability (μ_i) during the annealing process developed recently [5]. Consequently, the optimum accumulative annealing time (t_{Acc}) of 7.63 ks at the isothermal annealing temperature (T_{IA}) of 550 °C to attain the best μ_i have been worked out by this in-situ measurement during thermal-cycles [6].

EXPERIMENT

The $Fe_{73.5}Cu_1Nb_3Si_{13.5}B_9$ amorphous ribbons were prepared by the single-roller melt spinning method. For the annealing process, the ribbon specimens were subjected to repeated thermal-cycles. The specimens were heating up (at a rate of 5 °C/min) to an isothermal annealing temperature (T_{IA}) of 520 and 550 °C, respectively for a certain period of soaking time, furnace cooling. This was repeatedly carried out until certain accumulative annealing time (t_{Acc}) was reached, as shown in Fig. 1.

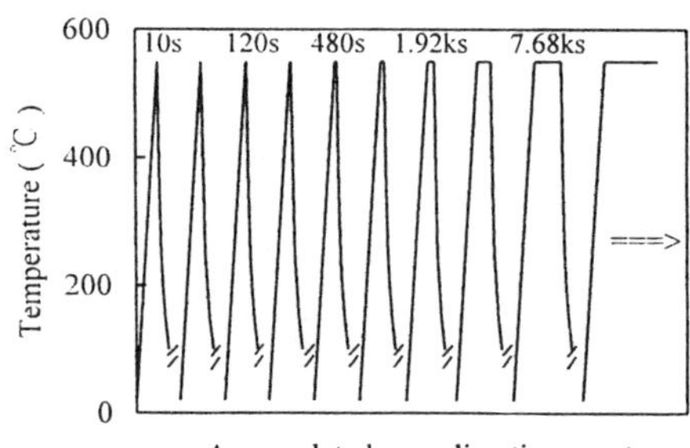

Fig. 1. The temperature and time profiles for the core specimen with T_{IA} at 550 °C; the annealing time shown at each thermal cycle is accumulated and discussed in the text.

Two flakes, which were annealed at 520 and 550 °C for t_{Acc} of about 910 s, respectively, were selected for TEM investigations and thinned directly by ion milling until electron transparency. The morphological observations were performed in a JEOL JEM-200CX electron microscope operated at 200 kV. The population of grain size is analyzed by a commercial software "Optimal". The data associated with the above annealing conditions were collected from at least three photographs and calculated by average.

Thermal analyses were conducted on tiny pieces taken from as-cast and annealed ribbons, respectively, using a Du Pont 1600 differential temperature analysis (DTA) and 2100 thermal analyzer with a heating rate of 20 °C/min in an argon atmosphere. The crystallization temperatures of $Fe_{73.5}Cu_1Nb_3Si_{13.5}B_9$ alloy ribbons were evaluated from the exothermic peaks of DSC and DTA curves during continuous heating from room temperature to 700 and 850 °C, respectively. Crystallization enthalpy was integrated from the DTA peak for each of the studied specimens.

RESULTS AND DISCUSSIONS

The TEM nanostructure of the annealed flake specimens corresponding to T_{IA} of 520 and 550 °C with the t_{Acc} of 910 s is shown in Figures 2 and 3, respectively. Though the T_{IA} is only about 520 °C, quantity of α-Fe(Si) crystals has been precipitated in the amorphous matrix as shown in Fig. 2. As T_{IA} is increased to 550 °C, both volume fraction and grain size of α-Fe(Si) crystals have been increased slightly, as shown in Fig. 3.

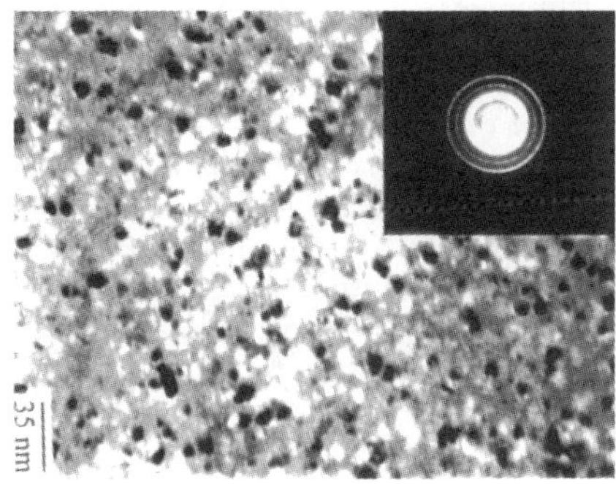

Fig. 2. TEM micrograph and electron diffraction pattern of the FeCuNbSiB alloy annealed at 520 °C with t_{Acc} of 910 s.

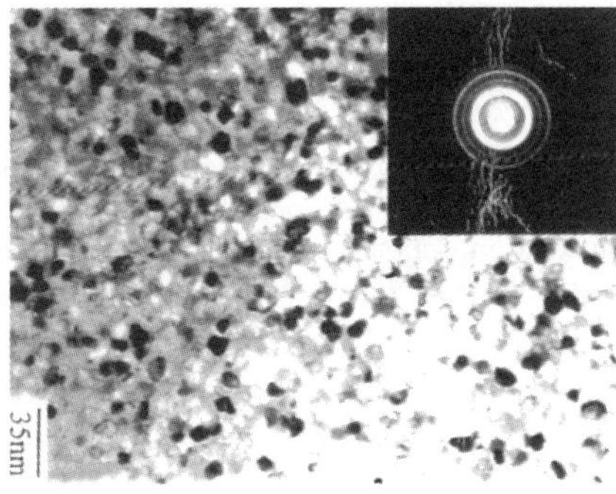

Fig. 3. TEM micrograph and electron diffraction pattern of the FeCuNbSiB alloy annealed at 550 °C with t_{Acc} of 910 s.

The software "Optimal" was used to analyze the population of grain size. The data associated with the above annealing conditions are collected from at least three photographs and averaged. The histograms of grain size of α-Fe(Si) crystal related to the microstructure shown in Figs. 2 and 3, are presented in Figs. 4 and 5, respectively. The dominant grain size is in the range of 3-9 and 3-11 nm, corresponding to 520 and 550 °C for t_{Acc} of about 910 s, respectively. The average grain size associated with the above two annealing conditions was estimated to be 7 and 8.5 nm, respectively. The portion of nanocrystals with grain size smaller than 10 nm is about 80 % for the annealing at 520 °C×910 s and 72 % for the annealing at 550 °C×910 s.

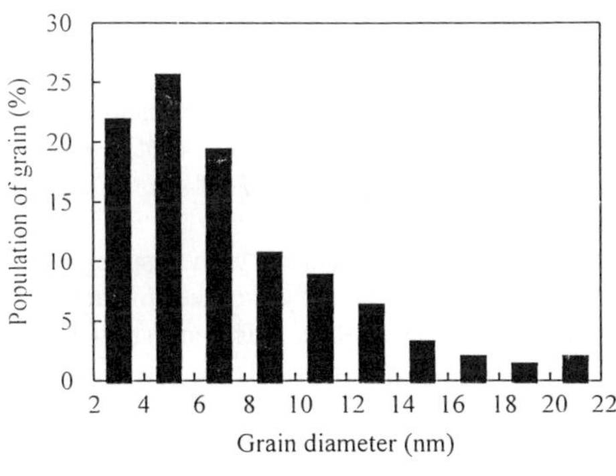

Fig. 4. The grain size distribution is attached from Fig. 2 for annealed at 520 °C with t_{Acc} of 910 s.

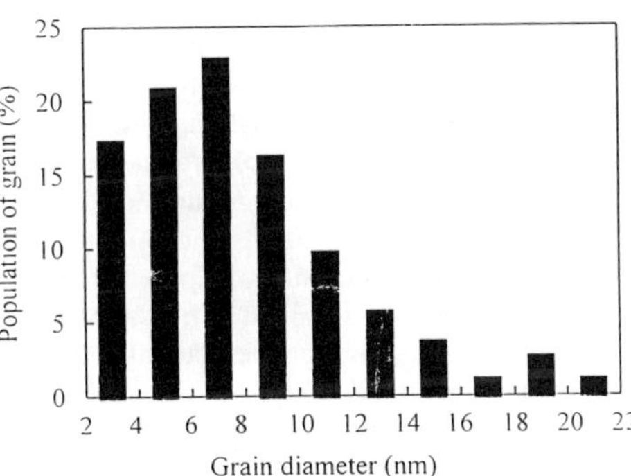

Fig. 5. The grain size distribution is attached from Fig. 3 for annealed at 550 °C with t_{Acc} of 910 s.

However, there is only one α-Fe(Si) phase evidenced in the attached TEM diffraction patterns on both Figures 2 and 3. The grain size (D) of α-Fe(Si) crystals appearing in the initial annealing stage at 520 °C×910 s and 550 °C×910 s is very small, mainly 3-9 nm, in good agreement with earlier observations [7,8].

A typical plot in Figure 6 shows the DSC curve of a flake taken from the as-cast $Fe_{73.5}Cu_1Nb_3Si_{13.5}B_9$ amorphous ribbon. The amorphous Curie temperature and crystallization temperatures were identified from the weak endothermic peak and sharp exothermic peaks, respectively. The Curie temperature of the amorphous phase can be identified directly from the weak endothermic peak at 328.4 °C. Two different intensive exothermic peaks appear at 544.8 and 699.5 °C, respectively. But it should be noted that the exothermic reaction associated with higher temperature occurs up to 725 °C and is still undergoing. The first peak (T_{X1}) starts at about 500 °C with a higher exothermic intensity represents the precipitation of α-Fe(Si) phase from the amorphous matrix, as proved by an earlier XRD study [2,7-9]. The lower intensity exothermic peak (T_{X2}) at the higher temperature is associated with the transformation of the remaining amorphous phase to other intermetallic compounds, such as Fe_2Si, Fe_2B etc. [2,7,8].

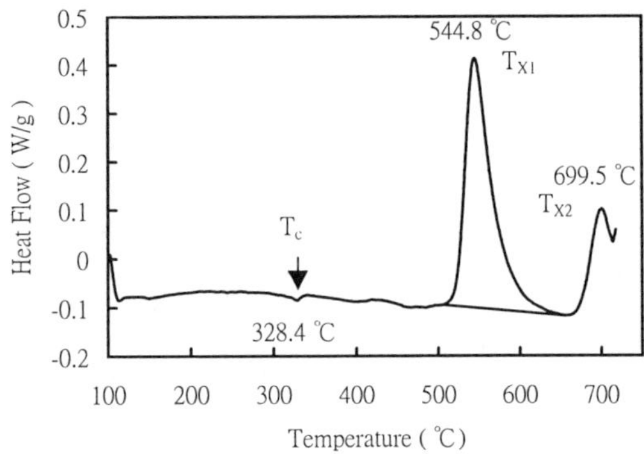

Fig. 6. DSC curve of an amorphous $Fe_{73.5}Cu_1Nb_3Si_{13.5}B_9$ alloy

The DTA measurements show three exothermic crystallization peaks at around 544.1, 697.7 and 732.9 °C, respectively, for an as-spun flake, as depicted in Fig. 7. The measured results of the first two peak-temperatures are in agreement with those of DSC measurement as shown in Fig. 6. In order to demonstrate the variation in crystallization response of a partially crystallized specimen, a DTA curve of the pre-annealed specimen (550 °C×7.63 ks) is shown in Figure 8.

The "crystallization enthalpy" ΔH is defined as the integrated area with respect to the exothermic peaks, as indicated by the hatched areas in Figure 7. The crystallization fraction of the amorphous alloy can be estimated from the decrement in crystallization enthalpy by the following equation:

$$X_f = (\Delta H_a - \Delta H_t)/\Delta H_a \qquad (1)$$

where ΔH_a and ΔH_t (joules/g) are, respectively, the crystallization enthalpy of the as-cast specimen (eg., Figure 7) and that of the isothermal-annealed for t ks (kilo-seconds) specimen (eg., Figure 8). The crystallization enthalpies associated with the main crystallization peaks T_{X1} and T_{X2} can be derived from the figure as shown in Table 1.

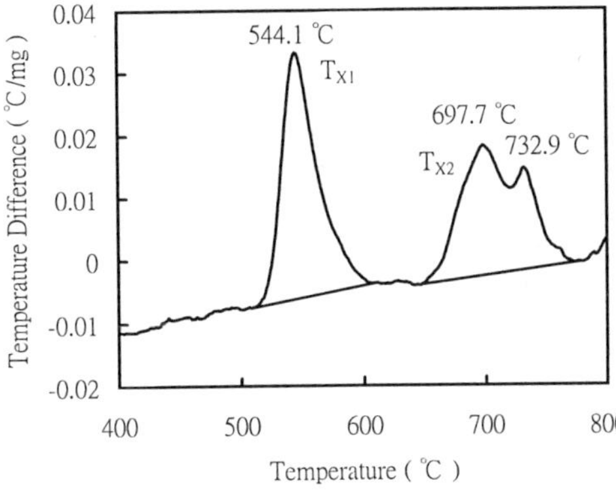

Fig. 7. DTA curve of an amorphous $Fe_{73.5}Cu_1Nb_3Si_{13.5}B_9$ alloy.

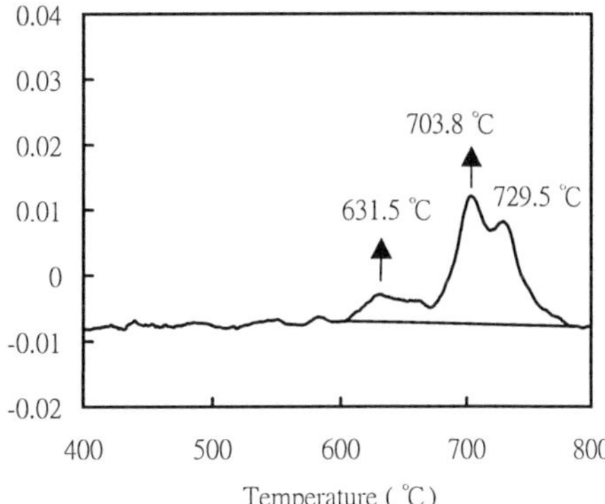

Fig. 8. DTA curve of $Fe_{73.5}Cu_1Nb_3Si_{13.5}B_9$ alloy for annealed at 550 °C with t_{Acc} of 910 s.

Table 1. The crystallization peak temperatures and enthalpies corresponding to the two main peaks of the specimens are obtained from DTA measurement.

Annealing condition	T_{X1} (°C)	T_{X2} (°C)	ΔH_1 (J/mg·min)	ΔH_2 (J/mg·min)
Nil (As-cast)	544.1	697.7	0.06370	0.06166
550 °C × 7.63 ks	631.5	703.8	0.00912	0.04707

Although there are two adjacent exothermic peaks spanning from 523 °C to 780 °C on Figures 7 and 8, only the first crystallization peak was taken into account, since it was the α-Fe(Si) phase which was concerned for the optimal crystallization of interest. So the two close exothermic peaks were deconvoluted to calculate the ΔH_1's. The crystallization fraction (X_f) of α-Fe(Si) phase, for the flake annealed at 550 °C×7.63 ks, can be calculated:

$$X_f^{Fe(Si)} = 1 - (\Delta H_1^{annealed}/\Delta H_1^{as\text{-}cast}) \simeq 86\ \% \qquad (2)$$

That is to say, the first crystallization (of α-FeSi) is 86 % completed at the annealing temperature of 550 °C for 7.63 ks.

It remains a problem how accurate this method will be. It should be noted that the total crystallization fraction obtained from DTA measurement includes the transformation from the original amorphous matrix into α-Fe(Si) nanocrystalline phase in the first crystallization and the transformation of the original amorphous phase to the remaining 'new' amorphous phase which is usually more stable than the original amorphous phase. The stabilization of the amorphous phase during the annealing can be manifest from the fact that the first crystallization peak shifts to a much higher temperature (631 versus 544 °C) after the annealing (Comparing Fig. 8 to Fig.7). The error might come from the estimation of $\Delta H_1^{annealed}$ which is obtained from the crystallization of α-Fe(Si) from the 'stabilized amorphous phase' during the 550 °C×7.63 ks annealing.

The calculated crystallization fraction, 86 %, after the annealing at 550 °C×7.63 ks, is higher than the estimation of 70 – 80 % by Herzer through a close analysis of the magnetization curves of the specimen annealed for 3.6 ks at 540 °C [4]. However, it is justifiable due to the 10 °C higher and doubled duration of annealing in our case. The crystallization rate decreases at extended annealing due to its sigmoidal reaction characteristics.

CONCLUSIONS

Thermomagnetic behaviors related to the annealing conditions can be observed for the $Fe_{73.5}Cu_1Nb_3Si_{13.5}B_9$ alloy by, as summarized below:

(1) The histograms of the grain size are dominant at 3-9 and 3-11 nm, with an average of 7 and 8.5 nm, corresponding to T_{IA} at 520 and 550 °C, respectively with t_{Acc} of 910 s by TEM examination.

(2) The crystallization fraction of the specimen corresponding to 550 °C×7.63 ks annealing is estimated to be about 86 % by the comparison of integrated first crystallization enthalpies of the DTA curves taken before and after the crystallization annealing.

ACKNOWLEDGMENTS

The authors are grateful for the support of this work partially from the Ministry of Economic Affairs, Taiwan, the Republic of China, and partially from the National Science Council of the Republic of China under the grant NSC 87-2216-E-007-043.

REFERENCES

1. Y. Yoshizawa, S. Oguma, and K. Yamauchi, J. Appl. Phys., 64, 6044 (1988).
2. Y. Yoshizawa and K. Yamauchi, Mat. Res. Soc. Symp. Proc., 183 (1991).
3. G. Herzer, Mater. Sci. Eng. A133, 1 (1991).
4. G. Herzer, IEEE Trans. Magn., vol. 25, 3327 (1989).
5. M.S. Leu and T. S. Chin, J. Appl. Phys. 81, 1 (1997).
6. M.S. Leu, T. S. Chin, I-C. Tung and C. M. Lee, Jpn. J. Appl. Phys. 38, 4083 (1999).
7. Manfred Müller, N. Mattern, and L. IIlgen, Z. Metallkde, 82, 895 (1991).
8. Tadeusz Kulik, Mater. Sci. Eng., A159, 95 (1992).
9. K. Hono, K. Hiraga, Q. Wang, A. Inoue and T. Sakurai, Acta Metall. Mater., 40, 2137 (1992).

RF-MÖSSBAUER STUDY OF THE COMPOSITIONAL DEPENDENCE OF SHORT RANGE ORDER IN AMORPHOUS Fe-M-B-Cu ALLOYS

M. KOPCEWICZ *, A. GRABIAS * and B. IDZIKOWSKI **
*Institute of Electronic Materials Technology, Wólczyńska 133, PL 01-919 Warsaw, Poland
**Institute of Molecular Physics, Polish Academy of Sciences,
M. Smoluchowskiego 17, PL 60-179 Poznań, Poland

ABSTRACT

The radio-frequency collapse of the magnetic hyperfine structure in the Mössbauer spectra is employed to study directly the short range order in amorphous Fe-M-B-Cu (M: Zr, Ti, Ta, Nb, Mo) alloys via the electric quadrupole hyperfine interaction. The shape of the quadrupole splitting distributions obtained from the rf collapsed spectra strongly suggests the existence of two short range orders with a high and low local content of iron. The decrease of boron content in FeZrB alloys leads to a more symmetric local structure experienced by Fe atoms.

INTRODUCTION

The short range order (SRO) in amorphous alloys is studied mostly by using X-ray and neutron diffraction. Another very powerful technique which can be used for the determination of the local structure in amorphous alloys is the Mössbauer spectroscopy. The most direct probe of the local structure is provided by the electric quadrupole hyperfine interaction which is completely determined by the nearest environment of the Mössbauer nuclei. However, most amorphous iron-based metal-metalloid alloys are ferromagnetic at room temperature and the magnetic dipole hyperfine interaction is usually much stronger than the electric quadrupole one, so the quadrupole interaction cannot be studied directly. To obtain information on the local structure from the electric quadrupole interaction it is best to separate this interaction from the magnetic one. This can be done by an unconventional rf-Mössbauer technique which makes use of the rf collapse effect induced in the magnetic material by a radio-frequency (rf) magnetic field [1-3]. The collapse of the magnetic hyperfine splitting observed in the rf-Mössbauer spectra is due to fast magnetization reversal induced by the external rf field. The magnetic hyperfine field can be averaged to zero if the frequency of the rf field considerably exceeds the Larmor precession frequency and if the rf field is sufficiently strong to overcome the magnetic anisotropy in the material. Such conditions can be fulfilled in many soft ferromagnets including amorphous alloys [4,5]. Then in place of the complex magnetically split six-line pattern the quadrupole split doublet appears in the rf-Mössbauer spectrum though the sample is in the ferromagnetic state. Therefore the distributions of the quadrupole splittings (QS) and the information regarding changes of the local structure can be obtained.

In this study we applied the rf-Mössbauer technique to investigate the structural changes in amorphous Fe-M-B-Cu (M: Zr, Ti, Ta, Nb, Mo) alloys related to alloy compositions. These alloys have been studied extensively because they serve as amorphous precursors in which the magnetically soft nanocrystalline phase can be formed due to controlled crystallization [6]. The SRO is studied as a function of the alloy composition via quadrupole splitting distributions, P(QS), derived from the rf collapsed Mössbauer spectra. In order to check the influence of boron content and of various metallic components on the SRO two series of alloys were studied: (i) $Fe_{93-x-y}Zr_7B_xCu_y$ (x = 4, 6, 8 or 12 and y = 0 or 2), and (ii) $Fe_{80}M_7B_{12}Cu_1$ (M: Ti, Ta, Nb, Mo). The results obtained by the rf-Mössbauer technique are compared with the conventional

Mat. Res. Soc. Symp. Proc. Vol. 577

Mössbauer measurements performed at low temperature from which information on the SRO was obtained from the magnetic hyperfine field distributions, P(H).

EXPERIMENT

Amorphous $Fe_{80}M_7B_{12}Cu_1$ (M: Ti, Nb, Ta, Mo) and $Fe_{93-x-y}Zr_7B_xCu_y$ (x = 4, 6, 8, 12; y = 0 or 2) alloys were prepared by the melt spinning technique. The amorphous structure of the as-quenched ribbons was confirmed by X-ray diffraction measurements. The rf-Mössbauer measurements were performed for the samples exposed to the rf field of 16 Oe at 60.9 MHz, the field sufficient enough to average to zero the magnetic hyperfine field and to induce the complete rf collapse effect. The P(QS) distributions were extracted from the rf collapsed spectra by using the histogram method [7,8] in the same way as in the earlier studies [1-3,5]. The conventional Mössbauer measurements without rf field applied were performed at 295 and 78K. The P(H) distributions were calculated from the magnetically split spectra using the same method [7,8]. The NORMOS program was used [9].

RESULTS

All amorphous FeZrBCu alloys investigated, except for $Fe_{89}Zr_7B_4$, are ferromagnetic at room temperature. Since the rf collapse effect occurs only in the ferromagnetic materials the $Fe_{89}Zr_7B_4$ alloy is not affected by the rf field. The Mössbauer spectra recorded for this alloy with and without the rf field are identical. As an example the rf collapsed spectra obtained for different compositions of the $Fe_{93-x}Zr_7B_x$ (x = 4, 6, 8, 12) alloys are shown in Figs.1a-1d. They consist of an asymmetric broadened quadrupole doublet. Similar spectra were obtained for Cu-containing alloys. The asymmetry of the QS doublet was taken into account in the fitting procedure by introducing a linear correlation between the isomer shift and the quadrupole splitting. The shape of the spectra changes considerably with alloy composition. For lower boron content (x = 4 or 6) the QS doublet becomes less resolved (compare Figs.1a and 1b with Fig.1d). This change is clearly reflected in the P(QS) distributions which consist of at least two well defined peaks: one at low QS values with the corresponding average quadrupole splitting of about 0.12 mm/s, and the second, much broader peak at about 0.55 mm/s (Figs.1a'-1d'). The relative contributions of these peaks to the total P(QS) distribution vary with boron content. The peak at low QS values increases substantially with decreasing boron content at the expense of the peak at high QS. Hence, the average quadrupole splitting <QS> determined from the entire P(QS) distribution also decreases with decreasing boron content (Table I). Addition of Cu does not affect the <QS> value, regardless of boron content. The width of P(QS) distribution remains almost unchanged for all alloy compositions (Table I).

The shape of P(QS) distributions observed for all alloy compositions strongly suggests that the structure of the amorphous FeZrBCu alloys is not homogeneous. There are two preferential short range orders: one in which Fe atoms experience a more symmetrical local atomic arrangement (small QS values) and the other with a more distorted local structure (high QS values). The relative content of iron atoms contained in the more symmetric SRO increases with increasing Fe abundance (decreasing boron content). Higher boron content induces stronger distortion of the structure due to the strong repulsive interaction between metalloid atoms. Two clearly different QS values strongly suggest composition fluctuations in the amorphous phase. Small QS values correspond to iron-rich FeZrB regions. The decrease of the local iron content leads to the increase of QS value in agreement with earlier systematic studies of the rf collapse effect in various iron-based amorphous alloys [1,2,5].

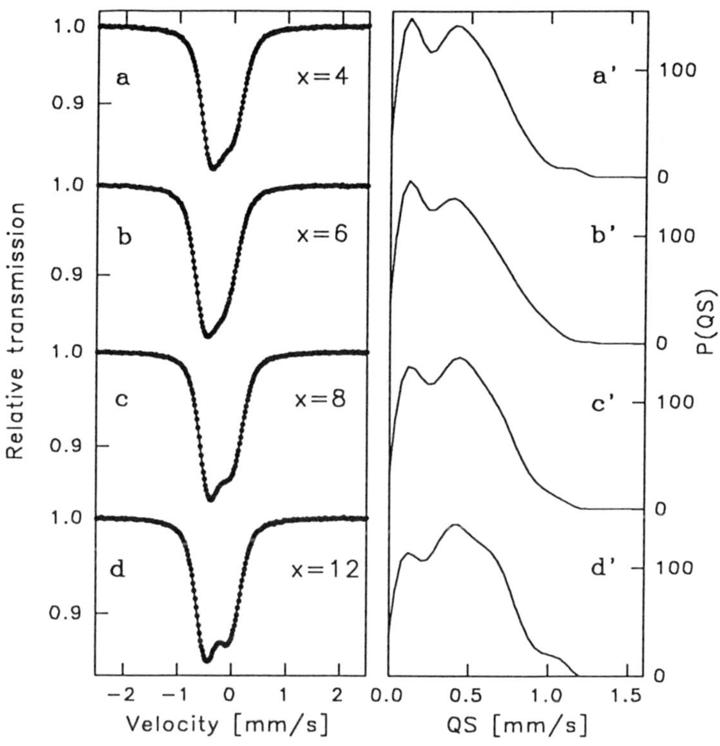

Fig.1. The rf-collapsed Mössbauer spectra of the amorphous $Fe_{93-x}Zr_7B_x$ alloys (a-d) and the corresponding quadrupole splitting distributions (a'-d').

Table I. The average quadrupole splittings <QS> and widths of the P(QS) distributions determined from the rf collapsed Mössbauer spectra for different alloy compositions:

Alloy composition	<QS> [mm/s] ±0.005 mm/s	ΔP(QS) [mm/s] ±0.01 mm/s
$Fe_{89}Zr_7B_4$	0.412	0.25
$Fe_{87}Zr_7B_6$	0.413	0.26
$Fe_{85}Zr_7B_6Cu_2$	0.420	0.26
$Fe_{85}Zr_7B_8$	0.428	0.25
$Fe_{83}Zr_7B_8Cu_2$	0.430	0.25
$Fe_{81}Zr_7B_{12}$	0.449	0.26
$Fe_{79}Zr_7B_{12}Cu_2$	0.457	0.26
$Fe_{80}Ti_7B_{12}Cu_1$	0.451	0.25
$Fe_{80}Ta_7B_{12}Cu_1$	0.453	0.26
$Fe_{80}Nb_7B_{12}Cu_1$	0.448	0.25
$Fe_{80}Mo_7B_{12}Cu_1$	0.452	0.25

The rf-Mössbauer results for amorphous Fe-M-B-Cu alloys, in which as M the Ti, Ta, Nb and Mo atoms were introduced, are shown in Fig.2. In distinction to the FeZrBCu case, the P(QS) distributions are almost identical for all alloy compositions, indicating that such substitutions do not affect significantly the short range order in Fe-M-B-Cu alloys. This result is not surprising because the atomic radii of the alloying elements are very similar: 1.47, 1.49, 1.46 and 1.39Å for Ti, Ta, Nb and Mo, respectively. Similarly to the FeZrBCu alloys the amorphous structure of Fe-M-B-Cu alloys is inhomogeneous as revealed by the P(QS) distributions consisting of two peaks (Fig.2), what suggests the presence of iron-rich and iron-poor regions.

In addition to the rf-Mössbauer studies of the amorphous alloys the conventional Mössbauer measurements (without rf field) were performed. The room temperature measurements performed for both series of alloys reveal that the hyperfine fields vary significantly with alloy compositions. Most of the alloys are ferromagnetic at room temperature, except for $Fe_{89}Zr_7B_4$ and $Fe_{80}Mo_7B_{12}Cu_1$. Figure 3 shows the spectra recorded at room temperature for the FeZrB alloys with various boron content. The changes of the hyperfine fields reflect the differences in the Curie temperatures (T_C) of the alloys studied, and are not related to the SRO. In order to extract information regarding the local structure from the conventional Mössbauer measurements data have to be collected at temperatures much lower than T_C of all alloys studied to avoid the influence of T_C on the hyperfine field distributions. Therefore the Mössbauer measurements were performed at 78K (Fig.4).

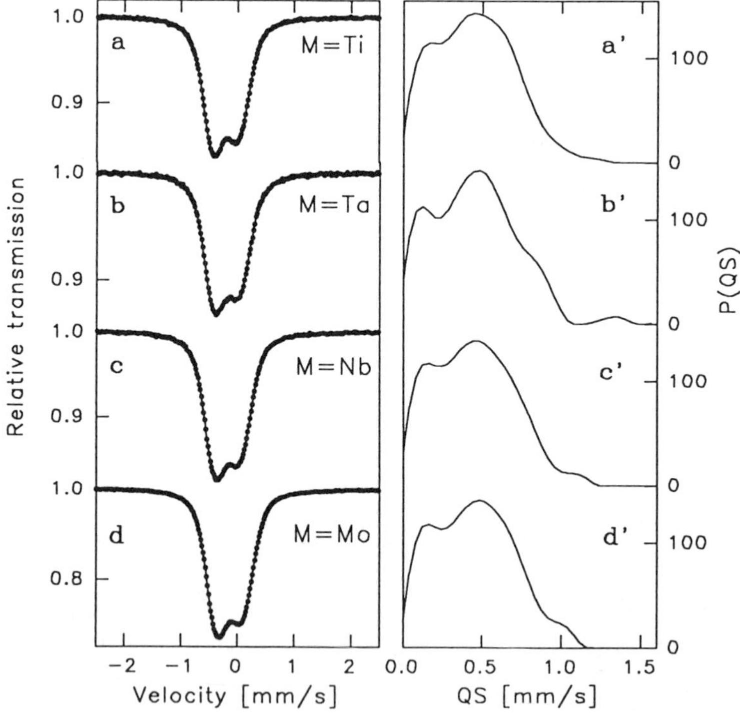

Fig.2. The rf-collapsed Mössbauer spectra of the amorphous $Fe_{80}M_7B_{12}Cu_1$ alloys (a-d) and the corresponding quadrupole splitting distributions (a'-d').

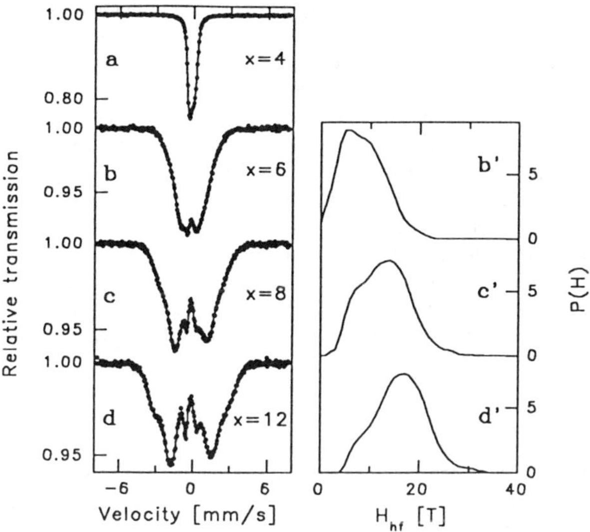

Fig.3. Mössbauer spectra recorded at room temperature for the amorphous $Fe_{93-x}Zr_7B_x$ alloys (a-d) and the corresponding hyperfine field distributions (b'-d').

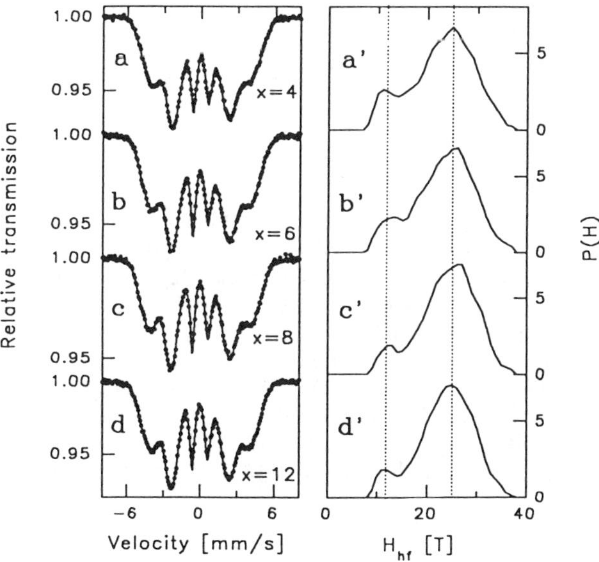

Fig.4. Mössbauer spectra recorded at T = 78K for the amorphous $Fe_{93-x}Zr_7B_x$ alloys (a-d) and the corresponding hyperfine field distributions (a'-d').

Figure 4 shows the Mössbauer spectra of $Fe_{93-x}Zr_7B_x$ alloys collected at 78K. As can be seen all spectra consist of a broadened sextet characteristic for amorphous alloys. The hyperfine field distributions, P(H), extracted using the same method as for the P(QS) distributions [7,8], consist of two well defined peaks: the main one at about 25T and the smaller one at low hyperfine fields (about 12T). The positions of both peaks in P(H) does not change with alloy compositions (Fig.4). The relative contribution of these peaks changes with boron content; the peak at 12T decreases slightly as boron content increases, what may suggests the change in the SRO as the alloy composition changes. However, the changes in the P(H) distributions are less distinct and less credible than those observed in P(QS).

The low temperature Mössbauer measurements performed for Fe-M-B-Cu alloys show that the P(H) distributions are very similar regardless of M atom substitution. This finding is in good agreement with the rf-Mössbauer results discussed above, and confirms that Ti, Ta, Nb and Mo substitutions do not affect significantly the short range order in the $Fe_{80}M_7B_{12}Cu_1$ amorphous alloys.

CONCLUSIONS

The rf-Mössbauer results are much more reliable for the determination of the changes in the short range order vs. alloy composition because the electric quadrupole hyperfine interaction is directly determined by the local structure, while the magnetic hyperfine field is only indirectly related to the symmetry experienced by Fe atoms, and may be significantly affected by the relation between the temperature of the measurement and the Curie point for a given material. The rf-Mössbauer technique provides more accurate and sensitive probe of the SRO than conventional diffraction methods.

ACKNOWLEDGEMENT

The financial support from the Grant No. KBN 2 P03B 06815 from the Polish Committee of Scientific Research is gratefully acknowledged.

REFERENCES

1. M. Kopcewicz, H.G. Wagner, U. Gonser, Hyperfine Inter. **27**, 413 (1986).
2. M.Kopcewicz, H.G. Wagner, U.Gonser, J. Phys. **F16**, 929 (1986).
3. M. Kopcewicz, B. Idzikowski, A. Wrzeciono, J. Magn. Magn. Mater. **125**, 290 (1993).
4. L. Pfeiffer, „Mössbauer Effect Methodology", Vol. 7; ed. I.J. Gruvermann, Plenum Press: New York, 1972, p. 263.
5. M. Kopcewicz, Struct. Chem. **2**, 313 (1991).
6. M. Kopcewicz, A. Grabias, P. Nowicki, D.L. Williamson, J. Appl. Phys. **79**, 993 (1996).
7. J. Hesse, A. Rübartsch, J. Phys. **E 7**, 526 (1974).
8. G. LeCaer, J.M. Dubois, J. Phys. **E 12**, 1083 (1979).
9. R.A. Brand, J. Lauer, D.M. Herlach, J. Phys. **F 13**, 675 (1983).

INFLUENCE OF NANOCRYSTALLIZATION ON THE HIGH-FREQUENCY MAGNETOIMPEDANCE IN ULTRATHIN $Fe_{83}Cu_1Nb_7B_9$ MELT-SPUN RIBBONS.

L. Brunetti, O. Rampado, P. Tiberto, F. Vinai
Istituto Elettrotecnico Nazionale Galileo Ferraris, Corso M. d'Azeglio 42, I-10125 Torino, Italy and INFM Torino Politecnico Unit

ABSTRACT

Ultrathin ribbons of amorphous $Fe_{83}Cu_1Nb_7B_9$ obtained by planar flow casting, have been submitted to dc Joule heating in vacuum to develop a nanocrystalline phase. Variations in the initial magnetic permeability as high as 1000 times the as-quenched value have been measured in a wide range of nanocrystalline samples. The magnetoimpedance behavior in the microwave region has been studied as a function of the nanocrystalline state. In this work the intensity of the impedance variation has been related to the samples degree of softness.

INTRODUCTION

Nanocrystalline magnetic phases obtained by suitable annealing of amorphous metallic alloys are being extensively studied owing to their peculiar soft magnetic properties [1]. The formation and structure of nanosized materials is strictly related not only to the chemical composition, but also to the details of the heat treatment. In particular, it is possible to develop off-equilibrium phases, possibly characterised by different and better physical properties, simply varying the heating rate at which the amorphous-to-crystalline transformation occurs. The dc Joule-heating technique, exploiting the heat released by an electrical current to the amorphous sample, is effective in obtaining a mean grain dimension slightly lower than in conventionally annealed samples [2]. In order to exploit nanocrystalline systems as new core materials, a substantial reduction of core losses in the high frequency range is strictly required. To minimise the eddy current losses, a relevant reduction of the sample thickness is to be achieved. To this aim, ultrathin Fe-Nb-Cu-B nanocrystalline materials (thickness $\approx$ 10 μm) have been recently developed. As expected, the development of a nanocrystalline phase is marked by a dramatic increase of the saturation induction and effective magnetic permeability with respect to the as-quenched state [3].

The discovery of large variations in the transport properties of soft magnetic materials in presence of a magnetic field deserved much attention for perspective technological applications as magnetic sensors. The search for materials displaying suitable working conditions for magnetic devices started with the finding of the so-called giant magnetoresistance effect (GMR) characterized by large changes in the electrical resistivity in multi-layered or granular systems when a dc magnetic field is applied [4]. More recently, much larger changes of the ac impedance have been discovered in Co-based amorphous materials when submitted to a dc magnetic field [5]. The basic mechanism responsible for impedance variations, usually called GMI, has an electromagnetic origin and it is related to the changes in the magnetic penetration depth associated with an ac current flowing through the conductor. The penetration depth turns out to depend on the conductivity of the material, the current frequency and the magnetic permeability μ [5,6]. The external dc field acts to modify the value of μ, therefore changing the effective conduction region and the conductor impedance. This effect has been measured in all soft magnetic alloys, either in wire or ribbon form, although to a lesser extent with respect to Co-based compositions. Its intensity has been related to the alloy magnetostriction constant

Mat. Res. Soc. Symp. Proc. Vol. 577

and, consequently, to the magnetic domain structure. However, it should be noted that generally the GMI ratio has turned out to be higher in wire-shaped samples with respect to ribbon ones, due to the fact that ribbons are characterised by lower transversal permeability values compared to the circumferential permeability measured in wires. The GMI intensity in melt-spun nanocrystalline alloys, ribbon and wire-shaped, has been found to be strictly related to the combination of high permeability and low electrical resistance values typical of the the nanocrystalline state [7].

Magnetoimpedance measurements have usually been extensively performed only up to 10 MHz using the voltage-current measurement technique, which gives a direct measurement of the transport properties [5,6]. However, in the high frequency range, i.e. in the microwave region, voltage and current are not always well defined quantities [8] and the classic measurement technique fails. Furthermore, transport properties at high frequency are not easily measurable in magnetic materials, mainly because the electromagnetic wave propagation does not occur in them. Indeed, the electromagnetic field attenuates very quickly inside the magnetic material, especially if this is also a good conductor. A technique, based on the S-parameter measurement of a transmission line section by means of an automatic vectorial network analyzer, has been applied to measure the GMI effect in soft amorphous ferromagnetic samples in the microwave region overcoming the previous problem [9]. In this work the high-frequency magnetoimpedance behavior in ultra-thin FeCuNbB alloys will be studied as a function of the nanocrystalline state.

EXPERIMENT

Ultrathin $Fe_{83}Cu_1Nb_7B_9$ amorphous ribbons were produced by the single-roll melt-spinning method in vacuum ($10^{-3} \div 10^{-5}$ Torr). Different strips cut from the same ribbon (width 3×10^{-3} m, length 9×10^{-2} m, thickness less than 10 μm) were submitted to dc Joule Heating in vacuum under selected electrical current densities in the range $2\ 10^7 \leq j \leq 7\ 10^7$ A/m^2 for a fixed time (t = 60 s). Room temperature measurements of the ac initial magnetic permeability were performed on magnetically open samples by means of a conventional fluxmetric technique with air flux compensation (driving field amplitude < 40 A/m). The magnetising frequency was 100 Hz.

The GMI effect has been studied through the S-parameter measurements performed by an automatic vectorial network analyser. The experimental set-up provides direct measurements of the transmission and the reflection coefficients of a dual-port device containing the sample as a function of frequency in the range $0.1 \div 6$ GHz. These parameters are complex wave amplitude ratios (S_{11}, S_{22} reflection parameters and S_{12}, S_{21} transmission parameters) that characterised the two-port device. The reflection parameters S_{11}, S_{22} are a direct measurement of the line characteristic impedance Z_∞. The dc magnetic field is given by a solenoid containing the measurement cell ($H_{max} = 7$ kA/m). A scheme of the experimental set-up has been described in details elsewhere [9].

As the magnetoimpedance measured in the high frequency can not be directly measured, an alternative method relating it to the characteristic impedance of a transmission line (a well known parameter that depends on geometrical structure of the line, dielectric permettivity, conductor conductivity and permeability [8]) is used. The measurement cell is simply a magnetic ribbon suspended on a ground plate. This typical transverse electromagnetic transmission line has the propagation vector coincident with the longitudinal axis of the line. In this condition the characteristic impedance Z_∞ of the measurement line turns out to be given by the following equation [8]:

$$Z_{\infty} = \frac{Z_s}{2\pi} \ln \frac{2D}{d} \qquad d << D \tag{1}$$

where Z_S is the intrinsic impedance of the line dielectric (approximately equal to the free space), D the distance of the ground plate from the symmetry center of the suspended conductor of diameter d, in the present case the magnetic sample. An analogous expression holds if the suspended conductor has a rectangular cross section, providing to substitute d with 4/w, being w the larger dimension of the conductor cross section.

Equation 1 is only valid for ideal conductors with d (w) negligible with respect to the distance D. In a real case the effect of the skin depth $\delta = (1/\mu\sigma\pi f)^{1/2}$, that reduces the dimension d of 2δ, must be taken into account. After this correction, an explicit relation between the characteristic impedance of the TEM line and the circular magnetic permeability of its suspended conductor (or, in the case of ribbons, transversal permeability) may be obtained.

The device behaves as an airline whose propagation constant γ and characteristic impedance Z_{∞} depend on the magnetic properties of the suspended conductor through the skin depth [8]. In our conditions the reflection parameters S_{11} and S_{22} are equal, being the measurement cell symmetric. On the other hand, S_{12} may be considered equal to S_{21} only in first approximation. These assumptions are supported by measurements in the limits of the experimental errors. Explicit analytical expressions relating the measured scattering parameters S_{ii} and S_{ij} (i,j=1or2) to propagation constant γ and characteristic impedance Z_{∞} are given by the transmission line theory [8].

The measured reflection coefficient is defined by the relation:

$$S_{ii} = \frac{Z - Z_0}{Z + Z_0} \tag{2}$$

where Z is the input impedance of the cell and Z_0 the reference impedance of the network analyser, which is 50 Ω in this experimental set-up.

Furthermore, it must be observed that the cell behaves as a transmission line terminated both sides by a load impedance Z_0 when connected between the test ports of the network analyzer. Therefore, if one side of the cell is regarded as input side, the associated input impedance Z is given by:

$$Z = Z_{\infty} \frac{Z_0 - Z_{\infty} \tan(\gamma L)}{Z_{\infty} + Z_0 \tan(\gamma L)} \tag{3}$$

This expression is basically the transformation of a load impedance Z_0 along a transmission line of length L, propagation constant γ and characteristic impedance Z_{∞}.

In order to obtain a simple relation between the characteristic impedance and the S parameters, two sets of measurements have been performed by modifying the experimental set-up: the measuring cell was terminated on one side alternatively with a standard short and an open circuit. By calculating Eq. 2 in these particular experimental conditions, one can obtain:

$$Z_0^2 \cdot \left(\frac{1 + S_{11}^c}{1 - S_{11}^c} \right) \left(\frac{1 + S_{11}^o}{1 - S_{11}^o} \right) = Z_{\infty}^2 \tag{4}$$

where S_{11}^c and S_{11}^o are the measured reflection coefficient of the standard short and open circuit respectively. All the parameters appearing in Equation 4 are complex. It should be noted that by extracting Z_∞ from Eq. 4, it is possible to obtain the sample magnetic permeability through the skin-depth expression.

RESULTS

The behaviour of the low-field magnetic permeability divided by the as-quenched value, μ/μ_{aq}, is reported in Fig. 1 as a function of the electrical current density for f = 100 Hz. No appreciable differences have been observed in the magnetic permeability ratio μ/μ_{aq} by varying the measuring frequency up to 1 kHz. For $j < 3.5\ 10^7\ A/m^2$ the permeability remains substantially the same as in the as-quenched material, indicating that no nanocrystalline fraction is developed. An increase in the ratio μ/μ_{aq} is observed for $j \geq 3.5\ A/m^2$, indicating the start of precipitation of α-Fe bcc grains. Massive nanocrystallization occurs in the interval $4 \leq j \leq 6\ 10^7\ A/m^2$, leading to an increase of μ up to about 1000 times the as-quenched value. Such a permeability enhancement is observed in a wide current density interval, indicating a substantial independence of the nanocrystallization from the Joule-heating current. The development of non-magnetic boride phases for $j > 6\ 10^7\ A/m^2$ is detrimental to the magnetic permeability, as previously observed [10].

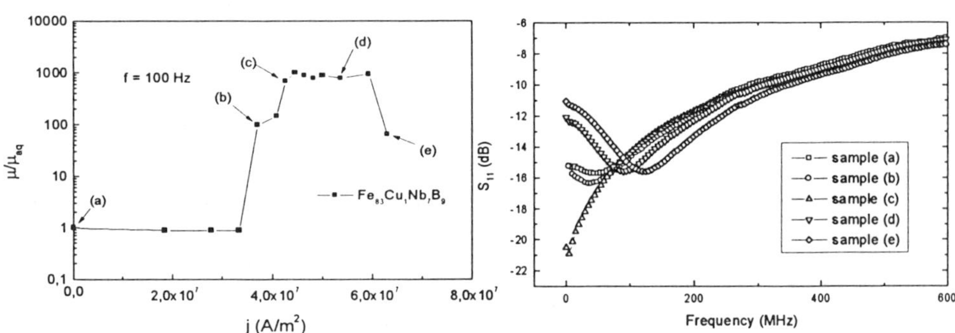

Fig. 1 Ratio of the low-field permeability to the as-quenched value measured at f = 100 Hz for all Joule-heated samples.

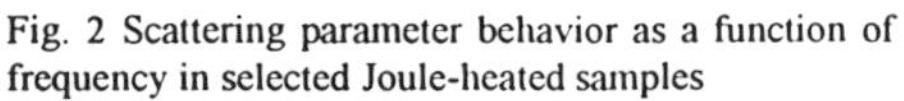

Fig. 2 Scattering parameter behavior as a function of frequency in selected Joule-heated samples

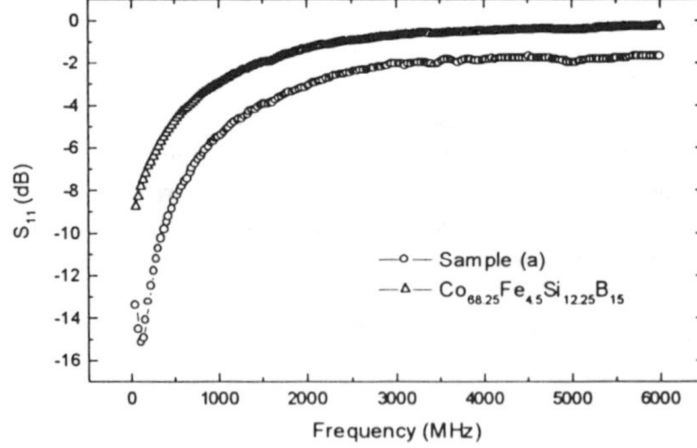

Fig. 3 S_{11} as a function of frequency in as-cast $Fe_{83}Cu_1Nb_7B_9$ and $Co_{68.25}Fe_{4.5}Si_{12.25}B_{15}$.

Selected Joule-heated samples (see arrows in Fig. 1: sample (a): as cast; (b) $j = 3.7\ 10^7\ A/m^2$; (c) $j = 4.26\ 10^7\ A/m^2$; (d) $j = 5.37\ 10^7\ A/m^2$; (e) $j = 6.3\ 10^7\ A/m^2$) characterized by different nanocrystalline state and magnetic permeability values have been chosen to study the high frequency magneto-impedance behavior. The frequency behavior of the scattering parameter S_{11} is reported up to 600

MHz in Fig. 2 for all the studied samples. In these measurements the magnetic field was set to zero. All the displayed curves are characterized by the presence of a minimum displacing towards higher frequency for samples treated with high electrical current values [sample (d), (e)]. By reporting the frequency behavior of the scattering parameter measured at selected magnetic field value no displacement at higher frequency values of the minimum has been observed. Consequently, this feature can not be related to a ferromagnetic resonance but more likely to intrinsic microstructural configuration of the material, as the length of the electromagnetic line has been properly chosen in order to avoid resonance phenomena [8]. It should be noted that in low magnetostrictive Co-based alloys [9,11] the S_{11} curve is seen to monotonically increase with increasing frequency. For comparison, the scattering parameter behavior as a function of frequency has been reported in Fig. 3 for sample (a) together with as-cast $Co_{68.25}Fe_{4.5}Si_{12.25}B_{15}$ amorphous ribbon.

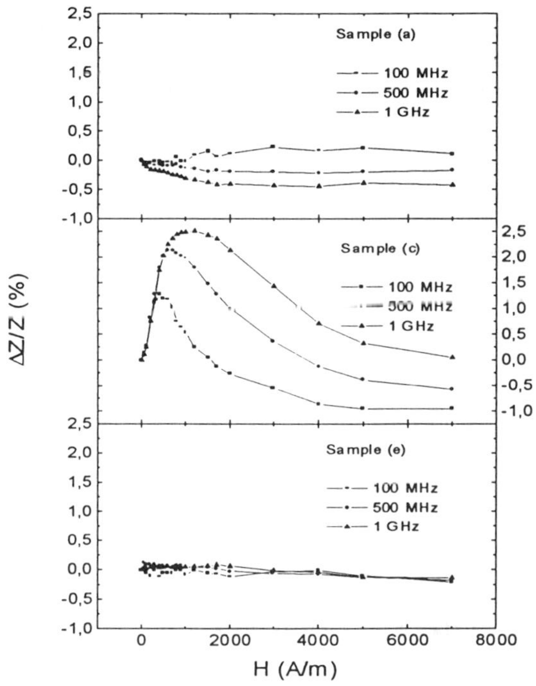

Fig. 4 $\Delta Z/Z$ = [Z(H)-Z(H=0)]/Z(H=0) behavior as a function of the applied magnetic field for (a): as-cast sample; (c) Joule-heated with j = 4.26 10^7 A/m^2; (e) Joule-heated with j =6.3 10^7 A/m^2.

It is possible to extract from the measured S_{11} data the corresponding impedance value Z by simply using Eq. 2. In this way the value of $\Delta Z/Z$ ratio is easily obtained (where $\Delta Z/Z$ = [Z(H)-Z(H=0)]/Z(H=0) x 100). The impedance variation as a function of the applied magnetic field at selected frequencies is reported in Fig. 4. It should be firstly noted that, as expected, the $\Delta Z/Z$ ratio in all the studied samples are very low compared to the ones usually observed in Co-based alloys. This fact has been usually related to the higher magnetostriction value of the Fe-based alloy and consequentely to the magnetic domain structure with respect to Co-based compositions [5,6]. The $\Delta Z/Z$ behavior as a function of the magnetic field in Sample (a) is shown not to have a well definite trend. By increasing the Joule-Heating current a massive nanocrystallization is induced and, as reported in Fig. 1, the presence of the nanocrystalline phase is reflected in a strong increase of the magnetic permeability ratio. The crystallization process is seen also to influence and improve the magnetoimpedance variation, as clearly evident from the $\Delta Z/Z$ behavior of Sample (c), which is seen to monotonically decrease after reaching a maximum value at low field (around 1000 A/mand depending on the measuring frequency). The field value at which the maximum appears is usually related to an anisotropy field, indicating that some transversal anisotropy has been induced during the electrical current treatment [12,13]. On the other hand, the development of the boride phase, occuring at higher electrical currents and inducing a dramatic decrease of μ (see Fig. 1), turns out to be detrimental also to the magnetotransport properties.

CONCLUSIONS

The GMI effect in the high frequency region has been studied in $Fe_{83}Cu_1Nb_7B_9$ melt-spun ribbons. Although characterized by low values of the magnetoimpedance ratio compared to the one observed in Co-based alloy, the nanocrystallization is seen to strongly increase the $\Delta Z/Z$ variation as a function of the applied field. This improvement has been related to the magnetic softness of the material. Moreover, the current-annealing dc Joule-heating technique turns out to induce a strong magnetic anisotropy evidenced by the maximum in the $\Delta Z/Z(H)$ behavior. The occurrence of the boride phase is detrimental either to the soft magnetic properties or magnetotransport ones.

REFERENCES

1. Y. Yoshizawa, S. Oguma, and Y. Yamauchi, J. Appl. Phys. **64**, 6044 (1988)
2. P. Allia, M. Baricco, M. Knobel, P. Tiberto, and F. Vinai, J. Magn. Magn. Mater. **133**, 243 (1994)
3. J.S. Lee, K. Y. Kim, T. H. Noh, I. K. Kang, and Y. C. Yoo, IEEE Trans on Magn **31**, 3901 (1995)
4. C.L. Chien, J.Q. Xiao, and J. S. Jiang, J. Appl. Phys. **73**, 5309 (1993).
5. K. Mohri, K. Kawashima, T. Kohzawa, Y. Yoshida, and L. V. Panina, IEEE Trans on Magn. **28**, 3150 (1992)
6. L.V. Panina, K. Mohri, K. Bushida, and M. Noda, J. Appl. Phys. **76**, 6198 (1994)
7. M. Knobel, J. Phys. IV France **8**, 213 (1998)
8. J.D.Kraus, *Electromagnetics*, (McGraw-Hill, New York, 1991), pp. 460-470.
9. L. Brunetti, P. Tiberto, and F. Vinai, Sensors and Actuators A **67**, 84 (1998)
10. P. Tiberto, F. Vinai, P. Allia, M. Baricco, J.Y. Park, and K.Y. Kim, J. Phys. IV France **8**, 39 (1998)
11. L. Brunetti, P. Tiberto, F. Vinai, H. Chiriac, and T. A. Ovari in *Non-Crystalline and Nanoscale Materials*, edited R. Rivas and M.A. Lopez-Quintela (World Scientific, Singapore 1998) p. 132
12. L.V. Panina, K. Mohri, T. Uchiyama, and M. Noda , IEEE Trans on Magn **31**, 1249. (1995)
13. P. Tiberto, F. Vinai, O. Rampado, H. Chiriac, and T.A. Ovari, J. Magn. Magn. Mater., in press (1999)

ELEMENT-SPECIFIC MAGNETIZATION REVERSAL IN FE/CE MULTILAYERS

M. MÜNZENBERG*, M. AREND*, W. FELSCH*, S. PIZZINI**, A. FONTAINE**, T. NEISIUS***, S. PASCARELLI***
*I. Physikalisches Institut, Universität Göttingen, Bunsenstr.9, 37077 Göttingen, Germany
**Laboratoire Louis Néel, CNRS, BP 166, 38042 Grenoble Cedex, France
***European Synchrotron Radiation Facility, BP 220, 38043 Grenoble Cedex, France

Corresponding author, electronic address: mmuenze@gwdg.de

ABSTRACT

Fe/Ce multilayers are magnetically soft with coercive fields of a few Oersteds. In these structures the Ce-5d states are magnetically polarized by hybridization with the 3d states of Fe. The element selectivity of X-ray magnetic circular dichroism (XMCD) is used to measure the magnetization reversal of the Ce-5d states at room temperature. Comparison with the macroscopic magnetization curves studied by the magnetooptical Kerr effect (MOKE) reveal that the ordered Ce-5d moment is not directly coupled to the overall Fe-layer magnetization: the coercivity of the Ce hysteresis is reduced a factor of two. In view of previous ^{57}Fe Mössbauer spectra this is attributed to an effect of the interface where nominally 1 atomic layer of Fe and Ce are intermixed. Magnetization reversal is initiated in this region due to a reduced magnetocrystalline anisotropy and then propagates into the Fe layers. This interpretation is supported by an experiment on a hydrided multilayer Fe/$CeH_{2-\delta}$: In this case, there is no intermixing at the interfaces and the magnetization of the Ce-5d electrons follows the Fe-layer magnetization directly.

INTRODUCTION

The magnetic properties of thin-film structures composed of ferromagnetic layers separated by nonmagnetic metallic layers are largely governed by the interlayers and by interaction at the interfaces. Interlayer exchange coupling and perpendicular magnetic anisotropy are well kown effects. The contribution of the interlying material to the total magnetic response of a composite structure can be measured separately by making use of the atomic selectivity intrinsic to X-ray magnetic circular dichroism (XMCD). We report on measurements of the 5d electron magnetization of cerium in Fe/Ce multilayers which is induced by iron. By studying the XMCD signal at the Ce-L_2 ($2p_{1/2}\rightarrow 5d$) absorption edge as a function of an applied magnetic field a hysteresis curve was obtained which differs considerably from the one measured by the magneto-optical Kerr effect (MOKE). The result reveals a new insight into the magnetization reversal process next to the interface and into the complex magnetic interaction in these multilayers.

The remarkable properties of the multilayer system Fe/Ce, which are closely related to the unconventional behavior of cerium, were reported in previous papers [1-3]. In Ce metal, the 4f-electron states can be localized (in the γ phase) or delocalized (in the α phase), depending on the degree of mixing with the conduction-electron states. Dramatic differences in the physical properties reflect the unlike electronic ground-state configurations [4]. In multilay-

Mat. Res. Soc. Symp. Proc. Vol. 577 © 1999 Materials Research Society

ers with Fe, detailed studies of X-ray absorption (XA) and XMCD have revealed that strong electronic correlations are effective. Ce adopts the electronic configuration of the α phase on a considerable length scale near the interfaces (up to ~18 Å). This has been attributed to the constraints resulting from the large mismatch between the interactomic spacings at the interfaces of Fe and Ce [1]. Complex hybridization effects at the interfaces, involving the Fe-3d and the Ce-5d and 4f states, induce magnetic order on the normally nonmagnetic 5d and 4f states of α-phase-like Ce. While the magnetic polarization of the 5d states affects the entire α-like part in the Ce layers the polarization of the 4f states is restricted to the direct interface with Fe [2,3]. The absorption of hydrogen in the Ce layers leads to a profound modification of the properties. This is demonstrated here for the MOKE and Ce-5d XMCD hysteresis loops measured on a Fe/$CeH_{2-\delta}$ multilayer, for comparison with the nonhydrided system. The differences are related to the relocalization of the 4f states, leading to a γ-phase-like electronic configuration of the hydride [5,6] and to an altered interface.

EXPERIMENT and RESULTS

For this study, multilayers [Fe(30Å)/Ce(XÅ)]×n with fixed Fe-layer and varying Ce-layer thicknesses were grown by ion-beam sputtering using argon gas in an ultra high vacuum chamber. A hydrided sample [Fe(30Å)/$CeH_{2-\delta}$(30Å)]×67 ($\delta \approx 0.5$) was prepared by reactive ion-beam sputtering in a hydrogen atmosphere of 8×10^{-6} mbar. The samples were deposited on Kapton foil precoated with a 40Å thick Cr buffer to permit measurements of XA and XMCD in transmission mode. For the basic structural characterization by X-ray diffraction Si(100) substrates were used. Because of the same buffer layer, the structural properties of the layered samples do not depend on the substrate. While the hydrided multilayer was deposited at room temperature, it was important to keep the sample holder at 90 K during deposition of the Fe/Ce multilayers to minimize diffusion. A 80-Å-thick Cr cap provided protection against oxidation. More details can be found in references [1] and [5].

X-ray reflectometry in Θ - 2Θ geometry revealed that both multilayer systems maintain a well defined layered structure with a sharp composition profiles. Modeling the small angle spectra by employing an optical theory [7] yielded rms roughnesses (full width at half maximum) of 3-4 Å for the Fe/Ce interface and 1-2Å for the Fe/$CeH_{2-\delta}$ interface. These differences are reflected in the ^{57}Fe Mössbauer spectra [8] addressed below. The high-angle diffraction spectra of the Fe/Ce multilayers show bcc Fe layers with (110) texture and small crystallites (of about 30 Å in the growth direction). For Ce, the spectra show a broad signature around the (111) reflection of the α phase which points to a nanocristalline or amorpous structure. At a thickness t_{Ce} near 40 Å, the (111) reflection of γ-Ce appears, superposed to that of the α-Ce. The high-angle diffraction spectra of the hydrided multilayer Fe/$CeH_{2-\delta}$ show good cristallinity of both components. Fe (bcc) and Ce (fcc) exhibit a marked (111) texture.

The differences in the electronic configuration of the Fe/Ce multilayers and their hydrided counterpart appear in the isotropic X-ray absorption (XA) spectra measured at the L_2 edge of Ce which involves the dipole-allowed transition from the $2p_{1/2}$ core level to the 5d continuum state (E_{L2}=6164 eV). These spectra were measured, together with XMCD spectra, in transmission mode at the energy-dispersive beamline ID24 of the European Synchrotron Radiation Facility (ESRF) in Grenoble, France. The energy resolution was between 0.3 and 0.5 eV. In the case of Fe/Ce multilayers, the XA spectra exhibit the typical double-peak structure indicating the presence of α-like Ce [1,2]: the profile is ascribed to the final states $2p^5 4f^1 5d^n$ and $2p^5 4f^0 5d^{n+1}$ and the relative intensities of the two lines is interpreted as an 'effective occupation number n_f of the Ce-4f states'[9]. The values deduced for the multilayers increase with increasing Ce-layer thickness t_{Ce}, from n_f =0.73 to 0.80 . As outlined previously [1,2], this reflects essentially the evolution from the α-like electronic configuration in Ce toward a laminated two-phase structure with γ-like Ce (n_f≈1) in the center of the Ce layers, as it can be seen in the X-ray diagrams by the appearance of the γ-like Ce signature, also. At t_{Ce}=45 Å, we estimate this part to ~10 Å as previously [1,2]. In the case of the hydrided multilayer the XA spectrum at the Ce-L_2 edge exhibits only a 'white-line' profile. This indicates that the electronic configuration of Ce is γ-phase like throughout the whole sublayer. As in the nonhydrided system, Fe induces a magnetic polarization of the 4f and 5d states of Ce [6].

For the measurements of the XMCD spectra circularly polarized light with a rate of 97.5% was produced by inserting a diamond quarter-wave plate in Laue geometry into the linearly polarized X-ray beam of the spectrometer [10]. A magnetic field was selected and fixed in the propagation direction of the beam at an angle of 30° with respect to the multilayer plane. The helicity of the light was changed by flipping the quarter wave plate between two angular 'offset' positions on either side of a diamond Bragg reflection profile suitable for the energies near the Ce-L_2 edge [11]. The difference signal corrected for a background provided the XMCD spectrum. This technique permits dispose of the external magnetic field (here up to a few 100 Oe) as a free parameter, which is indispensable for a measurement of the hysteresis (Fig. 1). These data measure the magnetic polarization of the 5d electrons of Ce. Errors given in the graphs are based on the subtraction of a reference background which was calculated from the spectrum of the demagnetized sample and the average of all spectra collected for a complete hysteresis curve (which includes the variation of the background with time). For all samples a comparison with the MOKE loops is made, using the same sample on Kapton substrate. They were recorded in longitudinal geometry and measure the total magnetization of the multilayers within the penetration depth of the light (~300 Å). But since according to the previous measurements the ordered 5d moment of Ce is at least 20 times smaller than the 3d moment of Fe [1,2,6] the MOKE intensity represents essentially the Fe-layer magnetization. The comparison of the hysteresis curves takes into account that the ordered Ce-5d and Fe-3d moments are directed oppositely [1,6].

It is immediately apparent, that the Fe/Ce multilayer is magnetically soft. In addition, the coercivity of the Ce-5d hysteresis of this sample is at least a factor of two smaller than that of the MOKE loop. This relation is observed for all samples investigated, differing in their Ce-layer thickness.

In contrast, the coercive fields of the Ce-5d and MOKE cycles are identical in the case of the hydrided multilayer.

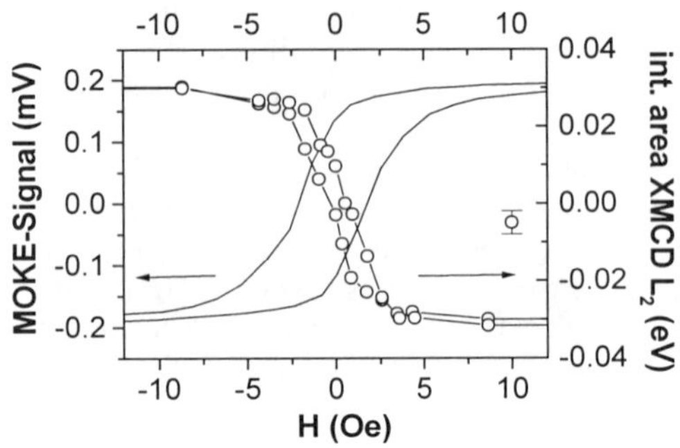

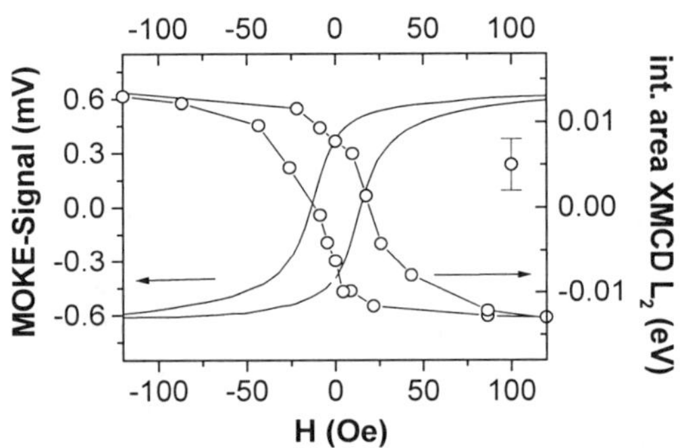

Fig. 1 Hysteresis curves resulting from the integrated Ce-L_2 XMCD spectra (open circles) of the multilayer [Fe(30Å)/Ce(35Å)]x62 (top) and the hydrided multilayer [Fe(30Å)/$CeH_{2-\delta}$(30Å]x67. Continuous curves: MOKE loops. Note the different scale of the magnetic field *H*.

DISCUSSION

The unequal reversal of the XMCD and MOKE signals for the Fe/Ce multilayers in the external magnetic field is remarkable since 5d magnetism on Ce is induced by Fe. In the present paper, we aim to focus on a simple model illustrated in Fig. 2 to explain the differences. We have previously shown, that substantial information on the interface in these structures can be derived by ^{57}Fe Moessbauer spectroscopy [8]. This study reveals the existence of an intermixed region at the Fe/Ce interface involving Fe atoms within a nominal depth of 2.5 Å which is in the order of the structural roughness. This region is only weakly magnetic as compared to Fe in the center of the sublayers (the magnetic hyperfine field is reduced by a factor of seven). Considering the nanocrystalline or amorphous structure of the Ce sublayers, the intermixed region must be highly disordered. As a result, it exhibits a ran-

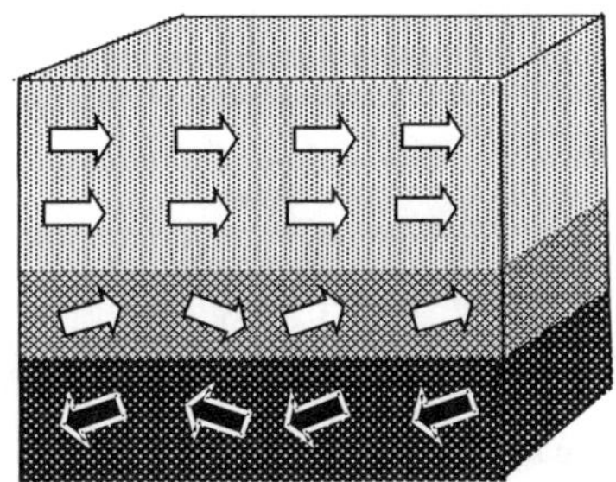

Fig. 2 On top: magnetic domain. Middle: disturbed interface region with reduced crystalline anisotropy. At bottom: Ce layer. The magnetic moments are coupled to the interface magnetization.

dom magnetocrystalline anisotropy <K> which can be expected to be small since the (α-phase-like) Ce-4f states are delocalized. This favors reversal of the magnetization at the interface in a smaller applied magnetic field as compared to the bulk of the Fe layer. Magnetic order of the Ce-5d electron spins is induced by Fe at the interface via 3d-5d hybridization.

This means that the interfacial Fe-3d and Ce-5d magnetic moments are rigidly coupled. Hence reversal of the Ce-5d magnetization by an applied field, as it is probed here by the XMCD measurement, mirrors in fact reversal of the Fe magnetic moments at the interface, i.e. in the intermixed layer. In other words, the observation of a smaller coercivity of the Ce-5d electron magnetization cycles indicates that magnetization reversal in the Fe/Ce multilayers is initiated at the interfaces and then propagates into the Fe layers. The smaller remanent signals of the dichroic loop as compared to the MOKE loop indicates that the magnetic moments of Fe at the interface are not collinear with the volume magnetization in zero magnetic field, supporting the picture of a random interfacial anisotropy.

To probe the proposed model we carried out an experiment on a sample, in which a 10-Å-thick La-layer was interleaved between the Ce and Fe layers. In this multilayer structure, magnetic polarization of the Ce-5d states by Fe is transmitted across the La interlayer [2]. However, the ^{57}Fe Mössbauer spectrum reveals that Fe and La are less intermixed at their interface and the perturbation of the magnetic hyperfine field at the interfacial Fe sites is considerably diminished as compared to the Fe/Ce interface [2]. Indeed, this sample shows the same coercive fields for the XMCD hysteresis cycle of the Ce-5d electrons and the MOKE loop.

An additional support for the model is the result obtained for the hydride multilayer Fe/$CeH_{2-\delta}$,. As is was shown above, the magnetization reversal of the Ce-5d electrons closely mirrors that of the Fe layers. Again, this is related to the different nature of the interfaces which are less rough and disordered and, according to the Mössbauer, not intermixed [8]. Magnetization reversal then occurs coherently through both sublayers. Furthermore, this multilayer is magnetically harder.

In conclusion, the present study illustrates the power of the XMCD technique for studying element-resolved details of magnetization process in thin-film heterostructures, which are not accessible by classical techniques.

ACKNOWLEDGEMENTS

This work was supported by the Deutsche Forschungsgemeinschaft within SFB 345.

REFERENCES

[11] S. Pizzini, A. Fontaine, L.M. Garcia, J.-F. Bobo, M. Piecuch, F. Baudelet, C. Malgrange, A. Alimoussa, E. Snoeck, M.J. and Casenove, J. Magn. Magn. Mater. **166**, 38 (1997).

[1] F. Klose, O. Schulte, F. Rose, W. Felsch, S. Pizzini, C. Giorgetti, F. Baudelet, E. Dartyge, G. Krill, and A. Fontaine, Phys. Rev. B **50**, 6174 (1994).

[2] M. Arend, M. Finazzi, O.Schulte, M. Münzenberg, A.-M. Dias, F. Baudelet, Ch. Giorgetti, E. Dartyge, P. Schaaf, J.-P. Kappler, G. Krill, and W. Felsch, Phys. Rev. B **57**, 2174 (1998).

[3] M. Finazzi, F.M.F. de Groot, A.-M. Dias, B. Kierren, F. Bertran, Ph. Sainctavit, J.-P. Kappler, O. Schulte, W. Felsch, and G. Krill, Phys. Rev. Lett. **75**, 4654 (1995).

[4] See *Handbook on the Physics and Chemistry of Rare Earths,* edited by K.A. Gschneidner, Jr., and L. Eyring, (North-Holland, Amsterdam, 1978), Vol. 1.

[5] O. Schulte, F. Klose, and W. Felsch, Phys. Rev. B **52**, 6480 (1995).

[6] M. Arend, W. Felsch, G. Krill, A. Delobbe, F. Baudelet, E. Dartyge, J.P. Kappler, M. Finazzi, S. Pizzini, A. Fontaine, A. San Miguel Fuster, Phys. Rev. B, in press.

[7] L.G. Parratt, Phys. Rev. B **95**, 359 (1954); L. Nevot and P. Croce, Rev. Phys. Appl. **15**, 761 (1980).

[10] C. Giles, C. Malgrange, J. Goulon, F. de Bergevin, C. Vettier, E. Dartyge, A. Fontaine, C. Giorgetti, and S Pizzini, J. Appl. Crystallogr. **27**, 232 (1994); C. Giles, C. Malgrange, J. Goulon, F. de Bergevin, C. Vettier, A. Fontaine, E. Dartyge, and S Pizzini, Nucl. Instr. and Meth. A **349**, 622 (1994).

[9] Röhler, in *Handbook on the Physics and Chemistry of Rare Earths,* edited by K.A. Gschneidner, Jr., L. Eyring, and S. Hüfner (North-Holland, Amsterdam, 1987), Vol. 10, p. 453.

[8] Ph. Bauer, F. Klose, O. Schulte, and W. Felsch, J. Magn. Magn. Mater.**138**, 163 (1994).

Part XIV

Nanocrystalline Magnetic Thin Films

PVD GROWTH OF FCC METAL FILMS ON SINGLE CRYSTAL Si AND Ge SUBSTRATES

K.H. Westmacott, S. Hinderberger, T. Radetic and U. Dahmen

National Center for Electron Microscopy
Lawrence Berkeley National Laboratory
1 Cyclotron Road
Berkeley, California 94720, USA

ABSTRACT

Epitaxial films of the fcc metals Al, Au, Ag and Ni were grown by physical vapor deposition on Si and Ge (111), (110) and (100) substrates at different deposition temperatures. The epitaxial relationships and morphological features of these films were characterized by transmission electron microscopy and diffraction in plan view and cross section. Ag formed single crystal films on all substrates at all temperatures. Au and Al could be grown as bicrystals, and under some conditions, Al and Ni grew as tricrystal films. The morphological effects of diffusion at the metal/substrate interface are ascribed to diffusion induced grain boundary migration.

1. INTRODUCTION

When a thin film grows on a crystalline substrate, the epitaxial relationship between the film and substrate lattices determines the type and degree of thin film texture. On a single crystal substrate, a film may grow as a single crystal if it forms in parallel epitaxy, as a bicrystal if it forms in two equivalent orientation variants, as a textured polycrystal if it grows in several equivalent orientation variants and as a general polycrystal if it grows with grains in many random orientations. The degree of film texture from a single crystal to fiber texture or even a texture-free polycrystal can be controlled through choice of thermodynamic, crystallographic, and physical parameters.

From the earliest studies it was believed that good atomic matching at the deposit/substrate interface was a necessary condition for epitaxy, on the reasonable assumption that such an interface would have a low energy. Bruce and Jaeger [1] studied fcc Ni, Cu, Au and Ag deposits on (100) Ag and W, and (110) W. The observed epitaxial relationships, which depended on the deposition temperature and the atomic misfit, agreed with earlier AES studies of the initial stages of deposition on W. For systems with a large disparity in the areal density of atoms at the interface, the strains required to bring primitive unit cells into coincidence become unrealistically large and the probable importance of supercell matching has been recognized [2,3]. As an example, primitive cell matching of Al/Si requires a 25% strain whereas in the vector ratio of 4:3 the strain is about 0.5%. A major difficulty in this approach is, however, the arbitrariness in the selection of the optimum supercell size and its associated strain, which undermines its use as a predictive tool.

Frequently the quality of thin films as characterized by X-ray diffraction, Rutherford backscattering, or TEM, is compared for different deposition conditions. Several studies have characterized the defect and grain boundary structure of such films and pointed out their unique topological and crystallographic properties. The faceting behavior, crystallography and grain boundary structure of (110) bicrystal films of Al have been studied in some detail [4,5,6] Such

Mat. Res. Soc. Symp. Proc. Vol. 577 © 1999 Materials Research Society

"mazed" bicrystal films have interesting crystallographic properties as they exhibit a maze of high angle grain boundaries with fixed misorientation but variable inclination, without the impeding effect of triple junctions. Similar studies on (111) bicrystal films of Au [7] and (100) tricrystal films of Al [8,9] have shown the importance of two- and three-color crystallography in the development of such microstructures. The multicolor crystallography of such films with Q^{th} order Potts model microstructures has been reviewed recently [10].

2. EXPERIMENTAL METHODS

Thin (usually 100nm) metal films were grown by thermal evaporation from a tungsten wire basket in a turbo-pumped Balzers evaporating unit. Single crystal wafers of silicon and germanium with (111), (100) and (110) orientation were used as substrates for the depositions. The silicon substrates were chemically oxidized in a solution of sulfuric acid and hydrogen peroxide (1:1) for about 5 minutes and subsequently rinsed in deionized water. They were then etched in a freshly-prepared 30% solution of hydrofluoric acid for 5 minutes, dried, and loaded immediately into the deposition chamber. Germanium substrates were given only the etching treatment.

Prior to all depositions the chamber was pumped to a vacuum better than 5×10^{-7} mbar. Depositions ranging in thickness from about 50nm to about 1μm were usually completed within a few seconds.

Specimens for transmission electron microscopy were prepared by standard procedures. Plan view specimens were mechanically dimpled from the substrate side to a thickness of approximately 10 μm and subsequently thinned to electron transparency by ion-beam milling. Cross section samples were prepared using the technique described by Sinclair. A JEOL 200CX microscope operated at 200 keV was usually used for electron microscopy.

3. RESULTS

Reproducible film growth was achieved on all three single crystal Si and Ge substrates so long as the optimum substrate preparation and growth conditions were maintained.

3.1 Single crystal films

On all three substrate orientations, for both Si and Ge, it was found that Ag always formed as a single crystal in parallel epitaxy with the substrate. On Si (111) and (100) substrates, this behavior was found to be independent of temperature over the range from 150-400°C. On Ge substrates, the dependence on temperature was not tested, but at 280°C parallel epitaxy was obtained in all three orientations. The deposition temperature appeared to affect mainly the internal defect structure of the films. This is illustrated in Figure 1, which shows the defect substructure of single crystal Ag films deposited on Si (111) at 150 and 400°C and on Ge (110) at 280°C.

Films deposited at the lower temperatures contained a high density of stacking faults (see Figure 1a). At the highest temperature (400°C) the stacking fault and dislocation densities were low, but a very high density of small defect clusters was now present (see Figure 1b). An intermediate defect density was found in films deposited at 280°C on Ge (110) (see Figure1c).

Single crystal films were also found for Al films on Si (111) substrates, but only at the highest growth temperature of 400°C. For comparison with earlier work [11] a film was also deposited at room temperature and subsequently annealed at 400°C without breaking vacuum. This treatment also produced a single crystal in parallel epitaxy but the film contained a very high density of randomly distributed dislocations.

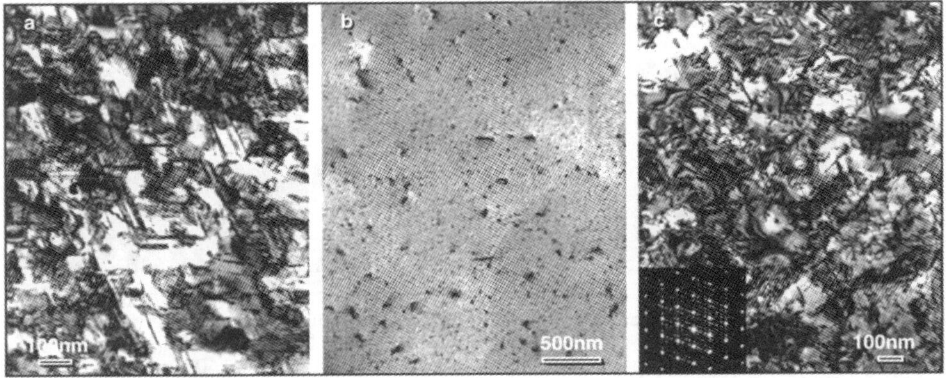

Figure 1: Single crystal films of silver formed on Si (111) at 150°C (a), and at 400°C (b). The film in (c) was grown on Ge (110) at 280°C. It is evident that with increasing temperature the density of stacking faults decreases but the density of point defect clusters increases.

3.2 Bicrystal films

Aluminum was found to form epitaxial bicrystal films on Si (100) surfaces over the temperature range 200-400°C. The observed structure, referred to as a mazed bicrystal, has unique properties which have been described in detail in earlier papers [4,5,8]. An epitaxial relationship, $(110)_{Al} \parallel (100)_{Si} : [\bar{1}10]_{Al} \parallel [011]_{Si}$, leads to two orientation variants rotated 90° to each other (see Figure 2a). A unique aspect of this microstructure is the fact that all grain boundaries form closed loops, and there are no triple junctions [12]. An increase in temperature led to a progressive increase in grain size.

A very similar bicrystal structure could be grown for Au on Ge (100) at temperatures between 200 and 280°C. This is illustrated in Figure 2b which shows that the boundaries are faceted and that the film contains occasional microtwins.

On the Ge (111) surface Au formed a bicrystal microstructure similar to that reported by Pashley et al. for Au on MoS_2 [13] and by Bauer et al. for Au on NaCl. Unlike the bicrystal structures on (100) substrates, which are generated by different variants of the same orientation relationship, this bicrystal is made of two different orientation relationships of the film, one in parallel epitaxy, the other rotated by 180° (equivalent to a 60° rotation, i.e. twin-related parallel to the substrate surface). This behavior has been called "double positioning" [13] because of the two different film positions with respect to the substrate, and has been attributed to the low stacking fault and twin boundary energies in Au. Figure 2c illustrates this structure near a <112> zone axis where the two grain orientations are not equivalent. The inset composite diffraction pattern was indexed as a [221] zone for one grain and near [112] for the other, consistent with their twin relationship. This type of microstructure has also been reported for silicide films on Si (111) [14].

Figure 2: Mazed bicrystal films of Al and Au on different substrates. The Al film in (a) and the Au film in (b) have the same crystallography and topography. Both were grown on (100) substrates and display a maze of 90° <110> grain boundaries. The Au film in (c) has the same topography but its crystallography is characterized by a misorientation of 60° about a <111> tilt axis.

These bicrystal films were also investigated in cross section. As illustrated in Figure 3, all three films exhibit uniformly flat surfaces and interfaces with the substrate. The <110> bicrystals of Al and Au (Figure 3a,b) have strictly columnar grains with straight grain boundaries that lie perpendicular to the interface. In sharp contrast, the <111> bicrystal in (c) exhibits many grain boundaries that are parallel to the interface. This explains the more complex appearance in the plan view micrograph (Figure 2c), where grains are often seen to overlap in projection.

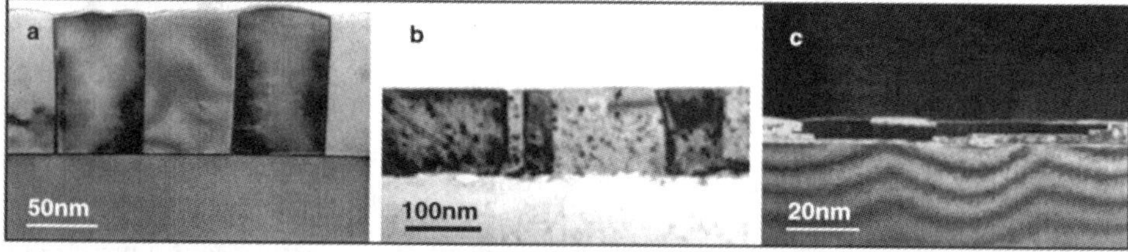

Figure 3: Cross sections of bicrystal films shown in Figure 2 showing uniform flat films that maintain relatively sharp interfaces with the substrate. In (a) and (b), grain boundaries are straight and perpendicular to the interface. In (c) grain boundaries are highly faceted and display a tendency to lie parallel to the substrate.

3.3 Tricrystal films

Tricrystal films of Al (100) could be grown Si (111) substrates at temperatures from 200-280°C. The special crystallographic properties of such tricrystals have been considered elsewhere [8,9] At higher temperatures, the epitaxial relationship changed to become parallel, and the films grew as single crystals. These microstructures are illustrated in Figure 4. Similar behavior was observed for Al on Ge (111) substrates, where a tricrystal structure could be grown at 220°C, but the structure changed to polycrystalline with a mixed (100) and (111) texture at higher substrate temperatures.

For a miscut Si substrate, the three orientation variants no longer occurred in equal proportion, but one variant was preferred over the other two. This leaves a single crystal of the majority (100) orientation variant, containing islands of the tricrystal structure, as seen in Figure 4b.

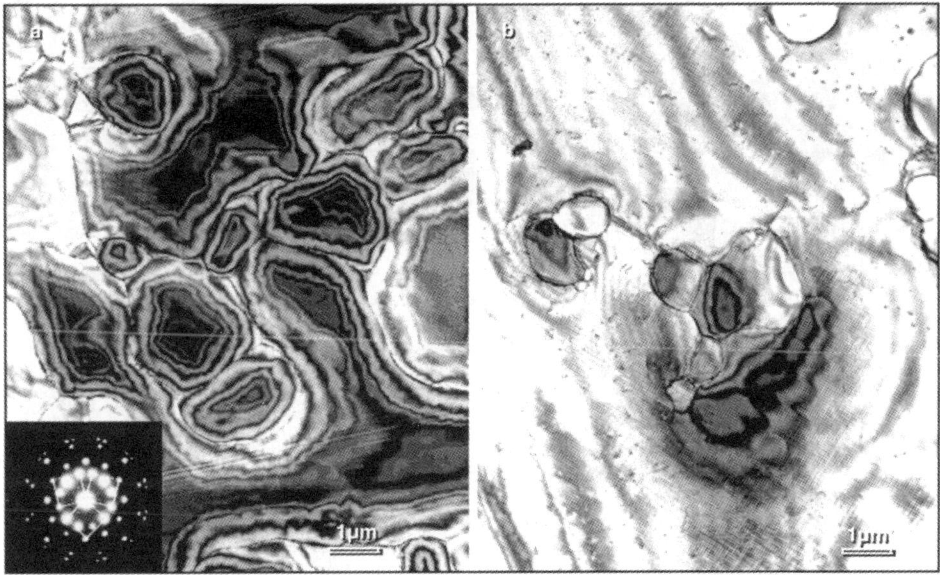

Figure 4: Al films grown on Si (111) substrates. A typical (100) tricrystal structure is seen in (a), exhibiting a random distribution of grains in only three orientations, rotated relative to each other by 30° around a common <100> axis. In (b) one of the three grain orientations is preferred over the other two due to a miscut of the substrate. The tricrystal structure now appears as islands in grain of the majority orientation.

Ni tricrystal films in (110) orientation could be grown on (111) Si if the substrate was held at a temperature of 280°C. The three grains were rotated relative to each other by 60° around a common <110> axis. As illustrated in Figure 5a the grains are smaller in size, and the crystallographic alignment less pronounced than for Al. The diffraction pattern in Figure 5b three interpenetrating <110> Ni patterns rotated by 120° to each other. For clarity, only one of the three patterns is marked.

When the growth temperature was raised to 350°C, a reaction with the Si substrate occurred. This is apparent from the cross section shown in Figure 5c. The reaction product was identified by electron diffraction as Ni_2Si.

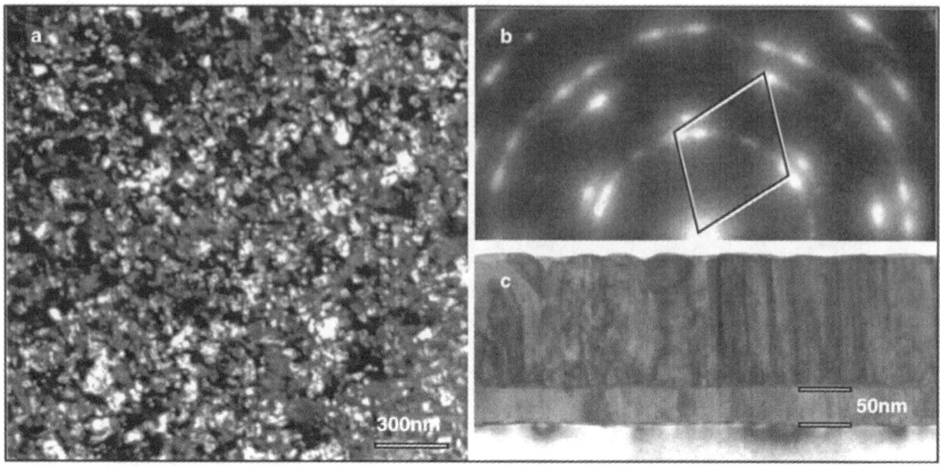

Figure 5: (110) tricrystal of nickel formed by deposition on Si (111) at 280°C (a). The diffraction pattern (b) consists of 3 superposed 110 spot patterns with relative rotations of 30°. For clarity only one of the variants has been marked on the pattern. The reaction layer formed during deposition at 350°C is apparent from the cross section in (c).

3.4. Post deposition annealing

To equilibrate the bicrystal structures, some films were annealed after deposition. In Al and Au films this led to faceting. However, extended annealing above the deposition temperature gave rise to some interesting instabilities. As seen from the weak beam micrograph in Figure 6a, the film is no longer flat but shows extensive variations in projected thickness. That this is not an artifact of specimen preparation is clear from the corresponding cross section in Figure 6b. By comparison with Figure 3b, which shows a similar cross section before annealing, it is clear that the interface undergoes substantial corrugation during the heat treatment. These corrugations have an amplitude of the order of the film thickness and are seen to affect the film surface as well as its interface with the Ge substrate.

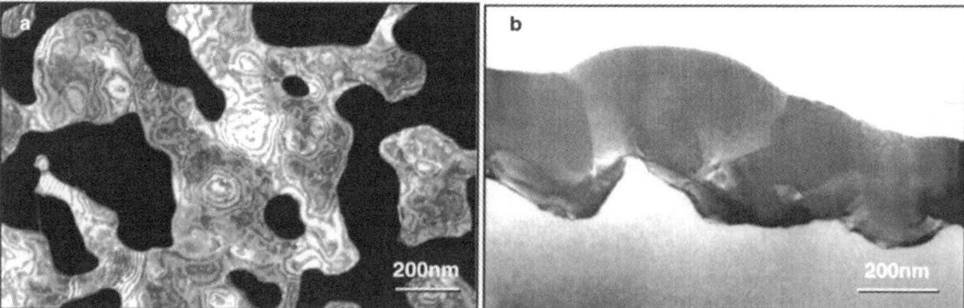

Figure 6: (110) Au bicrystal film on Ge substrate after annealing 1-3 days at 300°C. Weak beam image in plan view (a) shows thickness contours. Cross section (b) shows interface corrugation.

Although there appears to be a reacted layer at the interface in Figure 6b, diffraction and contrast analysis only showed Ge and Au. Initial energy dispersive X-ray analysis did not reveal any compositional differences between the Au film and the apparent interfacial phase. Thus, the layer is part of the Au film and contributes to the extensive thickness variations seen in plan view.

4. DISCUSSION

Ag, Al and Au form simple binary eutectics with Si and Ge. The eutectic temperatures and compositions are listed in Table 1. By comparison, the phase diagrams of Ni-Si and Ni-Ge are complex and display several intermediate phases. It is likely that the absence of intermediate phases inherent in a binary eutectic allows local equilibrium to be maintained during epitaxial growth. The temperature range for this type of growth is limited at the lower end by the atomic mobility of the metal and at the upper end by the eutectic reaction. Therefore, the largest temperature range for epitaxial growth is expected for Ag, which has by far the highest eutectic temperature with both Si and Ge. This is consistent with the observed behavior of Ag on all substrates and the increased stacking fault density at lower temperature.

Table 1: Eutectic temperatures and compositions

	Ag		Al		Au	
	T_{eu}	c_{eu}	T_{eu}	c_{eu}	T_{eu}	c_{eu}
Si	835	11	577	12.2	363	18.6
Ge	651	24.5	420	28.4	361	28

At the other extreme, Au has a very low eutectic with Si and Ge, and on some substrate orientations, epitaxial growth was not possible, even at growth temperatures well below the eutectic. It is likely that this difficulty is due to the same process that was observed during annealing of epitaxial films at 300°C (see Figure 6). This process leads to a substantial roughening of the interface and even the surface of the film.

Similar behavior, albeit much less pronounced has been observed during annealing of single crystal and bicrystal Al films on Si at 500°C [15]. It was found that during annealing of such films, craters of Al formed on the Si side of the interface, and after cooling to room temperature, Si crystallites were found on the film surface. This behavior is due to the asymmetry in mutual solubilities of Al and Si. While at elevated temperature a substantial amount of Si dissolves in the Al film, there is virtually no solubility for Al in the Si substrate. During cooling, the dissolved Si re-precipitates preferentially on the surface, rather than reversing the formation of interface craters or “spikes”. In-situ observations showed that this interdiffusion process initiates where the grain boundaries meet the interface. However, in the course of this process, the grain boundaries often move laterally as a result of diffusion induced grain boundary migration [e.g.16,17].

The same process is expected to operate for Au films on Ge. The eutectic temperature is substantially lower than for Al-Si (367 vs 577°C) and the maximum solubility of Ge in Au is about twice that of Si in Al. Hence, the process is expected to be more pronounced. Whereas in the case of Al on Si the film remained essentially flat, except for the location of the spikes at the interface and precipitates at the surfaces, in the case of Au on Ge, the interdiffusion was so extensive that the film became morphologically unstable, displaying corrugations with an amplitude about equal to the film thickness. This is could not be due simply to the volume of material transported from the interface (during annealing) to the surface (during cooling) because the equilibrium solubility at the annealing temperature is relatively small (<3at%). Hence another process must be responsible for the observed morphological change. It is likely that this process is due to the mechanism of interdiffusion, and is thus a manifestation of diffusion induced grain boundary migration.

Single crystal films that are free of grain boundaries and planar defects should not be subject to diffusion induced grain boundary migration and hence should be morphologically stable during annealing. However, it remains unclear why only Ag forms single crystal films whereas Au, Al and Ni generally form bi-, tri- or polycrystal films. It is difficult to understand the selection of epitaxial relationships for different metals on different semiconductor substrates.

Applying crystallographic concepts to metal-semiconductor interfaces it was found impossible to develop a coherent understanding of the observed epitaxial relationships. This problem remains even if the lattice correspondence between the metal and the substrate is based on a near-coincidence site lattice (n-CSL) [18,19,2,3]. The n-CSL is considered a more appropriate description of the match between two materials with a large difference in atomic density. Whereas in a primitive cell correspondence, all atoms are related to their counterparts across the interface by the same transformation rule, for an n-CL correspondence, the transformation rule applies only to the atoms at the lattice sites of the supercell. All other atoms must undergo shuffles and remain unaccounted for in the interface energy. Thus the larger the n-CSL cell the less significant it becomes as a measure of interface energy. On the other hand, larger cells allow smaller principal strains. This dilemma of choosing the appropriate balance of cell size and strain that will optimize the energy remains a serious drawback of this approach [20].

Regardless of the choice of lattice correspondence, if a thin film is under an elastic strain, elastic anisotropy is known to play an important role in epitaxial growth [21,22]. However, its effect cannot be used to rationalize the present results either. For example, large variations in epitaxy are observed for Al which is elastically isotropic, whereas no preference for the elastically softer (100) orientation is observed for Ag which exhibits a large anisotropy ratio (A=3). It is thus clear that elastic anisotropy does not play a dominant role in determining epitaxy in the present experiments.

Having ruled out crystallographic and elastic effects as determinants for the specific epitaxial relationships, we turn our attention to the state of the semiconductor surface. From the following selective comparison of the present results with recent work it becomes apparent that the choice of a metal-semiconductor epitaxial relationship depends on the state of the substrate surface and the degree of reaction at the interface. Because of the large body of literature on the subject, we limit ourselves to a selective review of Al film growth only on Si (111) surfaces.

It is well-known that a reduction in energy of a clean Si (100) surface can be effected by a 2x1 reconstruction that forms dimers and reduces the number of dangling bonds [23]. LEED measurements have also shown that this reconstruction is suppressed on a hydrogen-terminated surface [24]. Since the epitaxies found in the present work are likely to be due to H-passivation [25], it is not unreasonable to assume that replacement of the hydrogen by the Al leads to a form of 2x1 reconstruction. Evidence for Al-induced reconstruction of a Si (111) surface does exist [26], but no experiments of this type appear to have been done on Si (100). A consequence of a 2x1 reconstruction is to increase the effective spacing of a double set of (110) planes from its usual value of 0.768nm. In this circumstance the 2a[100]Al spacing of about 0.80 nm is readily accommodated without strain so that the (110) bicrystal structure remains the most favorable epitaxy over the entire temperature range. Support for this idea is drawn from the observations that the same bicrystal structure is found when Al is deposited on 2x1-reconstructed Si surfaces under the ultra-high vacuum conditions of ionized cluster beam deposition [27] and molecular beam epitaxy [28]. The observations reported in the present work are therefore consistent with an Al-induced reconstruction of the (100) surface.

5. SUMMARY

We have grown epitaxial films with single crystal, bicrystal or tricrystal structure using different fcc metals on Si and Ge single crystal substrates in (100), (110) and (111) orientation.

- Ag formed a single crystal film in parallel epitaxy on all substrates in all orientations over the full temperature range of the experiments.
- Au formed a (111) bicrystal on Ge (111) and a (110) bicrystal on both, Si and Ge (100).
- Al on Si and Ge (111) formed a (100) tricrystal at low temperatures but changed at higher temperature to a (111) single crystal on Si, or a textured polycrystal film on Ge. Al on Si and Ge (100) grew as a (110) bicrystal.
- Ni formed a tricrystal structure at low temperature and reacted to form a silicide interlayer at higher temperatures.

Interdiffusion during annealing led to diffusion induced grain boundary migration which caused Au films on Ge substrates to undergo large scale corrugations and was involved in the formation of spikes of Al on Si substrates.

ACKNOWLEDGMENTS

Thanks are due to Ute Salzberger for her expert assistance in some of the depositions and to John Cahn for helpful discussions. This work is supported by the Director, Office of Basic Energy Sciences, Materials Sciences Division of the U.S. Department of Energy under Contract No. DE-AC03-76SF00098.

REFERENCES

1. L.A. Bruce and H. Jaeger, Phil. Mag. A38, 223, (1978)
2. K.H. Westmacott and U. Dahmen, 40th EMSA Proc., G.W. Bailey (ed.), p. 620 (1982)
3. A. Zur. and T.C. McGill, J. Appl. Phys., 55 (2), 378 (1984)
4. U. Dahmen and K.H. Westmacott, Scripta metall., 22, 1673 (1988)
5. U. Dahmen, C.J.D. Hetherington, M.A. O'Keefe, K.H. Westmacott, M.J. Mills, M. Daw and V. Vitek, Phil. Mag. Lett., 62, 327 (1990)
6. U. Dahmen and N. Thangaraj, Mat. Sci. Forum, 126-128, 45 (1993)
7. C.J.D. Hetherington, U. Dahmen and J.M. Penisson, Mat. Res. Soc. Symp. Proc., 466, 215 (1997)
8. N. Thangaraj and U. Dahmen, Mat. Res. Soc. Symp. Proc., 238, 171 (1992)
9. J.W. Cahn and G. Kalonji, J. Phys. Chem. Solids 55, 1017 (1994)
10. U. Dahmen, in "Boundaries and Interfaces in Materials" , R.C. Pond, W.A.T. Clark, A.H. King and D.B. Williams, eds, TMS, Warrendale, p.225 (1998)
11. F.K. LeGoues, M. Liehr, M. Renier and W. Krakow, Phil. Mag.B, 57 (2), 179 (1988)
12. M.C. Madden, Appl. Phys. Lett. 55, 1077 (1989)
13. D.W. Pashley, M.J. Stowell, M.H. Jacobs and T.J. Law, Phil. Mag., 10, 127 (1964)
14. D. Cherns, J.C.H. Spence, G. Anstis, and J.L. Hutchison, Phil. Mag. A, 46, 849 (1982)
15. N. Thangaraj, S. Hinderberger, K.H. Westmacott and U. Dahmen, MRS Proc. ULSI in 1993, p. 247
16. D.Y. Yoon, Int. Mat. Rev. 40, 149 (1995)
17. A.H. King, Int. Mat. Rev. 32, 173 (1988)
18. W. Bollmann, Crystal Defects and Crystalline Interfaces, Springer, Berlin, (1970)
19. R.W. Balluffi, A. Brokman and A.H. King, Acta Met. 30, 1453 (1982)
20. Y. Gao, P. Shewmon and S.A. Dregia, Scr. Met. 22, 1521 (1988)
21. W.D. Nix, Metall. Trans. 20A, 2217 (1989)
22. J.A. Floro, C.V. Thompson, R. Carel and P.D. Bristowe, J. Mater. Res., 9, 2411 (1994)
23. R.E. Schlier. and H.F. Farnsworth, J. Chem. Phys. 30, 917 (1959)
24. P.O. Hahn, Res. Soc. Symp. Proc., 54, 645 (1986)
25. E. Yablonovitch, D.L. Allara, C.C. Chang, T. Gmitter and T.B. Bright, Phys. Rev. Lett., 57, 249 (1986)
26. J.J. Lander and J. Morrison, Surf. Sci., 2, 553 (1964)
27. I. Yamada, Mat. Res. Soc. Symp. Proc., 128, 113 (1989)
28. M.A. Hassan, G. Radnoczi, J.-E. Sundgren and G.V. Hansson, Surface Science, 236, 53 (1989)

CRYSTALLINE TEXTURE OF CoCrPt FILMS ON CrMn/NiAl AND Cr/NiAl UNDERLAYER STRUCTURES

J. ZOU *, B. LU *, A.E. BAYER **, D.E. LAUGHLIN **, and D.N. LAMBETH *
*Department of Electrical and Computer Engineering, **Department of Materials Science and Engineering, Data Storage Systems Center, Carnegie Mellon University, Pittsburgh, PA 15213, jzou@ece.cmu.edu

ABSTRACT

In this study, we examine the effects of CrMn/NiAl and Cr/NiAl underlayer structures on the crystalline texture and microstructure of CoCrPt magnetic films using x-ray and electron diffraction. The former underlayer structure was found to induce better CoCrPt $(10\bar{1}0)$ texture. The stress in the CoCrPt layer was measured. The effective in-plane anisotropy energy densities of the CoCrPt films on the above two underlayers were also measured by using out-of-plane torque and Miyajima methods and the result was consistent with the texture analysis.

INTRODUCTION

The underlayer structure has significant influence on Co alloy thin film magnetic media, because it can induce a certain texture in the Co alloy layer and control the Co alloy film grain size, via grain-to-grain epitaxy. NiAl, which has the B2 crystal structure and a lattice constant very similar to Cr, was found to be an excellent underlayer material [1,2]. It can induce unicrystalline $(10\bar{1}0)$ Co texture and smaller and more uniform Co grains. Inserting an intermediate layer of Cr, Cr alloys, or other materials, between the NiAl underlayer and Co alloy magnetic layer, was found to improve the Co in-plane texture and magnetic properties [3-6].

In this paper, the CoCrPt crystalline texture and effective in-plane anisotropy energy on two underlayer structures, CrMn/NiAl and Cr/NiAl, are compared.

EXPERIMENT

All films were deposited onto glass substrates by RF diode sputtering without substrate preheating. The Ar sputtering pressure was 10 mTorr and the base pressure was about 5×10^{-7} Torr. Deposition was performed at a fixed AC power density of 2.3 W/cm^2. All CoCrPt and NiAl layers were 300 and 1000 Å thick, respectively. The CrMn and Cr intermediate layer thickness was varied.

The effective in-plane anisotropy energy density of the samples was measured using torque magnetometry. Film textures were examined by x-ray diffractometry and electron microscopy (TEM).

Film stress was measured by a Tencor Alpha-Step 200 general metrology profiler. As shown in Fig. 1, after the NiAl underlayer, the CrMn or Cr intermediate layer, and the CoCrPt magnetic layer were sequentially deposited, the sample was scanned on the substrate side. Then the CoCrPt layer was removed by ion milling and the sample was scanned again on the substrate side. Using the two radii of curvature of the substrate with and without the CoCrPt layer, we calculated the stress in the CoCrPt magnetic layer by using the Stoney equation. The substrates were marked to ensure that the two scans would pass approximately the same location. The samples were supported by three steel balls to eliminate the possibility of undesired sample movement during scanning.

Mat. Res. Soc. Symp. Proc. Vol. 577 © 1999 Materials Research Society

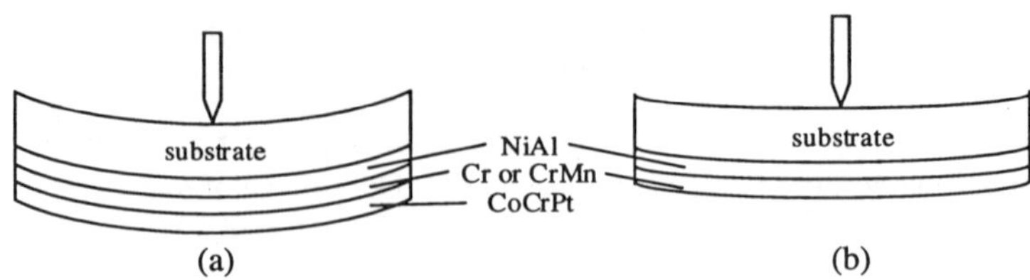

Fig. 1 Diagrammatic sketch of scanning the sample from the substrate side (a) with the CoCrPt magnetic layer and (b) with the CoCrPt layer removed.

RESULTS AND DISCUSSION

Fig. 2 shows the x-ray diffraction spectra of samples with various thicknesses of CrMn or Cr intermediate layers. Compared to the sample with no intermediate layer, adding a CrMn or Cr intermediate layer as thin as 25 Å increases the CoCrPt $(10\bar{1}0)$ peak. Further increasing the CrMn or Cr layer thickness did not improve this peak. Hence, only a very thin CrMn or Cr intermediate layer is needed to promote better in-plane texture in the CoCrPt magnetic layer. The samples with the CrMn layers showed slightly higher CoCrPt $(10\bar{1}0)$ peaks than the Cr samples with the same thickness.

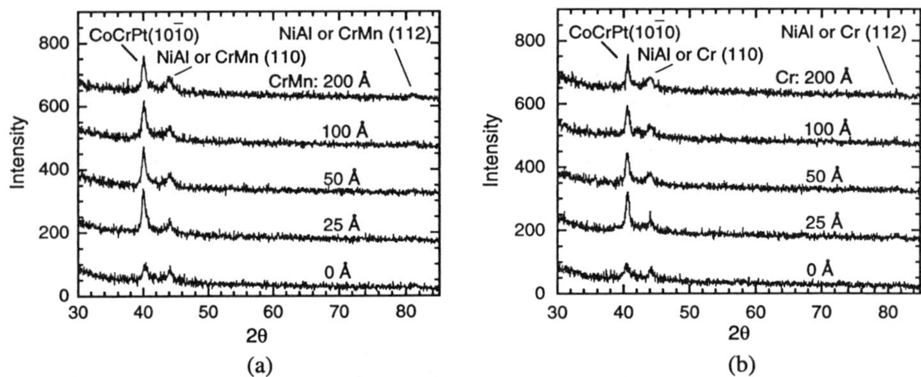

Fig. 2 X-ray spectra of (a) CoCrPt/CrMn/NiAl and (b) CoCrPt/Cr/NiAl films.

To compare the quality of the CoCrPt texture on top of the CrMn/NiAl and Cr/NiAl underlayer structures, the tilted electron diffraction technique was used [6,7]. Fig. 3 shows the 0°-tilt and 60°-tilt electron diffraction patterns of the CoCrPt layers on the above two underlayers. The CrMn and Cr intermediate layer thickness was 1000 Å. The arrows indicate the tilt axes and the three rings shown from inside to outside diameter, are $\{10\bar{1}0\}$, {0002}, and $\{10\bar{1}1\}$, respectively.

The $\{10\bar{1}0\}$ and $\{10\bar{1}1\}$ diffraction rings were present in the 0°-tilt diffraction patterns of both samples, which should not be seen in purely $(10\bar{1}0)$ textured hcp Co alloy samples. This implies that besides the $(10\bar{1}0)$ oriented CoCrPt grains, a portion of the grains were randomly oriented, for samples using both CrMn and Cr intermediate layers. This is not totally unanticipated, as there is substantial lattice mismatch between the high Pt content Co alloy and the CrMn or the Cr intermediate layer [4]. However, the intensities of these two rings were weaker for the samples using CrMn/NiAl underlayers than for the ones using the Cr/NiAl, which indicates that CoCrPt films on the former underlayer structure have a smaller portion of randomly oriented grains than the latter one. This is unanticipated as the lattice constant of Cr is changed little by the Mn addition.

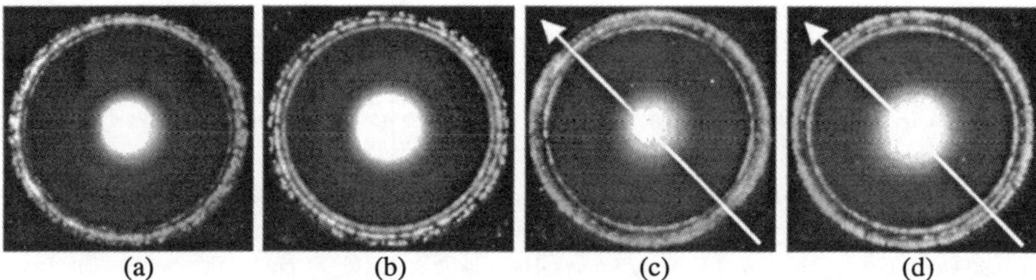

Fig. 3 The electron diffraction patterns of CoCrPt films. (a) & (c) were prepared on CrMn(1000Å)/NiAl(1000Å); (b) & (d) on Cr(1000Å)/NiAl(1000Å). (a) & (b) are 0° tilt; (c) & (d) are 60° tilt. The arrows indicate the tilt axes. The three rings shown, from inside to outside diameter, are $\{10\bar{1}0\}$, {0002}, and $\{10\bar{1}1\}$, respectively.

Tilting the sample changes the normal ring patterns into arc patterns and the length of the arcs is related to the texture axis distribution angle [6,7]. The arcs in the 60°-tilt electron diffraction pattern of the sample prepared on CrMn/NiAl underlayer structure were obviously shorter in length and more concentrated in intensity, indicating a smaller CoCrPt $(10\bar{1}0)$ texture distribution angle, than on Cr/NiAl.

Fig. 4 shows the plane-view TEM bright field images of the above two samples. The average grain size of the CoCrPt magnetic film on CrMn/NiAl underlayers was 16 nm, smaller than the 21 nm value for the sample on Cr/NiAl. The Co alloy grain size was also more evenly distributed for the former sample.

Fig. 4 Plane-view TEM bright field images of CoCrPt films on (a) CrMn(1000Å)/NiAl(1000Å) and (b) Cr(1000Å)/NiAl(1000Å) underlayers.

Fig. 5 reflects the film stress and shows the height of the sample substrate surfaces vs the scanning distance, before and after the CoCrPt magnetic layers were removed, for two samples using 1000 Å thick CrMn and Cr intermediate layers. The data points for the CoCrPt/Cr/NiAl sample in Fig. 5 (b) appear to be asymmetric, compared to the fit circle, which may be due to the substrate thickness unevenness across the scanning region. The sample using the Cr intermediate layer apparently had larger change in radius of curvature after the CoCrPt layer was removed, indicating a larger stress in the CoCrPt layer, than the sample using CrMn.

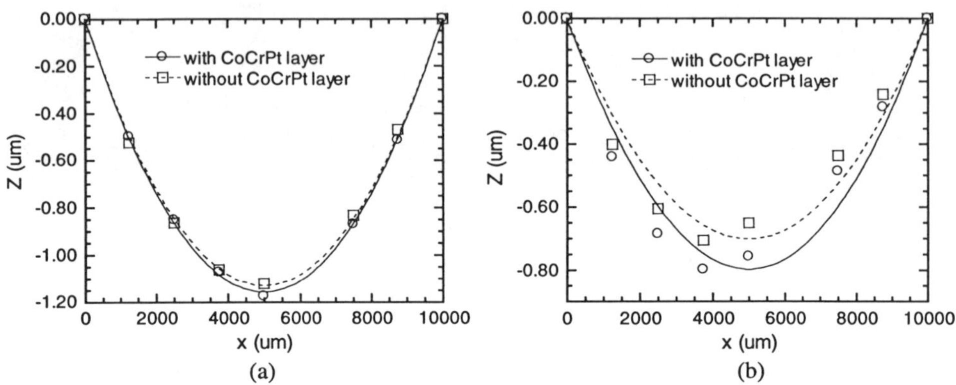

Fig. 5 The height of the substrate surface vs the scanning distance, for (a) CoCrPt/CrMn/NiAl and (b) CoCrPt/Cr/NiAl films, with and without the CoCrPt layer. The square and round dots are the experimental data and solid and dashed lines are the least squared error fit circular curves.

The stress in the CoCrPt magnetic layer was calculated using the Stoney equation:

$$\sigma = E_s d^2 (1/R - 1/R_0) / 6(1 - v_s)t \quad (1)$$

where σ is the film stress, E_s and v_s are Young's modulus and Poisson's ratio of the glass substrate, d is the substrate thickness, t is the film thickness, R and R_0 are the radii of curvature of the substrate with and without the CoCrPt layer, respectively. Table I shows the calculation results. Both samples showed compressive stress, which is consistent with the fact that Co alloys containing Pt have larger lattices than either CrMn or Cr. The CoCrPt layer had a larger compressive stress on top of the Cr/NiAl underlayers, than on the CrMn/NiAl. This is unexpected since incorporating Mn into Cr does not change the lattice constant much. However, the Mn addition into Cr may change the surface and/or the microstructure, causing the difference in the stress state in the sequentially deposited CoCrPt layer. The stress measurement results are consistent with the texture characterization – the less stressed film has the better texture.

TABLE I Stress in the CoCrPt layer for the CoCrPt/CrMn/NiAl and CoCrPt/Cr/NiAl films

Sample	R (m)	R_0 (m)	Stress (GPa)
CoCrPt/CrMn/NiAl	10.8	11.1	-0.3±0.1
CoCrPt/Cr/NiAl	15.7	17.8	-1.0±0.1

Out-of-plane torque method was used to examine the uniaxial anisotropy energy of the CoCrPt grains [8]. For a polycrystalline film with a 3D randomly distributed uniaxial crystalline anisotropy easy axes, the measured torque per unit volume by this method is:

$$L^{3D}/V = 2\pi Ms^2 \tag{2}$$

where the L^{3D} is the measured torque, V is the sample volume, and Ms is the magnetization. If the easy axes of the grains are 2D randomly distributed within the film plane, the measured torque per unit volume is:

$$L^{2D}/V = 2\pi Ms^2 + Ku/2 \tag{3}$$

where Ku is the uniaxial crystalline anisotropy energy density of the grains. We measured the out-of-plane torque using an applied magnetic field of 20 KOe, and used Eq. (3) to calculate the Ku^{eff}, which is the effective in-plane crystalline anisotropy energy density. The results are shown in Table II. The CoCrPt magnetic film had higher Ku^{eff} on CrMn/NiAl underlayers than on Cr/NiAl. This is consistent with the texture analysis results. There were more 3D randomly oriented CoCrPt grains on the Cr intermediate layer than on CrMn, which have no contribution to the Ku^{eff} obtained by this method. Furthermore, the CoCrPt $(10\bar{1}0)$ texture axis distribution angle, which is equal to the easy axis (c axis) distribution angle, was larger when prepared on Cr/NiAl than on CrMn/NiAl, giving rise to a smaller Ku^{eff} even if the crystalline anisotropy energy density of the grains was the same.

TABLE II The effective in-plane anisotropy energy density measurement results

Sample	Ku^{eff} (10^6 erg/cc)	
	Out-of-plane torque	Miyajima
CoCrPt/CrMn/NiAl	4.3	4.7
CoCrPt/Cr/NiAl	3.0	3.5

When the applied field is not much higher than the anisotropy field of the magnetic grains, the magnetization is not totally aligned with the applied field during the rotation of the sample and the out-of-plane method gives a torque value less than the ideal one when an infinitely large field is applied. Hence, this method gives the lower bound .of the anisotropy energy. The Miyajima method measures 45° out-of-plane torque at a series of applied fields and the torque at the infinitely large field can be extrapolated using the equation [9,10]:

$$(L/VH)^2 = Ms^2\,(L_\infty/V - L/V)\,/\,(2L_\infty/V) \tag{4}$$

where L_∞ is the torque at the infinitely large field. In Fig. 6, $(L/VH)^2$ is plotted vs L/V for both CoCrPt/CrMn/NiAl and CoCrPt/Cr/NiAl samples and the x-intercept of the extrapolated line using the high field values, gives the L_∞/V. The Ku^{eff} values calculated from L_∞/V using Eq. (3) are shown in Table II. Again, the CoCrPt film on CrMn/NiAl underlayers exhibited larger Ku^{eff} values than on Cr/NiAl.

The linear relation in Eq. (4) is strictly valid for the case that the easy axes are all aligned to one direction. For other situations such as 2D and 3D random cases, Eq. (4) is a very good approximation only when the applied field is much larger than the anisotropy field of the magnetic grains. For the field that is larger than the anisotropy field but still on the same order of magnitude, $(L/VH)^2$ is nearly a linear function of L/V, but the slope is slightly more gentle than the ideal case, causing an over-estimate of the extrapolated torque at the infinitely large field. Thus, the Miyajima method gives the upper bound of the anisotropy energy density.

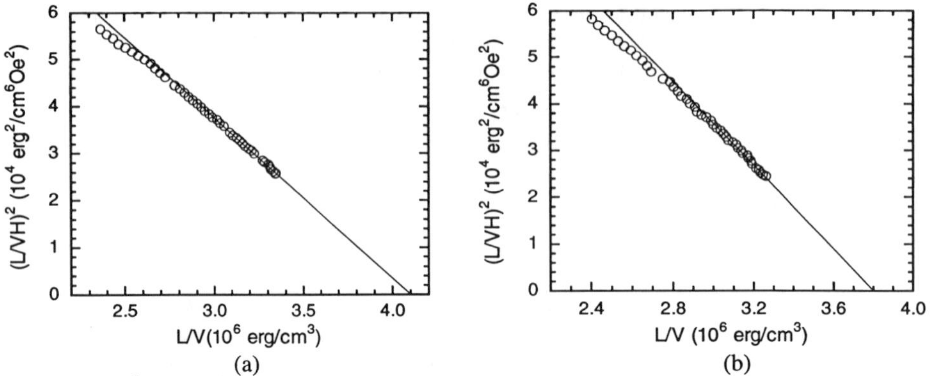

Fig. 6 $(L/VH)^2$ is plotted vs L/V for (a) CoCrPt/CrMn/NiAl and (b) CoCrPt/Cr/NiAl films.

CONCLUSIONS

Inserting a thin CrMn or Cr intermediate layer between the CoCrPt magnetic layer and the NiAl underlayer improves the CoCrPt ($10\bar{1}0$) texture. The CrMn/NiAl underlayer structure promotes better CoCrPt ($10\bar{1}0$) texture and less growth of randomly oriented CoCrPt grains than Cr/NiAl. CoCrPt films on CrMn/NiAl underlayers also have finer and more uniform grains, and less compressive stress, compared to Cr/NiAl. CoCrPt films on the former underlayer structure have higher effective in-plane anisotropy energy density, which is consistent with the texture analysis.

ACKNOWLEDGMENTS

This material is based upon the work supported in part by Seagate Technology, Inc. and by the Data Storage Systems Center of Carnegie Mellon University under a grant from the National Science Foundation # ECD-8907068. The U.S. government has certain rights to this material. A. E. Bayer was supported on an IBM Fellowship.

REFERENCES

1. L.-L. Lee, D. E. Laughlin, and D. N. Lambeth, IEEE Trans. Magn. **30**, 3951 (1994).
2. J. Li, M. Mirzamaani, X. Bian, M. Doerner, T. Arnoldussen, S. Duan, and K. Tang, to be published in J. Appl. Phys.
3. L.-L. Lee, D. E. Laughlin, L. Fang, and D. N. Lambeth, IEEE Trans. Magn. **31**, 2728 (1995).
4. J. Zou, D. E. Laughlin, and D. N. Lambeth, IEEE Trans. Magn. **34**, 1582 (1998).
5. J. Zou, D. E. Laughlin, and D. N. Lambeth, MRS Symp. Proc. **517**, 217 (1998).
6. B. Lu, D. E. Laughlin, D. N. Lambeth, S. Z. Wu, R. Ranjan, and G. C. Rauch, to be published in J. Appl. Phys.
7. L. Tang, Y. C. Feng, L.-L. Lee and D. E. Laughlin, J. Appl. Cryst. **29**, 419 (1996).
8. Y. Uesaka, K. Yoshida, Y. Nakatani, and N. Hayashi, J. Appl. Phys. **77** (10), 5303 (1995).
9. H. Miyajima and K. Sato, J. Appl. Phys. **47**, 4669 (1976).
10. H. N. Bertram and J.-G. Zhu, IEEE Trans. Magn. **27**, 5043 (1991).

MAGNETIC AND STRUCTURAL PROPERTIES OF LASER-ABLATED PLANAR AND CYLINDRICAL Co THIN FILMS

V. MADURGA*, J. VERGARA*, R. J. ORTEGA*, I. P. DE LANDAZÁBAL* and C. FAVIERES**.
*Departamento de Física. Universidad Pública de Navarra. Campus Arrosadía. E-31006 Pamplona. Spain. vmadurga@upna.es
**Departamento de Física de Materiales. Facultad de Ciencias Físicas. Universidad Complutense de Madrid. E-28040 Madrid. Spain.

ABSTRACT

Planar and cylindrical Co thin films have been obtained by pulsed laser ablation. X-ray diffractograms have shown no crystalline structure for the as-deposited samples, while the 450 °C annealed samples exhibit Co fcc crystalline peaks. The Scanning Tunnelling Microscopy has revealed a small increase of the surface roughness for the annealed films. The Hall effect has been used to determine the value of the spontaneous magnetization, M_s, at room temperature; $M_s \approx 14$ kgauss for the as-deposited sample and $M_s \approx 17.6$ kgauss for the 450 °C annealed sample. From transverse magnetooptic Kerr effect, it has been found that the as-deposited samples exhibit magnetic bistability, with a coercive field, H_c, ≈ 6 Oe. The annealed samples also show a bistable behavior until the annealing temperature is 450 °C. Besides, it has been observed an increase of H_c up to ≈ 50 Oe, when the annealing temperature increases. The vibrating sample magnetometry has confirmed these results, showing that the magnetization participating in the magnetooptic effect for these low fields is the total spontaneous magnetization of these samples. Moreover, the cylindrical films exhibit magnetoelastic behavior when they are subjected to angular deformation. It has been found that the saturation magnetostriction constant is negative.

INTRODUCTION

Amorphous and nanocrystalline phases of the ferromagnetic elements have been usually obtained by low temperature deposition techniques [1-3]. It has been shown that their saturation magnetization and Curie temperature values lie below the corresponding bulk values [4,5]. More recently, it has also been reported the possibility to obtain amorphous and nanocrystalline Co by Pulsed Laser Ablation Deposition, PLAD, at room temperatures [6].

In this work, we have studied the magnetic and structural properties of as-deposited PLAD Co thin films and the magnetoelastic behavior of PLAD Co cylindrical layers . We show moreover the evolution of these properties when the samples are annealed at progressively higher temperatures.

EXPERIMENTAL

A pulsed Nd: YAG laser, λ=1024 nm, 20 Hz repetition rate, with 4 ns pulses, energy 250 mJ per pulse, was used for the ablation process. The area of the laser beam on the targets was 2 mm^2. The pure Co single circular target, 20 mm in diameter, was rotated at 32 rpm. It is placed in a stainless-steel Neocera chamber under 10^{-5} mbar vacuum for 60 minutes. Planar and cylindrical samples were simultaneously produced, situated at a distance of 7.5 cm from the target. The planar films were deposited on glass substrates, and the cylindrical ones on capillary -glass as well as on copper wires substrates, rotating around their cylindrical axis of symmetry during the deposition, in order to achieve an homogeneous deposit. The wires are 88 mm long, 0.5 mm in diameter and the deposits are 40 mm long. The thickness of the samples is $\approx$ 140 nm. After deposition, the samples were annealed at temperatures between 150 °C and 525 °C for 10 minutes in an inert atmosphere.

The microstructure of the samples was studied with a XRD-3003 Seifert Diffractometer, using Mo anode and a grazing incidence geometry. A graphite secondary monochromator was placed just before the detector.

The surface topography of the samples was measured with a Metris-1000 STM from Burleigh and the thickness of the depositions, with a JEOL 6400 Scanning Electron Microscope.

The Hall effect has been used to determine the value of the spontaneous magnetization, M_s, of the films. Samples, 2 x 1 cm^2 have been used for these measurements. The maximum applied

Mat. Res. Soc. Symp. Proc. Vol. 577 © 1999 Materials Research Society

magnetic field is 2.3 Tesla. The 1 mA intensity has been provided by a one part in one million stable power supply.

Magnetic measurements were performed using a EG&G Vibrating Sample Magnetometer, at room temperature and in magnetic fields up to 1 Tesla.

Magnetooptical transverse Kerr effect, MOKE, was used for investigating the magnetization processes. A polarized beam laser was used for illuminating the samples and a photodiode to pick up the signal reflected by the samples. The diameter of the light spot is 0.5 mm.

The magnetoelastic properties exhibited by the cylindrical films were studied using the system described in reference 10. We have studied the Inverse Wiedemann effect, IWE. If an ac current is flowing through a magnetostrictive wire, creating an ac circular magnetic field, H_ϕ, the variation of the longitudinal magnetization, dM_z/dt, will produce an emf in a pick-up coil coaxial with the wire. Different angular deformation can modify the values of the non diagonal terms of the magnetic susceptibility tensor, through the magnetostriction constant, λ_s. We have used a pick-up coil of 4000 turns coaxial with the sample and an electronic fluxmeter (Walter Magnemetrics, model MF-3A) to detect M_z.

RESULTS AND DISCUSSION

Structure and Morphology

Fig.1 shows the X-ray diffractograms from a Co thin film. The as-deposited film shows a very broad peak centered at $2\theta = 20$ degrees, and two incipient peaks at higher angles. After annealing at 250 °C, the pattern is not significantly changed. This indicates that these two samples have no crystalline structure. For the 450 °C annealed sample, the diffraction maximum becomes sharper and also a new clear peak appears at $2\theta = 23$ degrees.

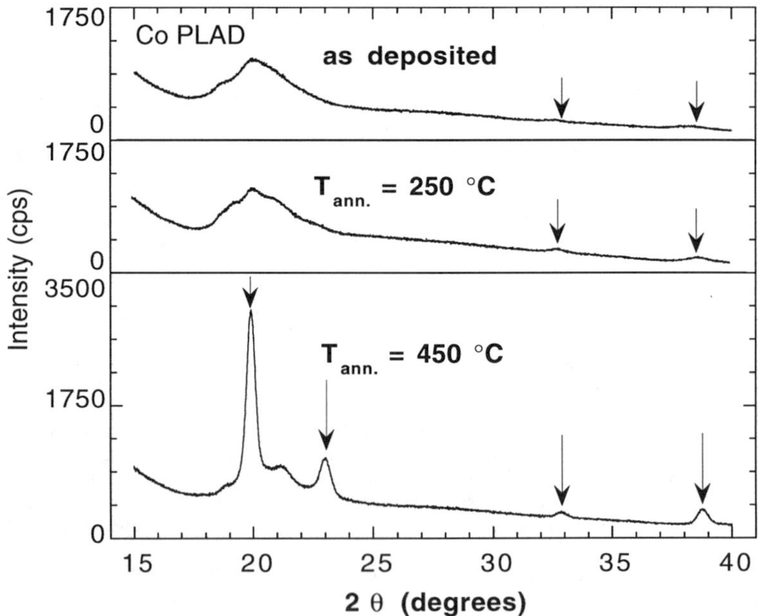

Fig.1. X-ray diffractograms corresponding to the as-deposited Co thin film and after annealing at 250 °C and 450 °C. The arrows represent the positions of the diffraction peaks of the Co fcc.

This last diffractogram corresponds to a Co fcc structure. It is well known that whereas at room temperature the hcp structure is stable for bulk Co, fine grain Co structures may show a fcc one at room temperature [7].

The topography of the samples indicates that a granular morphology is developed during the deposition process, see Fig.2. They show the presence of ≈ 60 nm size entities and a surface roughness of ≈ 6 nm. Although the influence of the heat treatment is increasing this size, these entities remain in this scale, so the strong structural changes observed by the X-ray diffraction must be attributed to intra-nano entities changes.

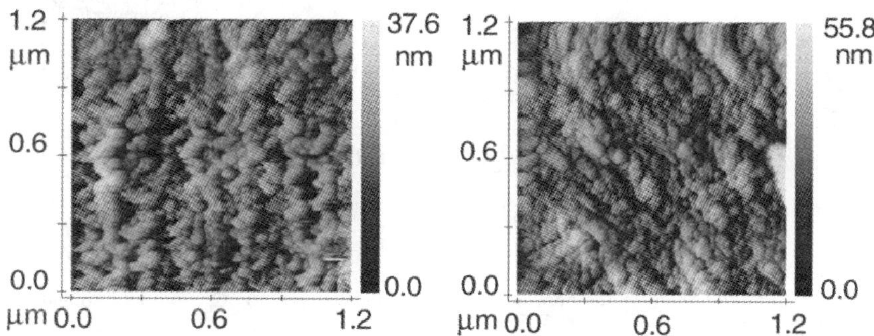

Fig.2. STM topographic images of two Co thin films: left, as-deposited; right, after annealing at 450 °C.

Magnetization

A.- Spontaneous magnetization

From the Hall effect measurements, see Fig.3, it is obtained the value of M_s, equal to the value of the applied magnetic field for which the extraordinary Hall effect has its end [8]. For the as-deposited sample, M_s is 14 kgauss, while for the 450 °C annealed sample, M_s is 17.6 kgauss. This result confirms the previous assumption of a non-crystalline structure for the as-deposited sample. Also, the estimated value of the extraordinary Hall coefficient, $R_s \approx 10^{-10}$ Ωcm/gauss for the as-deposited sample, decreases by more than one order of magnitude for this annealed sample [9].

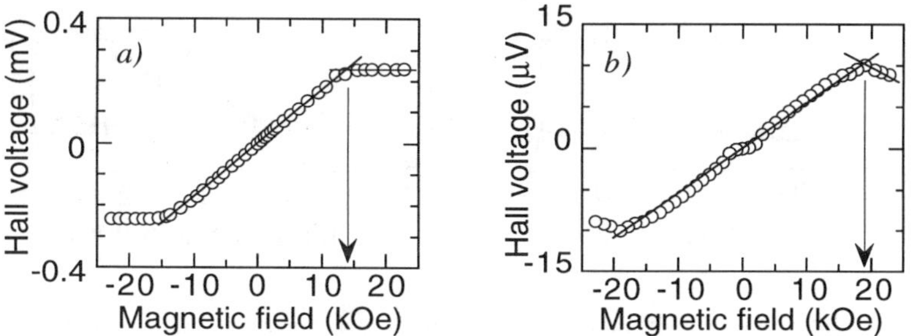

Fig.3. Hall effect measurements in (a) as-deposited film and (b) annealed at 450 °C. The value of M_s is equal to the value of the perpendicular to the plane applied magnetic field for which the extraordinary Hall effect is completed, as indicate the arrows in both cases.

B.- Magnetooptic behavior

Fig.4(a) shows the transverse Kerr effect, intensity of the reflected light versus applied magnetic field, for an as-deposited planar film. It exhibits a bistability, with a coercive field of ≈ 6 Oe. After annealing, it has been observed an increase of the coercive field, becoming ≈ 45 Oe, for the 450 °C annealed sample, over which the magnetic bistability is lost, see Fig.4(b).

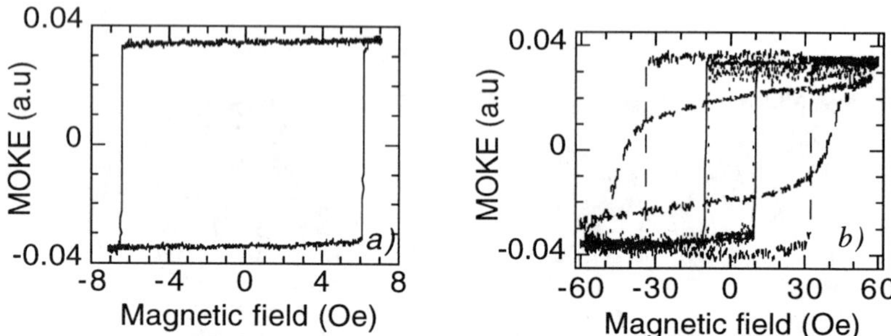

Fig.4. Kerr magnetic hysteresis loops corresponding to the planar samples, (a) as-deposited and (b) ——— as-deposited, - - - - -annealed at 150 °C, — —annealed at 450 °C, —— —— annealed at 525 °C.

C.- VSM

From the VSM measurements, it has been observed that the coercive field corresponding to the as-deposited samples is ≈ 8 Oe, increasing up to 50 Oe for the 525 °C annealed sample, as shown in Fig.5. It is seen that the as-deposited sample is saturated for this low magnetic field, indicating that the magnetization participating in the bistable MOKE is the total spontaneous magnetization of these samples.

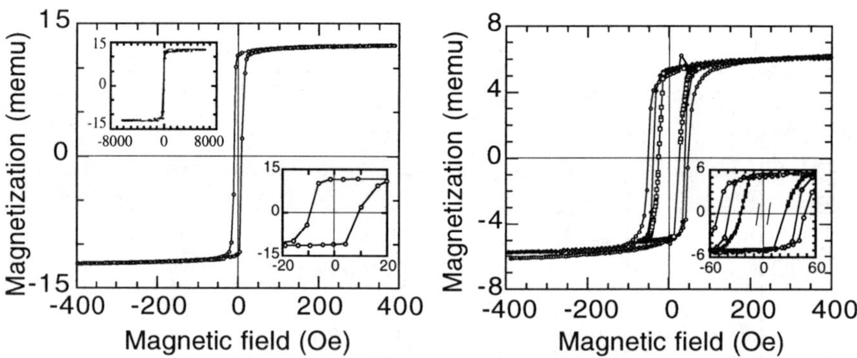

Fig. 5. VSM measurements from the planar samples, (a) as-deposited; the two insets show the hysteresis loop for a magnetic field of 8000 Oe and the value of the coercive field, ≈ 8 Oe. Note that the saturation is reached at 20 Oe; (b) After annealing at 350 °C, 420 °C and 525 °C. The inset shows how the coercive field increases to ≈ 25 and ≈ 40 and ≈ 50 Oe, respectively. The two parallel lines mark the H_c of the as-deposited sample.

The dependence with the annealing temperature of the coercive field from both, the magnetooptic and the VSM measurements, is shown in Fig.6. Note the change in the slope at around 420 °C, that can be associated with the growth of the Co fcc structure.

The as-deposited cylindrical layers show an alike behavior of that exhibited by the planar ones. They present a magnetic bistability, with coercive fields ≈ 10 Oe. The hysteresis loop obtained from the VSM is represented in Fig.7.

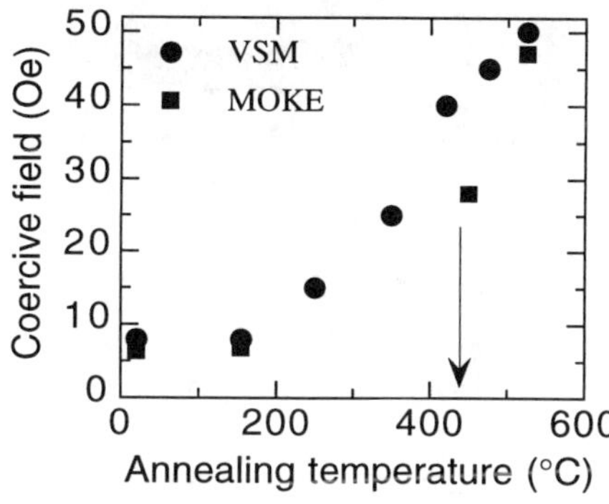

Fig.6. Dependence with the annealing temperature of the coercive field from both, the magnetooptic and the VSM measurements.

Fig.7. VSM hysteresis loop from a Co cylindrical layer.

For these cylindrical layers, we also show their magnetoelastic behavior. When they are subjected to a torsion, a magnetostrictive helical anisotropy is generated, and then a longitudinal component of the magnetization, M_z, appears if a circular magnetic field, H_ϕ, is applied. In Fig.8 we show the electromotive force, ~ dM_z/dt, (bottom signal in all photos), coming from the pick-up coil coaxial with the cylindrical layer, for different values of the torsion, ζ, and two values of H_ϕ created by an ac current flowing through the wire substrate, (top trace in all photos).

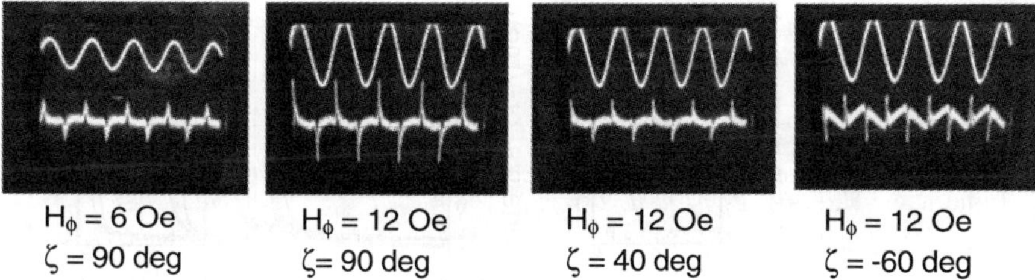

Fig.8. Dependence of dM_z/dt, (bottom signal in all photos), with the torsion, ζ, and with the ac circular magnetic field, H_ϕ, produced by an ac current (ν= 2 kHz) flowing through the wire substrate, (top trace in all photos).

When integrating the dM_z/dt and monitoring it versus H_ϕ we have obtained the Inverse Wiedemann Effect, IWE, or M_z-H_ϕ hysteresis loops, shown in Fig.9 for two different values of H_ϕ. From these measurements it has been found that the saturation magnetostriction constant of these samples is negative.

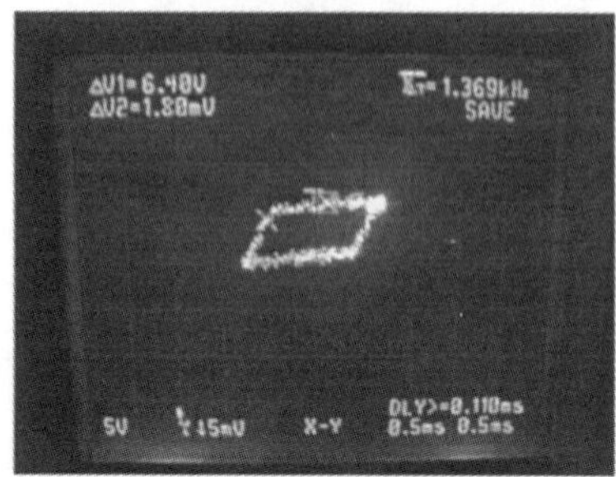

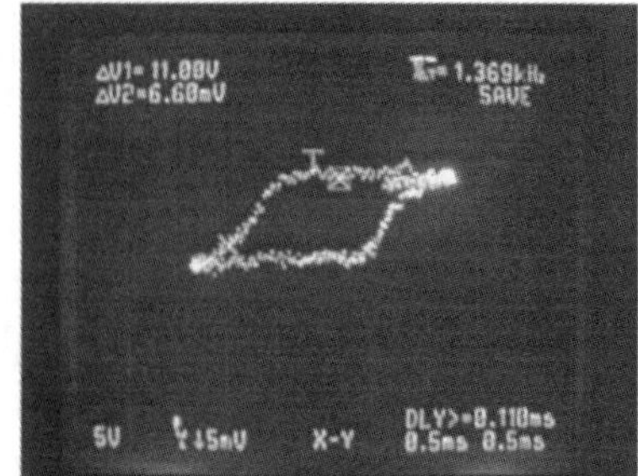

Fig.9. IWE M_z-H_ϕ hysteresis loops, (ν= 200 Hz), for a torsion of + 90 degrees, when two ac maximum circular magnetic field are applied: left, H_ϕ =18 Oe, right H_ϕ = 32 Oe.

CONCLUSIONS

Co planar and cylindrical thin films have been elaborated by pulsed laser ablation. X-ray diffractograms have shown no crystalline structure for the as-deposited samples, whereas after annealing at 450 °C, Co fcc crystalline peaks appear. It has been found that the spontaneous magnetization at room temperature, M_s is ≈ 14 kgauss for the as-deposited sample while it increases to ≈ 17.6 kgauss for the 450 °C annealed sample. The MOKE, 0.5 mm light spot diameter, has shown that the planar as-deposited samples have two stable magnetically saturated states; the irreversible change of the magnetization direction from one to another stable state is produced when the applied magnetic field achieves the value of ≈ 6 Oe. This value increases up to ≈ 50 Oe after annealing. This magnetic bistability is lost when the annealing temperature is over 450 °C. The VSM has confirmed these results, showing that the magnetization participating in the MOKE for low fields is the total spontaneous magnetization of these samples. The cylindrical layers show also bistable behavior. Moreover, when they are subjected to a controlled torsion, they exhibit magnetoelastic properties. From IWE measurements it has been found that the saturation magnetostriction constant of these samples is negative.

ACKNOWLEDGMENTS

This work has been supported by the Spanish CICYT, project MAT98-0404 and by the Gobierno de Navarra.

REFERENCES

1. C.W.B. Grigson, D.B. Dove, and G.R. Stilwell, Nature **204**, 173 (1965).
2. S. Fujime, Jap. J. Appl. Phys. **5**, 1029 (1966).
3. H.A. Davies, J. Aucote, and J.B. Hull, Nature Physical Science, **246**,13 (1973).
4. K. Tamura and H. Endo, Phys. Lett. **29A**, 52 (1969)
5. R.Z. Valiev, G.F. Korznikova, K.Y. Mulyukov, R.S. Mishra and A.K. Mukherjee, Phil. Mag. B **75**, 803 (1997).
6. Y. Hiroshima, T. Ishiguro, I. Urata, H. Makita, H. Ohta, M. Tohogi, and Y. Ichinose, J. Appl. Phys. **79,** 3572 (1996).
7. E. A. Owen, and D. Madoc Jones, Proc. Phys. Soc. **B67**, 456 (1954).
8. E.D. Dahlberg, K. Riggs, and G.A. Prinz, J. Appl. Phys. **63** (8) 1988.
9. C.L. Chien, and C.R. Westgate, in*The Hall effect and its applications*, edited by Plenum Press, New York (1980) and references therein.
10. V. Madurga, A. Hernando, and O.V. Nielsen. J. Phys. E: Sci. Instrum., **17**, 813 (1984).

C-AXIS PERPENDICULARLY ORIENTED BARIUM FERRITE THIN FILM MEDIA ON SILICON SUBSTRATE

Zailong Zhuang, Mark H. Kryder, Robert M. White, and David E. Laughlin*
Department of Electrical and Computer Engineering
*Department of Materials Science and Engineering
Data Storage Systems Center, Carnegie Mellon University, Pittsburgh, PA 15213-3890

ABSTRACT

Barium ferrite thin films with excellent perpendicular c-axis orientation were successfully fabricated on Si substrate without any buffer layer. To compensate for possible barium deficiency due to the inter-diffusion between films and substrate, a barium-rich target was used. For a 900 Å-thick film, the perpendicular remanent squareness is about 0.9, while the in-plane remanent squareness is about 0.3. The saturation magnetization (Ms) is about 220 emu/cc, while the coercivity is around 3500 Oe. X-ray diffraction (XRD) results show the (00l) perpendicular c-axis texture in the films. It was also found that the rapid thermal annealing conditions greatly affect the magnetic properties of barium ferrite films. With a certain flow rate of oxygen gas in the rapid thermal annealer (RTA), barium ferrite films generally crystallize with good perpendicular c-axis texture. Without oxygen gas, the hexagonal barium ferrite phase fails to develop; instead spinel Fe_3O_4 phase forms. The reason for the collapsing of hexagonal barium ferrite texture is thought to be an oxygen deficiency in the barium ferrite films due to the reduction of oxygen in the films during the high temperature annealing.

INTRODUCTION

Hexagonal barium ferrite thin films are promising candidates as future high-density magnetic recording media due to their excellent magnetic properties and mechanical and chemical stability. The as-deposited barium ferrite thin films without substrate heating are amorphous in nature. Crystalline barium ferrite thin films have to be obtained by either applying *in situ* substrate heating during deposition or through a post-deposition annealing. As the magnetic easy axis in barium ferrite thin films is parallel to the c-axis, much effort has been made to develop suitable substrates and underlayers for obtaining thin films with a high degree in c-axis perpendicular orientation. Many different kinds of underlayers and substrates have been used including ZnO, Pt, AlN, Si_3N_4, Si and thermally oxidized silicon(SiO_2)[1-5]. For the *in situ* heating during deposition case, barium ferrite thin films with excellent perpendicular c-axis orientation were achieved on the underlayer ZnO, thermally oxidized silicon either by target-facing type of sputtering or by conventional DC magnetron type of sputtering. For the post-deposition annealing case, barium ferrite thin films with perpendicular c-axis orientation were successfully grown on the Pt and AlN underlayers.

In this paper, we study the magnetic and structural properties of post-deposition annealed barium ferrite thin films deposited on Si substrate without any buffer layer. The lack of buffer layer between the films and substrate may cause the deficiency of barium content due to the inter-diffusion during high temperature annealing. Thus, a barium-rich barium ferrite target was used to compensate for barium deficiency in the thin films. The effect of rapid thermal annealing conditions on the magnetic and structural properties of barium ferrite thin films will be also explored.

Mat. Res. Soc. Symp. Proc. Vol. 577 © 1999 Materials Research Society

EXPERIMENTAL

Thin films were deposited by RF diode sputtering in a Leybold-Haereus Z-400 sputtering system. The sputtering power density and Ar gas pressure were set to 2.25W/cm^2 and 5.7mTorr, respectively. The base vacuum pressure was below 1x10^{-7} Torr. Silicon (100) substrates were used for all samples. The as-deposited barium ferrite thin films are amorphous and nonmagnetic. After deposition, all the samples were annealed at 800 °C in either the RTA oven for 60s or annealed at 800 °C in a thermal oven for 20 min. Argon gas was used to purge the RTA oven before annealing, and the oxygen gas was added into the oven during the RTA oven annealing. The flow rate of oxygen gas was varied as specified later.

The magnetic properties of the films were studied using either an alternating gradient magnetometer (AGM) or a vibrating sample magnetometer (VSM). The crystalline structures of the films were characterized by a Cu K_α X-ray diffractometer. An energy dispersive X-ray fluorescence system (EDXRF) was used to determine the chemical composition of the barium ferrite targets as well as of the deposited films

RESULTS AND DISCUSSION

Magnetic and structural properties of barium ferrite thin films

The magnetic properties of stoichiometric and barium rich barium ferrite thin films deposited on Si substrate are listed in Table I. All of the samples are post-deposition annealed at 800 °C in a thermal oven for 20 min. The 900 Å–thick stoichiometric barium ferrite thin film (sample A) shows a perpendicular c-axis orientation, as indicated by a perpendicular remanent squareness of 0.75 and in-plane remanent squareness of 0.44. The perpendicular Hc is around 1460 Oe, while the in-plane coercivity is around 1900 Oe. The MH loop of sample A is shown in Fig.1 Although sample A shows a perpendicular c-axis orientation, it is not suitable for use as a perpendicular recording medium, due to its smaller Hc, and lower perpendicular remanent squareness. The barium deficiency in the film is thought to be reason for the poor magnetic properties.

TABLE I FILM STRUCTURA AND MAGNETIC PROPERTIES OF BARIUM FERRITE THIN FILMS

Sample	Film Structure	Hc(⊥) (Oe)	Hc(//) (Oe)	Perpendicular Squareness	In-plane Squareness	δM	Ms (emu/cc)
A	Stoichiometric barium ferrite 900 Å/Si	1460	1400	0.78	0.32	-0.35	170
B	Stoichiometric barium ferrite 1500 Å/Si	1900	200	0.74	0.63	-0.40	
C	Ba-rich barium ferrite 900 Å/Si	3300	2600	0.88	0.28	-0.65	220
D	Ba-rich barium ferrite 1500Å/Si	4970	4940	0.70	0.66	-0.35	

To improve the magnetic properties, 10wt% barium rich barium ferrite thin films were deposited on Si substrate to compensate for the barium deficiency due to the inter-diffusion. The magnetic properties of a 900 Å–thick barium rich thin film (sample C) are shown in Table I, and the MH loop of sample C is shown in Fig. 2. Significant changes in magnetic properties were observed. Compared to the properties of sample A, sample C has a higher perpendicular remanent squareness of 0.88 and smaller in-plane remanent squareness of 0.28. The perpendicular Hc of sample C is around 3300 Oe, which is much larger than that of sample A. The increase of coercivity may be due to a decrease of grain size for barium rich thin films[6]. The much lower Hc observed in the stoichiometric barium ferrite films may be caused by the incoherent rotation of magnetization, which is more pronounced in larger grains. The Ms for barium-rich thin film samples C and D is around 220 emu/cc, while the Ms is only around 170 emu/cc for stoichiometric thin film samples A and B. It means that enriching barium content is effective to compensate for the barium loss, which tend to decrease of the saturation magnetization of barium ferrite thin films

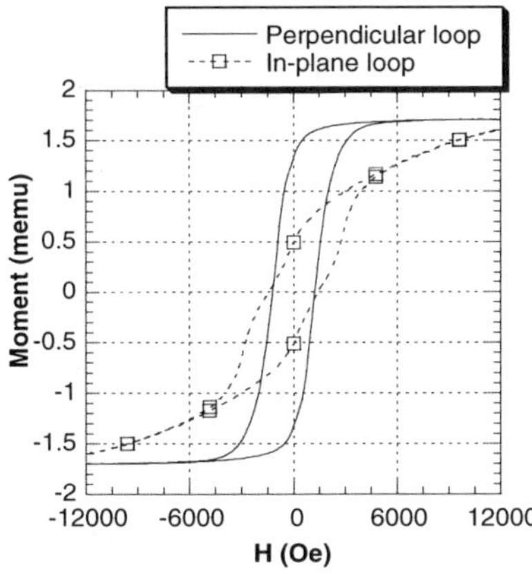

Fig. 1 In-plane and perpendicular MH loops for sample A

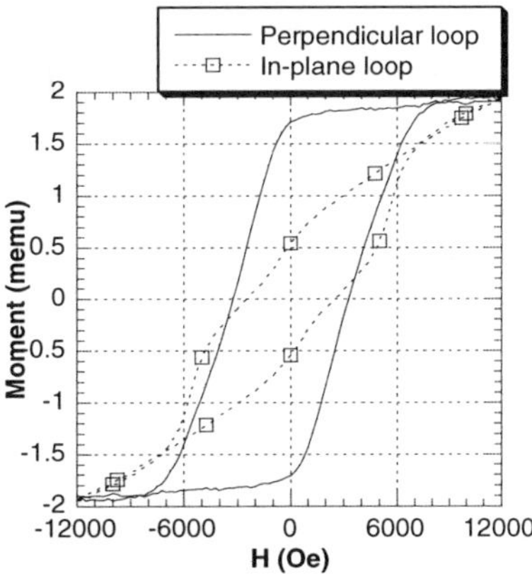

Fig. 2 In-plane and perpendicular MH loops for sample C

The magnetic properties of sample B and D, which have thickness of 1500 Å, are shown in Table I. There is no preferred c-axis orientation for sample D, while sample B shows a very weak perpendicular orientation. The similar values in the perpendicular and in-plane remanent squareness of sample D indicate random c-axis orientation of the film. The deterioration of perpendicular c-axis orientation with the increase in thickness is due to the increase of bulk nucleation sites, which favors to crystallize into random oriented barium ferrite grains[7]. δM curves were also measured for all samples and are also shown in Table I. There are only negative peaks for all barium ferrite films, which indicates magnetostatic interactions among grains. δM value is roughly the same for sample A and B. A significant difference in the δM values was observed for sample C and D. The smaller δM value for sample D with respect to sample C suggests a weaker magnetostatic interaction among barium ferrite grains for sample D. The coercivity Hc of sample D is much bigger than that of sample C. The increase in Hc is consistent with a reduction in magnetostatic interaction among barium ferrite grains, as shown by a decrease in absolute value of δM from a value of 0.65 to 0.35. Due to the directional nature of magnetic interaction, the magnetostatic interactions in the easy-axis randomly oriented films are weaker than in the easy-axis well-aligned films.

XRD was also used to characterize the structural properties for all the samples. XRD curves for sample A and sample C are shown in Fig 3. The dominant reflection peaks are (006) (008) and (0014), which indicate excellent c-axis orientation. The XRD results are consistent with the magnetic measurement results shown above. With the same thickness of 900 Å, the (00l) peaks of sample C are bigger than those of sample A, which indicates a better c-axis perpendicular orientation for barium rich film sample C. Sample B and D have weak (006) (008) and (0014) peaks, as shown in Fig. 4. In addition to the (00l) peaks, there are also weak peaks of (107) and (114) peaks. The existence of all (00l), (107) and (114) diffraction peaks indicates a random orientation of the c-axis for sample D. Sample B has relatively bigger (001) peaks over (107) and (104) peaks, as indicated in Fig.4. So sample B has a weak perpendicular c-axis orientation, which is consistent with magnetic properties shown in Table I.

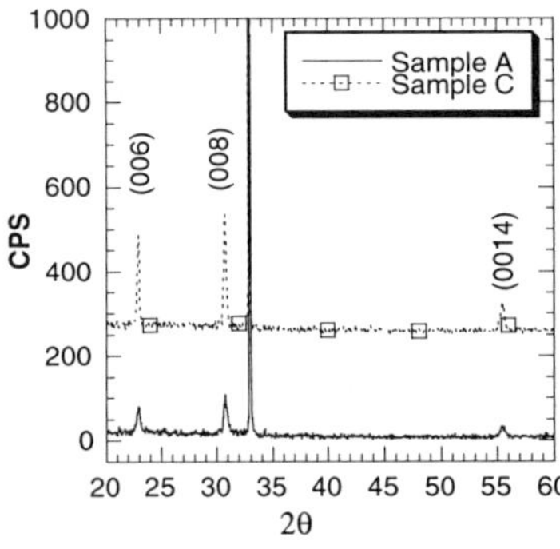

Fig. 3 The XRD curves for sample A and C

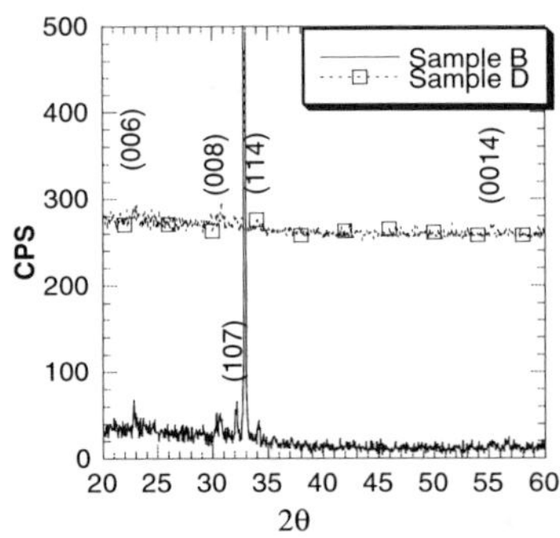

Fig. 4 The XRD curves for sample B and D

RAPID THERMAL ANNEALING EFFECT

Several samples with the same film structural as sample C were used to study the rapid thermal annealing effect on the c-axis texture and the magnetic properties of barium ferrite films. Sample E and F were annealed with an oxygen gas flow rate of 2 standard liter per minute (SLPM) and 5 SLPM respectively in a RTA at 800 °C for 60 s, while sample G was annealed in RTA without supplying oxygen gas at 800 °C for about 20 min.

Both samples E and F have a perpendicular c-axis orientation. The MH loops for sample F is shown in Fig.5. With the oxygen gas in the annealing process, barium ferrite films show perpendicular c-axis texture after annealing. The XRD curve for the sample E and F are shown in Fig. 7. The perpendicular c-axis texture of sample F is better than that of sample E, indicated by the larger (00l) peak intensity for sample F, as shown in Fig. 7. So a larger oxygen gas flow rate during annealing is beneficial to get better c-axis texture for barium ferrite thin films.

When annealing barium ferrite thin films in RTA without supplying oxygen gas, both film magnetic properties and film texture display a dramatic change. Sample G has an in-plane easy axis orientation, with a much higher Ms of 320 emu/cc, and a much lower coercivity of 600 Oe, as shown in Fig. 6. The XRD curve for sample G, as shown in Fig. 8, indicates spinel Fe_3O_4 (111), (222), and (511) reflections, and no hexagonal barium ferrite (006) (008) and (0014) reflections are observed. So the hexagonal barium ferrite crystallographic texture has collapsed with the formation of the spinel Fe_3O_4 (111) texture during annealing without oxygen gas. The reason for failing to form hexagonal barium ferrite texture is thought to be due to an oxygen deficiency in barium ferrite thin films. Without an oxygen atmosphere, oxygen atoms in the films will more likely diffuse to the Si substrate and react with Si during the high temperature annealing process, which makes the films oxygen deficient. Fe^{2+} ions may be produced by the reduction from Fe^{3+} ions due to the deficiency of oxygen in the film. So the spinel Fe_3O_4 phase is formed with the reduction of oxygen in the films during annealing.

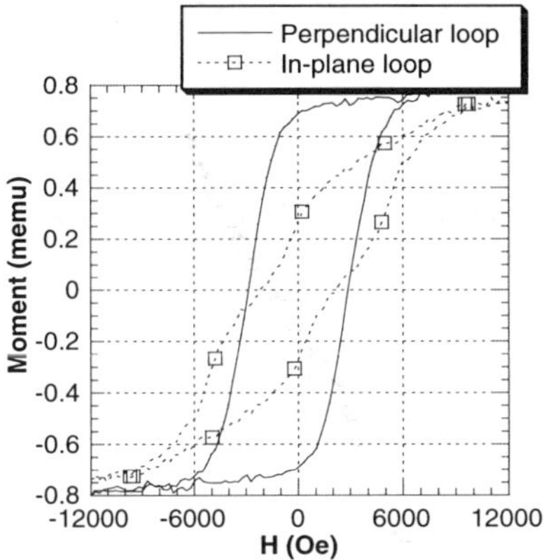

Fig.5 In-plane and perpendicular MH loops for sample E

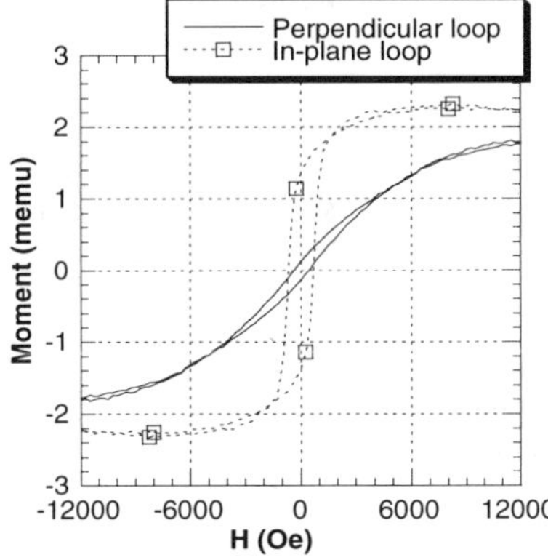

Fig. 6 In-plane and perpendicular MH loops for sample G

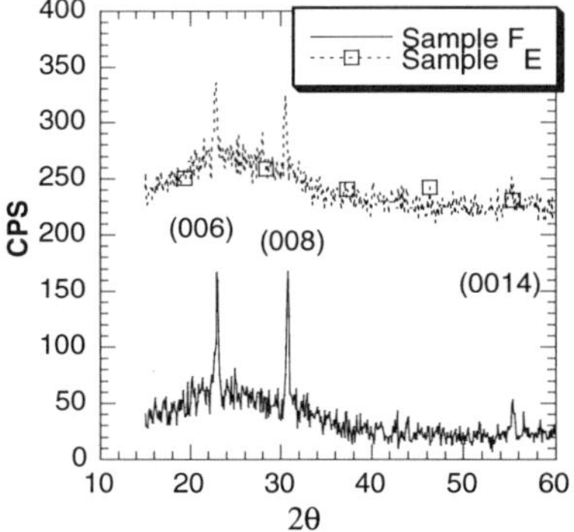

Fig. 7 The XRD curves for sample E and F

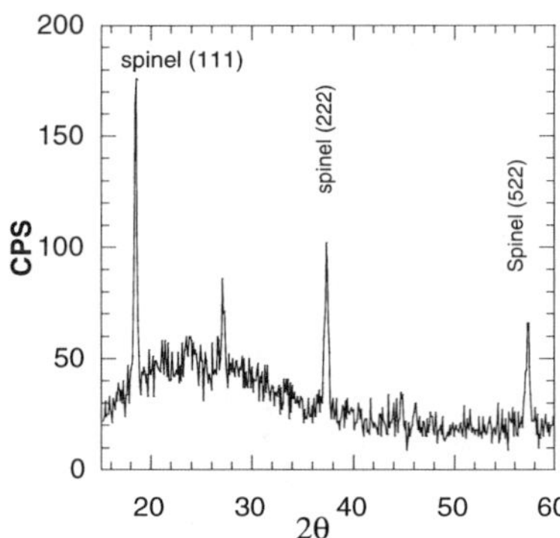

Fig. 8 The XRD curve for sample G

CONCLUSIONS

In this work we have found that barium ferrite thin films with excellent perpendicular c-axis orientation can be fabricated on Si substrate without using any buffer layer. Barium-rich thin films were deposited to compensate for the deficiency of barium content due to the inter-diffusion. For a 900 Å -thick film, the perpendicular remanent squareness for barium ferrite thin films is about 0.9, while the in-plane remanent squareness is about 0.3. It was also found that the rapid thermal annealing conditions greatly affect the magnetic properties of barium ferrite films. The perpendicular c-axis structure tends to collapse with the formation of the spinel Fe_3O_4 phase during annealing without oxygen gas flowing during annealing. The collapse of the hexagonal barium ferrite phase can be attributed to an oxygen deficiency in barium ferrite films due to the oxygen reduction in high temperature annealing. To avoid the reduction of oxygen during the annealing, an oxygen-rich annealing atmosphere is necessary.

ACKNOWLEDGMENTS

This work is supported by National Science Foundation under Grant No. ECD-8907068. The United State Government has certain rights in this material.

REFERENCES

[1]M. Matsuoka, Y. Hoshi, M. Naoe and S. Y. Yamanada, IEEE Trans. Magn., MAG-**18**, 1119(1982)
[2]Y. Hoshi, Y. Kubota, IEEE Trans. Magn., **31**, 2782 (1995)
[3]Y. Hoshi, Y. Kubota, J. Appl. Phys., **81,** 4667 (1997)
[4]X. Sui, M. H. Kryder, B. Y. Wong, and D. E. Laughlin, IEEE Trans. Magn., **29**, 3751 (1993).
[5]N. Hiratsuka, E. Niizuma, K. Kakizaki, J. Magn. Soc. Jpn, **21**, 77 (1997)
[6]Y. J. Chen, D. E. Laughlin, X. Ma, M. H. Kryder, J. Appl. Phys., **81**, 4380 (1997)
[7]Z. Zhuang, Maithri Rao, David Laughlin, Mark Kryder, submitted to the 43rd annual conference on magnetism & magnetic materials, 1998.
[8] J. Smit, H.P.J. Wijn, Ferrites; physical properties of ferrimagnetic oxides in relation to their technical applications, New York, Wiley, 1959.

STRUCTURE AND MAGNETIC PROPERTIES OF Fe-N THIN FILMS GROWN BY ECR DEPOSITION

Š. NÉMETH*, H. AKINAGA**, H. BOEVE*, H. BENDER*, J. DE BOECK*, G. BORGHS*
*IMEC, NMC/NM, Kapeldreef 75, B-3001 Leuven, Belgium
**JRCAT : NAIR, 1-1-4 Higashi, Tsukuba, Ibaraki 305-8562, Japan

ABSTRACT

The growth of Fe_xN_y thin films on GaAs, $In_{0.2}Ga_{0.8}As$, and SiO_2/Si substrates using an ultra high-vacuum (UHV) deposition chamber equipped with electron cyclotron resonance (ECR) microwave plasma source is presented. The structural properties of the deposited films have been measured using various techniques as x-ray diffraction (XRD), Auger electron spectroscopy (AES), and transmission electron microscopy (TEM). The results of XRD measurements show that the films consist of a combination of α-Fe, α'-Fe, γ-Fe_4N, and α''-$Fe_{16}N_2$ phases. The depth profiles, calculated from the Auger peak intensities, show a uniform nitrogen concentration through the films. The TEM reveals a columnar structure of these films. The properties of the different Fe-N layers have been exploited in the fabrication of Fe(N) / Fe_xN_y / Fe trilayer structures, where Fe(N) means a slightly nitrogen doped Fe film. The magneto-transport properties of this trilayer structure grown on $In_{0.2}Ga_{0.8}As$ substrates are presented.

INTRODUCTION

Fe-N thin films have received great attention in recent years because they are very attractive materials for magnetic recording heads for high-density recording. In addition, iron nitrides have significantly improved corrosion resistance over pure iron. The $Fe_{16}N_2$ was discovered and its crystal structure determined by Jack in 1951 [1]. Kim and Takahashi [2] reported in 1972 that Fe-N films deposited onto a glass substrate in low-pressure N_2 gas atmosphere, have a high saturation magnetic flux density (B_S) of 2.58T, which is higher than that of pure Fe by 17%. Furthermore, they confirmed that the Fe-N films are polycrystalline and mixtures of Fe and $Fe_{16}N_2$ crystallites. From the volume ratio of $Fe_{16}N_2$ in the films they attributed the high B_S to $Fe_{16}N_2$, of which the B_S is estimated to be 2.83T. Since then, many researchers have fabricated Fe-N films with high B_S. Up to present various techniques, like evaporation [2], N ion implantation [3], molecular beam epitaxy (MBE) [4,5], conventional sputtering and plasma evaporation [6,7] have been used to produce $Fe_{16}N_2$ single crystal films on various single crystal substrates such are GaAs, $In_{0.2}Ga_{0.8}As$, MgO and glass substrates.

In this paper we present our attempt to fabricate Fe-N thin films using an UHV deposition chamber equipped with ECR microwave plasma source on GaAs, $In_{0.2}Ga_{0.8}As$ grown on GaAs, and SiO_2/Si substrates. The grown Fe-N thin films were characterized using XRD (Copper $K_{\alpha 1}$, λ=1.540598 Å), AES, and TEM. Magnetotransport properties of Fe(N) / Fe_xN_y / Fe trilayer structures were investigated using a conventional room temperature electromagnet set-up.

EXPERIMENTAL PROCEDURES

Experiments were performed in an UHV system with ECR microwave plasma source. Figure 1 shows the schematic diagram of the equipment. A base pressure of $2x10^{-7}$ Pa is obtained using with the pumping systems consisting of a combination of a turbo-molecular pump with variable rotation speed and a rotary vane pump. The UHV chamber contains of the following

Mat. Res. Soc. Symp. Proc. Vol. 577 © 1999 Materials Research Society

parts: 2" substrate holder and heater with temperatures up to 800°C, high temperature Fe effusion cell with shutter, and high purity N_2 gas inlet controlled by a mass flow controller. The ECR microwave plasma source is composed of the following parts: two side flanges containing the water cooled magnet coils generating a magnetic field distribution (875 Gauss), and a coaxial antenna in the center of one of these flanges emitting 2.45 GHz microwave radiation. The magnetron radiating at 2.45 GHz is connected to the antenna by matched wave-guide impedance with mobile circuit terminators [8].

Figure 2 shows the typical emission spectrum of the ECR microwave plasma source operated at a microwave power of ≈200 Watt, magnet current of 50 A, and N_2 pressure of $9.5x10^{-2}$ Pa. The obtained emission spectrum reveals a series of atomic nitrogen emission peaks in the near-IR spectral region in addition to various species of molecular nitrogen. The relative low intensity of atomic nitrogen emission peaks at 745, 821, and 869 nm is due to the high pressure of the N_2 [9]. At a pressure of $1x10^{-1}$ Pa the mean free path of nitrogen atoms is greater than 5 cm. In our UHV ECR system the plasma is transported over the substrate along the magnetic field lines. As a consequence, nitrogen atoms impinge directly onto the grown film, and can form Fe-N.

For all the experiments three different substrate types such as GaAs, $In_{0.2}Ga_{0.8}As$ grown on GaAs, and SiO_2/Si were used. For all the experiments the Fe effusion cell temperature (1350°C), microwave power (≈200 W), and magnet current (49.9 A) were kept at identical values.

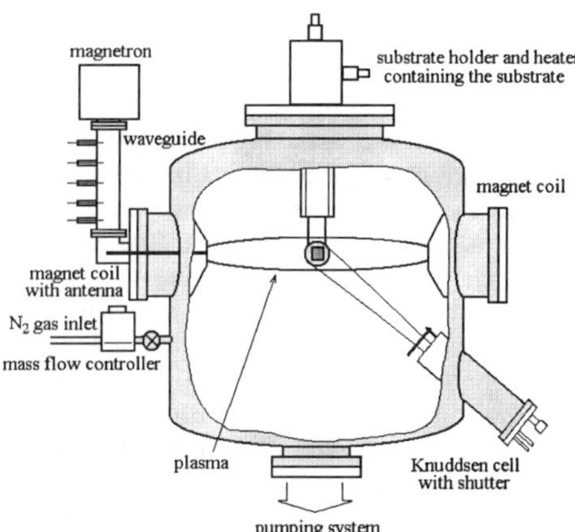

Figure 1: UHV deposition chamber containig an ECR microwave plasma source.

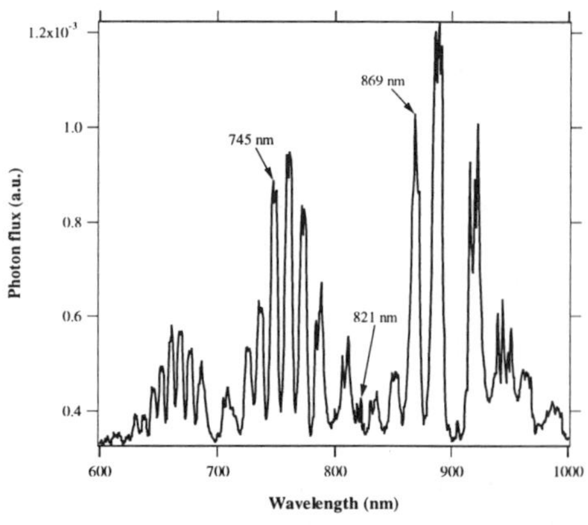

Figure 2: Nitrogen-plasma emission spectrum from an ECR microwave plasma source operated at microwave power of ≈ 200 Watt, magnet current of 50 A and N_2 pressure of $9.5x10^{-2}$ Pa.

RESULTS AND DISCUSSION

In this paper the results of various growth runs (S5, S6, S7, and S9) are discussed. The TEM cross-section image of Figure 3 (S7) shows the structure of a Fe-N layer grown on $In_{0.20}Ga_{0.80}As$. The thickness of the layer is 90 nm (± 8nm) for a deposition time of 7.5 hours, the growth temperature was 70 °C, and the N_2 pressure was set to $8.8x10^{-2}$ Pa. A columnar structure is clearly visible with an average column width of about 10 nm. At the interface a 2 nm thick interface layer between the grown Fe-N thin film and the $In_{0.20}Ga_{0.80}As$ substrate is visible.

The nature of this interface layer is unknown at the moment. We believe, that this is a As-rich layer, which is originating from an incomplete removal of the As-cap that was used to protect the $In_{0.20}Ga_{0.80}As$ epitaxial layer during transfer from the MBE machine to the ECR plasma deposition chamber.

Figure 3: TEM cross-section of Fe_xN_y film grown with ECR plasma deposition chamber (S7 grown on $In_{0.2}Ga_{0.8}As$).

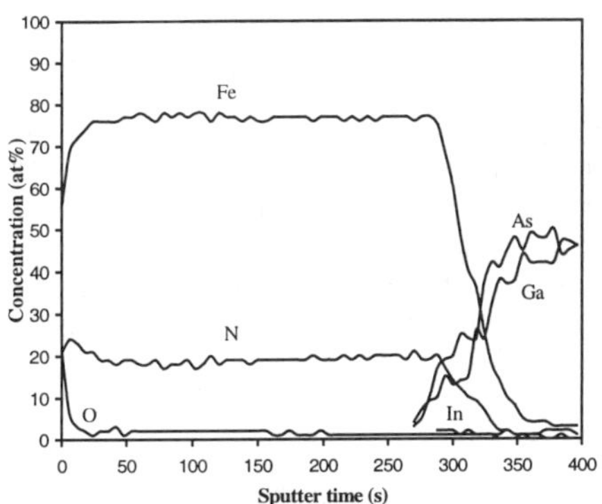

Figure 4: AES depth profile of FexNy film (S7 grown on GaAs), the Fe:N ratio is 80:20.

AES depth profile data (Figure 4) of (S7) grown on GaAs substrate show that the concentration of Fe to N is 80:20. No oxygen or carbon is present in the layer. Figure 5 shows the XRD spectra of (S7) grown on $In_{0.2}Ga_{0.8}As$. The presence of α-Fe, α'-Fe, and α''-$Fe_{16}N_2$ phases is clearly visible (film called Fe_xN_y in the remainder of this paper).

The AES spectrum of (S6) shows the Fe:N ratio of the three different regimes during the growth (Figure 6). The total thickness of this layer is 58 nm, the growth temperature was set to 70 °C, and the N_2 pressure was varied as described below. In section I the plasma was on with a N_2 pressure of 6.2×10^{-2} Pa. In section II, we suffered from an instability of the plasma during Fe deposition under the same N_2 pressure. During the final section III the plasma was again stabilized with a higher N_2 content (8.8×10^{-2} Pa). Comparing regions I and III in figure 6, it is clear that a modulation of the Fe-N composition is possible by altering the N_2 pressure during the film growth. This, in principle, enables the growth of heterostructures with different magnetic properties.

In region II, it was expected that a clear dip in the N-content would be seen, but only a gradual decrease to the interface - over section II and I - was found. Given the fact that the condition of the plasma was not very well defined in section II, we decided to prepare a separate sample without the plasma. Hence, we performed a growth (S5) under N_2 pressure of 8.8×10^{-2}

Pa, but without igniting the plasma. The thickness of the layer is 12 nm, the growth temperature was set to 104°C. From AES data (Figure 7) it is clear that no nitrogen was incorporated into the grown layer.

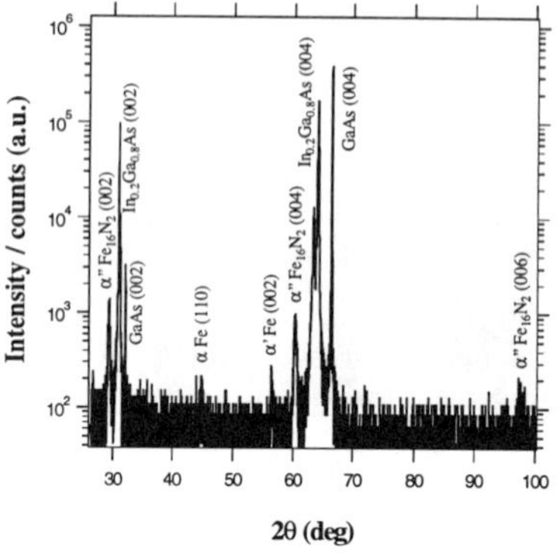

Figure 5: XRD spectrum of an ECR-deposited Fe_xN_y film (S7 on $In_{0.2}Ga_{0.8}As$).

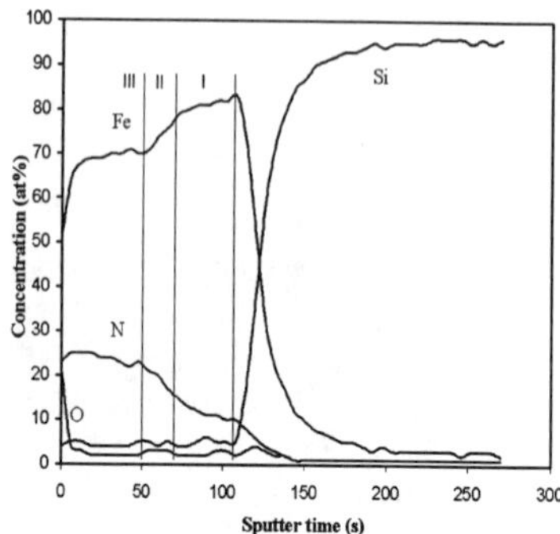

Figure 6: AES of a layer with varying plasma conditions. The plasma was on in regions I and III and off in II.

Our Fe_xN_y films show a metallic conductivity. However, the magnetization measurements by alternating gradient force magnetometer (AGFM) showed that the Fe_xN_y films were not ferromagnetic at room temperature. The reason is not clear so far. We fabricated the magnetic trilayer with the non-magnetic Fe_xN_y layer as a spacer. The stacking structure is Fe(N) (12 nm) / Fe_xN_y (1.2 nm) / Fe (12 nm), where Fe(N) means the Fe layer grown under N_2 pressure without igniting the plasma. The configuration of the magnetoresistance (MR) measurement is shown in Fig. 8 (a). The magnetic field was applied parallel and perpendicular to the trilayer plane. When the direction of the magnetic field is parallel to the plane, the MR curves showed two sharp peaks at around 0.01 T. Since the easy magnetization axis of the Fe single layer is in-plane and the coercive field of the Fe layer measured by AGFM corresponds to the magnetic field of 0.01 T, the peaks are attributed to the magnetization reversal of the bottom Fe layer. A more detailed analysis of the coupling behavior is in progress.

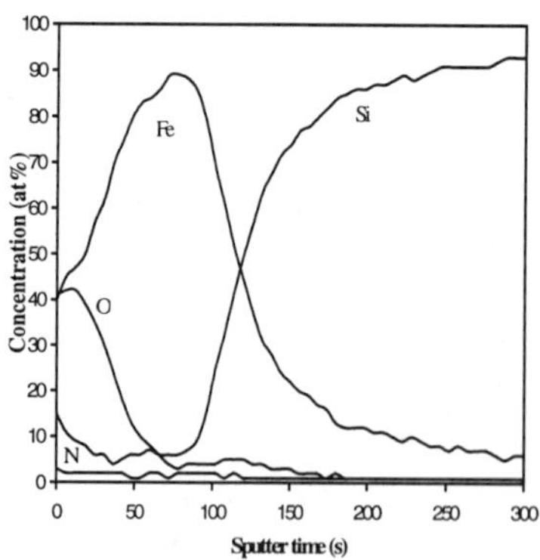

Figure 7. AES depth profile of the sample S5.

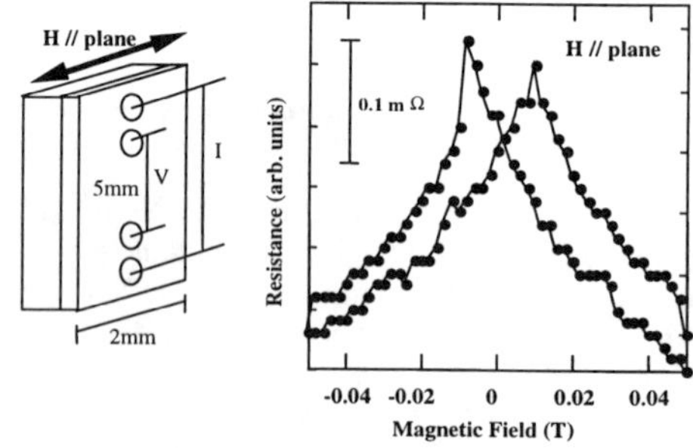

Figure 8. Preliminary MR data from sample S9.

CONCLUSION

The use of an ECR microwave plasma for the growth of Fe_xN_y films was studied. The structural analysis shows that the films were columnar, consisting of a combination of α-Fe, α'-Fe, γ-Fe_4N, and α''-$Fe_{16}N_2$ phases. The modulation of the plasma conditions leads to a variation in N content, enabling the fabrication of compositionally modulated layers where the magnetic properties can be varied during the deposition. The first magneto-transport data of a compositionally modulated structure grown on $In_{0.2}Ga_{0.8}As$ substrates are presented.

ACKNOWLEDGEMENTS

H. Boeve acknowledges the IWT for financial support. This work was performed in the framework of JRCAT - IMEC collaborative research agreement for spin-dependent devices.

REFERENCES

1. K. H. Jack, Proc R. Soc. London, Ser. **A 208**, p. 200 (1951)

2. T. K. Kim and M. Takahashi, Appl. Phys. Lett. **20**, p. 492 (1972).

3. K. Nakajima and S. Okamoto, Appl. Phys. Lett. **56**, p. 92 (1990).

4. M. Komuro, Y. Kozono, M. Hanazozo, and Y. Sugita, J. Appl. Phys. **67**, p. 5126 (1989).

5. Y. Sugita, K. Mitsuoka, M. Komuro, H. Hoshiya, Y. Kozono, and Hanazono, J. Appl. Phys. **70**, p. 5977 (1991).

6. M. A. Russak, C. V. Jahnes, E. Klokholm, J. W. Lee, M. E. Re, and B. C. Webb, J. Appl. Phys. **70**, p. 6427 (1991).

7. M. Takahashi, H. Shoji, H. Takahashi , T. Wakiyama, M. Kinoshita, and W. Ohta, IEEE Trans. Magn. **29**, p. 3040 (1993).

8. K. Deneffe, P. Van Mieghem, B. Brijs, W. Van Der Vorst, R. Mertens, and G. Borghs, Journal of the Less Common Metals, **164&165**, p. 307, (1990).

9. R. P. Vaudo, J. W. Cook, Jr., and J. F. Schetzina, J. Vac. Sci. Technol. **B 12(2)**, p. 1232, (1994).

INVESTIGATIONS ON IN SITU NANOCRYSTALLIZATION AND MAGNETIC PROPERTIES FOR AMORPHOUS $Fe_{78}Si_9B_{13}$ RIBBONS

Xiangcheng Sun[*a], **A. Cabral-Prieto**[*], and **M. Jose Yacaman**[*b]
[*]Instituto Nacional de Investigaciones Nucleares (ININ), Km.36.5, Carr. Mexico-Toluca, C.P.52045 Ocoyoacac, Edo. de Mexico, Mexico [a]E-mail: sunxc@nuclear.inin.mx
[b]Instituto de Fisica, National University of Mexico , Mexico D.F. Mexico

Wensheng Sun
State Key Lab for *RSA*, Institute of Metal Research, Chinese Academy of Science, Shenyang, 110015, P. R. China

ABSTRACT

The amorphous state of ferromagnetic $Fe_{78}Si_9B_{13}$ (Metglas 2605S-2) and its nanocrystallization were investigated by *in situ* transmission electron microscope (TEM), X-ray diffraction (XRD), Mössbauer spectroscopy (MS), differential scanning calorimeters (DSC) and magnetic moment measurements. The Mössbauer spectrum exhibited an essentially symmetric hyperfine field pattern of 259KOe in as-quenched amorphous state at room temperature. The Curie and crystallization temperature were determined to be T_C=708K and T_x= 803K, respectively. The T_x value was in good agreement with DSC measurement results. The occupied fraction of the nanocrystalline phases of α-Fe(Si) and Fe_2B at *in situ* optimum annealing temperature was about 57% and 43%, respectively. It is notable that the magnetization of the amorphous phase decreases more rapidly with increasing temperature than those of nanocrystalline ferromagnetism, suggesting the presence of the distribution of exchange interaction in the amorphous phase or high metalloid contents.

INTRODUCTION

Nanocrystalline magnetic materials have been attracted growing interest in the few past decedes[1-5], due to their excellent soft magnetic properties. Nanocrystalline alloys are obtained by the controlled crystallization of amorphous alloys, like Fe-Si-B [6] and Fe-B-Cu-(Zr,Nb)[7,8]. The ferromagnetic properties of these amorphous alloys can be improved by annealing them above the crystallization temperature T_x. This effect is caused by the homogeneous precipitation of bcc α-Fe(*M*) nanocrystals embedded in the ferromagnetic amorphous matrix. In general, the as-quenched amorphous ribbons are exposed to high temperatures ($T>T_x$) under inert gas to produce nanocrystalline alloys samples[1-8]. In this work, a study of the transformation process of the $Fe_{78}Si_9B_{13}$ amorphous alloy using *in situ* transmission electron microscopy (TEM), differential scanning calorimeters (DSC), and X-ray diffraction (XRD) will be presented. Furthermore, the magnetic properties were studied using Mössbauer spectroscopy (MS) and vibrating sample magnetometer (VSM) techniques.

Mat. Res. Soc. Symp. Proc. Vol. 577 © 1999 Materials Research Society

EXPERIMENT

Amorphous $Fe_{78}Si_9B_{13}$ (at%) alloy ribbons of 30mm width and about 25 μm thickness, were prepared by the melt-spinning techniques [6]. The *in situ* transmission electron micrographs (TEM) were obtained with a JEOL-JEM100 and JEOL 2010 EX electron microscopes, equipped with an *in situ* heating holder inside the electron microscope. The Mössbauer spectroscopy were recorded using a conventional spectrometer of the electromechanical type with a $^{57}Co/Rh$ source. Calorimetric measurements of the as-quenched ribbons were performed on a differential scanning calorimeter (DuPont 2000 Thermal Analysis DSC-2) using different heating rates under an nitrogen gas atmosphere. X-ray diffraction (XRD) were carried out on a Simiens x-ray diffactometer (Cu, Kα radiation, λ=0.15418nm). The magnetic moment measurements were carried out using a vibrating sample magnetometer (VSM) operating with applied magnetic fields up to 3600Oe from 50K to room temperature.

RESULTS

Fig.1 shows the differential scanning calorimeters (DSC) thermal analysis for as-quenched ribbons samples at a different heating rate of 5, 10, 15, 20K/min. Two exothermic peaks are observed at around 786K and 807K, indicating this is typical primary crystallization process, in other words, two crystallization processes (nucleation and growth) takes place, the maximum of the second peak around 807K reaches the crystallization temperature T_x for this $Fe_{78}Si_9B_{13}$ amorphous alloy.

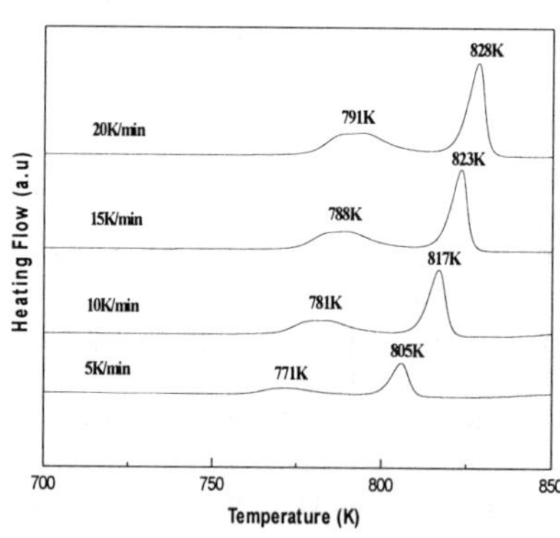

Fig. 1 The DSC curves for $Fe_{78}Si_9B_{13}$ amorphous ribbons at different heating rates.

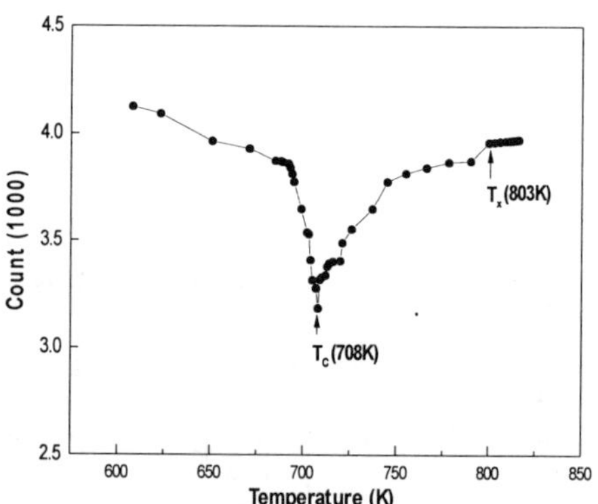

Fig. 2 Counts measured for 10s at zero Doppler velocity as a function of temperature. The heating is 1 k/min.

In order to determine the Curie temperature T_c, the velocity transducer of the Mössbauer spectrometer was set at zero velocity and counts were recorded for a fixed counting time, 15s, while the temperature was raised at a rate of 1K/min from 393K to 813K. The results are shown in Fig.2 . The Curie temperature is determined to be 708K. As the temperature was further increased, the counts rate increased slowly due to the second-order Doppler effect[10], and then increased suddenly at 803K. After the amorphous $Fe_{78}Si_9B_{13}$alloy has reached the crystallization temperature T_x = 803K, the amorphous state gradually transformed into microstructure of the nanocrystalline phases. Both T_x and T_c are in good agreement to previous measurements[3-6].

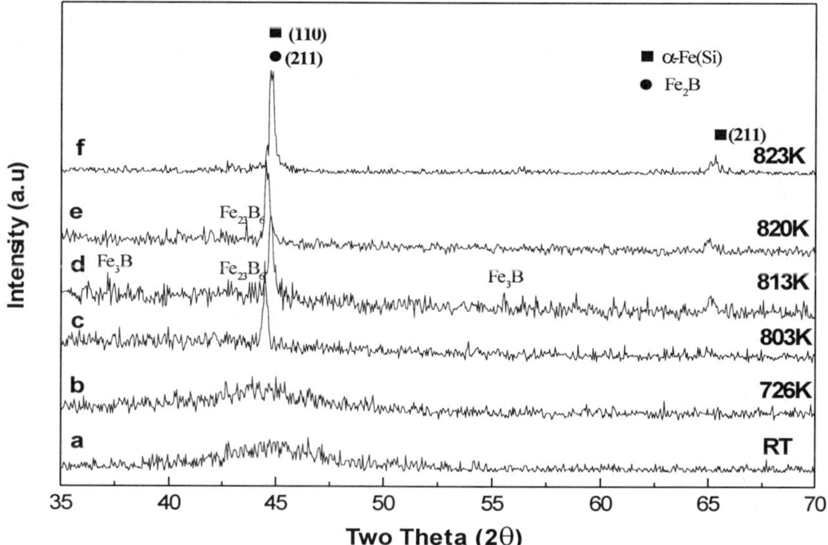

Fig. 3 The XRD patterns of $Fe_{78}Si_9B_{13}$ amorphous ribbon after different annealing temperatures.

Fig.3(a, b, c, d, e, f) shows the XRD patterns for amorphous and crystalline phases after thermal analysis treatment. Fig.3 (a) reveals that the as-quenched sample was in the amorphous state. On the heating the as-quenched sample at T_C<T< T_x, almost a similar XRD pattern to Fig.3 (a) appeared, as shown in Fig.3(b). Although minor changes are better observed from the Mössbauer spectra , as will be shown later. This phenomenon is mainly due to the internal strain relaxation which is induced by thermal treatment. However, when the annealing temperature is greater than T_x (T> T_x), crystalline phases appeared in the XRD patterns of Fig.3 (c, d, e). Precipitation of α-Fe(Si) (bcc) solid solution, and metastable $Fe_{23}B_6$ (bct), Fe_3B (bct), Fe_2B (bct) borides phases took place[1, 2,5,6]. Two main nanophases, α-Fe(Si) (bcc) and Fe_2B (bct), can be clearly identified from Fig.3 (c, d, e, f).

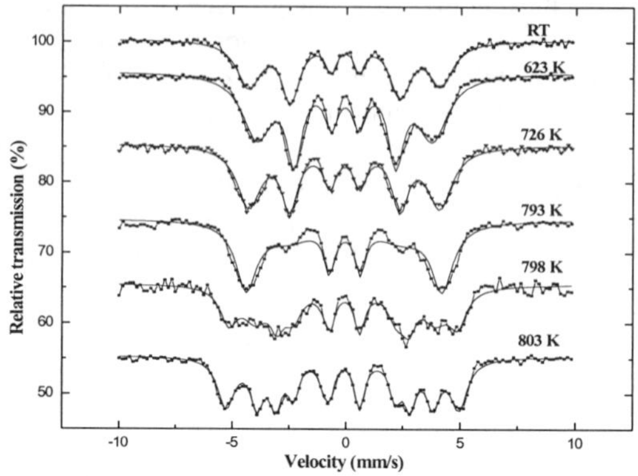

Fig. 4 Mössbauer spectra of amorphous $Fe_{78}Si_9B_{13}$ at various annealing temperature.

The Mössbauer spectra of the as-quenched amorphous samples from room temperature (RT) to above T_x exhibited magnetic patterns shown in Fig.4. The broad lines are to be expected in view of the disordered atomic arrangements, in which the strength of the hyperfine interactions change from site to site due to the structurally inequivalent Fe environment. The Mössbauer spectrum of amorphous $Fe_{78}Si_9B_{13}$ at RT exhibits a magnetic pattern with a hyperfine magnetic field of 259 KOe and relative intensity ratios close to 3 :4 :1 :1 :4 :3, indicating that the direction of the magnetic moment is close to 70° from the γ-ray direction. This spin direction is typical in ferromagnetic amorphous alloys, which has been induced by magnetic anisotropy due to the rapid quenching technique. However, this spin orientation can be changed by treating the sample at $T_C<T<T_x$, where the intensity ratios in close to 3 :1 :1 :1 :1 :3, which indicates that the spin orientation is perpendicular to the plane, *i.e*, the spin orientation is almost aligned to the γ-ray direction. Nevertheless, when $T > T_x$, the Mössbauer spectrum only consisted of two magnetic patterns due to two nanophases, which confirmed by former XRD patterns of Fig.3 (e, f). According to the literature reports [9, 10], the magnetic hyperfine field pattern of 317KOe and 240KOe belongs to bcc α-Fe(Si) solid solutions and bct Fe_2B boride, respectively. Table.1 shows the fitted Mössbauer parameters for the different thermal treatment samples, which giving the information of these two nanophases and the relative intensity variations of the treated samples.

Table 1. The fitted Mössbauer parameter for amorphous $Fe_{78}Si_9B_{13}$ alloy after different annealing treatments.

Temperature	**H_{hf} (KOe)**	**ΔE_Q (mm/s)**	**δ (mm/s)**	**area ratio %**
RT	**259**	**-0.037**	**0.107**	**100**
726 K	**267**	**-0.36**	**0.12**	**100**
803 K	**317**	**-0.028**	**0.05**	**57.65**
	240	**0.022**	**0.14**	**42.35**

*H_{hf} is the magnetic hyperfine field, ΔE_Q is the quadrupole splitting, δ is the isomer shift and area ratio (%) is area of various phases relative to total area, δ referred to α-Fe.

In particular, the magnetic moments of the amorphous $Fe_{78}Si_9B_{13}$ have been measured at H_{app} = 3.6 KOe at various temperatures from 50K to RT. The results are shown in Fig. 5. It is clear that the magnetization of the amorphous phase decreases more rapidly with increasing temperature than those of nanocrystalline ferromagnetism. And the saturation magnetization of the as-quenched amorphous materials extrapolated to 0K is found to be 2.01 μ_B/Fe atom. This value is smaller than the 2.22 μ_B of bcc Fe [11], suggesting that electron transfer from the metalloid atoms to the *d* band of the Fe atoms occur.

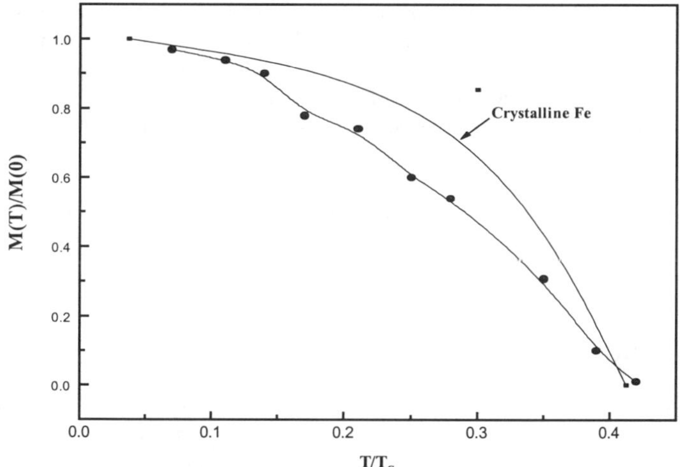

Fig. 5 Reduced magnetization vs reduced temperature of the amorphous $Fe_{78}Si_9B_{13}$ and the crystalline phases after crystallization.

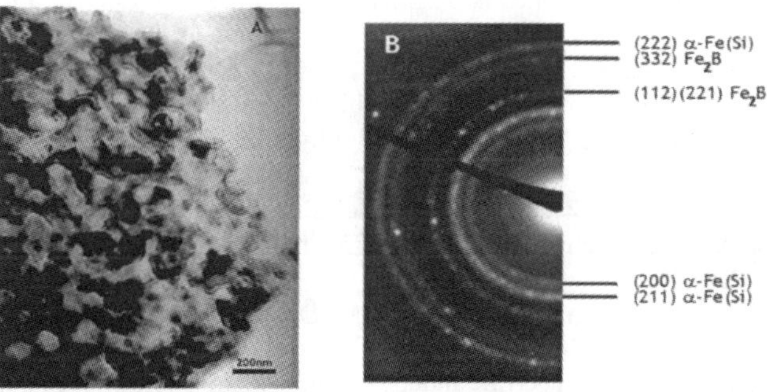

Fig. 6 (a, b). The typical *in situ* TEM morphologies and selected area electron diffraction for Fe-Si-B nanocrystalline phases

The typical *in situ* TEM images of the Fe-Si-B nanocrystalline alloy at annealing temperature, $T_a = 813K$ ($T > T_x$), is shown in Fig.6 (a, b). From Fig.6 (a), small almost spherical nanometer grains precipitated homogeneously is observed. The selected area electron diffraction (Fig.6 (b)) patterns confirms that the randomly distributed nanocrystalline phases were crystalline α-Fe(Si) (bcc) and Fe_2B (bct) with a random orientation. This is in good agreement with above XRD and Mössbauer analysis results.

CONCLUSIONS

Polycrystalline Fe-Si-B alloys with different nanometer grain sizes were successfully prepared by means of the *in situ* amorphous crystallization method. Two main nanophases are observed after full crystallization of the amorphous $Fe_{78}Si_9B_{13}$ alloy, which is consistent with previous reports. The XRD pattern, Mossbauer spectra and in situ TEM observations suggested that the formed nanophases were the α-Fe(Si) (bcc) solid solution and the Fe_2B (bct).

The Curie temperature (T_c) and crystallization temperature (T_x) have been determined to be 708K and 803K using DSC thermal analysis and Mössbauer spectroscopy measurements. The Mössbauer spectrum at different annealing temperature shown that the disordered atomic arrangements, in which the strength of the hyperfine interactions change from site to site due to the structurally inequivalent Fe environment.

ACKNOWLEDGMENTS

The present work is financially supported by CONACyT projects through "Quantum Colliods" in Mexico. We would like to thank Prof. J.R.Gasga for *in situ* TEM measurements; Mr. C.Rodriguez for DSC measurements ; Mr. R.Reglero for kind help of paper preparation.

REFERENCES:

1. J.C.Rawers, R, A.McCune and A.Adams, J. Mater. Sci. Lett., **7**, 958 (1988)
2. J.Y.Bang and R.Y. Lee, J. Mater. Sci., 26, 4961 (1991); C.F.Chang and J. Marti J. Mater. Sci., **18**, 2297, 1983.
3. Y.Yoshizawa, S.oguma and K.Yamauchi J. Appl. Phys., **64**(10), 6044(1988)
4. K.Lu, J.T.Wang, and W.D.Wei, J. Appl. Phys., **69**, 522(1991)
5. H.Y.Tong, J.T.Wang, H.G.Jiang and K.Lu J. Non-Cryst. Solids **150**, 444(1992)
6. X.D.Liu, K.Lu, and Z.Q Hu Mater. Sci. Eng. **A179/180,** 386(1994)
7. K.Suzuki, M.Kikuchi, A. Makino, A.Inoue and T.Masumoto, Mater. Trans. Jpn. Inst. Met., **32**, 961(1991)
8. H.nakabayashi, M.Doi, M.Matsui, and M. Doyama Mat. Res. Soc. Symp. Proc. **286**, 227 (1993)
9. C.S. Kim, S. B. Kim, J.S. Lee and T.H. Noh J. Appl. Phys. **79**(8) 5459 (1996)
10. M.B. Sterns Phys. Rev., **129**, 1136 (1963)
11. R.S. Tebble and D.J. Craik, Magnetic Materials (Wiley-Interscience, NewYork, 1969), P. 51

AUTHOR INDEX

SUBJECT INDEX